W9-AMT-637

THE ROLE OF DEGENERATE STATES IN CHEMISTRY

A SPECIAL VOLUME OF ADVANCES IN CHEMICAL PHYSICS

VOLUME 124

EDITORIAL BOARD

THE ROLE OF DEGENERATE STATES IN CHEMISTRY

ADVANCES IN CHEMICAL PHYSICS
VOLUME 124

Edited by

MICHAEL BAER and GERT DUE BILLING

Series Editors

I. PRIGOGINE

Center for Studies in Statistical Mechanics
and Complex Systems
The University of Texas
Austin, Texas
and
International Solvay Institutes
Université Libre de Bruxelles
Brussels, Belgium

STUART A. RICE

Department of Chemistry
and
The James Franck Institute
The University of Chicago
Chicago, Illinois

A JOHN WILEY & SONS, INC., PUBLICATION

Published by John Wiley & Sons, Inc., Hoboken, New Jersey.
Published simultaneously in Canada.

Library of Congress Catalog Number: 58-9935

ISBN 0-471-43817-0

Printed in the United States of America

10 9 8 7 6 5 4 3 2 1

CONTRIBUTORS TO VOLUME 124

RAVINDER ABROL, Arthur Amos Noyes Laboratory of Chemical Physics, Division of Chemistry and Chemical Engineering, California Institute of Technology, Pasadena, CA

SATRAJIT ADHIKARI, Department of Chemistry, Indian Institute of Technology, Guwahati, India

MICHAEL BAER, Applied Physics Division, Soreq NRC, Yavne, Israel

GERT D. BILLING, Department of Chemistry, Ørsted Institute, University of Copenhagen, Copenhagen, Denmark

MARK S. CHILD, Physical & Theoretical Chemistry Laboratory, South Parks Road, Oxford, United Kingdom

ERIK DEUMENS, Department of Chemistry and Physics, University of Florida Quantum Theory Project, Gainesville, FL

ROBERT ENGLMAN, Soreq NRC, Yavne, Israel

YEHUDA HAAS, Department of Physical Chemistry and the Farkas Center for Light-Induced Processes, Hebrew University of Jerusalem, Jerusalem, Israel

M. HAYASHI, Center for Condensed Matter Sciences, National Taiwan University, Taipei, Taiwan, ROC

J. C. JIANG, Institute of Atomic and Molecular Sciences, Academia Sinica, Taipei, Taiwan, ROC

V. V. KISLOV, Institute of Atomic and Molecular Sciences, Academia Sinica, Taipei, Taiwan, ROC

ARON KUPPERMAN, Arthur Amos Noyes Laboratory of Chemical Physics, Division of Chemistry and Chemical Engineering, California Institute of Technology, Pasadena, CA

K. K. LIANG, Institute of Atomic and Molecular Sciences, Academia Sinica, Taipei, Taiwan, ROC

S. H. LIN, Institute of Atomic and Molecular Sciences, Academia Sinica, Taipei, Taiwan, ROC

SPIRIDOULA MATSIKA, Department of Chemistry, Johns Hopkins University, Baltimore, MD

A. M. MEBEL, Institute of Atomic and Molecular Sciences, Academia Sinica, Taipei, Taiwan, ROC

N. YNGVE OHRN, Department of Chemistry and Physics, University of Florida Quantum Theory Project, Gainesville, FL

MILJENKO PERIĆ, Institut für Physikalische und Theoretische Chemie, Universitaet Bonn, Bonn, Germany

SIGRID D. PEYERIMHOFF, Institut für Physikalische und Theoretische Chemie, Universitaet Bonn, Bonn, Germany

MICHAEL ROBB, Chemistry Department, King's College London, Strand London, United Kingdom

A. J. C. VARANDAS, Departamento de Quimica, Universidade de Coimbra, Coimbra, Portugal

G. A. WORTH, Chemistry Department, King's College London, Strand London, United Kingdom

Z. R. XU, Departamento de Quimica, Universidade de Coimbra, Coimbra, Portugal

ASHER YAHALOM, The College of Judea and Samaria, Faculty of Engineering, Ariel, Israel

DAVID R. YARKONY, Department of Chemistry, Johns Hopkins University, Baltimore, MD

SHMUEL ZILBERG, Department of Physical Chemistry and the Farkas Center for Light-Induced Processes, Hebrew University of Jerusalem, Jerusalem, Israel

INTRODUCTION

Few of us can any longer keep up with the flood of scientific literature, even in specialized subfields. Any attempt to do more and be broadly educated with respect to a large domain of science has the appearance of tilting at windmills. Yet the synthesis of ideas drawn from different subjects into new, powerful, general concepts is as valuable as ever, and the desire to remain educated persists in all scientists. This series, *Advances in Chemical Physics*, is devoted to helping the reader obtain general information about a wide variety of topics in chemical physics, a field that we interpret very broadly. Our intent is to have experts present comprehensive analyses of subjects of interest and to encourage the expression of individual points of view. We hope that this approach to the presentation of an overview of a subject will both stimulate new research and serve as a personalized learning text for beginners in a field.

I. PRIGOGINE
STUART A. RICE

INTRODUCTION TO THE ADVANCES OF CHEMICAL PHYSICS VOLUME ON:

THE ROLE OF DEGENERATE STATES IN CHEMISTRY

The study of molecular systems is based on the Born–Oppenheimer *treatment*, which can be considered as one of the most successful theories in physics and chemistry. This treatment, which distinguishes between the fast-moving electrons and the slow-moving nuclei leads to electronic (adiabatic) eigenstates and the non-adiabatic coupling terms. The existence of the adiabatic states was verified in numerous experimental studies ranging from photochemical processes through photodissociation and unimolecular processes and finally bimolecular interactions accompanied by exchange and/or charge-transfer processes. Having the well-established adiabatic states many studies went one step further and applied the Born–Oppenheimer *approximation*, which assumes that for low enough energies the dynamics can be carried out on the lower surface only, thus neglecting the coupling to the upper states. Although on numerous occasions, this approximation was found to yield satisfactory results, it was soon realized that the relevance of this approximation is quite limited and that the interpretation of too many experiments whether based on spectroscopy or related to scattering demand the inclusion of several electronic states. For a while, it was believed that perturbation theory may be instrumental in this respect but this idea was not found in many cases to be satisfactory and therefore was only rarely employed.

In contrast to the successful introduction, of the electronic adiabatic states into physics and mainly into chemistry, the incorporation of the complementary counterpart of the Born–Oppenheimer treatment, that is, the electronic non-adiabatic coupling terms, caused difficulties (mainly due to their being "extended" vectors) and therefore were ignored. The non-adiabatic coupling terms are responsible for the coupling between the adiabatic states, and since for a long time most studies were related to the ground state, it was believed that the Born–Oppenheimer approximation always holds due to the weakness of the non-adiabatic coupling terms. This belief persisted although it was quite early recognized, due to the Hellmann–Feynman theorem, that non-adiabatic coupling terms are not necessarily weak, on the contrary, they may be large and eventually become infinite. They become infinite (or singular) at those instances when two successive adiabatic states turn out to be degenerate. Having singular non-adiabatic coupling terms not only leads to the breakdown of the Born–Oppenheimer approximation but also rules out the possibility of keeping it while applying perturbation theory. Nevertheless the Born–Oppenheimer approximation can be partly "saved," in particular while studying low-energy processes, by extending it to include the relevant non-adiabatic coupling terms. In this way, a new equation is obtained, for which novel methods to solve it were developed—some of them were discussed in this volume.

This volume in the series of *Advances of Chemical Physics* centers on studies of effects due to electronic degenerate states on chemical processes. However, since the degenerate states affect chemical processes via the singular non-adiabatic coupling terms, a major part of this volume is devoted to the study of features of the non-adiabatic coupling terms. This is one aspect related to this subject. Another aspect is connected with the Born–Oppenheimer Schrödinger equation which, if indeed degenerate states are common in molecular systems, frequently contains singular terms that may inhibit the possibility of solving this equation within the original Born–Oppenheimer adiabatic framework. Thus, an extensive part of this volume is devoted to various transformations to another framework—the diabatic framework—in which the adiabatic coupling terms are replaced by potential coupling—all analytic smoothly behaving functions.

In Chapter I, Child outlines the early developments of the theory of the geometric phase for molecular systems and illustrates it primarily by application to doubly degenerate systems. Coverage will include applications to given to $(E \times \epsilon)$ Jahn–Teller systems with linear and quadratic coupling, and with spin–orbit coupling. The origin of vector potential modifications to the kinetic energy operator for motion on well-separated lower adiabatic potential surfaces is also be outlined.

In Chapter II, Baer presents the transformation to the diabatic framework via a matrix—the adiabatic-to-diabatic transformation matrix—calculated employing a line-integral approach. This chapter concentrates on the theoretical–mathematical aspects that allow the rigorous derivation of this transformation matrix and, following that, the derivation of the diabatic potentials. An interesting finding due to this treatment is that, once the non-adiabatic coupling terms are arranged in a matrix, this matrix has to fulfill certain *quantization* conditions in order for the diabatic potentials to be single valued. Establishing the quantization revealed the existence of the topological matrix, which contains the topological features of the electronic manifold as related to closed contours in configuration space. A third feature fulfilled by the non-adiabatic coupling matrix is the curl equation, which is reminiscent of the Yang–Mills field. This suggests, among other things, that *pseudomagnetic* fields may "exist" along *seams* that are the lines formed by the singular points of the non-adiabatic coupling terms. Finally, having the curl equation leads to the proposal of calculating non-adiabatic coupling terms by solving this equation rather than by performing the tedious ab initio treatment. The various theoretical derivations are accompanied by examples that are taken from *real* molecular systems.

In Chapter III, Adhikari and Billing discuss chemical reactions in systems having conical intersections. For these situations they suggest to incorporate the effect of a geometrical phase factor on the nuclear dynamics, even at energies well below the conical intersection. It is suggested that if this phase factor is incorporated, the dynamics in many cases, may still be treated within a one-surface approximation. In their chapter, they discuss the effect of this phase factor by first considering a model system for which the two-surface problem can also easily be solved without approximation. Since many calculations involving heavier atoms have to be considered using approximate dynamical theories such as classical or quantum classical, it is important to be able to include the geometric phase factor into these theories as well. How this can be achieved is discussed for the three-particle problem. The connection between the so-called extended Born–Oppenheimer approach and the phase angles makes it possible to move from two-surface to multisurface problems. By using this approach a three-state model system is considered. Finally, the geometric phase effect is formulated within the so-called quantum dressed classical mechanics approach.

In Chapter IV, Englman and Yahalom summarize studies of the last 15 years related to the Yang–Mills (YM) field that represents the interaction between a set of nuclear states in a molecular system as have been discussed in a series of articles and reviews by theoretical chemists and particle physicists. They then take as their starting point the theorem that when the electronic set is complete so that the Yang–Mills field intensity tensor vanishes and the field is a pure gauge, and extend it to obtain some new results. These studies throw light on the nature of the Yang–Mills fields in the molecular and other contexts, and on the interplay between diabatic and adiabatic representations.

In Chapter V, Kuppermann and Abrol present a detailed formulation of the nuclear Schrödinger equation for chemical reactions occurring on multiple potential energy surfaces. The discussion includes triatomic and tetraatomic systems. The formulation is given in terms of hyperspherical coordinates and accordingly the scattering equations are derived. The effect of first and second derivative coupling terms are included, both in the adiabatic and the diabatic representations. In the latter, the effect of the non-removable (transverse) part of the first derivative coupling vector are considered. This numerical treatment led, finally, to the potential energy surfaces that are then employed for the scattering calculations. The coverage includes a detailed asymptotic analysis and expressions for the reactive scattering matrices, the associated scattering amplitudes and differential cross-sections. The inclusion of the geometric phase in these equations is discussed, as well as results of representative calculations.

In Chapter VI, Ohrn and Deumens present their electron nuclear dynamics (END) time-dependent, nonadiabatic, theoretical, and computational approach to the study of molecular processes. This approach stresses the analysis of such processes in terms of dynamical, time-evolving states rather than stationary molecular states. Thus, rovibrational and scattering states are reduced to less prominent roles as is the case in most modern wavepacket treatments of molecular reaction dynamics. Unlike most theoretical methods, END also relegates electronic stationary states, potential energy surfaces, adiabatic and diabatic descriptions, and nonadiabatic coupling terms to the background in favor of a dynamic, time-evolving description of all electrons.

In Chapter VII, Worth and Robb discuss techniques known as direct, or on-the-fly, molecular dynamics and their application to non-adiabatic processes. In contrast to standard techniques, which require a predefined potential energy surfaces, here the potential function, is provided by explicit evaluation of the electronic wave function for the states of interest. This fact makes the method very general and powerful, particularly for the study of polyatomic systems where the calculation of a multidimensional potential function is expected to be a complicated task. The method, however, has a number of difficulties that need to be solved. One is the sheer size of the problem—all nuclear and electronic degrees of freedom are treated explicitly. A second is the restriction placed on the form of the nuclear wave function as a local- or trajectory-based representation is required. This introduces the problem of including quantum effects into methods that are often based on classical mechanics. For non-adiabatic processes, there is the additional complication of the treatment of the non-adiabatic coupling. In this chapter these authors show how progress has been made in this new and exciting field, highlighting the different problems and how they are being solved.

In Chapter VIII, Haas and Zilberg propose to follow the phase of the total electronic wave function as a function of the nuclear coordinates with the aim of locating conical intersections. For this purpose, they present the theoretical basis for this approach and apply it for conical intersections connecting the two lowest singlet states (S_1 and S_0). The analysis starts with the Pauli principle and is assisted by the permutational symmetry of the electronic wave function. In particular, this approach allows the selection of two coordinates along which the conical intersections are to be found.

In Chapter IX, Liang et al. present an approach, termed as the "crude Born–Oppenheimer approximation," which is based on the Born–Oppenheimer approximation but employs the straightforward perturbation method. Within their chapter they develop this approximation to become a practical method for computing potential energy surfaces. They show that to carry out different orders of perturbation, the ability to calculate the matrix elements of the derivatives of the Coulomb interaction with respect to nuclear coordinates is essential. For this purpose, they study a diatomic molecule, and by doing that demonstrate the basic skill to compute the relevant matrix elements for the Gaussian basis sets. Finally, they apply this approach to the H_2 molecule and show that the calculated equilibrium position and force constant fit reasonable well those obtained by other approaches.

In Chapter X, Matsika and Yarkony present an algorithm for locating points of conical intersection for odd electron molecules. The nature of the singularity at the conical intersection is determined and a transformation to *locally* diabatic states that eliminates the singularity is derived. A rotation of the degenerate electronic states that represents the branching plane in terms of mutually orthogonal vectors is determined, which will enable us to search for confluences intersecting branches of a single seam.

In Chapter XI, Perić and Peyerimhoff discuss the Renner–Teller coupling in triatomic and tetraatomic molecules. For this purpose, they describe some of their theoretical tools to investigate this subject and use the systems FeH_2, CNC, and HCCS as adequate examples.

In Chapter XII, Varandas and Xu discuss the implications of permutational symmetry on the total wave function and its various components for systems having sets of identical particles. By generalizing Kramers' theorem and using double group theory, some drastic consequences are anticipated when the nuclear spin quantum number is one-half and zero. The material presented may then be helpful for a detailed understanding of molecular spectra and collisional dynamics. As case studies, they discuss, in some detail, the spectra of trimmeric species involving ^2S atoms. The effect of vibronic interactions on the two conical intersecting adiabatic potential energy surfaces will then be illustrated and shown to have an important role. In particular, the implications of the Jahn–Teller instability on the calculated energy levels, as well as the involved dynamic Jahn–Teller and geometric phase effects, will be examined by focusing on the alkali metal trimmers. This chapter was planned to be essentially descriptive, with the mathematical details being gathered on several appendixes.

MICHAEL BAER
GERT DUE BILLING

CONTENTS

EARLY PERSPECTIVES ON GEOMETRIC PHASE

M. S. CHILD

Physical and Theoretical Chemistry Laboratory
Oxford, United Kingdom

CONTENTS

I. INTRODUCTION

Subsequent chapters deal largely with developments in the theory of geometric phase and non-adiabatic coupling over the past 10 years, but the editors agreed with me that there would be some value in including a chapter on early contributions to the field, to provide a historical perspective. No doubt the choice of material will seem subjective to some. Others will find it redundant to repeat well-established results in an "Advances" volume, but this chapter is not

The Role of Degenerate States in Chemistry: A Special Volume of Advances in Chemical Physics, Volume 124, Edited by Michael Baer and Gert Due Billing. Series Editors I. Prigogine and Stuart A. Rice. ISBN 0-471-43817-0.

addressed to the experts; it is primarily intended for students seeking a pedagogical introduction to the subject. Discussion is limited to what is now known as the quantal adiabatic (Longuet-Higgins or Berry) phase, associated with motion on a single adiabatic electronic surface, on the assumption that the nuclear motion occurs far from any points of electronic degeneracy. The geometric phase and an associated vector potential term in the nuclear kinetic energy operator will be seen to arise from the presence of singularities in the non-adiabatic coupling operator, at so-called conical intersection points, but the wave function will appear as a product of a single electronic and a single nuclear factor.

The story begins with studies of the molecular Jahn–Teller effect in the late 1950s [1–3]. The Jahn–Teller theorems themselves [4,5] are 20 years older and static Jahn–Teller distortions of electronically degenerate species were well known and understood. Geometric phase is, however, a dynamic phenomenon, associated with nuclear motions in the vicinity of a so-called conical intersection between potential energy surfaces.

The simplest and most widely studied case is the $E \times \epsilon$ Jahn–Teller model [2,6,7] for which a double degeneracy at say an equilateral triangular geometry is relieved *linearly* by nuclear distortions in a doubly degenerate nuclear vibration. In the language of later discussions [8], the nuclear coordinates Q define a two-dimensional (2D) parameter space containing the intersection point Q_0, and the geometric phase is associated with evolution of the real adiabatic electronic eigenstates, say $|x_+(Q)\rangle$ and $|x_-(Q)\rangle$, on parameter paths around Q_0. The important points are that $|x_\pm(Q)\rangle$ are undefined at Q_0, but that they can be taken elsewhere as smooth functions of Q, in the sense that $\langle x_\pm(Q)|x_\pm(Q + \delta Q)\rangle \rightarrow 1$ as $\delta Q \rightarrow 0$, over any region free of other degeneracies. It is then a simple matter to demonstrate that the linearity of the separation between the two adiabatic potential surfaces, say $W_\pm(Q)$, also requires a sign change in $|x_\pm(Q)\rangle$, as they are transported around Q_0 [2,6,7]. Note that there is no corresponding geometric phase associated with symmetry determined electronic degeneracies in linear molecules for which the degeneracy is relieved quadratically in the bending coordinate [9]; in other words the two linear molecule adiabatic potential surfaces touch at Q_0 but do not intersect. Conical intersections, with associated geometric phase, do, however, arise at accidental degeneracies in linear molecules, between, for example, Σ and Π electronic states [6]; they can also occur in quite general geometries for nonsymmetric species, such as NaKRb. The latter were taken by Longuet-Higgins [7] as test cases to resolve a controversy over the "noncrossing rule" in polyatomics.

The next significant development in the history of the geometric phase is due to Mead and Truhlar [10]. The early workers [1–3] concentrated mainly on the spectroscopic consequences of localized non-adiabatic coupling between the upper and lower adiabatic electronic eigenstates, while one now speaks of the geometric phase associated with a well-separated lower adiabatic surface, such that the nuclear motions revolve around the intersection point Q_0, without passing close to it. Longuet-Higgins et al. [2] treat this situation in a linear coupling approximation, but Mead and Truhlar [10] were the first to provide a systematic formulation. Any treatment must recognize that the nature of the nuclear wave function is necessarily affected by the electronic sign change, since the total wave function must be a single-valued function of Q. This means either that the boundary conditions on the nuclear wave function must incorporate a compensating sign change for circuits around Q_0 or that the real adiabatic eigenstates, $|x_\pm\rangle$, must be defined with compensating phase factors, such that

$$|n_\pm\rangle = e^{i\psi_\pm(Q)}|x_\pm\rangle$$

is single valued around Q_0. Ham [11] analyses the ordering of vibronic eigenvalues in the presence of geometric phase from the former standpoint, while Mead and Truhlar [10] adopt the latter formulation, which leads to a *vector potential* contribution to the nuclear kinetic energy, dependent on the form of the chosen phase factor $\psi(Q)$. Residual arbitrariness in the choice of $\psi_\pm(Q)$, subject to the single valuedness of $|n_\pm\rangle$, must cancel out in any consistent treatment of the nuclear dynamics.

Berry [8] set the theory in a wider context, by defining a "gauge invariant" geometric phase, which is specific to the system in question and to the geometry of the chosen encircling path, but is also independent of the above residual arbitrariness. The resulting integrated geometric phase applies to quite general situations, provided there is a single isolated point of degeneracy. The degeneracy need not be twofold, nor need the encircling path lie in the plane containing Q_0, as demonstrated by Berry's [8] explicit treatment of angular momentum precession, with arbitrary $2J + 1$ degeneracy, in a slowly rotating magnetic field.

Macroscopic physical manifestations of the above adiabatic geometric phase may be found in the Aharonov–Bohm effect [12] and in nuclear magnetic resonance (NMR) systems subject to slowly rotating magnetic fields [13]. Their observation in molecular systems is less straightforward. Books have been written about the multisurface dynamics of Jahn–Teller systems [14,15], but effects attributable to the geometric phase on the lowest adiabatic potential surface are quite elusive. One example is an observed energy level dependence on the square of a half-odd quantum number, j, in Na_3 [16,17], as first predicted by Longuet-Higgins et al. [2]. It depends, however, on the assumption of strictly linear Jahn–Teller coupling, because j is conserved only in the absence of corrugations on the lower surface arising from the inclusion of quadratic and higher Jahn–Teller coupling terms (see Sections V.A and V.C). The strongest general prediction, for C_3 point groups, is that geometric phase causes a systematic inversion in the vibronic tunneling splitting associated with the above corrugations [11]; thus the levels of the lowest vibronic triplet are predicted in the order $E(E) < E(A)$, an order that is successively reversed and restored in the higher triplets. The possible observation of similar geometric phase related effects in molecular scattering situations is discussed in several of the following chapters.

Section II begins with a general discussion of conical intersections, including deductions from the point group and time-reversal symmetries, concerning connections between the nuclear coordinate dependencies of different electronic Hamiltonian matrix elements. Section III is concerned with the nature of electronic adiabatic eigenstates close to a conical intersection. The crucial result for later sections is that an $E \times \epsilon$ conical intersection gives rise to an adiabatic eigenvector sign change regardless of the size and shape of the encircling loop, provided that no other degenerate points are enclosed. Specifically, geometrical aspects of adiabatic eigenvector evolution are discussed in Section IV, along the lines of papers by Berry [8] and Aharonov et al. [18]. Different expressions for its evaluation are also outlined. Various aspects of the $E \times \epsilon$ Jahn–Teller problem, with linear and quadratic coupling, including and excluding spin–orbit coupling, are outlined in Section V. More general aspects of the nuclear dynamics on the lower potential sheet arising from a conical intersection are treated in Section VI, from two viewpoints. Section VI.A expounds Ham's general conclusions about the order of vibronic tunneling levels from a band theory standpoint [11], with sign-reversing boundary conditions on the nuclear wave functions. There is also an appendix for readers unfamiliar with Floquet theory arguments. By contrast, Section VI.B outlines the elements of Mead and Truhlar's theory [10], with normal boundary conditions on the nuclear wave function and a vector potential contribution to the nuclear kinetic energy, arising from the compensating phase factor $\psi(Q)$, which was discussed above. The relationship between the contributions of Aharonov et al. [18] and Mead and Truhlar [10] are described. Aspects of the symmetry with respect to nuclear spin exchange in the presence of geometric phase are also discussed. Section VII collects the main conclusions and draws attention to related early work on situations with greater complexity than the simple $E \times \epsilon$ problem.

II. CONICAL INTERSECTIONS

Molecular aspects of geometric phase are associated with *conical intersections* between electronic energy surfaces, $W(Q)$, where Q denotes the set of say k vibrational coordinates. In the simplest two-state case, the $W(Q)$ are eigensurfaces of the nuclear coordinate dependent Hermitian electronic Hamiltonian matrix,

$$H(Q) = \begin{pmatrix} H_{\mathrm{AA}}(Q) & H_{\mathrm{AB}}(Q) \\ H_{\mathrm{BA}}(Q) & H_{\mathrm{BB}}(Q) \end{pmatrix} \tag{1}$$

namely,

$$W_{\pm}(Q) = \frac{1}{2}[H_{\mathrm{AA}}(Q) + H_{\mathrm{BB}}(Q)] \pm \frac{1}{2}\sqrt{[H_{\mathrm{AA}}(Q) - H_{\mathrm{BB}}(Q)]^2 + 4|H_{\mathrm{AB}}(Q)|^2} \tag{2}$$

Strict degeneracy between the electronic energy surfaces therefore requires the existence of points Q_0 at which $H_{\mathrm{AA}}(Q) = H_{\mathrm{BB}}(Q)$ and $H_{\mathrm{AB}}(Q) = 0$. These two independent conditions will rarely occur by variation of a single coordinate Q [unless $H_{\mathrm{AB}}(Q) = 0$ by symmetry]—hence the diatomic "noncrossing rule." There is, however, no such prohibition in polyatomics. In the common case of a real representation, degeneracies can clearly lie on a surface of dimensionality $k - 2$, where k is the number of vibrations [6,7,19,20]. They are termed *conical* if $H_{\mathrm{AA}}(Q) - H_{\mathrm{BB}}(Q)$ and $H_{\mathrm{AB}}(Q)$ vanish linearly in Q. Such points are symmetry determined for Jahn–Teller systems [4], which include all electronically degenerate nonlinear polyatomics. They also occur as a result of bending at, say a $\Sigma - \Pi$ intersection in a linear molecule [6], and at more general configurations of nonsymmetrical species. For example, Longuet-Higgins [7] shows that Heitler–London theory for a system of three dissimilar H-like atoms, such as LiNaK, has a pair of doublet states with eigensurfaces governed by the Hamiltonian matrix

$$H = \begin{pmatrix} W - \alpha + \frac{1}{2}(\beta + \gamma) & \sqrt{\frac{3}{2}}(\beta - \gamma) \\ \sqrt{\frac{3}{2}}(\beta - \gamma) & W + \alpha - \frac{1}{2}(\beta + \gamma) \end{pmatrix} \tag{3}$$

where α, β, and γ are exchange integrals for the three interatomic bonds. A conical intersection therefore occurs at geometries such that $\alpha = \beta = \gamma$, which again implies two independent constraints.

Aspects of the Jahn–Teller symmetry argument will be relevant in later sections. Suppose that the electronic states are n-fold degenerate, with symmetry Γ_e at some symmetrical nuclear configuration Q_0. The fundamental question concerns the symmetry of the nuclear coordinates that can split the degeneracy linearly in $Q - Q_0$, in other words those that appear linearly in Taylor series for the n^2 matrix elements $\langle A|H|B\rangle$. Since the bras $\langle A|$ and kets $|B\rangle$ both transform as Γ_e and H are totally symmetric, it would appear at first sight that the Jahn–Teller active modes must have symmetry $\Gamma_Q = \Gamma_e \times \Gamma_e$. There are, however, further restrictions, dependent on whether the number of electrons is even or odd. The following argument [4,5] uses the symmetry of the electronic states under the time-reversal operator $\hat{T}$ to establish general relations between the various matrix elements. The essential properties are that $\hat{T}$ commutes with the Hamiltonian

$$\hat{H}\hat{T} = \hat{T}\hat{H}; \tag{4}$$

that any state $|A\rangle$ has a time-reverse $\hat{T}|A\rangle$, such that

$$\langle \hat{T}\beta|\hat{T}\alpha\rangle = \langle \beta|\alpha\rangle^*; \tag{5}$$

and that states with even and odd electrons are symmetric and antisymmetric under $\hat{T}^2$, respectively. It therefore follows that

$$\begin{aligned} \langle A|H|\hat{T}B\rangle &= \langle \hat{T}A|\hat{T}H\hat{T}B\rangle^* = \langle \hat{T}A|H|\hat{T}^2\beta\rangle^* = \pm\langle B|H|\hat{T}A\rangle \\ &= \frac{1}{2}\left(\langle A|H|\hat{T}B\rangle \pm \langle B|H|\hat{T}A\rangle\right) \end{aligned} \tag{6}$$

where the upper and lower signs apply for even and odd electron systems, respectively.

Suppose now that $|A\rangle$ and $|B\rangle$ belong to an electronic representation Γ_e. Since H is totally symmetric, Eq. (6) implies that the matrix elements $\langle A|H|\hat{T}B\rangle$ belong to the representation of symmetrized or anti-symmetrized products of the bras $\{\langle A|\}$ with the kets $\{|\hat{T}A\rangle\}$. However, the set $\{|\hat{T}A\rangle\}$ is, however, simply a reordering of the set $\{|A\rangle\}$. Hence, the symmetry of the matrix elements in the even- and odd-electron cases is given, respectively, by the symmetrized $[\Gamma_e \times \Gamma_e]$ and antisymmetrized $\{\Gamma_e \times \Gamma_e\}$ parts of the direct product of Γ_e with itself. A final consideration is that coordinates belonging to the totally symmetric representation, Γ_0, cannot break any symmetry determined degeneracy. The symmetries of the Jahn–Teller active modes are therefore given by

$$\begin{aligned} &\Gamma_Q \subset [\Gamma_e \times \Gamma_e] - \Gamma_0 \qquad \text{for even electron systems} \\ &\Gamma_Q \subset \{\Gamma_e \times \Gamma_e\} - \Gamma_0 \qquad \text{for odd electron systems} \end{aligned}$$

This is the central Jahn–Teller [4,5] result. Three important riders should be noted. First, $\Gamma_Q = 0$ for spin-degenerate systems, because $\{\Gamma_e \times \Gamma_e\} = \Gamma_0$. This is a particular example of the fact that Kramer's degeneracies, arising from spin alone can only be broken by magnetic fields, in the presence of which H and $\hat{T}$ no longer commute. Second, a detailed study of the molecular point groups reveals that all degenerate nonlinear polyatomics, except those with Kramer's degeneracy, have at least one vibrational coordinate covered by the above rules. Finally, no linear polyatom has such coordinates. Hence, there are no symmetry determined conical intersections in linear molecules. The leading vibronic coupling terms are quadratic in the nuclear coordinates, giving rise to a Renner–Teller [9] rather than a Jahn–Teller effect.

The symmetry argument actually goes beyond the above determination of the symmetries of Jahn–Teller active modes, the coefficients of the matrix element expansions in different coordinates are also symmetry determined. Consider, for simplicity, an electronic state of symmetry E in an even-electron molecule with a single threefold axis of symmetry, and choose a representation in which two complex electronic components, $|e_{\pm}\rangle = 1/\sqrt{2}(|e_{\mathrm{A}}\rangle \pm i|e_{\mathrm{B}}\rangle)$, and two degenerate complex nuclear coordinate combinations $Q_{\pm} = re^{\pm i\phi}$ each have character $\tau^{\pm 1}$ under the C_3 operation, where $\tau = e^{2\pi i/3}$. The bras $\langle e_{\pm}|$ have character $\tau^{\mp 1}$. Since the Hamiltonian operator is totally symmetric, the diagonal matrix elements $\langle e_{\pm}|H|e_{\pm}\rangle$ are totally symmetric, while the characters of the off-diagonal elements $\langle e_{\mp}|H|e_{\pm}\rangle$ are $\tau^{\pm 2}$. Since $\tau^3 = 1$, it follows that an expansion of the complex Hamiltonian matrix to quadratic terms in $Q_{\pm}$ takes the form

$$H = \begin{pmatrix} 0 & kQ_- + lQ_+^2 \\ kQ_+ + lQ_-^2 & 0 \end{pmatrix} \tag{7}$$

The corresponding expression in the real basis $(|e_{\mathrm{A}}\rangle, |e_{\mathrm{B}}\rangle)$ is

$$H = \begin{pmatrix} kr\cos\phi + lr^2\cos 2\phi & kr\sin\phi - lr^2\sin 2\phi \\ kr\sin\phi - lr^2\sin 2\phi & -kr\cos\phi - lr^2\cos 2\phi \end{pmatrix} \tag{8}$$

after substitution for (Q_+, Q_-) in terms of (r, ϕ). Equation (8) defines what is known as the $E \times \epsilon$ Jahn–Teller problem, which is discussed in Section V.

More general situations have also been considered. For example, Mead [21] considers cases involving degeneracy between two Kramers doublets involving four electronic components $|\alpha\rangle$, $|\alpha'\rangle$, $|\beta\rangle$, and $|\beta'\rangle$. Equations (4) and (5), coupled with antisymmetry under $\hat{T}^2$ lead to the following identities between the various matrix elements

$$\langle\alpha|\hat{H}|\alpha\rangle = \langle\hat{T}\alpha|\hat{T}\hat{H}|\alpha\rangle^* = \langle\hat{T}\alpha|\hat{H}|\hat{T}\alpha\rangle^* = \langle\alpha'|\hat{H}|\alpha'\rangle^* = \langle\alpha'|\hat{H}|\alpha'\rangle \tag{9}$$

$$\langle\alpha|\hat{H}|\alpha'\rangle = \langle\alpha|\hat{H}|\hat{T}\alpha\rangle = \langle\hat{T}\alpha|\hat{T}\hat{H}|\hat{T}\alpha\rangle^* = \langle\hat{T}\alpha|\hat{H}|\hat{T}^2\alpha\rangle^* = -\langle\alpha'|\hat{H}|\alpha\rangle^* \tag{10}$$

$$\langle\alpha|\hat{H}|\beta\rangle = \langle\hat{T}\alpha|\hat{T}\hat{H}|\beta\rangle^* = \langle\hat{T}\alpha|\hat{H}|\hat{T}\beta\rangle^* = \langle\alpha'|\hat{H}|\beta'\rangle^* \tag{11}$$

$$\langle\alpha|\hat{H}|\beta'\rangle = \langle\hat{T}\alpha|\hat{T}\hat{H}|\beta'\rangle^* = \langle\hat{T}\alpha|\hat{H}|\hat{T}\beta'\rangle^* = -\langle\alpha'|\hat{H}|\beta\rangle^* \tag{12}$$

The conclusion is therefore that the 4×4 Hamiltonian matrix, which is assumed to have zero trace, takes the form

$$H(Q) = \begin{pmatrix} w(Q) & 0 & u(Q) & v(Q) \\ 0 & w(Q) & -v^*(Q) & u^*(Q) \\ u^*(Q) & -v(Q) & -w(Q) & 0 \\ v^*(Q) & u(Q) & 0 & -w(Q) \end{pmatrix} \qquad (13)$$

where $w(Q)$ is real. Consequently, there are five independent conditions for a strict conical intersection between two Kramers doublets, although $v(Q)$ may, for example, vanish in model situations (see Section V.B). Moreover, there is no certainty that the intersection will lie in a dynamically accessible region of the coordinate space.

III. ADIABATIC EIGENSTATES NEAR A CONICAL INTERSECTION

Suppose that $|x_n(Q)\rangle$ is the adiabatic eigenstate of the Hamiltonian $H(q;Q)$, dependent on internal variables q (the electronic coordinates in molecular contexts), and parameterized by external coordinates Q (the nuclear coordinates). Since $|x_n(Q)\rangle$ must satisfy

$$H(q;Q)|x_n(Q)\rangle = E_n(Q)|x_n(Q)\rangle \qquad \langle x_m|x_n\rangle = \delta_{mn} \qquad (14)$$

it follows by the Hellman–Feynman theorem that

$$[H(q;Q) - E_n(Q)]\nabla_Q|x_n(Q)\rangle = [\nabla_Q E_n - \nabla_Q H]|x_n(Q)\rangle \qquad (15)$$

Thus, on expanding

$$\nabla_Q|x_n(Q)\rangle = \sum_m |x_m(Q)\rangle\langle x_m|\nabla_Q|x_n\rangle \qquad (16)$$

the off-diagonal matrix elements of ∇_Q may be derived from Eq. (15) in the form

$$\langle x_m|\nabla_Q|x_n\rangle = \frac{\langle x_m|\nabla_Q H|x_n\rangle}{E_n(Q) - E_m(Q)} \qquad (17)$$

The adiabatic approximation involves neglect of these off-diagonal terms, on the basis that $|E_n(Q) - E_m(Q)| \gg |\langle x_m|\nabla_Q H|x_n\rangle|$. The diagonal elements $\langle x_n|\nabla_Q|x_n\rangle$ are undetermined by this argument, but the gradient of the normalization integral, $\langle x_n|x_n\rangle = 1$, shows that

$$\nabla_Q\langle x_n|x_n\rangle = \langle x_n|\nabla_Q x_n\rangle + \langle\nabla_Q x_n|x_n\rangle = \langle x_n|\nabla_Q x_n\rangle + \langle x_n|\nabla_Q x_n\rangle^* = 0 \qquad (18)$$

Consequently,

$$\langle x_n|\nabla_Q x_n\rangle = -\langle x_n|\nabla_Q x_n\rangle^* \tag{19}$$

from which $\langle x_n|\nabla_Q x_n\rangle = 0$, for real $|x_n\rangle$.

Equations (16)–(20) show that the real adiabatic eigenstates are everywhere smooth and continuously differentiable functions of Q, except at degenerate points, such that $E_n(Q) - E_m(Q) = 0$, where, of course, the $|x_n\rangle$ are undefined. There is, however, no requirement that the $|x_n\rangle$ should be real, even for a real Hamiltonian, because the solutions of Eq. (14) contain an arbitrary Q dependent phase term, $e^{i\psi(Q)}$ say. Second, as we shall now see, the choice that $|x_n\rangle$ is real raises a different type of problem. Consider, for example, the model Hamiltonian in Eq. (8), with $l = 0$;

$$H = \begin{pmatrix} kr\cos\phi & kr\sin\phi \\ kr\sin\phi & -kr\cos\phi \end{pmatrix} \tag{20}$$

with a degeneracy at $r = 0$ and real eigenvectors

$$|x_+\rangle = \begin{pmatrix} \cos\frac{\phi}{2} \\ \sin\frac{\phi}{2} \end{pmatrix} \qquad |x_-\rangle = \begin{pmatrix} -\sin\frac{\phi}{2} \\ \cos\frac{\phi}{2} \end{pmatrix} \tag{21}$$

It is readily verified that

$$\langle x_+|\frac{\partial}{\partial\phi}|x_+\rangle = \langle x_-|\frac{\partial}{\partial\phi}|x_-\rangle = 0 \tag{22}$$

but the new problem is that

$$|x_\pm(\phi + 2\pi)\rangle = -|x_\pm(\phi)\rangle \tag{23}$$

which means that $|x_\pm(\phi)\rangle$ is double valued with respect to encirclement of the degeneracy at $r = 0$. In the molecular context, the assumption of a real adiabatic electronic eigenstate therefore requires boundary conditions such that the associated nuclear wave function also changes sign on any path around the origin, because the total wave function itself must be single valued. A more convenient alternative, for practical calculations, is often to add a phase modification, such that the modified eigenstates, $|n_\pm\rangle$, are single valued [2,10].

$$|n_+\rangle = e^{i\psi(Q)}\begin{pmatrix} \cos\frac{\phi}{2} \\ \sin\frac{\phi}{2} \end{pmatrix} \qquad |n_-\rangle = e^{i\psi(Q)}\begin{pmatrix} -\sin\frac{\phi}{2} \\ \cos\frac{\phi}{2} \end{pmatrix} \tag{24}$$

with $\psi(Q_f) - \psi(Q_i) = \pm\pi$. The simplest choice in the present context is $\psi(Q) = \phi/2$ but any phase factor, $e^{i\psi(\phi)}$, that changes sign around a circuit of ϕ is equally acceptable. Nevertheless, the geometric phase defined in Section IV and the associated vector potential theory outlined in Section VI.B are gauge invariant (i.e., independent of this phase ambiguity).

We should also notice explicitly that [22]

$$\langle x_-|\nabla_Q|x_+\rangle = \frac{\mathbf{e}_\phi}{2r} \tag{25}$$

where $\mathbf{e}_\phi$ is a unit vector in the direction of increasing ϕ. Equation (25) shows that the non-adiabatic coupling diverges at the conical intersection point, which is of course a manifestation of the fact that $|x_\pm\rangle$ are undefined at an exact degeneracy. It is readily verified that $\langle n_-|\nabla_Q|n_+\rangle$ and $\langle n_+|\nabla_Q|n_+\rangle$ also diverge in a similar way.

In turn, this leads to an important conclusion, for the general discussion, that the above sign change, for real eigenstates such that $\langle x_\pm(Q + \delta Q)|x_\pm(Q)\rangle \to 1$ as $\delta Q \to 0$, arises solely from the electronic degeneracy—not from the linearity of Eq. (20), because the adiabatic eigenstates were seen above to be smooth continuously differentiable functions of the nuclear coordinates Q, except at the conical intersection Q_0, where the divergence occurs. To reverse a famous argument of Longuet-Higgins [7], suppose that a sign change were observed for an arbitrarily small path C around Q_0, on which the linear approximation (20) is valid, but not around some larger loop L, which excludes other degeneracies. Now, imagine a continuous expansion and deformation that takes C into L, parameterized by a monotonically increasing parameter λ. There must be some point λ_0, at which $|x_-(Q)\rangle$, say, is sign reversing on $C(\lambda_0)$ but sign preserving on $C(\lambda_0 + d\lambda)$. In other words, the change from sign reversing to sign preservation on the larger loop requires the smoothly continuous function $|x_-(Q)\rangle$ to undergo a discontinuous change at λ_0—a logical impossibility.

Longuet-Higgins [7] actually uses the argument in reverse to infer the logical existence of conical intersections, from the observation of sign changes around arbitrary loops, a test that is now widely used to detect the existence of conical intersections between ab initio potential energy surfaces [23]. A generalization of the Longuet-Higgins argument to the case of a spin–orbit coupled doublet has been given by Stone [24]. As discussed above [see Eq. (13)] the Hamiltonian matrix is then intrinsically complex, and there are no real adiabatic eigenstates. Nevertheless one can still find "parallel transported" states $|x_\pm\rangle$, with vanishing diagonal elements, as in Eq. (22), which acquire a variable phase change, according to the radius of the encircling loop. The conical intersection is now removed by spin–orbit coupling, but it's influence is still apparent in simple sign changes of $|x_\pm\rangle$ around very large loops. The difference from the Longuet-Higgins case is that the phase change falls to zero on very small circles around the maximum on the lower adiabatic surface. This situation is further discussed in Section V.B.

Longuet-Higgins [7] also reinforces the discussion by the following qualitative demonstration of a cyclic sign change for the LiNaK like system subject to Eq. (3), in which rows and columns are labeled by the basis functions

$$
\begin{aligned}
{}^2\Psi_1 &= \frac{1}{\sqrt{2}}(\psi_B - \psi_C) \\
{}^2\Psi_2 &= \frac{1}{\sqrt{6}}(-2\psi_A + \psi_B + \psi_C)
\end{aligned}
\tag{26}
$$

where $\psi_A = (\bar{a}bc)$, and so on, with the β spin on atom A. Thus ${}^2\Psi_1$ may be recognized as the Heitler–London ground state of BC in the "reactant" A + BC geometry, at which $\beta = \gamma = 0$. Second, there is also a "transition state" geometry B–A–C at which $\alpha < \beta = \gamma$, where the lower eigenstate goes over to ${}^2\Psi_2$. The table below follows changes in the ground-state wave function as the system proceeds through various permutations of the three possible reactant and transition state geometries, subject to the constraint that the overlap from one step to the next is positive.

Geometry	Ground-State Wave Function
A + BC	$\frac{1}{\sqrt{2}}(\psi_B - \psi_C)$
A–B–C	$\frac{1}{\sqrt{6}}(2\psi_B - \psi_A - \psi_C)$
AB + C	$\frac{1}{\sqrt{2}}(\psi_B - \psi_A)$
B–A–C	$\frac{1}{\sqrt{6}}(-2\psi_A + \psi_B + \psi_C)$
B + AC	$\frac{1}{\sqrt{2}}(-\psi_A + \psi_C)$
B–C–A	$\frac{1}{\sqrt{6}}(-\psi_A - \psi_B + 2\psi_C)$
BC + A	$\frac{1}{\sqrt{2}}(-\psi_B + \psi_C)$

Comparison between the first and last lines of the table shows that the sign of the ground-state wave function has been reversed, which implies the existence of a conical intersection somewhere inside the loop described by the table.

IV. GEOMETRIC PHASE

While the presence of sign changes in the adiabatic eigenstates at a conical intersection was well known in the early Jahn–Teller literature, much of the discussion centered on solutions of the coupled equations arising from non-adiabatic coupling between the two or more nuclear components of the wave function in a spectroscopic context. Mead and Truhlar [10] were the first to focus on the consequences for both scattering and spectroscopy on a single adiabatic electronic energy surface, influenced by, but well separated from a conical intersection (see Section VI). Berry [8], who coined the term *geometric phase*, set the argument in a more general context. Given the existence of an infinity of phase modified adiabatic eigenstates of any given problem, the questions at issue are

1. Whether there are any physical invariants of the system, independent of phase modifications.
2. How such invariants can be computed.

Berry [8] starts by assuming the existence of a *single-valued* adiabatic eigenstate $|n(Q)\rangle$, such as that in Eq. (24), subject to

$$H(Q)|n(Q)\rangle = E_n(Q)|n(Q)\rangle \qquad \langle m|n\rangle = \delta_{nm} \tag{27}$$

Solutions of the time-dependent Schrödinger equation

$$i\hbar \frac{d|\Psi(Q(t))\rangle}{dt} = H(Q(t))|\Psi(Q(t))\rangle \tag{28}$$

are sought then in the form

$$|\Psi(Q(t))\rangle = |n(Q(t))\rangle e^{i\gamma(t)-(i/\hbar)\int E_n(Q(t))dt} \qquad \gamma(0) = 0 \tag{29}$$

as the system is taken slowly round a time dependent path $Q(t)$. It readily follows from Eq. (28) and (29) that

$$\nabla_{\mathbf{Q}}|n(Q)\rangle \cdot \dot{\mathbf{Q}} + i\frac{d\gamma}{dt}|n(Q)\rangle = 0 \tag{30}$$

from which it follows by integrating around a closed path C in parameter space that

$$\gamma_C = \gamma(T) - \gamma(0) = i\oint_C \langle n|\nabla_{\mathbf{Q}} n\rangle \cdot \dot{\mathbf{Q}}\, dt = i\oint_C \langle n|\nabla_{\mathbf{Q}} n\rangle \cdot \mathbf{dQ} \tag{31}$$

It should be noted, by taking the gradient of the normalization identity that

$$\langle n|\nabla_{\mathbf{Q}} n\rangle = -\langle \nabla_{\mathbf{Q}} n|n\rangle = -\langle n|\nabla_{\mathbf{Q}} n\rangle^*. \tag{32}$$

In other words, $\langle n|\nabla_{\mathbf{Q}} n\rangle$ is imaginary, making γ_C real. As an illustrative example, $|n\rangle$ may assumed to be given by Eq. (24), in which case

$$\gamma_C = i\oint_C \langle n|\nabla_{\mathbf{Q}} n\rangle \cdot \mathbf{dQ} = -\oint_C \nabla_{\mathbf{Q}}\psi \cdot \mathbf{dQ} = -[\psi(T) - \psi(0)] = \mp\pi \tag{33}$$

The sign of γ_C is actually indeterminate for this particular model because the quantity of physical interest is $e^{i\gamma_C} = e^{\mp i\pi} = -1$.

Equation (31) is the fundamental geometric phase formula. It is termed geometric for two reasons. First, the combination of $|\nabla_{\mathbf{Q}} n\rangle$ and $\dot{\mathbf{Q}}\,dt$ in the central term ensures that γ_C depends only on the geometry of the path C—not on the rate at which it is traversed. Second, it is gauge invariant, in the sense that multiplication of any single-valued eigenstate $|n\rangle$ by a phase factor $e^{i\Delta\psi}$, such that $\Delta\psi(T) = \Delta\psi(0)$ leaves γ_C invariant. All single-valued solutions of Eq. (27) have the same geometric phase γ_C. The arbitrariness in $\psi(Q)$ allows, however, for different manifestations of Eq. (31). For example, the choice $\psi = -\phi/2$, coupled with Eq. (25) for the linear $E \times \epsilon$ model allows the identity

$$\langle n|\nabla_{\mathbf{Q}}|n\rangle = i\nabla_{\mathbf{Q}}\psi = -i\frac{\mathbf{e}_\phi}{2r} = -i\langle x_-|\nabla_{\mathbf{Q}}|x_+\rangle \tag{34}$$

so that Eq. (31) may be expressed as

$$\gamma_C = \oint_C \langle x_-|\nabla_{\mathbf{Q}}|x_+\rangle \cdot \mathbf{dQ} \tag{35}$$

Phase factors of this type are employed, for example, by the Baer group [25,26]. While Eq. (34) is strictly applicable only in the immediate vicinity of the conical intersection, the continuity of the non-adiabatic coupling, discussed in Section III, suggests that the integrated value of $\langle x_-|\nabla_{\mathbf{Q}}|x_+\rangle$ is independent of the size or shape of the encircling loop, provided that no other conical intersection is encountered. The mathematical assumption is that there exists some phase function, $\psi(Q)$, such that

$$\nabla_{\mathbf{Q}}\psi = -\langle x_-|\nabla_{\mathbf{Q}}|x_+\rangle \tag{36}$$

a condition that requires that $\nabla_{\mathbf{Q}} \times \langle x_-|\nabla_{\mathbf{Q}}|x_+\rangle = 0$, because $\nabla_{\mathbf{Q}} \times \nabla_{\mathbf{Q}}\psi = 0$ for any function $\psi(Q)$. Equation (34) ensures that this curl condition is satisfied for the linear $E \times \epsilon$ model, but it would not be valid for the spin $\frac{1}{2}$ model discussed below, for example (or for the isomorphous ${}^2E \times \epsilon$ model discussed in Section V.B), for which the adiabatic eigenstates are intrinsically complex.

Other forms for γ_C are also available. Consider, for example, the phase modification

$$|n(Q)\rangle = e^{i\psi(Q)}|x(Q)\rangle \tag{37}$$

so that $\psi[Q(T)] - \psi[Q(0)] = \arg\langle x[Q(0)]|x[Q(T)]\rangle$, because $\langle n[Q(0)]|n[Q(T)]\rangle = 1$ [i.e., $|n(Q)\rangle$ is single valued]. It follows from Eq. (31) that

$$\gamma_C = \arg\langle x[Q(0)]|x[Q(T)]\rangle + i\oint_C \langle x|\nabla_{\mathbf{Q}}x\rangle \cdot \mathbf{dQ} \tag{38}$$

which applies to a quite general adiabatic eigenstate $|x(Q)\rangle$. At one extreme, $|x\rangle$ is single-valued and Eq. (38) reverts to Eq. (31), while at the opposite limit $|x\rangle$ is real, $\langle x|\nabla_{\mathbf{Q}}x\rangle$ vanishes and γ_C takes values 0 or π according to whether $|x\rangle$ evolves to $\pm|x\rangle$.

Another form

$$\gamma_C = i\int\int \nabla_{\mathbf{Q}} \times \langle n|\nabla_{\mathbf{Q}}n\rangle \cdot \mathbf{dS} \tag{39}$$

with the integral taken over an area enclosed by the contour C, was obtained by Berry [8] by applying Stokes theorem [27] to the line integral in Eq. (31). Care is, however, required to ensure that the chosen surface excludes the conical intersection point, Q_0, where $\langle n|\nabla_{\mathbf{Q}}n\rangle$ diverges, because Stokes theorem requires that $\langle n|\nabla_{\mathbf{Q}}n\rangle$ should be continuously differentiable over the surface.

A variant of Eq. (39), with the integral taken over a surface bounded by a path C' that excludes Q_0, is illuminating in situations where $|n\rangle$ is given by Eq. (37), with $|x\rangle$ real. One then finds that

$$\gamma_{C'} = i\int\int \nabla_{\mathbf{Q}} \times \nabla_{\mathbf{Q}}\psi \cdot \mathbf{dS'} = \mathbf{0} \tag{40}$$

because $\nabla_{\mathbf{Q}} \times \nabla_{\mathbf{Q}}\psi = 0$ for any function $\psi(Q)$. This means that the value of γ_C is independent of the size and shape of the path C, provided that no degenerate points, other than Q_0 are enclosed, because any distortion of C can be interpreted as taking in an additional area over which the integrals in Eqs. (31) and (39) both vanish. This Stokes theorem argument confirms the earlier topological conclusions applicable to real adiabatic eigenstates $|x\rangle$.

A third expression may be obtained by taking the curl inside the bracket in Eq. (39) and using the identities

$$|\nabla_{\mathbf{Q}}n\rangle = \sum |m\rangle\langle m|\nabla_{\mathbf{Q}}n\rangle \tag{41}$$

$$\langle m|\nabla_{\mathbf{Q}}n\rangle = -\frac{\langle m|\nabla_{\mathbf{Q}}H|n\rangle}{E_m(Q) - E_n(Q)} \tag{42}$$

to yield [8,10]

$$\gamma_C = i \iint \sum_{m \neq n} \frac{\langle n|\nabla_{\mathbf{Q}} H|m\rangle \times \langle m|\nabla_{\mathbf{Q}} H|n\rangle}{[E_m(Q) - E_n(Q)]^2} \cdot \mathbf{dS} \tag{43}$$

The sum excludes $m = n$, because the derivation involves the vector product of $\langle n|\nabla_{\mathbf{Q}} H|n\rangle$ with itself, which vanishes. The advantage of Eq. (43) over Eq. (31) is that the numerator is independent of arbitrary phase factors in $|n\rangle$ or $|m\rangle$; neither need be single valued. On the other hand, Eq. (43) is inapplicable, for the reasons given above if the degenerate point lies on the surface S.

Consequently, Eq. (43) is inapplicable to the model of Eq. (20), because the eigenstates, given by Eq. (21) or (24), are only defined in the (x, y) plane, which contains the degeneracy. On the other hand, Berry [8] extends the model in the form

$$H = \frac{1}{2}\begin{pmatrix} z & x - iy \\ x + iy & -z \end{pmatrix} = x\sigma_x + y\sigma_y + z\sigma_z \tag{44}$$

where the components of the vector σ are the Pauli spin matrices. Thus

$$\nabla_{\mathbf{Q}} H = (\sigma_x, \sigma_y, \sigma_z) \tag{45}$$

Moreover, because there are only two eigenstates, it follows from the completeness property, the vanishing of $\langle n|\nabla_{\mathbf{Q}} H|n\rangle$ and the angular momentum commutation relations that

$$\langle n|\nabla_{\mathbf{Q}} H|m\rangle \times \langle m|\nabla_{\mathbf{Q}} H|n\rangle = \langle n|\nabla_{\mathbf{Q}} H \times \nabla_{\mathbf{Q}} H|n\rangle = i\langle n|\sigma|n\rangle \tag{46}$$

The level splitting for this model is $E_m(Q) - E_n(Q) = \sqrt{x^2 + y^2 + z^2} = r$ and the eigenstates may be taken as

$$|n_+\rangle = \begin{pmatrix} \cos\frac{\theta}{2} \\ \sin\frac{\theta}{2} e^{i\phi} \end{pmatrix} \qquad |n_-\rangle = \begin{pmatrix} -\sin\frac{\theta}{2} \\ \cos\frac{\theta}{2} e^{i\phi} \end{pmatrix} \tag{47}$$

It is readily verified that

$$\langle n_\pm|\sigma|n_\pm\rangle = \pm\frac{\hat{\mathbf{r}}}{2} \tag{48}$$

where $\hat{\mathbf{r}}$ is a unit vector perpendicular to the surface of the sphere, used for evaluation of the surface integral. One readily finds, by use of Eqs. (43)–(47), that

$$\gamma_C = \pm i \iint_S \frac{i\langle n|\sigma|n\rangle}{r^2} \cdot \mathbf{dS} = -\frac{1}{2}\iint_S d\Omega \tag{49}$$

where $d\Omega$ is the solid angle element. The choice of a contour at constant z or θ therefore yields

$$\gamma_C = \pm(1 - \cos\theta)\pi \tag{50}$$

which reduces to Eq. (33) for $\theta = \pi/2$, a comparison that is justified by noting that the model (44) with $\theta = \pi/2$ reduces to that in Eq. (20), in a complex rather than a real representation. The factors $e^{i\gamma_C}$ for the two states, which could be obtained more directly by substitution from Eq. (47) in Eq. (31), now take different values for $\theta \neq \pi/2$.

There is also an interesting alternative approach by Aharonov et al. [18], who start by using projection operators, $\Pi_n = |n\rangle\langle n|$ to partition the Hamiltonian

$$H = \frac{\mathbf{P}^2}{2m} + H_1(\mathbf{q}; \mathbf{Q}) \tag{51}$$

between the adiabatic eigenstates $|n\rangle$ of $H_1(q; r)$, rather than immediately assuming an adiabatic representation. Since $\mathbf{P}$ does not commute with the Π_n, products such as $\Pi_n \mathbf{P}^2 \Pi_n$ must be interpreted as

$$\Pi_n \mathbf{P}^2 \Pi_n = \Pi_n \mathbf{P} \cdot \sum_m \Pi_m \mathbf{P} \Pi_n \tag{52}$$

an expression that can be simplified by decomposing $\mathbf{P}$ into two parts; a part $\mathbf{P} - \mathbf{A}$ that acts only within an adiabatic subspace and a part $\mathbf{A}$ that causes non-adiabatic transitions. Thus

$$[(\mathbf{P} - \mathbf{A}), \Pi_n] = 0 \tag{53}$$

while ambiguity in $\mathbf{A}$ can be removed by requiring that

$$\Pi_n \mathbf{A} \Pi_n = 0 \tag{54}$$

In other words, $\mathbf{A}$ is a strictly off-diagonal operator that can be evaluated as the difference between $\mathbf{P}$ and its diagonal parts

$$\mathbf{A} = \mathbf{P} - \sum_m \Pi_m \mathbf{P} \Pi_m = \frac{1}{2} \sum_m [\Pi_m, [\Pi_m, \mathbf{P}]] \tag{55}$$

The operation of $(1/2m)\mathbf{P}^2$ within any particular subspace may therefore be represented as

$$\begin{aligned}
\frac{1}{2m} \Pi_n \mathbf{P}^2 \Pi_n &= \frac{1}{2m} \Pi_n (\mathbf{P} - \mathbf{A} + \mathbf{A})^2 \Pi_n \\
&= \frac{1}{2m} \left[\Pi_n (\mathbf{P} - \mathbf{A}) \Pi_n (\mathbf{P} - \mathbf{A}) \Pi_n + \sum_m \Pi_n \mathbf{A} \Pi_m \mathbf{A} \Pi_n \right] \\
&= \frac{1}{2m} [\Pi_n (\mathbf{P} - \mathbf{A})^2 \Pi_n + \Pi_n \mathbf{A}^2 \Pi_n]
\end{aligned} \tag{56}$$

Equations (53) and (54) are used to perform the manipulations in (56).

The strengths of this approach are that the operator **A** is gauge invariant and that equation (55) can be employed for its computation, regardless of the number of components. To see the connection with geometric phase, arguments given by Stone and Goff [28] show that **A** determines a field strength operator with components

$$F_{ij} = -i[P_i - A_i, P_j - A_j] \tag{57}$$

the diagonal elements of which determine the phase change $\Delta\gamma$ over a loop $\Delta r^i \Delta r^j$ in parameter space, in the form [18]

$$\Delta\gamma_n = \Delta r^i \Delta r^j \langle n|F_{ij}|n\rangle = i\Delta r^i \Delta r^j \langle n|[A_i, A_j]|n\rangle \tag{58}$$

In three-dimensional (3D) applications the overall phase change over a cycle may therefore be expressed as a surface integral, analogous to Eq. (43), namely,

$$\gamma_n = i\int\!\!\int_S \langle n|[A_i, A_j]|n\rangle \, dS^{ij} \tag{59}$$

Comparison with Eq. (43) is illuminating. By the method of construction, the matrix elements of **A** are identical with the off-diagonal elements of **P**; thus, with the help of Eqs. (41) and (42)

$$\mathbf{A}|n\rangle = \sum_{m\neq n} |m\rangle\langle m|\mathbf{A}|n\rangle = -i\sum_{m\neq n} |m\rangle\langle m|\nabla_Q|n\rangle = i\sum_{m\neq n} \frac{|m\rangle\langle m|\nabla_{\mathbf{Q}}H|n\rangle}{E_m(Q) - E_n(Q)} \tag{60}$$

Consequently, Eqs. (43) and (59) are identical, for applications in a 3D parameter space, except that the vector product in the former is expressed as a commutator in the latter. Both are computed as diagonal elements of combinations of strictly off-diagonal operators; and both give gauge independent results. Equally, however, both are subject to the limitations with respect to the choice of surface for the final integration that are discussed in the sentence following Eq. (43).

Equations (31)–(43) assume a 3D parameter space, Q, although the gradient $\nabla_{\mathbf{Q}}|n\rangle$ has an obvious generalization to higher dimensions. Further generalizations, to include the curl, transform equation (38) into the integral of a two form over the surface bounded by C, this two form being obtained by replacing ∇ by the exterior derivative d and $\times$ by the wedge product $\wedge$ of the theory of differential forms [29].

V. THE $E \times \epsilon$ JAHN–TELLER PROBLEM

The $E \times \epsilon$ Jahn–Teller problem, described by Eq. (7) or (8), plus an additional nuclear term $h_0(Q)$, common to the two electronic states, is the prototype for all subsequent discussions. In its linear variant, with $l = 0$ in the above equations, it provided the first example of geometric phase, plus a less familiar half-integral quantum number [2]. The effects of spin–orbit coupling on geometric phase [21,24] are also conveniently illustrated. Addition of the quadratic terms in Eq. (7) or (8) is of interest in introducing a threefold corrugation on the lower adiabatic potential surface leading to an "inverted" pattern of vibronic multiplets (E levels below A, in the lowest triplet), which is one of the clearest experimental manifestations of geometric phase [11]. There is also an interesting question concerning the relative magnitudes of the linear and quadratic terms in Eqs. (7) and (8). We shall find that there is no geometric phase effect unless $k \neq 0$, which raises questions as to the nature of its disappearance as $k/l \to 0$.

A. The Linear Jahn–Teller Effect

It is convenient to discuss the linear Jahn–Teller model in the scaled complex representation

$$H = \begin{pmatrix} h_0 & kre^{-i\phi} \\ kre^{i\phi} & h_0 \end{pmatrix} \tag{61}$$

where

$$h_0 = -\frac{1}{2r}\frac{\partial}{\partial r}\left(r\frac{\partial}{\partial r}\right) - \frac{1}{2r^2}\frac{\partial^2}{\partial \phi^2} + \frac{1}{2}r^2 \tag{62}$$

rather than in the real representation in Eq. (20). It is readily verified, by ignoring the kinetic energy terms, that the eigensurfaces take the form

$$W_{\pm} = \frac{1}{2}r^2 \pm kr \tag{63}$$

with single-valued eigenstates

$$|n_{\pm}\rangle = \frac{1}{\sqrt{2}}\begin{pmatrix} 1 \\ \pm e^{i\phi} \end{pmatrix} \tag{64}$$

Substitution in Eq. (33) therefore yields $\gamma_C = -\pi$, in agreement with the result obtained from the real representation of the Hamiltonian in Eq. (20).

Figure 1a shows that the eigensurfaces form an interconnected double sheet, the lower member of which has a ring of equivalent minima at $r = k$ and $W_- = -\frac{1}{2}k^2$. As expected angular momentum is conserved, but with the complication that it is vibronic, rather than purely vibrational in character,

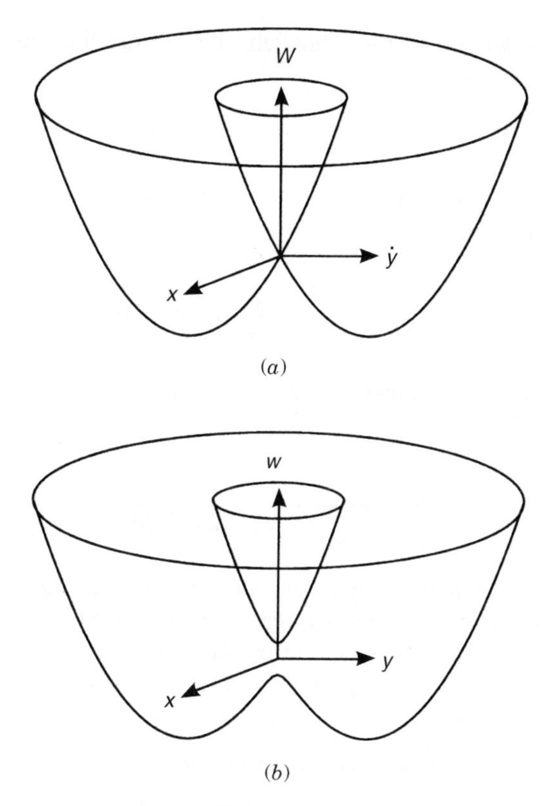

Figure 1. Adiabatic potential surfaces (*a*) for the linear $E \times \epsilon$ case and (*b*) for a ^2E state with linear Jahn–Teller coupling and spin–orbit coupling to a ^2A state.

because it may be confirmed that the operator

$$\hat{j} = \hat{l}_z + \sigma_z \qquad \hat{l}_z = -i\frac{\partial}{\partial\phi} \tag{65}$$

commutes with H; and $\hat{j}$ includes an electronic component, σ_z, as well as the vibrational term $\hat{l}_z$. The single valued eigenstates of $\hat{j}$, belonging to the upper and lower potential surfaces, may be obtained by multiplying Eq. (64) by $e^{i(j-1/2)\phi}$; thus

$$|u_{j\pm}(\phi)\rangle = \frac{1}{\sqrt{2}}\begin{pmatrix} e^{i(j-1/2)\phi} \\ \pm e^{i(j+1/2)\phi} \end{pmatrix} \qquad j = \frac{1}{2}, \frac{3}{2}, \frac{5}{2} \cdots \tag{66}$$

They must be coupled by separate radial factors in a full calculation [2] but, to the extent that non-adiabatic coupling between the upper and lower surfaces is neglected, the total lower adiabatic wave function may be expressed as

$$|\Psi_-\rangle = r^{-1/2}|u_{j-}(\phi)\rangle|v_-(r)\rangle \tag{67}$$

with the radial wave function on the lower adiabatic surface, $|v_-(r)\rangle$, taken as an eigenstate of the radial operator

$$\hat{h}_r = -\frac{1}{2r}\frac{\partial^2}{\partial r^2} + \frac{j^2}{2r^2} + \frac{1}{2}r^2 - kr \tag{68}$$

For large k, the approximate potential minimum lies at $r = k$ and the lower vibronic eigenvalues are given by [2]

$$E_{vj} = -\frac{k^2}{2} + \left(v + \frac{1}{2}\right) + \frac{j^2}{2k^2} \tag{69}$$

The presence of the half-odd quantum number j in Eq. (69) is potentially a physically measurable consequence of geometric phase, which was first claimed to have been detected in the spectrum of Na_3 [16]. The situation is, however, quite complicated and the first unambiguous evidence for geometric phase in Na_3 was reported only in 1999 [17].

B. Spin–Orbit Coupling in a ^2E State

The effects of spin–orbit coupling on geometric phase may be illustrated by imagining the vibronic coupling between the two Kramers doublets arising from a ^2E state, spin–orbit coupled to one of symmetry ^2A. The formulation given below follows Stone [24]. The four ^2E components are denoted by $|e_+\alpha\rangle$, $|e_-\alpha\rangle$, $|e_+\beta\rangle$, $|e_-\beta\rangle$, and those of ^2A by $|e_0\alpha\rangle$, $|e_0\beta\rangle$. The spin–orbit coupling operator has nonzero matrix elements

$$\langle e_+\beta|H_{so}|e_0\alpha\rangle = \langle e_-\alpha|H_{so}|e_0\beta\rangle \tag{70}$$

giving rise to a second-order splitting, of say 2Δ, between one Kramers doublet, $|e_+\alpha\rangle$, $|e_-\beta\rangle$, and the other, $|e_-\alpha\rangle$, $|e_+\beta\rangle$. There is also a spin-preserving vibronic coupling term, of the form in Eq. (61), giving rise to a Hamiltonian of the form

$$H = \begin{pmatrix} h_0 + \Delta & kre^{-i\phi} \\ kre^{i\phi} & h_0 - \Delta \end{pmatrix} \tag{71}$$

for one coupled pair and the complex conjugate form for the other. Notice that Eq. (71) conforms to Eq. (13) with $w = \Delta$, $u = kre^{-i\phi}$, and $v = 0$. The eigensurfaces now take the forms

$$W_{\pm} = \frac{1}{2}r^2 \pm \sqrt{k^2r^2 + \Delta^2} \tag{72}$$

with an avoided conical intersection, as shown in Figure 1*b*.

It is convenient, for comparison with Section V.A, to employ substitutions

$$\Delta = \rho(r)\cos\theta(r) \qquad kr = \rho(r)\sin\theta(r) \qquad \tan\theta(r) = \frac{kr}{\Delta} \tag{73}$$

which convert the Hamiltonian in Eq. (71) to the form in Eq. (44). Comparison with Eq. (50) shows that the geometric phase, for a cycle of constant radius, r, is given by

$$\gamma_C = -(1 - \cos\theta(r))\pi \tag{74}$$

It reverts to the unspin–orbit modified value, $\gamma_C = -\pi$, for paths such that $kr \gg \Delta$, but vanishes as $r \to 0$.

Reverting to the vibronic structure, the operator $\hat{\jmath}$ again commutes with $\hat{H}$, and the analogue of the lower adiabatic eigenstate of $\hat{\jmath}$ in Eq. (66) becomes

$$|u_{j-}(r,\phi)\rangle = \frac{1}{\sqrt{2}}\begin{pmatrix} -\sin\frac{\theta}{2}e^{i(j-1/2)\phi} \\ \cos\frac{\theta}{2}e^{i(j+1/2)\phi} \end{pmatrix} \qquad j = \frac{1}{2}, \frac{3}{2}, \frac{5}{2} \cdots \tag{75}$$

where the r dependence of $|u_{j-}(r,\phi)\rangle$ comes from that of $\theta(r)$. There is also an equivalent complex conjugate eigenstate of the complex conjugate Hamiltonian to that in Eq. (71). One finds after some manipulation that

$$\begin{aligned} &\langle u_{j\_}(r,\phi)|h_0|r^{-1/2}u_{j\_}(r,\phi)\rangle \\ &\quad = r^{-1/2}\left\{-\frac{1}{2}\frac{\partial^2}{\partial r^2} + \frac{j^2 + j\cos\theta}{2r^2} + W_-(r) + \frac{1}{8}\left(\frac{d\theta}{dr}\right)^2\right\} \end{aligned} \tag{76}$$

The radial factor in the total wave function

$$|\Psi_-\rangle = r^{-1/2}|u_{j-}(\phi)\rangle|v_-(r)\rangle \tag{77}$$

must therefore be an eigenstate of

$$\hat{h}_r = -\frac{1}{2}\frac{\partial^2}{\partial r^2} + \frac{j^2 + j\cos\theta}{2r^2} + W_-(r) + \frac{1}{8}\left(\frac{d\theta}{dr}\right)^2 \tag{78}$$

The principal differences from Eq. (68) lie in the form of the potential $W_{-}(r)$ and in the presence of the term $j\cos\theta$, of which the latter arises from the dependence of the geometric phase on the radius of the encircling path. The eigenvalues of $\hat{h}_r$ are no longer doubly degenerate, but a precisely equivalent Kramer's twin radial Hamiltonian may be derived from the complex conjugate of Eq. (71).

C. The Quadratic Jahn–Teller Effect

The quadratic Jahn–Teller effect is switched on by including the quadratic terms in Eq. (7); thus, with the inclusion of the additional diagonal Hamiltonian h_0,

$$H = \begin{pmatrix} h_0 & kre^{-i\phi} + lr^2 e^{2i\phi} \\ kre^{i\phi} + lr^2 e^{-2i\phi} & h_0 \end{pmatrix} \tag{79}$$

The eigensurfaces are given by

$$W_{\pm}(r,\phi) = \frac{1}{2}r^2 \pm r\sqrt{k^2 + 2klr\cos 3\phi + l^2 r^2} \tag{80}$$

with a threefold corrugation around the minimum of $W_{-}(r)$, in place of the line of continuous minima in Figure 1. The three absolute minima in Figure 2, at $\phi = 0, \pm 2\pi/3$, correspond to three equivalent isosceles distortions of an initially equilateral triangular molecule.

There is no simple general form for the adiabatic eigenvectors, except in the limits, $k = 0$ and $l = 0$, when, for example,

$$\begin{aligned} |x_{-}\rangle &= \begin{pmatrix} e^{-i\phi/2} \\ -e^{i\phi/2} \end{pmatrix} \qquad l = 0 \\ &= \begin{pmatrix} e^{-i\phi} \\ -e^{i\phi} \end{pmatrix} \qquad k = 0 \end{aligned} \tag{81}$$

In the first case, $|x_{-}\rangle$ changes sign as ϕ increases to $\varphi + 2\pi$, while in the second, $|x_{-}\rangle$ is single valued. There is therefore a geometric phase of $\pm\pi$, for $lr \ll k$, but no geometric phase in the opposite limit, $lr \gg k$. The interesting questions concern (1) the effects of the corrugations on the vibronic eigenvalues; and (2) the origin of the change in geometric phase behavior as the ratio lr/k increases.

The first of these questions is deferred to Section VI. The second is addressed by considering the degeneracy condition $W_{+}(r,\phi) = W_{-}(r,\phi)$. One solution lies at $r = 0$, and there are three others at $r = k/l$ and $\phi = \pi, \pm\pi/3$ [30,31]. A circuit of ϕ with $r < k/l$ therefore encloses a single degenerate point, which accounts for the "normal" sign change, $e^{\pm i\pi} = -1$, whereas as circuit with $r > k/l$ encloses four degenerate points, with no sign change because $e^{\pm 4i\pi} = 1$.

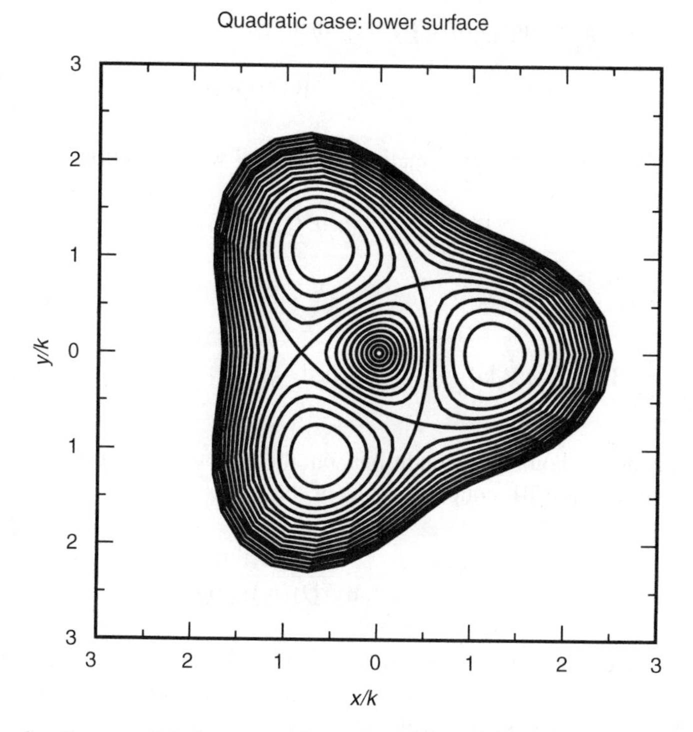

Figure 2. Contours of the lower potential surface in the quadratic $E \times \epsilon$ Jahn–Teller case.

Any proper treatment of the dynamics, including motion in the r variable therefore requires knowledge of the position of the minima of $W_{-}(r, \phi)$, which are found to lie at $r = k/(1 - 2l^2)$ [units are dictated by the form of the scaled restoring term $r^2/2$ in Eq. (80)]. The potential minima therefore lie inside the critical circle $r = k/l$ if $l < 1/\sqrt{3}$ and outside it if the sense of the inequality is reversed. Single surface dynamics, in the sense to be discussed below, may therefore be assumed to apply with a geometric phase of π if $l \ll 1/\sqrt{3}$, and with no geometric phase if $l \gg 1/\sqrt{3}$. Cases with $l \simeq 1/\sqrt{3}$, with significant wave function amplitude at the degenerate points with $r = k/l$, cannot be validly treated in an adiabatic approximation.

VI. SINGLE-SURFACE NUCLEAR DYNAMICS

Given the full-Hamiltonian

$$H(q, Q) = \sum \frac{\hat{P}_i^2}{2m_i} \nabla_{Q_i}^2 + H_{\text{el}}(q; Q) \tag{82}$$

and adiabatic eigenstates $|n(q;Q)\rangle$, such that

$$H_{\rm el}(q;Q)|n(q;Q)\rangle = W(Q)|n(q;Q)\rangle \tag{83}$$

the Born–Oppenheimer approximation to the total wave function,

$$|\Psi(q,Q)\rangle = |n(q;Q)\rangle|v(Q)\rangle \tag{84}$$

requires that

$$\left(\sum \frac{1}{2m_i}[\hat{\mathbf{P}}_i^2 + 2\langle n|\hat{\mathbf{P}}_i|n\rangle \cdot \hat{\mathbf{P}}_i + \langle n|\hat{\mathbf{P}}_i^2|n\rangle] + W(Q)\right)|v(Q)\rangle = E|v(Q)\rangle \tag{85}$$

with appropriate boundary conditions on the vibrational factors $|v(Q)\rangle$. As discussed in Section III, coupling terms of the form

$$\langle n|\nabla_Q|m\rangle = \frac{\langle n|\nabla_Q H_{\rm el}|m\rangle}{W_n(Q) - W_m(Q)} \tag{86}$$

have been neglected in the derivation of Eq. (85). The assumption is that the wave function has negligible amplitude in the vicinity of any points at which $W(Q)$ has a close degeneracy with any other eigensurface.

Geometric phase complications necessarily arise, however, whenever the nuclear wave function has significant amplitude on a loop around an isolated degeneracy. They can be treated in two ways, according to whether the adiabatic eigenstate $|n(q;Q)\rangle$ is taken to be multivalued or single-valued around the loop in nuclear coordinate space Q. Illustrations are given below for the two different approaches. The first concerns the energy ordering of the vibronic eigenstates arising from a strong quadratic Jahn–Teller effect [11]. The second outlines the vector potential approach, due to Mead and Truhlar [10], with applications to the above $E \times \epsilon$ linear Jahn–Teller problems and to scattering problems involving identical nuclei.

A. The Ordering of Vibronic Multiplets

It was seen in Section V.C that quadratic Jahn–Teller coupling terms result in a threefold corrugation around the minimum energy path on the lower potential surface $W_-(Q)$ and that there is a geometric phase, $\gamma_C = \pi$, provided that the radius of the minimum energy path satisfies $r < k/l$. We now consider the influence of geometric phase on the relative energies of the (A, E) symmetry levels in different tunneling triplets. The solution, due to Ham [11], applies band theory arguments to assess the influence of antiperiodic, $\psi(\phi + 2\pi) = -\psi(\phi)$, boundary conditions on solutions of the threefold periodic angular equation

$$\left\{\frac{\hbar^2}{2m}\frac{d^2}{d\phi^2} + E - V(\phi)\right\}\psi(\phi) = 0 \qquad V\left(\phi + \frac{2\pi}{3}\right) = V(\phi) \tag{87}$$

Note that there is no first derivative term in Eq. (87), because the first line of Eq. (81) ensures that $\langle x|\partial/\partial\phi|x\rangle = 0$.

The first strand of Ham's argument [11] is that $V(\phi)$ supports continuous bands of Floquet states, with wave functions of the form

$$\psi_k(\phi) = e^{ik(E)\phi}\xi(\phi) \tag{88}$$

where $\xi(\phi)$ has the same periodicity as $V(\phi)$ [32]. Elements of Floquet theory, collected in the appendix, show that the spectrum is bounded by $-\frac{3}{2} < k \leqslant \frac{3}{2}$, and that the dispersion curves, $E(k)$ obtained by inversion of $k(E)$ in Eq. (88), have turning points at $k = 0$ and $k = \frac{3}{2}$.

A second constraint is that the relative order of the critical energies at $k = 0$ and $k = \frac{3}{2}$ is invariant to the presence or absence of the potential $V(\phi)$ [11]. Equation (A.6) shows that the free motion band structure can be folded onto the interval $-\frac{3}{2} < k \leqslant \frac{3}{2}$. Consequently, preservation of relative energy orderings at $k = 0$ and $k = \frac{3}{2}$ implies a band structure for $V(\phi) \neq 0$, with the form shown in Figure 3.

The question of vibronic energy ordering, with and without geometric phase, now turns on the appropriate values of k in Eq. (88), bearing in mind that all energy levels are doubly degenerate except those at $k = 0$ and $k = \frac{3}{2}$. Normal periodic boundary conditions require integral k, with $E(0) < E(\pm 1)$, in the lowest energy band. However, introduction of a sign change in $\psi_k(\phi)$, to compensate the electronic geometric phase factor, introduces half odd-integral values of k, with $E(\pm\frac{1}{2}) < E(\frac{3}{2})$. This ordering is seen from Figure 3 to be reversed and restored in the successively excited bands. It may also be noted that an explicit calculation of the lowest 89 vibrational levels of Na_3 [33] confirms that the ordering of vibronic energy levels is the clearest observable molecular manifestation of geometric phase.

B. Vector-Potential Theory: The Molecular Aharonov–Bohm Effect

Mead and Truhlar [10] broke new ground by showing how geometric phase effects can be systematically accommodated in scattering as well as bound state problems. The assumptions are that the adiabatic Hamiltonian is real and that there is a single isolated degeneracy; hence the eigenstates $|n(q;Q)\rangle$ of Eq. (83) may be taken in the form

$$|n(q;Q)\rangle = e^{i\psi(Q)}|x(q;Q)\rangle \tag{89}$$

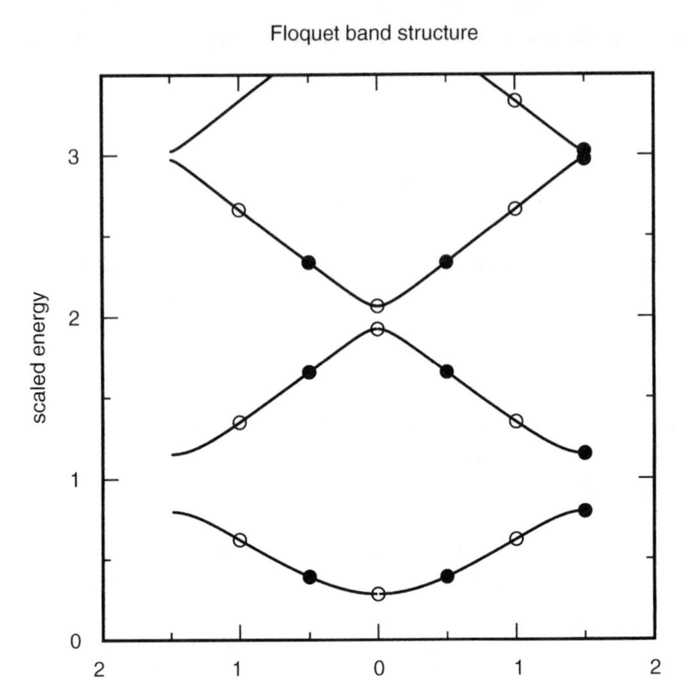

Figure 3. Floquet band structure for a threefold cyclic barrier (*a*) in the plane wave case after using Eq. (A.11) to fold the band onto the interval $-\frac{3}{2} < k \leqslant \frac{3}{2}$; and (*b*) in the presence of a threefold potential barrier. Open circles in case (*b*) mark the eigenvalues at $k = 0, \pm 1$, consistent with periodic boundary conditions. Closed circles mark those at $k = \pm\frac{1}{2}, \frac{3}{2}$, consistent with sign-changing boundary conditions. The point $k = -\frac{3}{2}$ is assumed to be excluded from the band.

where $|x(q;Q)\rangle$ is real, and $\psi(Q)$ is designed to ensure that $|n(q;Q)\rangle$ is single valued around the degeneracy. Consequently, Eq. (85) takes the form

$$\left(\sum \frac{1}{2m_i}[\{\hat{\mathbf{P}}_i - \mathbf{a}_i\}^2 + \langle x|\hat{\mathbf{P}}_i^2|x\rangle] + W(Q)\right)|v(Q)\rangle = E|v(Q)\rangle \tag{90}$$

where

$$\mathbf{a}_i = -\hbar\nabla_{Q_i}\psi \tag{91}$$

The term $\mathbf{a}_i$ therefore plays the role of a vector potential in electromagnetic theory, with a particularly close connection with the Aharonov–Bohm effect, associated with adiabatic motion of a charged quantal system around a magnetic flux line [12], a connection that has led to the phrase molecular Aharonov–Bohm effect [34,35] for the influence of $\mathbf{a}_i$ on the nuclear dynamics. Note also that single valuedness of $|n(q;Q)\rangle$ allows considerable ambiguity in the definition of $\psi(Q)$, but it is easily verified that the substitution of $\tilde{\psi}(Q)$ for $\psi(Q)$ merely alters the phase of $|v(Q)\rangle$ by a factor $e^{i(\psi-\tilde{\psi})}$, without altering the essential dynamics. The simplest choice $\psi(Q) = \mu\phi$, where μ is a half-odd integer and ϕ is an angle measured around the degeneracy, is therefore normally employed in molecular Aharonov–Bohm theory.

An advantage of Eq. (90) for computational purposes is that the solutions are subject to single-valued boundary conditions. It is also readily verified that inclusion of an additional factor $e^{i\Delta\psi(Q)}$ on the right-hand side of Eq. (89) adds a term $\Delta\mathbf{a}_i = -\hbar\nabla_{Q_i}\Delta\psi$ to the vector potential, which leads in turn to a compensating factor $e^{-i\Delta\psi(Q)}$ in the nuclear wave function. The total wave function is therefore invariant to changes in such phase factors.

We now consider the connection between the preceding equations and the theory of Aharonov et al. [18] [see Eqs. (51)–(60)]. The tempting similarity between the structures of Eqs. (56) and (90), hides a fundamental difference in the roles of the vector operator $\mathbf{A}$ in Eq. (56) and the vector potential $\mathbf{a}$ in Eq. (90). The former is defined, in the adiabatic partitioning scheme, as a strictly off-diagonal operator, with elements $\langle m|\mathbf{A}|n\rangle = \langle m|\mathbf{P}|n\rangle$, thereby ensuring that $(\mathbf{P} - \mathbf{A})$ is diagonal. By contrast, the Mead–Truhlar vector potential $\mathbf{a}$ arises from the influence of nonzero diagonal elements, $\langle n|\mathbf{P}|n\rangle$ on the nuclear equation for $|v\rangle$, an aspect of the problem not addressed by Arahonov et al. [18]. Suppose, however, that Eq. (56) was contracted between $\langle n|$ and $|n\rangle|v\rangle$ in order to handle the adiabatic nuclear dynamics within the Aharonov scheme. The result becomes

$$\frac{1}{2m}\langle n|\mathbf{P}^2|n\rangle|v\rangle = \frac{1}{2m}\left[\langle n|(\mathbf{P} - \mathbf{A})^2|n\rangle|v\rangle + \langle n|\mathbf{A}^2|n\rangle|v\rangle\right] \tag{92}$$

Given a real electronic Hamiltonian, with single-valued adiabatic eigenstates of the form $|n\rangle = e^{i\psi(Q)}|x_n\rangle$ and $|x_m\rangle$, the matrix elements of $\mathbf{A}$ become

$$\langle m|\mathbf{A}|n\rangle = \langle x_m|\mathbf{A}|x_n\rangle = \langle x_m|\mathbf{P}|x_n\rangle \tag{93}$$

so that

$$\langle n|\mathbf{A}^2|n\rangle = \sum_m \langle n|\mathbf{A}|m\rangle\langle m|\mathbf{A}|n\rangle = \sum_m \langle x_n|\mathbf{P}|x_m\rangle\langle x_m|\mathbf{P}|x_n\rangle = \langle x_n|\mathbf{P}^2|x_n\rangle \tag{94}$$

The sum over all m is justified by the fact that the diagonal elements $\langle x_n|\mathbf{P}|x_n\rangle$ vanish in a real representation. It is also evident from the factorization of $|n\rangle$ and the absence of diagonal elements of $\mathbf{A}$ that

$$\langle n|\mathbf{P} - \mathbf{A}|n\rangle = \langle n|\mathbf{P}|n\rangle = \hbar\nabla_Q\psi = -\mathbf{a} \tag{95}$$

Consequently, inclusion of the nuclear derivatives of $|v\rangle$ leads to

$$\langle n|(\mathbf{P} - \mathbf{A})^2|n\rangle|v\rangle = (\mathbf{P} - \mathbf{a})^2|v\rangle \tag{96}$$

The upshot is that Eq. (95) goes over precisely to the kinetic energy part of Eq. (90). Despite some phrases in the introduction to Aharonov et al. [18] there is therefore no fundamental contradiction with Mead and Truhlar [10].

Some final comments on the relevance of non-adiabatic coupling matrix elements to the nature of the vector potential $\mathbf{a}$ are in order. The above analysis of the implications of the Aharonov coupling scheme for the single-surface nuclear dynamics shows that the off-diagonal operator $\mathbf{A}$ provides nonzero contributions only via the term $\langle n|\mathbf{A}^2|n\rangle$. There are therefore no necessary contributions to $\mathbf{a}$ from the non-adiabatic coupling. However, as discussed earlier, in Section IV [see Eqs. (34)–(36)] in the context of the $E \times \epsilon$ Jahn–Teller model, the phase choice $\psi = -\phi/2$ coupled with the identity

$$\nabla_Q\psi = -\langle x_-|\nabla_Q|x_+\rangle = -\frac{\mathbf{e}_\phi}{2r} \tag{97}$$

close to the degeneracy, allows a representation for $\mathbf{a}$ in terms of $\langle x_-|\nabla_Q|x_+\rangle$, without recourse to arguments [36,37] that have aroused some controversy [38]. The resulting ADT form for the vector potential may have computational advantages in avoiding the need to identify the precise conical intersection point; a number of successful applications have been reported [25,26,39]. Notice that adoption of the first equality in Eq. (97) implies a new phase choice ($\psi \neq \pm\frac{1}{2}\phi$) which must, by the continuity argument in Section III, still ensure a single-valued adiabatic eigenstate $|n\rangle$. It must be emphasized, however, that such an ADT representation for the vector potential is subject to the same restrictions as those that apply to the corresponding representation for the geometric phase in Eq. (35).

1. Symmetry Considerations

It is beyond the scope of these introductory notes to treat individual problems in fine detail, but it is interesting to close the discussion by considering certain, geometric phase related, symmetry effects associated with systems of identical particles. The following account summarizes results from Mead and Truhlar [10] for three such particles. We know, for example, that the fermion statistics for H atoms require that the vibrational–rotational states on the ground $^1A_1$ electronic energy surface of NH_3 must be antisymmetric with respect to binary exchange and symmetric with respect to cyclic permutations; that is, they must belong to the A_2 representation of the C_{3v} point group. We now consider how similar symmetry constraints are introduced in a scattering context, in the presence of geometric phase. It is convenient to formulate the theory in a symmetrical coordinate system, which is here taken, for historical reasons, in the form employed by Mead and Truhlar [10]. An alternative hyperspherical formulation is also available in the literature [40].

The three internal coordinates are expressed as combinations of squares of the interparticle distances;

$$\begin{aligned} Q &= R_{\rm AB}^2 + R_{\rm BC}^2 + R_{\rm CA}^2 \\ u &= R_{\rm BC}^2 + R_{\rm CA}^2 - 2R_{\rm AB}^2 = S\cos\phi \\ v &= \sqrt{3}(R_{\rm BC}^2 - R_{\rm CA}^2) = S\sin\phi \\ S^2 &= u^2 + v^2 = 2[(R_{\rm AB}^2 - R_{\rm BC}^2)^2 + (R_{\rm BC}^2 - R_{\rm CA}^2)^2 + (R_{\rm CA}^2 - R_{\rm AB}^2)^2] \end{aligned} \tag{98}$$

Note that Mead and Truhlar [10] employ the symbol θ in place of the present ϕ, which is preferred here for consistency with the previous text.

There is a line of degeneracies at the equilateral geometries, $S = 0$, and deviations from the degeneracy line are expressed in terms of u and v, subject at a given value of Q to $u^2 + v^2 \leq Q$, this bounding circle being the locus of linear geometries. The properties of Eqs. (98) are summarized in Figure 4, from which

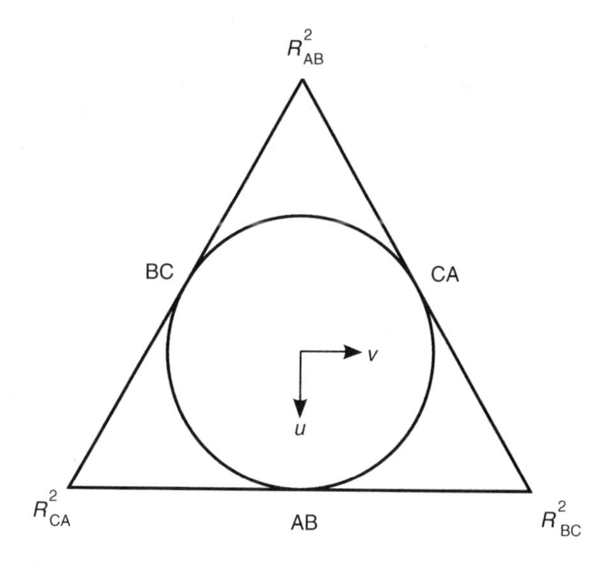

Figure 4. Triangular phase diagram, showing partitions of $R_{\rm AB}^2$, $R_{\rm BC}^2$, and $R_{\rm CA}^2$, at fixed $Q > R_{\rm AB}^2 + R_{\rm BC}^2 + R_{\rm CA}^2$. Physically allowed combinations lie inside the circle, with the conical intersection, corresponding to equilateral triangular ABC, at the center.

it is evident that there three equivalent specifications for u and v, according to whether AB, BC, or CA is taken as the unique particle pair.

Mead and Truhlar [10] further demonstrate that the real adiabatic eigenstates close to $S = 0$ behave in the AB representation as

$$\begin{pmatrix} |+\rangle \\ |-\rangle \end{pmatrix} = \begin{pmatrix} \sin\frac{\phi}{2} & \cos\frac{\phi}{2} \\ \cos\frac{\phi}{2} & -\sin\frac{\phi}{2} \end{pmatrix} \begin{pmatrix} |X_{\mathrm{AB}}\rangle \\ |Y_{\mathrm{AB}}\rangle \end{pmatrix} \tag{99}$$

where the reference kets $|X_{\mathrm{AB}}\rangle$ and $|Y_{\mathrm{AB}}\rangle$ are, respectively, symmetric and antisymmetric with respect to exchange of particles A and B. The geometry also dictates the existence of alternative basis kets $(|X_{\mathrm{BC}}\rangle, |Y_{\mathrm{BC}}\rangle)$ and $(|X_{\mathrm{CA}}\rangle, |Y_{\mathrm{CA}}\rangle)$, related to $(|X_{\mathrm{AB}}\rangle, |Y_{\mathrm{AB}}\rangle)$ by

$$\begin{aligned} \begin{pmatrix} |X_{\mathrm{AB}}\rangle \\ |Y_{\mathrm{AB}}\rangle \end{pmatrix} &= \begin{pmatrix} -\frac{1}{2} & \frac{\sqrt{3}}{2} \\ -\frac{\sqrt{3}}{2} & -\frac{1}{2} \end{pmatrix} \begin{pmatrix} |X_{\mathrm{BC}}\rangle \\ |Y_{\mathrm{BC}}\rangle \end{pmatrix} \\ &= \begin{pmatrix} -\frac{1}{2} & -\frac{\sqrt{3}}{2} \\ \frac{\sqrt{3}}{2} & -\frac{1}{2} \end{pmatrix} \begin{pmatrix} |X_{\mathrm{CA}}\rangle \\ |Y_{\mathrm{CA}}\rangle \end{pmatrix} \end{aligned} \tag{100}$$

To see the implications of Eqs. (98)–(100) for the reaction

$$\mathrm{AB} + \mathrm{C} \rightarrow \mathrm{A} + \mathrm{BC}$$

where A, B, and C are equivalent atoms, we note first that the reactant geometry, $R_{\mathrm{BC}} \simeq R_{\mathrm{CA}} \gg R_{\mathrm{AB}}$ corresponds to $\phi \rightarrow 0$, for which $|-\rangle \rightarrow |X_{\mathrm{AB}}\rangle$ and $|+\rangle \rightarrow |Y_{\mathrm{AB}}\rangle$. It follows from the definitions of $|X_{\mathrm{AB}}\rangle$ and $|Y_{\mathrm{AB}}\rangle$ that diatomics in electronic states that are symmetric or antisymmetric with respect to nuclear exchange have $|-\rangle$ or $|+\rangle$ as the ground adiabatic eigenstate, respectively. The former possibility (applicable to Σ_g^+ or Σ_u^- rather than Σ_g^- or Σ_u^+ symmetry [41]) is assumed in what follows. Thus attention is focused on the state $|-\rangle$.

The next step is to note that the permutation ABC $\rightarrow$ CAB corresponds to an increase in the angle ϕ by $2\pi/3$ [10]. As a result

$$\begin{aligned} |-\rangle \rightarrow \cos\left(\frac{\phi}{2} + \frac{\pi}{3}\right)|X_{\mathrm{AB}}\rangle - \sin\left(\frac{\phi}{2} + \frac{\pi}{3}\right)|Y_{\mathrm{AB}}\rangle \\ = \cos\frac{\phi}{2}|X_{\mathrm{CA}}\rangle + \sin\frac{\phi}{2}|Y_{\mathrm{CA}}\rangle \end{aligned} \tag{101}$$

where the second line follows from Eq. (100). The result is the negative of $|-\rangle$ as given by Eq. (99) in the CA representation. On the other hand, repetition of the argument, with an additional phase factor $e^{\pm 3i\phi/2}$, shows that the four functions

$e^{\pm 3i\phi/2}|\pm\rangle$ are all symmetric under cyclic permutations, as required by both bose and fermi statistics. Moreover, these phase modified eigenstates are also single valued in ϕ.

Finally, following Mead and Truhlar [10], it may be seen that an interchange of A and B is equivalent to a sign reversal of ϕ followed by a rotation perpendicular to the AB bond, under the latter of which $|X_{\mathrm{AB}}\rangle$ is invariant and $|Y_{\mathrm{AB}}\rangle$ changes sign. The net effect is therefore to induce the transitions $e^{\pm 3i\phi/2}|-\rangle \rightarrow e^{\mp 3i\phi/2}|-\rangle$ and $e^{\pm 3i\phi/2}|+\rangle \rightarrow -e^{\mp 3i\phi/2}|+\rangle$.

The upshot of these considerations is that total solutions associated, for example, with the state $|-\rangle$ must be taken in one or other of the symmetrized forms

$$|\Psi\rangle = [|v_+(Q)\rangle e^{3i\phi/2} \pm |v_-(Q)\rangle e^{-3i\phi/2}]|-\rangle \tag{102}$$

where $|v_\pm(Q)\rangle$ are complex functions satisfying the nuclear equations

$$\hat{H}_\pm |v_\pm(Q)\rangle = E|v_\pm(Q)\rangle \tag{103}$$

in which $\hat{H}_\pm$ differ from the normal nuclear Hamiltonian by the substitution $\hat{p}_\phi \rightarrow \hat{p}_\phi \pm 3\hbar/2$. Equation (102) assumes that the electronic states of the fragment diatomics are symmetric with respect to binary exchange (e.g., Σ_g^+ or Σ_u^-), using the upper and lower signs for bose and fermi statistics, respectively. A corresponding form with $|+\rangle$ in place of $|-\rangle$ applies when the fragments electronic states are antisymmetric with respect to nuclear exchange (e.g., Σ_g^- or Σ_u^+), using lower and upper signs for the bose and fermi cases, respectively, in view of the substitution $e^{\pm 3i\phi/2}|+\rangle \rightarrow -e^{\mp 3i\phi/2}|+\rangle$ under binary exchange.

VII. CONCLUSIONS AND EXTENSIONS

The above discussion centers around the seminal contributions of Longuet-Higgins [2,6,7], Mead and Truhlar [10], Berry [8], and Ham [11], supplemented by symmetry arguments due to Jahn and Teller [4,5] and Mead [21]. Topics covered concerned the conditions required for a conical intersection between adiabatic potential energy surfaces (Section II); the behavior of adiabatic electronic eigenstates near a double degeneracy (Section III); the definition and computation of geometric phase (Section IV); and the influence of geometric phase on the nuclear dynamics on a well-separated adiabatic potential surface (Section VI). Illustrations were provided by the simplest and most widely studied $E \times \epsilon$ Jahn–Teller model.

First, the starting point for the discussion is that the real smoothly varying electronic eigenstates $|x(Q)\rangle$ close to a double degeneracy, Q_0, are found to change sign around any path in a nuclear coordinate plane, Q, containing the degeneracy. Second, this electronic sign change must be compensated by a suitable choice of nuclear wave function, such that the total wave function is single valued with respect to any circuit around Q_0. Two possibilities are therefore open for the nuclear dynamics; either the nuclear wave functions must change sign on paths around Q_0, or the theory must be formulated in terms phase modified adiabatic eigenstates

$$|n(Q)\rangle = e^{\pm i\psi(Q)}|x(Q)\rangle$$

such that $|n(Q)\rangle$ is single valued. There is, however, considerable ambiguity in the choice of the phase function $\psi(Q)$. Berry's first contribution [8] was to show that the integrated geometric phase

$$\gamma_C = i\oint_C \langle n|\nabla_{\mathbf{Q}} n\rangle \cdot \mathbf{dQ}$$

depends only on the geometry of the encircling path, regardless of the choice of $\psi(Q)$, provided that $|n(Q)\rangle$ is single valued. Moreover, there is no requirement that the path C should lie in a plane containing Q_0. In addition, Berry derived an alternative expression for γ_C that relaxes the single valuedness condition on $|n(Q)\rangle$.

Simple aspects of the theory were discussed in Section V by reference to the simplest and most widely studied $E \times \epsilon$ Jahn–Teller model. They include the existence a half-odd quantum number j in the linear coupling model, which has been detected in the spectroscopy of Na_3 [16]. However, j is no longer conserved in the presence of quadratic and higher coupling terms, due to the presence of corrugations on the potential energy surface. Next, complications due to an avoided conical intersection were illustrated by the case of a ^2E state with spin–orbit coupling, which may also be viewed as the case of a circuit in a plane from which the intersection point is excluded. The geometric phase is then no longer independent of the size and shape of the encircling path; it takes the "normal" value of π on large circuits far from the avoided intersection, but is quenched to zero as the radius of the circuit decreases.

The two basic approaches to the influence of geometric phase on the nuclear dynamics were outlined in Section VI. The first follows Ham [11] in using band theory arguments to demonstrate that the nuclear sign change, characteristic of the $E \times \epsilon$ problem with real eigenstates $|x\rangle$, causes a reversal in the ordering of vibronic tunneling triplets arising from threefold potential surface corrugations; the normal order $E(A) < E(E)$, $E(E) < E(A)$, and so on in successive triplets is replaced by $E(E) < E(A)$, $E(A) < E(E)$, and so on. The second follows Mead and Truhlar [10] in replacing $|x\rangle$ by the above single-valued functions $|n\rangle$, in which case the modifying phase $\psi(Q)$ contributes a vector potential term to the nuclear kinetic energy operator. There was also shown to be a geometric phase related contribution to the nuclear spin statistics.

The above results mainly apply to the Longuet-Higgins $E \times \epsilon$ problem, but this historical survey would be incomplete without reference to early work on the much more challenging problems posed by threefold or higher electronic degeneracies in molecules with tetrahedral or octahedral symmetry [3]. For example, tetrahedral species, with electronic symmetry T_1 or T_2, have at least five Jahn–Teller active vibrations belonging to the representations E and T with individual coordinates (Q_a, Q_b) and (Q_x, Q_y, Q_z) say. The linear terms in the nine Hamiltonian matrix elements were shown in 1957 [3] to be

$$H = \begin{pmatrix} \frac{1}{2}k_E(Q_a + \sqrt{3}Q_b) & k_TQ_z & k_TQ_y \\ k_TQ_z & \frac{1}{2}k_E(Q_a - \sqrt{3}Q_b) & k_TQ_x \\ k_TQ_y & k_TQ_x & -k_EQ_a \end{pmatrix} \qquad (104)$$

and the corresponding quadratic terms are also well established [42] (see also Appendix IV of [14]). The cubic group, vector coupling coefficients in Griffith's book [43] are very helpful for calculations of this kind. Mead's recent review [44] is largely devoted to the geometric phase aspects of this complicated case, in which one is now concerned with possible circuits in at least a five-dimensional (5D) parameter space (recall that CH_4^+ has two vibrations with symmetry t_2), some of which encircle lines of degeneracy, while others do not. There is also no readily tractable means to determine the adiabatic eigenvectors at arbitrary nuclear geometries, except in the remarkable O'Brien d model [46,47] with $k_E = k_T$, which seems to be relatively little known in molecular physics. The interesting findings, in this special case, are that the Hamiltonian (104) may be shown to commute with the three components of a vibronic angular momentum, somewhat analogous to the operator $\hat{\jmath}$ in Eq. (65) for the linear $E \times \epsilon$ case. Consequently, the eigenvectors at arbitrary nuclear geometries can be expressed in terms of Wigner matrix elements [48] and an explicit expression for the vector potential in the Mead and Truhlar formalism has been worked out [47]. The model is of restricted practical interest, but anyone interested in the complexities of geometric phase, in its more challenging contexts, is strongly advised to study these interesting papers. The review by Judd [49] adds useful mathematical detail.

APPENDIX A: ELEMENTS OF FLOQUET THEORY

Floquet solutions of the periodic second-order equation (taken here to be threefold periodic)

$$\left\{\frac{\hbar^2}{2m}\frac{d^2}{d\phi^2} + E - V(\phi)\right\}\psi(\phi) = 0 \qquad V\left(\phi + \frac{2\pi}{3}\right) = V(\phi) \qquad (A.1)$$

are defined to propagate from ϕ to $\phi + 2\pi/3$ in the form

$$\psi_k\left(\phi + \frac{2\pi}{3}\right) = e^{2\pi i k(E)/3}\psi_k(\phi) \tag{A.2}$$

so that $|\psi_k(\phi + 2\pi/3)| = |\psi_k(\phi)|$. The factor $k/3$ is introduced so that, after threefold repetition of Eq. (A.2),

$$\psi_k(\phi + 2\pi) = e^{2\pi i k(E)}\psi_k(\phi) \tag{A.3}$$

It also follows that the function

$$\xi(\phi) = e^{-ik\phi}\psi_k(\phi) \tag{A.4}$$

is periodic, because on combining Eqs. (A.2) and (A.4)

$$\xi\left(\phi + \frac{2\pi}{3}\right) = e^{-ik(\phi+2\pi/3)}\psi_k\left(\phi + \frac{2\pi}{3}\right) = \xi(\phi) \tag{A.5}$$

Consequently, Floquet solutions may be expressed as

$$\psi_k(\phi) = e^{ik(E)\phi}\xi(\phi) \tag{A.6}$$

where $\xi(\phi)$ has the same periodicity as $V(\phi)$. Equation (A.6) defines the energy dependent wavevector $k(E)$, which is the inverse of the dispersion function $E(k)$ for the band in question. Different bands have increasingly many nodes in the periodic factor $\xi(\phi)$.

The existence of such Floquet states, and the nature of the resulting band structure, is explained by the following argument, due to Whittaker and Watson [32]. Consider a pair of independent solutions of Eq. (A.1), say $f_1(\phi)$ and $f_2(\phi)$, and allow ϕ to increase by $2\pi/3$. In view of the periodicity of $V(\phi)$, the propagated solutions $f_i(\phi + 2\pi/3)$ must be expressible as linear combinations of the $f_i(\phi)$ themselves;

$$\begin{pmatrix} f_1(\phi + 2\pi/3) \\ f_2(\phi + 2\pi/3) \end{pmatrix} = \begin{pmatrix} u_{11} & u_{12} \\ u_{21} & u_{22} \end{pmatrix} \begin{pmatrix} f_1(\phi) \\ f_2(\phi) \end{pmatrix} \tag{A.7}$$

Now, continuity requires that the wronskian $f_1 f_2' - f_2 f_1'$ is preserved, from which it may be verified that

$$\det u = 1 \tag{A.8}$$

Moreover, the trace of the matrix u, $t(E) = u_{11} + u_{22}$, is invariant to a similarity transformation; that is to an alternative choice of $f_1(\phi)$ and $f_2(\phi)$. Consequently, the eigenvalue equation,

$$\lambda^2 - t(E)\lambda + 1 = 0 \tag{A.9}$$

is also independent of this choice. The solutions $\lambda_\pm$, which have product unity, take the Floquet form

$$\lambda_\pm = e^{\pm 2\pi i k(E)/3} \tag{A.10}$$

implied by Eq. (A.2), if $-2 < t(E) < 2$; otherwise $\lambda_\pm$ are real and different from unity, which means, via the analogue of Eq. (A.2) that the corresponding solutions $\psi(\phi)$ increase or decrease progressively as ϕ increases by multiples of $2\pi/3$.

Values of λ on the unit circle restrict $k(E)$ to $-\frac{3}{2} < k \leqslant \frac{3}{2}$, with the band edges at the special points $k = 0$ and $k = \frac{3}{2}$, where the two roots coincide. The repeated root condition means that the corresponding dispersion curve $E(k)$ has turning points at its edges, while every other level is doubly degenerate. We also note in passing that plane wave solutions can be expressed in the Floquet form of Eq. (A.6),

$$e^{ik\phi} = e^{i(k-3n)\phi} e^{3in\phi} \tag{A.11}$$

and that n can always be chosen such that $-\frac{3}{2} < k - 3n \leqslant \frac{3}{2}$. Consequently, the free motion dispersion curve, $E = k^2\hbar^2/2m$, can always be folded onto the above interval.

As a concrete illustration of the Floquet band structure for a threefold barrier, Section 3.4 of Child [50] contains an explicit analytical form for the matrix u;

$$u = \begin{pmatrix} \sqrt{1 + \varkappa^2(E)} e^{i\sigma'(E)} & -i\varkappa(E) e^{-i\sigma(E)} \\ i\varkappa(E) e^{i\sigma(E)} & \sqrt{1 + \varkappa^2(E)} e^{-i\sigma'(E)} \end{pmatrix} \tag{A.12}$$

where $\sigma(E)$ and $\sigma'(E)$ increase monotonically with E, while $\varkappa(E)$ decreases monotonically to zero as $E \to \infty$. Consequently, the trace $t(E)$ varies as

$$t(E) = 2\sqrt{1 + \varkappa^2(E)} \cos \sigma'(E) = 2 \cos \frac{2\pi k(E)}{3} \tag{A.13}$$

A proper calculation requires that $\sigma'(E)$ and $\varkappa(E)$ should be evaluated in terms of semiclassical phase integrals, but it is sufficient for illustrative purposes to

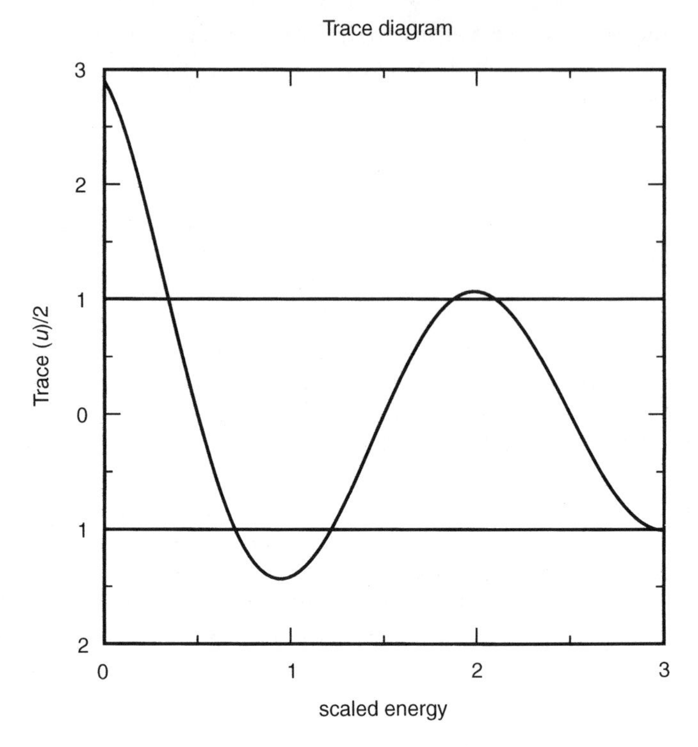

Figure 5. Variation of *trace*(u) with scaled energy, $E/\hbar\omega$, derived from Eq. (A.13) with $E_0 = 0.5\,\hbar\omega$. The Floquet bands in Figure 3*b* cover energy ranges such that $|\mathrm{trace}(u)| < 2$.

employ the approximations $\sigma'(E) = E/\hbar\omega$ and $\varkappa(E) = e^{-(E-E_0)/\hbar\omega}$, where $\hbar\omega$ is an appropriate energy quantum. Successive Floquet bands cover the energy ranges for which $|t(E)| < 2$ in Figure 5. The corresponding dispersion curves shown in Figure 3 were obtained by inversion of the function $k(E)$, determined by Eq. (A.13).

Acknowledgments

The author is grateful for discussions with D. E. Manolopoulos. The hospitality of the Joint Institute for Astrophysics at the University of Colorado during the final preparation of the manuscript is gratefully acknowledged.

References

1. W. Moffitt and A. D. Liehr, *Phys. Rev.* **106**, 1195 (1956).
2. H. C. Longuet-Higgins, U. Opik, M. H. L. Pryce, and R. A. Sack, *Proc. R. Soc. London, Ser. A* **244**, 1 (1958).

3. W. Moffitt and W. R. Thorson, *Phys. Rev.* **108**, 1251 (1957).
4. H. A. Jahn and E. Teller, *Proc. R. Soc. London, Ser. A* **161**, 220 (1937).
5. H. A. Jahn, *Proc. R. Soc. London, Ser. A* **164**, 117 (1938).
6. G. Herzberg and H. C. Longuet-Higgins, *Discuss. Faraday Soc.* **35**, 77 (1963).
7. H. C. Longuet-Higgins, *Proc. R. Soc. London, Ser. A* **344**, 147 (1975).
8. M. V. Berry, *Proc. R. Soc. London, Ser. A* **392**, 45 (1984).
9. R. Renner, *Z. Physik* **92**, 172 (1934).
10. C. A. Mead and D. G. Truhlar, *J. Chem. Phys.* **70**, 2284 (1979).
11. F. S. Ham, *Phys. Rev. Lett.* **58**, 725 (1987).
12. Y. Aharonov and D. Bohm, *Physical Rev.* **115**, 485 (1959).
13. D. Suter, G. C. Chingas, R. A. Harris, and A. Pines, *Molec. Phys.* **61**, 1327 (1987).
14. R. Englman, *The Jahn–Teller effect in Molecules and Crystals*, Wiley-Interscience, New York, 1972.
15. I. B. Bersuker, *The Jahn–Teller effect and Vibronic Interactions in Modern Chemistry*, Plenum Press, New York, 1984.
16. G. Delacrétaz, E. R. Grant, R. L. K. Whetton, L. Wöste, and J. W. Zwanziger, *Phys. Rev. Lett.* **56**, 2598 (1986).
17. H. van Bosch, V. Dev, H. A. Eckel, J. Wang, W. Demtroder, P. Sebald, and W. Meyer, *Phys. Rev. Lett.* **82**, 3560 (1999).
18. Y. Aharonov, E. Ben-Reuven, S. Popescu, and D. Rohrlich, *Nuclear Phys.* **B350**, 818 (1990).
19. J. von Neumann and E. P. Wigner, *Physik Z.* **30**, 467 (1929).
20. E. Teller, *J. Phys. Chem.* **41**, 109 (1937).
21. C. A. Mead, *J. Chem. Phys.* **70**, 2276 (1979).
22. M. Baer and R. Englman, *Mol. Phys.* **75**, 293 (1992).
23. D. R. Yarkony, *Rev. Mod. Phys.* **68**, 985 (1996).
24. A. J. Stone, *Proc. R. Soc. London, Ser. A* **351**, 141 (1976).
25. Z.-R. Xu, M. Baer, and A. J. C. Varandas, *J. Chem. Phys.* **112**, 2746 (2000).
26. A. M. Mebel, A. Yahalom, R. Englman, and M. Baer, *J. Chem. Phys.* **115**, 3673 (2001).
27. G. Arfken, *Mathematical methods for Physicists*, 3rd ed., Academic Press, San Diego, 1985.
28. M. Stone and W. Goff, *Nucl. Phys.* **B295** [FS21], 243 (1988).
29. V. I. Arnold, *Mathematical methods of classical dynamics*, Chap. 7, Springer, New York, 1978.
30. J. W. Zwanziger and E. R. Grant, *J. Chem. Phys.* **87**, 2954 (1987).
31. H. Koizumi and I. B. Bersuker, *Phys. Rev. Lett.* **83**, 3009 (1999).
32. E. T. Whittaker and G. N. Watson, *Modern Analysis*, Cambridge University Press, 1962.
33. B. Kendrick, *Phys. Rev. Lett.* **79**, 2431 (1997).
34. C. A. Mead, *Chem. Phys.* **49**, 23 (1980).
35. C. A. Mead, *Chem. Phys.* **49**, 33 (1980).
36. M. Baer and R. Englman, *Chem. Phys. Lett.* **265**, 105 (1997).
37. M. Baer, *J. Chem. Phys.* **107**, 2694 (1997).
38. B. K. Kendrick, C. A. Mead, and D. G. Truhlar, *J. Chem. Phys.* **110**, 7594 (1999).
39. R. Baer, D. M. Charutz, R. Kosloff, and M. Baer, *J. Chem. Phys.* **105**, 9141 (1996).
40. B. Kendrick and R. T. Pack, *J. Chem. Phys.* **104**, 7457 (1996); **104**, 7502 (1996); **106**, 3519 (1997).

41. G. Herzberg, *Spectra of Diatomic Molecules*, Van Nostrand, New York, 1950.
42. D. Feller and Davidson, *J. Chem. Phys.* **80**, 1006 (1983).
43. J. S. Griffith, *The theory of transition metal ions*, Cambridge University Press, 1961.
44. C. A. Mead, *Rev. Mod. Phys.* **64**, 51 (1992).
45. H. Koizumi, I. B. Bersuker, J. E. Boggs, and V. Z. Polinger, *J. Chem. Phys.* **112**, 8470 (2000).
46. M. C. M. O'Brien, *J. Phys.* **C4**, 2554 (1971).
47. C. C. Chancey, *J. Phys.* **A21**, 3347 (1988).
48. D. M. Brink and G. R. Satchler, *Angular Momentum*, 2nd ed, Oxford University Press, 1968.
49. B. R. Judd, *Adv. Chem. Phys.* **57**, 247 (1984).
50. M. S. Child, *Semiclassical mechanics with molecular applications*, Oxford University Press, 1991.

THE ELECTRONIC NON-ADIABATIC COUPLING TERM IN MOLECULAR SYSTEMS: A THEORETICAL APPROACH

MICHAEL BAER

Applied Physics Division, Soreq NRC, Yavne, Israel

CONTENTS

The Role of Degenerate States in Chemistry: A Special Volume of Advances in Chemical Physics, Volume 124, Edited by Michael Baer and Gert Due Billing. Series Editors I. Prigogine and Stuart A. Rice.
ISBN 0-471-43817-0.

I. INTRODUCTION

Electronic non-adiabatic effects are an outcome of the Born–Oppenheimer–Huang treatment and as such are a result of the distinction between the fast moving electrons and the slow moving nuclei [1,2]. The non-adiabatic coupling terms, together with the adiabatic potential energy surfaces (which are also an outcome of this treatment) form the origin for the driving forces that govern the motion of the atoms in the molecular system. Still they differ from the potential energy surfaces because they are, as we shall show, *derivative coupling* and as such they are vectors, in contrast to adiabatic potential energy surfaces, which are scalars.

The ordinary way to get acquainted with objects like the non-adiabatic coupling terms is to derive them from first principles, via ab initio calculations [4–6], and study their spatial structure—somewhat reminiscent of the way potential energy surfaces are studied. However, this approach is not satisfactory because the non-adiabatic coupling terms are frequently singular (in addition to being vectors), and therefore theoretical means should be applied in order to understand their role in molecular physics. During the last decade, we followed both courses but our main interest was directed toward studying their physical–mathematical features [7–13]. In this process, we revealed (1) the necessity to form sub-Hilbert spaces [9,10] in the region of interest in configuration space and (2) the fact that the non-adiabatic coupling *matrix* has to be *quantized* for this sub-space [7–10].

In the late 1950s and the beginning of the 1960s Longuet-Higgins and colleagues [14–17] discovered one of the more interesting features in molecular physics related to the Born–Oppenheimer–Huang electronic adiabatic eigenfunctions (which are parametrically dependent on the nuclear coordinates). They found that these functions, when surrounding a point of degeneracy in configuration space, may acquire a phase that leads to a flip of sign of these functions. Later, this feature was explicitly demonstrated by Herzberg and Longuet-Higgins [16] for the Jahn–Teller conical intersection model [18–29] (see also Appendix A). This interesting observation implies that if a molecular system possesses a conical intersection at a point in configuration space the relevant electronic eigenfunctions that are parametrically dependent on the nuclear coordinates, are multivalued (this finding was later confirmed for a real case following ab initio calculations [30]). No hints were given to the fact that this phenomenon is associated with the electronic non-adiabatic coupling terms. In this chapter, we not only discuss this connection but also extend the two-state case to the multistate cases.

In molecular physics, one distinguishes between (1) the adiabatic framework that is characterized by the adiabatic surfaces and the above-mentioned non-adiabatic coupling terms [31–46] and (2) the diabatic framework that is characterized by the smoothly behaving potential couplings (and the nonexistence of non-adiabatic couplings) [31–53]. The adiabatic framework is in most cases inconvenient for treating the nuclear Schrödinger equation because of two reasons. (1) The non-adiabatic coupling terms are usually spiky [3,54] and frequently singular [3,36,55,56] so that any numerical recipe for solving this equation becomes unstable. (2) The singular feature of the non-adiabatic coupling terms dictates certain boundary conditions that may not be easily implemented for deriving the solution within this framework [56]. Therefore, transforming to the diabatic framework, to be termed *diabatization*, is essentially enforced when treating the multistate problem as created by the Born–Oppenheimer–Huang approach.

The diabatization can be carried out in various ways—and we discuss some of them here—but the way recommended in the present composition is based on the following two-step process: (1) Forming the Schrödinger equation within the adiabatic framework (which also includes the non-adiabatic coupling terms) and (2) employing a unitary transformation that eliminates these terms from the adiabatic Schrödinger equation and replacing them with the relevant potential coupling terms [34–36,57]. This two-step process creates, as will be shown, an opportunity to study the features of the non-adiabatic coupling terms and the results of this study constitute the main subject of this chapter.

The theoretical foundations for this study were laid in a 1975 publication [34] in which this transformation matrix, hence to be termed the adiabatic-to-diabatic transformation matrix, was shown to be a solution of an integral equation defined along a given contour. In what follows, this equation will be termed as a *line integral*. The line integral reduces, for the two-state case, to an ordinary integral over the corresponding non-adiabatic coupling term, and yields the adiabatic-to-diabatic transformation angle [34–36]. In addition, the sufficient conditions that guarantee the existence and the single-valuedness of the integral-equation solution (along a contour in a given region in configuration space) were derived. In this context, it was shown that these conditions, hence termed the *curl* conditions, are fulfilled by the system of Born–Oppenheimer–Huang eigenfunctions that span a full-Hilbert space [34], and sometimes, under certain conditions, also span a sub-Hilbert space [8–10].

These two findings form a connection of the theory of the electronic non-adiabatic coupling terms with the Yang–Mills isotopic gauge transformation [58,59]. The existence of the *curl* conditions may lead to nonzero Yang–Mills fields as will be proposed in Section XIV. Still, it is important to emphasize that the curl condition as it emerges from our theory and the Yang–Mills field that is a quantum mechanical extension of the classical electromagnetic theory are far from being identical or of the same origin.

In 1992, Baer and Englman [55] suggested that Berry's topological phase [60–62], as derived for molecular systems, and likewise the Longuet-Higgins phase [14–17], should be related to the adiabatic-to-diabatic transformation angle as calculated for a two-state system [56] (see also [63]). Whereas the Baer–Englman suggestion was based on a study of the Jahn–Teller conical intersection model, it was later supported by other studies [11,12,64–75]. In particular, it can be shown that these two angles are related by comparing the "extended" Born–Oppenheimer *approximation*, once expressed in terms of the gradient of the Longuet-Higgins phase (see Appendix A) and once in terms of the two-state non-adiabatic coupling term [75].

Although the two angles seem to serve the same purpose, there is one fundamental difference between the two: The Longuet-Higgins phase (or the molecular Berry phase), when followed along a closed contour becomes, due to an *ansatz*, a multiple of π. Contrary to this ansatz, the situation with respect to the adiabatic-to-diabatic transformation angle is much more fundamental, because of the close relationship between the non-adiabatic coupling terms and the *diabatic* potentials. It was proved that the corresponding non-adiabatic coupling *matrix* has to be "quantized" (see Section IV) in order to yield single-valued diabatic potentials. This "quantized" non-adiabatic coupling matrix yields, in the case of a two-state isolated system, an adiabatic-to-diabatic transformation angle, with features as demanded by the Longuet-Higgins ansatz. In other words, the adiabatic-to-diabatic transformation angle when calculated along closed contours becomes, just like the Longuet-Higgins phase, a multiple of π (or zero).

The next question asked is whether there are any indications, from ab initio calculations, to the fact that the non-adiabatic transformation angles have this feature. Indeed such a study, related to the H_3 system, was reported a few years ago [64]. However, it was done for circular contours with exceptionally small radii (at most a few tenths of an atomic unit). Similar studies, for circular and noncircular contours of much larger radii (sometimes up to five atomic units and more) were done for several systems showing that this feature holds for much more general situations [11,12,74]. As a result of the numerous numerical studies on this subject [11,12,64–75] the quantization of a quasi-isolated two-state non-adiabatic coupling term can be considered as established for realistic systems.

Like the curl condition is reminiscent of the Yang–Mills field, the quantization just mentioned is reminiscent of a study by Wu and Yang [76] for the quantization of Dirac's magnetic monopole [77–78]. As will be shown, the present quantization conditions just like the Wu and Yang conditions result from a *phase factor*, namely, the exponential of a phase and not just from a *phase*.

As mentioned above, the starting point in this field is the Born–Oppenheimer–Huang treatment. However in the first derivations [34] it was always assumed that the corresponding Born–Oppenheimer–Huang eigenfunctions form a full-Hilbert space. Here, the derivation is repeated for a finite sub-Hilbert space, which is defined by employing features of the non-adiabatic coupling terms. It will be shown that this particular sub-space behaves, for all practical purposes, as a full-Hilbert space [8–10]. These subjects are treated in Sections II and III. The connection between the non-adiabatic coupling matrix and the uniqueness of the relevant diabatic potential matrix is presented in Section IV; the quantization of the non-adiabatic coupling matrix is discussed in Section V and the conditions for breaking up the complete

Hilbert space into sub-Hilbert spaces and sub-sub-Hilbert spaces are given in Section VI. Three subjects related to topological effects are presented in Sections VII–IX, and multidegeneracy at a point is further (briefly) discussed in Section X. Section XI is devoted to a practical aspect of the theory, namely, how and when one may/can *diabatize* an electronic adiabatic framework. An interesting relationship between the adiabatic-to-diabatic transformation matrix and Wigner's rotation matrix is discussed in Section XII. Two "exotic" subjects—one related to pseudomagnetic fields in molecular systems and the other related to the possibility of calculating the non-adiabatic coupling terms from the curl equations—are presented in Sections XIII and XIV, respectively. Throughout the review, we show results as derived from ab initio calculations. However, more situations and examples are given in Section XV. A summary and conclusions are presented in Section XVI.

II. THE BORN–OPPENHEIMER–HUANG TREATMENT

A. The Born–Oppenheimer Equations for a Complete Hilbert Space

The Hamiltonian, $\mathbf{H}$, of the nuclei and the electrons is usually written in the following form:

$$\mathbf{H} = \mathbf{T}_{\mathrm{n}} + \mathbf{H}_{\mathrm{e}}(\mathrm{e} \mid \mathrm{n}) \tag{1}$$

where $\mathbf{T}_{\mathrm{n}}$ is the nuclear kinetic energy, $\mathbf{H}_{\mathrm{e}}(\mathrm{e} \mid \mathrm{n})$ is the electronic Hamiltonian that also contains the nuclear Coulombic interactions and depends parametrically on the nuclei coordinates, and e and n stand for the electronic and the nuclear coordinates, respectively.

The Schrödinger equation to be considered is of the form:

$$(\mathbf{H} - E)\Psi(\mathrm{e}, \mathrm{n}) = 0 \tag{2}$$

where E is the total energy and $\Psi(\mathrm{e,n})$ is the complete wave function that describes the motions of both the electrons and the nuclei. Next, we employ the Born–Oppenheimer–Huang expansion:

$$\Psi(\mathrm{e}, \mathrm{n}) = \sum_{i=1}^{N} \psi_i(\mathrm{n})\zeta_i(\mathrm{e} \mid \mathrm{n}) \tag{3}$$

where the $\psi_i(\mathrm{n})$, $i = 1, \ldots, N$ are nuclear-coordinate dependent coefficients (recognized later as the nuclear wave functions) and $\zeta_i(\mathrm{e} \mid \mathrm{n}), i = 1, \ldots, N$ are the electronic eigenfunctions of the above introduced electronic Hamiltonian:

$$[\mathbf{H}_{\mathrm{e}}(\mathrm{e} \mid \mathrm{n}) - u_i(\mathrm{n})]\zeta_i(\mathrm{e} \mid \mathrm{n}) = 0 \qquad i = 1, \ldots, N \tag{4}$$

Here $u_i(\mathrm{n}), i = 1, \ldots, N$ are the electronic eigenvalues recognized, later, as the (adiabatic) potential energy surfaces (PES) that governs the motion of the nuclei. In this treatment, we assume that the Hilbert space is of dimension N. Substituting Eq. (3) in Eq. (2), multiplying it from the left by $\zeta_j(\mathrm{e} \mid \mathrm{n})$, and integrating over the electronic coordinates while recalling Eqs. (1) and (4), yields the following set of coupled equations:

$$\sum_{i=1}^{N} \langle \zeta_j | \mathbf{T}_\mathrm{n} \psi_i(\mathrm{n}) | \zeta_i \rangle + (u_j(\mathrm{n}) - E)\psi_j(\mathrm{n}) = 0 \qquad j = 1, \ldots, N \tag{5}$$

where the bra–ket notation means integration over electronic coordinates. To continue, we recall that the kinetic operator $\mathbf{T}_\mathrm{n}$ can be written (in terms of mass-scaled coordinates) as

$$\mathbf{T}_\mathrm{n} = -\frac{1}{2m}\boldsymbol{\nabla}^2 \tag{6}$$

where m is the mass of the system and $\boldsymbol{\nabla}$ is the gradient (vector) operator. By substituting Eq. (6) in Eq. (5) yields the more explicit form of the Born–Oppenheimer–Huang system of coupled equations:

$$-\frac{1}{2m}\nabla^2\psi_j + (u_j(\mathrm{n}) - E)\psi_j - \frac{1}{2m}\sum_{i=1}^{N}(2\boldsymbol{\tau}_{ji}^{(1)} \cdot \nabla\psi_i + \tau_{ji}^{(2)}\psi_i) = 0$$
$$j = 1, \ldots, N \tag{7}$$

where $\boldsymbol{\tau}^{(1)}$ is the non-adiabatic (vector) matrix of the first kind with the elements:

$$\boldsymbol{\tau}_{ji}^{(1)} = \langle \zeta_j | \nabla \zeta_i \rangle \tag{8a}$$

and $\boldsymbol{\tau}^{(2)}$ is non-adiabatic (scalar) matrix of the second kind, with the elements:

$$\boldsymbol{\tau}_{ji}^{(2)} = \langle \zeta_j | \nabla^2 \zeta_i \rangle \tag{8b}$$

For a system of real electronic wave functions, $\boldsymbol{\tau}^{(1)}$ is an antisymmetric matrix.

Equation (7) can also be written in a matrix form as follows:

$$-\frac{1}{2m}\nabla^2\boldsymbol{\Psi} + (\mathbf{u} - E)\boldsymbol{\Psi} - \frac{1}{2m}(2\boldsymbol{\tau}^{(1)} \cdot \nabla + \boldsymbol{\tau}^{(2)})\boldsymbol{\Psi} = 0 \tag{9}$$

where $\boldsymbol{\Psi}$ is column vector that contains nuclear functions.

B. The Born–Oppenheimer–Huang Equation for a (Finite) Sub-Hilbert Space

Next, the full-Hilbert space is broken up into two parts—a finite part, designated as the P space, with dimension M, and the complementary part, the Q space (which is allowed to be of an infinite dimension). The breakup is done according to the following criteria [8–10]:

$$\boldsymbol{\tau}_{ij}^{(1)} \cong 0 \qquad \text{for} \qquad i \leq M \qquad j > M \tag{10}$$

In other words, the non-adiabatic coupling terms between P and Q states are all assumed to be zero. These requirements will later be reconsidered for a relaxed situation where these coupling terms are assumed to be not necessarily identically zero but small, that is, of the order ε in regions of interest.

To continue, we define the following two relevant Feshbach projection operators [79], namely, P_M, the projection operator for the P space

$$\mathrm{P}_M = \sum_{j=1}^{M} |\zeta_j\rangle\langle\zeta_j| \tag{11a}$$

and Q_M, the projection operator for the Q space

$$\mathrm{Q}_M = I - P_M \tag{11b}$$

Having introduced these operators, we are now in a position to express the P part of the $\boldsymbol{\tau}^{(2)}$ matrix (to be designated as $\boldsymbol{\tau}_M^{(2)}$) in terms of the P part of $\boldsymbol{\tau}^{(1)}$ (to be designated as $\boldsymbol{\tau}_M^{(1)}$). To do that, we consider Eq. (8a) and derive the following expression:

$$\nabla\boldsymbol{\tau}_{ji}^{(1)} = \nabla\langle\zeta_j|\nabla\zeta_i\rangle = \langle\nabla\zeta_j\nabla\zeta_i\rangle + \langle\zeta_j|\nabla^2\zeta_i\rangle$$

or, by recalling Eq. (8b), we get

$$\boldsymbol{\tau}_{ji}^{(2)} = -\langle\nabla\zeta_j|\nabla\zeta_i\rangle + \nabla\boldsymbol{\tau}_{ji}^{(1)} \tag{12}$$

The first term on the right-hand side can be further treated as follows:

$$\langle\nabla\zeta_j|\nabla\zeta_i\rangle = \langle\nabla\zeta_j P_M + Q_M|\nabla\zeta_i\rangle$$

which for $i, j \leq M$ becomes

$$\langle\nabla\zeta_j|\nabla\zeta_i\rangle|_M = \langle\nabla\zeta_j|P_M|\nabla\zeta_i\rangle = \sum_{k=1}^{M}\langle\nabla\zeta_j|\zeta_k\rangle\langle\zeta_k|\nabla\zeta_i\rangle \tag{13}$$

(the contribution due to Q_M can be shown to be zero), or also:

$$\langle \nabla\zeta_j | \nabla\zeta_i \rangle |_M = (\boldsymbol{\tau}_M^{(1)})^2_{ij} \qquad i,j \leq M \tag{13a}$$

where $\boldsymbol{\tau}_M^{(1)}$ is, as mentioned above, of dimension M. Therefore within the Pth subspace the matrix $\boldsymbol{\tau}_M^{(2)}$ can be presented in terms of $\boldsymbol{\tau}_M^{(1)}$ in the following way:

$$\boldsymbol{\tau}_M^{(2)} = (\boldsymbol{\tau}_M^{(1)})^2 + \nabla\boldsymbol{\tau}_M^{(1)} \tag{14}$$

Substituting the matrix elements of Eq. (14) in Eq. (7) yields the final form of the Born–Oppenheimer–Huang equation for the P subspace:

$$-\frac{1}{2m}\nabla^2\boldsymbol{\Psi} + \left(u - \frac{1}{2m}\boldsymbol{\tau}_M^2 - E\right)\boldsymbol{\Psi} - \frac{1}{2m}(2\boldsymbol{\tau}_M \cdot \nabla + \nabla\boldsymbol{\tau}_M)\boldsymbol{\Psi} = 0 \tag{15}$$

where the dot designates the scalar product, $\boldsymbol{\Psi}$ is a column matrix that contains the nuclear functions $\{\psi_i;\ i = 1, \ldots, M\}$, $\mathbf{u}$ is a diagonal matrix that contains the adiabatic potentials, and $\boldsymbol{\tau}_M$, for reasons of convenience, replaces $\boldsymbol{\tau}_M^{(1)}$. Equation (15) can also be written in the form [9]:

$$-\frac{1}{2m}(\nabla + \boldsymbol{\tau}_M)^2\boldsymbol{\Psi} + (u - E)\boldsymbol{\Psi} = 0 \tag{16}$$

which is writing the Schrödinger equation more compactly. (A similar Hamiltonian was employed by Pacher et al. [41] within their block-diagonalized approach to obtain quasidiabatic states.)

III. THE ADIABATIC-TO-DIABATIC TRANSFORMATION

A. The Derivation of the Adiabatic-to-Diabatic Transformation Matrix

The aim in performing what is termed the adiabatic-to-diabatic transformation is to eliminate from Eq. (16) the eventually problematic matrix, $\boldsymbol{\tau}_M$, which is done by replacing the column matrix $\boldsymbol{\Psi}$ in Eq. (16) by another column matrix $\boldsymbol{\Phi}$ where the two are related as follows:

$$\boldsymbol{\Psi} = A\boldsymbol{\Phi} \tag{17}$$

At this stage, we would like to emphasize that the same transformation has to be applied to the electronic adiabatic basis set in order not to affect the total wave function of both the electrons and the nuclei. Thus if ξ is the electronic basis set that is attached to $\boldsymbol{\Phi}$ then ζ and ξ are related to each other as

$$\xi = \zeta\mathbf{A}^\dagger \tag{18}$$

Here, $\mathbf{A}$ is an undetermined matrix of the coordinates ($\mathbf{A}^{\dagger}$ is its Hermitian conjugate). Our next step is to obtain an $\mathbf{A}$ matrix, which will eventually simplify Eq. (16) by eliminating the $\boldsymbol{\tau}_M$ matrix. For this purpose, we consider the following expression:

$$\begin{aligned}(\nabla + \boldsymbol{\tau}_M)^2 \mathbf{A}\boldsymbol{\Phi} &= (\nabla + \boldsymbol{\tau}_M)(\nabla + \boldsymbol{\tau}_M)\mathbf{A}\boldsymbol{\Phi} \\ &= (\nabla + \boldsymbol{\tau}_M)(\mathbf{A}\nabla\boldsymbol{\Phi} + (\nabla\mathbf{A})\Phi + \boldsymbol{\tau}_M \mathbf{A}\boldsymbol{\Phi}) \\ &= 2(\nabla\mathbf{A}) \cdot \nabla\boldsymbol{\Phi} + \mathbf{A}\nabla^2\boldsymbol{\Phi} + (\nabla^2\mathbf{A})\boldsymbol{\Phi} + (\nabla\boldsymbol{\tau}_M)\mathbf{A}\boldsymbol{\Phi} \\ &\quad + 2\boldsymbol{\tau}_M(\nabla\mathbf{A})\boldsymbol{\Phi} + 2\boldsymbol{\tau}_M\mathbf{A}(\nabla\boldsymbol{\Phi}) + \boldsymbol{\tau}_M^2\mathbf{A}\boldsymbol{\Phi}\end{aligned}$$

which can be further developed to become

$$\therefore = \mathbf{A}\nabla^2\boldsymbol{\Phi} + 2(\nabla\mathbf{A} + \boldsymbol{\tau}_M\mathbf{A}) \cdot \nabla\boldsymbol{\Phi} + \{(\boldsymbol{\tau}_M + \nabla) \cdot (\nabla\mathbf{A} + \boldsymbol{\tau}_M\mathbf{A})\}\boldsymbol{\Phi}$$

where the ∇ parameters, in the third term, do not act beyond the curled parentheses $\{\}$. Now, if $\mathbf{A}$ (henceforth to be designated as $\mathbf{A}_M$ in order to remind us that it belongs to the M-dimensional P subspace) is chosen to be a solution of the following equation:

$$\nabla\mathbf{A}_M + \boldsymbol{\tau}_M\mathbf{A}_M = 0 \tag{19}$$

then the above (kinetic energy) expression takes the simplified form:

$$(\nabla + \boldsymbol{\tau}_M)^2\mathbf{A}\boldsymbol{\Phi} = \mathbf{A}_M\nabla^2\boldsymbol{\Phi} \tag{20}$$

and therefore Eq. (16) becomes

$$-\frac{1}{2m}\nabla^2\boldsymbol{\Phi} + (W_M - E)\boldsymbol{\Phi} = 0 \tag{21}$$

where we used the fact that $\mathbf{A}_M$ is a unitary matrix (seen Appendix B) and $\mathbf{W}_M$, the diabatic potential matrix, is given in the form:

$$\mathbf{W}_M = (\mathbf{A}_M)^{\dagger}\mathbf{u}_M\mathbf{A}_M \tag{22}$$

Equation (21) is the diabatic Schrödinger equation.

In what follows, the $\mathbf{A}$ matrix (or the $\mathbf{A}_M$ matrix) will be called the adiabatic-to-diabatic transformation matrix.

B. The Necessary Condition for Having a Solution for the Adiabatic-to-Diabatic Transformation Matrix

The $\mathbf{A}$ matrix has to fulfill Eq. (19). It is obvious that all features of $\mathbf{A}$ are dependent on the features of the $\boldsymbol{\tau}$-matrix elements. Thus, for example, if we want the adiabatic-to-diabatic transformation matrix to have second derivatives or more in a given region, the $\boldsymbol{\tau}$-matrix elements have to be analytic functions in this region, namely, they have to have well-defined derivatives. However, this is not enough to guarantee the analyticity of **A**. In order for it to be analytic, there are additional conditions that the elements of this matrix have to fulfill, namely, that the result of two (or more) mixed derivatives should not depend on the order of the differentiation. In other words, if p and q are any two coordinates then the following condition has to hold:

$$\frac{\partial^2}{\partial p \partial q}\mathbf{A} = \frac{\partial^2}{\partial q \partial p}\mathbf{A} \tag{23}$$

The conditions for that to happen are derived in Appendix B (under Analyticity) and are given here:

$$\frac{\partial}{\partial p}\boldsymbol{\tau}_q - \frac{\partial}{\partial q}\boldsymbol{\tau}_p - [\boldsymbol{\tau}_q, \boldsymbol{\tau}_p] = 0 \tag{24}$$

which can also be written more compact as a vector equation:

$$\text{curl}\,\boldsymbol{\tau} - [\boldsymbol{\tau} \times \boldsymbol{\tau}] = 0 \tag{25}$$

For a two-state system Eq. (25) simplifies significantly to become

$$\text{curl}\,\boldsymbol{\tau} = 0 \tag{26}$$

In what follows, Eq. (25) [and Eq. (26)] will be referred to as the *curl* condition. In Appendix C it is proved, employing the integral representation [see Eq. (27)], that the fulfillment of this condition at every point throughout a given region, guarantees the *single valuedness* of the **A** matrix throughout this region.

The importance of the adiabatic-to-diabatic transformation matrix is in the fact that given the adiabatic potential matrix it yields the diabatic potential matrix. Since the potentials that govern the motion of atomic species have to be analytic and *single valued*, and since the adiabatic potentials usually have these features, we expect the adiabatic-to-diabatic transformation to yield diabatic potentials with the same features. Whereas the analyticity feature is guaranteed because the adiabatic-to-diabatic transformation matrix is usually analytic it is more the uniqueness requirement that is of concern. The reason being that in cases where the electronic eigenfunctions become degenerate in configuration space the corresponding non-adiabatic coupling terms become singular (as is well known from the Hellmann–Feynman theorem [3,36]) and this as is proved in Appendix C, may cause the adiabatic-to-diabatic transformation matrix to become multivalued. Thus we have to make sure that the relevant diabatic potentials will also stay single-valued in cases where the adiabatic-to-diabatic transformation matrix is not. All these aspects will be discussed in Section IV.

By returning to the diabatic potentials as defined in Eq. (22), the condition expressed in Eq. (25) also guarantees well-behaved (namely, single-valued) diabatic potentials. However, it is known (as was already discussed above) that the $\boldsymbol{\tau}$-matrix elements are not always well behaved because they may become singular, implying that in such regions Eq. (25) is not satisfied at *every point*. In such a situation the analyticity of the adiabatic-to-diabatic transformation matrix may still be guaranteed (except at the close vicinity of these singular points) but no longer its single-valuedness. The question is to what extent this "new" difficulty is going to affect the single-valuedness of the diabatic potentials (which have to be single valued if a solution for the corresponding Schrödinger equation is required). Section IV is devoted to this issue.

IV. THE ADIABATIC-TO-DIABATIC TRANSFORMATION MATRIX AND THE LINE INTEGRAL APPROACH

From now on, the index M will be omitted and it will be understood that any subject to be treated will refer to a finite sub-Hilbert space of dimension M.

Equation (19) can also be written as an integral equation along a contour in the following way [34–36]:

$$\mathbf{A}(s, s_0 \mid \Gamma) = \mathbf{A}(s_0 \mid \Gamma) - \int_{s_0}^{s} \mathbf{ds}' \cdot \boldsymbol{\tau}(s' \mid \Gamma)\mathbf{A}(s', s_0 \mid \Gamma) \tag{27}$$

where Γ is the given contour in the multidimensional configuration space, the points s and s_0 are located on this contour, $\mathbf{ds}'$ is a differential vector along this contour, and the dot is a scalar product between this differential vector and the (vectorial) non-adiabatic coupling matrix $\boldsymbol{\tau}$. Note that the $\boldsymbol{\tau}$ matrix is the kernel of this equation and since, as mentioned above, some of the non-adiabatic coupling terms may be singular in configuration space (but not necessarily along the contour itself), it has implication on the multivaluedness of both the $\mathbf{A}$ matrix and the diabatic potentials.

A. The Necessary Conditions for Obtaining Single-Valued Diabatic Potentials and the Introduction of the Topological Matrix

The solution of Eq. (19) can be written in the form [57]:

$$\mathbf{A}(s, s_0) = \wp \exp\left(-\int_{s_0}^{s} ds \cdot \boldsymbol{\tau}\right)\mathbf{A}(s_0) \tag{28}$$

where the symbol $\wp$ is introduced to indicate that this integral has to be carried out in a given order [57,80]. In other words, $\wp$ is a path ordering operator. The solution in Eq. (28) is well defined as long as $\boldsymbol{\tau}$, along Γ, is well defined. However, as mentioned earlier, the solution may not be *uniquely* defined at every point in configuration space. Still, we claim that under certain conditions such a solution is of importance because it will lead to uniquely defined diabatic potentials. This claim brings us to formulate the necessary condition for obtaining uniquely defined diabatic potentials.

Let us consider a closed path Γ defined in terms of a continuous parameter λ so that the starting point s_0 of the contour is at $\lambda = 0$. Next, β is defined as the value attained by λ once the contour completes a full cycle and returns to its starting point. For example, in the case of a circle, λ is an angle and $\beta = 2\pi$.

With these definitions we can now look for the necessary condition(s). Thus, we assume that at *each point* s_0 in configuration space the diabatic potential matrix $\mathbf{W}(\lambda)$ $[\equiv \mathbf{W}(s, s_0)]$ fulfills the condition:

$$\mathbf{W}(\lambda = 0) = \mathbf{W}(\lambda = \beta) \tag{29}$$

Following Eq. (22), this requirement implies that for every point s_0 we have

$$\mathbf{A}^{\dagger}(0)u(0)\mathbf{A}(0) = \mathbf{A}^{\dagger}(\beta)u(\beta)\mathbf{A}(\beta) \tag{30}$$

Next, we introduce another transformation matrix, $\mathbf{B}$, defined as

$$\mathbf{B} = \mathbf{A}(\beta)\mathbf{A}^{\dagger}(0) \tag{31}$$

which, for every s_0 and a given contour Γ, connects $u(\beta)$ with $u(0)$:

$$u(\beta) = \mathbf{B}u(0)\mathbf{B}^{\dagger} \tag{32}$$

The $\mathbf{B}$ matrix is, by definition, a unitary matrix (it is a product of two unitary matrices) and at this stage except for being dependent on Γ and, eventually, on s_0, it is rather arbitrary. In what follows, we shall derive some features of $\mathbf{B}$.

Since the electronic eigenvalues (the adiabatic PESs) are uniquely defined at each point in configuration space we have $u(0) \equiv u(\beta)$, and therefore Eq. (32) implies the following commutation relation:

$$[\mathbf{B}, u(0)] = 0 \tag{33}$$

or more explicitly:

$$\sum_{j=1} (\mathbf{B}^{*}_{kj}\mathbf{B}_{kj} - \delta_{kj})u_j(0) = 0 \tag{34}$$

Equation (34) has to hold for every arbitrary point s_0 ($\equiv \lambda = 0$) on the path Γ and for an essential, arbitrary set of nonzero adiabatic eigenvalues, $u_j(s_0)$; $j = 1, \ldots, M$. Due to the arbitrariness of s_0, and therefore also of the $u_j(s_0)$ values, Eqs. (34) can be satisfied if and only if the **B**-matrix elements fulfill the relation:

$$\mathbf{B}_{kj}^{*}\mathbf{B}_{kj} = \delta_{kj} \qquad j, k \leq M \tag{35}$$

or

$$\mathbf{B}_{jk} = \delta_{jk}\exp(i\chi_k) \tag{36}$$

Thus **B** is a diagonal matrix that contains in its diagonal (complex) numbers whose norm is 1 (this derivation holds as long as the adiabatic potentials are nondegenerate along the path Γ). From Eq. (31), we obtain that the **B**-matrix transforms the **A**-matrix from its initial value to its final value while tracing a closed contour:

$$\mathbf{A}(\beta) = \mathbf{BA}(0) \tag{37}$$

Let us now return to Eq. (28) and define the following matrix:

$$\mathbf{D} = \wp \exp\left(-\oint_{\Gamma} ds \cdot \tau\right) \tag{38}$$

Notice from Eq. (28) that if the contour Γ is a closed loop (which returns to s_0) the **D** matrix transforms $\mathbf{A}(s_0)$ to its value $\mathbf{A}(s = s_0 | s_0)$ at the end of the closed contour, namely;

$$\mathbf{A}(s = s_0 \mid s_0) = \mathbf{DA}(s_0) \tag{39}$$

Now, by comparing Eq. (37) with Eq. (39) it is noticed that **B** and **D** are identical, which implies that all the features that were found to exist for the **B** matrix also apply to the matrix **D** as defined in Eq. (38).

Returning to the beginning of this section, we established the following: The *necessary* condition for the **A** matrix to yield single-valued diabatic potentials is that the **D** matrix, defined in Eq. (38), be diagonal and has, in its diagonal, numbers of norm 1. Since we consider only real electronic eigenfunctions these numbers can be either (+1) or (−1) established. By following Eq. (39), it is also obvious that the **A** matrix is not necessarily single-valued because the **D** matrix, as was just proved, is not necessarily a unit matrix. In what follows, the number of (−1) values in a given matrix **D** will be designated as **K**.

The **D** matrix plays an important role in the forthcoming theory because it contains all topological features of an electronic manifold in a region surrounded by a contour Γ as will be explained next.

That the electronic adiabatic manifold can be multivalued is a well-known fact, going back to Longuet-Higgins et al. [14–17]. In this section, we just proved that the same applies to the adiabatic-to-diabatic transformation matrix and for this purpose we introduced the diabatic framework. The diabatic manifold is, by definition, a manifold independent of the nuclear coordinates and therefore single-valued in configuration space. Such a manifold always exists for a complete Hilbert space [36b] (see Appendix D). Next, we assume that an *approximate* (partial) diabatic manifold like that can be found for the present sub-Hilbert space defined with respect to a certain region in configuration space. This approximate diabatic manifold is, by definition, single valued. Then, we consider Eq. (18), in which the electronic diabatic manifold is presented in terms of the product $\zeta \mathbf{A}^{\dagger}$, where ζ is the adiabatic electronic manifold. Since this product is singled valued in configuration space (because it produces a diabatic manifold) it remains single valued while tracing a closed contour. In order for this product to remain single valued, the number of wave functions that flip sign in this process has to be identical to the topological number K. Moreover the positions of the (-1)s in the **D** matrix have to match the electronic eigenfunctions that flip their sign. Thus, for example, if the third element in the **D** matrix is (-1) this implies that the electronic eigenfunction that belongs to the third state flips sign.

It is known that multivalued adiabatic electronic manifolds create topological effects [23,25,45]. Since the newly introduced **D** matrix contains the information relevant for this manifold (the number of functions that flip sign and their identification) we shall define it as the *Topological Matrix*. Accordingly, K will be defined as the *Topological Number*. Since **D** is dependent on the contour Γ the same applies to K thus: $K = K(\Gamma)$.

B. The Quasidiabatic Framework

In Section IV.A, the adiabatic-to-diabatic transformation matrix as well as the diabatic potentials were derived for the relevant sub-space without running into theoretical conflicts. In other words, the conditions in Eqs. (10) led to a *finite* sub-Hilbert space which, for all practical purposes, behaves like a full (infinite) Hilbert space. However it is inconceivable that such strict conditions as presented in Eq. (10) are fulfilled for real molecular systems. Thus the question is to what extent the results of the present approach, namely, the adiabatic-to-diabatic transformation matrix, the curl equation, and first and foremost, the diabatic potentials, are affected if the conditions in Eq. (10) are replaced by more realistic ones? This subject will be treated next.

The quasidiabatic framework is defined as the framework for which the conditions in Eqs. (10) are replaced by the following less stricked ones [81]:

$$\tau_{ij}^{(1)} \cong O(\varepsilon) \qquad \text{for} \qquad i \leq M \qquad j > M \tag{40}$$

Thus, we still relate to the same sub-space but it is now defined for P-states that are weakly coupled to Q-states. We shall prove the following *lemma*: If the interaction between any P- and Q-state measures like $O(\varepsilon)$, the resultant P-diabatic potentials, the P-adiabatic-to-diabatic transformation matrix elements and the P-curl τ equation are all fulfilled up to $O(\varepsilon^2)$.

1. The Adiabatic-to-Diabatic Transformation Matrix and the Diabatic Potentials

We prove our statement in two steps: First, we consider the special case of a Hilbert space of three states, the two lowest of which are coupled strongly to each other but the third state is only weakly coupled to them. Then, we extend it to the case of a Hilbert space of N states where M states are strongly coupled to each other, and L $(= N - M)$ states, are only loosely coupled to these M original states (but can be strongly coupled among themselves).

We start with the first case where the components of two of the $\boldsymbol{\tau}$-matrix elements, namely, τ_{13} and τ_{23}, are of the order of $O(\varepsilon)$ [see Eq. (40)].

The 3×3 $\mathbf{A}$ matrix has nine elements of which we are interested in only four, namely, a_{11}, a_{12}, a_{21}, and a_{22}. However, these four elements are coupled to a_{31} and a_{32} and, therefore, we consider the following *six* line integrals [see Eq. (27)]:

$$a_{ij}(s) = a_{ij}(s_0) - \sum_{k=1}^{3} \int_{s_0}^{s} ds \cdot \tau_{ik}(s) a_{kj}(s) \qquad i = 1,2,3 \qquad j = 1,2 \tag{41}$$

Next, we estimate the magnitudes of a_{31} and a_{32} and for this purpose we consider the equations for a_{31} and a_{32}. Thus, assuming a_{1j} and a_{2j} are given, the solution of the relevant equations in Eq. (41), is

$$a_{3j}(s) = a_{3j}(s_0) - \int_{s_0}^{s} ds' \cdot (\tau_{31} a_{1j} + \tau_{32} a_{2j}) \tag{42}$$

For obvious reasons, we assume $a_{3j}(s_0) = 0$. Since both, a_{1j} and a_{2j}, are at most (in absolute values) unity, it is noticed that the magnitude of a_{31} and a_{32} are of the order of $O(\varepsilon)$ just like the assumed magnitude of the components of τ_{i3} for $i = 1, 2$. Now, returning to Eq. (41) and substituting Eq. (42) in the last term in each summation, one can see that the integral over $\tau_{i3} a_{3j}$; $j = 1, 2$ is of the second order in ε, which can be specified as $O(\varepsilon^2)$. In other words, ignoring the coupling between the two-state system and a third state introduces a second-order error in the calculation of each of the elements of the two-state $\mathbf{A}$ matrix.

To treat the general case, we assume $\mathbf{A}$ and $\boldsymbol{\tau}$ to be of the following form:

$$\mathbf{A} = \begin{pmatrix} \mathbf{A}^{(M)} & \mathbf{A}^{(M,L)} \\ \mathbf{A}^{(L,M)} & \mathbf{A}^{(L)} \end{pmatrix} \tag{43a}$$

and

$$\boldsymbol{\tau} = \begin{pmatrix} \boldsymbol{\tau}^{(M)} & \boldsymbol{\tau}^{(M,L)} \\ \boldsymbol{\tau}^{(L,M)} & \boldsymbol{\tau}^{(L)} \end{pmatrix} \tag{43b}$$

where we recall that M is the dimension of the P sub-space. As before, the only parts of the $\mathbf{A}$ matrix that are of interest for us are $\mathbf{A}^{(M)}$ and $\mathbf{A}^{(L,M)}$. By substituting Eqs. (43) in Eq. (27), we find for $\mathbf{A}^{(M)}$ the following integral equation:

$$\mathbf{A}^{(M)} = \mathbf{A}_0^{(M)} - \int_{s_0}^{s} ds \cdot \boldsymbol{\tau}^{(M)}\mathbf{A}^{(M)} - \int_{s_0}^{s} ds \cdot \boldsymbol{\tau}^{(M,L)}\mathbf{A}^{(L,M)} \tag{44}$$

where $\mathbf{A}$ stands for $\mathbf{A}(s)$ and $\mathbf{A}_0$ for $\mathbf{A}(s_0)$. Our next task is to get an estimate for $\mathbf{A}^{(L,M)}$. For this purpose, we substitute Eqs. (43) in Eq. (19) and consider the first-order differential equation for this matrix:

$$\nabla\mathbf{A}^{(L,M)} + \boldsymbol{\tau}^{(L,M)}\mathbf{A}^{(M)} + \boldsymbol{\tau}^{(L)}\mathbf{A}^{(L,M)} = 0 \tag{45a}$$

which will be written in a slightly different form:

$$\nabla\mathbf{A}^{(L,M)} + \boldsymbol{\tau}^{(L)}\mathbf{A}^{(L,M)} = -\boldsymbol{\tau}^{(L,M)}\mathbf{A}^{(M)} \tag{45b}$$

in order to show that it is an inhomogeneous equation for $\mathbf{A}^{(L,M)}$ (assuming the elements of $\mathbf{A}^{(M)}$ are known). Equation (45b) will be solved for the initial conditions where the elements of $\mathbf{A}^{(L,M)}$ are zero (this is the obvious choice in order for the isolated sub-space to remain as such in the diabatic framework as well). For these initial conditions, the solution of Eq. (45a) can be shown to be

$$\mathbf{A}^{(L,M)} = \exp\left(-\int_{s_0}^{s} ds' \cdot \boldsymbol{\tau}^{(L)}\right)\left\{\int_{s_0}^{s} \exp\left(\int_{s_0}^{s'} ds'' \cdot \boldsymbol{\tau}^{(L)}\right) ds' \cdot \boldsymbol{\tau}^{(L,M)}\mathbf{A}^{(M)}\right\} \tag{46}$$

In performing this series of integrations, it is understood that they are carried out in the correct order and always for consecutive infinitesimal sections along the given contour Γ [57]. Equation (46) shows that all elements of $\mathbf{A}^{(L,M)}$ are linear combinations of the (components of the) $\tau^{(L,M)}$ elements, which are all assumed to be of first order in ε. We also reiterate that the absolute values of all elements of $\mathbf{A}^{(M)}$ are limited by the value of the unity.

Now, by returning to Eq. (44) and replacing $\mathbf{A}^{(L,M)}$ by the expression in Eq. (46) we find that the line integral to solve $\mathbf{A}^{(M)}$ is perturbed to the second order, namely,

$$\mathbf{A}^{(M)} = \mathbf{A}_0^{(M)} - \int_{s_0}^{s} ds \cdot \boldsymbol{\tau}^{(M)} \mathbf{A}^{(M)} + O(\varepsilon^2) \tag{47}$$

This concludes our derivation regarding the adiabatic-to-diabatic transformation matrix for a finite N. The same applies for an infinite Hilbert space (but finite M) if the coupling to the higher Q-states decays fast enough.

Once there is an estimate for the error in calculating the adiabatic-to-diabatic transformation matrix it is possible to estimate the error in calculating the *diabatic potentials*. For this purpose, we apply Eq. (22). It is seen that the error is of the second order in ε, namely, of $O(\varepsilon^2)$, just like for the adiabatic-to-diabatic transformation matrix.

2. *The Curl Condition*

Next, we analyze the P-curl condition with the aim of examining to what extent it is affected when the weak coupling is ignored as described in Section IV.B.1 [81]. For this purpose, we consider two components of the (unperturbed) $\boldsymbol{\tau}$ matrix, namely, the matrices $\boldsymbol{\tau}_q$ and $\boldsymbol{\tau}_p$, which are written in the following form [see Eq. (43)]:

$$\boldsymbol{\tau}_x = \begin{pmatrix} \boldsymbol{\tau}_x^{(M)} & \boldsymbol{\tau}_x^{(M,L)} \\ \boldsymbol{\tau}_x^{(L,M)} & \boldsymbol{\tau}_x^{(L)} \end{pmatrix} \qquad x = q, p \tag{48}$$

Here, $\boldsymbol{\tau}_x^{(M)}$ (and eventually $\boldsymbol{\tau}_x^{(L)}$); $x = p, q$ are the matrices that contain the strong non-adiabatic coupling terms, whereas $\boldsymbol{\tau}_x^{(M,L)}$ [and $\boldsymbol{\tau}_x^{(L,M)}$]; $x = p, q$ are the matrices that contain the weak non-adiabatic coupling terms, all being of the order $O(\varepsilon)$. Employing Eqs. (24) and (25) and by substituting Eq. (48) for $\boldsymbol{\tau}_q$ and $\boldsymbol{\tau}_p$, it can be seen by algebraic manipulations that the following relation holds:

$$\frac{\partial \boldsymbol{\tau}_p^{(M)}}{\partial q} - \frac{\partial \boldsymbol{\tau}_q^{(M)}}{\partial p} - [\boldsymbol{\tau}_p^{(M)}, \boldsymbol{\tau}_q^{(M)}] = \{\boldsymbol{\tau}_p^{(M,L)} \boldsymbol{\tau}_q^{(L,M)} - \boldsymbol{\tau}_q^{(M,L)} \boldsymbol{\tau}_p^{(L,M)}\} \tag{49}$$

Notice, all terms in the curled parentheses are of $O(\varepsilon^2)$, which implies that the curl condition becomes

$$\text{curl}\, \boldsymbol{\tau}^{(M)} - [\boldsymbol{\tau}^{(M)} \times \boldsymbol{\tau}^{(M)}] = O(\varepsilon^2) \tag{50}$$

namely, the curl condition within the sub-space, is fulfilled up to $O(\varepsilon^2)$.

Obviously, the fact that the solution of the adiabatic-to-diabatic transformation matrix is only perturbed to second order makes the present approach rather attractive. It not only results in a very efficient approximation but also yields an estimate for the error made in applying the approximation.

V. THE QUANTIZATION OF THE NON-ADIABATIC COUPLING MATRIX

One of the main outcomes of the analysis so far is that the topological matrix **D**, presented in Eq. (38), is identical to an adiabatic-to-diabatic transformation matrix calculated at the end point of a closed contour. From Eq. (38), it is noticed that **D** does not depend on any particular point along the contour but on the contour itself. Since the integration is carried out over the non-adiabatic coupling matrix, $\boldsymbol{\tau}$, and since **D** has to be a diagonal matrix with numbers of norm 1 for *any* contour in configuration space, these two facts impose severe restrictions on the non-adiabatic coupling terms.

In Section V.A, we present a few analytical examples showing that the restrictions on the $\boldsymbol{\tau}$-matrix elements are indeed quantization conditions that go back to the early days of quantum theory. Section V.B will be devoted to the general case.

A. The Quantization as Applied to Model Systems

In this section, we intend to show that for a certain type of models the above imposed "restrictions" become the ordinary well-known Bohr–Sommerfeld quantization conditions [82]. For this purpose, we consider the following non-adiabatic coupling matrix $\boldsymbol{\tau}$:

$$\boldsymbol{\tau}(s) = \mathbf{g}\mathbf{t}(s) \tag{51}$$

where $\mathbf{t}(s)$ is a vector whose components are functions in configuration space and $\mathbf{g}$ is a *constant* antisymmetric matrix of dimension M. For this case, one can evaluate the ordered exponential in Eq. (38). Thus substituting Eq. (51) in Eq. (38) yields the following solution for the **D** matrix:

$$\mathbf{D} = \mathbf{G}\exp\left(-\boldsymbol{\omega}\oint_{\Gamma} ds \cdot \mathbf{t}(s)\right)\mathbf{G}^{\dagger} \tag{52}$$

where $\boldsymbol{\omega}$ is a diagonal matrix that contains the eigenvalues of the $\mathbf{g}$ matrix and $\mathbf{G}$ is a matrix that diagonalizes g ($\mathbf{G}^{\dagger}$ is the Hermitian conjugate of $\mathbf{G}$). Since $\mathbf{g}$ is an antisymmetric matrix all its eigenvalues are either imaginary or zero.

Next, we concentrate on a few special cases.

1. The Two-State Case

The **g** matrix in this case is given in the form:

$$\mathbf{g} = \begin{pmatrix} 0 & 1 \\ -1 & 0 \end{pmatrix} \tag{53}$$

The matrix **G** that diagonalizes it is

$$\mathbf{G} = \frac{1}{\sqrt{2}} \begin{pmatrix} 1 & 1 \\ i & -i \end{pmatrix} \tag{54}$$

and the corresponding eigenvalues are $\pm i$. Substituting Eq. (54) in Eq. (52) and replacing the two ω parameters by $\pm i$ yields the following **D** matrix:

$$\mathbf{D} = \begin{pmatrix} \cos\left(\oint_{\Gamma} \mathbf{t}(s) \cdot ds\right) & -\sin\left(\oint_{\Gamma} \mathbf{t}(s) \cdot ds\right) \\ \sin\left(\oint_{\Gamma} \mathbf{t}(s) \cdot ds\right) & \cos\left(\oint_{\Gamma} \mathbf{t}(s) \cdot ds\right) \end{pmatrix} \tag{55}$$

Next, we refer to the requirements to be fulfilled by the matrix **D**, namely, that it is diagonal and that it has the diagonal numbers that are of norm 1. In order for that to happen, the vector-function $\mathbf{t}(s)$ has to fulfill along a given (closed) path Γ the condition:

$$\oint_{\Gamma} \mathbf{t}(s) \cdot ds = n\pi \tag{56}$$

where n is an *integer*. These conditions are essentially the Bohr–Sommerfeld quantization conditions [82] (as applied to the single term of the two-state $\boldsymbol{\tau}$ matrix).

Equation (56) presents the condition for the extended conical intersection case. It is noticed that if n is an odd integer the diagonal of the **D** matrix contains two (-1) terms, which means that the elements of the adiabatic-to-diabatic transformation matrix flip sign while tracing the closed contour in Eq. (56) [see Eq. (39)]. This case is reminiscent of what happened in the simplified Jahn–Teller model as was studied by Herzberg–Longuet–Higgins [16] in which they showed that if two eigenfunctions that belong to the two states that form a conical intersection, trace a closed contour around that conical intersection, both of them flip sign (see Appendix A).

If the value of n in Eq. (56), is an even integer, the diagonal of the **D** matrix contains two $(+1)$ terms, which implies that in this case none of the elements of the adiabatic-to-diabatic transformation matrix flip sign while tracing the closed contour. This situation will be identified as the case where the above mentioned two eigenfunctions trace a closed contour but do not flip sign—the case known as the Renner–Teller model [15,83]. Equation (56) is the extended version of the Renner–Teller case.

In principle, we could have a situation where one of the diagonal elements is (+1) and one (−1) but from the structure of the **D** matrix one can see that this case can never happen.

In our introductory remarks, we said that this section would be devoted to model systems. Nevertheless it is important to emphasize that although this case is treated within a group of model systems this *model* stands for the general case of a two-state sub-Hilbert space. Moreover, this is the only case for which we can show, analytically, for a nonmodel system, that the restrictions on the **D** matrix indeed lead to a quantization of the relevant non-adiabatic coupling term.

2. *The Three-State Case*

The non-adiabatic coupling matrix $\boldsymbol{\tau}$ will be defined in a way similar to that in the Section V.A [see Eq. (51)], namely, as a product between a vector-function $\mathbf{t}(s)$ and a constant antisymmetric matrix $\mathbf{g}$ written in the form

$$\mathbf{g} = \begin{pmatrix} 0 & 1 & 0 \\ -1 & 0 & \eta \\ 0 & -\eta & 0 \end{pmatrix} \tag{57}$$

where η is a (constant) parameter. By employing this form of $\mathbf{g}$, we assumed that g_{13} and g_{31} are zero (the more general case is treated elsewhere [80]). The eigenvalues of this matrix are

$$\omega_{1,2} = \pm i\omega \qquad \omega_3 = 0 \qquad \omega = \sqrt{1+\eta^2} \tag{58}$$

and the corresponding matrix, $\mathbf{G}$, that diagonalizes the matrix $\mathbf{g}$ is

$$\mathbf{G} = \frac{1}{\omega\sqrt{2}} \begin{pmatrix} 1 & 1 & \eta\sqrt{2} \\ i\omega & -i\omega & 0 \\ -\eta & -\eta & \sqrt{2} \end{pmatrix} \tag{59}$$

By again employing Eq. (52), we find the following result for the **D** matrix

$$\mathbf{D} = \omega^{-2} \begin{pmatrix} \eta^2 + C & \omega S & \eta(1-C) \\ \omega S & \omega^2 C & -\eta\omega S \\ \eta(1-C) & \eta\omega S & 1+\eta^2 C \end{pmatrix} \tag{60}$$

where

$$C = \cos\left(w\oint_\Gamma \mathbf{t}(s)\cdot ds\right) \qquad \text{and} \qquad S = \sin\left(w\oint_\Gamma \mathbf{t}(s)\cdot ds\right) \tag{61}$$

Notice that the necessary and sufficient condition for this matrix to become diagonal is that the following condition:

$$\omega\oint_\Gamma \mathbf{t}(s)\cdot ds = \sqrt{1+\eta^2}\oint_\Gamma \mathbf{t}(s)\cdot ds = 2n\pi \tag{62}$$

be fulfilled. Moreover, this condition leads to a **D** matrix that contains in its diagonal numbers of norm 1 as required. However, in contrast to the previously described two-state case, they, all three of them, are positive, namely, (+1). In other words the "quantization" of the matrix $\boldsymbol{\tau}$ as expressed in Eq. (62) leads to a **D** matrix that is a unit matrix, and therefore will maintain the adiabatic-to-diabatic transformation matrix single valued along any contour that fulfills this quantization. This is, to a certain extent, an unexpected result but, as we shall see in the Section V.A.3, it is not the typical result. Still it is an interesting result and we shall return to it in Sections X and XII.

3. *The Four-State Case*

The **g** matrix in this case will be written in the form

$$\mathbf{g} = \begin{pmatrix} 0 & 1 & 0 & 0 \\ -1 & 0 & \eta & 0 \\ 0 & -\eta & 0 & \sigma \\ 0 & 0 & -\sigma & 0 \end{pmatrix} \tag{63}$$

where η and σ are two parameters. The matrix **G** that diagonalizes **g** is

$$\mathbf{G} = \frac{1}{\sqrt{2}}\begin{pmatrix} i\lambda_q & i\lambda_q & -i\lambda_p & -i\lambda_p \\ p\lambda_q & -p\lambda_q & -q\lambda_p & q\lambda_p \\ i\lambda_p & i\lambda_p & i\lambda_q & i\lambda_q \\ q\lambda_p & -q\lambda_p & p\lambda_q & -p\lambda_q \end{pmatrix} \tag{64}$$

where p and q are defined as

$$\begin{aligned} p &= \frac{1}{\sqrt{2}}(\varpi^2 + \sqrt{\varpi^4 - 4\sigma^2})^{(1/2)} \\ q &= \frac{1}{\sqrt{2}}(\varpi^2 - \sqrt{\varpi^4 - 4\sigma^2})^{(1/2)} \end{aligned} \tag{65}$$

and λ_p and λ_q are defined as

$$\lambda_p = \sqrt{\frac{p^2 - 1}{p^2 - q^2}} \qquad \lambda_q = \sqrt{\frac{1 - q^2}{p^2 - q^2}} \tag{66}$$

and ϖ as

$$\varpi = \sqrt{(1 + \eta^2 + \sigma^2)} \tag{67}$$

From Eq. (65), it is obvious that $p > q$. The four eigenvalues are

$$(\omega_1, \omega_2, \omega_3, \omega_4) \equiv (ip, -ip, iq, -iq) \tag{68}$$

Again, by employing Eq. (52) we find the following expressions for the **D**-matrix elements:

$$\begin{aligned}
&\mathbf{D}_{11}(\alpha) = \lambda_q^2 C_p + \lambda_p^2 C_q && \mathbf{D}_{12}(\alpha) = p\lambda_q^2 S_p + q\lambda_p^2 S_q \\
&\mathbf{D}_{13}(\alpha) = \lambda_p\lambda_q(-C_p + C_q) && \mathbf{D}_{14}(\alpha) = \lambda_p\lambda_q(-qS_p + pS_q) \\
&\mathbf{D}_{22}(\alpha) = p^2\lambda_q^2 C_p + q^2\lambda_p^2 C_q && \mathbf{D}_{23}(\alpha) = \lambda_p\lambda_q(pS_p - qS_q) \\
&\mathbf{D}_{24}(\alpha) = pq\lambda_p\lambda_q(C_p - C_q) && \mathbf{D}_{33}(\alpha) = (\lambda_p^2 C_p + \lambda_q^2 C_q \\
&\mathbf{D}_{34}(\alpha) = -(q\lambda_p^2 S_p + p\lambda_q^2 S_q) && \mathbf{D}_{44}(\alpha) = q^2\lambda_p^2 C_p + p^2\lambda_q^2 C_q \\
&\mathbf{D}_{21}(\alpha) = -\mathbf{D}_{12}(\alpha) \quad \mathbf{D}_{31}(\alpha) = \mathbf{D}_{13}(\alpha) && \mathbf{D}_{32}(\alpha) = -\mathbf{D}_{23}(\alpha) \\
&\mathbf{D}_{41}(\alpha) = -\mathbf{D}_{14}(\alpha) \quad \mathbf{D}_{42}(\alpha) = \mathbf{D}_{24}(\alpha) && \mathbf{D}_{43}(\alpha) = -\mathbf{D}_{34}(\alpha)
\end{aligned} \tag{69}$$

where

$$C_p = \cos(p\alpha) \qquad \text{and} \qquad S_p = \sin(p\alpha) \tag{70}$$

and similar expressions for C_q and S_q. Here α stands for

$$\alpha = \oint_\Gamma \boldsymbol{t}(s') \cdot ds' \tag{71}$$

Next, we determine the conditions for this matrix to become diagonal (with numbers of norm 1 in the diagonal), which will happen if and only if when p and q fulfill the following relations:

$$p\alpha = p\oint_\Gamma \boldsymbol{t}(s') \cdot ds' = 2\pi n \tag{72a}$$

$$q\alpha = q\oint_\Gamma \boldsymbol{t}(s') \cdot ds' = 2\pi\ell \tag{72b}$$

where n (>1) and ℓ, defined in the range $n > \ell \geq 0$, are allowed to be either integers or half-integers but m $(= n - \ell)$ can only attain *integer* values. The difference between the case where n and ℓ are integers and the case where both are half-integers is as follows: By examining the expressions in Eq. (69), notice that in the first case all diagonal elements of **D** are $(+1)$, so that, **D** is, in fact, the unit matrix and therefore the elements of the adiabatic-to-diabatic transformation matrix are single valued in configuration space. In the second case, we get from Eq. (69), that all four diagonal elements are (-1). In this case, when the adiabatic-to-diabatic transformation traces a closed contour all its elements flip sign.

Since p and q are directly related to the non-adiabatic coupling terms η and σ [see Eqs. (65) and (66)] the two conditions in Eqs. (72) imply, again, "quantization" conditions for the values of the τ-matrix elements, namely, for η and σ, as well as for the vectorial function $\mathbf{t}(s)$.

It is interesting to note that this is the first time that in the present framework the quantization is formed by two quantum numbers: a number n to be termed the principal quantum number and a number ℓ, to be termed the secondary quantum number. This case is reminiscent of the two quantum numbers that characterize the hydrogen atom.

4. *Comments Concerning Extensions*

In Sections V.A.1–V.A.3, we treated one particular group of τ matrices as presented in Eq. (51), where $\mathbf{g}$ is an antisymmetric matrix with constant elements. The general theory demands that the matrix **D** as presented in Eq. (52) be diagonal and that as such it contains $(+1)$ and (-1) values in its diagonal. In the three examples that were worked out, we found that for this particular class of τ matrices the corresponding **D** matrix contains either $(+1)$ or (-1) terms but never a mixture of the two types. In other words, the **D** matrix can be represented in the following way:

$$\mathbf{D} = (-1)^k \mathbf{I} \tag{73}$$

where k is either even or odd and **I** is the unit matrix. Indeed, for the two-state case k was found to be either odd or even, for the three-state case it was found to be only even, and for the four-state case it was again found to be either odd or even. It seems to us (without proof) that this pattern applies to any dimension. If this really is the case, then we can make the following two statements:

1. In case the dimension of the τ matrix is an odd number, the **D** matrix will always be the unit matrix **I**, namely, k must be an even number. This is so because an odd dimensional $\mathbf{g}$ matrix, always has the zero as an eigenvalue and this eigenvalue produces the $(+1)$ in the **D** matrix that dictates the value of k in Eq. (73).

2. In case the dimension of the $\boldsymbol{\tau}$ matrix is an even number, the **D** matrix will (always) be equal either to **I** or to $(-\mathbf{I})$.
3. These two facts imply that in case of an odd dimension the quantization is characterized by (a series of) integers only [as in Eq. (62)] but in case of an even dimension it is characterized either by (a series of) integers or by (a series of) half-integers [as in Eqs. (72)].

B. The Treatment of the General Case

The derivation of the **D** matrix for a given contour is based on first deriving the adiabatic-to-diabatic transformation matrix, **A**, as a function of s and then obtaining its value at the end of the arbitrary closed contours (when s becomes s_0). Since **A** is a real unitary matrix it can be expressed in terms of cosine and sine functions of given angles. First, we shall consider briefly the two special cases with $M = 2$ and 3.

The case of $M = 2$ was treated in Section V.A.4. Here, this treatment is repeated with the aim of emphasizing different aspects and also for reasons of completeness. The matrix $\mathbf{A}^{(2)}$ takes the form:

$$\mathbf{A}^{(2)} = \begin{pmatrix} \cos\gamma_{12} & \sin\gamma_{12} \\ -\sin\gamma_{12} & \cos\gamma_{12} \end{pmatrix} \tag{74}$$

where γ_{12}, the adiabatic-to-diabatic transformation angle, can be shown to be [34]

$$\gamma_{12} = \int_{s_0}^{s} \tau_{12}(s') \cdot ds' \tag{75}$$

Designating α_{12} as the value of γ_{12} for a closed contour, namely,

$$\alpha_{12} = \oint_{\Gamma} \tau_{12}(s') \cdot ds' \tag{76}$$

the corresponding $\mathbf{D}^{(2)}$ matrix becomes accordingly [see also Eq. (55)]:

$$\mathbf{D}^{(2)} = \begin{pmatrix} \cos\alpha_{12} & \sin\alpha_{12} \\ -\sin\alpha_{12} & \cos\alpha_{12} \end{pmatrix} \tag{77}$$

Since for any closed contour $\mathbf{D}^{(2)}$ has to be a diagonal matrix with $(+1)$ and (-1) terms, it is seen that $\alpha_{12} = n\pi$ where n is either odd or even (or zero) and therefore the only two possibilities for $\mathbf{D}^{(2)}$ are as follows:

$$\mathbf{D}^{(2)} = (-1)^k \mathbf{I} \tag{78}$$

where **I** is the unit matrix and k is either even or odd.

The case of $M = 3$ is somewhat more complicated because the corresponding orthogonal matrix is expressed in terms of three angles, namely, γ_{12}, γ_{13}, and γ_{23} [36,84,85]. This case was recently studied by us in detail [85] and here we briefly repeat the main points.

The matrix $\mathbf{A}^{(3)}$ is presented as a product of three rotation matrices of the form:

$$\mathbf{Q}_{13}^{(3)}(\gamma_{13}) = \begin{pmatrix} \cos\gamma_{13} & 0 & \sin\gamma_{13} \\ 0 & 1 & 0 \\ -\sin\gamma_{13} & 0 & \cos\gamma_{13} \end{pmatrix} \tag{79}$$

[the other two, namely, $\mathbf{Q}_{12}^{(3)}(\gamma_{12})$ and $\mathbf{Q}_{23}^{(3)}(\gamma_{23})$, are of a similar structure with the respective cosine and sine functions in the appropriate positions) so that $\mathbf{A}^{(3)}$ becomes:

$$\mathbf{A}^{(3)} = \mathbf{Q}_{12}^{(3)}\mathbf{Q}_{23}^{(3)}\mathbf{Q}_{13}^{(3)} \tag{80}$$

or, following the multiplication, the more explicit form:

$$\mathbf{A}^{(3)} = \begin{pmatrix} c_{12}c_{13} - s_{12}s_{23}s_{13} & s_{12}s_{23} & c_{12}s_{13} + c_{12}s_{23}c_{13} \\ -s_{12}c_{13} - c_{12}s_{23}s_{13} & c_{12}c_{23} & -s_{12}s_{13} + c_{12}s_{23}c_{13} \\ -c_{23}s_{13} & -s_{23} & c_{23}c_{13} \end{pmatrix} \tag{81}$$

Here, $c_{ij} = \cos(\gamma_{ij})$ and $s_{ij} = \sin(\gamma_{ij})$. The three angles are obtained by solving the following three coupled first-order differential equations, which follow from Eq. (19) [36,84,85]:

$$\begin{aligned} \nabla\gamma_{12} &= \tau_{12} - \tan\gamma_{23}(-\tau_{13}\cos\gamma_{12} + \tau_{23}\sin\gamma_{12}) \\ \nabla\gamma_{23} &= -(\tau_{23}\cos\gamma_{12} + \tau_{13}\sin\gamma_{12}) \\ \nabla\gamma_{13} &= -(\cos\gamma_{23})^{-1}(-\tau_{13}\cos\gamma_{12} + \tau_{23}\sin\gamma_{12}) \end{aligned} \tag{82}$$

These equations were integrated as a function of φ (where $0 \le \varphi \le 2\pi$), for a *model* potential [85] along a circular contour of radius ρ (for details see Appendix E). The φ-dependent γ angles, that is, $\gamma_{ij}(\varphi \mid \rho)$, for various values of ρ and $\Delta\varepsilon$ ($\Delta\varepsilon$ is the potential energy shift defined as the shift between the two original coupled adiabatic states and a third state, at the origin, i.e., at $\rho = 0$.) are presented in Figure 1. Thus for each φ we get, employing Eq. (81), the $\mathbf{A}^{(3)}(\varphi)$ matrix elements. The relevant $\mathbf{D}^{(3)}$ matrix is obtained from $\mathbf{A}^{(3)}$ by substituting $\varphi = 2\pi$. If α_{ij} are defined as

$$\alpha_{ij} = \gamma_{ij}(\varphi = 2\pi) \tag{83}$$

then, as is noticed from Figure 1, the values of α_{ij} are either zero or π. A simple analysis of Eq. (81), for these values of α_{ij}, shows that $\mathbf{D}^{(3)}$ is a diagonal matrix with two (-1) terms and one $(+1)$ in the diagonal.

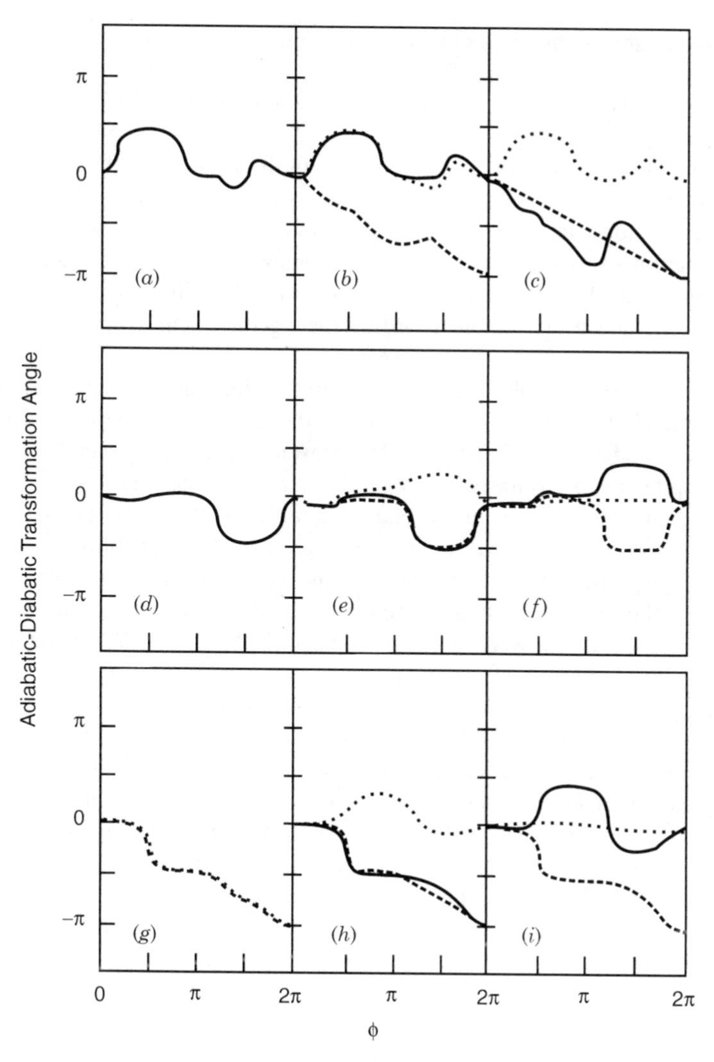

Figure 1. The three adiabatic–diabatic transformation angles [obtained by solving Eqs. (77) for a 3 × 3 diabatic model potential presented in Section XIII.B] $\gamma_{12}(\varphi)$, $\gamma_{23}(\varphi)$, $\gamma_{13}(\varphi)$ as calculated for different values of ρ and $\Delta\varepsilon$: (*a*) $\gamma = \gamma_{12}$, $\Delta\varepsilon = 0.0$; (*b*) $\gamma = \gamma_{12}$, $\Delta\varepsilon = 0.05$; (*c*) $\gamma = \gamma_{12}$, $\Delta\varepsilon = 0.25$; (*d*) $\gamma = \gamma_{23}$, $\Delta\varepsilon = 0.0$; (*e*) $\gamma = \gamma_{23}$, $\Delta\varepsilon = 0.05$; (*f*) $\gamma = \gamma_{23}$, $\Delta\varepsilon = 0.25$; (*g*) $\gamma = \gamma_{13}$, $\Delta\varepsilon = 0.0$; (*h*) $\gamma = \gamma_{13}$, $\Delta\varepsilon = 0.05$; (*i*) $\gamma = \gamma_{13}$, $\Delta\varepsilon = 0.25$. ——— $\rho = 0.01$; - - - - - - - $\rho = 0.1$; .... $\rho = 0.5$.

This result will now be generalized for an arbitrary $\mathbf{D}^{(3)}$ matrix in the following way: Since a general $\mathbf{A}^{(3)}$ matrix can always be written as in Eq. (81) the corresponding $\mathbf{D}^{(3)}$ matrix becomes diagonal if and only if:

$$\alpha_{ij} = \gamma_{ij}(\varphi = 2\pi) = n_{ij}\pi \tag{84}$$

the diagonal terms can, explicitly, be represented as

$$\mathbf{D}_{ij}^{(3)} = \delta_{ij} \cos\alpha_{jn} \cos\alpha_{jm} \qquad j \neq n \neq m \quad j = 1, 2, 3 \tag{85}$$

This expression shows that the $\mathbf{D}^{(3)}$ matrix, in the most general case, can have either three $(+1)$ terms in the diagonal or two (-1) terms and one $(+1)$. In the first case, the contour does not surround any conical intersection, whereas in the second case it surrounds either one or two conical intersections (a more general discussion related to the solution of the corresponding line integral is given in Section VIII and a discussion regarding the "geometrical" aspect is given in Section IX).

It is important to emphasize that this analysis, although it is supposed to hold for a general three-state case, contradicts the analysis we performed of the three-state model in Section V.A.2. The reason is that the "general (physical) case" applies to an (arbitrary) aggregation of conical intersections whereas the previous case applies to a special (probably unphysical) situation. The discussion on this subject is extended in Section X. In what follows, the cases for an aggregation of conical intersections will be termed the "breakable" situations (the reason for choosing this name will be given later) in contrast to the type of models that were discussed in Sections V.A.2 and V.A.3 and that are termed as the "unbreakable" situation.

Before discussing the general case, we would like to refer to the present choice of the rotation angles. It is well noticed that they differ from the ordinary *Euler* angles that are routinely used to present three-dimensional (3D) orthogonal matrices [86]. In fact, we could apply the Euler angles for this purpose and get identical results for $\mathbf{A}^{(3)}$ (and for $\mathbf{D}^{(3)}$). The main reason we prefer the "democratic" choice of the angles is that this set of angles can be extended to an arbitrary number of dimensions as will be done next.

The M-dimensional adiabatic-to-diabatic transformation matrix $\mathbf{A}^{(M)}$ will be written as a product of elementary rotation matrices similar to that given in Eq. (80) [9]:

$$\mathbf{A}^{(M)} = \prod_{i=1}^{M-1} \prod_{j>i}^{M} \mathbf{Q}_{ij}^{(M)}(\gamma_{ij}) \tag{86}$$

where $\mathbf{Q}_{ij}^{(M)}(\gamma_{ij})$ [like in Eq. (79)] is an $M \times M$ matrix with the following terms: In its (ii) and (jj) positions (along the diagonal) are located the two relevant cosine functions and at the rest of the (M − 2) positions are located $(+1)s$; in the (ij) and (ji) off-diagonal positions are located the two relevant ±sine functions and at all other remaining positions are zeros. From Eq. (86), it can be seen that the number of matrices contained in this product is $M(M-1)/2$ and that this is also the number of independent γ_{ij} angles that are needed to describe an $M \times M$

unitary matrix (we recall that the missing $M(M+1)/2$ conditions follow from the ortho-normal conditions). The matrix $\mathbf{A}^{(M)}$ as presented in Eq. (86) is characterized by two important features: (1) Every *diagonal* element contains at least one term that is a product of cosine functions only. (2) Every *off-diagonal* element is a summation of products of terms where each product contains at least one sine function. These two features will lead to conditions to be imposed on the various γ_{ij} angles to ensure that the topological matrix, $\mathbf{D}^{(M)}$, is diagonal as discussed in the Section IV.A.

To obtain the γ_{ij} angles one usually has to solve the relevant first-order differential equations of the type given in Eq. (82). Next, like before, the α_{ij} angles are defined as the γ_{ij} angles at the end of a closed contour. In order to obtain the matrix $\mathbf{D}^{(M)}$, one has to replace, in Eq. (86), the angles γ_{ij} by the corresponding α_{ij} angles. Since $\mathbf{D}^{(M)}$ has to be a *diagonal* matrix with $(+1)$ and (-1) terms in the diagonal, this can be achieved if and only if *all* α_{ij} angles are zero or multiples of π. It is straightforward to show that with this structure the elements of $\mathbf{D}^{(M)}$ become [9]:

$$\mathbf{D}_{ij}^{(M)} = \delta_{ij} \prod_{k \neq i}^{M} \cos \alpha_{ik} = \delta_{ij} (-1)^{\sum_{k \neq i}^{M} n_{ik}} \qquad i = 1, \ldots, M \tag{87}$$

where n_{ik} are integers that fulfill $n_{ik} = n_{ki}$. From Eq. (87), it is noticed that along the diagonal of $\mathbf{D}^{(M)}$ we may encounter K numbers that are equal to (-1) and the rest that are equal to $(+1)$. It is important to emphasize that in case a contour does not surround any conical intersection the value of K is zero.

VI. THE CONSTRUCTION OF SUB-HILBERT SPACES AND SUB-SUB-HILBERT SPACES

In Section II.B, it was shown that the condition in Eq. (10) or its relaxed form in Eq. (40) enables the construction of sub-Hilbert space. Based on this possibility we consider a prescription first for constructing the sub-Hilbert space that extends to the full configuration space and then, as a second step, constructing of the sub sub-Hilbert space that extends only to (a finite) portion of configuration space.

In the study of (electronic) curve crossing problems, one distinguishes between a situation where two electronic curves, $E_j(R)$, $j = 1, 2$, approach each other at a point $R = R_0$ so that the difference $\Delta E(R = R_0) = E_2(R = R_0) - E_1$ is relatively small and a situation where the two electronic curves interact so that $\Delta E(R) \sim$ Const is relatively large. The first case is usually treated by the Landau–Zener formula [87–92] and the second is based on the Demkov approach [93]. It is well known that whereas the Landau–Zener type interactions are strong enough to cause transitions between two adiabatic states, the Demkov-type interactions are usually weak and affect the motion of the interacting molecular species relatively slightly. The Landau–Zener situation is the one that may become the Jahn–Teller conical intersection in two dimensions [15–21]. We shall also include the Renner–Teller parabolic intersection [15,22,26,83], although it is characterized by two interacting potential energy surfaces that behave quadratically (and not linearly as in the Landau–Zener case) in the vicinity of the above mentioned degeneracy point.

A. The Construction of Sub-Hilbert Spaces

By following Section II.B, we shall be more specific about what is meant by "strong" and "weak" interactions. It turns out that such a criterion can be assumed, based on whether two *consecutive* states do, or do not, form a conical intersection or a parabolical intersection (it is important to mention that only consecutive states can form these intersections). The two types of intersections are characterized by the fact that the nonadiabatic coupling terms, at the points of the intersection, become infinite (these points can be considered as the "black holes" in molecular systems and it is mainly through these "black holes" that electronic states interact with each other.). Based on what was said so far we suggest breaking up complete Hilbert space of size N into L sub-Hilbert spaces of varying sizes $N_P, P = 1, \ldots, L$ where

$$N = \sum_{P=1}^{L} N_P. \tag{88}$$

(L may be finite or infinite.)

Before we continue with the construction of the sub-Hilbert spaces, we make the following comment: Usually, when two given states form conical intersections, one thinks of isolated points in configuration space. In fact, conical intersections are not points but form (finite or infinite) seams that "cut" through the molecular configuration space. However, since our studies are carried out for planes, these planes usually contain isolated conical intersection points only.

The criterion according to which the break-up is carried out is based on the non-adiabatic coupling term $\boldsymbol{\tau}_{ij}$ as were defined in Eq. (8a). In what follows, we distinguish between two kinds of non-adiabatic coupling terms: (1) The intra-non-adiabatic coupling terms $\boldsymbol{\tau}_{ij}^{(P)}$, which are formed between two eigenfunctions belonging to a given sub-Hilbert space, namely, the Pth sub-space:

$$\boldsymbol{\tau}_{ij}^{(P)} = \langle \zeta_i^{(P)} | \nabla \zeta_j^{(P)} \rangle \qquad i,j = 1, \ldots, N_P \tag{89}$$

and (2) Inter-non-adiabatic coupling terms $\boldsymbol{\tau}_{ij}^{(P,Q)}$, which are formed between two eigenfunctions, the first belonging to the Pth sub-space and the second to the Qth sub-space:

$$\boldsymbol{\tau}_{ij}^{(P,Q)} = \langle \zeta_i^{(P)} | \nabla \zeta_j^{(Q)} \rangle \qquad i = 1, \ldots, N_P \qquad j = 1, \ldots, N_Q \tag{90}$$

The Pth sub-Hilbert space is defined through the following two requirements:

1. All N_p states belonging to the Pth sub-space interact strongly with each other in the sense that each pair of *consecutive* states have at least one point where they form a Landau–Zener type interaction. In other words, all $\boldsymbol{\tau}_{jj+1}^{(P)}$; $j = 1, \ldots, N_P - 1$ form at least at one point in configuration space, a conical (parabolical) intersection.
2. The range of the Pth sub-space is defined in such a way that the lowest (or the first) state and the highest (the N_Pth) state that belong to this sub-space form Demkov-type interactions with the highest state belonging to the lower $(P-1)$th sub-space and with the lowest state belonging to the upper $(P+1)$th sub-space, respectively (see Fig. 2). In other words, the two non-adiabatic coupling terms fulfill the conditions:

$$\boldsymbol{\tau}_{N_{P-1}1}^{(P-1,P)} \sim O(\varepsilon) \qquad \text{and} \qquad \boldsymbol{\tau}_{N_P 1}^{(P,P+1)} \sim O(\varepsilon) \tag{91}$$

At this point, we make two comments: (a) Conditions (1) and (2) lead to a well-defined sub-Hilbert space that for any further treatments (in spectroscopy or scattering processes) has to be treated as a whole (and not on a "state by state" level). (b) Since all states in a given sub-Hilbert space are adiabatic states, strong interactions of the Landau–Zener type can occur between two consecutive states only. However, Demkov-type interactions may exist between any two states.

B. The Construction of Sub-Sub-Hilbert Spaces

As we have seen, the sub-Hilbert spaces are defined for the whole configuration space and this requirement could lead, in certain cases, to situations where it will be necessary to include the complete Hilbert space. However, it frequently happens that the dynamics we intend to study takes place in a given, isolated, region that contains only part of the conical intersection points and the question is whether the effects of the other conical intersections can be ignored?

The answer to this question can be given following a careful study of these effects employing the line integral approach presented in terms of Eq. (27). For this purpose, we analyze what happens along a certain line Γ that surrounds

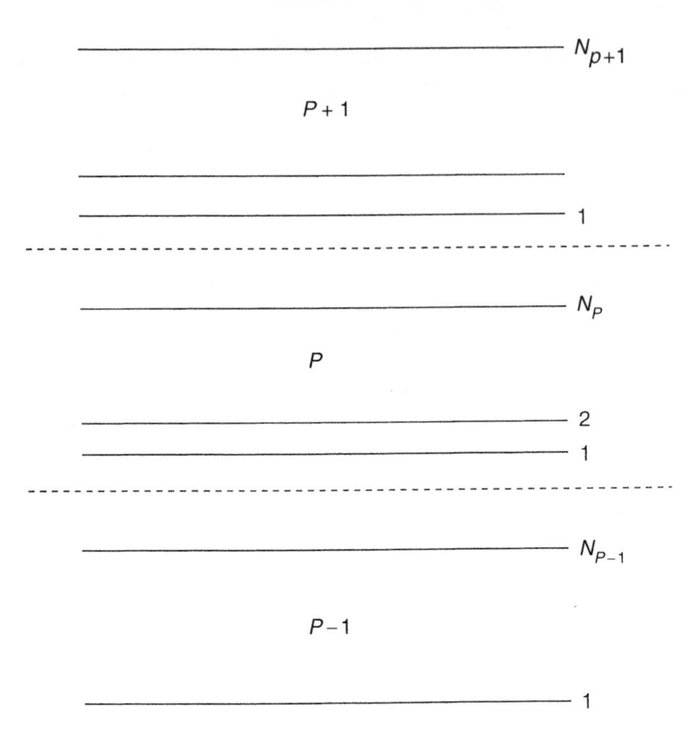

Figure 2. A schematic picture describing the three consecutive sub-Hilbert spaces, namely, the $(P-1)$th, the Pth, and the $(P+1)$th. The dotted lines are separation lines.

one or several conical intersections. To continue, we employ the same procedure as discussed in Section IV.B: We break up the adiabatic-to-diabatic transformation matrix **A** and the $\boldsymbol{\tau}$ matrix as written in Eq. (43). In this way, we can show that if, along the particular line Γ, the noninteresting parts of the $\boldsymbol{\tau}$ matrix are of order ε the error expected for the interesting part in the **A** matrix is of order $O(\varepsilon^2)$ [81]. If this happens for any contour in this region, then we can ignore the effects of conical intersection that are outside this region and carry out the dynamic calculations employing the reduced set of states.

VII. THE TOPOLOGICAL SPIN

Before we continue and in order to avoid confusion, two matters have to be clarified: (1) We distinguished between two types of Landau–Zener situations, which form (in two dimensions) the Jahn–Teller conical intersection and the Renner–Teller parabolical intersection. The main difference between the two is that the parabolical intersections do not produce topological effects and therefore, as far as this subject is concerned, they can be ignored. Making this distinction leads to the conclusion that the more relevant magnitude to characterize topological effects, for a given sub-space, is not its dimension M but N_J, the number of conical intersections. (2) In general, one may encounter more than one conical intersection between a pair of states [12,22,26,66,74]. However, to simplify the study, we assume one conical intersection for a pair of states so that $(N_J + 1)$ stands for the *number of states* that form the conical intersections.

So far, we introduced three different integers M, N_J, and K. As mentioned earlier, M is a characteristic number of the sub-space (see Section VI.B) but is not relevant for topological effects; instead N_J, as just mentioned, is a characteristic number of the sub-space and relevant for topological effects, and K, the number of (-1) terms in the diagonal of the topological matrix $\mathbf{D}$ (or the number of eigenstates that flip sign while the electronic manifold traces a closed contour) is relevant for topological effects but may vary from one contour to another, and therefore is not a characteristic feature for a given sub-space.

Our next task is to derive all possible K values for a given N_J. First, we refer to a few special cases: It was shown before that in case of $N_J = 1$ the $\mathbf{D}$ matrix contains two (-1) terms in its diagonal in case the contour surrounds the conical intersection and no (-1) terms when the contour does not surround the conical intersection. Thus the allowed values of K are either 2 or 0. The value $K = 1$ is not allowed. A similar inspection of the case $N_J = 2$ reveals that K, as before, is equal either to 2 or to 0 (see Section V.B). Thus the values $K = 1$ or 3 are not allowed. From here, we continue to the general case and prove the following statement:

In any molecular system, K can attain only *even* integers in the range [9]:

$$K = \{0, 2, \ldots, K_J\} \quad \begin{cases} K_J = N_J & N_J = 2p \\ K_J = (N_J + 1) & N_J = 2p + 1 \end{cases} \tag{92}$$

where p is an integer.

The proof is based on Eq. (87). Let us assume that a certain closed contour yields a set of α_{ij} angles that produce a number K. Next, we consider a slightly different closed contour, along which one of these α_{ij} parameters, say α_{st}, changed its value from zero to π. From Eq. (87), it can be seen that only two $\mathbf{D}$ matrix elements contain $\cos(\alpha_{st})$, namely, $\mathbf{D}_{ss}$ and $\mathbf{D}_{tt}$. Now, if these two matrix elements were positive following the first contour, then changing α_{st} from $0 \to \pi$ would produce two additional (-1) terms, thus increasing K to $K + 2$; if these two matrix elements were negative, this change would cause K to decrease to $K - 2$; and if one of these elements was positive and the other negative, then changing α_{st} from $0 \rightarrow \pi$ would not affect K. Thus, immaterial to the value of N_J, the various K values *differ* from each other by even integers only. Now, since any set of K values also contains the value $K = 0$ (the case when the closed loop does not surround any conical intersections), this implies that K can attain only even integers. The final result is the set of values presented in Eq. (92).

The fact that there is a one-to-one relation between the (-1) terms in the diagonal of the topological matrix and the fact that the eigenfunctions flip sign along closed contours (see discussion at the end of Section IV.A) hints at the possibility that these sign flips are related to a kind of a spin quantum number and in particular to its magnetic components.

The spin in quantum mechanics was introduced because experiments indicated that individual particles are not completely identified in terms of their three spatial coordinates [87]. Here we encounter, to some extent, a similar situation: A system of items (i.e., distributions of electrons) in a given point in configuration space is usually described in terms of its set of eigenfunctions. This description is incomplete because the existence of conical intersections causes the electronic manifold to be multivalued. For example, in case of two (isolated) conical intersections we may encounter at a given point in configuration space four different sets of eigenfunctions (see Section VIII).

$$\begin{aligned} &\text{(a)} \quad (\zeta_1, \zeta_2, \zeta_3) \\ &\text{(b)} \quad (-\zeta_1, -\zeta_2, \zeta_3) \\ &\text{(c)} \quad (\zeta_1, -\zeta_2, -\zeta_3) \\ &\text{(d)} \quad (-\zeta_1, \zeta_2, -\zeta_3) \end{aligned} \tag{93}$$

In case of three conical intersections, we have as many as eight different sets of eigenfunctions, and so on. Thus we have to refer to an additional characterization of a given sub-sub-Hilbert space. This characterization is related to the number N_J of conical intersections and the associated possible number of sign flips due to different contours in the relevant region of configuration space, traced by the electronic manifold.

In [7,8,80], it was shown that in a two-state system the nonadiabatic coupling term, τ_{12}, has to be "quantized" in the following way:

$$\oint_\Gamma \boldsymbol{\tau}_{12}(s') \cdot ds' = n\pi \tag{94}$$

where n is an integer (in order to guarantee that the 2×2 diabatic potential be single valued in configuration space). In case of conical intersections, this number has to be an odd integer and for our purposes it is assumed to be $n = 1$. Thus each conical intersection can be considered as a "spin." Since in a given sub-space N_J conical intersections are encountered, we could define the spin J of this sub-space as $(N_J/2)$. However, this definition may lead to more sign flips than we actually encounter (see Section VIII). In order to make a connection between J and N_J as well as with the "magnetic components" M_J of J and the number of the actual sign flips, the spin J has to be defined as [9]:

$$J = \frac{1}{2}\frac{K_J}{2} \qquad \begin{cases} K_J = N_J & N_J = 2p \\ K_J = (N_J + 1) & N_J = 2p + 1 \end{cases} \tag{95}$$

and, accordingly, the various M_J values are defined as

$$M_J = J - K/2 \qquad \text{where} \qquad K = \{0, 2, \ldots, K_J\} \tag{96}$$

For the seven lowest N_J values, we have the following assignments:

$$\begin{array}{lll} \text{For } N_J = 0 & \{J = 0 & M_J = 0\} \\ \text{For } N_J = 1 & \{J = 1/2 & M_J = 1/2, -1/2\} \\ \text{For } N_J = 2 & \{J = 1/2 & M_J = 1/2, -1/2\} \\ \text{For } N_J = 3 & \{J = 1 & M_J = 1, 0, -1\} \\ \text{For } N_J = 4 & \{J = 1 & M_J = 1, 0, -1\} \\ \text{For } N_J = 5 & \{J = 3/2 & M_J = 3/2, 1/2, -1/2, -3/2\} \\ \text{For } N_J = 6 & \{J = 3/2 & M_J = 3/2, 1/2, -1/2, -3/2\} \\ \text{For } N_J = 7 & \{J = 2 & M_J = 2, 1, 0, -1, -2\} \end{array} \tag{97}$$

The general formula and the individual cases as presented in Eq. (97) indicate that indeed the number of conical intersections in a given sub-space and the number of possible sign flips within this sub-sub-Hilbert space are interrelated, similar to a spin J with respect to its magnetic components M_J. In other words, each decoupled sub-space is now characterized by a spin quantum number J that connects between the number of conical intersections in this system and the topological effects which characterize it.

VIII. AN ANALYTICAL DERIVATION FOR THE POSSIBLE SIGN FLIPS IN A THREE-STATE SYSTEM

In Section IX, we intend to present a geometrical analysis that permits some insight with respect to the phenomenon of sign flips in an M-state system ($M > 2$). This can be done without the support of a parallel mathematical study [9]. In this section, we intend to supply the mathematical foundation (and justification) for this analysis [10,12]. Thus employing the line integral approach, we intend to prove the following statement:

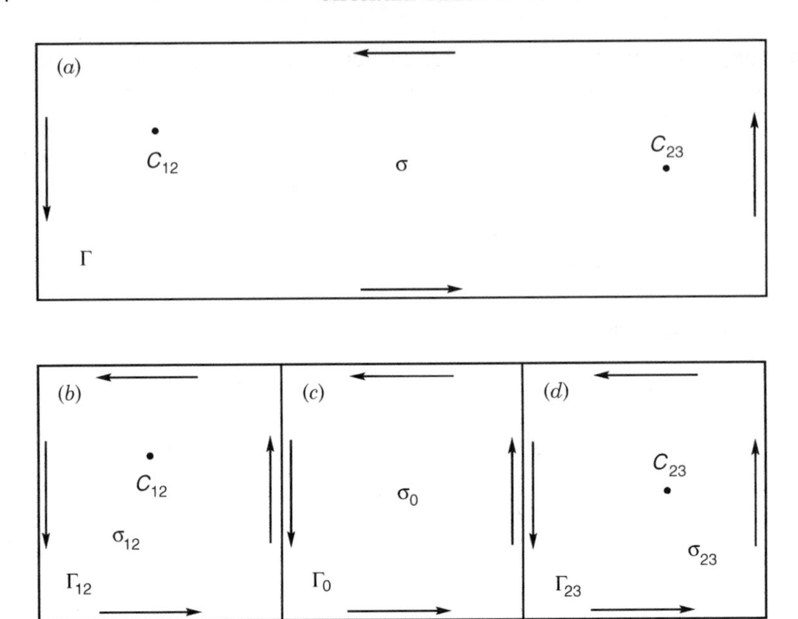

Figure 3. The breaking up of a region σ, which contains two conical intersections (at C_{12} and C_{23}), into three subregions: (*a*) The full region σ defined in terms of the closed contour Γ. (*b*) The region σ_{12}, which contains a conical intersection at C_{12} and is defined by the closed contour Γ_{12}. (*c*) The region σ_0, which is defined by the closed contour Γ_0 and does not contain any conical intersection. (*d*) The region σ_{23}, which contains a conical intersection at C_{23} and is defined by the closed contour Γ_{23}. It can be seen that $\Gamma = \Gamma_{12} + \Gamma_0 + \Gamma_{23}$.

If a contour in a given plane surrounds two conical intersections belonging to two different (adjacent) pairs of states, only two eigenfunctions flip sign—the one that belongs to the lowest state and the one that belongs to the highest one.

To prove this, we consider the following three regions (see Fig. 3): In the first region, designated σ_{12}, is located the main portion of the interaction, t_{12}, between states 1 and 2 with the point of the conical intersection at C_{12}. In the second region, designated as σ_{23}, is located the main portion of the interaction, t_{23}, between states 2 and 3 with the point of the conical intersection at C_{23}. In addition, we assume a third region, σ_0, which is located in-between the two and is used as a buffer zone. Next, it is assumed that the intensity of the interactions due to the components of t_{23} in σ_{12} and due to t_{12} in σ_{23} is ~ 0. This situation can always be achieved by shrinking $\sigma_{12}(\sigma_{23})$ toward its corresponding center $C_{12}(C_{23})$. In σ_0, the components of both t_{12} and t_{23} may be of arbitrary magnitude but no conical intersection of any pair of states is allowed to be there.

To prove our statement, we consider the line integral [see Eq. (27)]:

$$\mathbf{A} = \mathbf{A}_0 - \oint_{\Gamma} ds \cdot \boldsymbol{\tau}\mathbf{A} \tag{98}$$

where the integration is carried out along a closed contour Γ, $\mathbf{A}$ is the 3×3 adiabatic-to-diabatic transformation matrix to be calculated, the dot stands for a scalar product, and $\boldsymbol{\tau}$ is the matrix of 3×3 that contains the two non-adiabatic coupling terms, namely,

$$\boldsymbol{\tau}(s) = \begin{pmatrix} 0 & t_{12} & 0 \\ -t_{12} & 0 & t_{23} \\ 0 & -t_{23} & 0 \end{pmatrix} \tag{99}$$

Note the components of $t_{13} \sim 0$. This assumption is not essential for the proof, but simplifies the derivation.

The integral in Eq. (98) will now be presented as a sum of three integrals (for a detailed discussion on that subject: see Appendix C), namely,

$$\mathbf{A} = \mathbf{A}_0 - \oint_{\Gamma_{12}} ds \cdot \boldsymbol{\tau}\mathbf{A} - \oint_{\Gamma_0} ds \cdot \boldsymbol{\tau}\mathbf{A} - \oint_{\Gamma_{23}} ds \cdot \boldsymbol{\tau}\mathbf{A} \tag{100}$$

Since there is no conical intersection in the buffer zone, σ_0, the second integral is zero and can be deleted so that we are left with the first and the third integrals. In general, the calculation of each integral is independent of the other; however, the two calculations have to yield the same result, and therefore they have to be interdependent to some extent. Thus we do each calculation separately but for different (yet unknown) boundary conditions: The first integral will be done for $\mathbf{G}_{12}$ as a boundary condition and the second for $\mathbf{G}_{23}$. Thus $\mathbf{A}$ will be calculated twice:

$$\mathbf{A} = \mathbf{G}_{ij} - \oint_{\Gamma_{ij}} ds \cdot \boldsymbol{\tau}\mathbf{A} \tag{101}$$

Next are introduced the topological matrices $\mathbf{D}$, $\mathbf{D}_{12}$, and $\mathbf{D}_{23}$, which are related to $\mathbf{A}$ in the following way [see Eq. (39)]:

$$\mathbf{A} = \mathbf{D}\mathbf{A}_0 \tag{102a}$$

$$\mathbf{A} = \mathbf{D}_{12}\mathbf{G}_{12} \tag{102b}$$

$$\mathbf{A} = \mathbf{D}_{23}\mathbf{G}_{23} \tag{102c}$$

The three equalities can be fulfilled if and only if the two $\mathbf{G}$ matrices, namely, $\mathbf{G}_{12}$ and $\mathbf{G}_{23}$, are chosen to be

$$\mathbf{G}_{12} = \mathbf{D}_{23}\mathbf{A}_0 \qquad \text{and} \qquad \mathbf{G}_{23} = \mathbf{D}_{12}\mathbf{A}_0 \tag{103}$$

Since the $\mathbf{D}$ matrices are diagonal the same applies to $\mathbf{D}_{12}$ and $\mathbf{D}_{23}$ so that $\mathbf{D}$ becomes

$$\mathbf{D} = \mathbf{D}_{13} = \mathbf{D}_{12}\mathbf{D}_{23} \tag{104}$$

Our next task will be to obtain $\mathbf{D}_{12}$ and $\mathbf{D}_{23}$. For this purpose, we consider $\boldsymbol{\tau}_{12}$ and $\boldsymbol{\tau}_{23}$—the two partial $\boldsymbol{\tau}$ matrices—defined as follows:

$$\boldsymbol{\tau}_{12}(s) = \begin{pmatrix} 0 & t_{12} & 0 \\ -t_{12} & 0 & 0 \\ 0 & 0 & 0 \end{pmatrix} \qquad \text{and} \qquad \boldsymbol{\tau}_{23}(s) = \begin{pmatrix} 0 & 0 & 0 \\ 0 & 0 & t_{23} \\ 0 & -t_{23} & 0 \end{pmatrix} \tag{105}$$

so that

$$\boldsymbol{\tau} = \boldsymbol{\tau}_{12} + \boldsymbol{\tau}_{23} \tag{106a}$$

We start with the first of Eqs. (101), namely,

$$\mathbf{A} = \mathbf{G}_{12} - \oint_{\Gamma_{ij}} ds \cdot \boldsymbol{\tau}_{12}\mathbf{A} \tag{107}$$

where $\boldsymbol{\tau}_{12}$ replaces $\boldsymbol{\tau}$ because $\boldsymbol{\tau}_{23}$ is assumed to be negligibly small in σ_{12}. The solution and the corresponding $\mathbf{D}$ matrix, namely, $\mathbf{D}_{12}$ are well known (see discussion in Sections V.A.1 and V.B). Thus

$$\mathbf{D}_{12} = \begin{pmatrix} -1 & 0 & 0 \\ 0 & -1 & 0 \\ 0 & 0 & 1 \end{pmatrix} \tag{108}$$

which implies (as already explained in Section IV.A) that the first (lowest) and the second functions flip sign. In the same way, it can be shown that $\mathbf{D}_{23}$ is equal to

$$\mathbf{D}_{23} = \begin{pmatrix} 1 & 0 & 0 \\ 0 & -1 & 0 \\ 0 & 0 & -1 \end{pmatrix} \tag{109}$$

which shows that the second and the third (the highest) eigenfunctions flip sign. Substituting Eqs. (108) and (109) in Eq. (104) yields the following result for $\mathbf{D}_{13}$:

$$\mathbf{D}_{13} = \begin{pmatrix} -1 & 0 & 0 \\ 0 & 1 & 0 \\ 0 & 0 & -1 \end{pmatrix} \tag{110}$$

In other words, surrounding the two conical intersections indeed leads to the flip of sign of the first and the third eigenfunctions, as was claimed.

This idea can be extended, in a straightforward way, to various situations as will be done in Section IX.

IX. THE GEOMETRICAL INTERPRETATION FOR SIGN FLIPS

In Sections V and VII, we discussed the possible K values of the $\mathbf{D}$ matrix and made the connection with the number of signs flip based on the analysis given in Section IV.A. Here, we intend to present a geometrical approach in order to gain more insight into the phenomenon of signs flip in the M-state system ($M > 2$).

As was already mentioned, conical intersections can take place only between two adjacent states (see Fig. 4). Next, we make the following definitions:

1. Having two consecutive states j and $(j+1)$, the two form the conical intersection to be designated as C_j as shown in Figure 4, where N_J conical intersection are presented.

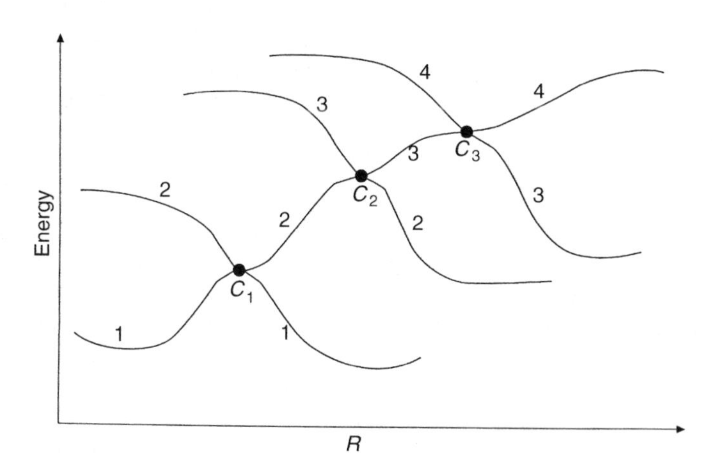

Figure 4. Four interacting adiabatic surfaces presented in terms of four adiabatic curves. The points C_j; $j = 1,2,3$, stand for the three conical intersections.

2. The contour that surrounds a conical intersection at C_j will be designated as Γ_{jj+1} [see Fig. 5(a)].
3. A contour that surrounds two consecutive conical intersections that is, C_j and C_{j+1} will be designated as Γ_{jj+2} [see Fig. 5(b)]. In the same way a contour that surrounds n consecutive conical intersections namely C_j, $C_{j+1} \ldots C_{j+n}$ will be designated as Γ_{jj+n} [see Fig. 5(c) for $N_J = 3$].
4. In case of three conical intersections or more, a contour that surrounds C_j and C_k but not the in-between conical intersections will be designated as $\Gamma_{j,k}$. Thus, for example, $\Gamma_{1,3}$ surrounds C_1 and C_3 but not C_2 (see Fig. 5d).

We also introduce an algebra of closed contours based on the analysis given in Section VIII (see also Appendix C):

$$\Gamma_{jn} = \sum_{k=j}^{n-1} \Gamma_{kk+1} \tag{111}$$

and also

$$\Gamma_{j,k} = \Gamma_{jj+1} + \Gamma_{kk+1} \qquad \text{where} \qquad (k > j + 1) \tag{112}$$

This algebra implies that in case of Eq. (111) the only two functions (out of n) that flip sign are ζ_1 and ζ_n because all in-between ζ functions get their sign flipped twice. In the same way, Eq. (112) implies that all four electronic functions mentioned in the expression, namely, the jth and the $(j+1)$th, the kth and the $(k+1)$th, all flip sign. In what follows, we give a more detailed explanation based on the mathematical analysis of the Section VIII.

In Sections VII and VIII, it was mentioned that K yields the number of eigenfunctions that flip sign when the electronic manifold traces certain closed paths. In what follows, we shall show how this number is formed for various N_J values.

The situation is obvious for $N_J = 1$. Here, the path either surrounds or does not surround a C_1. In case it surrounds it, two functions, that is, ζ_1 and ζ_2, flip sign so that $K = 2$ and if it does not surround it no ζ function flips sign and $K = 0$. In case of $N_J = 2$, we encounter two conical intersections, namely, C_1 and the C_2 (see Fig. 5a and 5b). Moving the electronic manifold along the path Γ_{12} will change the signs of ζ_1 and ζ_2, whereas moving it along the path Γ_{23} will change the signs ζ_2 and ζ_3. Next, moving the electronic manifold along the path, Γ_{13} (and Fig. 5b) causes the sign of ζ_2 to be flipped twice (once when surrounding C_1 and once when surrounding C_2) and therefore, altogether, its sign remains unchanged. Thus in the case of $N_J = 2$ we can have either no change of sign (when the path does not surround any conical intersection) or three cases where two different functions change sign.

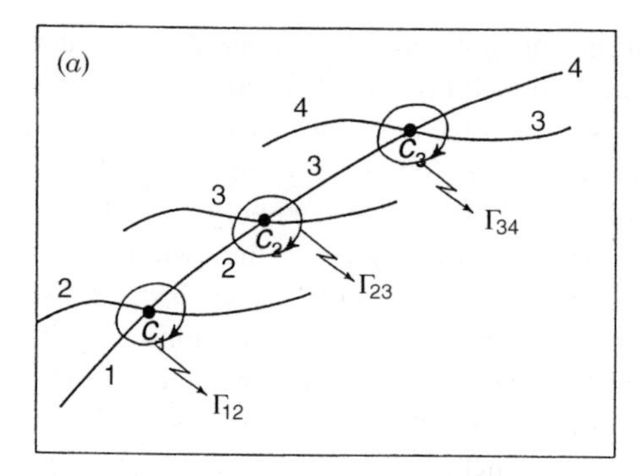

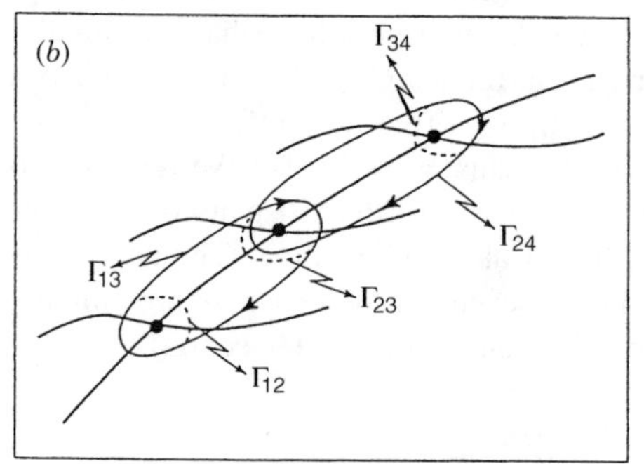

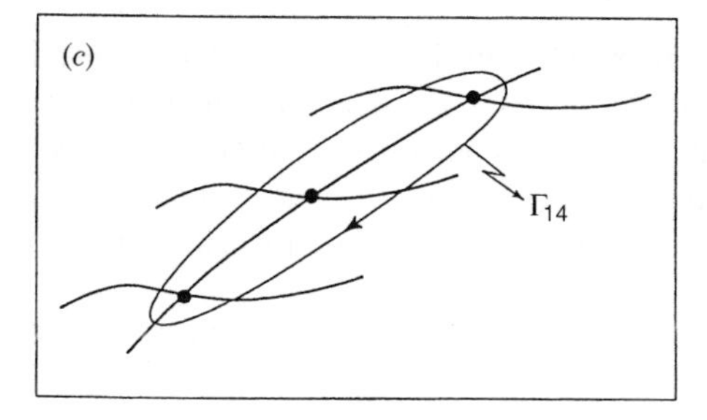

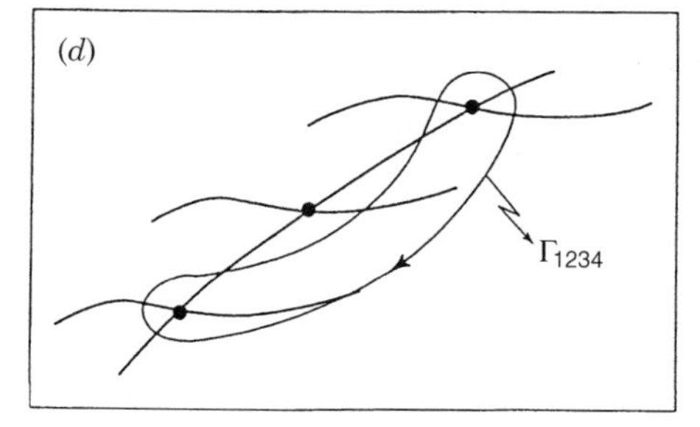

Figure 5. The four interacting surfaces, the three points of conical intersection and the various contours leading to sign conversions: (*a*) The contours Γ_{jj+1} surrounding the respective C_j; $j = 1, 2, 3$ leading to the sign conversions of the *j*th and the $(j + 1)$th eigenfunctions. (*b*) The contours Γ_{jj+2} surrounding the two (respective) conical intersections namely C_j and C_{j+1}; $j = 1, 2$ leading to the sign conversions of the *j*th and the $(j + 2)$th eigenfunctions but leaving unchanged the sign of the middle, the $(j + 1)$, eigenfunction. Also shown are the contours Γ_{jj+1} surrounding the respective C_j; $j = 1, 2, 3$ using partly dotted lines. It can be seen that $\Gamma_{jj+2} = \Gamma_{jj+1} + \Gamma_{j+1j+2}$. (*c*) The contour Γ_{14} surrounding the three conical intersections, leading to the sign conversions of the first and the fourth eigenfunctions but leaving unchanged the signs of the second and the third eigenfunctions. Based on Figure (5*b*) we have $\Gamma_{14} = \Gamma_{12} + \Gamma_{23} + \Gamma_{34}$. (*d*) The contour $\Gamma_{1,3}$ surrounding the two external conical intersections but not the middle one, leading to the sign conversions of all four eigenfunctions, that is, $(\zeta_1, \zeta_2, \zeta_3, \zeta_4); \rightarrow (-\zeta_1, -\zeta_2, -\zeta_3, -\zeta_4)$. Based on Figure (5*b*) we have $\Gamma_{1,3} = \Gamma_{12} + \Gamma_{34}$.

A somewhat different situation is encountered in case of $N_J = 3$, and therefore we shall briefly discuss it as well (see Fig. 5*d*). It is now obvious that contours of the type $\Gamma_{jj+1}; j = 1, 2, 3$ surround the relevant C_j (see Fig. 5*a*) and will flip the signs of the two corresponding eigenfunctions. From Eq. (111), we get that surrounding two consecutive conical intersections, namely, C_j and C_{j+1}, with Γ_{jj+2}; $j = 1, 2$ (see Fig. 5*b*), will flip the signs of the two external eigenfunctions, namely, ζ_j and ζ_{j+2}, but leave the sign of ζ_{j+1} unchanged. We have two such cases—the first and the second conical intersections and the second and the third ones. Then we have a contour Γ_{14} that surrounds all three conical intersections (see Fig. 5*c*) and here, like in the previous where $N_J = 2$ [see also Eq. (111)], only the two external functions, namely, ζ_1 and ζ_4 flip sign but the two internal ones, namely, ζ_2 and ζ_3, will be left unchanged. Finally, we have the case where the contour $\Gamma_{1,3}$ surrounds C_1 and C_3 but not C_2 (see Fig. 5*d*). In this case, all four functions flip sign [see Eq. (112)].

We briefly summarize what we found in this $N_J = 3$ case: We revealed six different contours that led to the sign flip of six (different) pairs of functions and one contour that leads to a sign flip of all four functions. The analysis of Eq. (87) shows that indeed we should have seven different cases of sign flip and one case without sign flip (not surrounding any conical intersection).

X. THE MULTIDEGENERATE CASE

The emphasis in our previous studies was on isolated two-state conical intersections. Here, we would like to refer to cases where at a given point three (or more) states become degenerate. This can happen, for example, when two (line) seams cross each other at a point so that at this point we have three surfaces crossing each other. The question is: How do we incorporate this situation into our theoretical framework?

To start, we restrict our treatment to a tri-state degeneracy (the generalization is straightforward) and consider the following situation:

1. The two lowest states form a conical intersection, presented in terms of $\tau_{12}(\rho)$, located at the origin, namely, at $\rho = 0$.
2. The second and the third states form a conical intersection, presented in terms of $\tau_{23}(\rho, \varphi \mid \rho_0, \varphi_0)$, located at $\rho = \rho_0$, $\varphi = \varphi_0$ [24].
3. The tri-state degeneracy is formed by letting $\rho_0 \to 0$, namely,

$$\lim_{\rho_0 \to 0} \tau_{23}(\rho, \varphi \mid \rho_0, \varphi_0) = \tau_{23}(\rho, \varphi) \tag{113}$$

so that the two conical intersections coincide. Since the two conical intersections are located at the same point, every closed contour that surrounds one of them will surround the other so that this situation is the case of one contour $\Gamma(= \Gamma_{13})$

surrounding two conical intersections (see Fig. 5*b*). According to the discussion of Section IX, only two functions will flip signs (i.e., the lowest and the highest one). Extending this case to an intersection point of n surfaces will not change the final result, namely, only two functions will flip signs, the lowest one and the highest one.

This conclusion contradicts the findings discussed in Sections V.A.2 and V.A.3. In Section V.A.2, we treated a three-state model and found that functions can *never* flip signs. In Section V.A.3, we treated a four-state case and found that either all four functions flip their sign or none of them flip their sign. The situation where two functions flip signs is not allowed under any conditions.

Although the models mentioned here are of a very specialized form (the non-adiabatic coupling terms have identical spatial dependence), still the fact that such contradictory results are obtained for the two situations could hint to the possibility that in the transition process from the nondegenerate to the degenerate situation, in Eq. (113), something is not continuous.

To date, this contradiction has not been resolved but we still would like to make the following suggestion. In molecular physics, we may encounter two types of multidegeneracy situations: (1) The one described above is formed from an aggregation of two-state conical intersections and depends on external coordinates (the coordinates that form the seam). Thus this multidegeneracy is created by varying these external coordinates in a proper way. In the same way, the multidegeneracy can be removed by varying these coordinates. Note that this kind of a degeneracy is not an essential degeneracy because the main features of the individual conical intersections are unaffected while assembling or disassembling this degeneracy. We shall term this degeneracy as a *breakable* multidegeneracy. (2) The other type mentioned above is the one that is not formed from an aggregation of conical intersections and therefore will not breakup under any circumstances. Therefore, this degeneracy is termed the *unbreakable* multidegeneracy.

XI. THE NECESSARY CONDITIONS FOR A RIGOROUS MINIMAL DIABATIC POTENTIAL MATRIX

This Section considers one of the more important dilemmas in molecular physics: Given a Born–Oppenheimer–Huang system, what is the minimal sub-Hilbert space for which diabatization is still valid.

A. Introductory Comments

When studying molecular systems one encounters two almost insurmountable difficulties: (1) That of numerically treating the non-adiabatic coupling terms that are not only spiky—a feature that is in itself a "recipe" for numerical instabilities—but also, singular. (2) That of having to consider large portions of the Hilbert space. As we will show in this section, the two apparently unrelated difficulties are strongly interrelated. Moreover, we will show that resolving the first difficulty may, in many cases, also settle the second.

As discussed earlier, one distinguishes between (1) the adiabatic framework that is characterized by the adiabatic surfaces and the non-adiabatic (derivative) coupling terms and (2) the diabatic framework that is characterized by the fact that derivative couplings are eliminated and replaced by (smoothly behaving) potential couplings. Because of the unpleasant features of the non-adiabatic coupling terms the dynamics is expected to be more easily carried out within the diabatic framework. Therefore, transforming to the diabatic framework (also to be termed *diabatization*) is the right thing to do when treating the multistate problem as created by the Born–Oppenheimer–Huang approach [1,2]. However, because the non-adiabatic coupling terms are frequently singular functions may cause difficulties and therefore the diabatization becomes more of a theoretical–mathematical problem rather than a numerical one.

In 1975, Baer suggested that the diabatic arrangement be reached by first forming the adiabatic framework and then transforming it to the diabatic one by employing the non-adiabatic coupling terms [34]. This approach becomes particularly simple when applied to two states, because it amounts to the calculation of an angle (related to a 2×2 orthogonal matrix), which is formed by integration over the (1,2) non-adiabatic coupling term along a given contour [34–36]. This approach was successfully employed to treat charge-transfer processes [54,94–97], which until that time were solely carried out using classical trajectories [3,98,99], reactive exchange processes between neutrals [100–102] and photodissociation processes [103,104].

Because of difficulties in calculating the non-adiabatic coupling terms, this method did not become very popular. Nevertheless, this approach, was employed extensively in particular to simulate spectroscopic measurements, with a "modification" introduced by Macias and Riera [47,48]. They suggested looking for a symmetric operator that behaves "violently" at the vicinity of the conical intersection and use it, instead of the non-adiabatic coupling term, as the integrand to calculate the adiabatic-to-diabatic transformation. Consequently, a series of operators such as the electronic dipole moment operator, the transition dipole moment operator, the quadrupole moment operator, and so on, were employed for this purpose [49,52,53,105]. However, it has to be emphasized that immaterial to the success of this approach, it is still an ad hoc procedure.

For example, there are also other approaches by Pacher et al. [106], Romero et al. [107], Sidis [40], and Domcke and Stock [42], which developed recipes for construction ab initio diabatic states. These methods can be efficient as long as one encounters, *at most*, one isolated conical intersection in a given region in configuration space but have to be further developed, if several conical intersections are located at the region of interest.

Now we intend to present the purpose of this section. In order to do this in a comprehensive way, we need to explain what is meant, within the *present* framework, by the statement that "the diabatization is non-physical." The procedure discussed above is based on a transformation matrix of a dimension M derived within a sub-Hilbert space of the same dimension. The statement "a diabatization is non-physical" implies that some of the elements of the diabatic matrix formed in this process are *multivalued* in configuration space. (In this respect, it is important to emphasize that the nuclear Schrödinger equation cannot be solved for multivalued potentials.) We show that if an M-dimensional sub-Hilbert space is not large enough, some elements of the diabatic potential matrix will not be single valued. Thus a resolution to this difficulty seems to be in increasing the dimension of the sub-Hilbert space, that is, the value of M. However, increasing M indefinitely will significantly increase the computational volume. Therefore it is to everyone's interest to keep M as small as possible. Following this explanation, we can now state the purpose of this section:

We intend to show that an adiabatic-to-diabatic transformation matrix based on the non-adiabatic coupling matrix can be used not only for reaching the diabatic framework but also as a mean to determine the *minimum* size of a sub-Hilbert space, namely, the minimal M value that still guarantees a valid diabatization.

For example: one forms, within a two-dimensional (2D) sub-Hilbert space, a 2×2 diabatic potential matrix, which is not single valued. This implies that the 2D transformation matrix yields an invalid diabatization and therefore the required dimension of the transformation matrix has to be at least three. The same applies to the size of the sub-Hilbert space, which also has to be at least three. In this section, we intend to discuss this type of problems. It also leads us to term the conditions for reaching the minimal relevant sub-Hilbert space as "the necessary conditions for diabatization."

In this section, diabatization is formed employing the adiabatic-to-diabatic transformation matrix **A**, which is a solution of Eq. (19). Once **A** is calculated, the diabatic potential matrix **W** is obtained from Eq. (22). Thus Eqs. (19) and (22) form the basis for the procedure to obtain the diabatic potential matrix elements.

Note that since the *adiabatic* potentials are single valued by definition, the single valuedness of **W** (viz, the single valuedness of each of its terms) depends on the features of the **A** matrix [see Eq. (22)]. It is also obvious that if **A** is single valued, the same applies to **W** [the single valuedness of the **A** matrix in a given region is guaranteed if Eq. (25) is fulfilled throughout this region]. However, in Section (IV.A) we showed that **A** does not have to be single valued in order to guarantee the single valuedness of **W**. In fact, it was proved that the necessary condition for having single-valued diabatic potentials along a given contour Γ is that the **D** matrix introduced in Eq. (38) is *diagonal* with numbers of norm 1, namely, numbers that are either $(+1)$ or (-1). If this condition is fulfilled for every contour in the region of interest, then we may say that for this particular dimension M the diabatic potential is single valued or in other words the diabatization is valid. However, if this condition is fulfilled for a give value M but not for $(M-1)$, then M is the minimal value for which diabatization is valid in this particular case.

According to Section VI, the size M of the sub-Hilbert space is determined whether the respective M states form an isolated set of states fulfilling Eqs. (91). In this case, diabatization is always valid for this subsystem. However, it can happen that under certain geometrical situations the size of the sub-Hilbert space for which diabatization is valid is even smaller than this particular M

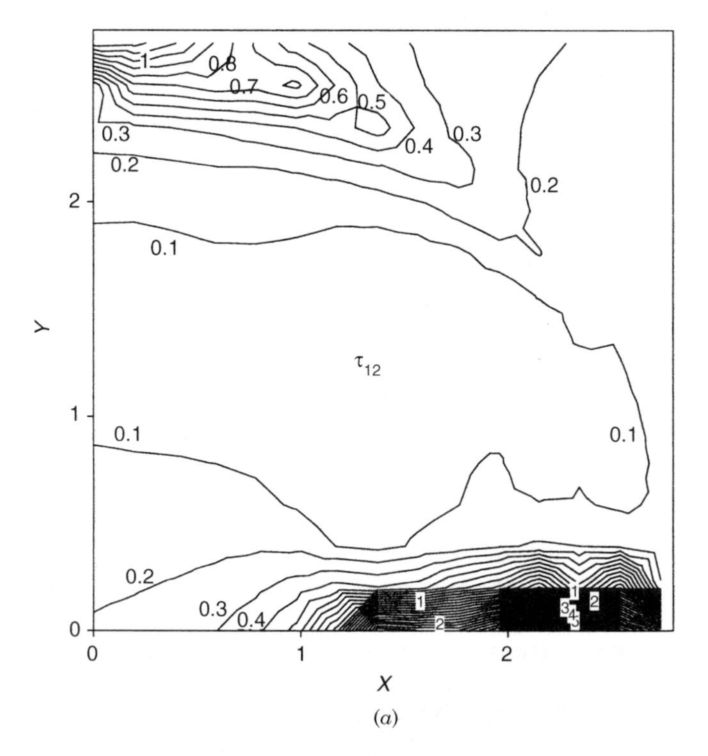

Figure 6. Equi-nonadiabatic coupling lines for the terms $\tau_{12}(x, y)$ and $\tau_{23}(x, y)$ as calculated for the C_2H molecule for a fixed C–C distance, that is, $r_{CC} = 1.35$ Å. (*a*) Equi-non-adiabatic coupling term lines for the $\tau_{12}(x, y)$. (*b*) Equi-non-adiabatic coupling term lines for $\tau_{23}(x, y)$. The Cartesian coordinates (x, y) are related to (q, θ) as follows: $x = q\cos\theta$; $y = q\sin\theta$, where q and θ are measured with respect to the midpoint between the two carbons.

value. These geometrical situations are considered in Section XI.B, and to simplify the discussion we refer to $M = 3$.

B. The Noninteracting Conical Intersections

Let us consider a system of three states where the two lower states are coupled by $\tau_{12}(s)$ and the two upper ones by $\tau_{23}(s)$ (see Section VIII for details). By the concept "noninteracting conical intersections" we mean the case where the spatial distribution of $\tau_{12}(s)$ and $\tau_{23}(s)$ is such that they overlap only slightly at the region of interest. As an example we may consider a case where the main intensity of $\tau_{12}(s)$ is concentrated along one ridge and the main intensity of $\tau_{23}(s)$ is concentrated along another ridge. Next, we assume that these two ridges are approximately parallel and located far enough apart so that the overlap between $\tau_{12}(s)$ and $\tau_{23}(s)$ is minimal (see Figs. 6 and 7).

We are interested in calculating the diabatic potentials for a region in configuration space, that contains the two conical intersections. According

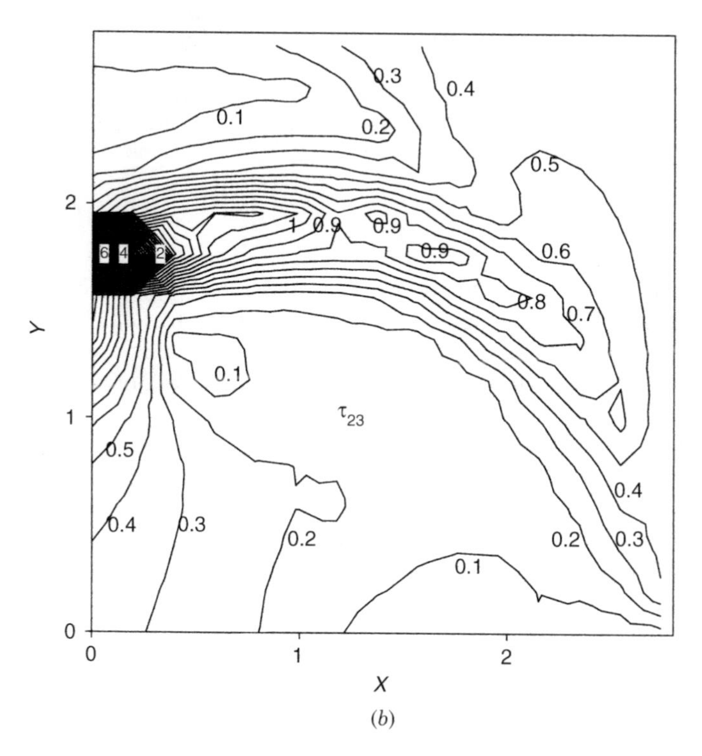

(*b*)

Figure 6 (*Continued*)

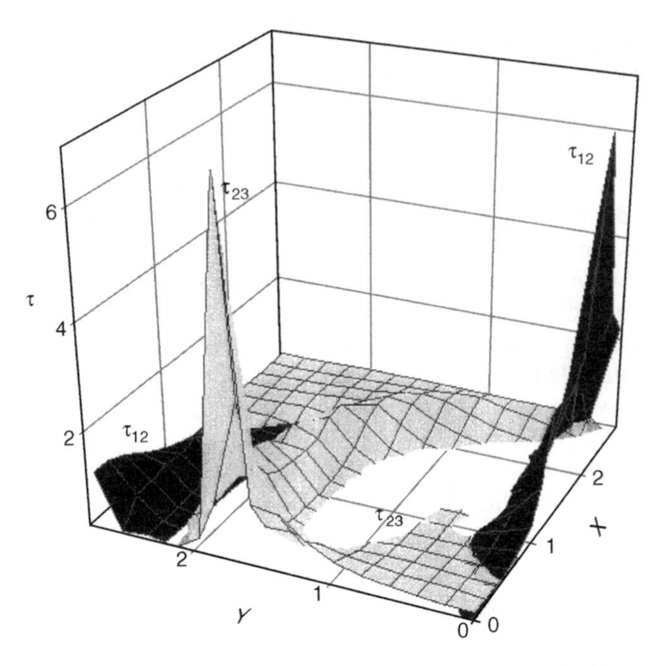

Figure 7. The geometrical positions (with respect to the CC axis) of $\tau_{12}(x, y)$ and $\tau_{23}(x, y)$. All distances are in angstroms (Å). The Cartesian coordinates (x, y) are related to (q, θ) as follows: $x = q\cos\theta$; $y = q\sin\theta$, where q and θ are measured with respect to the midpoint between the two carbons.

to Section VI, for this purpose we need the *three* states because the two lowest states (1 and 2) and the two highest states (2 and 3) are *strongly* coupled to each other. Next, we examine to see if under these conditions this is really necessary or could the diabatization be achieved with only the two lowest states.

For this purpose, we consider Figure 8 with the intention of examining what happens along the contour Γ, in particularly when it gets close to C_{23}. Note that some segments of the contour Γ are drawn as full lines and others are as dashed lines. The full lines denote segments along which $\boldsymbol{\tau}_{12}(s)$ is of a strong intensity but $\boldsymbol{\tau}_{23}(s)$ is negligibly weak. The dashed lines denote segments along which $\boldsymbol{\tau}_{12}(s)$ is negligibly weak.

Next, consider the following line integral [see Eq. (27)]:

$$\mathbf{A}(s) = \mathbf{A}(s_0) - \int_{s_0}^{s} ds \cdot \boldsymbol{\tau}\mathbf{A} \tag{114}$$

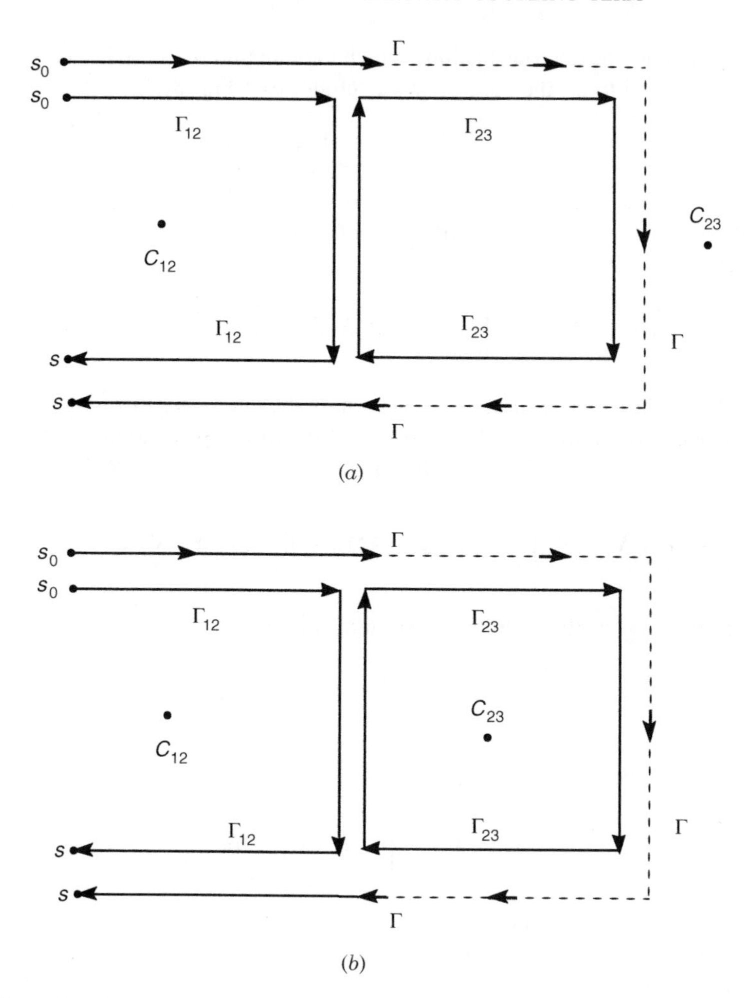

Figure 8. The representation of an open contour Γ in terms of an open contour Γ_{12} in the vicinity of the conical intersection at C_{12} and a closed contour Γ_{23} at the vicinity of a conical intersection at C_{23}: $\Gamma = \Gamma_{12} + \Gamma_{23}$. It is assumed that the intensity of $\boldsymbol{\tau}_{12}$ is strong along Γ_{12} (full line) and weak along Γ_{23} (dashed line). (*a*) The situation when C_{23} is outside the closed contour Γ_{23}. (*b*) The situation when C_{23} is inside the closed contour Γ_{23}.

where $\mathbf{A}(s)$ and $\boldsymbol{\tau}(s)$ are of dimensions 3×3 [the explicit form of $\boldsymbol{\tau}(s)$ is given in Eq. (99)]. We also consider the two other $\boldsymbol{\tau}$ matrices, namely, $\boldsymbol{\tau}_{12}(s)$ and $\boldsymbol{\tau}_{23}(s)$ (see Eq. (105)] and it is easy to see that

$$\boldsymbol{\tau}(s) = \boldsymbol{\tau}_{12}(s) + \boldsymbol{\tau}_{23}(s) \tag{106b}$$

To continue, the contour Γ, along which the integration in Eq. (114) is carried out, is assumed to be the sum of two contours [see Fig. 8]:

$$\Gamma = \Gamma_{12} + \Gamma_{23} \tag{115}$$

where Γ_{23} is a closed contour in the vicinity of C_{23} that may or may not surround it and Γ_{12} is an open contour near C_{12}. By substituting Eq. (115) in (114) we get

$$\mathbf{A}(s) = \mathbf{A}(s_0) - \int_{\Gamma_{12}} ds' \cdot \boldsymbol{\tau}(s')\mathbf{A}(s') - \oint_{\Gamma_{23}} ds' \cdot \boldsymbol{\tau}(s')\mathbf{A}(s') \tag{116}$$

Next, by recalling the assumptions concerning the intensities of $\boldsymbol{\tau}_{12}(s)$ and $\boldsymbol{\tau}_{23}(s)$ we replace $\boldsymbol{\tau}(s)$, in the second term of Eq. (116) with $\boldsymbol{\tau}_{12}(s)$ and in the third term with $\boldsymbol{\tau}_{23}(s)$. As a result Eq. (116) becomes

$$\mathbf{A}(s) = \mathbf{A}(s_0) - \int_{\Gamma_{12}} ds' \cdot \boldsymbol{\tau}_{12}(s')\mathbf{A}(s') - \oint_{\Gamma_{23}} ds' \cdot \boldsymbol{\tau}_{23}(s')\mathbf{A}(s') \tag{117}$$

By defining $\mathbf{A}_{23}$ as the following (constant) matrix,

$$\mathbf{A}_{23} = \mathbf{A}(s_0) - \oint_{\Gamma_{23}} ds' \cdot \boldsymbol{\tau}_{23}(s')\mathbf{A}(s') \tag{118}$$

Eq. (117) becomes

$$\mathbf{A}(s) = \mathbf{A}_{23} - \int_{\Gamma_{12}} ds' \cdot \boldsymbol{\tau}_{12}(s')\mathbf{A}(s') \tag{119}$$

where the matrix $\mathbf{A}_{23}$ is the corresponding "boundary" value matrix. As for $\mathbf{A}_{23}$, it is noticed to be the solution of Eq. (118), namely, the outcome of an integration performed along a closed contour (Γ_{23}) where $\boldsymbol{\tau}_{23}$ is the kernel. Consequently, this matrix can be presented as [see Eq. (39)]:

$$\mathbf{A}_{23} = \mathbf{D}_{23}\mathbf{A}(s_0) \tag{120}$$

where from the analysis in Section VIII we get that $\mathbf{D}_{23}$ will have a $(+1)$ at position (1,1) and a (-1) at positions (2,2) and (3,3) when it surrounds the conical intersection at C_{23}.

Now, by returning to Eq. (119) it can be shown that if $\mathbf{A}_{12}$ is the solution of the equation

$$\mathbf{A}_{12}(s) = \mathbf{A}_0 + \int_{\Gamma_{12}} ds' \cdot \boldsymbol{\tau}_{12}(s')\mathbf{A}_{12}(s') \tag{121}$$

where the contour Γ_{12} can be any contour, then

$$\mathbf{A}(s) = \mathbf{D}_{23}\mathbf{A}_{12}(s) \tag{122}$$

To summarize: Following the theory presented above, we have to distinguish between two situations. (1) As long as Γ does not surround C_{23} the matrix $\mathbf{A}(s)$ is given in the form

$$\mathbf{A}(s) = \begin{pmatrix} \cos\gamma_{12} & \sin\gamma_{12} & 0 \\ -\sin\gamma_{12} & \cos\gamma_{12} & 0 \\ 0 & 0 & 1 \end{pmatrix} \tag{123}$$

where $\gamma_{12}(s)$ is given in the form

$$\gamma_{12}(s) = \gamma_{12}(s_0) - \int_{s_0}^{S} ds \cdot \boldsymbol{\tau}_{12}(s) \tag{124}$$

(2) In the case where Γ surrounds the C_{23} conical intersection, the value of $\gamma_{12}(s)$ may change its sign (for more details see Ref. [108]).

Since for any assumed contour the most that can happen, due to C_{23}, is that $\gamma_{12}(s)$ flips its sign, the corresponding 2×2 diabatic matrix potential, $\mathbf{W}(s)$, will not be affected by that as can be seen from the following expressions:

$$\begin{aligned} W_{11}(s) &= u_1(s)\cos^2\gamma_{12}(s) + u_2(s)\sin^2\gamma_{12}(s) \\ W_{22}(s) &= u_1(s)\sin^2\gamma_{12}(s) + u_2(s)\cos^2\gamma_{12}(s) \\ W_{12}(s) &= W_{21}(s) = \frac{1}{2}(u_2(s) - u_1(s))\sin(2\gamma_{12}(s)) \end{aligned} \tag{125}$$

In other words, the calculation of $\mathbf{W}$ can be carried out by ignoring C_{23} [or $\boldsymbol{\tau}_{23}(s)$] altogether.

XII. THE ADIABATIC-TO-DIABATIC TRANSFORMATION MATRIX AND THE WIGNER ROTATION MATRIX

The adiabatic-to-diabatic transformation matrix in the way it is presented in Eq. (28) is somewhat reminiscent of the Wigner rotation matrix [109a] (assuming that $\mathbf{A}(s_0) \equiv I$). In order to see this, we first present a few well-known facts related to the definition of ordinary angular momentum operators (we follow the presentation by Rose [109b] and the corresponding Wigner matrices and then return to discuss the similarities between Wigner's $\mathbf{d}^j(\beta)$ matrix and the adiabatic-to-diabatic transformation matrix.

A. Wigner Rotation Matrices

The ordinary angular rotation operator R(**k**,θ) in the limit $\theta \to 0$ is written as

$$\mathrm{R}(\mathbf{k}, \theta) = \exp(-i\mathrm{S}(\mathbf{k}, \theta)) \tag{126}$$

where **k** is a unit vector in the direction of the axis of rotation, θ is the angle of rotation, and S(**k**,θ) is an operator that has to fulfill the condition $\mathrm{S}(\mathbf{k}, \theta) \to 0$ for $\theta \to 0$ to guarantee that in this situation (i.e., when $\theta \to 0$) $\mathrm{R}(\mathbf{k}, \theta) \to \mathbf{I}$. Moreover, since $\mathrm{R}(\mathbf{k}, \theta)$ has to be unitary, the operator $\mathrm{S}(\mathbf{k}, \theta)$ has to be Hermitian. Next, it is shown that $\mathrm{S}(\mathbf{k}, \theta)$ is related to the total angular momentum operator, **J**, in the following way:

$$\mathrm{S}(\mathbf{k}, \theta) = (\mathbf{k} \cdot \mathbf{J})\theta \tag{127}$$

where the dot stands for scalar product. By substituting Eq. (127) in Eq. (126) we get the following expression for $\mathrm{R}(\mathbf{k}, \theta)$:

$$\mathrm{R}(\mathbf{k}, \theta) = \exp(-i(\mathbf{k} \cdot \mathbf{J})\theta) \tag{128}$$

It has to be emphasized that in this framework **J** is the angular momentum operator in ordinary coordinate space (i.e., configuration space) and θ is a (differential) ordinary angular polar coordinate.

Next, Euler's angles are employed for deriving the outcome of a general rotation of a system of coordinates [86]. It can be shown that $\mathrm{R}(\mathbf{k}, \theta)$ is accordingly presented as

$$\mathrm{R}(\mathbf{k}, \theta) = e^{-i\alpha J_z} e^{-i\beta J_y} e^{-i\gamma J_z} \tag{129}$$

where J_y and J_z are the y and the z components of **J** and α, β, and γ are the corresponding three Euler angles. The explicit matrix elements of the rotation operator are given in the form:

$$\mathbf{D}^j_{m'm}(\theta) = \langle jm'|\mathrm{R}(k, \theta)|jm\rangle = e^{-i(m'\alpha+m\gamma)}\langle jm'|e^{-i\beta J_y}|jm\rangle \tag{130}$$

where m and m' are the components of **J** along the J_z and $J_{z'}$ axes, respectively, and $|jm\rangle$ is an eigenfunction of the Hamiltonian of J^2 and of J_z. Equation (130) will be written as

$$\mathbf{D}^j_{m'm}(\theta) = e^{-i(m'\alpha+m\gamma)}\mathbf{d}^j_{m'm}(\beta) \tag{131}$$

The $\mathbf{D}^j$ matrix as well as the $\mathbf{d}^j$ matrix are called the Wigner matrices and are the subject of this section. Note that if we are interested in finding a relation between the adiabatic-to-diabatic transformation matrix and Wigner's matrices, we should mainly concentrate on the $\mathbf{d}^j$ matrix. Wigner derived a formula for these matrix elements [see [109b], Eq. (4.13)] and this formula was used by us to obtain the explicit expression for $j = \frac{3}{2}$ (the matrix elements for $j = 1$ are given in [109b], p. 72).

B. The Adiabatic-to-Diabatic Transformation Matrix and the Wigner $\mathbf{d}^j$ Matrix

The obvious way to form a similarity between the Wigner rotation matrix and the adiabatic-to-diabatic transformation matrix defined in Eqs. (28) is to consider the (unbreakable) multidegeneracy case that is based, just like Wigner rotation matrix, on a single axis of rotation. For this sake, we consider the particular set of $\boldsymbol{\tau}$ matrices as defined in Eq. (51) and derive the relevant adiabatic-to-diabatic transformation matrices. In what follows, the degree of similarity between the two types of matrices will be presented for three special cases, namely, the two-state case which in Wigner's notation is the case, $j = \frac{1}{2}$, the tri-state case (i.e., $j = 1$) and the tetra-state case (i.e., $j = \frac{3}{2}$).

However, before going into a detail comparison between the two types of matrices it is important to remind the reader what the elements of the $\mathbf{J}_y$ matrix look like. By employing Eqs (2.18) and (2.28) of [109b] it can be shown that

$$\langle jm | J_y | jm + k \rangle = \boldsymbol{\delta}_{1k} \frac{1}{2i} \sqrt{(j + m + 1)(j - m)} \tag{132a}$$

$$\langle jm + k | J_y | jm \rangle = -\boldsymbol{\delta}_{1k} \frac{1}{2i} \sqrt{(j - m + 1)(j + m)} \tag{132b}$$

Now, by defining $\tilde{\mathbf{J}}_y$ as

$$\tilde{\mathbf{J}}_y = \mathbf{i} J_y \tag{133}$$

it is seen that the $\tilde{\mathbf{J}}_y$ matrix is an antisymmetric matrix just like the $\boldsymbol{\tau}$ matrix. Since the $\mathbf{d}^j$ matrix is defined as

$$\mathbf{d}^j(\beta) = \exp(-i\beta J_y) = \exp(\beta \tilde{\mathbf{J}}_y) \tag{134}$$

It is expected that for a certain choice of parameters (that define the $\boldsymbol{\tau}$ matrix) the adiabatic-to-diabatic transformation matrix becomes identical to the corresponding Wigner rotation matrix. To see the connection, we substitute Eq. (51) in Eq. (28) and assume $\mathbf{A}(s_0)$ to be the unity matrix.

The three matrices of interest were already derived and presented in Section V.A. There they were termed the $\mathbf{D}$ (topological) matrices (not related to the above mentioned Wigner $\mathbf{D}^j$ matrix) and were used to show the kind of quantization one should expect for the relevant non-adiabatic coupling terms. The only difference between these topological matrices and the adiabatic-to-diabatic transformation matrices requested here, is that in Eqs. (55), (61), and (72) the closed-line integral [see Eq. (76)] is replaced by $\gamma(s)$ defined along an (open) contour [see Eq. (75)]:

For the three cases studied in Section V.A, the similarity to the three corresponding Wigner matrices is achieved in the following way:

1. For the two-state case (i.e., $j = \frac{1}{2}$), $\mathbf{d}^{1/2}(\beta)$ is identified with the corresponding adiabatic-to-diabatic transformation matrix [see Eq. (74)] for which $\beta = \gamma$.
2. For the tri-state case ($j = 1$), we consider Eq. (60). The corresponding $\mathbf{d}^1(\beta)$ matrix is obtained by assuming $\eta = 1$ [see Eq. (57)], and therefore $\omega = \sqrt{2}$. From Eq. (61) or (62), it is seen that $\beta = \gamma\sqrt{2}$. For the sake of completeness we present the corresponding $\mathbf{d}^1(\beta)$ matrix [109b]:

$$\mathbf{d}^1(\beta) = \frac{1}{2}\begin{pmatrix} 1 + C(\beta) & \sqrt{2}S(\beta) & 1 - C(\beta) \\ \sqrt{2}S(\beta) & 2C(\beta) & -\sqrt{2}S(\beta) \\ 1 - C(\beta) & \sqrt{2}S(\beta) & 1 + C(\beta) \end{pmatrix} \tag{135}$$

   where $C(\beta) = \cos\beta$ and $S(\beta) = \sin\beta$.
3. For the tetra-state case ($j = \frac{3}{2}$), we consider Eq. (69). The corresponding $\mathbf{d}^{3/2}(\beta)$ matrix is obtained by assuming $\eta = \sqrt{4/3}$ and $\sigma = 1$ [see Eq. (63)]. This will yield for ϖ the value $\varpi = \sqrt{10/3}$ [see Eq. (67)]. Since $\beta = p\gamma$ [see Eqs. (72)] we have to determine the value of p, which can be shown to be $p = \sqrt{3}$ [see Eq. (65)] and therefore $\beta = \gamma\sqrt{3}$. For the sake of completeness, we present the $\mathbf{d}^{3/2}(\beta)$ matrix:

$$\mathbf{d}^{3/2}(\beta') = \begin{pmatrix} C^3 & -\sqrt{3}C^2S & -\sqrt{3}S^2C & S^3 \\ \sqrt{3}C^2S & C(1 - 3S^2) & -S(1 - 3C^2) & -\sqrt{3}S^2C \\ -\sqrt{3}S^2C & S(1 - 3C^2) & C(1 - 3S^2) & -\sqrt{3}C^2S \\ -S^3 & -\sqrt{3}S^2C & \sqrt{3}C^2S & C^3 \end{pmatrix} \tag{136}$$

   where $C = \cos(\beta/2)$ and $S = \sin(\beta/2)$.

The main difference between the adiabatic-to-diabatic transformation and the Wigner matrices is that whereas the Wigner matrix is defined for an ordinary spatial coordinate the adiabatic-to-diabatic transformation matrix is defined for a rotation coordinate in a different space.

XIII. CURL CONDITION REVISITED: INTRODUCTION OF THE YANG–MILLS FIELD

In this section, the curl condition is extended to include the points of singularity as discussed in Appendix C. The study is meant to shed light as to the origin of the non-adiabatic coupling terms and to connect them with pseudomagnetic fields.

A. The Non-Adiabatic Coupling term as a Vector Potential

In Section III.B, and later in Appendix C, it was shown that the sufficient condition for the adiabatic-to-diabatic transformation matrix **A** to be single valued in a given region in configuration space is the fulfillment of the following "curl" condition [8,34]:

$$\operatorname{curl} \boldsymbol{\tau} - [\boldsymbol{\tau} \times \boldsymbol{\tau}] = 0 \tag{137}$$

This condition is fulfilled as long as the components of τ are analytic functions at the point under consideration (in case part of them become singular at this point, curl τ is not defined).

The expression in Eq. (137) is reminiscent of the Yang–Mills field, however, it is important to emphasize that the Yang–Mills field was introduced for a different physical situation [58,59]. In fact, what Eq. (137) implies is that for molecular systems the Yang–Mills field is zero if the following two conditions are fulfilled:

1. The group of states, for which Eq. (137) is expected to be valid, forms a sub-Hilbert space that is isolated with respect to other portions of the Hilbert space following the definition in Eqs. (40).
2. The $\boldsymbol{\tau}$-matrix elements are analytic functions (vectors) in the above-mentioned region of configuration space.

In what follows, we assume that indeed the group of states form an isolated sub-Hilbert space, and therefore have a Yang–Mills field that is zero or not will depend on whether or not the various elements of the $\boldsymbol{\tau}$ matrix are singular.

In order to extend the existence of Eq. (137) for the singular points as well we write it as follows:

$$\operatorname{curl} \boldsymbol{\tau} - [\boldsymbol{\tau} \times \boldsymbol{\tau}] = \mathbf{H} \tag{138}$$

where **H** is zero at the regular points.

In order to get more insight, we return to the Born–Oppenheimer–Huang equation [1,2] as written in Eq. (16) and, for simplicity, limit ourselves to the two-state case:

$$-\frac{1}{2m}(\nabla + \tau)^2\Psi + (u - E)\Psi = 0 \tag{139}$$

so that $\boldsymbol{\tau}$ is given in the form

$$\boldsymbol{\tau} = \begin{pmatrix} 0 & \tau \\ -\tau & 0 \end{pmatrix} \tag{140}$$

Although Eq. (139) looks like a Schrödinger equation that contains a vector potential $\boldsymbol{\tau}$, it cannot be interpreted as such because $\boldsymbol{\tau}$ is an *antisymmetric* matrix (thus, having diagonal terms that are equal to zero). This "inconvenience" can be "repaired" by employing the following unitary transformation:

$$\Psi = \mathbf{G}\Phi \tag{141}$$

where $\mathbf{G}$ is the (constant) matrix

$$\mathbf{G} = \frac{1}{\sqrt{2}}\begin{pmatrix} 1 & 1 \\ i & -i \end{pmatrix} \tag{142}$$

By substituting Eq. (141) in Eq. (139) and multiplying it from the left by $G^{\dagger}$ yields

$$\frac{1}{2m}(-i\nabla + \mathbf{t})^2\Phi + (\mathbf{w} - E)\Phi = 0 \tag{143}$$

where $\mathbf{t}$ is now a diagonal matrix

$$\mathbf{t} = \begin{pmatrix} \boldsymbol{\tau} & 0 \\ 0 & -\boldsymbol{\tau} \end{pmatrix} \tag{144}$$

and $\mathbf{w}$ is an ordinary potential matrix of the kind

$$\mathbf{w} = \frac{1}{2}\begin{pmatrix} u_1 + u_2 & -(u_2 - u_1) \\ -(u_2 - u_1) & u_1 + u_2 \end{pmatrix} \tag{145}$$

The important outcome from this transformation is that now the non-adiabatic coupling term $\boldsymbol{\tau}$ is incorporated in the Schrödinger equation in the same way as a vector potential due to an external magnetic field. In other words, $\boldsymbol{\tau}$ behaves like a vector potential and therefore is expected to fulfill an equation of the kind [111a]

$$\text{curl}\,\boldsymbol{\tau} = \mathbf{H} \tag{146}$$

where $\mathbf{H}$ is a pseudomagnetic field. Equation (146) looks similar to Eq. (138) but is in fact identical to it because in the case of two states the cross-term $[\boldsymbol{\tau} \times \boldsymbol{\tau}]$ is zero. Now, by returning to the Yang–Mills field we recall that $\mathbf{H} \neq 0$ at the singular points of $\boldsymbol{\tau}$. In the present study, we consider a case of *one* singular point.

The question is if in reality such magnetic fields exist. It turns out that such fields can be formed by long and narrow solenoids [111b]. It is well known that in this case the magnetic fields are nonzero only inside the solenoid but zero outside it [111b]. Moreover, it has a nonzero component along the solenoid axis only. Thus simulating the molecular *seam* [36,54,110] as a solenoid we can identify the non-adiabatic coupling term as a vector potential produced by an infinitesimal narrow solenoid.

The quantum mechanical importance of a vector potential **A**, in regions where the magnetic field is zero, was first recognized by Aharonov and Bohm in their seminal 1959 paper [112].

B. The Pseudomagnetic Field and the Curl Equation

To continue, we assume the following situation: We concentrate on an x–y plane, which is chosen to be perpendicular to the seam. In this way, the pseudomagnetic field is guaranteed to be perpendicular to the plane and will have a nonzero component in the z direction only. In addition, we locate the origin at the point of the singularity, that is, at the crossing point between the plane and the seam. With these definitions the pseudomagnetic field is assumed to be of the form [113].

$$H = H_z = 2\pi \frac{\delta(q)}{q} f(\theta) \tag{147}$$

Here, $\delta(q)$ is the Dirac δ function and $f(\theta)$ is an arbitrary function to be determined [it can be shown that any function of the type $f(q, \theta)$ leads to the same result because of the $\delta(q)$ function]. By considering Eq. (146) for the z component, we obtain (employing polar coordinates):

$$\frac{1}{q}\left(\frac{\partial \tau_\theta}{\partial q} - \frac{\partial \tau_q}{\partial \theta}\right) = 2\pi \frac{\delta(q)}{q} f(\theta) \tag{148}$$

Here, (τ_q, τ_θ) are the radial and the angular components of $\boldsymbol{\tau}$ (the z component, i.e., the out-of-plane component, is by definition equal to zero). Equation (148) can be shown (by substitution) to have the following solution:

$$\tau_\theta(q, \theta) - \int_0^q dq \frac{\partial \tau_q}{\partial \theta} = \pi h(q) f(\theta) \tag{149}$$

where $h(q)$ is the Heaviside function

$$h(q) = \begin{cases} 1 & q \geq 0 \\ 0 & q < 0 \end{cases} \tag{150}$$

Since q is a radius it is always positive, and therefore Eq. (149) can be written, without loss of generality, as

$$\tau_\theta(q, \theta) - \int_0^q dq \frac{\partial \tau_q}{\partial \theta} = \pi f(\theta) \tag{151}$$

Next, we consider the "quantization" condition introduced earlier [see Eq. (94)]. Assuming Γ to be a circle with radius q, Eq. (94) implies

$$\int_0^{2\pi} \tau_\theta(q, \theta) d\theta = n\pi \tag{152}$$

A similar integration over θ along the $(0,2\pi)$ range can be carried out for Eq. (151). Thus, let us first consider the integration over the second term

$$\int_0^{2\pi} d\theta \int_0^q dq \frac{\partial \tau_q}{\partial \theta} = \int_0^q dq \int_0^{2\pi} \frac{\partial \tau_q}{\partial \theta} d\theta = \int_0^q dq(\tau_q(q, \theta = 2\pi) - \tau_q(q, \theta = 0))$$

In Section XIV.A, it is proved that $\tau_q(q, \theta)$ is, for every value of q, single valued with respect to θ so that we have

$$\int_0^{2\pi} d\theta \int_0^q dq \frac{\partial \tau_q}{\partial \theta} = 0 \tag{153}$$

Combining Eqs. (151)–(153) yields the following outcome:

$$\int_0^{2\pi} f(\theta)\, d\theta = n \tag{154}$$

In other words, the quantization that was encountered for the non-adiabatic coupling terms is associated with the "quantization" of the *intensity* of the "magnetic" field along the seam. Moreover, Eq. (154) reveals another feature, namely, that there are fields for which n is an odd integer, namely, conical intersections and there are fields for which n is an even integer, namely, parabolical intersections.

Equation (151) can be applied to obtain $f(\theta)$. Ab initio calculation for small enough q values will yield $\tau_\theta(\theta, q \sim 0)$ and these, as is seen from Eq. (151), can be directly related to $f(\theta)$:

$$f(\theta) \sim \frac{1}{\pi} \tau_\theta(q \sim 0, \theta) \tag{155}$$

where the contribution of the second term on the left-hand side (for small enough q values) is ignored.

C. Conclusions

This section is devoted to the idea that the electronic non-adiabatic coupling terms can be simulated as vector potentials. For this purpose, we considered a two-state system, shifted (rigorously) the off-diagonal non-adiabatic coupling terms to the diagonal and employed the relevant Maxwell equation. As is also noticed, the simulation created a connection between the "curl" condition as fulfilled by the non-adiabatic coupling terms and the Yang–Mills field.

As noticed, a pseudomagnetic field is assumed to exist along the seam formed by varying indirect coordinates (i.e., coordinates not related to the plane for which the vector potential is not zero) of a given molecular system. In this respect, we want to suggest that eventually the pseudomagnetic field is "formed," semiclassically, by the zero-point vibrational motion of the indirect coordinates. For this purpose, we consider a three-atom molecular system ABC and assume the AB distance to be the *indirect* coordinate. Varying the AB distance builds up, semiclassically, a motion along the seam. Consequently, the zero-point vibrational motion along the AB bond creates, semiclassically, a periodic motion along the seam. This motion eventually causes charges that are concentrated along the seam (or its vicinity) to oscillate and in this way to form a pseudoelectromagnetic field.

XIV. A THEORETIC-NUMERIC APPROACH TO CALCULATE THE ELECTRONIC NON-ADIABATIC COUPLING TERMS

In this section, we discuss the possibility that the electronic non-adiabatic coupling terms will be derived, not by ab initio treatments but, by solving the curl equations for a given set of boundary conditions obtained from ab initio calculations [114,115]. In other words, instead of performing an ab initio calculation at any point in configuration space we suggest solving the relevant differential equations for boundary conditions obtained from a (limited) ab initio calculation [64–74] or perturbation theory [66,67].

A. The Treatment of the Two-State System in a Plane

1. *The Solution for a Single Conical Intersection*

The curl equation for a two-state system is given in Eq. (26):

$$\text{curl}\,\boldsymbol{\tau} = 0 \tag{26}$$

Equation (26) is fulfilled at any point in configuration space for which the components of $\boldsymbol{\tau}$ are analytic functions.

Equation (26) is a set of partial first-order differential equations. Each component of the Curl forms an equation and this equation may or may not be "coupled" to the other equations. In general, the number of equations is equal to the number of components of the Curl equations. At this stage, to solve this set of equation in its most general case seems to be a formidable task.

In what follows, we shall limit ourselves to the following situation. Assuming a system of N coordinates $(z_1, z_2, \ldots, z_N)$, with the following components:

$$\tau_{z_j} = \tau_{z_j}(z_1, z_2, \ldots z_N) \qquad j = 1, 2, \ldots, N \tag{156a}$$

and assume that two of them, that is, $\boldsymbol{\tau}_{z_1}$ and $\boldsymbol{\tau}_{z_2}$ depend only on their own coordinates, namely, (z_1, z_2), thus

$$\boldsymbol{\tau}_{z_j} = \boldsymbol{\tau}_{z_j}(z_1, z_2) \qquad j = 1, 2 \tag{156b}$$

then the following partial curl equation

$$\frac{\partial \boldsymbol{\tau}_{z_1}}{\partial z_2} - \frac{\partial \boldsymbol{\tau}_{z_2}}{\partial z_1} = 0$$

is the only equation to be considered within the (z_1, z_2) space because due to Eq. (156b) all the other relevant components lead to the results

$$\frac{\partial \boldsymbol{\tau}_{z_n}}{\partial z_1} = \frac{\partial \boldsymbol{\tau}_{z_n}}{\partial z_2} 0 \qquad n = 3, \ldots, N$$

In what follows, the 2D space is assumed to be a plane, and therefore we apply either the polar coordinates (q, θ) or the Cartesian coordinates (x, y).

We start treating the curl equation expressed in terms of polar coordinates:

$$\frac{1}{q}\left(\frac{\partial \boldsymbol{\tau}_\theta}{\partial q} - \frac{\partial \boldsymbol{\tau}_q}{\partial \theta}\right) = 0 \Rightarrow \frac{\partial \boldsymbol{\tau}_\theta}{\partial q} - \frac{\partial \boldsymbol{\tau}_q}{\partial \theta} = 0 \tag{157}$$

Integrating the second equation with respect to q along the interval $[0, q]$ yields

$$\boldsymbol{\tau}_\theta(q, \theta) - \int_0^q dq \frac{\partial \boldsymbol{\tau}_q}{\partial \theta} = \boldsymbol{\tau}_\theta(q \sim 0, \theta) \tag{158a}$$

Next, Eq. (158a) is integrated with respect to θ along the interval $[0, 2\pi]$ and we get

$$\int_0^{2\pi} \boldsymbol{\tau}_\theta(q, \theta) d\theta - \int_0^{2\pi} \int_0^q dq \frac{\partial \boldsymbol{\tau}_q}{\partial \theta} d\theta = \int_0^{2\pi} \boldsymbol{\tau}_\theta(q \sim 0, \theta) d\theta \tag{158b}$$

which due to the fact that $\boldsymbol{\tau}_\theta(q, \theta)$ is quantized (for every value of q) in the following way [see Eq. (94)]:

$$\int_0^{2\pi} \boldsymbol{\tau}_\theta(q, \theta) d\theta = n\pi \tag{159}$$

yields the result:

$$\int_0^{2\pi} d\theta \int_0^q dq \frac{\partial \tau_q}{\partial \theta} = 0 \tag{160}$$

If we evaluate the integrand and change the order of integration we get

$$\int_0^{2\pi} d\theta \int_0^q dq \frac{\partial \tau_q}{\partial \theta} = \int_0^q dq \int_0^{2\pi} \frac{\partial \tau_q}{\partial \theta} d\theta = \int_0^q dq(\tau_q(q, \theta = 2\pi) - \tau_q(q, \theta = 0))$$

This result implies that $\tau_q(q, \theta)$ is, for every value of q, single valued with respect to θ.

In what follows, we assume that the second term in Eq. (158a) is *negligibly small* and as a result $\tau_\theta(q, \theta)$ becomes independent of q. Thus

$$\tau_\theta(q, \theta) = \tau_\theta(q = q_0, \theta) \tag{161a}$$

where q_0 is a fixed q value and $\tau_\theta(q = q_0, \theta)$ is a boundary value (at $q_0 \sim 0$) for $\tau_\theta(q, \theta)$ determined either by ab initio calculations or perturbation theory. We also recall that $\tau_\theta(q, \theta)$ fulfills the quantization condition as written in Eq. (159).

To examine our assumption regarding the dependence of $\tau_\theta(q, \theta)$ on q, we consider the well-known (collinear) conical intersection of the C_2H molecule formed by the two lowest states, namely, the $1^2A'$ and the $2^2A'$ states [12,72,105]. Figure 9 presents $\tau_\theta(q, \theta)$ as calculated for a fixed C—C distance, that is, $R_{CC} = 1.35$ Å and for different q values. It is seen that the basic shape of $\tau_\theta(q, \theta)$ is approximately preserved although q is varied along a relative large interval, that is, the [0.05, 1.0 Å] interval. It is noticed that the shape $\tau_\theta(q, \theta)$ is significantly affected only when $q = 1$ Å and $\theta \sim \pi$. The reason is that in this situation the point ($q = 1$ Å, $\theta = \pi$) gets very close to one of the carbons (the distance becomes ~0.3 Å) and therefore the ab initio values for $\tau_\theta(q, \theta)$ are not for an isolated conical intersection anymore as it should be [12].

In Section XIV.A.2, we intend to obtain the vector function $\boldsymbol{\tau}(q, \theta)$ for a given *distribution* of conical intersections. Thus, first we have to derive an expression for a conical intersection removed from the origin, namely, assumed to be located at some point, (q_{j0}, θ_{j0}), in the plane.

Combining Eqs. (151), (158a), and (161a) we get that $\tau_\theta(q, \theta)$ can be writtern as:

$$\tau_\theta(q, \theta) = \pi f(\theta) \tag{161b}$$

To shift it to some arbirtrary point (q_{j0}, θ_{j0}) we first express Eq. (161b) in terms of *Cartesian* coordinates, and then shift the solution to the point of interest, namely, to $(x_{j0}, y_{j0})[\equiv (q_{j0}, \theta_{j0})]$. Once completed, the solution is transformed back to polar coordinates (for details see Appendix F). Following

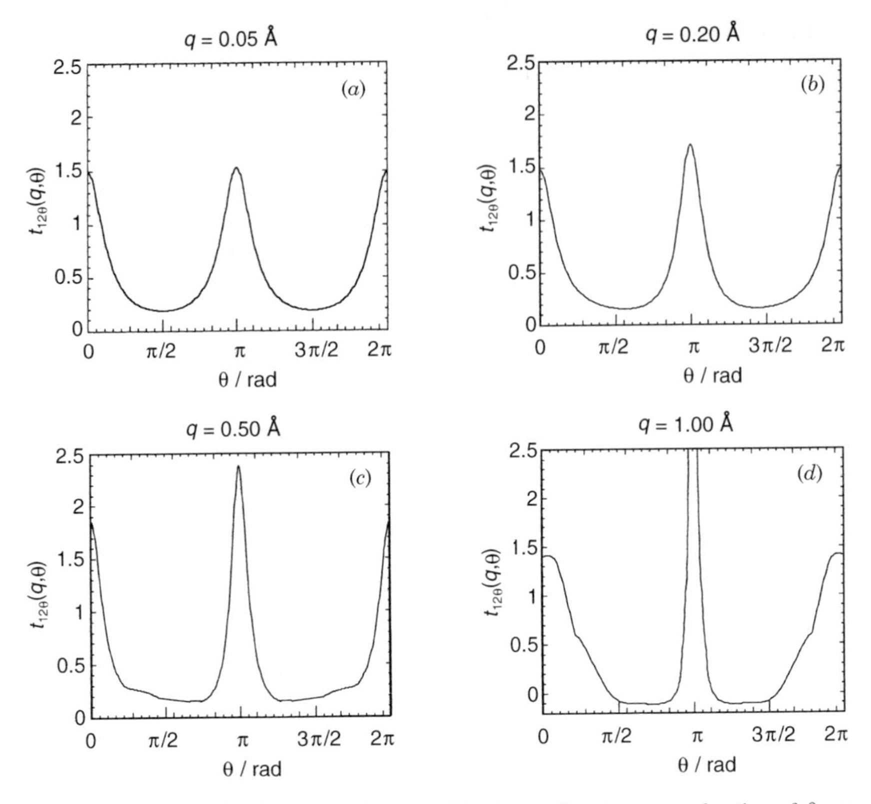

Figure 9. The $\tau_\theta(q, \theta)$—the angular non-adiabatic coupling term as a function of θ—as calculated for different q values. (*a*) $q = 0.05$ Å; (*b*) $q = 0.2$ Å; (*c*) $q = 0.5$ Å; (*d*) $q = 1.0$ Å.

this procedure, $\tau_\theta(q, \theta)$ and $\tau_q(q, \theta)$ (which is now different from zero) become

$$\begin{aligned} \tau_q(q, \theta) &= -f_j(\theta_j)\frac{1}{q_j}\sin(\theta - \theta_j) \\ \tau_\theta(q, \theta) &= f_j(\theta_j)\frac{q}{q_j}\cos(\theta - \theta_j) \end{aligned} \tag{162}$$

where q_j and θ_j are the coordinates of an arbitrary point, $P(q, \theta)$, with respect to the conical intersection position. The coordinates (q_j, θ_j) are related to (q, θ) as follows:

$$\begin{aligned} q_j &= \sqrt{(q\cos\theta - q_{j0}\cos\theta_{j0})^2 + (q\sin\theta - q_{j0}\sin\theta_{j0})^2} \\ \cos\theta_j &= \frac{q\cos\theta - q_{j0}\cos\theta_{j0}}{q_j} \end{aligned} \tag{163}$$

and $f_j(\theta_j)$ is defined as

$$f_j(\theta_j) \equiv \tau_{j\theta}(q_j \sim 0, \theta_j) \tag{164}$$

In Eq. (162) [as well as in Eq. (164)], we attached a subscript j to $f(\theta)$ to indicate that each conical intersection (in this case the jth one) may form a different spatial (angular) distribution.

Note that for $q_{j0} = 0$ the solution in Eq. (161b) is restored (and τ_q becomes zero).

2. *The Solution for a Distribution of Conical Intersections*

With the modified expression we can now extend the solution of Eq. (162) to any number of conical intersections. The solution in Eq. (162) stands for a single conical intersection located at an arbitrary point (q_{j0}, θ_{j0}). Since $\tau_\theta(q, \theta)$ and $\tau_q(q, \theta)$ are scalars the solution in case of N conical intersections located at the points (q_{j0}, θ_{j0}); $j = 1, \ldots, N$ are obtained by summing up the individual contributions [114]:

$$\begin{aligned} \tau_q(q, \theta) &= -\sum_{j=1}^{N} f_j(\theta_j) \frac{1}{q_j} \sin(\theta - \theta_j) \\ \tau_\theta(q, \theta) &= q \sum_{j=1}^{N} f_j(\theta_j) \frac{1}{q_j} \cos(\theta - \theta_j) \end{aligned} \tag{165}$$

Equation (165) yields the two components of $\tau(q, \theta)$, the vectorial non-adiabatic coupling term, for a distribution of two-state conical intersections expressed in terms of the values of the *angular* component of each individual non-adiabatic coupling term at the *closest* vicinity of each conical intersection. These values have to be obtained from ab initio treatments (or from perturbation expansions); however, all that is needed is a set of these values along a single closed circle, each surrounding one conical intersection.

To summarize our findings so far, we may say that if indeed the radial component of a single completely *isolated* conical intersection can be assumed to be negligible small as compared to the angular component, then we can present, almost fully analytically, the 2D "field" of the non-adiabatic coupling terms for a two-state system formed by any number of conical intersections. Thus, Eq. (165) can be considered as the non-adiabatic coupling field in the case of two states.

In Section XIV.B, this derivation is extended to a three-state system.

B. The Treatment of the Three-State System in a Plane

To study the three-state case, we consider two non-adiabatic coupling terms: one, between the lowest and the intermediate state, designated as τ_{12} with its origin located at $P_a(q_a, \theta_a)$, and the other between the intermediate and the highest state, designated as τ_{23} with its origin located at $P_b(q_b, \theta_b)$. As will be seen, in addition to τ_{12} and τ_{23} we also have to consider τ_{13}, although no degeneracy point exists between the lowest and the highest states. In other words, we shall show how the interaction between the above mentioned two conical intersections builds up τ_{13}, which does not have a source of its own. Thus the $\boldsymbol{\tau}$ matrix for the most general case has to be of the form:

$$\boldsymbol{\tau} = \begin{pmatrix} 0 & \tau_{12} & \tau_{13} \\ -\tau_{12} & 0 & \tau_{23} \\ -\tau_{13} & -\tau_{23} & 0 \end{pmatrix} \tag{166}$$

The curl equation for three (or more) states is given in Eq. (25) and is presented here again for the sake of completeness:

$$\operatorname{curl} \boldsymbol{\tau} - [\boldsymbol{\tau} \times \boldsymbol{\tau}] = 0 \tag{25}$$

It is well noted that, in contrast to the two-state equation [see Eq. (26)], Eq. (25) contains an additional, nonlinear term. This nonlinear term enforces a perturbative scheme in order to solve the required $\boldsymbol{\tau}$-matrix elements.

The derivation of the $\boldsymbol{\tau}$-matrix elements will be done in two steps: (1) first by considering each of the conical intersection as being isolated, namely, the one independent of the other; and (2) secondly by employing Eq. (20) to treat the two conical intersections as one complete system. Thus within the first step we obtain zeroth-order expressions for $\boldsymbol{\tau}_{12}$ and $\boldsymbol{\tau}_{23}$, that is, $\boldsymbol{\tau}_{012}$ and $\boldsymbol{\tau}_{023}$, respectively, whereas within the second step we not only correct these expressions so that Eq. (25), is ($\sim$) fulfilled for three states, but also derive the missing $\boldsymbol{\tau}_{13}$ term. The study is done, as before, for a plane in configuration space employing polar coordinates.

To study the two isolated conical intersections, we have to treat two-state curl equations that are given in Eq. (26). Here, the first 2×2 $\boldsymbol{\tau}$ matrix contains the (vectorial) element, that is, $\boldsymbol{\tau}_{012}$ and the second 2×2 $\boldsymbol{\tau}$ matrix contains $\boldsymbol{\tau}_{023}$. As before each of the non-adiabatic coupling terms, $\boldsymbol{\tau}_{012}$ and $\boldsymbol{\tau}_{023}$ has the following components:

$$\boldsymbol{\tau}_{0jj+1} = (\boldsymbol{\tau}_{0q\,jj+1}, \boldsymbol{\tau}_{0\theta jj+1}) \qquad j = 1, 2 \tag{167}$$

where $\boldsymbol{\tau}_{0qjj+1}$ and $\boldsymbol{\tau}_{0\theta jj+1}$; $j = 1, 2$, were derived in Section XIV.A [see Eqs. (165)], and therefore no further treatment is necessary.

In Section XI.B, we discussed situations (based on ab initio calculations) where the two non-adiabatic coupling terms τ_{12} and τ_{23} slightly overlap [12,108]. Based on ab initio calculations (as were carried out for the C_2H molecule) it was found that in many cases the non-adiabatic coupling is not evenly distributed around its point of degeneracy but rather is concentrated along a radial ridge that starts at the point of degeneracy (see Figs. 6 and 7). Therefore, in these cases, only slight overlaps are expected, in particular, when the two points of degeneracy $P_x(q_x, \theta_x)$; $x = a, b$ are located far enough from each other [108].

Thus if τ_{jj+1}—the full non-adiabatic coupling term—and the unperturbed non-adiabatic coupling term, τ_{0jj+1}, are assumed to be related to each other as

$$\boldsymbol{\tau}_{jj+1} = \boldsymbol{\tau}_{0jj+1} + \delta\boldsymbol{\tau}_{jj+1} \qquad j = 1, 2 \tag{168}$$

then it follows, from the above discussion, that the components of the two vectorial perturbations (i.e., $\delta\tau_{q\,jj+1}$ and $\delta\tau\theta_{jj+1}$) are likely to be (much) smaller than the corresponding components, namely, $\tau_{0q\,jj+1}$ and $\tau_{0\theta jj+1}$.

Next, we return to Eq. (25) and recall that we are interested only in the components of $\boldsymbol{\tau}_{jj+1}$ $j = 1, 2$ in a plane perpendicular to the z axis. It can be shown that if $\boldsymbol{\tau}_{0jj+1}$, $j = 1, 2$ do not posses a z component, the same applies to the perturbations $\delta\boldsymbol{\tau}_{jj+1}$ $j = 1, 2$, as well as to τ_{13}.

Substituting Eq. (168) in Eq. (166) and the result in Eq. (25) yields the (inhomogeneous) differential equations for the components of $\delta\boldsymbol{\tau}_{jj+1}$; $j = 1, 2$

$$\begin{aligned} \text{curl}(\delta_{12}) &= \frac{\partial(\delta\tau_{q12})}{\partial\theta} - \frac{\partial(\delta\tau_{\theta12})}{\partial q} = \tau_{\theta13}\tau_{0q23} - \tau_{q13}\tau_{0\theta23} \\ \text{curl}(\delta\tau_{23}) &= \frac{\partial(\delta\tau_{q23})}{\partial\theta} - \frac{\partial(\delta\tau_{\theta23})}{\partial q} = \tau_{q13}\tau_{0\theta12} - \tau_{\theta13}\tau_{0q12} \end{aligned} \tag{169}$$

where the second-order terms were deleted. In this derivation, we employed the fact that:

$$\text{curl}\,\boldsymbol{\tau}_{012} = \text{curl}\,\boldsymbol{\tau}_{023} = 0 \tag{170}$$

In the same way, with similar assumptions, we obtain the (inhomogeneous) differential equation for the components of $\boldsymbol{\tau}_{13}$

$$\text{curl}\,\tau_{13} = \frac{\partial\tau_{q13}}{\partial\theta} - \frac{\partial\tau_{\theta13}}{\partial q} = \tau_{0\theta12}\tau_{0q23} - \tau_{0q12}\tau_{0\theta23} \tag{171}$$

Equation (171) is the an explicit "curl" equation for a coupling that does not has a "source" of its own but is formed due to the interaction between two "real" conical intersection.

Equations (169) and (171), together with Eqs. (170), form the basic equations that enable the calculation of the non-adiabatic coupling matrix. As is noticed, this set of equations creates a hierarchy of approximations starting with the assumption that the cross-products on the right-hand side of Eq. (171) have small values because at any point in configuration space at least one of the multipliers in the product is small [115].

XV. STUDIES OF SPECIFIC SYSTEMS

In this section, we concentrate on a few examples to show the degree of relevance of the theory presented in the previous sections. For this purpose, we analyze the conical intersections of two *real* two-state systems and one *real* system resembling a tri-state case.

A. The Study of *Real* Two-State Molecular Systems

We start by mentioning the studies of Yarkony et al. [64] who were the first to apply the line integral approach to reveal the existence of a conical intersection for a "real" molecular system—the H_3 system—by calculating the relevant non-adiabatic coupling terms from first principles and then deriving the topological angle α [see Eq. (76)]. Later Yarkony and co-workers applied this approach to study other tri-atom system such as AlH_2 [65], CH_2 [66,69], H_2S [66], HeH_2 [68], and Li_3 [70].

Recently, Xu et al. [11] studied in detail the H_3 molecule as well as its two isotopic analogues, namely, H_2D and D_2H, mainly with the aim of testing the ability of the line integral approach to distinguish between the situations when the contour surrounds or does not surround the conical intersection point. Some time later Mebel and co-workers [12,72–74,116] employed *ab initio* non-adiabatic coupling terms and the line-integral approach to study some features related to the C_2H molecule.

Some results of these studies will be presented in Sections XV.A.1–XV.A.3.

1. *The H_3-System and Its Isotopic Analogues*

Although the study to be described is for a "real" system, the starting point was not the ab initio adiabatic potential energy surfaces and the ab initio non-adiabatic coupling terms but a diabatic potential [117], which has its origin in the LSTH potential [118] improved by including three-center terms [119]. These were used to calculate the adiabatic-to-diabatic transformation angle γ by employing the Hellmann–Feynman theorem [3,36]. However, we present our results in term of the diabatic-to-adiabatic transformation angle β, which is also know as the *mixing angle*. We start by proving, analytically that these two angles are identical up to an integration constant.

We consider a 2D diabatic framework that is characterized by an angle, $\beta(s)$, associated with the orthogonal transformation that diagonalizes the diabatic potential matrix. Thus, if $\mathbf{V}$ is the diabatic potential matrix and if $\mathbf{u}$ is the adiabatic one, the two are related by the orthogonal transformation matrix $\mathbf{A}$ [34]:

$$\mathbf{u} = \mathbf{A}^{\dagger}\mathbf{V}\mathbf{A} \tag{172}$$

where $\mathbf{A}^{\dagger}$ is the complex conjugate of the $\mathbf{A}$ matrix. For the present two-state case, $\mathbf{A}$ can be written in the form:

$$\mathbf{A} = \begin{pmatrix} \cos\beta & -\sin\beta \\ \sin\beta & \cos\beta \end{pmatrix} \tag{173}$$

where β—the above mentioned mixing angle—is given by [36a]:

$$\beta = \frac{1}{2}\tan^{-1}\frac{2V_{12}}{V_{11} - V_{22}} \tag{174}$$

Recalling γ(s), the adiabatic-to-diabatic transformation angle [see Eqs. (74) and (75)] it is expected that the two angles are related. The connection is formed by the Hellmann–Feynman theorem, which yields the relation between the s component of the non-adiabatic coupling term, $\boldsymbol{\tau}$, namely, τ_s, and the characteristic diabatic magnitudes [13]

$$\tau_s(s) = (u_2 - u_1)^{-1}\mathbf{A}_1^*\frac{\partial \mathbf{V}}{\partial s}\mathbf{A}_2 = \frac{\sin 2\beta}{2W_{12}}\mathbf{A}_1^*\frac{\partial \mathbf{V}}{\partial s}\mathbf{A}_2 \tag{175}$$

where $\mathbf{A}_i, i = 1, 2$ are the two columns of the $\mathbf{A}$ matrix in Eq. (173). By replacing the two $\mathbf{A}_i$ columns by their explicit expressions yields for τ_s the expression

$$\tau_s(s) = \frac{\sin 2\beta}{2V_{12}}\left[\frac{-\sin 2\beta}{2}\frac{\partial}{\partial s}(V_{11} - V_{22}) + \cos 2\beta\frac{\partial}{\partial s}V_{12}\right] \tag{176}$$

Next, by differentiating Eq. (174) with respect to s

$$\frac{\partial}{\partial s}(V_{11} - V_{22}) = 2\left(V_{12}\frac{\partial}{\partial s}\cot 2\beta + \cot 2\beta\frac{\partial}{\partial s}V_{12}\right) \tag{177}$$

and by substituting Eq. (177) in Eq. (176), yields the following result for $\tau_s(s)$:

$$\tau_s(s) = \frac{\partial \beta}{\partial s} \tag{178}$$

Comparing this equation with Eq. (75), it is seen that the mixing angle β is, up to an additive constant, identical to the relevant adiabatic-to-diabatic transformation—angle γ:

$$\gamma(s) = \beta(s) - \beta(s_0) \tag{179}$$

This relation will be used to study geometrical phase effects within the diabatic framework for the H_3 system and its two isotopic analogues. What is meant by this is that since our starting point is the 2×2 diabatic potential matrix, we do not need to obtain the adiabatic-to-diabatic transformation angle by solving a line integral; it will be obtained simply by applying Eqs. (174) and (178). The forthcoming study is carried out by presenting $\beta(\varphi)$ as a function of an angle φ to be introduced next.

In the present study, we are interested in finding the locus of the seam defined by the conditions $r_{AB} = r_{BC} = r_{AC}$ [14–17] where r_{AB}, r_{BC}, and r_{AC} are the interatomic distances. Since we intend to study the geometrical properties produced by this seam we follow a suggestion by Kuppermann and co-workers [29,120,121] and employ the hyperspherical coordinates (ρ, θ, φ) that were found to be suitable for studying topological effects for the H–H_2 (and its isotopic analogues) because one of the hyperspherical (angular) coordinates surrounds the seam in case of the pure-hydrogenic case. Consequently, following previous studies [29,122–124], we express the three above-mentioned distances in terms of these coordinates, that is,

$$\begin{aligned} r_{AB}^2 &= \frac{1}{2} d_C \rho^2 \left[1 + \sin\frac{\theta}{2} \cos(\varphi + \chi_{AC}) \right] \\ r_{BC}^2 &= \frac{1}{2} d_A \rho^2 \left[1 + \sin\frac{\theta}{2} \cos(\varphi) \right] \\ r_{AC}^2 &= \frac{1}{2} d_B \rho^2 \left[1 + \sin\frac{\theta}{2} \cos(\varphi - \chi_{AB}) \right] \end{aligned} \tag{180}$$

where

$$\begin{aligned} d_X^2 &= \frac{m_X}{\mu}\left(1 - \frac{m_X}{M}\right) \qquad & \chi_{XY} &= 2\tan^{-1}\left(\frac{m_Z}{\mu}\right) \\ \mu &= \sqrt{\frac{m_A m_B m_C}{M}} \qquad & M &= m_A + m_B + m_C \end{aligned} \tag{181}$$

Here X,Y,Z stand for A,B,C and

$$\rho = \sqrt{r_{AB}^2 + r_{AC}^2 + r_{BC}^2} \tag{182}$$

By equating the three interatomic distances with each other, we find that the seam is a straight line, for which ρ is arbitrary but φ and θ have fixed values φ_s and θ_s determined by the masses only.

$$\varphi_s = \tan^{-1}\left\{ \frac{\cos\chi_{AC} - t\cos\chi_{AB} - \left(\frac{d_A}{d_C}\right)^2 + t\left(\frac{d_A}{d_B}\right)^2}{\sin\chi_{AC} - t\sin\chi_{AB}} \right\} \tag{183}$$

and

$$\theta_s = 2\sin^{-1}\left\{\frac{\left(\frac{d_{\mathrm{A}}}{d_{\mathrm{B}}}\right)^2 - 1}{\cos(\varphi_s - \chi_{\mathrm{AB}}) - \left(\frac{d_{\mathrm{A}}}{d_{\mathrm{B}}}\right)^2 \cos\varphi_s}\right\} \tag{184}$$

where t is given in the form

$$t = \left[\left(\frac{d_{\mathrm{A}}}{d_{\mathrm{C}}}\right)^2 - 1\right]\left[\left(\frac{d_{\mathrm{A}}}{d_{\mathrm{B}}}\right)^2 - 1\right]^{-1} \tag{185}$$

Equations (182)–(185) are valid when all three masses are different. In case two masses are equal, namely, $m_{\mathrm{B}} = m_{\mathrm{C}}$, we get for θ_s the simplified expression

$$\theta_s = 2\sin^{-1}\left\{\left|\frac{m_{\mathrm{B}} - m_{\mathrm{A}}}{m_{\mathrm{B}} + 2m_{\mathrm{A}}}\right|\right\} \tag{186}$$

and for φ_s the value π when $m_{\mathrm{A}} > m_{\mathrm{B}}$ and the value zero when $m_{\mathrm{A}} < m_{\mathrm{B}}$. In case all three masses are equal (then $t = 1$), we get $\theta_s = 0$ and $\varphi_s = \pi$.

In what follows, we discuss the H_2D system. For this purpose Eq. (186) is employed for which it is obtained that the straight line seam is defined for the following values of θ_s and φ_s, namely, $\theta_s = 0.4023$ rad, and $\varphi_s = \pi$. In the H_3 case, the value of θ_s is zero and this guarantees that all the circles with constant ρ and θ encircle the seam. The fact that θ_s is no longer zero implies that not all the circles with constant ρ and θ encircle the seam; thus, circles for which $\theta > \theta_s$ will encircle the seam and those with $\theta < \theta_s$ will not.

In Figure 10 are presented $\beta(\varphi)$ curves for H_2D, all calculated for $\rho = 6a_0$. In this calculation, the hyperspherical angle φ, defined in along the $[0,2\pi]$ interval, is the independent angular variable. Figure 10*a* shows two curves for the case where the line integral does not encircle the seam, namely, for $\theta = 0.2$ and 0.4 rad and in Figure 10*b* for the case where the line integral encircles the seam, namely, for $\theta = 0.405$ and 2.0 rad. Notice that the curves in Figure 10*a* reach the value of zero and those in Figure 10*b* reach the value of π. In particular, two curves, that in Figure 10*a* for $\theta = 0.4$ rad and the other in Figure 10*b* for $\theta = 0.405$ rad, were calculated along very close contours (that approach the locus of the seam) and indeed their shapes are similar—they both yield an abrupt step—but one curve reaches the value of zero and the other the value π. Both types of results justify the use of the line integral to uncover the locus of the seam. More detailed results as well as the proper analysis can be found in [11].

These results as well as others presented in [11] are important because on various occasions it was implied that the line integral approach is suitable only

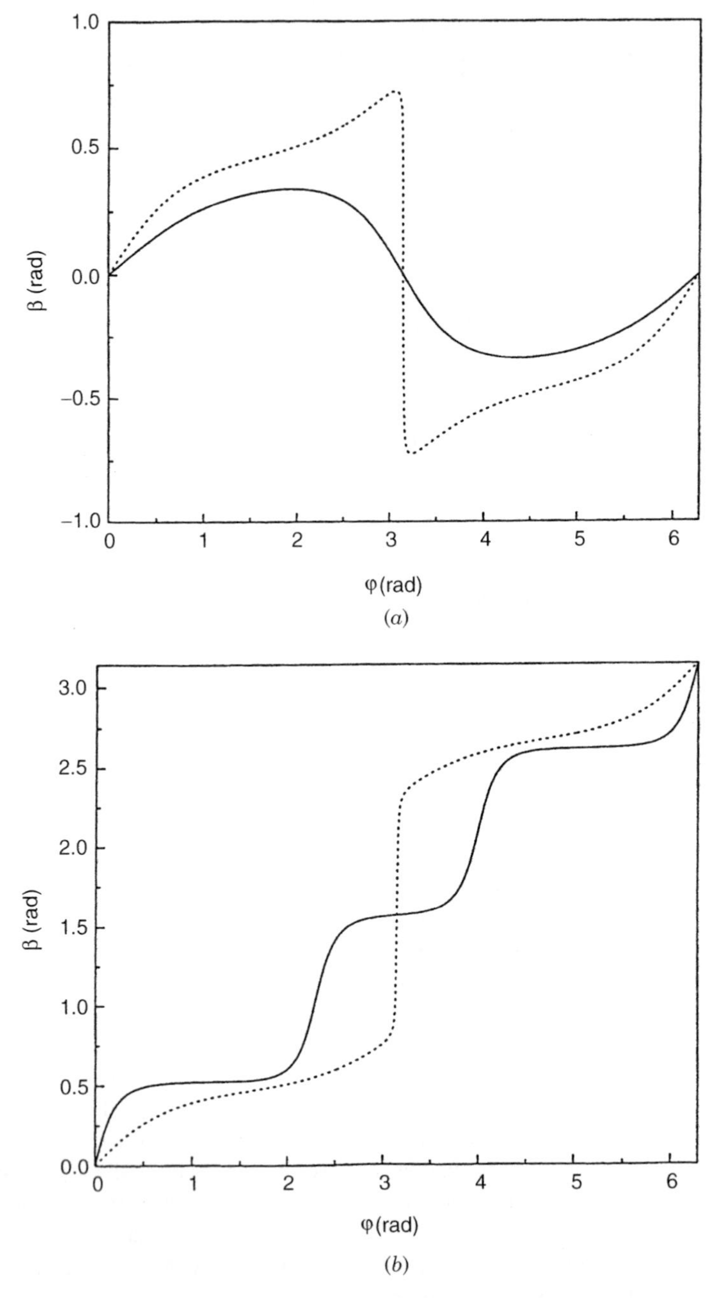

Figure 10. The mixing angle β, for the H_2D system, as a function of hyperspherical angle φ, calculated for hyperspherical radius $\rho = 6\ a_0$: (*a*) Results for $\theta = 0.2$ rad ———— and $\theta = 0.4$ rad .... (*b*) The same as (*a*) but for $\theta = 2.0$ rad ————; and $\theta = 0.405$ rad ....

for cases when relatively small radii around the conical intersection are applied [64]. In [11], it is shown for the first time that this approach can be useful even for large radii, which does not mean that it is relevant for any assumed contour surrounding a conical intersection (or for that matter a group of conical intersections) but means that we can always find contours with large radii that will reveal the conical intersection location for a given pair of states.

2. *The C_2H-Molecule: The Study of the (1,2) and the (2,3) Conical Intersections*

In the first part of this study, we were interested in non-adiabatic coupling terms between the $1^2A'$ and $2^2A'$ and between the $2^2A'$ and $3^2A'$ electronic states. The calculations were done employing MOLPRO [6], which yield the six relevant non-adiabatic coupling elements as calculated with respect to the Cartesian center-of-mass coordinates of each atom. These coupling terms were then transformed, employing chain rules [12,73], to non-adiabatic coupling elements with respect to the internal coordinates of the C_2H molecule, namely, $\langle\zeta_i|\partial\zeta_j/\partial r_1\rangle$ $(=\tau_{r_1})$, $\langle\zeta_i|\partial\zeta_j/\partial r_2\rangle(=\tau_{r_2})$, and $\langle\zeta_i|\partial\zeta_j/\partial\varphi\rangle(=\tau_\varphi)$. Here r_1 and r_2 are the C—C and C—H distances, respectively, and φ is the relevant CC$\cdots$CH angle. The adiabatic-to-diabatic transformation angle, $\gamma(\varphi|r_1,r_2)$, is derived next employing the following line integral [see Eq. (75)], where the contour is an arc of a circle with radius r_2:

$$\gamma(\varphi \mid r_1,r_2) = \int_0^{\varphi} d\varphi'\tau_\varphi(\varphi' \mid r_1,r_2) \tag{187}$$

The corresponding topological phase, $\alpha(r_1,r_2)$ [see Eq. (76)] defined as $\gamma(\varphi = 2\pi \mid r_1,r_2)$, was also obtained for various values of r_1 and r_2.

First, we refer to the (1,2) conical intersection. A detailed inspection of the non-adiabatic coupling terms revealed the existence of a conical intersection between these two states, for example, at the point $\{\varphi = 0, r_1 = 1.35\,\text{Å}, r_2 = 1.60\,\text{Å})$ as was established before [105]. More conical intersections of this kind are expected at other r_1 values. Next, were calculated the $\gamma(\varphi \mid r_1,r_2)$ angles as a function of φ for various r_2. The $\tau_\varphi(\varphi \mid r_1,r_2)$ functions as well as the adiabatic-to-diabatic transformation angles are presented in Figure 11 for three different r_2 values, namely, $r_2 = 1.8, 2.0, 3.35\,\text{Å}$. Mebel et al. [12] also calculated the topological angle $\alpha(r_1,r_2)$ for these three r_2 values employing Eq. (76) and got, for the first two r_2 values, the values 3.136 and 3.027 rad, respectively—thus, in both cases, values close to the expected π value. A different situation is encountered in the third case when the circle surrounds the two (symmetrical) CIs as can be seen from the results presented in the third panel of Figure 11*e* and *f*. In such a case, the angle α is expected to be either an even multiple of π or zero. The integration according to Eq. (76) yields the value

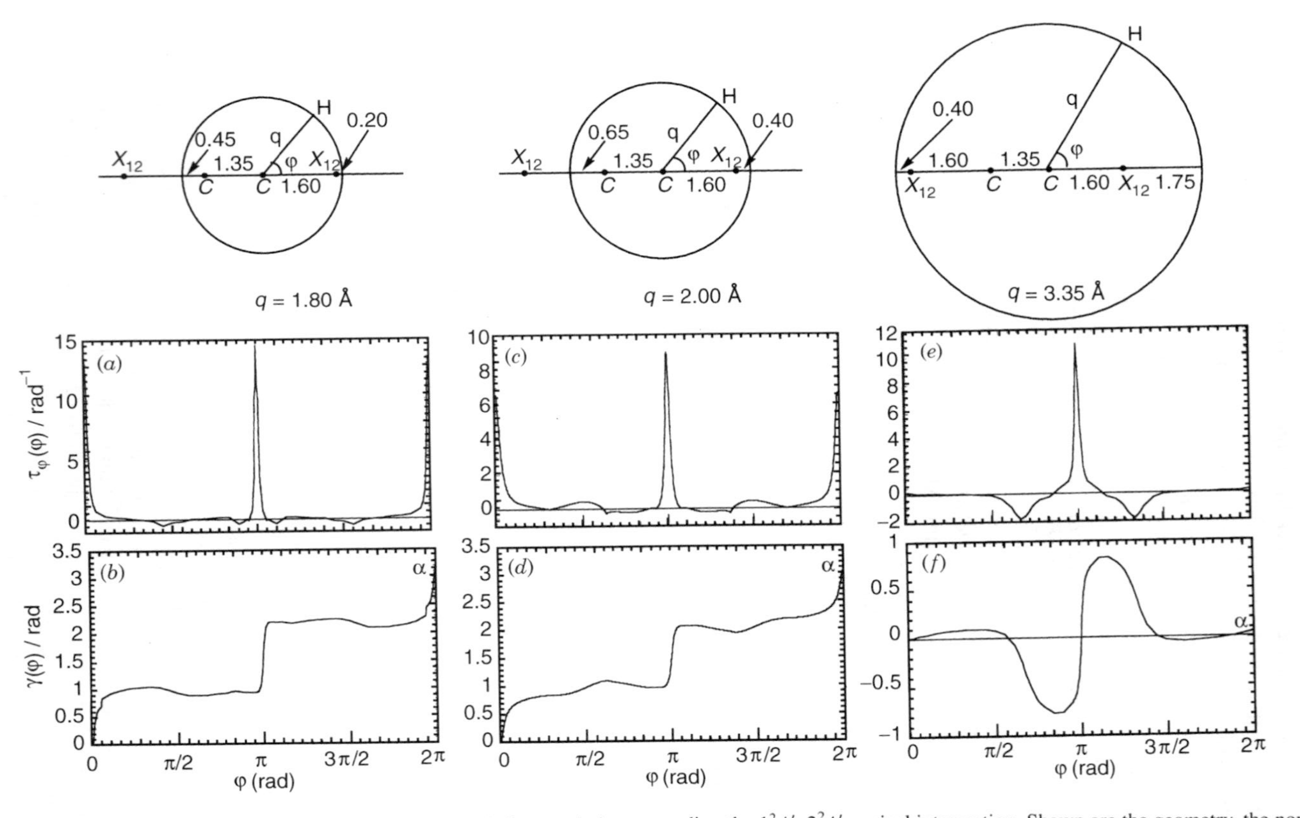

Figure 11. Results for the C_2H molecule as calculated along a circle surrounding the $1^2A'$–$2^2A'$ conical intersection. Shown are the geometry, the nonadiabatic coupling matrix elements $\tau\varphi(\varphi \mid r_2)$ and the adiabatic-to-diabatic transformation angles $\gamma(\varphi \mid r_2)$ as calculated for r_1 (=CC distance) = 1.35 Å and for three r_2 values (r_2 is the CH distance): (*a*) and (*b*) $r_2 = 1.80$ Å; (*c*) and (*d*) $r_2 = 2.00$ Å; (*e*) and (*f*) $r_2 = 3.35$ Å. (Note that $q \equiv r_2$.)

of 0.048 rad, namely, a value close to zero. It is important to mention that we also performed integrations along closed circles that do not surround any conical intersections and got the value zero as was proved in Appendix C (for more details about these calculations see [12]).

In this series of results, we encounter a somewhat unexpected result, namely, when the circle surrounds two conical intersections the value of the line integral is zero. This does not contradict any statements made regarding the general theory (which asserts that in such a case the value of the line integral is either a multiple of 2π or *zero*) but it is still somewhat unexpected, because it implies that the two conical intersections behave like *vectors* and that they arrange themselves in such a way as to reduce the effect of the non-adiabatic coupling terms. This result has important consequences regarding the cases where a pair of electronic states are coupled by more than one conical intersection.

On this occasion, we want also to refer to an incorrect statement that we made more than once [72], namely, that the (1,2) conical intersection results indicate "that for any value of r_1 and r_2 the two states under consideration form an *isolated* two-state sub-Hilbert space." We now know that in fact they do not form an isolated system because the second state is coupled to the third state via a conical intersection as will be discussed next. Still, the fact that the series of topological angles, as calculated for the various values of r_1 and r_2, are either multiples of π or zero indicates that we can form, for this adiabatic two-state system, *single-valued* diabatic potentials. Thus if for some numerical treatment only the two lowest adiabatic states are required, the results obtained here suggest that it is possible to form from these two adiabatic surfaces single-valued diabatic potentials employing the line-integral approach. Indeed, recently Billing et al. [104] carried out such a photodissociation study based on the two lowest adiabatic states as obtained from ab initio calculations. The complete justification for such a study was presented in Section XI.

Reference [73] presents the first line-integral study between two excited states, namely, between the second and the third states in this series of states. Here, like before, the calculations are done for a fixed value of r_1 (results are reported for $r_1 = 1.251$ Å) but in contrast to the previous study the origin of the system of coordinates is located at the point of this particular conical intersection, that is, the (2,3) conical intersection. Accordingly, the two polar coordinates (φ, q) are defined. Next is derived the φ-th non-adiabatic coupling term i.e. τ_φ $(= \langle\zeta_2|\partial\zeta_3/\partial\varphi\rangle)$ again employing chain rules for the transformation $(\tau_\gamma, \tau_{r2}) \rightarrow \tau_\varphi(\tau_q$ is not required because the integrals are performed along a circle with a fixed radius q—see Fig. 12).

Figure 12 presents $\tau_\varphi(\varphi \mid q)$ and $\gamma(\varphi \mid q)$ for three values of q, that is, $q = 0.2, 0.3, 0.4$ Å. The main features to be noticed are (1) The function $\tau_\varphi(\varphi \mid q)$ exhibits the following symmetry properties: $\tau_\varphi(\varphi) = \tau_\varphi(\pi - \varphi)$ and

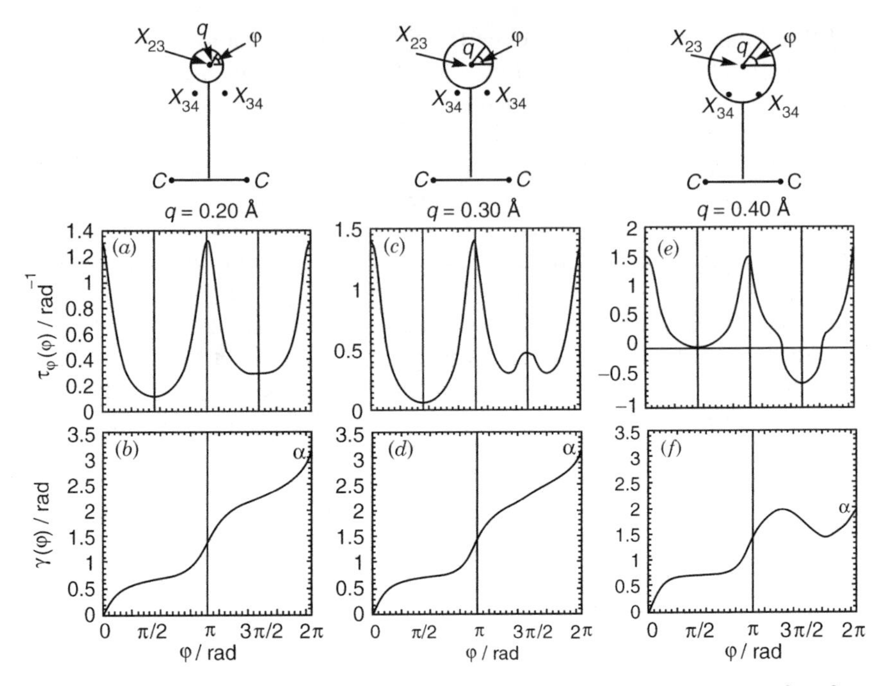

Figure 12. Results for the C_2H molecule as calculated along a circle surrounding the $2^2A'$–$3^2A'$ conical intersection. The conical intersection is located on the C_{2v} line at a distance of 1.813 Å from the CC axis, where r_1 (=CC distance) = 1.2515 Å. The center of the circle is located at the point of the conical intersection and defined in terms of a radius q. Shown are the non-adiabatic coupling matrix elements $\tau\varphi(\varphi|q)$ and the adiabatic-to-diabatic transformation angles $\gamma(\varphi|q)$ as calculated for (a) and (b) where $q = 0.2$ Å; (c) and (d) where $q = 0.3$ Å; (e) and (f) where $q = 0.4$ Å. Also shown are the positions of the two close-by (3,4) conical intersections (designated as X_{34}).

$\tau_\varphi(\pi + \varphi) = \tau_\varphi(2\pi - \varphi)$, where $0 \leq \varphi \leq \pi$. In fact, since the origin is located on the C_{2v} axis we should expect only $|\tau_\varphi(\varphi)| = |\tau_\varphi(\pi - \varphi)|$ and $|\tau_\varphi(\pi + \varphi)| = |\tau_\varphi(2\pi - \varphi)|$, where $0 \leq \varphi \leq \pi$ but due to continuity requirements these relations also have to be satisfied without the absolute signs. (2) It is seen that the adiabatic-to-diabatic transformation angle, $\gamma(\varphi \mid q)$, increases, for the two smaller q-values, monotonically to become $\alpha(\Gamma \mid q)$, with the value of π (in fact, we got 0.986π and 1.001π for $q = 0.2$ and 0.3 Å, respectively). The two-state assumption seems to break down in case $q = 0.4$ Å because the calculated value of $\alpha(\Gamma \mid q)$ is not anymore π but only 0.63π. The reason being that the $q = 0.4$-Å circle not only passes too close to two (3,4) conical intersections—the distances at the closest points are $\sim$0.04 Å—and so the (2,3) system can not be considered anymore as an isolated sub-Hilbert space but in fact surrounds these two conical intersections (see third panel of Fig. 12). More details are given in Section XV.B [116].

B. The Study of a *Real* Three-State Molecular System: Strongly Coupled (2,3) and (3,4) Conical Intersections

We ended Section XV.A by claiming that the value $\alpha(\Gamma \mid q = 0.4\,\text{Å})$ is only 0.63π instead of π (thus damaging the two-state quantization requirement) because, as additional studies revealed, of the close locations of two (3,4) conical intersections. In this section, we show that due to these two conical intersections our sub-space has to be extended so that it contains three states, namely, the second, the third, and the fourth states. Once this extension is done, the quantization requirement is restored but for the three states (and not for two states) as will be described next.

In Section IV, we introduced the topological matrix **D** [see Eq. (38)] and showed that for a sub-Hilbert space this matrix is diagonal with (+1) and (−1) terms a feature that was defined as quantization of the non-adiabatic coupling matrix. If the present three-state system forms a sub-Hilbert space the resulting **D** matrix has to be a diagonal matrix as just mentioned. From Eq. (38) it is noticed that the **D** matrix is calculated along contours, Γ, that surround conical intersections. Our task in this section is to calculate the **D** matrix and we do this, again, for circular contours.

The numerical part is based on two circles, C_3 and C_4, related to two different centers (see Fig. 13). Circle C_3, with a radius of 0.4 Å, has its center at the position of the (2,3) conical intersection (like before). Circle C_4, with a radius 0.25 Å, has its center (also) on the C_{2v} line, but at a distance of 0.2 Å from the (2,3) conical intersection and closer to the two (3,4) conical intersections. The computational effort concentrates on calculating the exponential in Eq. (38) for the given set of ab initio 3×3 τ matrices computed along the above mentioned two circles. Thus, following Eq. (28) we are interested in calculating the following expression:

$$\mathbf{A}(\varphi \mid q) = \wp \exp\left(-\int_0^{\varphi} \tau_\varphi(\varphi' \mid q)d\varphi'\right) \tag{188}$$

where value of q determines the circular contour. The matrix $\mathbf{D}(q)$ is, accordingly:

$$\mathbf{D}(q) = \mathbf{A}(\varphi = 2\pi \mid q) \tag{189}$$

To calculate $\mathbf{A}(\varphi \mid q)$ the angular interval $\{0, \varphi\}$ is divided into n (small enough) segments with $\{\varphi_0(=0), \varphi_1, \ldots, \varphi_n(=\varphi)\}$ as division points, so that the **A** matrix can be presented as

$$\mathbf{A}(\varphi = \varphi_n) = \prod_{k=1}^{n} \exp\left(-\int_{\varphi_{k-1}}^{\varphi_k} \tau(\varphi')d\varphi'\right) \tag{190}$$

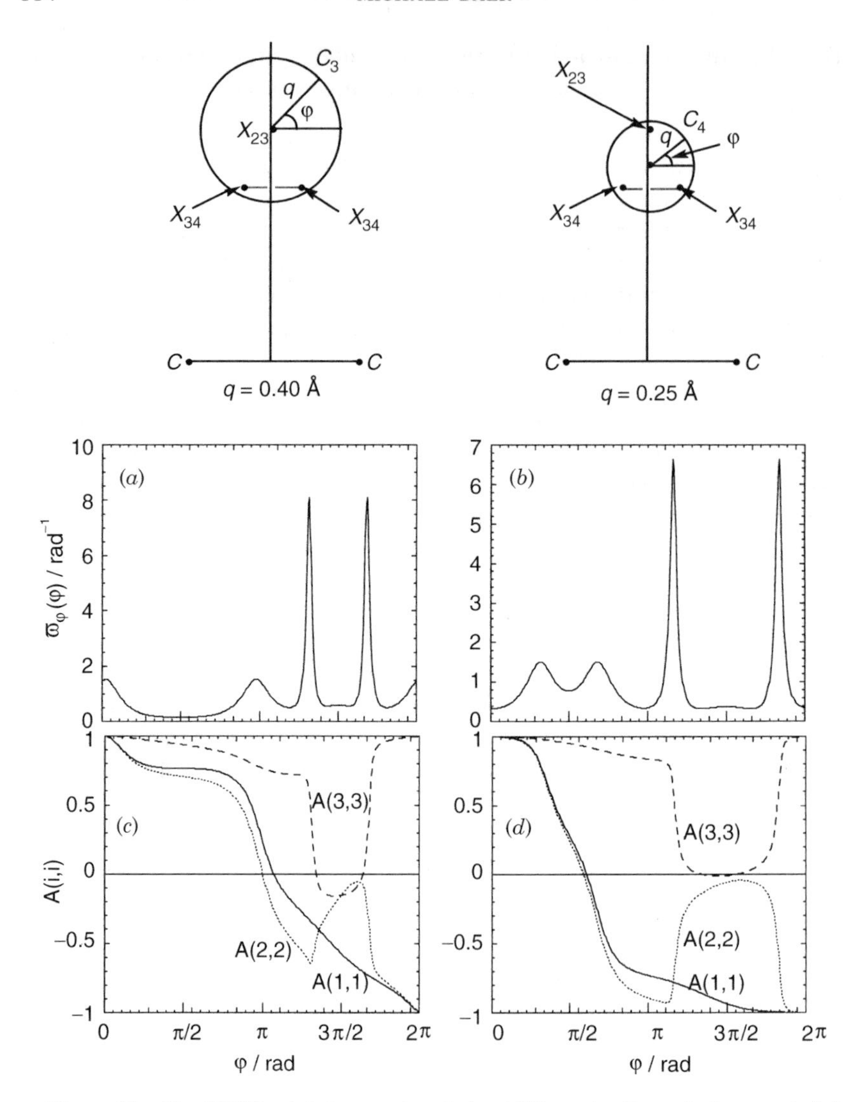

Figure 13. The NACT $\varpi(\varphi)$ (see text) and the ADT matrix diagonal elements $A_{ii}(\varphi)$; $i = 1,2,3$, as calculated for two contours surrounding all three CIs: (*a*) and (*c*) Results for the C_3 contour ($q = 0.4$ Å). (*b*) and (*d*) Results for the C_4 contour ($q = 0.25$ Å). The upper panels present the geometrical situation for each case: The contour C_3 has its center at the point of the (2,3) CI and its radius is $q = 0.4$ Å. The contour C_4 has its center (at a distance of 0.2 Å) in-between the (2,3) CI point and the two (3,4) CIs axis and its radius is $q = 0.25$ Å.

where the variable q is deleted. By following the procedure described in [57], one presents $\mathbf{A}(\varphi_n)$ as

$$\mathbf{A}(\varphi_n) = \prod_{k=1}^{n} \mathbf{G}_k^{\dagger}\mathbf{E}(\varphi_k)\mathbf{G}_k \tag{191}$$

where $\mathbf{G}_k$ is a unitary matrix that diagonalizes $\tau(\varphi)$ at the mid-point of the kth segment: $\tilde{\varphi}_k = (\varphi_k + \varphi_{k-1})/2$ and $\mathbf{E}(\varphi_k)$ is a diagonal matrix with elements $(m = 1, 2, \ldots, M)$:

$$\mathbf{E}_m(\varphi_k) = \exp\left(-\int_{\varphi_k}^{\varphi_{k+1}} t_m(\varphi)d\varphi\right) = \exp(-t_m(\tilde{\varphi}_k)\Delta\varphi) \tag{192}$$

Here, $t_m(\tilde{\varphi})$; $m = 1, 2, \ldots, M$ are the eigenvalues of $\tau(\tilde{\varphi})$ and $\Delta\varphi$ is the angular grid size. The order of the multiplication in Eq. (191) is such that the $k = 0$ term is the first term from the right-hand side in the product. With these definitions the matrix $\mathbf{D}$ is defined as $\mathbf{D}(q) = A(\varphi = \varphi_N \mid q)$, where $\varphi_N = 2\pi$ [see Eq. (189)].

Going back to our case and recalling that $\tau(\varphi \mid q)$ is a 3×3 antisymmetric matrix it can be shown that one of its eigenvalues is always zero and the others are two imaginary conjugate functions, namely, $\pm i\varpi(\varphi)$ where $\varpi(\varphi) = \sqrt{\tau_{12}^2 + \tau_{23}^2 + \tau_{13}^2}$. In Figure 13$a$ and b we present $\varpi(\varphi)$ functions as calculated for the two circles C_3 and C_4 (see the relevant upper panels of Fig. 13). The two strong spikes are due to the two (3,4) conical intersections and they occur at points where the circles cross their axis line.

To perform the product in Eq. (191) we need the $\mathbf{G}$ matrices and, for this 3×3 matrix, these can be obtained analytically [7,80]. Thus

$$\mathbf{G} = \frac{1}{\varpi\lambda\sqrt{2}} \begin{pmatrix} i\tau_{13}\varpi - \tau_{23}\tau_{12} & -i\tau_{13}\varpi - \tau_{23}\tau_{12} & \tau_{23}\lambda\sqrt{2} \\ i\tau_{23}\varpi + \tau_{13}\tau_{12} & -i\tau_{23}\varpi + \tau_{13}\tau_{12} & -\tau_{13}\lambda\sqrt{2} \\ \lambda^2 & \lambda^2 & \tau_{12}\lambda\sqrt{2} \end{pmatrix} \tag{193}$$

where $\lambda = \sqrt{\tau_{23}^2 + \tau_{13}^2}$.

In Figure 13c and d we present the three diagonal elements of the corresponding adiabatic-to-diabatic transformation matrices $\mathbf{A}(\varphi \mid q)$ as calculated for the two circles. Note that $\mathbf{A}_{11}(\varphi \mid q)$, in both cases, behaves smoothly while varying essentially undisturbed, from $(+1)$ to (-1). The second diagonal term in each case, that is, $\mathbf{A}_{22}(\varphi \mid q)$, follows the relevant $\mathbf{A}_{11}(\varphi \mid q)$, until the contour enters the region of the (3,4) conical intersections. There the $\mathbf{A}_{22}(\varphi \mid q)$ terms start to increase like they would do if only one (3,4) conical intersection were present. However, once they have reached the region of the second (3,4) conical intersection this conical intersection pushes the curve down again so that finally the $\mathbf{A}_{22}(\varphi \mid q)$ terms become (-1), instead of $(+1)$. The third term, $\mathbf{A}_{33}(\varphi \mid q)$ in each case, proceeds undisturbed as long as it is out of the range of the two (3,4) conical intersections. Once it enters the region of the first conical intersection, the curve starts to decrease and eventually becomes (-1) as it should if only one conical intersection was present. However, as the contour reaches the region of the second conical intersection, $\mathbf{A}_{33}(\varphi \mid q)$ is pushed back and ends up with the value of $(+1)$, instead of (-1). The value of each term $\mathbf{A}_{ii}(\varphi = 2\pi \mid q), i = 1, 2, 3$ constitutes the diagonal of the $\mathbf{D}$ matrix for the particular contour:

The results for $C_4(q = 0.25\,\text{Å})$ are as follows:

$$\mathbf{D}_{11} = -0.9998; \quad \mathbf{D}_{22} = -0.9999; \quad \mathbf{D}_{33} = 0.9997.$$

The results for $C_3(q = 0.4\,\text{Å})$ are as follows:

$$\mathbf{D}_{11} = -0.990; \quad \mathbf{D}_{22} = -0.988; \quad \mathbf{D}_{33} = 0.997.$$

While studying these results we have to pay attention to two features: (1) In each case, these numbers must, in absolute value, be as close as possible to 1; and (2) two of these numbers have to be negative. Then, we also have to be able to justify the fact that it is the first two diagonal elements that have to be negative and it is the third one that must be positive. Note that these $\mathbf{D}_{ii}$ terms are reasonably close to fulfilling the expected features just mentioned:

For the three relevant (absolute) numbers, the two different calculations yielded $\mathbf{D}_{jj}$ values (three for each case) all in the range $0.99 \leq |\mathbf{D}_{jj}| \leq 0.9999$—thus the quantization is fulfilled to a very high degree.

The values due to the two separate calculations are of the same quality we usually get from (pure) two-state calculations, that is, very close to 1.0 but two comments have to be made in this respect: (1) The quality of the numbers are different in the two calculations: The reason might be connected with the fact that in the second case the circle surrounds an area about three times larger than in the first case. This fact seems to indicate that the deviations are due "noise" caused by CIs belonging to neighbor states [e.g., the (1,2) and the (4,5) CIs]. (2) We would like to remind the reader that the diagonal element in case of the two-state system was only (–)0.39 [73] [instead of (–)1.0] so that incorporating the third state led, indeed, to a significant improvement.

The requirement of having two negative values and one positive is also fulfilled. Since this subject has been treated several times before (see Sections VIII and IX) it will be discussed within the next subject, related to the locations of the negative terms, that requires some analysis.

The positions of the (-1) terms in the diagonal indicate which of the electronic eigenfunction flips sign upon tracing the closed contour under consideration [see Section (IV.A)]. The results of this study show that in both cases the eigenfunctions of the two lower states (i.e., $2^2A'$ and $3^2A'$) flip sign, whereas the sign of the third function (i.e., $4^2A'$) remains the same. In situations where we have a single conical intersection between *each* consecutive pair of states it is the first and the third eigenfunctions that flip sign (see Section VIII). Here, we encounter the situation of one conical intersection between the lower pair of states but *two* (not one) conical intersections between the upper pair of states.

To analyze this case, we employ, as before, "contour algebra" (see Section IX): From Figure 14, it is noticed that Γ_{23} is a contour that surrounds the (2,3) conical intersection, Γ_{34} is the contour that surrounds the two (3,4) conical intersections, and Γ_{24} is a contour that surrounds all three conical intersections. According to "contour algebra" the event that "takes place" along Γ_{24} is the sum of the events along each individual contour. Thus,

$$\Gamma_{24} = \Gamma_{23} + \Gamma_{34} \tag{194}$$

Next, we are aware of the fact that if the system traces Γ_{23} it will be the two lower eigenfunctions that flip signs. If the system traces Γ_{34}, then no function flips its sign because two such conical intersections cancel each other [12,22,26,74,125]. Now, if the system traces Γ_{24} then, from Eq. (194) it follows that again, only the two lowest functions flip their sign, so that the effect due to the single lower CI will be preserved. In other words, the *two-state* topological effects are not disturbed along those contours that surround all three CIs. The results will be different once we choose a contour that surrounds, in addition to the lower CI, only one of the two upper CIs [see Ref. (117b)].

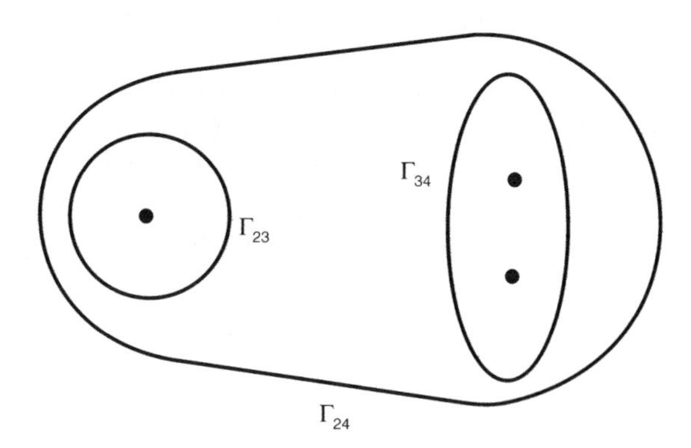

Figure 14. The three contours for the three situations discussed in the text: Γ_{23} surrounds the (2,3) CI, Γ_{34} surrounds the two (3,4) CIs, and Γ_{24} surrounds all three CIs.

XVI. SUMMARY AND CONCLUSIONS

This field currently differs from others fields in molecular physics mainly in two ways: (1) It is a highly theoretical field and as such it requires chemical–physical intuition and mathematical skill. (2) This field is still open to new developments that could significantly affect chemistry when treated on the molecular level. In this chapter, we tried to summarize the various findings related to this field and to give the reader its state of the art. Some of the subjects presented here were already discussed in previous reviews [8,13]. Still, due to last year's intensive efforts, we managed to include several new issues—some of them may open new venues for more research in this field. Since, as mentioned, part of the subjects presented here were already summarized in a previous review [13], in this section we will mainly concentrate on the implications of new subjects thus avoiding unnecessary repetition. We distinguish between two kinds of topics: (1) Practical ones that are associated with the possibility of treating dynamical processes related to excited states, namely, the *diabatization* process. (2) Less practical ones, which are interesting from a theoretical point of view but with potential prospects.

We start summarizing our findings regarding diabatization. There is no doubt that diabatization is essential for any dynamical study that involves electronically excited states. Diabatization is applied (on and off) for almost three decades mainly for studying charge-transfer processes between ion and molecules [54,94–97,125,127–131] and sporadically for other purposes [100–104]. However, only recently the conditions for a *correct* diabatization, subject to minimal numerical efforts, were formulated [108]. This subject is discussed in Section XI. The diabatization as presented here is shown to be closely connected with the fact that the non-adiabatic coupling matrix has to be quantized to guarantee single-valued diabatic potentials. One of the more fundamental answers regarding the quantization of the nonadiabatic coupling matrix were given in a series of ab initio calculations for different molecules [64–74]. The quantization for two-state systems for real systems was discussed in our previous reviews [8,13] but here, in Section XV.B, we extended the discussion to a three-state case found to exist for the second, third, and fourth states of the C_2H molecule [117]. This study is particularly important because it produces, for first time, the proof that the quantization is a general feature that goes beyond the two-state systems.

The two other subjects, as we already mentioned, are more theoretical but eventually may lead to interesting practical findings.

In Section XIII, we made a connection between the *curl* condition that was found to exist for Born–Oppenheimer–Huang systems and the Yang–Mills field. Through this connection we found that the non-adiabatic coupling terms can be considered as vector potentials that have their source in pseudomagnetic fields defined along *seams*. We speculated that these fields could be, semiclassically, associated with the zero-point vibrational motion [113].

Another subject with important potential application is discussed in Section XIV. There we suggested employing the *curl* equations (which any Bohr–Oppenheimer–Huang system has to obey for the for the relevant sub-Hilbert space), instead of ab initio calculations, to derive the non-adiabatic coupling terms [113,114]. Whereas these equations yield an analytic solution for any two-state system (the abelian case) they become much more elaborate due to the nonlinear terms that are unavoidable for any realistic system that contains more than two states (the non-abelian case). The solution of these equations is subject to boundary conditions that can be supplied either by ab initio calculations or perturbation theory.

This chapter centers on the mathematical aspects of the non-adiabatic coupling terms as single entities or when grouped in matrices, but were it not for the available ab initio calculation, it would have been almost impossible to proceed thus far in this study. Here, the ab initio results play the same crucial role that experimental results would play in general, and therefore the author feels that it is now appropriate for him to express his appreciation to the groups and individuals who developed the numerical means that led to the necessary numerical outcomes.

APPENDIX A: THE JAHN–TELLER MODEL AND THE LONGUET–HIGGINS PHASE

We consider a case where in the vicinity of a point of degeneracy between two electronic states the diabatic potentials behave linearly as a function of the coordinates in the following way [16–21]

$$\mathbf{W} = k\begin{pmatrix} y & x \\ x & -y \end{pmatrix} \tag{A.1}$$

where (x, y) are some generalized nuclear coordinates and k is a force constant. The aim is to derive the eigenvalues and the eigenvectors of this potential matrix. The eigenvalues are the adiabatic potential energy states and the eigenvectors form the columns of the adiabatic-to-diabatic transformation matrix. In order to perform this derivation, we shall employ polar coordinates (q,φ), namely,

$$y = q\cos\varphi \qquad \text{and} \qquad x = q\sin\varphi \tag{A.2}$$

By substituting for x and y, we get φ-*independent* eigenvalues of the form

$$u_1 = kq \quad \text{and} \quad u_2 = -kq \qquad \text{where} \qquad q = \{0, \infty\} \quad \text{and} \quad \varphi = \{0, 2\pi\}$$

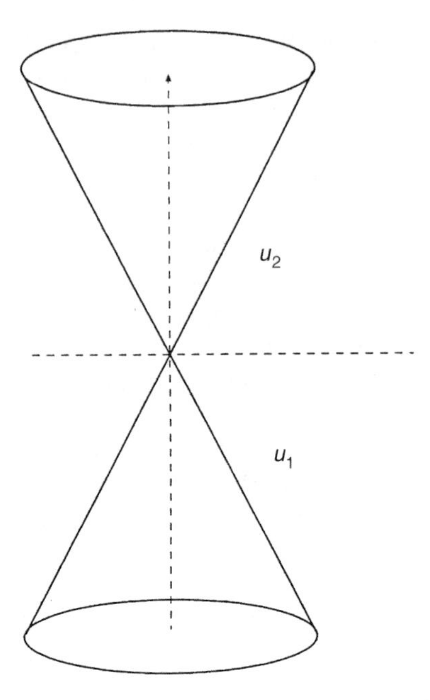

Figure 15. The two interacting cones within the Jahn–Teller model.

As noticed from Figure 15, the two surfaces u_1 and u_2 are conelike potential energy surfaces with a common apex. The corresponding eigenvectors are

$$\begin{aligned}\zeta_1 &= \left(\cos\frac{\varphi}{2}, \quad \sin\frac{\varphi}{2}\right)\\ \zeta_2 &= \left(\sin\frac{\varphi}{2}, \quad -\cos\frac{\varphi}{2}\right)\end{aligned} \tag{A.3}$$

The components of the two vectors (ξ_1, ξ_2), when multiplied by the electronic (diabatic) basis set ($|\phi_1\rangle, |\phi_2\rangle$), form the corresponding electronic adiabatic basis set ($|\eta_1\rangle, |\eta_2\rangle$):

$$\begin{aligned}|\eta_1\rangle &= \cos\frac{\varphi}{2}|\phi_1\rangle + \sin\frac{\varphi}{2}|\phi_2\rangle\\ |\eta_2\rangle &= \sin\frac{\varphi}{2}|\phi_1\rangle - \cos\frac{\varphi}{2}|\phi_2\rangle\end{aligned} \tag{A.4}$$

The adiabatic functions are characterized by two interesting features: (1) they depend only on the angular coordinate (but not on the radial coordinate) and (2) they are not single valued in configuration space because when φ is replaced by $(\varphi + 2\pi)$—a rotation that brings the adiabatic wave functions back to their initial position—*both* of them change sign. This last feature, which was revealed by Longuet-Higgins [14–17], may be, in certain cases, very crucial because multivalued electronic eigenfunctions cause the corresponding nuclear wave functions to be multivalued as well, a feature that has to be incorporated explicitly (through specific boundary conditions) while solving the nuclear Schrödinger equation. In this respect, it is important to mention that ab initio electronic wave functions indeed, possess the multivaluedness feature as described by Longuet–Higgins [30].

One way to eliminate the multivaluedness of the electronic eigenfunctions is by multiplying it by a phase factor [15], namely,

$$\zeta_j(\varphi) = \exp(i\vartheta)\eta_j(\varphi) \qquad j = 1, 2 \tag{A.5}$$

where a possible choice for ϑ is

$$\vartheta = \varphi/2 \tag{A.6}$$

Note that $\zeta_j(\varphi)$; $j = 1, 2$ are indeed single-valued eigenfunctions; however, instead of being real, they become complex.

The fact that the electronic eigenfunctions are modified as presented in Eq. (A.5) has a direct effect on the non-adiabatic coupling terms as introduced in Eqs. (8a) and (8b). In particular, we consider the term $\tau_{11}^{(1)}$ (which for the case of real eigenfunctions is identically zero) for the case presented in Eq. (A.5):

$$\boldsymbol{\tau}_{11}^{(1)} = \langle\zeta_1|\nabla\zeta_1\rangle = i\nabla\vartheta + \langle\eta_1|\nabla\eta_1\rangle$$

but since

$$\langle\eta_1|\nabla\eta_1\rangle = 0$$

it follows that $\boldsymbol{\tau}_{11}^{(1)}$ becomes

$$\boldsymbol{\tau}_{11}^{(1)} = i\nabla\vartheta \tag{A.7}$$

In the same way, we obtain

$$\boldsymbol{\tau}_{11}^{(2)} = i\nabla^2\vartheta - (\nabla\vartheta)^2 \tag{A.8}$$

The fact that now $\boldsymbol{\tau}_{11}^{(1)}$ is not zero will affect the ordinary Born–Oppenheimer approximation. To show that, we consider Eq. (15) for $M = 1$, once for a real eigenfunction and once for a complex eigenfunction. In the first case, we get from Eq. (15) the ordinary Born–Oppenheimer equation:

$$-\frac{1}{2m}\nabla^2\psi + (u - E)\psi = 0 \tag{A.9}$$

because for real electronic eigenfunctions $\boldsymbol{\tau}_{11}^{(1)} \equiv 0$ but in the second case for which $\boldsymbol{\tau}_{11}^{(1)} \neq 0$ the Born–Oppenheimer approximation becomes

$$-\frac{1}{2m}(\nabla + i\nabla\vartheta)^2\psi + (u - E)\psi = 0 \tag{A.10}$$

which can be considered as an 'extended' Born–Oppenheimer approximation for a case of a single isolated state expressed in terms of a complex electronic eigenfunction [132]. This equation was interpreted for some time as the adequate Schrödinger equation to describe the effect of the conical intersection that originate from the two interacting states. As it stands it contains an effect due to an ad hoc phase attached to a ground-state electronic eigenfunction [63].

The extended Born–Oppenheimer approximation based on the nonadiabatic coupling terms was discussed on several occasions [23,25,26,55,56,133,134] and is also presented here by Adhikari and Billing (see Chapter 3).

APPENDIX B: THE SUFFICIENT CONDITIONS FOR HAVING AN ANALYTIC ADIABATIC-TO-DIABATIC TRANSFORMATION MATRIX

The adiabatic-to-diabatic transformation matrix, $\mathbf{A}_p$, fulfills the following first-order differential vector equation [see Eq. (19)]:

$$\nabla\mathbf{A}_M + \tau_M\mathbf{A}_M = 0 \tag{B.1}$$

In order for $\mathbf{A}_M$ to be a regular matrix at every point in the assumed region of configuration space it has to have an inverse and its elements have to be analytic functions in this region. In what follows, we prove that if the elements of the components of τ_M are analytic functions in this region and have derivatives to any order and if the P subspace is decoupled from the corresponding Q subspace then, indeed, $\mathbf{A}_M$ will have the above two features.

I. ORTHOGONALITY

We start by proving that $\mathbf{A}_M$ is a unitary matrix and as such it will have an inverse (the proof is given here again for the sake of completeness). Let us consider the complex conjugate of Eq. (B.1):

$$\nabla \mathbf{A}_M^\dagger - \mathbf{A}_M^\dagger \boldsymbol{\tau}_M = 0 \tag{B.2}$$

where we recall that $\boldsymbol{\tau}_M$, the non-adiabatic coupling matrix, is a real anti-symmetric matrix. By multiplying Eq. (B.2) from the right by $\mathbf{A}_M$ and Eq. (B.1) from the left by $\mathbf{A}_M^\dagger$ and combining the two expressions we get

$$\mathbf{A}_M^\dagger \nabla \mathbf{A}_M + (\nabla \mathbf{A}_M^\dagger)\mathbf{A}_M = (\nabla \mathbf{A}_M^\dagger \mathbf{A}_M) = 0 \Rightarrow \qquad \mathbf{A}_M^\dagger \mathbf{A}_M = \text{constant}$$

For a proper choice of boundary conditions, the above mentioned constant matrix can be assumed to be the identity matrix, namely,

$$\mathbf{A}_M^\dagger \mathbf{A}_M = I \tag{B.3}$$

Thus $\mathbf{A}_P$ is a unitary matrix at any point in configuration space.

II. ANALYTICITY

From basic calculus, it is known that a function of a single variable is analytic at a given interval if and only if it has well-defined derivatives, to any order, at any point in that interval. In the same way, a function of several variables is analytic in a region if at any point in this region, in addition to having well-defined derivatives for all variables to any order, the result of the differentiation with respect to any *two* different variables does not depend on the order of the differentiation.

The fact that the $\mathbf{A}_M$ matrix fulfills Eq. (B.1) ensures the existence of derivatives to any order for any variable, at a given region in configuration space, if $\boldsymbol{\tau}_M$ is analytic in that region. In what follows, we assume that this is, indeed, the case. Next, we have to find the conditions for a mixed differentiation of the $\mathbf{A}_M$ matrix elements to be independent of the order.

For that purpose, we consider the p and the q components of Eq. (B.1) (the subscript M will be omitted to simplify notation):

$$\begin{aligned} \frac{\partial}{\partial p}\mathbf{A} + \boldsymbol{\tau}_p \mathbf{A} &= 0 \\ \frac{\partial}{\partial q}\mathbf{A} + \boldsymbol{\tau}_q \mathbf{A} &= 0 \end{aligned} \tag{B.4}$$

By differentiating the first equation according to q we find

$$\frac{\partial}{\partial q}\frac{\partial}{\partial p}\mathbf{A} + \left(\frac{\partial}{\partial q}\boldsymbol{\tau}_p\right)\mathbf{A} + \boldsymbol{\tau}_p \frac{\partial}{\partial q}\mathbf{A} = 0$$

or

$$\frac{\partial}{\partial q}\frac{\partial}{\partial p}\mathbf{A} + \left(\frac{\partial}{\partial q}\tau_p\right)\mathbf{A} - \tau_p\tau_q\mathbf{A} = 0 \tag{B.5a}$$

In the same way, we get from the second equation the following expression:

$$\frac{\partial}{\partial p}\frac{\partial}{\partial q}\mathbf{A} + \left(\frac{\partial}{\partial p}\tau_q\right)\mathbf{A} - \tau_q\tau_p\mathbf{A} = 0 \tag{B.5b}$$

Requiring that the mixed derivative is independent of the order of the differentiation yields

$$\left(\frac{\partial}{\partial p}\tau_q - \frac{\partial}{\partial q}\tau_p\right)\mathbf{A} = (\tau_q\tau_p - \tau_p\tau_q)\mathbf{A} \tag{B.6}$$

or (since **A** is a unitary matrix):

$$\frac{\partial}{\partial p}\tau_q - \frac{\partial}{\partial q}\tau_p = [\tau_q, \tau_p] \tag{B.7}$$

Thus, in order for the **A** matrix to be analytic in a region, any two components of $\boldsymbol{\tau}$, locally, have to fulfill Eq. (B.7). Equation (B.7) can also be written in a more compact way

$$\text{curl}\,\boldsymbol{\tau} - [\boldsymbol{\tau} \times \boldsymbol{\tau}] = 0 \tag{B.8}$$

where $\times$ stands for a vector product.

The question to be asked is: Under what conditions (if at all) do the components of $\boldsymbol{\tau}$ fulfill Eq. (B.8)? In [34] it is proved that this relation holds for any full Hilbert space. Here, we shall show that this relation holds also for the P sub-Hilbert space of dimension M, as defined by Eq. (10). To show that we employ, again, the Feshbach projection operator formalism [79] [see Eqs. (11)].

We start by considering the pth and the qth components of Eqs. (8a)

$$\left(\frac{\partial \tau_q}{\partial p}\right)_{jk} = \left\langle \frac{\partial \zeta_j}{\partial p} \middle| \frac{\partial \zeta_k}{\partial q} \right\rangle + \left\langle \zeta_j \middle| \frac{\partial^2 \zeta_k}{\partial p \partial q} \right\rangle \qquad j,k \leq M \tag{B.9a}$$

and

$$\left(\frac{\partial \tau_p}{\partial q}\right)_{jk} = \left\langle \frac{\partial \zeta_{kj}}{\partial q} \middle| \frac{\partial \zeta_k}{\partial p} \right\rangle + \left\langle \zeta_j \middle| \frac{\partial^2 \zeta_k}{\partial q \partial p} \right\rangle \qquad j,k \leq M \tag{B.9b}$$

Subtracting Eq. (B.9b) from Eq. (B.9a) and assuming that the electronic eigenfunctions are *analytic* functions with respect to the nuclear coordinates yields the following result:

$$\left(\frac{\partial}{\partial p}\tau_q - \frac{\partial}{\partial q}\tau_p\right)_{jk} = \left\langle \frac{\partial \zeta_{kj}}{\partial p} \middle| \frac{\partial \zeta_k}{\partial q} \right\rangle - \left\langle \frac{\partial \zeta_j}{\partial q} \middle| \frac{\partial \zeta_k}{\partial_p} \right\rangle \qquad j,k \leq M \qquad \text{(B.10)}$$

Equation (B.10) stands for the (j,k) matrix element of the left-hand side of Eq. (B.7). Next, we consider the (j,k) element of the first term on the right-hand side of Eq. (B.7), namely,

$$(\boldsymbol{\tau}_q \boldsymbol{\tau}_p)_{jk} = \sum_{i=1}^{M} \left\langle \zeta_j \middle| \frac{\partial \zeta_i}{\partial q} \right\rangle \left\langle \zeta_i \middle| \frac{\partial \zeta_k}{\partial p} \right\rangle$$

Since for real functions

$$\left\langle \zeta_j \middle| \frac{\partial \zeta_i}{\partial q} \right\rangle = -\left\langle \frac{\partial \zeta_j}{\partial q} \middle| \zeta_i \right\rangle$$

we get for this matrix element the result

$$(\boldsymbol{\tau}_q \boldsymbol{\tau}_p)_{jk} = -\sum_{i=1}^{M} \left\langle \frac{\partial \zeta_j}{\partial q} \middle| \zeta_i \right\rangle \left\langle \zeta_i \middle| \frac{\partial \zeta_k}{\partial p} \right\rangle = -\left\langle \frac{\partial \zeta_j}{\partial q} \middle| \left(\sum_{i=1}^{M} |\zeta_i\rangle\langle\zeta_i| \right) \middle| \frac{\partial \zeta_k}{\partial p} \right\rangle$$

Recalling that the summation within the round parentheses can be written as $[1 - Q_M]$, where Q_M is the projection operator for Q subspace, we obtain

$$(\boldsymbol{\tau}_q \boldsymbol{\tau}_p)_{jk} = -\left\langle \frac{\partial \zeta_j}{\partial q} \middle| \frac{\partial \zeta_k}{\partial p} \right\rangle - \sum_{i=M+1}^{N} \left\langle \frac{\partial \zeta_j}{\partial q} \middle| \zeta_i \right\rangle \left\langle \zeta_i \middle| \frac{\partial \zeta_k}{\partial p} \right\rangle \qquad j,k \leq M$$

Since under the summation sign each term is zero (no coupling between the inside and the outside subspaces) [see Eq. (10)] we finally get

$$(\boldsymbol{\tau}_q \boldsymbol{\tau}_p)_{jk} = -\left\langle \frac{\partial \zeta_j}{\partial q} \middle| \frac{\partial \zeta_k}{\partial p} \right\rangle \qquad \text{(B.11a)}$$

A similar result will be obtained for Eq. (B.7), namely,

$$(\boldsymbol{\tau}_p \boldsymbol{\tau}_q)_{jk} = -\left\langle \frac{\partial \zeta_j}{\partial p} \middle| \frac{\partial \zeta_k}{\partial q} \right\rangle \qquad \text{(B.11b)}$$

Subtracting Eq. (B.11b) from Eq. (B.11a) yields Eq. (B.10), thus proving the existence of Eq. (B.7).

Summary: In a region where the $\boldsymbol{\tau}_M$ elements are analytic functions of the coordinates, A_M is an orthogonal matrix with elements that are analytic functions of the coordinates.

APPENDIX C: ON THE SINGLE/MULTIVALUEDNESS OF THE ADIABATIC-TO-DIABATIC TRANSFORMATION MATRIX

In this appendix, we discuss the case where two components of $\boldsymbol{\tau}_M$, namely, $\boldsymbol{\tau}_{Mp}$ and $\boldsymbol{\tau}_{Mq}$ (p and q are Cartesian coordinates) are singular in the sense that at least one element in each of them is singular at the point $B(p = a, q = b)$ located on the plane formed by p and q. We shall show that in such a case the adiabatic-to-diabatic transformation matrix *may* become multivalued.

We consider the integral representation of the two relevant first-order differential equations [namely, the p and the q components of Eq. (19)]:

$$\begin{aligned} \frac{\partial}{\partial p}\mathbf{A}_M + \boldsymbol{\tau}_{Mp}\mathbf{A}_M &= 0 \\ \frac{\partial}{\partial q}\mathbf{A}_M + \boldsymbol{\tau}_{Mq}\mathbf{A}_M &= 0 \end{aligned} \tag{C.1}$$

In what follows, the subscript M will be omitted to simplify the notations. If the initial point is $P(p_0, q_0)$ and we are interested in deriving the value of $\mathbf{A}(\equiv \mathbf{A}_M)$ at a final point $Q(p, q)$ then one integral equation to be solved is

$$\mathbf{A}(p,q) = \mathbf{A}(p_0,q_0) - \int_{p0}^{p} dp'\boldsymbol{\tau}_p(p',q_0)\mathbf{A}(p',q_0) - \int_{q0}^{q} dq'\boldsymbol{\tau}_q(p,q')\mathbf{A}(p,q') \tag{C.2a}$$

Another way of obtaining the value of $\mathbf{A}(p,q)$ [we shall designate it as $\tilde{\mathbf{A}}(p,q)$] is by solving the following integral equation:

$$\tilde{\mathbf{A}}(p,q) = \mathbf{A}(p_0,q_0) - \int_{q_0}^{q} dq'\boldsymbol{\tau}_q(p_0,q')\tilde{\mathbf{A}}(p_0,q') - \int_{p_0}^{p} dp'\boldsymbol{\tau}_p(p',q)\tilde{\mathbf{A}}(p',q) \tag{C.2b}$$

In Eq. (C.2a), we derive the solution by solving it along the path Γ' characterized by two straight lines and three points (see Fig. 16*a*):

$$\Gamma': \quad P(p_0,q_0) \rightarrow P'(p_0,q) \rightarrow Q(p,q) \tag{C.3a}$$

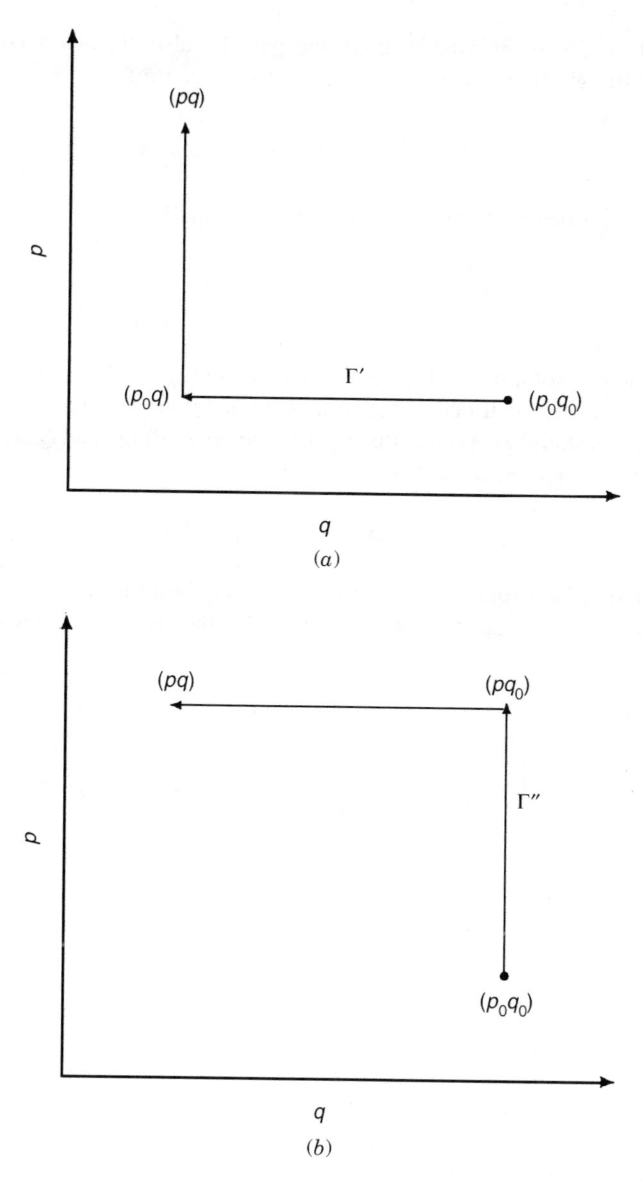

Figure 16. The rectangular paths Γ' and Γ'' connecting the points (p_0, q_0) and (p, q) in the (p, q) plane.

and in Eq. (C.2b) by solving it along the path Γ'' also characterized by two (different) straight lines and the three points (see Fig. 16*b*)

$$\Gamma'': \quad P(p_0, q_0) \rightarrow Q'(p, q_0) \rightarrow Q(p, q) \tag{C.3b}$$

Note that Γ, formed by Γ' and Γ'' written schematically as

$$\Gamma = \Gamma' - \Gamma'' \tag{C.4}$$

is a closed path.

Since the two solutions of Eq. (C.1) presented in Eqs. (C.2a) and (C.2b) may not be identical we shall derive the sufficient conditions for that to happen.

To start this study, we assume that the four points P, P', Q', and Q are at small distances from each other so that if

$$p = p_0 + \Delta p \qquad q = q_0 + \Delta q$$

then Δp and Δq are small enough distances as required for the derivation.

Subtracting Eq. (C.2b) from Eq. (C.2a) yields the following expression:

$$\begin{aligned}\Delta\mathbf{A}(p,q) = &-\int_{q_0}^{q_0+\Delta q} dq'(\boldsymbol{\tau}_q(p_0,q')\mathbf{A}(p_0,q') - \boldsymbol{\tau}_q(p,q')\mathbf{A}(p,q')) \\ &+ \int_{p_0}^{p_0+\Delta p} dp'(\boldsymbol{\tau}_p(p',q_0)\mathbf{A}(p',q_0) - \boldsymbol{\tau}_p(p',q)\mathbf{A}(p',q))\end{aligned} \tag{C.5}$$

where

$$\Delta\mathbf{A}(p,q) = \mathbf{A}(p,q) - \tilde{\mathbf{A}}(p,q) \tag{C.6}$$

Next, we consider two cases.

1. The case where the point $B(a,b)$ is not surrounded by the path Γ (see Fig. 17*a*). In this case, both $\boldsymbol{\tau}_p$ and $\boldsymbol{\tau}_q$ are *analytic* functions of the coordinates in the region enclosed by Γ, and therefore the integrands of the two integrals can be replaced by the corresponding derivatives calculated at the respective intermediate points, namely,

$$\begin{aligned}\Delta\mathbf{A}(p,q) = &\Delta p\int_{q_0}^{q_0+\Delta q} dq' \frac{\partial(\boldsymbol{\tau}_q(\tilde{p},q')\mathbf{A}(\tilde{p},q'))}{\partial p} \\ &- \Delta q\int_{p_0}^{p_0+\Delta p} dp' \frac{\partial(\boldsymbol{\tau}_p(p',\tilde{q})\mathbf{A}(p',\tilde{q}))}{\partial q}\end{aligned} \tag{C.7}$$

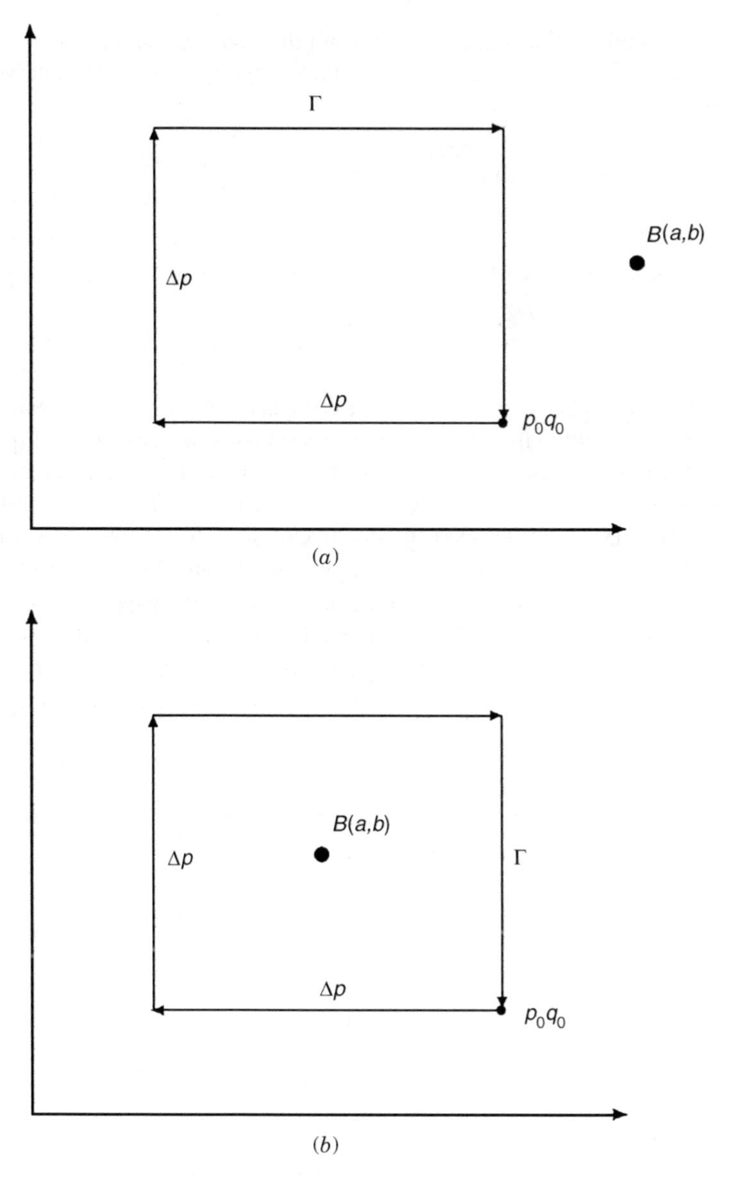

Figure 17. The differential closed paths Γ and the singular point $B(a, b)$ in the (p, q) plane: (a) The point B is not surrounded by Γ. (b) The point B is surrounded by Γ.

To continue the derivation, we recall that Δp and Δq are small enough so that the two integrands vary only slightly along the interval of integration so, that ΔA becomes

$$\Delta \mathbf{A}(p,q) = \Delta p \Delta q \left\{ \frac{\partial(\boldsymbol{\tau}_q(\tilde{p},\tilde{\tilde{q}})\mathbf{A}(\tilde{p},\tilde{\tilde{q}}))}{\partial p} - \frac{\partial(\boldsymbol{\tau}_p(\tilde{\tilde{p}},\tilde{q})\mathbf{A}(\tilde{\tilde{p}},\tilde{q})}{\partial q} \right\} \tag{C.8}$$

If we assume again that all relevant functions are smooth enough, the expression in the curled parentheses can be evaluated further to become

$$\Delta \mathbf{A}(p,q) = \left\{ \left(\frac{\partial \boldsymbol{\tau}_q(p,q)}{\partial p} - \frac{\partial \boldsymbol{\tau}_p(p,q)}{\partial q} \right) - [\boldsymbol{\tau}_q, \boldsymbol{\tau}_p] \right\} \mathbf{A}(p,q) \Delta p \Delta q \tag{C.9}$$

where Eqs. (C.1) were used to express the derivatives of $\mathbf{A}(p,q)$. Since the expression within the curled parentheses is identically zero due to Eq. (24), ΔA becomes identically zero or in other words the two infinitesimal paths Γ' and Γ'' yield identical solutions for the $\mathbf{A}$ matrix. The same applies to ordinary (viz., not necessarily small) closed paths because they can be constructed by "integrating" over closed infinitesimal paths (see Fig. 18).

2. The case when one of the differential closed paths surrounds the point $B(a,b)$ (see Fig. 17*b*). Here the derivation breaks down at the transition from Eqs. (C.5)–(C.7) and later, from Eqs. (C.7)–(C.8), because $\boldsymbol{\tau}_p$ and $\boldsymbol{\tau}_q$ become infinitely large in the close vicinity of $B(a,b)$, and therefore their

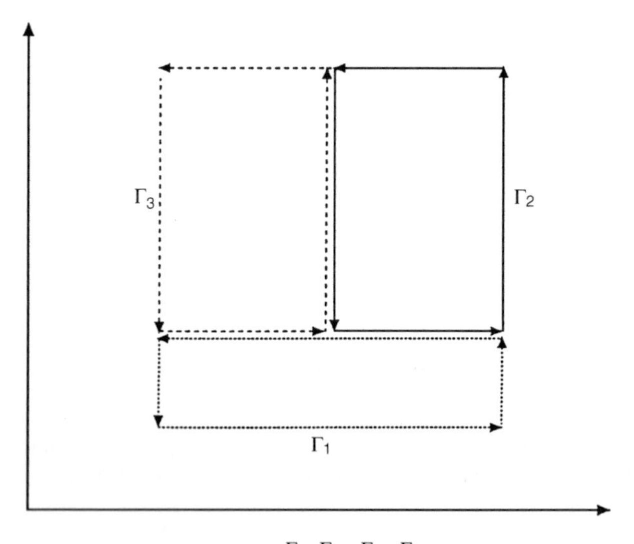

Figure 18. The closed (rectangular) path Γ as a sum of three partially closed paths Γ_1, Γ_2, Γ_3.

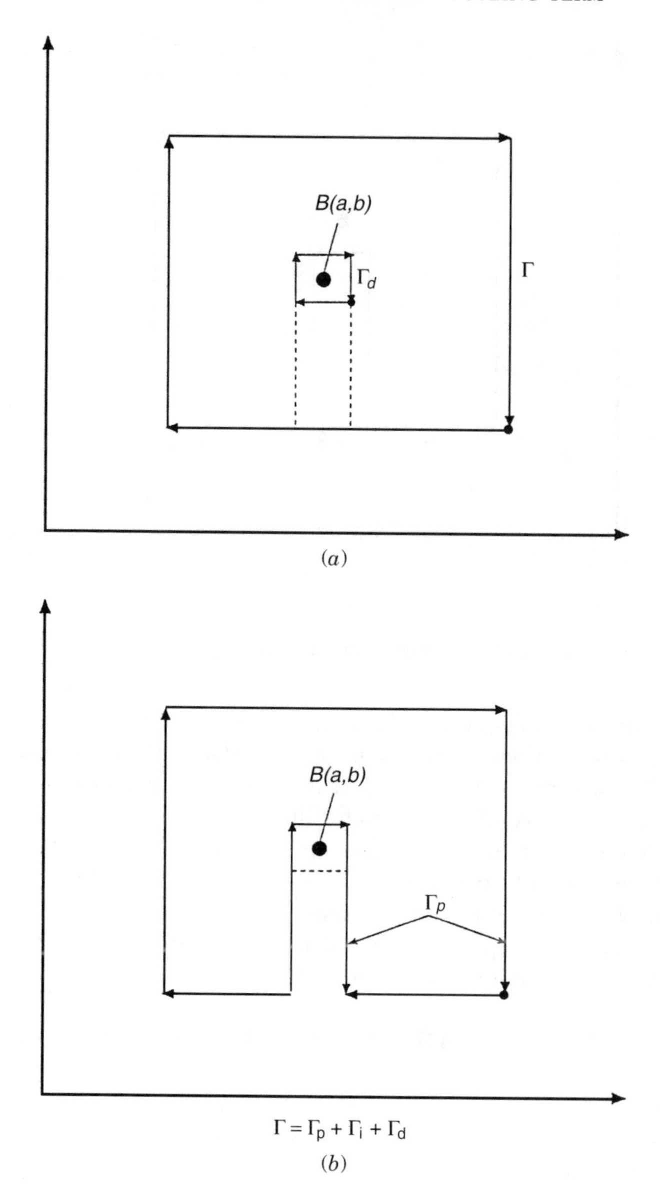

Figure 19. The closed path Γ as a sum of three closed paths Γ_d, Γ_p, Γ_i. (*a*) The closed (rectangular) paths, that is, the large path Γ and the differential path Γ_d both surrounding the singular point $B(a, b)$. (*b*) The closed path Γ_p that does not surround the point $B(a, b)$. (*c*) The closed path Γ_i that does not surround the point $B(a, b)$.

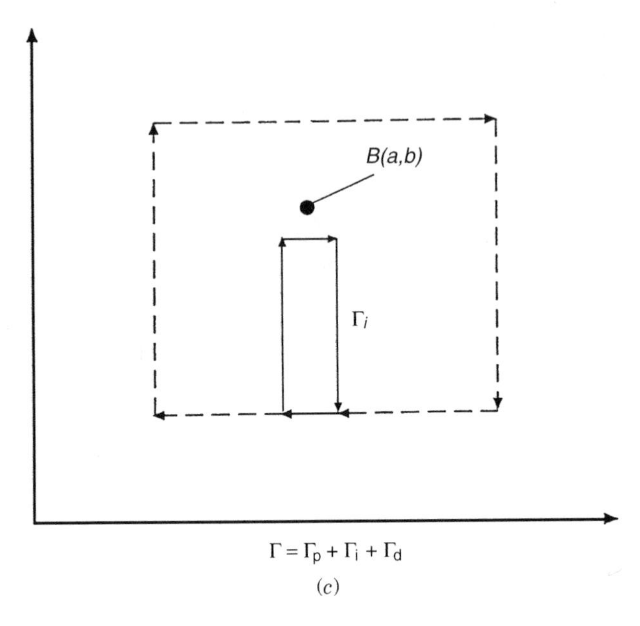

(*c*)

Figure 19 (*Continued*)

intermediate values cannot be estimated. As a result it is not clear whether the two solutions of the **A** matrix calculated along the two different differential paths are identical or not. The same applies to a regular size (i.e., not necessarily small) path Γ that surrounds the point $B(a, b)$. This closed path can be constructed from a differential path Γ_d that surrounds $B(a, b)$, a path Γ_p that does not surround $B(a, b)$, and a third, a connecting path Γ_i, which, also, does not surround $B(a, b)$ (see Fig. 19). It is noted that the small region surrounded by Γ_d governs the features of the **A** matrix in the entire region surrounded by Γ, immaterial of how large Γ is.

APPENDIX D: THE DIABATIC REPRESENTATION

Our starting equation is Eq. (3) in Section II.A with one difference, namely, we replace $\zeta_i(\mathrm{e} \mid \mathrm{n})$ by $\zeta_i(\mathrm{e} \mid \mathrm{n}_0)$; $i = 1, \ldots, N$, where n_0 stands for a fixed set of nuclear coordinates. Thus

$$\Psi(\mathrm{e}, \mathrm{n} \mid \mathrm{n}_0) = \sum_{i=1}^{N} \psi_i(\mathrm{n})\zeta_i(\mathrm{e} \mid \mathrm{n}_0) \tag{D.1}$$

Here, $\zeta_i(\mathrm{e} \mid \mathrm{n}_0)$, like $\zeta_i(\mathrm{e} \mid \mathrm{n})$, is an eigenfunction of the following Hamiltonian

$$(\mathbf{H}_\mathrm{e}(\mathrm{e} \mid \mathrm{n}_0) - u_i(\mathrm{n}_0))\zeta_i(\mathrm{e} \mid \mathrm{n}_0) = 0 \qquad i = 1, \ldots, N \tag{D.2}$$

where $u_i(\mathrm{n}_0), i = 1, \ldots, N$ are the electronic eigenvalues as calculated for this (fixed) set of nuclear coordinates. Substituting Eqs. (1) and (D.1) in Eq. (2) yields the following expression:

$$\sum_{i=1}^{N} \mathbf{T}_n \psi_i(\mathrm{n})|\zeta_i(\mathrm{e} \mid \mathrm{n}_0)\rangle + \sum_{i=1}^{N} \psi_i(\mathrm{n})[\mathbf{H}_\mathrm{e}(\mathrm{e} \mid \mathrm{n}) - E]|\zeta_i(\mathrm{e} \mid \mathrm{n}_0)\rangle = 0 \qquad \text{(D.3)}$$

It has to be emphasized that whereas n_0 is fixed, n is a variable. Substituting Eq. (6) for T_n, multiplying Eq. (D.3) by $\langle\zeta_j(\mathrm{e} \mid \mathrm{n}_0)|$, and integrating over the electronic coordinates yields the following result:

$$\left(-\frac{1}{2m}\nabla^2 - E\right)\psi_j(\mathrm{n}) + \sum_{i=1}^{N} \langle\zeta_j(\mathrm{e} \mid \mathrm{n}_0)|\mathbf{H}_\mathrm{e}(\mathrm{e} \mid \mathrm{n})|\zeta_i(\mathrm{e} \mid \mathrm{n}_0)\rangle\psi_i(\mathrm{n}) = 0 \qquad \text{(D.4)}$$

Recalling

$$\mathbf{H}_\mathrm{e}(\mathrm{e} \mid \mathrm{n}) = \mathbf{T}_\mathrm{e} + \mathbf{u}(\mathrm{e} \mid \mathrm{n}) \qquad \text{(D.5a)}$$

and, therefore, also

$$\mathbf{H}_\mathrm{e}(\mathrm{e} \mid \mathrm{n}_0) = \mathbf{T}_\mathrm{e} + \mathbf{u}(\mathrm{e} \mid \mathrm{n}_0) \qquad \text{(D.5b)}$$

where $u(\mathrm{e} \mid \mathrm{n})$ is the Coulombic field, we can replace $H_\mathrm{e}(\mathrm{e} \mid \mathrm{n})$ in Eq. (D.4) by the following expression:

$$\mathbf{H}_\mathrm{e}(\mathrm{e} \mid \mathrm{n}) = \mathbf{H}_\mathrm{e}(\mathrm{e} \mid \mathrm{n}_0) + \{\mathbf{u}(\mathrm{e} \mid \mathrm{n}) - \mathbf{u}(\mathrm{e} \mid \mathrm{n}_0)\} \qquad \text{(D.6)}$$

Equation (D.6) is valid because the electronic coordinates are independent of the nuclear coordinates. Having this relation, we can calculate the following matrix element:

$$\langle\chi_j(\mathrm{e} \mid \mathrm{n}_0)|\mathbf{H}_\mathrm{e}(\mathrm{e} \mid \mathrm{n})|\chi_i(\mathrm{e} \mid \mathrm{n}_0)\rangle = u_j(\mathrm{n}_0)\delta_{ji} + v_{ij}(\mathrm{n} \mid \mathrm{n}_0) \qquad \text{(D.7)}$$

where

$$v_{ij}(\mathrm{n} \mid \mathrm{n}_0) = \langle\chi_j(\mathrm{e} \mid \mathrm{n}_0)|\mathbf{u}(\mathrm{e} \mid \mathrm{n}) - \mathbf{u}(\mathrm{e} \mid \mathrm{n}_0)|\chi_i(\mathrm{e} \mid \mathrm{n}_0)\rangle \qquad \text{(D.8)}$$

Defining

$$\mathbf{V}_{ij}(\mathrm{n} \mid \mathrm{n}_0) = v_{ij}(\mathrm{n} \mid \mathrm{n}_0) + u_j(\mathrm{n}_0)\delta_{ji} \qquad \text{(D.9)}$$

and recalling Eq. (D.7), we get for Eq. (D.4) the expression

$$\left(-\frac{1}{2m}\nabla^2 - E\right)\psi_j(\mathrm{n}) + \sum_{i=1}^{N} V_{ji}(\mathrm{n} \mid \mathrm{n}_0)\psi_i(\mathrm{n}) = 0 \tag{D.10}$$

This equation can be also written in matrix form

$$-\frac{1}{2m}\nabla^2\Psi + (\mathbf{V} - E)\Psi = 0 \tag{D.11}$$

Here, $\mathbf{V}$ is the diabatic potential matrix and, in contrast to $\mathbf{u}$ in Eq. (9) of Section (II.A), it is a full matrix. Thus Eq. (D.11) is the Schrödinger equation within the diabatic representation.

APPENDIX E: A NUMERICAL STUDY OF A THREE-STATE MODEL

In Section V.B, we discussed to some extent the 3×3 adiabatic-to-diabatic transformation matrix $\mathbf{A}(\equiv \mathbf{A}^{(3)})$ for a tri-state system. This matrix was expressed in terms of three (Euler-type) angles $\gamma_{ij}, i = 1,2,3$ [see Eq. (81)], which fulfill a set of three coupled, first-order, differential equations [see Eq. (82)].

In what follows, we treat a tri-state model system defined in a plane in terms of two polar coordinates (ρ, φ) [85]. In order to guarantee that the non-adiabatic matrix $\boldsymbol{\tau}$, yields single-valued diabatic potentials we shall start with a 3×3 *diabatic* potential matrix and form, employing the Hellmann–Feynman theorem [3,36,85], the corresponding non-adiabatic coupling matrix $\boldsymbol{\tau}$. The main purpose of studying this example is to show that the $\mathbf{A}$ matrix may not be uniquely defined in configuration space although the diabatic potentials are all single valued.

The tri-state diabatic potential that is employed in this study is closely related (but not identical) to the one used by Cocchini et al. [39,135] to study the excited states of Na_3. It is of the following form (for more details see [85]):

$$\mathbf{V} = \begin{pmatrix} \varepsilon_E + U_1 & U_2 & W_1 - W_2 \\ U_2 & \varepsilon_E - U_1 & W_1 + W_2 \\ W_1 - W_2 & W_1 + W_2 & \varepsilon_A \end{pmatrix} \tag{E.1}$$

Here ε_E and ε_A are the values of two electronic states (an E-type state and an A-type state, respectively), $U_i; i = 1,2$ are two potentials defined as

$$U_1 = k\rho\cos\varphi + \frac{1}{2}g\rho^2\cos(2\varphi) \tag{E.2a}$$

and

$$U_2 = k\rho \sin \varphi - \frac{1}{2} g\rho^2 \sin (2\varphi) \tag{E.2b}$$

W_i; $i = 1, 2$ are potentials of the same functional form as the U_i parameters but defined in terms of a different set of parameters f and p, which replace g and k, respectively. The numerical values for these four parameters are [135]

$$k = \sqrt{2}p = 5.53\,\text{a.u.} \qquad \text{and} \qquad g = \sqrt{2}f = 0.152\,\text{a.u.}$$

Equation (82) is solved, for fixed ρ values, but for a varying angular coordinate, φ, defined along the interval $(0,2\pi)$. Thus ρ serves as a parameter and the results will be presented for different ρ values. A second parameter that will be used is the potential energy shift, $\Delta\varepsilon(= \varepsilon_E - \varepsilon_A)$, defined as the shift between the two original coupled adiabatic states and the third state at the origin, that is, at $\rho = 0$ (in case $\Delta\varepsilon = 0$, all three states are degenerate at the origin). The results will be presented for several of its values. In Figure 20 are shown the three non-adiabatic coupling terms $\tau_{\varphi ij}(\varphi)$; $i,j = 1,2,3(i > j)$ as calculated for different values of ρ and $\Delta\varepsilon$. The main feature to be noticed is the well-defined (sharp) tri-peak structure of $\tau_{\varphi 12}$ and $\tau_{\varphi 23}$ as a function of φ. There are other interesting features to be noticed but these are of less relevance to the present study (for a more extensive discussion see [85]).

Figure 1 presents the three γ angles as a function of φ for various values of ρ and $\Delta\varepsilon$. The two main features that are of interest for the present study are (1) following a full cycle, all three angles in all situations obtain the values either of π or of zero. (2) In each case (viz., for each set of ρ and $\Delta\varepsilon$), following a full cycle, two angles become zero and one becomes π. From Eq. (81) notice that the $\mathbf{A}$ matrix is diagonal at $\varphi = 0$ and $\varphi = 2\pi$ but in the case of $\varphi = 0$ the matrix $\mathbf{A}$ is the unit matrix and in the second case it has two (-1) terms and one $(+1)$ in its diagonal. Again recalling Eq. (39), this implies that the $\mathbf{D}$ matrix is indeed diagonal and has in its diagonal numbers of norm 1. However, the most interesting fact is that $\mathbf{D}$ is not the unit matrix. In other words, the adiabatic-to-diabatic transformation matrix presented in Eq. (81) is not single valued in configuration space although the corresponding diabatic *potential* matrix is single valued, by definition [see Eqs. (E.1) and (E.2)]. The fact that $\mathbf{D}$ has two (-1) terms and one $(+1)$ in its diagonal implies that the present $\boldsymbol{\tau}$ matrix produces topological effects, as was explained in the last two paragraphs of Section IV.A: Two electronic eigenfunctions flip sign upon tracing a closed path and one electronic function remains with its original sign.

As much as the results in the last section (Appendix D) are interesting the rather more interesting case is the one for $\Delta\varepsilon = 0$, namely, the case where the three states degenerate at one point. Here we find that even in this case $\mathbf{D}$ is *not*

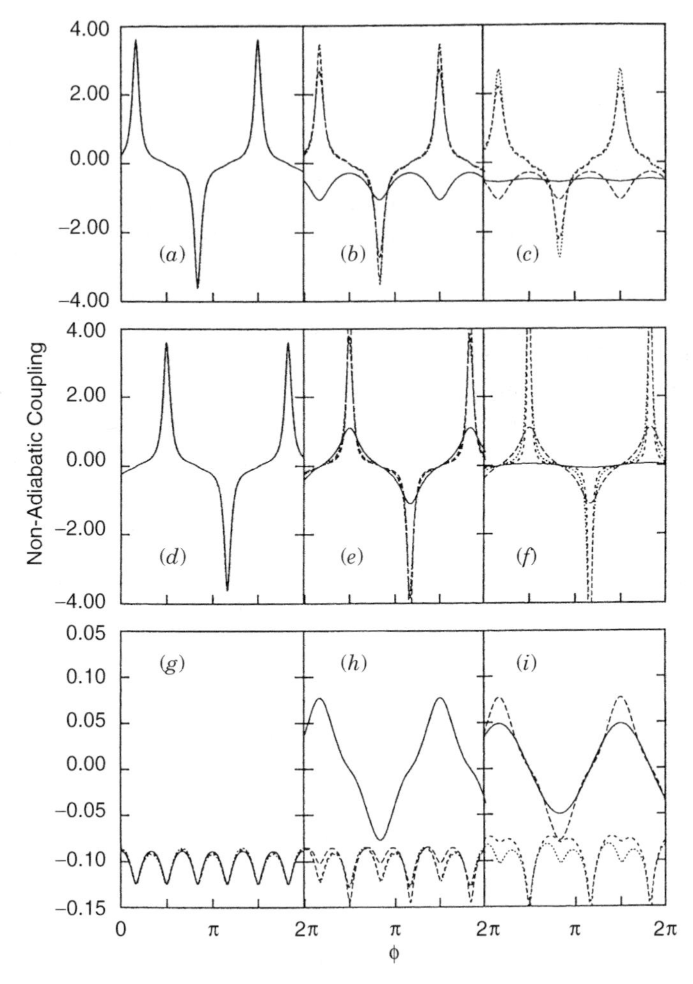

Figure 20. The three non-adiabatic coupling terms (obtained for the model potential described in Appendix E, see also Section V.B) $\boldsymbol{\tau}_{12\varphi}(\varphi), \boldsymbol{\tau}_{23\varphi}(\varphi), \boldsymbol{\tau}_{13\varphi}(\varphi)$ as a function φ calculated for different values of ρ and $\Delta\varepsilon$: (*a*) $\tau = \tau_{12}$, $\Delta\varepsilon = 0.0$; (*b*) $\tau = \tau_{12}$, $\Delta\varepsilon = 0.05$; (*c*) $\tau = \tau_{12}$, $\Delta\varepsilon = 0.5$; (*d*) $\tau = \tau_{23}$, $\Delta\varepsilon = 0.0$; (*e*) $\tau = \tau_{23}$, $\Delta\varepsilon = 0.05$; (*f*) $\tau = \tau_{23}$, $\Delta\varepsilon = 0.5$; (*g*) $\tau = \tau_{13}$, $\Delta\varepsilon = 0.0$; (*h*) $\tau = \tau_{13}$, $\Delta\varepsilon = 0.05$; (*i*) $\tau = \tau_{13}$, $\Delta\varepsilon = 0.5$. ———— $\rho = 0.01$; – – – – – – $\rho = 0.1$; - - - - - - - $\rho = 0.5$; .... $\rho = 1.0$.

the unit matrix but it keeps the features it encountered for $\Delta\varepsilon \neq 0$. In other words, the transition from the $\Delta\varepsilon \neq 0$ situation to the $\Delta\varepsilon = 0$ situation is continuous as was discussed in Section X. However, the present $\Delta\varepsilon = 0$ **D** matrix is in contradiction with the **D** matrix in Section V.A.2, which was derived for a particular type of a 3×3 $\boldsymbol{\tau}$ matrix that also refers to a trifold degeneracy at a single point. In this case, as we may recall, it was proved that it has to be a unit matrix if it is expected to yield single-valued diabatic potentials. These two examples support the finding of Section X where we distinguished between breakable and unbreakable multidegeneracy. The Cocchini et al. [135] model belongs, of course, to those models that yield the breakable degeneracy.

APPENDIX F: THE TREATMENT OF A CONICAL INTERSECTION REMOVED FROM THE ORIGIN OF COORDINATES

We start by writing the curl equation in Eq. (157) for a vector $\mathbf{t}(x, y)$ in Cartesian coordinates.

$$\frac{\partial t_x}{\partial y} - \frac{\partial t_y}{\partial x} = 0 \tag{F.1}$$

The solution to Eq. (F.1)

$$\mathbf{t}(x, y) = f\left(\frac{y}{x}\right)\frac{-y\mathbf{i_x} + x\mathbf{i_y}}{x^2 + y^2} \tag{F.2}$$

where $\mathbf{i_x}$ and $\mathbf{i_y}$ are unit vectors along the x and the y axes, respectively. To shift this solution from the origin to some given point (x_{j0}, y_{j0}) the variable x is replaced by $(x - x_{j0})$ and the variable y by $(y - y_{j0})$ so that the solution of Eq. (F.1) is given in the form

$$\mathbf{t}(x, y) = f\left(\frac{y - y_{j0}}{x - x_{j0}}\right)\frac{-(y - y_{j0})\mathbf{i_x} + (x - x_{j0})\mathbf{i_y}}{(x - x_{j0})^2 + (y - y_{j0})^2} \tag{F.3}$$

Next, we are interested in expressing this equation in terms of polar coordinates (q, θ). For this purpose, we recall the following relations:

$$x = q\cos\theta \qquad y = q\sin\theta \tag{F.4}$$

and introduce the following definitions:

$$x - x_{j0} = q_j\cos\theta_j \qquad y - y_{j0} = q_j\sin\theta_j \tag{F.5}$$

Since we are interested in the polar components of $\mathbf{t}(q, \theta)$, that is, t_q and t_θ, we need to know their relation with t_x and t_y as well, which was derived sometime ago [84].

$$\begin{aligned} t_q &= \left\langle \chi_1 \middle| \frac{\partial}{\partial q} \chi_2 \right\rangle = \cos\theta t_x + \sin\theta t_y \\ t_\theta &= \left\langle \chi_1 \middle| \frac{\partial}{\partial \theta} \chi_2 \right\rangle = q(-\sin\theta t_x + \cos\theta t_y) \end{aligned} \tag{F.6}$$

where χ_1 and χ_2 are the two lowest electronic adiabatic wave functions. By employing Eqs. (F.3), (F.5), and (F.6), we finally get

$$\begin{aligned} t_q(q,\theta) &= -f(\theta_j)\frac{1}{q_j}\sin(\theta - \theta_j) \\ t_\theta(q,\theta) &= \frac{q}{q_j} f(\theta_j)\cos(\theta - \theta_j) \end{aligned} \tag{F.7}$$

Equations (F.7) are the equations employed in the text [see Eqs. (164)].

Acknowledgments

The author would like to thank: Dr. Y. T. Lee, Dr. S. H. Lin, and Dr. A. Mebel for their warm hospitality at the Institute of Atomic and Molecular Science, Taipei, where the first sections of this chapter were written, and the Academia Sinica of Taiwan for partly supporting this research; Dr. Gert D. Billing for his exceptional hospitality during his stay at the Department of Chemistry, University of Copenhagen where the new sections of this chapter were written and to the Danish Research Training Council for partly supporting this research; Dr. A. Dalgarno and Dr. K. Kirby for their warm hospitality at the Institute for Theoretical Atomic and Molecular Physics at the Harvard Smithsonian Center for Astrophysics where this chapter was completed and to the National Science Foundation for partly supporting this work through a grant for the Institute for Theoretical Atomic and Molecular Physics; Dr. R. Englman for many years of scientific collaboration and for his continuous encouragement, to Dr. A. Mebel, for the more recent intensive fruitful collaboration that among other things led to important applications for *real* molecular systems, to Dr. A. Alijah, Dr. S. Adhikari, and Dr. A. Yahalom for making significant contributions at various stages of the research and for illuminating discussions, and finally to Dr. Gert D. Billing and Dr. John Avery for intensive discussions in particular with regard to Sections XI, XIII, and XIV.

References

1. M. Born and J. R. Oppenheimer, *Ann. Phys. (Leipzig)* **84**, 457 (1927).
2. M. Born and K. Huang, *Dynamical theory of Crystal Lattices*, Oxford University Press, New York, 1954.
3. R. K. Preston and J. C. Tully, *J. Chem. Phys.* **54**, 4297 (1971).
4. A. Kuppermann and R. Abrol (Chapter 5 in this volume).
5. R. Krishnan, M. Frisch, J. A. Pople, *J. Chem. Phys.* **72**, 4244 (1980).
6. MOLPRO is a package of ab initio programs written by H.-J. Werner and P. J. Knowles, with contributions from J. Almlöf, R. D. Amos, M. J. O. Deegan, S. T. Elbert, C. Hampel, W. Meyer, K. Peterson, R. Pitzer, A. J. Stone, P. R. Taylor, and R. Lindh.

7. M. Baer and A. Alijah, *Chem. Phys. Lett.* **319**, 489 (2000).
8. M. Baer,*Chem. Phys.* **259**, 123 (2000).
9. M. Baer,*Chem. Phys. Lett.* **329**, 450 (2000).
10. M. Baer, *J. Chem. Phys. Chem. A* **105**, 2198 (2001).
11. Z. R. Xu, M. Baer, and A. J. C. Varandas, *J. Chem. Phys.* **112**, 2746 (2000).
12. A. Mebel, A. Yahalom, R. Englman, and M. Baer, *J. Chem. Phys.* **115**, 3673 (2001).
13. M. Baer, *Phys. Reps.* **358**, 75 (2002).
14. H. C. Longuet-Higgins, U. Opik, M. H. L. Pryce, and R. A. Sack, *Proc. R. Soc. London, Ser. A* **244**, 1 (1958).
15. H. C. Longuet-Higgins, *Adv. Spectrosc.* **2**, 429 (1961).
16. G. Herzberg and H. C. Longuet-Higgins, *Discuss. Faraday Soc.* **35**, 77 (1963).
17. H. C. Longuet-Higgins, *Proc. R. Soc. London, Ser. A* **344**, 147 (1975).
18. H. A. Jahn and E. Teller, *Proc. R. Soc. London, Ser. A* **161**, 220 (1937).
19. E. Teller, *J. Phys. Chem.* **41**, 109 (1937).
20. E. Teller, *Isr. J. Chem.* **7**, 227 (1969).
21. R. Englman, *The Jahn–Teller Effect in Molecules and Crystals*, John Wiley & Sons, Inc., Interscience, New York, 1972; I. B. Bersucker and V. Z. Polinger, *Vibronic Interactions in Molecules and Crystals*, Springer Verlag, 1989.
22. J. W. Zwanziger and E.R. Grant, *J. Chem. Phys.* **87**, 2954 (1987).
23. R. Baer, D. Charutz, R. Kosloff, and M. Baer, *J. Chem. Phys.* **105**, 9141 (1996).
24. M. Baer, A. Yahalom, and R. Englman, *J. Chem. Phys.* **109**, 6550 (1998).
25. S. Adhikari and G. D. Billing, *J. Chem. Phys.* **111**, 40 (1999).
26. M. Baer, *J. Chem. Phys.* **107**, 10662 (1997).
27. M. Baer, R. Englman, and A. J. C Varandas, *Mol. Phys.* **97**, 1185 (1999).
28. R. Englman, A. Yahalom, and M. Baer, *Eur. Phys. J* **D8**, 1 (2000).
29. A. Kuppermann, in *Dynamics of Molecules and Chemical Reactions*, R. E. Wyatt and J. Z. H. Zhang, eds., Marcel, Dekker, New York, 1996, p. 411.
30. A. J. C. Varandas, J. Tennyson, and J. N. Murrell, *Chem. Phys. Lett.* **61**, 431 (1979).
31. W. D. Hobey and A. D. Mclachlan, *J. Chem. Phys.* **33**, 1695 (1960).
32. W. Lichten, *Phys. Rev.* **164**, 131 (1967).
33. F. T. Smith, *Phys. Rev.* **179**, 111 (1969).
34. M. Baer, *Chem. Phys. Lett.* **35**, 112 (1975).
35. M. Baer, *Chem. Phys.* **15**, 49 (1976).
36. (a) M. Baer, in *Theory of Chemical Reaction Dynamics*, M. Baer, ed., CRC, Boca Raton, FL, 1985, Vol. II, Chap. 4; (b) M. Baer, *Adv. Chem. Phys.* **49**, 191 (1982) (see Chap. VI).
37. H. Köppel, W. Domcke, and L. S. Cederbaum, *Adv. Chem. Phys.* **57**, 59 (1984).
38. T. Pacher, L. S. Cederbaum, and H. Köppel, *J. Chem. Phys.* **89**, 7367 (1988).
39. R. Meiswinkel and H. Köppel, *Chem. Phys.* **144**, 117 (1990).
40. V. Sidis, in *State-to-State Ion Molecule Reaction Dynamics*, M. Baer and C. Y. Ng, eds., Vol. II p. 73; *Adv. Chem. Phys.* **82**, 73 (1992).
41. T. Pacher, L. S. Cederbaum, and H. Köppel, *Adv. Chem. Phys.* **84**, 293 (1993).
42. W. Domcke and G. Stock, *Adv. Chem. Phys.* **100**, 1 (1997).
43. A. Thiel and H. Köppel, *J. Chem. Phys.* **110**, 9371 (1999).

44. D. A. Micha, *Adv. Chem. Phys.* **30**, 7 (1975).
45. B. Lepetit, and A. Kuppermann, *Chem. Phys. Lett.* **166**, 581 (1990); Y.-S. M. Wu, B. Lepetit, and A. Kuppermann, *Chem. Phys. Lett.* **186**, 319 (1991); Y.-S. M. Wu and A. Kuppermann, *Chem. Phys. Lett.* **201**, 178 (1993); A. Kuppermann and Y.-S. M. Wu, *Chem. Phys. Lett.* **205**, 577 (1993); Y.-S. M. Wu and A. Kuppermann, *Chem. Phys. Lett.* **235**, 105 (1995).
46. R. Abrol, A. Shaw, A. Kuppermann, and D. R. Yarkony, *J. Chem. Phys.* **115**, 4640 (2001); R. Abrol and A. Kuppermann, *J. Chem. Phys.* **116**, 1035 (2002).
47. A. Macias and A. Riera, *J. Chem. Phys. B* **11**, L489 (1978).
48. A. Macias and A. Riera, *Int. J. Quantum Chem.* **17**, 181 (1980).
49. H.-J. Werner and W. Meyer, *J. Chem. Phys.* **74**, 5802 (1981).
50. X. Chapuisat, A. Nauts, and D. Dehareug-Dao, *Chem. Phys. Lett.* **95**, 139 (1983).
51. D. Hehareug-Dao, X. Chapuisat, J. C. Lorquet, C. Galloy, and G. Raseev, *J. Chem. Phys.* **78**, 1246 (1983).
52. C. Petrongolo, G. Hirsch, and R. Buenker, *Mol. Phys.* **70**, 825 (1990).
53. C. Petrongolo, G. Hirsch, and R. Buenker, *Mol. Phys.* **70**, 835 (1990).
54. Z. H. Top and M. Baer, *Chem. Phys.* **25**, 1 (1977).
55. M. Baer and R. Englman, *Mol. Phys.* **75**, 293 (1992).
56. M. Baer and R. Englman, *Chem. Phys. Lett.* **265**, 105 (1997).
57. M. Baer, *Mol. Phys.* **40**, 1011 (1980).
58. C. N. Yang and R. L. Mills, *Phys. Rev.* **96**, 191 (1954).
59. L. O'Raifeartaigh, *The Dawning of Gauge Theory*, Princeton University Press, Princeton, N. J., 1997.
60. M. V. Berry, *Proc. R. Soc. London. Ser. A* **392**, 45 (1984).
61. B. Simon, *Phys. Rev. Lett.* **51**, 2167 (1983).
62. Y. Aharonov and J. Anandan, *Phys. Rev. Lett.* **58**, 1593 (1987).
63. Y. Aharonov, E. Ben-Reuven, S. Popescu, and D. Rohrlich, *Nucl. Phys. B* **350**, 818 (1991).
64. D. R. Yarkony, *J. Chem. Phys.* **105**, 10456 (1996).
65. G. Chaban, M. S. Gordon, and D. R. Yarkony, *J. Phys. Chem.* **101A**, 7953 (1997).
66. N. Matsunaga and D. R. Yarkony, *J. Chem. Phys.* **107**, 7825 (1997).
67. N. Matsunaga and D. R. Yarkony, *Mol. Phys.* **93**, 79 (1998).
68. R. G. Sadygov and D. R. Yarkony, *J. Chem. Phys.* **109**, 20 (1998).
69. D. R. Yarkony, *J. Chem. Phys.* **110**, 701 (1999).
70. R. G. Sadygov and D. R. Yarkony, *J. Chem. Phys.* **110**, 3639 (1999).
71. D. R. Yarkony, *J. Chem. Phys.* **112**, 2111 (2000).
72. A. Mebel, M. Baer, and S. H. Lin, *J. Chem. Phys.* **112**, 10703 (2000).
73. A. Mebel, M. Baer, V. M. Rozenbaum, and S. H. Lin, *Chem. Phys. Lett.* **336**, 135 (2001).
74. A. Mebel, M. Baer, and S. H. Lin, *J. Chem. Phys.* **114**, 5109 (2001).
75. M. Baer, *J. Chem. Phys.* **107**, 2694 (1997).
76. T. T. Wu and C. N. Yang, *Phys. Rev. D* **12**, 3845 (1975).
77. P. A. M. Dirac, *Proc. R. Soc., London Ser. A* **133**, 60 (1931).
78. J. D. Jackson, *Classical Electrodynamics*, 3rd ed., John Wiley & Sons, Inc., 1998, p. 275.
79. H. Feshbach, *Ann. Phys. (N.Y.)* **5**, 357 (1958).
80. M. Baer, *J. Phys. Chem. A* **104**, 3181 (2000).

81. M. Baer and R. Englman, *Chem. Phys. Lett.* **335**, 85 (2001).
82. D. Bohm, *Quantum Theory*, Dover Publications, Inc., N.Y., 1989, p. 41.
83. R. Renner, *Z. Phys.* **92**, 172 (1934).
84. Z. H. Top and M. Baer, *J. Chem. Phys.* **66**, 1363 (1977).
85. A. Alijah and M. Baer, *J. Phys. Chem. A* **104**, 389 (2000).
86. H. Goldstein, *Classical Mechanics*, Addison-Wesley Publishing Company, 1966, p.107.
87. L. D. Landau and E. M. Lifshitz, *Quantum Mechanics*, Pergamon Press, Oxford, U.K., 1965, p. 188.
88. C. Zener, *Proc. R. Soc. London Ser. A* **137**, 696 (1932).
89. L. D. Landau, *Phys. Z. Sowjetunion* **2**, 46 (1932).
90. H. Nakamura and C. Zhu, *Comments At. Mol. Phys.* **32**, 249 (1996).
91. D. Elizaga, L. F. Errea, A. Macias, L. Mendez, A. Riera, and A. Rojas, *J. Phys. B* **32**, L697 (1999).
92. A. Alijah and E. E. Nikitin, *Mol. Phys.* **96**, 1399 (1999).
93. Yu. N. Demkov, *Sov. Phys. JETP* **18**, 138 (1964).
94. M. Baer and A. J. Beswick, *Phys. Rev. A* **19**, 1559 (1979).
95. M. Baer, G. Niedner-Schatteburg, and J. P. Toennies, *J. Chem. Phys.* **91**, 4169 (1989).
96. M. Baer, C-L. Liao, R. Xu, G. D. Flesch, S. Nourbakhsh, and C. Y. Ng, *J. Chem. Phys.* **93**, 4845 (1990); M. Baer and C. Y. Ng, *J. Chem. Phys.* **93**, 7787 (1990).
97. I. Last, M. Gilibert, and M. Baer, *J. Chem. Phys.* **107**, 1451 (1997).
98. A. Bjerre and E. E. Nikitin, *Chem. Phys. Lett.* **1**, 179 (1967).
99. S. Chapman, in *State-to-State Ion Molecule Reaction Dynamics*, M. Baer and C. Y. Ng, eds., Vol. II; *Adv. Chem. Phys.* **82** (1992).
100. P. Halvick and D. G. Truhlar, *J. Chem. Phys.* **96**, 2895 (1992).
101. G. J. Tawa, S. L. Mielke, D. G. Truhlar, and D. W. Schwenke, in *Advances in Molecular Vibrations Collision Dynamics*, J. M. Bowman, ed., JAI Press, Greenwich, CT, 1993, Vol. 2B, p. 45.
102. S. L. Mielke, D. G. Truhlar, and D. W. Schwenke, *J. Phys. Chem.* **99**, 16210 (1995).
103. T. Suzuki, H. Katayanagi, S. Nanbu, and M. Aoyagi, *J. Chem. Phys.* **109**, 5778 (1998).
104. G. D. Billing, M. Baer, and A. M. Mebel, *Chem. Phys. Lett.* (in press).
105. H. Thümmel, M. Peric, S. D. Peyerimhoff, and R. J. Buenker, *Z. Phys. D: At. Mol. Clusters* **13**, 307 (1989).
106. T. Pacher, H. Koppel, and L. S. Cederbaum, *J. Chem. Phys.* **95**, 6668 (1991).
107. T. Romero, A. Aguilar, and F. X. Gadea,, *J. Chem. Phys.* **110**, 6219 (1999).
108. M. Baer, A. M. Mebel, and G. D. Billing, *J. Phys. Chem. A* (in press).
109a. E. P. Wigner, *Gruppentheorie*, FriedrichVieweg und son, Braunschweig, 1931.
109b. 109M. E. Rose, *Elementary Theory of Angular Momentum*, John Wiley & Sons, Inc., New York, 1957.
110. M. Baer, *Chem. Phys. Lett.* **347**, 149 (2001).
111. (a) R. P. Feynman, R. B Leighton, and M. Sands, *The Feynman Lectures on Physics*, Addison-Wesley Publishing Co., 1964, Vol. II, Sect. 14.1; (b) *ibid.* Sect. 14.4.
112. Y. Aharonov and D. Bohm, *Phys. Rev.* **115**, 485 (1959).
113. M. Baer, *Chem. Phys. Lett.* **349**, 84 (2001).

114. J. Avery, M. Baer, and G. D. Billing, *Mol. Phys.* **100**, 1011 (2002).
115. M. Baer, A. M. Mebel, and G. D. Billing (submitted for publication).
116. M. Baer and A. M. Mebel, *Int. J. Quant. Chem.* (in press); A. M. Mebel, G. Halasz, A. Vibok, and M. Baer (submitted for publication).
117. A. J. C. Varandas, F. B. Brown, C. A. Mead, D. G. Truhlar, and N. C. Blaise, *J. Chem. Phys.* **86**, 6258 (1987).
118. S. Liu and P. Siegbahn, *J. Chem. Phys.* **68**, 2457 (1978); D. G. Truhlar and C. J. Horowitz, *J. Chem. Phys.* **68**, 2466 (1978).
119. I. Last and M. Baer, *J. Chem. Phys.* **75**, 288 (1981); ibid. **80**, 3246 (1984).
120. Y.-S. M. Wu, B. Lepetit, and A. Kuppermann, *Chem. Phys. Lett.* **186**, 319 (1991).
121. A. Kuppermann and Y.-S. M. Wu, *Chem. Phys. Lett.* **241**, 229 (1995).
122. R. C. Whitten and F. T. Smith, *J. Math. Phys.* **9**, 1103 (1968).
123. B. R. Johnson, *J. Chem. Phys.* **73**, 5051 (1980).
124. G. D. Billing and N. Markovic, *J. Chem. Phys.* **99**, 2674 (1993).
125. M. Baer, in *State Selected and State-to-Sate Ion-Molecule Reaction Dynamics: Theory*, M. Baer and C. Y. Ng, eds., Vol. II, p. 187; *Adv. Chem. Phys.* **82**, 1992.
126. R. Englman, A. Yahalom, A. M. Mebel, and M. Baer, *Int. J. Quant. Chem.* (in press).
127. M. Chajia and R. D. Levine, *Phys. Chem. Chem. Phys.* **1**, 1205 (1999).
128. T. Takayanki, Y. Kurasaki, and A. Ichihara, *J. Chem. Phys.* **112**, 2615 (2000).
129. L. C. Wang, *Chem. Phys.* **237**, 305 (1998).
130. C. Shin and S. Shin, *J. Chem. Phys.* **113**, 6528 (2000).
131. T. Takayanki and Y. Kurasaki, *J. Chem. Phys.* **113**, 7158 (2000).
132. C. A. Mead, *Chem. Phys.* **49**, 23 (1980).
133. M. Baer, S. H. Lin, A. Alijah, S. Adhikari, and G. D. Billing. *Phys. Rev. A* **62**, 032506-1 (2000).(*)
134. S. Adhikari and G. D. Billing, A. Alijah, S. H. Lin, and M. Baer, *Phys. Rev. A* **62**, 032507-1 (2000).
135. F. Cocchini, T. H. Upton, and W. J. Andreoni, *Chem. Phys.* **88**, 6068 (1988).

NON-ADIABATIC EFFECTS IN CHEMICAL REACTIONS: EXTENDED BORN–OPPENHEIMER EQUATIONS AND ITS APPLICATIONS

SATRAJIT ADHIKARI

Department of Chemistry, Indian Institute of Technology, Guwahati, India

GERT DUE BILLING

Department of Chemistry, University of Copenhagen, Copenhagen, Denmark

CONTENTS

The Role of Degenerate States in Chemistry: A Special Volume of Advances in Chemical Physics, Volume 124, Edited by Michael Baer and Gert Due Billing. Series Editors I. Prigogine and Stuart A. Rice.
ISBN 0-471-43817-0.

I. INTRODUCTION

One of the most interesting observations in molecular physics was made by Herzberg and Longuet-Higgins (HLH) [1] when they were investigating the Jahn–Teller (JT) conical intersection (CI) problem [2–15]. These authors found that in the presence of a CI located at some point in configuration space (CS), the adiabatic electronic wave functions that are parametrically dependent on the nuclear coordinates became multivalued and proposed to correct the "deficiency" by multiplying the adiabatic wave functions of the two states with a unique phase factor (see Appendix A). More specifically, in the theory of molecular dynamics the Born–Oppenheimer (BO) treatment [16] (see Appendix B) is based on the fact that the fast-moving electrons are distinguishable from the slow-moving nuclei in a molecular system. The BO approximation [16,17] (see Appendix B) has been made with this distinction and once the electronic eigenvalue problem is solved, the nuclear Schrödinger equation employing the BO approximation should be properly modified in order to avoid wrong observations. The BO approximation implies that the non-adiabatic coupling terms (see Appendix B) [18–30] are negligibly small, that is, it has been assumed that particularly at low-energy processes, the nuclear wave function on the upper electronic surface affect the corresponding lower wave function very little. As a consequence of this approximation, the product of the nuclear wave function on the upper electronic state and the non-adiabatic coupling terms are considered to be very small and will have little effect on the dynamics. On the other hand, when the non-adiabatic coupling terms are sufficiently large or infinitely large, the use of the ordinary BO approximation becomes invalid even at very low energies. Even though the components of the upper state wave function in the total wave function are small enough, their product with large or infinitely large non-adiabatic coupling terms may not be. The reason for having large non-adiabatic coupling terms is that the fast-moving electron may, in certain situations, create exceptionally large forces, causing the nuclei in some regions of CS to be strongly accelerated so that their velocities are no longer negligibly small. In this situation, when these terms responsible for this accelerated motion are ignored within the ordinary BO approximation, the relevance of the ordinary

BO approximation vanishes even at low energies, hence the formulation of generalized BO equations become worth while considering.

The aim of the generalized BO treatment is to avoid multivaluedness of the total wave function. The Longuet-Higgins suggestion [1] of obtaining a generalized BO treatment for the JT model [2–5] by multiplying a complex phase factor with the adiabatic wave functions of the two states responsible for forming the CI, was reformulated by Mead and Truhlar [31–33] by introducing a vector potential into the nuclear Schrödinger equation (SE) in order to ensure a single valued and continuous total wave function. In their approach, the adiabatic wave function is multiplied by the Longuet-Higgins phase and by operating with the nuclear kinetic energy operator (KEO) on this product function, the KEO acquire some additional terms. Terms, that appear as a vector potential. Thus, when the nuclear coordinates travel through a closed path around the CI, the vector potential can introduce the required sign change and make the total wave function continuous and single valued. For general coordinate systems and complicated molecules where the point of CI does not coincide with any special symmetry of the coordinate system, the introduction of a vector potential so as to obtain the extended BO equations is a more general approach than the one that multiplies the adiabatic wave function with an HLH phase.

For systems with three identical atoms, the JT effect is the best known phenomena [34–37] and well investigated in bound systems [38–43]. Significant differences in the reaction cross-section of the $H + H_2$ system (and its isotopic variants) obtained by theoretical calculations and experimental measurements indicate the complication due to the consideration of the ordinary BO separation in the theory of electronic and nuclear motion of a molecular system having a CI between the electronic states. In this respect, we would like to mention the pioneering studies of Kuppermann and co-workers [44–46] and others [47–48] who incorporated the required sign change by multiplying the adiabatic wave function of the $D + H_2$ reactive system with the HLH phase. Kuppermann and co-workers identified the effect of this geometric (or topological) phase for the first time in a chemical reaction. Their theoretically calculated integral cross-sections agreed well with experimental data at different energies [49–52]. In particular, they found that such effects are noticeable in differential cross-sections. This series of studies renewed interest in this subject.

As the CI of the ground and the excited states of the H_3 system occurs at the symmetric triangular configuration, it is possible to incorporate the HLH phase directly in the basis functions as Kuppermann et al. did so that the nuclear SE does not require any extra term through a vector potential. Even though this approch could be a reasonably good approximation for the isotopic variants of X_3 with the dynamics expressed in hyperspherical coordinates, the vector potential approach, as we mentioned earlier, will be more rigorous for general coordinate and reactive systems. We have formulated [53,54] the general form of the vector potential in hyperspherical coordinates for the $A + B_2$ type of reactive system even in cases where the position of the CI is arbitrary. The influence of the vector potential on the integral and differential cross-sections of the $D + H_2$ reactive system has been estimated [53–55] by quasiclassical trajectory calculations. We found qualitatively the same relative shift of the rotational state distribution and the change of scattering angle distributions in the presence of the vector potential as indicated by Kuppermann et al. through directly introducing the HLH phase change. We also performed semiclassical calculations [56] and include either a vector potential in the nuclear SE or incorporated a phase factor in the basis functions and again obtained the same relative shift of the rotational state distribution.

The effect of singularities on scattering processes has also been investigated by extending the JT model [57,58]. The geometric phase effect on the proper symmetry allowed vibrational transition probabilities in the nonreactive and reactive channels of a simple two-dimensional (2D) quasi-JT model is an interesting topic. The ordinary BO equations can be extended either by including the HLH phase [1] or by adding extra terms through a vector potential [59,60]. Quantum mechanical calculations indicated that in the case of the quasi-JT model, ordinary BO equations could not give the proper symmetry allowed transitions, whereas the extended BO equations could. Finally, a two surface diabatic calculation on the quasi-JT model confirmed the validity of the extended BO equations. It is also important to point out that calculations were done both in the time-independent [59] and time-dependent framework [60]. The findings were the same.

The generalization of BO equations based on the HLH phase seemed to be the right thing to do so far, but generally two questions arise: (1) Is it really necessary to incorporate an ad hoc correction of the HLH type into the quantum theory of an atom and a molecule? (2) Is it guaranteed that such a treatment can offer correct results in all cases or not? In this context, we would like to mention the work by Baer and Englman [57]. As the non-adiabatic coupling terms appear in the off-diagonal positions in the SE, in order to construct a single approximated BO equation the non-adiabatic coupling terms must be shifted from their original off-diagonal position to the diagonal position. In the first attempt, it was shown that such a possibility may exist and an approximate version of the extended BO equations for the two-state case has been derived. In a subsequent article, Baer [58] derived a new set of coupled BO–SEs from first principles (and without approximations) for the 2D Hilbert space where all the non-adiabatic coupling terms are shifted from the off-diagonal to the diagonal position. These two equations remain coupled but the coupling term become potential coupling. As this potential coupling term is multiplied by the original adiabatic wave function associated with the upper electronic state, which is small particularly at low energy processes, two decoupled extended BO equations are obtained by deleting this product.

The adiabatic-to-diabatic transformation (ADT) matrix (see Appendix B) is responsible for the transformation from the electronic adiabatic (see Appendix B) eigenfunctions to the diabatic framework (see Appendix B). The adiabatic framework describes the functions that govern the motion of the nuclei, namely, on the potential energy surface (PES) and the non-adiabatic coupling terms. It is the non-adiabatic terms that cause difficulties when studying nuclear dynamics of a system having a CI. These terms are abruptly behaving—sometime even spiky—functions of the coordinate [19,20,61] and therefore cause numerical instabilities when solution of the corresponding nuclear SE is attempted. It has been well known for quite some time that the only way to overcome this numerical difficulty is to move from the adiabatic to diabatic framework where the non-adiabatic coupling terms are replaced by the potential coupling terms that are much smoother functions of the coordinates [20,62]. Recently, a direct connection has been found between a given non-adiabatic coupling matrix and the uniqueness of the relevant diabatic potential matrix [63,64]. It has been proven that in order to produce a uniquely defined diabatic potential energy matrix from the non-adiabatic coupling matrix, the ADT matrix has to fulfill quantization-type requirements. In simple cases, these requirements become ordinary quantizations of the eigenvalues of the non-adiabatic coupling matrix. As, for example, for systems having a CI between two states, the average values of the non-adiabatic coupling over a closed path is only allowed to have the value $n/2$, where n is an integer. This value is the same as that given by the HLH phase factor. Similarly, for systems having a CI among three states this average becomes n, where n is now an integer. The main advantage of this new derivation is that it can be extended to any N-state system. Baer and others, along with the present authors, proved an "existence theorem" that shows the possibility of a derivation of the extended BO equation for an N-state system [65] having a CI at a particular point. We obtained extended BO equations for a tri-state JT model [66] using quantization-type requirement of the ADT matrix and these extended BO equations are different from those obtained by using the HLH phase. Finally, we perform numerical calculation on the ground adiabatic surface of the tri-state JT model using those extended BO equations obtained by considering that three states are coupled and found that the results agreed well with the diabatic results.

Finally, in brief, we demonstrate the influence of the upper adiabatic electronic state(s) on the ground state due to the presence of a CI between two or more than two adiabatic potential energy surfaces. Considering the HLH phase, we present the extended BO equations for a quasi-JT model and for an $A + B_2$ type reactive system, that is, the geometric phase (GP) effect has been introduced either by including a vector potential in the system Hamiltonian or by incorporating a phase factor into the adiabatic nuclear wave function. The effect of a topological phase on reactive and non-reactive transition probabilities were obtained by using a time-dependent wavepacket approach in a 2D quasi-JT model. Even when we replace the operators in the Hamiltonian (with or without introducing a vector potential) of the $D + H_2$ reactive system with the corresponding classical variables and calculate integral and differential cross-sections, we can clearly identify the signature of the GP effect. Semiclassical results on the same system also indicate an effect. We also present the results obtained by quasiclassical trajectory calculations for the $H + D_2$ reaction. In case of a two-state isolated system (a Hilbert space of dimension 2), the formulation of extended BO equations to perform scattering calculations on a quasi-JT model and $A + B_2$ type reactive systems is based on the idea of a Longuet-Higgins phase. If more than one excited state is coupled with the ground state, the phase factor could be different from the Longuet-Higgins phase factor as shown by Baer et al. [65], where the phase angle is defined through the ADT matrix. It indicates that even for reaction dynamic studies on the ground adiabatic surface one needs to know the number of excited states coupled with the ground state and depending on this number, the phase factor changes, hence the form of extended BO equations will be modified. We present the outline of the derivation of the extended version of the BO approximate equations and perform scattering calculation on a two-arrangement–two-coordinate tri-state model system. These calculations were done three times for each energy: Once without any approximations, that is, a diabatic calculation; next with those extended BO equations derived by using the HLH phase; and finally with those extended BO equations derived by using the new phase factor due to tri-state coupling. The state-to-state (reactive and nonreactive) transition probabilities obtained indicate that only the new approximate BO equations can yield the relevant results for a tri-state system. In Section V, we introduce a new formulation of quantum molecular dynamics (so-called quantum dressed classical mechanics) and give the form of the vector potential needed for incorporating topological phase effects if the dynamics is solved using this approach.

II. LONGUET-HIGGINS PHASE-BASED TREATMENT

As mentioned in the introduction, the simplest way of approximately accounting for the geometric or topological effects of a conical intersection incorporates a phase factor in the nuclear wave function. In this section, we shall consider some specific situations where this approach is used and furthermore give the vector potential that can be derived from the phase factor.

A. The Geometric Phase Effect in a 2D Two Surface System

The non-adiabatic effect on the ground adiabatic state dynamics can as mentioned in the introduction be incorporated either by including a vector potential into the system Hamiltonian derived by considering the HLH phase or by multiplying the HLH phase directly on the basis functions. We have studied a two-coordinate quasi-"JT scattering" model [37] where the nuclear kinetic energy operator in Cartesian coordinates can be written as,

$$T_n(R,r) = -\frac{\hbar^2}{2m}\left[\frac{\partial^2}{\partial r^2} + \frac{\partial^2}{\partial R^2}\right] \tag{1}$$

or in terms of polar coordinates we have,

$$T_n(q,\phi) = -\frac{\hbar^2}{2m}\left[\frac{\partial^2}{\partial q^2} + \frac{1}{q}\frac{\partial}{\partial q} + \frac{1}{q^2}\frac{\partial^2}{\partial \phi^2}\right] \tag{2}$$

R and r are defined in the intervals, $-\infty \leq R \leq \infty$ and $-\infty \leq r \leq \infty$ and these are related to q and ϕ in the following way:

$$r = q\sin\phi,\ R = q\cos\phi,\ \text{and}\ \phi = \arctan(r/R)$$

The effective nuclear kinetic energy operator due to the vector potential is formulated by multiplying the adiabatic eigenfunction of the system, $\psi(R,r)$ with the HLH phase $\exp(i/2\arctan(r/R))$, and operating with $T_n(R,r)$, as defined in Eq. (1), on the product function and after little algebraic simplification, one can obtain the following effective kinetic energy operator,

$$T'_n(R,r) = -\frac{\hbar^2}{2m}\left[\frac{\partial^2}{\partial r^2} + \frac{\partial^2}{\partial R^2} + \left(\frac{R}{r^2+R^2}\right)i\frac{\partial}{\partial r} - \left(\frac{r}{r^2+R^2}\right)i\frac{\partial}{\partial R} - \frac{1}{4(r^2+R^2)}\right] \tag{3}$$

Similarly, the expression for the effective kinetic energy operator in polar coordinates will be,

$$T'_n(q,\phi) = -\frac{\hbar^2}{2m}\left[\frac{\partial^2}{\partial q^2} + \frac{1}{q}\frac{\partial}{\partial q} + \frac{1}{q^2}\frac{\partial^2}{\partial \phi^2} - i\frac{1}{q^2}\frac{\partial}{\partial \phi} - \frac{1}{4q^2}\right] \tag{4}$$

If the position of the conical intersection is shifted from the origin of the coordinate system to $(r_0,\ R_0)$, the relation between Cartesian and polar coordinates for the present system can be written as, $r \pm r_0 = q\sin\phi$, $R \pm R_0 = q\cos\phi$ and $\phi = \arctan(r \pm r_0/R \pm R_0)$. Consequently, the effective nuclear kinetic energy operator will be [68],

$$\begin{aligned} T''_n(R,\ r) = -\frac{\hbar^2}{2m}\Bigg[\frac{\partial^2}{\partial r^2} + \frac{\partial^2}{\partial R^2} &+ \left(\frac{R \pm R_0}{(r \pm r_0)^2 + (R \pm R_0)^2}\right)i\frac{\partial}{\partial r} \\ &- \left(\frac{(r \pm r_0)}{(r \pm r_0)^2 + (R \pm R_0)^2}\right)i\frac{\partial}{\partial R} - \left(\frac{1}{4((r \pm r_0)^2 + (R \pm R_0)^2)}\right)\Bigg] \end{aligned} \tag{5}$$

Hence, the expression of Eq. (5) indicates that, in a polar coordinate system, Eq. (4) will remain unchanged even if the position of the conical intersection is shifted from the origin of the coordinate system.

The ordinary BO equations in the adiabatic representation can also be used for single surface calculations where the geometrical phase effect is incorporated by an HLH phase change in ϕ. The correct phase treatment of the ϕ coordinate has been introduced by using a special technique [44–48] when the kinetic energy operators are evaluated numerically. More specifically, the geometrical phase effect has been introduced by modifying the fast Fourier transformation (FFT) procedure when evaluating the kinetic energy terms. The wave function $\psi(q, \phi)$ is multiplied with $\exp(i\phi/2)$, then after doing a forward FFT the coefficients are multiplied with a slightly different frequency factor containing $(k + \frac{1}{2})$ instead of k and finally after completing the backward FFT [69], the wave function is multiplied with $\exp(-i\phi/2)$. The procedure needs to be repeated in each time step of the propagation.

The transition probabilities obtained due to the above two modified treatments of single-surface calculations need to be compared with those transition probabilities obtained by two surface calculations that confirms the validity of these former treatments.

1. Scattering Calculation with the Quasi-Jahn–Teller Model

The two adiabatic potential energy surfaces that we will use in the present calculations, are called a reactive double-slit model (RDSM) [59] where the first surface is the lower and the second is the upper surface, respectively,

$$\begin{aligned} u_1(R, r) &= \frac{1}{2} m(\omega_0 - \tilde{\omega}_1(R))^2 r^2 + Af(R, r) + g(R)u_2(R, r) \\ u_2(R, r) &= \frac{1}{2} m\omega_0^2 r^2 - (D - A)f(R, r) + D \end{aligned} \tag{6}$$

with

$$\tilde{\omega}_1(R) = \omega_1 \exp\left(-\left(\frac{R}{\sigma}\right)^2\right)$$

$$f(R, r) = \exp\left(-\left(\frac{R^2 + r^2}{\sigma^2}\right)\right)$$

and

$$g(R) = 0 \tag{7}$$

The parameters used in the above expressions for the potential energy surfaces and the calculations are given in Table 1 of [60].

The two surface calculations by using the following Hamiltonian matrix are rather straightforward in the diabatic representation

$$\mathbf{H} = \mathbf{T} + \mathbf{W} = T_n \begin{pmatrix} 1 & 0 \\ 0 & 1 \end{pmatrix} + \begin{pmatrix} W_{11} & W_{12} \\ W_{21} & W_{22} \end{pmatrix} \tag{8}$$

where the diabatic potential matrix elements,

$$\begin{aligned} W_{11} &= \frac{1}{2}[u_1 + u_2 + (u_1 - u_2)\cos\phi] \\ W_{22} &= \frac{1}{2}[u_1 - u_2 + (u_1 - u_2)\cos\phi] \\ W_{12} = W_{21} &= \frac{1}{2}(u_1 - u_2)\sin\phi \end{aligned} \tag{9}$$

are obtained by the following orthogonal transformation:

$$\mathbf{W} = \mathbf{T}^{\dagger}\mathbf{U}\mathbf{T} \tag{10}$$

$$\text{with} \quad \mathbf{T} = \begin{pmatrix} \cos\phi/2 & -\sin\phi/2 \\ \sin\phi/2 & \cos\phi/2 \end{pmatrix} \quad \text{and} \quad \mathbf{U} = \begin{pmatrix} u_1 & 0 \\ 0 & u_2 \end{pmatrix}$$

Single surface calculations with a vector potential in the adiabatic representation and two surface calculations in the diabatic representation with or without shifting the conical intersection from the origin are performed using Cartesian coordinates. As in the asymptotic region the two coordinates of the model represent a translational and a vibrational mode, respectively, the initial wave function for the ground state can be represented as,

$$\Psi_{\text{ad}}(R, r, t_0) = \psi_{\text{GWP}}^{k_0}(R, t_0)\Phi_v(r, t_0) \tag{11}$$

where $\psi_{\text{GWP}}^{k_0}(R, t_0)$ is a Gaussian wavepacket and $\Phi_v(r, t_0)$ a harmonic oscillator wave function.

It is important to note that the two surface calculations will be carried out in the diabatic representation. One can get the initial diabatic wave function matrix for the two surface calculations using the above adiabatic initial wave function by the following orthogonal transformation,

$$\begin{pmatrix} \Psi_{\text{di}}^1(R, r, t_0) \\ \Psi_{\text{di}}^2(R, r, t_0) \end{pmatrix} = \begin{pmatrix} \cos\phi/2 & \sin\phi/2 \\ -\sin\phi/2 & \cos\phi/2 \end{pmatrix} \begin{pmatrix} \Psi_{\text{ad}}(R, r, t_0) \\ 0 \end{pmatrix} \tag{12}$$

Single surface calculations with proper phase treatment in the adiabatic representation with shifted conical intersection has been performed in polar coordinates. For this calculation, the initial adiabatic wave function $\Psi_{\mathrm{ad}}(q, \phi, t_0)$ is obtained by mapping $\Psi_{\mathrm{ad}}(R, r, t_0)$ into polar space using the relations, $r \pm r_0 = q \sin\phi$ and $R \pm R_0 = q \cos\phi$. At this point, it is necessary to mention that in all the above cases the initial wave function is localized at the positive end of the R coordinate where the negative and positive ends of the R coordinate are considered as reactive and nonreactive channels.

The kinetic energy operator evaluation and then, the propagation of the R, r, or q, ϕ degrees of freedom have been performed by using a fast Fourier transformation FFT [69] method for evaluating the kinetic energy terms, followed by Lanczos reduction technique [70] for the time propagation. A negative imaginary potential [48]

$$V_{\mathrm{Im}}(R) = -\frac{iV_{\mathrm{Im}}^{\max}}{\cosh^2[(R_{\max}^{\pm} - R)/\beta]} \tag{13}$$

has been used to remove the wavepacket from the grid before it is reflected from the negative and positive ends of the R grid boundary. The parameters used in the above expression and other data are given in Table II of [60].

The transition probability at a particular total energy (E_n) from vibrational level i to f may be expressed as the ratio between the corresponding outgoing and incoming quantities [71]

$$P_{i\to f}^{\pm}(E_n) = \frac{\int_{-\infty}^{\infty} \xi_{k_{f,n}}^{\pm}(t)\, dt}{\int_{k_{\min,i,n}}^{k_{\max,i,n}} |\eta(k_{i,n})|^2\, dk_{i,n}} \tag{14}$$

where the (+) and (−) signs in the above expression indicate nonreactive and reactive transition probabilities. If we propagate the system from the initial vibrational state, i, and are interested in projecting at a particular energy, E_n, and final vibrational state, f, the following equation can dictate the distribution of energy between the translational and vibrational modes,

$$\frac{\hbar^2 k_{i,n}^2}{2m} + \hbar\omega_0\left(i + \frac{1}{2}\right) = E_n = \frac{\hbar^2 k_{f,n}^2}{2m} + \hbar\omega_0\left(f + \frac{1}{2}\right) \tag{15}$$

One can obtain the explicit expressions for $\xi_{k_{f,n}}^{+}$ and $\xi_{k_{f,n}}^{-}$ as defined in Eq. (13) considering the following outgoing fluxes in the nonreactive and reactive channels

$$\xi_f^{+}(t) = \mathrm{Re}\left\{ \psi_f^{\star}(R_0^{+}, t) \times (-i\hbar/m) \times \left(\frac{\partial\psi_f(R, t)}{\partial R}\right)_{R_0^{+}} \right\} \tag{16}$$

and

$$\xi_f^-(t) = \mathrm{Re}\left\{ \psi_f^\star(R_0^-, t) \times (i\hbar/m) \times \left(\frac{\partial \psi_f(R, t)}{\partial R}\right)_{R_0^-} \right\} \tag{17}$$

where

$$\psi_f(R, t) = \int_{-\infty}^{\infty} \Psi_{\mathrm{ad}}^\star(R, r, t)\Phi_f(r)dr \tag{18}$$

The discrete Fourier expansion of $\psi_f(R, t)$ can be written as

$$\begin{aligned} \psi_f(R, t) &= \sum_{n=-(N_R/2)+1}^{N_R/2} C_n(t)\exp\left[2i\pi(n-1)\left(\frac{R - R_{\min}}{R_{\max} - R_{\min}}\right)\right] \\ &= \sum_n C_n(t)\exp\left[\frac{2i\pi(n-1)(i-1)}{N_R}\right] \end{aligned} \tag{19}$$

where, $R = R_{\min} + (i-1)(R_{\max} - R_{\min})/N_R$ and N_R is total number of grid points in R space. By substituting Eq. (18) into Eqs. (15) and (16), one can easily arrive at

$$\begin{aligned} \xi_f^+(t) = \mathrm{Re}\Bigg\{ \psi_f^\star(R_0^+, t) \sum_{n=0}^{N_R/2} \Bigg\{ \left(\frac{2\pi\hbar(n-1)}{m(R_{\max} - R_{\min})}\right) \\ \times \exp\left[\frac{2i\pi(n-1)(i_0^+ - 1)}{N_R}\right] C_n(t) \Bigg\}\Bigg\} \end{aligned} \tag{20}$$

and

$$\begin{aligned} \xi_f^-(t) = \mathrm{Re}\Bigg\{ \psi_f^\star(R_0^-, t) \sum_{n=-(N_R/2)+1}^{-1} \Bigg\{ \left(\frac{2\pi\hbar(n-1)}{m(R_{\max} - R_{\min})}\right) \\ \times \exp\left[\frac{2i\pi(n-1)(i_0^- - 1)}{N_R}\right] C_n(t) \Bigg\}\Bigg\} \end{aligned} \tag{21}$$

where $i_0^\pm = [(R_0^\pm - R_{\min})/(R_{\max} - R_{\min})] + 1$. It is important to note that in $\xi_f^+(t)$ only positive and in $\xi_f^-(t)$ only negative values of n have been considered. It has been numerically verified that negative components of n in $\xi_f^+(t)$ and positive components of n in $\xi_f^-(t)$ actually have negligible contribution.

We are now in a position to write from Eqs. (19) and (20) that,

$$\xi_f^+(t) = \sum_{n=0}^{N_R/2} \xi_{k_{f,n}}^+ \tag{22}$$

and

$$\xi_f^-(t) = \sum_{n=-(N_R/2)+1}^{-1} \xi_{k_{f,n}}^- \tag{23}$$

respectively.

The denominator in Eq. (13) can be interpreted as an average value over the momentum distribution from the initial wavepacket, that is,

$$\eta(k_{i,n}) = \frac{1}{2\pi}\int_0^{\infty} \psi_{\mathrm{GWP}}^{k_0}(R, t_0)\exp(ik_{i,n}R)\, dR \tag{24}$$

and the limits $(k_{\mathrm{min},i,n}, k_{\mathrm{max},i,n})$ of the integral in the denominator of Eq. (13) over the variable $k_{i,n}$ can be obtained if we consider the wavenumber interval of the corresponding final f channel,

$$k_{\mathrm{min},f,n} = k_{f,n} - \frac{\pi}{R_{\mathrm{max}} - R_{\mathrm{min}}}, \qquad k_{\mathrm{max},f,n} = k_{f,n} + \frac{\pi}{R_{\mathrm{max}} - R_{\mathrm{min}}} \tag{25}$$

These values are related to the initial wavenumber intervals by the following equations:

$$\begin{aligned} \frac{\hbar^2}{2m}(k_{\mathrm{min},i,n})^2 + \hbar\omega_0\left(i + \frac{1}{2}\right) &= \frac{\hbar^2}{2m}(k_{\mathrm{min},f,n})^2 + \hbar\omega_0\left(f + \frac{1}{2}\right) \\ \frac{\hbar^2}{2m}(k_{\mathrm{max},i,n})^2 + \hbar\omega_0\left(i + \frac{1}{2}\right) &= \frac{\hbar^2}{2m}(k_{\mathrm{max},f,n})^2 + \hbar\omega_0\left(f + \frac{1}{2}\right) \end{aligned} \tag{26}$$

We have used the above analysis scheme for all single- and two-surface calculations. Thus, when the wave function is represented in polar coordinates, we have mapped the wave function, $\Psi_{\mathrm{ad}}(q, \phi, t)$ to $\Psi_{\mathrm{ad}}(R, r, t)$ in each time step to use in Eq. (17) and as the two surface calculations are performed in the diabatic representation, the wave function matrix is back transformed to the adiabatic representation in each time step as

$$\begin{pmatrix} \Psi_{\mathrm{ad}}^1(R, r, t) \\ \Psi_{\mathrm{ad}}^2(R, r, t) \end{pmatrix} = \begin{pmatrix} \cos\phi/2 & -\sin\phi/2 \\ \sin\phi/2 & \cos\phi/2 \end{pmatrix} \begin{pmatrix} \Psi_{\mathrm{di}}^1(R, r, t) \\ \Psi_{\mathrm{di}}^2(R, r, t) \end{pmatrix} \tag{27}$$

and used in Eq. (17) for analysis.

For all cases we have propagated, the system is started in the initial vibrational state, $i = 0$, with total average energy 1.75 eV and projected at four selected energies, 1.0, 1.5, 2.0, and 2.5 eV, respectively.

2. *Results and Discussion*

In Table I we present vibrational state-to-state transition probabilities on the ground adiabatic surface obtained by two-surface calculations and compare with those transition probabilities obtained by single-surface calculations with or without including the vector potential in the nuclear Hamiltonian. In these calculations, the position of the conical intersection coincides with the origin of the coordinates. Again shifting the position of the conical intersection from the origin of the coordinates, two-surface results and modified single-surface results obtained either by introducing a vector potential in the nuclear Hamiltonian or by incorporating a phase factor in the basis set, are also presented.

At this point, it is important to note that as the potential energy surfaces are even in the vibrational coordinate (r), the same parity, that is, even $\rightarrow$ even and odd $\rightarrow$ odd transitions should be allowed both for nonreactive and reactive cases but due to the conical intersection, the diabatic calculations indicate that the allowed transition for the reactive case are odd $\rightarrow$ even and even $\rightarrow$ odd whereas in the case of nonreactive transitions even $\rightarrow$ even and odd $\rightarrow$ odd remain allowed.

In Table I(a), various reactive state-to-state transition probabilities are presented for four selected energies where calculations have been performed assuming that the point of conical intersection and the origin of the coordinate system are at the same point. The numbers of the first row of this table have been obtained from two-surface diabatic calculations and we notice that only odd $\rightarrow$ even and even $\rightarrow$ odd transitions are allowed. Single-surface results including a vector potential not only give the correct parity for the transitions but also good agreement between the first- and the second-row numbers for all energies. The third row of Table I(a) indicates the numbers from a single-surface calculation without a vector potential. We see that the parity as well as the actual numbers are incorrect.

Again, in Table I(b), we present reactive state-to-state transition probabilities at the four selected energies where the position of the conical intersection is shifted from the origin of the coordinates. The first row of this table indicates results from a two-surface diabatic calculation where in the nonreactive case the same parity and in the reactive case opposite parity transitions appear as allowed transitions. Calculated numbers shown in the second row came from single-surface calculations with a vector potential and the results not only follow the parity (same parity for the nonreactive case and different parity for the reactive case) but also agree well for all energies with the numbers shown in the first row of Table I(b). Results from single-surface calculations incorporating a

TABLE I(a)
Reactive State-to-State Transition Probabilities when Calculations are Performed Keeping the Position of the Conical Intersection at the Origin of the Coordinates

E (eV)	0 → 0	0 → 1	0 → 2	0 → 3	0 → 4	0 → 5	0 → 6	0 → 7	0 → 8	0 → 9
1.0	0.0000[a]	0.0033	0.0000	0.0220						
	0.0001[b]	0.0101	0.0008	0.0345						
	0.0094[c]	0.0000	0.0361	0.0000						
1.5	0.0000	0.1000	0.0000	0.0342	0.0000	0.0764				
	0.0001	0.1046	0.0001	0.0370	0.0004	0.0582				
	0.0719	0.0000	0.0664	0.0000	0.0827	0.0000				
2.0	0.0000	0.1323	0.0000	0.0535	0.0000	0.0266	0.0000	0.2395		
	0.0002	0.1323	0.0000	0.0583	0.0001	0.0267	0.0007	0.2383		
	0.1331	0.0000	0.0208	0.0000	0.0300	0.0000	0.1963	0.0000		
2.5	0.0000	0.0987	0.0000	0.0858	0.0000	0.0901	0.0000	0.0248	0.0000	0.2529
	0.0001	0.0983	0.0001	0.0903	0.0005	0.0870	0.0010	0.0297	0.0007	0.2492
	0.2116	0.0000	0.0382	0.0000	0.0121	0.0000	0.1783	0.0000	0.1119	0.0000

[a] Two-surface calculation.
[b] Single-surface calculation with vector potential.
[c] Single-surface calculation without vector potential.

TABLE I(b)
Reactive State-to-State Transition Probabilities when Calculations are Performed by Shifting the Position of Conical Intersection from the Origin of the Coordinate System

E (eV)	0 → 0	0 → 1	0 → 2	0 → 3	0 → 4	0 → 5	0 → 6	0 → 7	0 → 8	0 → 9
1.0	0.0000[a]	0.0119	0.0000	0.0090						
	0.0001[b]	0.0113	0.0004	0.0060						
	0.0003[c]	0.0363	0.0004	0.0271						
1.5	0.0000	0.1043	0.0000	0.0334	0.0000	0.0571				
	0.0000	0.1084	0.0001	0.0346	0.0002	0.0592				
	0.0001	0.1390	0.0000	0.0183	0.0001	0.0050				
2.0	0.0000	0.1281	0.0000	0.0561	0.0000	0.0365	0.0000	0.2443		
	0.0001	0.1286	0.0002	0.0604	0.0001	0.0319	0.0001	0.2609		
	0.0000	0.1040	0.0001	0.0853	0.0004	0.0526	0.0002	0.2185		
2.5	0.0000	0.0869	0.0000	0.0909	0.0000	0.0788	0.0000	0.0211	0.0000	0.2525
	0.0002	0.0864	0.0002	0.0981	0.0007	0.0750	0.0002	0.0342	0.0018	0.2387
	0.0000	0.0711	0.0002	0.0877	0.0006	0.0932	0.0009	0.0479	0.0019	0.2611

[a] Two-surface calculation.
[b] Single-surface calculation with vector potential.
[c] Single-surface calculation with phase change.

phase factor into the basis set are shown in the third row of Table I(b) for the reactive channel. Though the phase treatment can offer proper parity allowed transitions, these numbers have for all energies less agreement with those presented in the first and second rows of Table I(b).

In this model calculation, using a time-dependent wavepacket approach, we studied the extended JT model in order to investigate the symmetry effects on the scattering processes. First, we performed two surface diabatic calculations, which are considered to be the exact ones as they can follow interference effects due to the conical intersection. We see that the ordinary BO approximation has failed to treat the symmetry effect but the modified single-surface calculations, either by including a vector potential into the nuclear Hamiltonian or by incorporating a phase factor in the basis set, can reproduce the two-surface results for different situations. Though the transition probabilities calculated by Baer et al. [59] using the same model are qualitatively the same as the present numbers, small quantitative differences are present, particularly, at higher energies. We believe that some of these deviations could be due to the dynamic effects of the potential, the vector potential, or the phase changes in the wave function. We may therefore conclude that if the energy is below the conical intersection, then the effect of it is well described by simply adding a vector potential to the Hamiltonian or by the simple phase change in the angle ϕ, which when increased by 2π makes the system encircle the intersection and appear to work well even in cases where the intersection is shifted away from the origin.

B. Three-Particle Reactive System

We derive the effective Hamiltonian considering the HLH phase change for any reaction involving three atoms and discuss integral and differential cross-sections obtained either classically or semiclassically. An easy way of incorporating the geometric phase effect is to use the hyperspherical coordinates in which the encircling of the intersection is connected with a phase change by 2π of one of the hyperangles (ϕ).

In the presence of a phase factor, the momentum operator ($\hat{P}$), which is expressed in hyperspherical coordinates, should be replaced [53,54] by ($\hat{P} - \hbar\nabla\eta$) where $\nabla\eta$ creates the vector potential in order to define the effective Hamiltonian (see Appendix C). It is important to note that the angle entering the vector potential is strictly only identical to the hyperangle ϕ for an A_3 system.

The general form of the effective nuclear kinetic energy operator ($\hat{T}'$) can be written as

$$
\begin{aligned}
\hat{T}' &= \frac{1}{2\mu}(\hat{P} - \hbar\nabla\eta)^2 \\
&= \frac{1}{2\mu}(\hat{P}^2 - \hbar^2\nabla^2\eta - 2\hbar\nabla\eta\hat{P} + \hbar^2\nabla\eta\,\nabla\eta)
\end{aligned} \tag{28}
$$

It is now convenient to introduce hyperspherical coordinates (ρ, θ, and ϕ), which specify the size and shape of the ABC molecular triangle and the Euler angles α, β, and γ describing the rotation of the molecular shape in space. If the Euler angles are treated as classical variables and the coordinate system is such that the z axis is aligned [67] with the total angular momentum J, the semi-classical kinetic energy operator $\hat{T}_{\rm scl}$ for a three-particle system can be expressed in a modified form of Johnson's hyperspherical coordinates [72] as below,

$$\hat{T}_{\rm scl} = \frac{\hat{P}^2}{2\mu} = \frac{1}{2\mu}\left[\hat{P}_\rho^2 + \frac{4}{\rho^2}\hat{L}^2(\theta,\phi)\right] + \frac{P_\gamma[P_\gamma - 4\cos\theta\hat{P}_\phi]}{2\mu\rho^2\sin^2\theta} + \frac{(J^2 - P_\gamma^2)(1+\sin\theta\cos 2\gamma)}{\mu\rho^2\cos^2\theta} - \frac{\hbar^2}{2\mu\rho^2}\left[\frac{1}{4} + \frac{1}{\sin^2 2\theta}\right] \tag{29}$$

where ρ is the hyperradius, and θ and ϕ are the hyperangles with

$$\hat{L}^2 = -\hbar^2\left[\frac{\partial^2}{\partial\theta^2} + \frac{1}{\sin^2\theta}\frac{\partial^2}{\partial\phi^2}\right]$$

Due to the special choice of coordinates, the momenta conjugate to α and β are constants of motion, that is, $P_\alpha = J$, $P_\beta = 0$, and $P_\gamma = J\cos\beta$.

When we wish to replace the quantum mechanical operators with the corresponding classical variables, the well-known expression for the kinetic energy in hyperspherical coordinates [73] is

$$T_{\rm cl} = \frac{1}{2\mu}\left[P_\rho^2 + \frac{4}{\rho^2}\left(P_\theta^2 + \frac{1}{\sin^2\theta}P_\phi^2\right)\right] + \frac{P_\gamma[P_\gamma - 4\cos\theta P_\phi]}{2\mu\rho^2\sin^2\theta} + \frac{(P_\alpha^2 - P_\gamma^2)(1+\sin\theta\cos 2\gamma)}{\mu\rho^2\cos^2\theta} \tag{30}$$

The explicit expressions of the other terms in Eq. (27) can be evaluated in terms of hyperspherical coordinates using the results of Appendix C,

$$-\frac{\hbar^2}{2\mu}\nabla^2\eta = -\frac{\hbar^2}{2\mu}\sum_i\frac{\partial^2\eta}{\partial X_i^2} \qquad \text{where} \qquad X_i \equiv (r_x, r_y, r_z, R_x, R_y, R_z)$$

$$\begin{aligned}
&= \frac{4\hbar^2}{\mu\rho^2\sin\theta}\{[\sin\theta_0\cos\theta\sin\phi/2[\sin^2\theta\sin^2\phi + (\cos\theta_0\sin\theta\cos\phi \\
&+ \sin\theta_0\cos\theta)^2]] + [\sin\theta_0\sin\theta\sin\phi\{\sin^2\phi\cos\theta\sin\theta \\
&+ (\cos\theta_0\sin\theta\cos\phi + \sin\theta_0\cos\theta)(\cos\theta_0\cos\theta\sin\phi - \sin\theta_0\sin\theta)\} \\
&+ (\cos\theta_0\sin\theta + \sin\theta_0\cos\theta\cos\phi)\{\sin^2\theta\sin\phi\cos\phi \\
&- \cos\theta_0\sin\theta\sin\phi(\cos\theta_0\sin\theta\cos\phi + \sin\theta_0\cos\theta)\}]/ \\
&[\sin^2\theta\sin^2\phi + (\cos\theta_0\sin\theta\cos\phi + \sin\theta_0\cos\theta)^2]^2\}
\end{aligned} \tag{31}$$

$$\frac{\hbar^2}{2\mu}\nabla\eta\nabla\eta = \frac{\hbar^2}{2\mu}\sum_i \frac{\partial\eta}{\partial X_i}\frac{\partial\eta}{\partial X_i} \qquad \text{where} \qquad X_i \equiv (r_x, r_y, r_z, R_x, R_y, R_z)$$

$$= \frac{2\hbar^2}{\mu\rho^2}\frac{[\sin^2\theta_0 \sin^2\phi + (\cos\theta_0 \sin\theta + \sin\theta_0 \cos\theta \cos\phi)^2]}{[\sin^2\theta \sin^2\phi + (\cos\theta_0 \sin\theta \cos\phi + \sin\theta_0 \cos\theta)^2]^2} \tag{32}$$

and

$$-\frac{\hbar}{\mu}\nabla\eta P = -\frac{\hbar}{\mu}\sum_i \frac{\partial\eta}{\partial X_i} P_{X_i} \qquad \text{where} \qquad X_i \equiv (r_x, r_y, r_z, R_x, R_y, R_z) \tag{33}$$

where the general form of the momenta $P^{\star}_{X_i}$ ($\star$ indicates that the Coriolis term is not included) in hyperspherical coordinates can be expressed as

$$P^{\star}_{X_i} = P_\rho \frac{\partial\rho}{\partial X_i} + P_\theta \frac{\partial\theta}{\partial X_i} + P_\phi \frac{\partial\phi}{\partial X_i}$$

It is to easy to evaluate $\partial\rho/\partial X_i$, $\partial\theta/\partial X_i$, and $\partial\phi/\partial X_i$ [for $X_i \equiv (r_x, r_y, r_z, R_x, R_y, R_z)$] using equation (C.2) and after introducing the Coriolis term [72], the momenta P_{X_i} become

$$\begin{aligned}
P_{r_x} &= \left(\frac{r_x}{\rho}P_\rho - \frac{2R_y}{\rho^2}P_\theta + \frac{2R_x}{\rho^2 \sin\theta}P_\phi - \omega_z r_y\right)\\
P_{r_y} &= \left(\frac{r_y}{\rho}P_\rho - \frac{2R_x}{\rho^2}P_\theta - \frac{2R_y}{\rho^2 \sin\theta}P_\phi + \omega_z r_x\right)\\
P_{r_z} &= (\omega_x r_y - \omega_y r_x)\\
P_{R_x} &= \left(\frac{R_x}{\rho}P_\rho + \frac{2r_y}{\rho^2}P_\theta - \frac{2r_x}{\rho^2 \sin\theta}P_\phi - \omega_z R_y\right)\\
P_{R_y} &= \left(\frac{R_y}{\rho}P_\rho + \frac{2r_x}{\rho^2}P_\theta + \frac{2r_y}{\rho^2 \sin\theta}P_\phi + \omega_z R_x\right)\\
P_{R_z} &= (\omega_x R_y - \omega_y R_x)
\end{aligned} \tag{34}$$

where ω_x, ω_y, and ω_z are the components of instantaneous angular velocity of the rotating axes XYZ with respect to the stationary axes $X'Y'Z'$.

By substituting $\partial\eta/\partial X_i$ and P_{X_i} in Eq. (32), after some simplification we get,

$$-\frac{\hbar}{\mu}\nabla\eta P = -\frac{4\hbar[\sin\theta_0 \sin\theta \sin\phi P_\theta + (\cos\theta_0 \sin\theta + \sin\theta_0 \cos\theta \cos\phi)P_\phi]}{\mu\rho^2 \sin\theta[\sin^2\theta \sin^2\phi + (\cos\theta_0 \sin\theta \cos\phi + \sin\theta_0 \cos\theta)^2]} \tag{35}$$

It is important to note that Eq. (34) becomes independent of the Coriolis term because the symmetrical components of P and $\nabla\eta$ cancel it identically.

Thus, the total effective Hamiltonian (H) in the presence of a vector potential is now defined and it is for an X_3 type reactive system ($\theta_0 = 0$) given by

$$H_{\mathrm{scl(cl)}} = T_{\mathrm{scl(cl)}} + \frac{2\hbar^2}{\mu\rho^2 \cos^2\theta} - \frac{4\hbar P_\phi}{\mu\rho^2 \sin^2\theta} + V(\rho, \theta, \phi) \tag{36}$$

Thus the inclusion of the geometric phase in this case adds two terms to the Hamiltonian. The first is an "additional" potential term and the second has the effect that $2\hbar$ is added to P_γ in the Coriolis coupling term [see Eq. (35)].

1. *Quasiclassical Trajectory (QCT) Calculation on* $D + H_2$

The total effective Hamiltonian H, in the presence of a vector potential for an $A + B_2$ system is defined in Section II.B and the coupled first-order Hamilton equations of motion for all the coordinates are derived from the new effective Hamiltonian by the usual prescription [74], that is,

$$\begin{aligned} \dot{q}_i &= \frac{\partial H}{\partial p_i} \\ \dot{p}_i &= -\frac{\partial H}{\partial q_i} \end{aligned} \tag{37}$$

During initialization and final analysis of the QCT calculations, the numerical values of the Morse potential parameters that we have used are given as $D_e = 4.580$ eV, $r_e = 0.7416$ Å, and $\beta = 1.974$ Å^{-1}. Moreover, the potential energy as a function of internuclear distances obtained from the analytical expression (with the above parameters) and the LSTH [75,76] surface asymptotically agreed very well.

In the final analysis of the QCT calculations, j' is uniquely defined. By using the final coordinate (r') and the momentum (p'), the rotational angular momentum ($L = r' \times p'$) and j' [setting $L^2 = j'(j' + 1)\hbar^2$] can be determined. Once the rotational angular momentum (L) is obtained, we can find the vibrational energy ($E_{\mathrm{vib}} = E_{\mathrm{int}} - E_{\mathrm{rot}}$). From the vibrational energy, the final vibrational quantum number, v', is obtained using the expression of the energy levels of a Morse oscillator. However, at higher values of v' the energy level expression of the Morse oscillator may not be accurate and the following semiclassical formula based on the Bohr–Sommerfeld quantization

$$v' = -\frac{1}{2} + \frac{1}{h}\oint p_r\, dr \tag{38}$$

can be used instead. We have performed QCT calculations for obtaining integral cross-sections of the $D + H_2$ $(v = 1, j) \rightarrow DH(v', j') + H$ reaction at the total energy of 1.8 eV (translational energy 1.0 eV) with the LSTH [75,76] potential energy parameters. These studies have been done with or without inclusion of the geometric phase and starting from initial states $(v = 1, j = 1)$. For this case, 1.2×10^5 trajectories are taken noting that convergence was actually obtained with $\sim 5 \times 10^4$ trajectories. The distribution of integral cross-sections with $(\theta_0 = 11.5°)$ or without inclusion of the geometric phase as a function of j' $(v' = 1)$ has been shown in Figure 1 and compared with those QCT results obtained by using $\theta_0 = 0$.

In Figure 1, we see that there are relative shifts of the peak of the rotational distribution toward the left from $j' = 12$ to $j' = 8$ in the presence of the geometric phase. Thus, for the $D + H_2$ $(v = 1, j) \rightarrow DH$ $(v', j') + H$ reaction with the same total energy 1.8 eV, we find qualitatively the same effect as found quantum mechanically. Kuppermann and Wu [46] showed that the peak of the rotational state distribution moves toward the left in the presence of a geometric phase for the process $D + H_2$ $(v = 1, j = 1) \rightarrow DH\,(v' = 1, j') + H$. It is important to note the effect of the position of the conical intersection (θ_0) on the rotational distribution for the $D + H_2$ reaction. Although the absolute position of the peak (from $j' = 10$ to $j' = 8$) obtained from the quantum mechanical calculation is different from our results, it is worthwhile to see that the peak

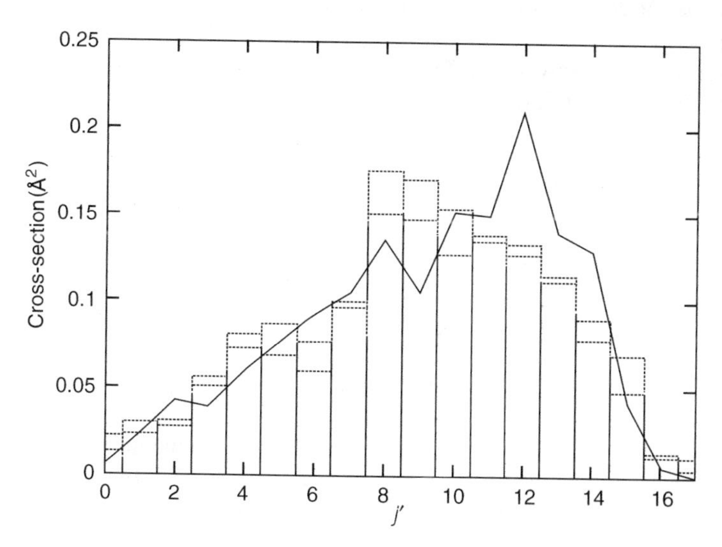

Figure 1. Quasiclassical cross-sections for the reaction $D + H_2$ $(v = 1, j = 1) \rightarrow DH$ $(v' = 1, j') + H$ at 1.8-eV total energy as a function of j'. The solid line indicates results obtained without including the geometric phase effect. Boxes show the results with the geometric phase included using either $\theta_0 = 0$ (dashed) or $\theta_0 = 11.5°$ (dotted).

position of the rotational distribution without a geometric phase effect using classical hyperspherical calculation comes at $j' = 12$ (the overestimation is due to the use of classical mechanics), which is different from the quantum calculation having the peak at $j' = 10$.

The relative shift of the peak position of the rotational distribution in the presence of a vector potential thus confirms the effect of the geometric phase for the $D + H_2$ system displaying conical intersections. The most important aspect of our calculation is that we can also see this effect by using classical mechanics and, with respect to the quantum mechanical calculation, the computer time is almost negligible in our calculation. This observation is important for heavier systems, where the quantum calculations are even more troublesome and where the use of classical mechanics is also more justified.

The effect of the GP is expected to be even more pronounced in differential cross-sections and the computation of differential cross-sections are again carried out by QCT calculations for the $D + H_2\,(v = 1, j = 1) \rightarrow DH\,(v' = 1, j') +$ H reaction at the total energy of 1.8 eV (initial kinetic energy 1.0 eV) with the London–Sato–Trūhlar–Horowitz (LSTH) [75,76] potential energy parameters. We calculated the scattering angle distributions for different final rotational states $(v' = 1, j')$ with or without inclusion of a geometric phase starting from the initial state $[(v = 1, j = 1)]$. The convergence of these distributions has appeared when there are a sufficient number of trajectories in each scattering angle. Nearly 1.0×10^6 number of trajectories have been computed to obtain converged distributions for all the final j' states.

The rotationally resolved differential cross-section are subsequently smoothened by the moments expansion (M) in cosines [77–79]:

$$
\begin{aligned}
\frac{d\sigma_R(j',\theta)}{d\omega} &= \frac{\sigma_R(j')}{4\pi}\left[1 + \sum_{k=1}^{M} c_k \cos\left(k\pi a(\theta)\right)\right] \\
c_k &= \frac{2}{N_R(j')}\sum_{s=1}^{N_R(j')} \cos\left(k\pi a(\theta_s)\right) \\
\sigma_R(j') &= \pi b_{\max}^2 N_R(j')/N \\
a(\theta) &= \frac{1}{2}(1 - \cos\theta)
\end{aligned}
\tag{39}
$$

where N is the total number of trajectories and $N_R(j')$ is the number of reactive trajectories leading to the DH(j') product. Also, θ is the scattering angle, s labels the reactive trajectories leading to the same product, and $b_{\max}$ is the impact parameter.

The calculations showed [54,55] significant effect of the GP on scattering angle resolved cross-sections for a particular final rotational state. It is interesting to see the change of these distributions due to the geometric phase compared to those obtained without the geometric phase. It appears that for lower final rotational states ($j' \leq 10$), scattering angle distributions in the presence of the geometric phase, have higher peaks compared to those without geometric-phase situations. Similarly, for higher final rotational states ($j' \geq 10$), nongeometric-phase cases have predominance over geometric-phase cases. Finally, there are crossings between these distributions at $j' = 10$. The rotationally resolved differential cross-section results as shown in [55] are quite expected when considering the integral cross-section distributions displayed in Figure 1. Kuppermann and Wu showed differential cross-sections at $E_{\text{tot}} = 1.8$ eV (initial kinetic energy 1.0 eV) for the $D + H_2$ ($v = 1, j = 1$) $\rightarrow$ $DH(v' = 1, j') + H$ reaction with or without considering the geometric phase. In their calculations, the differential cross-section distributions represented either with or without the geometric-phase cases have crossings at $j' = 8$, where for lower j' values the "with geometric phase" and for higher j' values the "without geometric phase" cases have predominance. Qualitatively, we have found the same features for differential cross-section distributions as they have obtained except that the crossing position is in our case $j' = 10$ as compared to theirs $j' = 8$. Again, this difference in crossing position comes about due to the use of classical mechanics. The scattering angle resolved differential cross-sections in the presence of a vector potential indicate and confirm the effect of the geometric phase in the $D + H_2$ reaction having a conical intersection. The fact that these effects can be seen using classical mechanics is the most important aspect of our calculations since the computational cost in this case is very small.

2. *Semiclassical Calculation on a $D + H_2$ Reaction*

Considering the semiclassical Hamiltonian from Eq. (28), one can expand the total wave function as,

$$\Psi(\rho, \theta, \phi, t) = \sum_k \psi_k(\theta, \phi, t)\Phi_k(\rho, t) \tag{40}$$

where ρ, θ, and ϕ are quantum degrees of freedom and $\Phi_k(\rho, t)$ are Hermite basis functions with expansion coefficients $\psi_k(\theta, \phi, t)$.

The Hermite basis functions $\Phi_k(\rho, t)$ have the following form:

$$\Phi_k(\rho, t) = \pi^{1/4} \exp\left(\frac{i}{\hbar}(\gamma(t) + P_\rho(t)(\rho - \rho(t)) + \text{Re}\, A(t)(\rho - \rho(t))^2)\right)\xi_k(x) \tag{41}$$

where

$$x = \sqrt{\frac{2\,\text{Im}\, A(t)}{\hbar}}(\rho - \rho(t))$$

and

$$\xi_k(x) = \frac{1}{\sqrt{k!2^k\sqrt{\pi}}} \exp(-x^2/2)H_k(x)$$

are the harmonic oscillator basis functions.

In this semiclassical calculation, we use only one wavepacket (the classical path limit), that is, a Gaussian wavepacket, rather than the general expansion of the total wave function. Equation (39) then takes the following form:

$$\Psi(\rho, \theta, \phi, t_0) = \psi_I(\theta, \phi, t_0)\Phi_{\text{GWP}}(\rho, t_0) \tag{42}$$

where $\Phi_{\text{GWP}}(\rho, t)$ is $\Phi_0(\rho, t)$ as defined in Eq. (39), and the expansion coefficient [80] is

$$\psi_I(\theta, \phi, t_0) = \rho\sqrt{\sin(-\eta)/\zeta g_v(\zeta)P_j(\cos\eta)} \tag{43}$$

where g_v and P_j are the Morse vibrational and normalized Legendre wave functions, respectively. The variables ζ and η can be expressed using the asymptotic representation of θ and ϕ,

$$\theta = \theta_0 + \zeta \sin\eta$$
$$\phi = \phi_0 + \zeta \cos\eta$$

The general hyperspherical formulation of the vector potential arising due to an arbitrary position of the conical intersection of the adiabatic potential energy hypersurfaces of an A + BC type reactive system has been formulated [54]. For the H_3 system, the location of the conical intersection is at $\theta_0 = \phi_0 = 0$ but for the $D + H_2$ system it is at $\phi_0 = 0$ and $\theta_0 = 11.5°$. As we wish to compare the results obtained by introducing a vector potential in the system Hamiltonian with those obtained by multiplying the wave function with a complex phase factor, we approximated the vector potential expression using $\theta_0 = 0$ and the corresponding extra terms are added to the Hamiltonian.

In hyperspherical coordinates, the wave function changes sign when ϕ is increased by 2π. Thus, the correct phase treatment of the ϕ coordinate can be obtained using a special technique [44–48] when the kinetic energy operators are evaluated: The wave function $f(\phi)$ is multiplied with $\exp(-i\phi/2)$, and after the forward FFT [69] the coefficients are multiplied with slightly different frequencies. Finally, after the backward FFT, the wave function is multiplied with $\exp(i\phi/2)$.

The kinetic energy operator evaluation and then the propagation of the θ, ϕ degrees of freedom have been performed using the FFT [69] method followed by the Lanczos iterative reduction technique [70] for the time propagation. In the classical path picture, the propagation of the ρ motion has some additional equations of motion for the width parameter $A(t)$. The variables γ and its conjugate momentum P_γ are propagated using classical equations of motion and a mean-field potential averaged over the ρ, θ, and ϕ dependence.

The energy and state resolved transition probabilities are the ratio of two quantities obtained by projecting the initial wave function on incoming plane waves (I) and the scattered wave function on outgoing plane waves (F)

$$P_{I\to F}(E) = \lim_{t\to\infty} \frac{k_F}{k_I} \frac{|\int\int d\theta\, d\phi \sum_k \int d\rho \exp(-ik_F\rho)\Phi_k(\rho, t)\psi_k(\theta, \phi, t)|^2}{|\int d\rho \exp(+ik_I\rho)\Phi_{\text{GWP}}(\rho, t_0)\psi_I(\theta, \phi, t_0)|^2} \tag{44}$$

where the total energy, $E = (\hbar^2 k_I^2/2\mu) + E_I = (\hbar^2 k_F^2/2\mu + E_F$, and μ is the reduced mass for the ρ motion.

The integral over ρ can be evaluated analytically due to use of a Hermite basis,

$$P_{I\to F}(E) = \frac{k_F}{k_i}\sqrt{\frac{g_F}{g_I}} \exp\{-g_F(P_\rho(t) - \hbar k_F)^2 + g_I(-P_\rho^0 + \hbar k_I)\}$$
$$\times \left| \sum_{k=0} \int\int d\theta\, d\phi \psi_k(\theta, \phi, t)\psi_I(\theta, \phi, t_0) \times (-1)^k \frac{H_k(\Delta)\exp(-ik\delta)}{\sqrt{k!2^k}} \right|^2 \tag{45}$$

where H_k is a Hermite polynomial, and

$$\delta = \arctan\left(\frac{\text{Im}\, A(t)}{\text{Re}\, A(t)}\right)$$
$$\Delta = \sqrt{g_F}(P_\rho(t) - \hbar k_F)$$
$$g_F = \frac{\text{Im}\, A(t)}{2\hbar|A(t)|^2}$$
$$g_I = \frac{\text{Im}\, A(t_0)^2}{2\hbar|A(t_0)|}$$

and t should be large enough for the interaction potential to vanish and g_F to approach a constant value.

With each random choice of γ and its conjugate momentum P_γ, one can have a separate trajectory with a different final wave function. After a series of calculations, the energy and state resolved cross-sections are obtained.

This semiclassical method, using one wavepacket only (GWP), has been applied for the reaction $D + H_2(v = 1, j = 1) \rightarrow DH\ (v' = 1, j') + H$ by using the LSTH potential energy surface [75,76], where in order to obtain integral cross-sections we have considered the total angular momentum vector J to be different from zero. We have performed all calculations with total energy 1.80 eV ($E_{tr} = 1.0$ eV) with or without introducing geometric phase effects. The transition probabilities as a function of total energy can be obtained by Eq. (43) for each trajectory and finally, series of trajectories can give state resolved cross-sections with good accuracy around the total energy 1.80 eV. The trajectories had randomly selected values of the total angular momentum in the range 0 to $J_{max} = 50$ in units of $\hbar$. The parameter P_γ as well as γ are also selected randomly. The propagation has been carried out with the initial values of width parameters, $\mathrm{Re}\,A(t_0) = 0$, $\mathrm{Im}\,A(t_0) = 0.5$ amu $\tau^{-1}(1\tau = 10^{-14}\,\mathrm{s})$ assuming that the quantum classical correlation will remain small during the entire collision, that is, the traditional classical path picture is valid [81].

In Figure 2, we present integral cross-sections as a function of rotational quantum number j', with or without including the geometric phase effect. Each calculation has been performed with a product-type wave function consisting of one wavepacket ($\Phi_0(\rho, t)$) and a grid size ($N_\theta \times N_\phi$) in (θ, ϕ) space equal to 256×64 has been used. Though in this result the peak of the rotational state distribution without including the geometric phase effect is at $j' = 8$ instead of

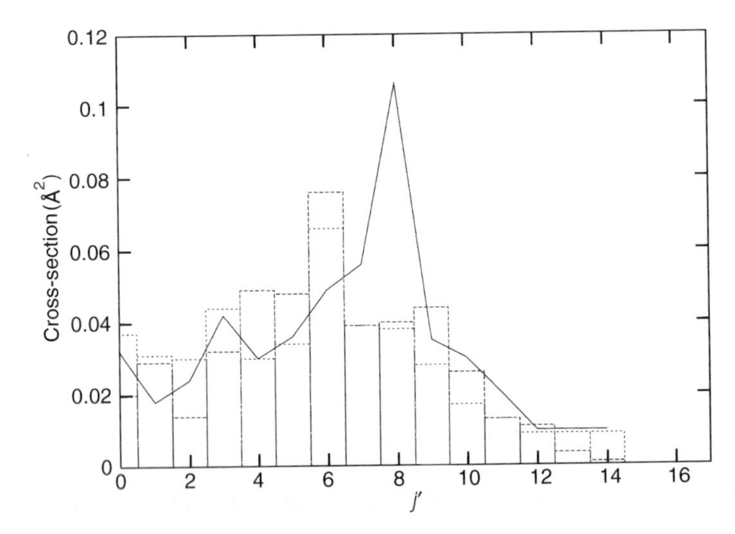

Figure 2. Quantum classical cross-sections for the reaction $D + H_2\ (v = 1, j = 1) \rightarrow DH\ (v' = 1, j') + H$ at 1.8-eV total energy as a function of j'. The solid line indicates results obtained without including the geometric phase effect. Boxes show the results with geometric phase effect included using either a complex phase factor (dashed) or a vector potential (dotted).

at $j' = 10$ (corresponding to accurate quantum mechanical calculation [46]), the result is still impressive because we have used only one basis function instead of a grid in the ρ coordinate. But the important point is that the peak position is again shifted to the left, from $j' = 8$–6 if the GP effect is considered either by including the vector potential into the system Hamiltonian or by incorporating a phase factor in the adiabatic nuclear wave function.

3. *Quasiclassical Trajectory Calculations on a* $H + D_2$ *Reaction at* 2.20 eV

Satisfactory agreement between experimentally measured and theoretically (without considering the GP effect) calculated results [80,82–87] for the reaction, $H + D_2$ $(v = 0, j = 0) \rightarrow HD(v', j') + D$, at a collisional energy of 2.20 eV has renewed theoretical interest in this area. As at this collisional energy, the CI is located at 2.7 eV, a significant contribution from the geometric phase is expected to appear. We studied the difference between results obtained with or without including the GP effect. We have calculated integral and differential cross-sections for the same reaction using the QCT approach with or without including the general expression of a vector potential into the system Hamiltonian. As we mentioned earlier, the simplest way of including the phase effect is to switch to hyperspherical coordinates, in which the HLH phase factor is $\exp(i\eta/2)$ where the hyperangle η increase by 2π as the conical intersection is encircled. When the nuclear kinetic energy operator operates on the wave function multiplied by the HLH phase factor, the Hamiltonian accumulate an additional potential (a vector potential). In this calculation, we wish to replace the quantum operators by classical variables. The reason for this is that the classical trajectories are easy to integrate to obtain reliable values of integral and differential cross-sections. In particular, our previous QCT calculations showed that the GP effect was predicted qualitatively correct. For each trajectory, the final vibrational quantum number (v') is calculated using the semiclassical formula based on the Bohr–Sommerfeld quantization rule,

$$v' = -\frac{1}{2} + \frac{1}{h}\oint p_r \, dr$$

Integral and differential cross-sections for the $H + D_2(v = 0, j = 0) \rightarrow DH$ $(v', j') + D$ reaction at total enery 2.3917 eV (collisional energy 2.20 eV) are computed by using QCTs on the LSTH potential energy surface and these calculations have been performed with or without including a vector potential into the system Hamiltonian. For each case (with or without GP) $\sim 1.2 \times 10^5$ QCTs are computed to get the product rotational state distributions of the final vibrational state (v') although convergence is nearly obtained with 5×10^4 QCTs. The scattering angle distributions for different final rotational state (j') are calculated from 1.0×10^6 QCTs for each case (with or without GP).

Differential cross-sections for particular final rotational states (j') of a particular vibrational state (v') are usually smoothened by the moment expansion (M) in cosine functions mentioned in Eq. (38). Rotational state distributions for the final vibrational state $v' = 0$ and 1 are presented in [88]. In each case, with or without GP results are shown. The peak position of the rotational state distribution for $v' = 0$ is slightly left shifted due to the GP effect, on the contrary for $v' = 1$, these peaks are at the same position. But both these figures clearly indicate that the absolute numbers in each case (with or without GP) are different.

We have also presented scattering angle distributions for $v' = 0$, $j' = 0$–12 and $v' = 1$, $j' = 0$–12 in [88] where in each figure results obtained with or without considering GP effect are shown. These figures clearly demonstrate that the differential cross-section as a function of scattering angle for with or without GP are rather different.

III. THE EXTENDED BORN–OPPENHEIMER APPROXIMATION

The BO coupled equations in the adiabatic representation (see Appendix B) are

$$-\frac{\hbar^2}{2m}\nabla^2\psi_j(n) + (u_j(n) - E)\psi_j(n) - \frac{\hbar^2}{2m}\sum_{i=1}^{N}\{2\tau_{ji}^{(1)}\nabla\psi_i(n) + \tau_{ji}^{(2)}\psi_i(n)\} = 0 \tag{46}$$

where $\psi_j(n)$ and $u_j(n)$, $j = 1, \ldots, N$ are the nuclear wave functions and the adiabatic potential energy surfaces, ∇ is the gradient (vector) operator, m is the reduced mass of the system, $\tau^{(1)}$ is the non-adiabatic vector matrix, and $\tau^{(2)}$ is non-adiabatic scalar matrix. Recalling their relation from Appendix B, Eq. (45), they can be written in the following matrix notation:

$$-\frac{\hbar^2}{2m}\nabla^2\Psi + \left[u - \frac{\hbar^2}{2m}\tau^2 - E\right]\Psi - \frac{\hbar^2}{2m}(2\tau\cdot\nabla + \nabla\tau)\Psi = 0 \tag{47}$$

where Ψ is a column matrix that contains the nuclear wave functions ψ_j, u is a diagonal potential (adiabatic) matrix, the dot product designates a scalar product, and τ replaces $\tau^{(1)}$ to simplify the notation.

If we consider the transformation $\Psi = A\Phi$, then Eq. (46) can be transformed into the following diabatic matrix equation:

$$-\frac{\hbar^2}{2m}\nabla^2\Phi + (\mathbf{A}^{-1}u\mathbf{A} - E)\Phi = 0 \tag{48}$$

where the transformation matrix (unitary) **A** has to satisfy the following matrix equation:

$$\nabla \mathbf{A} + \tau \mathbf{A} = 0 \tag{49}$$

and we are interested in exploring the detailed properties of the transformation matrix **A** when it satisfies Eq. (48).

As stated in the introduction, we present the derivation of an extended BO approximate equation for a Hilbert space of arbitary dimensions, for a situation where all the surfaces including the ground-state surface, have a degeneracy along a single line (e.g., a conical intersection) with the excited states. In a two-state problem, this kind of derivation can be done with an arbitary $\boldsymbol{\tau}$ matrix. On the contrary, such derivation for an $N \geq 2$ dimensional case has been performed with some limits to the elements of the $\boldsymbol{\tau}$ matrix. Hence, in this sence the present derivation is not general but hoped that with some additional assumptions it will be applicable for more general cases.

The $\boldsymbol{\tau}$ matrix is an antisymmetric vector matrix with the component $\boldsymbol{\tau}_p, p = x, y, z, X, Y, Z$, and so on, and $\boldsymbol{\tau}_p$ is assumed to be a product of a scalar <u>function</u> t_p and a <u>constant</u> antisymmetric matrix **g** (which does not depend on p). Thus,

$$\begin{aligned} \tau_p &= t_p g \\ t_p g_{ji} &= \langle \xi_j | \nabla \xi_i \rangle \end{aligned} \tag{50}$$

If we consider **G** as a unitary transformation matrix that diagonalizes the **g** matrix and $\mathbf{i\omega}$ is the diagonal matrix with elements $\mathbf{i\omega}_j$, $j = 1, \ldots, N$ as the corresponding eigenvalues, it can be shown that, following the unitary transformation performed with **G**, Eq. (46) becomes

$$-\frac{\hbar^2}{2m}(\nabla + it\omega)^2\chi + (\mathbf{W} - E)\chi = 0 \tag{51}$$

where χ is related to Ψ through the transformation $\Psi = G\chi$ and the nondiagonal diabatic potential matrix **W** is related to the adiabatic potential matrix **u** as $\mathbf{W} = \mathbf{G}^{\dagger}\mathbf{u}\mathbf{G}$. Due to the above transformation, the non-adiabatic coupling matrix $\boldsymbol{\tau}$ becomes a diagonal matrix $\boldsymbol{\omega}$ and a new off-diagonal potential matrix is formed that couples the various differential equations. It is important to note that so far the derivation is rigorous and no approximations have been imposed. Hence, the solution of Eq. (46) will be the same as the solution of Eq. (50), but it will be convenient to impose the BO approximation in Eq. (50). For low enough energies, all upper adiabatic states are assumed to be classically closed, that is, each of the corresponding adiabatic functions ψ_j, $j = 2, \ldots, N$ is expected to fulfill the condition

$$|\psi_1| \gg |\psi_j|; \qquad j = 2, \ldots, N \tag{52}$$

in those regions of configuration space (CS) where the lower surface is energetically allowed. This assumption has to be employed with great care and is found nicely fulfilled for two- or three-state systems although some risk is involved by extending this assumption to an arbitrary number of states. We can analyze the product $W\chi$ for the jth equation,

$$\begin{aligned}(W\chi)_j &= \{(G^\star uG)(G^\star\Psi)\}_j = (G^\star u\Psi)_j = \sum_{k=1}^{N} G^\star_{jk} u_k \Psi_k \\ &= u_1\chi_j - u_1 \sum_{k=1}^{N} G^\star_{jk}\Psi_k + \sum_{k=1}^{N} G^\star_{jk} u_k \Psi_k \\ &= u_1\chi_j + \sum_{k=2}^{N} G^\star_{jk}(u_j - u_1)\psi_k, \qquad j = 1, \ldots, N\end{aligned} \tag{53}$$

By substituting Eq. (52) in Eq. (50) and introducing the approximation

$$-\frac{\hbar^2}{2m}(\nabla + it\omega_j)^2\chi_j + (u_1 - E)\chi_j = 0, \qquad j = 1, \ldots, N \tag{54}$$

the N equations for the $N\chi$ functions are uncoupled and each equation stands on its own and can be solved independently. These equations are solved for the same adiabatic PES u_1 but for different ω_js.

Now, we assume that the functions, $t\omega_j$, $j = 1, \ldots, N$ are such that these uncoupled equations are gauge invariant, so that the various χ values, if calculated within the same boundary conditions, are all identical. Again, in order to determine the boundary conditions of the χ function so as to solve Eq. (53), we need to impose boundary conditions on the Ψ functions. We assume that at the given (initial) asymptote all ψ^i_i values are zero except for the ground-state function ψ^i_1 and for a low enough energy process, we introduce the approximation that the upper electronic states are closed, hence all final wave functions ψ^f_i are zero except the ground-state function ψ^f_1.

Hence, in order to contruct extended BO approximated equations for an N-state coupled BO system that takes into account the non-adiabatic coupling terms, we have to solve N uncoupled differential equations, all related to the electronic ground state but with different eigenvalues of the non-adiabatic coupling matrix. These uncoupled equations can yield meaningful physical solutions only when the eigenvalues of the **g** matrix fulfill certain requirement. For example, these eigenvalues produce gauge invariant equations, that is, its solution will be compatible with the assumption concerning the BO approximation.

A. The Quantization of the Non-Adiabatic Coupling Matrix Along a Closed Path

In this section, we prove that the non-adiabatic matrices have to be quantized (similar to Bohr–Sommerfeld quantization of the angular momentum) in order to yield a continous, uniquely defined, diabatic potential matrix $\mathbf{W}(s)$. In another way, the extended BO approximation will be applied only to those cases that fulfill these quantization rules. The ADT matrix $\mathbf{A}(s, s_0)$ transforms a given adiabatic potential matrix $\mathbf{u}(s)$ to a diabatic matrix $\mathbf{W}(s, s_0)$

$$\mathbf{W}(s, s_0) = \mathbf{A}^{\star}(s, s_0)\mathbf{u}(s)\mathbf{A}(s, s_0) \tag{55}$$

$\mathbf{A}^{\star}(s, s_0)$ is the complex conjugate matrix of $\mathbf{A}(s, s_0)$, s_0 is an initial point in CS, and s is another point. It is assumed that $\mathbf{W}(s, s_0)$ and $\mathbf{u}(s, s_0)$ are uniquely defined throughout the CS and to ensure the uniqueness of $\mathbf{W}(s, s_0)$ our aim is to derive the features to be fulfilled by the $\mathbf{A}(s, s_0)$.

We introduce a closed-path Γ defined by a parameter λ. At the starting point s_0, $\lambda = 0$ and when the path complete a full cycle, $\lambda = \beta(2\pi$, in case of circle).

We now express our assumption regarding the uniqueness of $\mathbf{W}(s, s_0)$ in the following way:

$$W(\lambda = 0) = W(\lambda = \beta) \tag{56}$$

By using Eq. (54), we can rewrite Eq. (55) as

$$\mathbf{A}^{\star}(0)\mathbf{u}(0)\mathbf{A}(0) = \mathbf{A}^{\star}(\beta)\mathbf{u}(\beta)\mathbf{A}(\beta) \tag{57}$$

Hence, $\mathbf{u}(\beta)$ and $\mathbf{u}(0)$ are connected as below

$$\mathbf{u}(\beta) = \mathbf{D}u(0)\mathbf{D}^{\star} \qquad \mathbf{D} = \mathbf{A}(\beta)\mathbf{A}^{\star}(0) \tag{58}$$

The **D** matrix is by definition a unitary matrix (it is product of two unitary matrices) and since the adiabatic eigenvalues are uniquely defined in CS, we have, $u(0) \equiv u(\beta)$. Then, Eq. (57) can be written as

$$\mathbf{u}(0) = \mathbf{D}\mathbf{u}(0)\mathbf{D}^{\star} \tag{59}$$

By performing the matrix multiplication, one can get the following relations between the adiabatic eigenvalues $u_j(0)$ and the $\mathbf{D}$ matrix elements

$$\sum_{j=1}(\mathbf{D}^{\star}_{kj}\mathbf{D}_{kj} - \delta_{kj})u_j(0) = 0 \qquad k = 1, \ldots, N \tag{60}$$

Equation (59) is valid for every arbitary point in CS and for an arbitary set of nonzero adiabatic eigenvalues, $u_j(0)$, $j = 1, \ldots, N$, hence the $\mathbf{D}$ matrix elements fulfill the relation

$$(\mathbf{D}_{jk})^{\star}\mathbf{D}_{jk} = \delta_{jk} \qquad j, k = 1, \ldots, N \tag{61}$$

Thus $\mathbf{D}$ is a diagonal matrix that contains diagonal complex numbers whose norm is 1. By recalling Eq. (57), we get

$$\mathbf{A}(\beta) = \mathbf{DA}(0) \tag{62}$$

Again, we already know that the ADT becomes possible only when the transformation matrix $\mathbf{A}$ satisfy Eq. (63)

$$\nabla\mathbf{A} + \boldsymbol{\tau}\mathbf{A} = 0 \tag{63}$$

where $\boldsymbol{\tau}$ is the non-adiabatic coupling matrix. A uniquely defined $\mathbf{A}$ matrix will be guaranteed if and only if the elements of the $\boldsymbol{\tau}$ matrix are regular functions of the nuclear coordinates at every point in CS.

However, in order to obtain a uniquely defined diabatic potential matrix, it is not necessary for the $\mathbf{A}$ matrix to be uniquely defined throughout CS. Still, we ignore this difficulty and go ahead to derive $\mathbf{A}$ by a direct integration of Eq. (62),

$$\mathbf{A}(s) = \exp\left[-\int_{s_0}^{s} \mathbf{ds} \cdot \tau\right]\mathbf{A}(s_0) \tag{64}$$

where the integration is performed along a closed-path Γ that combines s and s_0, $\mathbf{ds}$ is a differential vector length element along this path, and the dot stands for a scalar product. We already define the matrix $\mathbf{G}$ as the unitary transformation matrix that diagonalizes the $\boldsymbol{\tau}$ matrix,

$$\mathbf{A}(s) = \mathbf{G}\exp\left[-i\omega\int_{s_0}^{s} ds \cdot t(s)\right]\mathbf{G}^{\star}A(s_0) \tag{65}$$

Hence, the matrix $\mathbf{D}$ along a path Γ takes the following form:

$$\mathbf{D} = \mathbf{G} \exp\left(-i\omega \oint_{\Gamma} ds \cdot t(s) \right) \mathbf{G}^{\star} \tag{66}$$

As the $\mathbf{D}$ matrix is a diagonal matrix with a complex number of norm 1, the exponent of Eq. (65) has to fulfill the following quantization rule:

$$\frac{1}{2\pi} \omega_j \int_{\tau} ds \cdot t(s) = n_j \qquad j = 1, \ldots, N \tag{67}$$

where n_j is an integer and if the $\mathbf{D}$ matrix is multiplied by (-1) the values of all n_j parameters have to be one-half of an odd integer. This fact is the necessary conditions for Eq. (53) to be gauge invariant or this quantization requirement that is a necessary condition for having uniquely defined diabatic potentials also guarantees the extended BO equation. Thus, the effect of non-adiabatic coupling terms lead to a extended BO approximation.

B. The Quantization of the Three-State Non-Adiabatic Coupling Matrix

We concentrate on an adiabatic tri-state model in order to derive the quantization conditions to be fulfilled by the eigenvalues of the non-adiabatic coupling matrix and finally present the extended BO equation. The starting point is the 3×3 non-adiabatic coupling matrix,

$$\boldsymbol{\tau} = \begin{pmatrix} 0 & t_1 & t_2 \\ -t_1 & 0 & t_3 \\ -t_2 & t_3 & 0 \end{pmatrix} \tag{68}$$

where $t_j, j = 1, 2, 3$ are arbitary functions of the nuclear coordinates. The matrix $\mathbf{G}$ diagonalizes $\boldsymbol{\tau}$ at a given point in CS

$$\mathbf{G} = \frac{1}{\tilde{\omega}\lambda\sqrt{2}} \begin{pmatrix} it_2\tilde{\omega} - t_3t_1 & -it_2\tilde{\omega} - t_3t_1 & t_3\lambda\sqrt{2} \\ it_3\tilde{\omega} + t_2t_1 & -it_3\tilde{\omega} + t_2t_1 & -t_2\lambda\sqrt{2} \\ \lambda^2 & \lambda^2 & t_1\lambda\sqrt{(2)} \end{pmatrix} \tag{69}$$

where $\lambda = \sqrt{t_2^2 + t_3^2}$, $\tilde{\omega} = \sqrt{t_1^2 + t_2^2 + t_3^2}$, and the three eigenvalues $(0, \pm i\tilde{\omega})$.

We already assume that the $\boldsymbol{\tau}$ matrix fulfills the conditions in Eqs. (48) and (49). These conditions ensures that the matrix $\mathbf{G}$ diagonalizes $\boldsymbol{\tau}(s)$ along a close path is independent of s and by employing Eq. (65), we obtain the **D** matrix,

$$\mathbf{D} = \tilde{\omega}^{-2} \begin{pmatrix} t_3^2 + (t_1^2 + t_2^2)C_1 & t_1\tilde{\omega}S_1 - 2t_2t_3S_2 & -\tilde{\omega}t_2S_1 + 2t_1t_3S_2 \\ t_1\tilde{\omega}S_1 - 2t_2t_3S_2 & t_2^2 + (t_1^2 + t_3^2)C_1 & -t_3\tilde{\omega}S_1 + 2t_1t_2C_2 \\ \tilde{\omega}t_2S_1 + 2t_1t_3S_2 & t_3\tilde{\omega}S_1 + 2t_1t_2C_2 & t_1^2 + (t_2^2 + t_3^2)C_1 \end{pmatrix} \tag{70}$$

where $S_1 = \sin(\oint \tilde{\omega} \cdot ds)$; $C_1 = \cos(\oint \tilde{\omega} \cdot ds)$; $S_2 = \sin^2(\frac{1}{2}(\oint \tilde{\omega} \cdot ds))$; $C_2 = \cos^2(\frac{1}{2}(\oint \tilde{\omega} \cdot ds))$.

As the **D** matrix has to be a unit matrix in order to get a continuous, uniquely defined diabatic matrix, the following integral is quantized as:

$$\frac{1}{2\pi}\oint \tilde{\omega} \cdot ds = \frac{1}{2\pi}\oint \sqrt{t_1^2 + t_2^2 + t_3^2} \cdot ds = n \tag{71}$$

Thus from the **D** matrix, it is easy to say that for three states n will be an integer and for two states n will be one-half of an odd integer.

C. The Study of the Three-State System

The numerical calculations have been done on a two-coordinate system with q being a radial coordinate and ϕ the polar coordinate. We consider a 3×3 non-adiabatic (vector) matrix $\boldsymbol{\tau}$ in which τ_q and τ_ϕ are two components. If we assume $\tau_q = 0$, τ_ϕ takes the following form,

$$\tau_\phi = t_\phi \mathbf{g} = \frac{t_0}{q}\mathbf{g} \tag{72}$$

where t_0 is a constant and $\mathbf{g}$ is a 3×3 matrix of the form

$$\mathbf{g} = \begin{pmatrix} 0 & 1 & 0 \\ -1 & 0 & \eta \\ 0 & -\eta & 0 \end{pmatrix} \tag{73}$$

where η is a constant. The $\boldsymbol{\tau}$ matrix couples the ground adiabatic state to the first excited state and then the first excited state to the second excited state. There is no direct coupling between the ground and the second excited state.

The adiabatic coupled SE for the above 3×3 non-adiabatic coupling matrix are

$$\begin{aligned} \left(T + u_1 + \frac{t_0^2}{2mq^2} - E\right)\psi_1 + \frac{t_0}{mq}\frac{\partial}{\partial\phi}\psi_2 - \frac{\eta t_0^2}{2mq^2}\psi_3 &= 0 \\ \left(T + u_2 + \frac{t_0^2(1+\eta^2)}{2mq^2} - E\right)\psi_2 - \frac{t_0}{mq}\frac{\partial}{\partial\phi}\psi_1 + \frac{\eta t_0}{mq}\frac{\partial}{\partial\phi}\psi_3 &= 0 \\ (T + u_3 + \frac{\eta^2 t_0^2}{2mq^2} - E)\psi_3 - \frac{\eta t_0}{mq}\frac{\partial}{\partial\phi}\psi_2 - \frac{\eta t_0^2}{2mq^2}\psi_1 &= 0 \end{aligned} \tag{74}$$

where T is the nuclear kinetic energy operator

$$T = -\frac{1}{2m}\left(\frac{\partial^2}{\partial q^2} + \frac{1}{q}\frac{\partial}{\partial q} + \frac{1}{q^2}\frac{\partial^2}{\partial \phi^2}\right) \tag{75}$$

In the case of a coupled system of three adiabatic equations, η (which is a constant) is chosen such that the quantization condition is fulfilled. Inserting the following values for t_j, $j = 1, 2, 3$: $t_1 = 1/2q$, $t_2 = 0$, and $t_3 = \eta/2q$, we get the following η value:

$$\eta = \sqrt{4n^2 - 1} \qquad \text{for} \qquad n = 1 \rightarrow \eta = \sqrt{3} \tag{76}$$

Now, we are in a position to present the relevant extended approximate BO equation. For this purpose, we consider the set of uncoupled equations as presented in Eq. (53) for the $N = 3$ case. The function $i\omega_j$ that appears in these equations are the eigenvalues of the $\mathbf{g}$ matrix and these are $\omega_1 = 2$; $\omega_2 = -2$, and $\omega_3 = 0$. In this three-state problem, the first two PESs are u_1 and u_2 as given in Eq. (6) and the third surface u_3 is chosen to be similar to u_2 but with $D_3 = 10$ eV. These PESs describe a two arrangement channel system, the reagent-arrangement defined for $R \rightarrow \infty$ and a product—arrangement defined for $R \rightarrow -\infty$.

D. Results and Discussion

We present state-to-state transition probabilities on the ground adiabatic state where calculations were performed by using the extended BO equation for the $N = 3$ case and a time-dependent wave-packet approach. We have already discussed this approach in the $N = 2$ case. Here, we have shown results at four energies and all of them are far below the point of CI, that is, $E = 3.0$ eV.

In [66], we have reported inelastic and reactive transition probabilities. Here, we only present the reactive case. Five different types of probabilities will be shown for each transition: (a) Probabilities due to a full tri-state calculation carried out within the diabatic representation; (b) Probabilities due to a two-state calculation (for which $\eta = 0$) performed within the diabatic representation; (c) Probabilities due to a single-state extended BO equation for the $N = 3$ case ($\omega_j = 2$); (d) Probabilities due to a single-state extended BO equation for the $N = 2$ case ($\omega_j = 1$); (e) Probabilities due to a single-state ordinary BO equation when $\omega_j = 0$.

At this stage, we would like to mention that the model, without the vector potential, is constructed in such a way that it obeys certain selection rules, namely, only the even $\rightarrow$ even and the odd $\rightarrow$ odd transitions are allowed. Thus any deviation in the results from these selection rules will be interpreted as a symmetry change due to non-adiabatic effects from upper electronic states.

TABLE II
Reactive State-to-State Transition Probabilities when Calculations are Performed Keeping the Position of the Conical Intersection at the Origin of the Coordinate System

E (eV)	$0 \rightarrow 0$	$0 \rightarrow 1$	$0 \rightarrow 2$	$0 \rightarrow 3$	$0 \rightarrow 4$	$0 \rightarrow 5$	$0 \rightarrow 6$	$0 \rightarrow 7$	$0 \rightarrow 8$	$0 \rightarrow 9$
1.0	0.0044 [a]	0.0000	0.0063	0.0000						
	0.0000 [b]	0.0049	0.0000	0.0079						
	0.0047 [c]	0.0000	0.0195	0.0000						
	0.0000 [d]	0.0045	0.0000	0.0080						
	0.0094 [e]	0.0000	0.0362	0.0000						
1.5	0.0325	0.0000	0.0592	0.0000	0.0311	0.0000				
	0.0000	0.1068	0.0000	0.0256	0.0000	0.0068				
	0.0419	0.0000	0.0648	0.0000	0.0308	0.0000				
	0.0000	0.1078	0.0000	0.0248	0.0000	0.0075				
	0.0644	0.0000	0.0612	0.0000	0.0328	0.0000				
2.0	0.1110	0.0000	0.0279	0.0000	0.0319	0.0000	0.2177	0.0000		
	0.0000	0.1232	0.0000	0.0333	0.0000	0.0633	0.0000	0.1675		
	0.1068	0.0000	0.0172	0.0000	0.0274	0.0000	0.2277	0.0000		
	0.0000	0.1264	0.0000	0.0353	0.0000	0.0656	0.0000	0.1678		
	0.1351	0.0000	0.0217	0.0000	0.0304	0.0000	0.2647	0.0000		
2.5	0.1318	0.0000	0.0295	0.0000	0.0091	0.0000	0.1375	0.0000	0.2043	0.0000
	0.0000	0.0936	0.0000	0.0698	0.0000	0.1350	0.0000	0.0200	0.0000	0.2398
	0.1256	0.0000	0.0155	0.0000	0.0084	0.0000	0.1545	0.0000	0.1977	0.0000
	0.0000	0.0947	0.0000	0.0658	0.0000	0.1363	0.0000	0.0190	0.0000	0.2365
	0.1831	0.0000	0.0343	0.0000	0.0089	0.0000	0.1607	0.0000	0.1157	0.0000

[a] Tri-surface calculation.
[b] Two-surface calculation.
[c] Single-surface calculation ($\omega = 2$).
[d] Single-surface calculation ($\omega = 1$).
[e] Single-surface calculation ($\omega = 0$).

Effects due to the non-adiabatic coupling terms on reactive transition probabilities are given in Table II. The two-state results and the corresponding extended approximated BO equation results follow the odd $\rightarrow$ even selection rules instead of even $\rightarrow$ even or odd $\rightarrow$ odd transitions in case of an ordinary BO scheme. This symmetry change has been discussed at length in Section II.A.2. The more interesting results are those for the tri-state case that apparently does not show any GP effect. Diabatic calculations, extended, and ordinary adiabatic BO calculations show the same selection rules. We thought that the extended BO equation could be partially wrong and the GP effects would become apparent but they did not. The present calculation reveals two points: (1) That geometrical features do not necessarily show up where they are expected as in the present tri-state case. (2) The extended approximated BO equation contains the correct information regarding the geometric effects. So, due to the conical intersection, in the two-state case, it contains the GP effects, whereas in tri-state case it tells us that such effects do not exist.

IV. QUANTUM DRESSED CLASSICAL MECHANICS

It is possible to parametarize the time-dependent Schrödinger equation in such a fashion that the equations of motion for the parameters appear as classical equations of motion, however, with a potential that is in principle more general than that used in ordinary Newtonian mechanics. However, it is important that the method is still exact and general even if the trajectories are propagated by using the ordinary classical mechanical equations of motion.

Thus it is possible to obtain a very convenient formulation, which is appealing from a computational point of view and allows the blending of classical and quantum concepts in a new way, by a selection of the initial time-dependent variables as in ordinary classical mechanics and an application of Newtons mechanics for the propagation of these parameters. Thus the classical mechanical part of the problem can, for example, be used to decide on the branching ratio in a chemical reaction, whereas the quantum mechanical part, which consist of grid points with quantum amplitudes, is used to project onto asymptotic wave functions of the product channels. In this fashion, we avoid describing the whole of space quantum mechanically at the same time, but only locally around the classical trajectories. The consequence is a large saving in the number of grid points and since it is also possible to minimize the computing effort when propagating the equations of motion, the final theory is not only easy to program, it is also efficient from a numerical point of view.

A. Theory

We directly give the relevant equations of motion for the simplest but nevertheless completely general scheme that involves propagation of grid points in a discrete variable representation (DVR) of the wave function. The grid points are propagated by classical equations of motion in a so-called fixed width approach for the basis set. For a derivation of these equations the reader is referred to [81,89,90]. As mentioned, the theory generates classical equations of motion for the center of the basis set or in the DVR representation the center of the DVR grid points. Thus, the grid points follow the classical equations of motion in space and if an odd number of grid points is used the middle one is the classical trajectory. For a one-dimensional (1D) problem we therefore have the following equations of motion:

$$\dot{x}(t) = p_x(t)/m \tag{77}$$

$$\dot{p}_x(t) = -\left.\frac{dV(x)}{dx}\right|_{x=x(t)} \tag{78}$$

defining the trajectory. For the quantum amplitudes, we have the matrix equation

$$i\hbar\dot{\mathbf{d}}(t) = (\mathbf{W}(t) + \mathbf{K})\mathbf{d}(t) \tag{79}$$

where **W** is a diagonal matrix and **K** the "kinetic coupling" matrix. The elements of the kinetic matrix is for a 1D system given as

$$\mathbf{K}_{ij} = \frac{\hbar\alpha_0}{m}\sum_n \tilde{\phi}_n(z_i)(2n+1)\tilde{\phi}_n(z_j) \tag{80}$$

where m is the mass associated with the x degree of freedom and α_0 is the imaginary part of the width parameter, that is, $\alpha_0 = \mathrm{Im}\,A$ of the Gauss–Hermite (G–H) basis set [81]. Since the kinetic operators have already worked on the basis functions before the DVR is introduced, the above matrix is what is left of the kinetic coupling.

We also notice that in coordinates weighted by $\sqrt{\alpha_0/m}$ the kinetic matrix is universal, that is, independent of the system.

The zeros of the Nth Hermite polynomial are denoted z_i and

$$\tilde{\phi}_n(z_i) = \phi_n(z_i)/\sqrt{A_i} \tag{81}$$

$$A_i = \sum_n \phi_n(z_i)^2 \tag{82}$$

where

$$\phi_n(z) = \frac{1}{\sqrt{2^n n!\sqrt{\pi}}}\exp\left(-\frac{1}{2}z^2\right)H_n(z) \tag{83}$$

The elements of the diagonal matrix **W** are given as

$$\mathbf{W}(x_i) = V(x_i) - V(x(t)) - \left.\frac{dV}{dx}\right|_{x=x(t)}(x_i - x(t)) - \frac{2\alpha_0^2}{m}(x_i - x(t))^2 \tag{84}$$

that is, the actual potential $V(x)$ from which a "reference" potential defined by the forces evaluated at the trajectory is subtracted. In the fixed width approach, the second derivative term V'' is related to the imaginary part of the width, that is, by the equation $V'' = 4\,\mathrm{Im}\,A^2/m$. This relation secures that $\mathrm{Im}\,A(t) = \text{constant}$ if $\mathrm{Re}\,A(t_0) = 0$. In the simplest possible approach, the first derivative is furthermore taken as the classical force in the sense of Newton.

But we emphasize that more general forces may be applied [81,91]. The grid points follow the trajectory and are defined through

$$x_i = x(t) + \sqrt{\alpha_0/2\hbar}z_i \tag{85}$$

For an atom–diatom collision, it is convenient for the formulation of the time dependent Gauss–Hermite (TDGH) discrete variable representation (DVR) theory to use Cartesian coordinates. That is, the center-of-mass distance $\mathbf{R} = (X, Y, Z)$ and the three coordinates for the orientation of the diatomic molecule $\mathbf{r} = (x, y, z)$ in a space-fixed coordinate with origo in the center of mass of the diatomic molecules. Thus the dimension of the grid is 6 and will be denoted $(n_X, n_Y, n_Z, n_x, n_y, n_z)$, where n_i is the number of grid points in degrees of freedom i. Note that in this approach $n_i = 1$ is an acceptable number of grid points (the classical limit). The dimension of the quantum problem is then $\Pi_{i=1}^{6} n_i$. But since one grid point in each mode makes sense from a dynamical point of view it is possible to explore the simplest quantum corrections to the classical limit, namely, the corrections obtained by adding grid points in each dimension.

The initial amplitudes $d_i(t_0)$ are obtained by projecting the initial wave function on the DVR basis set. For the initial wave function, we use

$$\Psi(X, Y, Z, x, y, z) \sim \frac{1}{R}\Phi_{\mathrm{GWP}}(R)\frac{1}{r}g_n(r)Y_{jm}(\theta, \phi) \tag{86}$$

where $\Phi_{\mathrm{GWP}}(R)$ is a Gaussian wavepacket in R, $g_n(r)$ a Morse vibrational wave function, and Y_{jm} a spherical harmonics for the diatomic molecule. The GWP is projected on planewave functions $\exp(ikR)$ when energy is resolving the wavepacket.

We can pick the initial random variables for the classical coordinates and momenta in the way it is done in an ordinary classical trajectory program.

The projection on the final channel is done in the following manner. We let the trajectory decide on the channel—just as in an ordinary classical trajectory program. Once the channel is determined we project the wave function (in the DVR representation) on the appropriate wave function for that channel

$$\frac{1}{R'}\exp(ik'R')\frac{1}{r'}g_{n'}(r')Y_{j'm'}(\theta', \phi') \tag{87}$$

where R' is the center-of-mass distance between A and BC, B and AC, or C and AB according to the channel specification. Likewise r', θ', and ϕ' specify the orientation of the diatom in the reactive channel found by the trajectory. This projection determines the final state $(n'j'm')$ distribution and the amplitudes therefore. The final probability distribution is added for all the trajectories of the channel and normalized with the classical total reactive cross-section of that channel to get the cross-section.

B. The Geometric Phase Effect

As demonstrated in [53] it is convenient to incorporate the geometrical phase effect by adding the vector potential in hyperspherical coordinates. Thus we found that the vector potential gave three terms, the first of which was zero, the second is just a potential term

$$V_a = \frac{2\hbar^2}{\mu\rho^2 \sin^2\theta} \tag{88}$$

and the third term, V_b, contains first derivative operators. By adding these terms to the normal Hamiltonian operator, we can incorporate the geometric phase effect.

We can express V_b as

$$V_b = -\frac{\hbar}{\mu}\sum_i \frac{\partial\eta}{\partial X_i}\hat{P}_{X_i} \tag{89}$$

where $\mu = \sqrt{m_1 m_2 m_3}/(m_1 + m_2 + m_3)$, $\eta = \phi/2$, and $\partial\phi/\partial X_i$ is given in Appendix C.

In order to incorporate the geometric phase effect in a formulation based on an expansion in G–H basis functions we need to consider the operation of the momentum operator on a basis function, that is, to evaluate terms as

$$\frac{\hbar}{i}\frac{\partial}{\partial x}\pi^{1/4}\exp\left(\frac{i}{\hbar}(\gamma(t) + p_x(t)(x - x(t)) + \mathrm{Re}\,A_x(t)(x - x(t))^2)\right)\phi_n(x,t) \tag{90}$$

Since we will normally use the fixed-width approach we can simplify the calculation by using $\mathrm{Re}\,A(t) = 0$. Thus we have

$$\begin{aligned}&(2\,\mathrm{Im}\,A(t)/\hbar)^{1/4}\exp(ip_x(t)(x - x(t))/\hbar)(p_x(t)\phi_n(x,t)\\ &\quad + (\hbar/i)\sqrt{\mathrm{Im}\,A(t)/\hbar}(\sqrt{n}\phi_{n-1} - \sqrt{n+1}\phi_{n+1})\end{aligned} \tag{91}$$

where we have used

$$\mathrm{Im}\,\gamma(t) = -\frac{\hbar}{4}\ln\left(\frac{2\,\mathrm{Im}\,A(t)}{\pi\hbar}\right) \tag{92}$$

$$\phi_n(x,\,t) = \frac{1}{\sqrt{n!2^n\sqrt{\pi}}}\exp\left(\frac{-\xi^2}{2}\right)H_n(\xi) \tag{93}$$

$$\xi = \sqrt{2\,\mathrm{Im}\,A(t)/\hbar}(x - x(t)) \tag{94}$$

C. The DVR Formulation

In the basis set formulation, we need to evaluate matrix elements over the G–H basis functions. We can avoid this by introducing a discrete variable representation method. We can obtain the DVR expressions by expanding the time-dependent amplitudes $a_n(t)$ in the following manner:

$$a_n(t) = \sum_{i=1}^{N} c_i(t)\phi_n(z_i) \tag{95}$$

where z_i are zeros of the N'th Hermite polynomium and $n = 0, 1, \ldots, N-1$. Thus we can insert this expansion in the expression for $\dot{a}_n(t)$ and obtain equations for $\dot{c}_i(t)$ instead. In this operation, we need to use

$$\sum_n \phi_n(z_i)\phi_n(z_j) = A_i\delta_{ij} \tag{96}$$

$$\sum_n \phi_n(z_j)\phi_n(\xi) \sim \frac{\delta(\xi - z_j)}{A_j} \tag{97}$$

After a little manipulation, we obtain

$$i\hbar\dot{d}_j(t) = \sum_i d_i(t)(H_{ij}\delta_{ij} + M_{ij}^{(x)} + T_{ij}) \tag{98}$$

where

$$H_{ij} = W(x_i)\delta_{ij} \tag{99}$$

$$\begin{aligned} M_{ij}^{(x)} &= F_x(p_x(t)\delta_{ij} - i\hbar\sqrt{\operatorname{Im} A/\hbar}A_i^{-1/2}A_j^{-1/2} \\ &\quad \times \left(\sum_n \phi_n(z_i)(\sqrt{n}\phi_{n-1}(z_j) - \sqrt{n+1}\phi_{n+1}(z_j))\right) \end{aligned} \tag{100}$$

$$T_{ij} = \frac{\hbar \operatorname{Im} A(t)}{m} A_i^{-1/2}A_j^{-1/2} \sum_n \phi_n(z_i)(2n+1)\phi_n(z_j) \tag{101}$$

Thus, the matrix elements $\mathbf{M}_{ij}^{(x)}$ are those that should be added in order to incorporate the geometric phase effect.

Extension to six dimensions is now straightforward. We obtain similar expressions just with the y and z components and the index n running over the basis functions included in the particular degree of freedom. For the functions

F_x, and so on, we obtain

$$F_x = F(X + x\,\text{tg}\,\phi/d_1^2) \tag{102}$$

$$F_y = F(Y + y\,\text{tg}\,\phi/d_1^2) \tag{103}$$

$$F_z = F(Z + z\,\text{tg}\,\phi/d_1^2) \tag{104}$$

$$F_X = F(x - X\,\text{tg}\,\phi d_1^2) \tag{105}$$

$$F_Y = F(y - Y\,\text{tg}\,\phi d_1^2) \tag{106}$$

$$F_Z = F(z - Z\,\text{tg}\,\phi d_1^2) \tag{107}$$

where the function F is given as $F = -(\hbar/2\mu)\cos\phi\sin\phi/(\mathbf{r}\cdot\mathbf{R})$ and $d_1^2 = m_1(1 - m_1/M)/\mu$ with $M = m_1 + m_2 + m_3$ [72,73]. In six dimensions, the amplitudes $d_i(t)$ [in Eq. (97)] will be of dimension $N = \Pi_{i=1}^6 N_i$. Here, in mass scaled coordinates we have used [48]

$$r^2/d_1^2 = \frac{1}{2}\rho^2(1 + \sin\theta\cos\phi) \tag{108}$$

$$R^2 d_1^2 = \frac{1}{2}\rho^2(1 - \sin\theta\cos\phi) \tag{109}$$

$$\mathbf{r}\cdot\mathbf{R} = -\frac{1}{2}\rho^2\sin\theta\sin\phi \tag{110}$$

Since the geometric phase effect is related to the angle ϕ we express ϕ as

$$\text{tg}\,\phi = -\frac{\mathbf{r}\cdot\mathbf{R}}{r^2/d_1 - d_1 R^2} \tag{111}$$

and obtain

$$\frac{\partial\phi}{\partial x} = \frac{\cos\phi\sin\phi}{\mathbf{r}\cdot\mathbf{R}}(X + x\,\text{tg}\,\phi/d_1^2) \tag{112}$$

$$\frac{\partial\phi}{\partial X} = \frac{\cos\phi\sin\phi}{\mathbf{r}\cdot\mathbf{R}}(x - X\,\text{tg}\,\phi d_1^2) \tag{113}$$

plus similar expressions for the y and z components.

Note that in this TDGH–DVR formulation of quantum dynamics, the inclusion of the geometric phase effects through the addition of a vector potential is very simple and the calculations can be carried out with about the same effort as what is involved in the ordinary scattering case.

Figure 3 shows the results with and without including the geometric phase effect for the $D + H_2$ reaction. The basis set is taken as 1,1,1,15,15,15, that is,

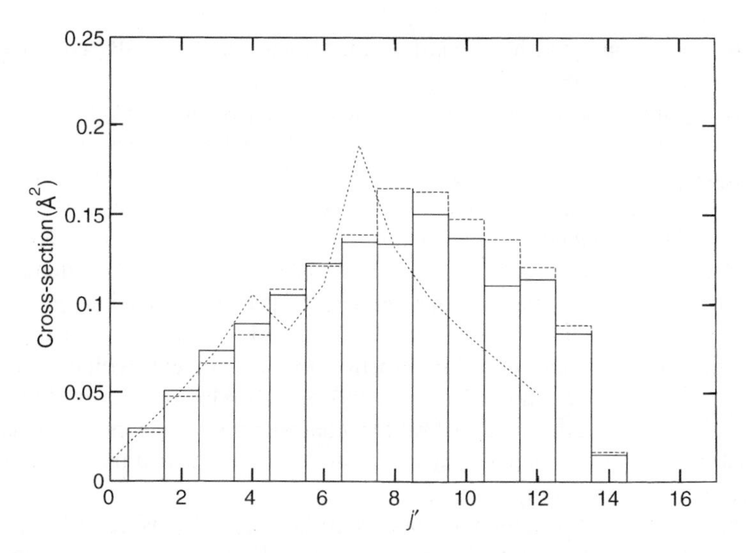

Figure 3. Cross-sections obtained with a (1,1,1,15,15,15) basis set and the TDGH–DVR method for the $D + H_2$ ($v = 1, j = 1$) $\rightarrow$ DH ($v' = 1, j'$) + H reaction at 1.8-eV total energy. The solid line indicates the values obtained without the vector potential and the dashed with a vector potential. The dashed line indicates the experimental results [49–52].

the X, Y, Z variables are treated classically. Altogether 200 trajectories were calculated. We notice that the branching ration, that is, the total reactive cross-section is obtained from the trajectories but the distribution is obtained by a projection of the DVR points on final rotational–vibrational states of the product. The maximum of the distribution is now $j' = 9$ (in better agreement with full quantum calculations). It is shifted to $j' = 8$ if the geometric phase is included. The agreement with experimental data is good for j' values <8 but overestimated at higher values. Since part of the system is still treated classically, we attribute this discrepancy to the lacking ability of classical trajectories to yield proper state-resolved reaction cross-sections (see also Fig. 1).

V. CONCLUSION

In this chapter, we discussed the significance of the GP effect in chemical reactions, that is, the influence of the upper electronic state(s) on the reactive and nonreactive transition probabilities of the ground adiabatic state. In order to include this effect, the ordinary BO equations are extended either by using a HLH phase or by deriving them from first principles. Considering the HLH phase due to the presence of a conical intersection between the ground and the first excited state, the general form of the vector potential, hence the effective kinetic energy operator, for a quasi-JT model and for an $A + B_2$ type reactive system were presented.

The ordinary BO approximate equations failed to predict the proper symmetry allowed transitions in the quasi-JT model whereas the extended BO equation either by including a vector potential in the system Hamiltonian or by multiplying a phase factor onto the basis set can reproduce the so-called exact results obtained by the two-surface diabatic calculation. Thus, the calculated transition probabilities in the quasi-JT model using the extended BO equations clearly demonstrate the GP effect. The multiplication of a phase factor with the adiabatic nuclear wave function is an approximate treatment when the position of the conical intersection does not coincide with the origin of the coordinate axis, as shown by the results of [60]. Moreover, even if the total energy of the system is far below the conical intersection point, transition probabilities in the JT model clearly indicate the importance of the extended BO equation and its necessity.

The integral and differential cross-section obtained by using QCT calculations on the ground adiabatic surface of the $D + H_2$ system at a total energy of 1.8 eV, clearly indicates the GP effect where the ground state of this system has a conical intersection with its' first excited state at a total energy of 2.7 eV. Similarly, semiclassical calculations on the same system with or without including a vector potential in the system Hamiltonian confirms this effect. Preliminary calculations with the new TDGH–DVR method also show a less dramatic effect. In the case of the $H + D_2$ reaction at total energy 2.4 eV, calculated rotational state and scattering angle distributions obtained from the QCT calculations on the LSTH surface demonstrate quantitative change due to the GP effect but the qualitative variation, at least in the integral cross-section, is not significant.

Formulation of the extended BO approximate equations using the HLH phase is based on the consideration of two coupled states. If the ground state of a system is coupled with more than one excited state, it has been demonstrated that the phase factor could be different from the HLH phase factor. In this formulation, we consider the BO coupled equations with the aim of deriving an approximate set of uncoupled equations that will contain the effect of non-adiabatic coupling terms. When the electronic states are degenerate, some of the non-adiabatic coupling terms may become infinite and affect the dynamics of the nuclei irrespective of how far it occurs from the point of the degeneracy. Hence, the importance of non-adiabatic coupling terms has been taken into account when deriving the uncoupled BO from the coupled ones. In this approch, the non-adiabatic coupling terms are not eliminated but shifted from the off-diagonal position to the diagonal one and the BO approximation has been introduced afterward. This shift has been done with the physical assumption that the non-adiabatic coupling matrix guarantees the continuous, single-valued diabatic potential matrix in the CS, that is, along a close path the non-adiabatic coupling matrix follows the Bohr–Sommerfeld type quantization rule. This quantization guarantees that all N decoupled equations obtained by deleting the potential coupling terms are invariant under gauge transformations and follow proper boundary conditions. These extended–approximated BO equations are tested for a tri-state system. First, we performed a so-called exact calculation in the diabatic representation to obtain reactive and nonreactive transition probabilities on the ground adiabatic surface and then the extended–approximated BO equations for the ground adiabatic surface are solved to get the relevant results. State-to-state transition probabilities obtained by both calculations indicate that the new approximate BO equations yield correct results for a tri-state system.

Hence, systems having conical intersections between two or more than two electronic states exhibit geometric phase effects. For two-states systems, the HLH phase factor is the same as that obtained by Baer et al. from first principles but the new phase factor appears to be different and depends on the number of electronic states coupled. Considering a conical intersection between the ground and first excited state of the $D + H_2$ reactive system, the extended BO equations are the same in both of the above-mentioned approaches and we found significant GP effect at a total energy of 1.8 eV. However, it has been possible to obtain good agreement between experiment and theory without including the effect for the $H + D_2$ system at a total energy 2.4 eV. At this point, it is worth noting that the calculations on the $H + D_2$ reaction were carried out on a different potential energy surface than the one we used in our calculations. May be the reactivity of one potential energy surface could hide the GP effect while another could expose it. At the same time, the importance of the GP effect is clearly understood in the quasi-JT model. The inclusion of a simple phase factor (HLH) or by using the extended BO equations can change the parity for vibrational transitions in the 2D two-surface model and give good agreement with results obtained by an exact two-state diabatic calculation. Again, calculations on a tri-state 2D quasi-JT model using the extended BO equations ($N \geq 2$) derived by Baer et al. not only exhibit geometric phase effects but also the new phase factor that changes with the number of electronic states coupled.

APPENDIX A: THE JAHN–TELLER MODEL AND THE HERZBERG–LONGUET–HIGGINS PHASE

When two electronic states are degenerate at a particular point in configuration space, the elements of the diabatic potential energy matrix can be modeled as a linear function of the coordinates in the following form:

$$W = k\begin{pmatrix} y & x \\ x & -y \end{pmatrix} \tag{A.1}$$

where k is the force constant and (x, y) are the nuclear coordinates. The eigenvalues and eigenvectors of the above matrix represent the adiabatic potential energy surfaces and the columns of the ADT matrix, respectively. In order to carry out this diabatization, we use the following transformations between the Cartesian (x, y) and polar (q, ϕ) coordinates: $x = q \sin\phi$ and $y = q \cos\phi$.

The eigenvalues and eigenfunctions of this model are

$$u_{1,2} = \pm kq \tag{A.2}$$

where $q = 0, \infty$ and $\phi = 0, 2\pi$, and

$$\begin{aligned} \xi_1 &= \left(\frac{1}{\sqrt{\pi}} \cos\phi/2, \quad \frac{1}{\sqrt{\pi}} \sin\phi/2 \right) \\ \xi_2 &= \left(\frac{1}{\sqrt{\pi}} \sin\phi/2, \quad -\frac{1}{\sqrt{\pi}} \cos\phi/2 \right) \end{aligned} \tag{A.3}$$

respectively.

These adiabatic eigenfunctions depend only on the angular coordinate ϕ and are not single valued in configuration space when ϕ changes to $\phi + 2\pi$—a rotation that brings the adiabatic wave functions back to their initial position. This multivaluedness of the adiabatic eigenfunctions was first revealed by Herzberg and Longuet-Higgins. In order to avoid multivalued electronic eigenfunctions they suggested that the corresponding nuclear wave function be treated with care. While solving the nuclear SE, this feature needs to be incorporated explicitly through specific boundary conditions. It is worth mentioning that in the HLH state realistic ab initio electronic wave functions posses the multivaluedness feature.

Longuet-Higgins corrected the multivaluedness of the electronic eigenfunctions by multiplying them with a phase factor, namely,

$$\eta_j(\phi) = \exp(i\alpha)\xi_j(\phi) \qquad j = 1, 2 \tag{A.4}$$

where $\alpha = \phi/2$. It is important to note that $\eta_j(\phi)$, $j = 1, 2$ are single-valued complex eigenfunctions.

APPENDIX B: THE BORN–OPPENHEIMER TREATMENT

The total electron–nuclear Hamiltonian of a molecular sytem is defined as

$$\hat{H} = \hat{T}_n + \hat{H}_e(e \mid n) \tag{B.1}$$

where $\hat{T}_n$ is the kinetic energy operator for the nuclei and $\hat{H}_e(e \mid n)$ is the electronic Hamiltonian and

$$\hat{H}_e = \hat{T}_e + \hat{V}(e \mid n) \tag{B.2}$$

with $\hat{T}_e$ being the kinetic energy operator of the electrons and $\hat{V}(e \mid n)$ the potential energy operator as a function of electronic coordinates(e) with nuclear coordinates(n).

The BO expansion of the molecular wave function

$$\Psi(e, n) = \sum_{i=1}^{N} \psi_i(n)\xi_i(e \mid n_0) \tag{B.3}$$

where the functions $\psi_i(n)$ are the nuclear coordinate-dependent coefficients, later considered as the nuclear wave function, and the $\xi_i(e \mid n_0)$s are the electronic eigenfunctions satisfying the equation

$$\hat{H}_e(n_0)\xi_i(e \mid n_0) = u_i(n_0)\xi_i(e \mid n_0) \qquad i = 1, \ldots, N \tag{B.4}$$

Here, the $u_i(n_0)$s are the electronic eigenvalues dependent on the nuclear coordinate n_0. Note that $n_0 \equiv n$ is defined as the adiabatic case and $n_0 \neq n$ is defined as the diabatic case.

Substituting Eqs. (B.1) and (B.3) into the time-independent Schrödinger equation $H\Psi(e, n) = E\Psi(e, n)$, one obtains

$$(\hat{T}_n + \hat{H}_e - E)\sum \psi_i(n)\xi_i(e \mid n_0) = 0 \tag{B.5}$$

Below, we apply the bra–ket notation to electronic coordinates only,

$$\langle \xi_j(n) | \xi_i(n_0) \rangle = \begin{cases} g_{ji}(n, n_0); for n \neq n_0 \\ \delta_{ji}; for n = n_0 \end{cases} \tag{B.6}$$

By returning back to Eq. (B.5), we have

$$\sum_{i=1}^{N} T_n \psi_i(n) |\xi_i(e \mid n_0)\rangle + \sum_{i=1}^{N} \psi_i(n)(H_e - E)|\xi_i(e \mid n_0)\rangle = 0 \tag{B.7}$$

If we consider the *ADIABATIC* ($n_0 \equiv n$) case, we get

$$\sum_{i=1}^{N} T_n \psi_i(n) |\xi_i(e \mid n)\rangle + \sum_{i=1}^{N} \psi_i(n)(u_i(n) - E)|\xi_i(e \mid n)\rangle = 0 \tag{B.8}$$

Multiplying by $\langle \xi_j|$ and integrating over electronic coordinates yields

$$\sum_{i=1}^{N} \langle \xi_j | T_n \psi_n(n) | \xi_i \rangle + (u_j(n) - E)\psi_j(n) = 0 \qquad j = 1, \ldots, N \tag{B.9}$$

where ∇ is the gradient operator and $T_n = -(1/2m)\nabla^2$.

Hence, the following matrix element becomes

$$\langle \xi_j | T_n \psi_i(n) | \xi_i \rangle = -\frac{1}{2m}\{\delta_{ij}\nabla^2\psi_i + 2\langle \xi_j|\nabla\xi_i\rangle + \langle \xi_j|\nabla^2\xi\rangle\psi_i\} \tag{B.10}$$

and the non-adiabatic coupling matrix elements are defined as below,

$$\tau_{ji}^{(1)} = \langle \xi_j|\nabla\xi_i\rangle \qquad \tau_{ji}^{(2)} = \langle \xi_j|\nabla^2\xi_i\rangle \tag{B.11}$$

For example, in the case of the x component of the nuclear coordinates we have

$$\tau_{xji}^{(1)} = \left\langle \xi_j \middle| \frac{\partial}{\partial x}\xi_i \right\rangle \qquad \tau_{xji}^{(2)} = \left\langle \xi_j \middle| \frac{\partial^2}{\partial x^2}\xi_i \right\rangle \tag{B.12}$$

Therefore, Eq. (B.10) in terms of this notation becomes

$$\langle \xi_j | T_n \psi_i(n) | \xi_i \rangle = -\frac{1}{2m}\{\delta_{ji}\nabla^2\psi_i + 2\tau_{ji}^{(1)} \cdot \nabla\psi_i + \tau_{ji}^{(2)}\psi_i\} \tag{B.13}$$

It is important to note that the non-adiabatic coupling terms have a direct effect on the momentum of the nuclei, which is the reason it is called a dynamic coupling. By substituting Eq. (B.13) in Eq. (B.9), we get

$$-\frac{1}{2m}\nabla^2\psi_j + (u_j(n) - E)\psi_j(n) - \frac{1}{2m}\sum_{i=1}^{N}(2\tau_{ji}^{(1)} \cdot \nabla\psi_i + \tau_{ji}^{(2)}\psi_i) = 0 \tag{B.14}$$

This is the electronic adiabatic Schrödinger equation and in the case of a single coordinate x Eq. (B.14) takes the following form:

$$-\frac{1}{2m}\frac{d^2}{dx^2}\psi_j + (u_j(n) - E)\psi_j(n) - \frac{1}{2m}\sum_{i=1}^{N}\left(2\tau_{xji}^{(1)} \cdot \frac{d}{dx}\psi_i + \tau_{xji}^{(2)}\psi_i\right) = 0 \tag{B.15}$$

When the non-adiabatic coupling terms $\tau^{(1)}$ and $\tau^{(2)}$ are considered negligibly small and dropped from Eq. (B.15), we get the uncoupled approximate Schrödinger equation

$$-\frac{1}{2m}\frac{d^2}{dx^2}\psi_j + (u_j(n) - E)\psi_j(n) = 0 \tag{B.16}$$

or more general,

$$\frac{1}{2m}\nabla^2\psi_j(n) + (u_j(n) - E)\psi_j(n) = 0 \tag{B.17}$$

The approximation involved in Eq. (B.17) is known as the Born–Oppenheimer approximation and this equation is called the Born–Oppenheimer equation.

By assuming the Hilbert space of dimension N, one can easily establish the relation between coupling matrices $\boldsymbol{\tau}^{(1)}$ and $\boldsymbol{\tau}^{(2)}$ by considering the (ij)th matrix element of $\nabla \cdot \tau^{(1)}$,

$$\begin{aligned}\nabla\tau_{ij}^{(1)} &= \nabla\langle\xi_i|\nabla\xi_j\rangle = \langle\nabla\xi_i|\nabla\xi_j\rangle + \langle\xi_i|\nabla^2\xi_j\rangle \\ &= \langle\nabla\xi_i|\nabla\xi_j\rangle + \tau_{ij}^{(2)}\end{aligned}$$

We can resolve the unity operator in the following way:

$$I = \sum_{k=1}^{N} |\xi_k\rangle\langle\xi_k|$$

and obtain,

$$\begin{aligned}\langle\nabla\xi_i|\nabla\xi_j\rangle &= \langle\nabla\xi_i|I|\nabla\xi_j\rangle = \langle\xi_i|\left(\sum_{k=1}^{N}|\xi_k\rangle\langle\xi_k\right)|\xi_j\rangle \\ &= \sum_{k=1}^{N}\langle\nabla\xi_i|\xi_k\rangle\langle\xi_k|\nabla\xi_j\rangle = -\sum_{k=1}^{N}\langle\xi_k|\nabla\xi_i\rangle\langle\xi_k|\nabla\xi_j\rangle \\ &= -\sum_{k=1}^{N}\tau_{ki}^{(1)}\tau_{kj}^{(1)} = -(\tau^{(1)})_{ij}^2\end{aligned}$$

Hence, the elements of $\tau^{(1)}$ and $\tau^{(2)}$ are related as below

$$\tau_{ij}^{(2)} = (\tau^{(1)})_{ij}^2 + \nabla\tau_{ij}^{(1)}$$

and finally in matrix notation

$$\boldsymbol{\tau}^{(2)} = (\tau^{(1)})^2 + \nabla\tau^{(1)} \tag{B.18}$$

Incorporating relation (B.18) in Eq. (B.14), we can write in matrix form,

$$-\frac{1}{2m}\nabla^2\psi + \left(u - \frac{1}{2m}\tau^{(1)2} - E\right)\psi - \frac{1}{2m}(2\tau^{(1)}\cdot\nabla + \nabla\tau^{(1)})\psi = 0 \tag{B.19}$$

which can be expressed in compact form as

$$-\frac{1}{2m}(\nabla + \tau)^2\psi + (u - E)\psi = 0 \tag{B.20}$$

So far, we have treated the case $n \equiv n_0$, which was termed the adiabatic representation. We will now consider the diabatic case where n is still a variable but n_0 is constant as defined in Eq. (B.3). By multiplying Eq. (B.7) by $\langle\xi_j(e \mid n_0)|$ and integrating over the electronic coordinates, we get

$$\left(-\frac{1}{2m}\nabla^2 - E\right)\psi_j(n) + \sum_{i=1}^{N}\langle\xi_j(e \mid n_0)|\hat{H}_e(e \mid n)|\xi_i(e \mid n_0)\rangle\psi_i(n) = 0 \tag{B.21}$$

We can rewrite the electronic Hamiltonian in the following form:

$$\begin{aligned} H_e(e \mid n) &= T_e + V(e \mid n) \\ H_e(e \mid n_0) &= T_e + V(e \mid n_0) \\ H_e(e \mid n) &= H_e(e \mid n_0) + \{V(e \mid n) - V(e \mid n_0)\} \end{aligned} \tag{B.22}$$

and by using Eq. (B.22), we can calculate the following matrix element:

$$\langle\xi_j(e \mid n_0)|H_e(e \mid n)|\xi_i(e \mid n_0)\rangle = u_j(n_0)\delta_{ji} + \tilde{\nu}_{ij}(n \mid n_0) \tag{B.23}$$

where

$$\begin{aligned} \tilde{\nu}_{ji}(n \mid n_0) &= \langle\xi_j(e \mid n_0)|V(e \mid n) - V(e \mid n_0)|\xi_i(e \mid n_0)\rangle \\ \nu_{ij}(n \mid n_0) &= \tilde{\nu}_{ij}(n \mid n_0) + u_j(n_0)\delta_{ji} \end{aligned} \tag{B.24}$$

By substituting the expression for the matrix elements in Eq. (B.21), we get the final form of the Schrödinger equation within the diabatic representation

$$\left(-\frac{1}{2m}\nabla^2 - E\right)\psi_j(n) + \sum_{i=1}^{N}\nu_{ji}(n \mid n_0)\psi_i(n) = 0 \tag{B.25}$$

where the coupling terms among the states are due to potential coupling.

By substituting the following transformation

$$\psi = A\Phi \tag{B.26}$$

into the adiabatic Schrödinger equation (B.20), we obtain the following expression,

$$-\frac{1}{2m}\{A\nabla^2\Phi + 2(\nabla A + \tau A)\cdot\nabla\Phi + \{(\tau + \nabla)\cdot(\nabla A + \tau A)\}\Phi\} + (u - E)A\Phi = 0 \tag{B.27}$$

If the transformation matrix **A** is chosen in such a way that $\nabla\mathbf{A} + \tau\mathbf{A} = 0$, Eq. (B.27) can be rearranged to the following form:

$$-\frac{1}{2m}\nabla^2\Phi + (A^{-1}uA - E)\Phi = 0 \tag{B.28}$$

which is basically in "diabatic" representation and identical looking with Eq. (B.25) and **A** is the adiabatic–diabatic transformation matrix.

APPENDIX C: FORMULATION OF THE VECTOR POTENTIAL

The vector potential is derived in hyperspherical coordinates following the procedure in [54], where the connections between Jacobi and the hyperspherical coordinates have been considered as below (see [67])

$$\begin{aligned}
r_x &= -\frac{\rho}{\sqrt{2}}\left(\cos\frac{\theta}{2} + \sin\frac{\theta}{2}\right)\cos\frac{\phi}{2} \\
r_y &= \frac{\rho}{\sqrt{2}}\left(\cos\frac{\theta}{2} - \sin\frac{\theta}{2}\right)\sin\frac{\phi}{2} \\
r_z &= 0 \\
R_x &= \frac{\rho}{\sqrt{2}}\left(\cos\frac{\theta}{2} + \sin\frac{\theta}{2}\right)\sin\frac{\phi}{2} \\
R_y &= \frac{\rho}{\sqrt{2}}\left(\cos\frac{\theta}{2} - \sin\frac{\theta}{2}\right)\cos\frac{\phi}{2} \\
R_z &= 0
\end{aligned} \tag{C.1}$$

The interatomic distances of the triangle ABC formed due to any A + BC type reactive system are as follows:

$$\begin{aligned}
\frac{R_{\mathrm{AB}}^2}{d_1^2} &= \frac{\rho^2}{2}(1 + \sin\theta\cos\phi) \\
\frac{R_{\mathrm{BC}}^2}{d_2^2} &= \frac{\rho^2}{2}(1 + \sin\theta\cos(\phi - \xi_2)) \\
\frac{R_{\mathrm{CA}}^2}{d_3^2} &= \frac{\rho^2}{2}(1 + \sin\theta\cos(\phi + \xi_3))
\end{aligned} \tag{C.2}$$

and these interatomic distances can also be expressed in terms of Jacobi coordinates

$$
\begin{aligned}
R_{\mathrm{AB}}^2 &= (r_x^2 + r_y^2)d_1^2 \\
R_{\mathrm{BC}}^2 &= (R_x^2 + R_y^2)\frac{d_2^2(1 - \cos\xi_2)}{2} + (r_x^2 + r_y^2)\frac{d_2^2(1 + \cos\xi_2)}{2} \\
&\quad - (r_xR_x + r_yR_y)d_2^2 \sin\xi_2 \\
R_{\mathrm{CA}}^2 &= (R_x^2 + R_y^2)\frac{d_3^2(1 - \cos\xi_3)}{2} + (r_x^2 + r_y^2)\frac{d_3^2(1 + \cos\xi_3)}{2} \\
&\quad + (r_xR_x + r_yR_y)d_3^2 \sin\xi_3
\end{aligned}
\tag{C.3}
$$

where $d_k^2 = (m_k/\mu)(1 - m_k/M)$, m_1 m_2 and m_3 are the masses of the atom A, B, and C, respectively, in the corners of the triangle ABC. The parameters $M = m_1 + m_2 + m_3$ and $\mu = \sqrt{m_1m_2m_3/M}$ and the angles are given by $\xi_2 = 2\arctan\,(m_3/\mu)$ and $\xi_3 = 2\arctan\,(m_2/\mu)$.

By using Eq. (C.2) one can write

$$
\tan\phi = \frac{\left(\frac{R_{\mathrm{CA}}^2}{d_3^2} - \frac{R_{\mathrm{AB}}^2}{d_1^2}\right)\cos\xi_2 - \left(\frac{R_{\mathrm{BC}}^2}{d_2^2} - \frac{R_{\mathrm{AB}}^2}{d_1^2}\right)\cos\xi_3 + \left(\frac{R_{\mathrm{BC}}^2}{d_2^2} - \frac{R_{\mathrm{CA}}^2}{d_3^2}\right)}{\left(\frac{R_{\mathrm{AB}}^2}{d_1^2} - \frac{R_{\mathrm{CA}}^2}{d_3^2}\right)\sin\xi_2 - \left(\frac{R_{\mathrm{BC}}^2}{d_2^2} - \frac{R_{\mathrm{AB}}^2}{d_1^2}\right)\sin\xi_3}
\tag{C.4}
$$

It would be convenient for obtaining the expressions of the gradient of the hyperangle ϕ with respect to Jacobi coordinates to introduce the physical region of the conical intersection in the following manner:

$$
\begin{aligned}
\frac{\partial\phi}{\partial r_i} &= \frac{\partial\phi}{\partial R_{\mathrm{AB}}}\frac{\partial R_{\mathrm{AB}}}{\partial r_i} + \frac{\partial\phi}{\partial R_{\mathrm{BC}}}\frac{\partial R_{\mathrm{BC}}}{\partial r_i} + \frac{\partial\phi}{\partial R_{\mathrm{CA}}}\frac{\partial R_{\mathrm{CA}}}{\partial r_i} \\
\frac{\partial\phi}{\partial R_i} &= \frac{\partial\phi}{\partial R_{\mathrm{AB}}}\frac{\partial R_{\mathrm{AB}}}{\partial R_i} + \frac{\partial\phi}{\partial R_{\mathrm{BC}}}\frac{\partial R_{\mathrm{BC}}}{\partial R_i} + \frac{\partial\phi}{\partial R_{\mathrm{CA}}}\frac{\partial R_{\mathrm{CA}}}{\partial R_i}
\end{aligned}
\tag{C.5}
$$

where $i \equiv x,\, y,\, z$. To obtain explicit expressions for $\nabla\phi$, we have used Eqs. (C.2–C.5) and after some algebra (!) it is interesting to note that $\nabla\phi$ becomes independent of d_k and ξ_k for any arbitrary A + BC type reactive system. We obtain

$$
\begin{aligned}
\frac{\partial\phi}{\partial r_i} &= -\frac{2}{\rho^2\sin\theta}(r_i\sin\phi + R_i\cos\phi) \\
\frac{\partial\phi}{\partial R_i} &= \frac{2}{\rho^2\sin\theta}(-r_i\cos\phi + R_i\sin\phi) \\
\frac{\partial\phi}{\partial r_z} &= 0 \\
\frac{\partial\phi}{\partial R_z} &= 0
\end{aligned}
\tag{C.6}
$$

where $i \equiv x, y$. Similarly, explicit expressions for $\nabla\theta$ are obtained using Eqs. (C.1)

$$\begin{aligned}
\frac{\partial\theta}{\partial r_x} &= -\frac{2R_y}{\rho^2} \\
\frac{\partial\theta}{\partial r_y} &= -\frac{2R_x}{\rho^2} \\
\frac{\partial\theta}{\partial R_x} &= \frac{2r_y}{\rho^2} \\
\frac{\partial\theta}{\partial R_y} &= \frac{2r_x}{\rho^2} \\
\frac{\partial\theta}{\partial r_z} &= 0 \\
\frac{\partial\theta}{\partial R_z} &= 0
\end{aligned} \tag{C.7}$$

The azimuthal angle (η) about the conical intersection is related with hyper-angles θ and ϕ as

$$\eta(\theta, \phi) \equiv \phi' = \arctan\left[\frac{\sin\theta\sin\phi}{\cos\theta_0\sin\theta\cos\phi + \sin\theta_0\cos\theta}\right] \tag{C.8}$$

where θ_0 indicates the position of the conical intersection.

The gradient of $\nabla\eta$ with respect to Jacobi coordinates (the vector potential) considering the physical region of the conical intersection, is obtained by using Eqs. (C.6–C.8) and after some simplification (!) we get,

$$\begin{aligned}
\frac{\partial\eta}{\partial r_x} &= -\frac{2}{\rho^2}\frac{[R_y\sin\theta_0\sin\phi + (\cos\theta_0\sin\theta + \sin\theta_0\cos\theta\cos\phi)(r_x\sin\phi + R_x\cos\phi)]}{[\sin^2\theta\sin^2\phi + (\cos\theta_0\sin\theta\cos\phi + \sin\theta_0\cos\theta)^2]} \\
\frac{\partial\eta}{\partial r_y} &= -\frac{2}{\rho^2}\frac{[R_x\sin\theta_0\sin\phi + (\cos\theta_0\sin\theta + \sin\theta_0\cos\theta\cos\phi)(r_y\sin\phi + R_y\cos\phi)]}{[\sin^2\theta\sin^2\phi + (\cos\theta_0\sin\theta\cos\phi + \sin\theta_0\cos\theta)^2]} \\
\frac{\partial\eta}{\partial r_z} &= 0 \\
\frac{\partial\eta}{\partial R_x} &= \frac{2}{\rho^2}\frac{[r_y\sin\theta_0\sin\phi + (\cos\theta_0\sin\theta + \sin\theta_0\cos\theta\cos\phi)(-r_x\cos\phi + R_x\sin\phi)]}{[\sin^2\theta\sin^2\phi + (\cos\theta_0\sin\theta\cos\phi + \sin\theta_0\cos\theta)^2]} \\
\frac{\partial\eta}{\partial R_y} &= \frac{2}{\rho^2}\frac{[r_x\sin\theta_0\sin\phi + (\cos\theta_0\sin\theta + \sin\theta_0\cos\theta\cos\phi)(-r_y\cos\phi + R_y\sin\phi)]}{[\sin^2\theta\sin^2\phi + (\cos\theta_0\sin\theta\cos\phi + \sin\theta_0\cos\theta)^2]} \\
\frac{\partial\eta}{\partial R_z} &= 0
\end{aligned} \tag{C.9}$$

For the H_3 isotopic variants, we can calculate the values of θ_0 and ϕ_0 by introducing $R_{AB} = R_{BC} = R_{CA}$. Moreover, we get $\theta_0 = \phi_0 = 0$ for an A_3 and $\phi_0 = 0$ for an AB_2 type reactive system. In case of an $A + B_2$ type reaction, one can use

$$\sin\theta_0 = \left|\frac{d_1^2 - d_2^2}{d_2^2 \cos\xi_2 - d_1^2}\right| \tag{C.10}$$

and obtain $\theta_0 = 11.5°$ for DH_2 and $\theta_0 = 14.5°$ for HD_2. The actual position of the CI on the PES is obtained through the equation, $V(\rho_0, \theta_0, \phi_0) = E_{CI}$ where E_{CI} is the potential energy at the point of the CI.

References

1. G. Herzberg and H. C. Longuet-Higgins, *Discuss. Faraday Soc.* **35**, 77 (1963).
2. H. A. Jahn and E. Teller, *Proc. R. Soc. London Ser. A* **161**, 220 (1937).
3. E. Teller, *J. Phys. Chem.* **41**, 109 (1937).
4. E. Teller, *Isr. J. Chem.* **7**, 227 (1969).
5. R. Englman, *The Jahn–Teller Effect in Molecules and Crystals*, Wiley-Interscience, New York, 1972.
6. Y. T. Lee, R. J. Gordon, and D. R. Herschbach, *J. Chem. Phys.* **54**, 2410 (1971).
7. J. D. McDonald, P. R. LeBreton, Y. T. Lee, and D. R. Herschbach, *J. Chem. Phys.* **56**, 769 (1972).
8. G. Herzbach, *Molecular Spectra and Molecular Structure III, Electronic Spectra and Electronic Structure of Polyatomic Molecules*, Van Nostrand Reinhold, New York, 1966, pp. 442–444.
9. T. Carrington, *Discuss. Faraday Soc.* **53**, 27 (1972).
10. J. O. Hirschfelder, *J. Chem. Phys.* **6**, 795 (1938).
11. R. N. Porter, R. M. Stevens, and M. Karplus, *J. Chem. Phys.* **49**, 5163 (1968).
12. J. L. Jackson and R. E. Wyatt, *Chem. Phys. Lett.* **18**, 161 (1973).
13. B. M. Smirnov, *Zh. Eksp. Teor. Fiz.* **46**, 578 (1964) (English Transl.: *Sov. Phys.* JETP **19**, 394 (1964)).
14. F. A. Matsen, *J. Phys. Chem.* **68**, 3283 (1964).
15. E. Frenkel and Z. Naturforsch, *Teil A* **25**, 1265 (1970).
16. M. Born and J. R. Oppenheimer, *Ann. Phy. (Leipzig)* **84**, 457 (1927).
17. M. Born and K. Huang, *Dynamical Theory of Crystal Lattices*, Oxford University Press, New York, 1954.
18. M. Baer, *Chem. Phys. Lett.* **35**, 112 (1975).
19. Z. H. Top and M. Baer, *J. Chem. Phys.* **66**, 1363 (1977).
20. Z. H. Top and M. Baer, *Chem. Phys.* **25**, 1 (1977).
21. M. Baer, *Mol. Phys.* **40**, 1011 (1980).
22. M. Baer, in *The theory of Chemical Reaction Dynamics*, M. Baer, ed., CRC, Boca Raton, FL, 1985, Vol. II., Chap. 4.
23. M. Baer, in *State-Selected and State-to-State Ion–Molecule Reaction Dynamics*, M. Baer and C. Y. Ng., eds., John Wiley & Sons, Inc., New York, 1992, Vol. II, Chap. 4.

24. V. Sidis, in *State-Selected and State-to-State Ion–Molecule Reaction Dynamics*, M. Baer and C. Y. Ng., eds., John Wiley & Sons, Inc., New York, 1992, Vol. II, Chap. 2.
25. T. Pacher, L. S. Cederbaum, and H. Köppel, *Adv. Chem. Phys.* **84**, 293 (1984).
26. H. Köppel, W. Domcke, and L. S. Cederbaum, *Adv. Chem. Phys.* **57**, 59 (1984).
27. C. A. Mead and D. G. Truhlar, *J. Chem. Phys.* **77**, 6090 (1982).
28. C. Petrongolo, R. J. Buekener and D. S. Peyerimhoff, *J. Chem. Phys.* **78**, 7284 (1983).
29. T. J. Gregory, M. L. Steven, D. G. Trulhar, and D. Schwenke, in *Advances in Molecular Vibrations and Collision Dynamics*, J. M. Bowman, ed., JAi, CT, 1994, Vol. 2B, Chap. III.
30. T. Pacher, C. A. Mead, L. S. Cederbaum, and H. Köppel, *J. Chem. Phys.* **91**, 7057 (1989).
31. C. A. Mead and D. G. Trulhar, *J. Chem. Phys.* **70**, 2284 (1979).
32. C. A. Mead, *Chem. Phys.* **49**, 23 (1980).
33. C. A. Mead, *J. Chem. Phys.* **72**, 3839 (1980).
34. A. J. C. Varandas and Z. R. Xu, *Int. J. Quant. Chem.* **75**, 89 (1999).
35. W. Moffitt and W. Thorson, *Phys. Rev.* **108**, 1251 (1957).
36. H. C. Longuet-Higgins, U. Opik, M. H. L. Pryce, and R. A. Sack, *Proc. R. Soc. London Ser. A* **244**, 1 (1958).
37. M. S. Child and H. C. Longuet-Higgins, *Philos. Trans. R. Soc. London Ser. A* **254**, 259 (1961).
38. W. H. Gerber and E. Schumacher, *J. Chem. Phys.* **69**, 1692 (1978).
39. W. Duch and G. A. Segal, *J. Chem. Phys.* **79**, 2951 (1983).
40. W. Duch and G. A. Segal, *J. Chem. Phys.* **82**, 2392 (1985).
41. T. C. Thompson, D. G. Truhlar, and C. A. Mead, *J. Chem. Phys.* **82**, 2392 (1985).
42. M. Baer and R. Englman, *Mol. Phys.* **75**, 293 (1992).
43. J. Schon and H. Köppol, *J. Chem. Phys.* **103**, 9292 (1995).
44. Y. M. Wu, B. Lepetit, and A. Kuppermann, *Chem. Phys. Lett.* **186**, 319 (1991).
45. Y. M. Wu and A. Kuppermann, *Chem. Phys. Lett.* **201**, 178 (1993).
46. A. Kuppermann and Y. M. Wu, *Chem. Phys. Lett.* **205**, 577 (1993).
47. X. Wu, R. E. Wyatt, and M. D'mello, *J. Chem. Phys.* **101**, 2953 (1994).
48. G. D. Billing and N. Markovic, *J. Chem. Phys.* **99**, 2674 (1993).
49. D. A. V. Kliner and R. N. Zare, *J. Chem. Phys.* **92**, 2107 (1990).
50. D. A. V. Kliner, D. E. Adelman, and R. N. Zare, *J. Chem. Phys.* **95**, 1648 (1991).
51. D. E. Adelman, H. Xu, and R. N. Zare, *Chem. Phys. Lett.* **203**, 573 (1993).
52. H. Xu, N. E. Shafar-Ray, F. Merkt, D. J. Hughes, M. Springer, R. P. Tuckett, and R. N. Zare, *J. Chem. Phys.* **103**, 5157 (1995).
53. S. Adhikari and G. D. Billing, *J. Chem. Phys.* **107**, 6213 (1997).
54. S. Adhikari and G. D. Billing, *Chem. Phys. Lett.* **284**, 31 (1998).
55. S. Adhikari and G. D. Billing, *Chem. Phys. Lett.* **289**, 219 (1998).
56. S. Adhikari and G. D. Billing, *Chem. Phys. Lett.* **305**, 109 (1999).
57. M. Baer and R. Englman, *Chem. Phys. Lett.* **265**, 105 (1996).
58. M. Baer, *J. Chem. Phys.* **107**, 10662 (1997).
59. R. Baer, D. M. Charutz, R. Kosloff, and M. Baer, *J. Chem. Phys.* **105**, 9141 (1996).
60. S. Adhikari and G. D. Billing, *J. Chem. Phys.* **111**, 40 (1999).
61. R. K. Preston and J. C. Tully, *J. Chem. Phys.* **54**, 4297 (1971).

62. M. Baer and A. J. Beswick, *Phys. Rev A* **19**, 1559 (1979).
63. M. Baer and A. Alijah, *Chem. Phys. Lett.* **319**, 489 (2000).
64. M. Baer, *J. Phys. Chem. A* **104**, 3181 (2000).
65. M. Baer, S. H. Lin, A. Alijah, S. Adhikari, and G. D. Billing, *Phys. Rev. A* **62**, 32506:1–8 (2000).
66. S. Adhikari, G. D. Billing, A. Alijah, S. H. Lin, and M. Baer, *Phys. Rev. A* **62**, 32507:1–7 (2000).
67. B. R. Johnson, *J. Chem. Phys.* **73**, 5051 (1980).
68. M. Baer, A. Yahalom, and R. Englman, *J. Chem. Phys.* **109**, 6550, (1998).
69. D. Kosloff and R. Kosloff, *J. Comput. Phys.* **52**, 35 (1983).
70. T. J. Park and J. C. Light, *J. Chem. Phys.* **85**, 5870 (1986).
71. G. Jolicard and G. D. Billing, *Chem. Phys.* **149**, 261 (1991).
72. B. R. Johnson, *J. Chem. Phys.* **79**, 1916 (1983).
73. B. R. Johnson, *J. Chem. Phys.* **79**, 1906 (1983).
74. H. Goldstein, *Classical Mechanics*, Addison-Wesley, Reading, MA, 1950.
75. P. Siegbahn and B. Liu, *J. Chem. Phys.* **68**, 2457 (1978).
76. D. G. Truhlar and C. J. Horowitz, *J. Chem. Phys.* **68**, 2466 (1978).
77. D. G. Truhlar and N. C. Blais, *J. Chem. Phys.* **67**, 1532 (1977).
78. A. Kosmas, E. A. Gislason and A. D. Jorgensen, *J. Chem. Phys.* **75**, 2884 (1981).
79. F. J. Aoiz, V. J. Herrero, and V. Saez Rabonos, *J. Chem. Phys.* **94**, 7991 (1991).
80. L. Schnieder, K. Seekamp-Rahn, J. Borkowski, E. Wrede, K. H. Welge, F. J. Aoiz, L. Banares, M. J. D'Mello, V. J. Herrero, V. Saez Rabanos, and R. E. Wyatt, *Science* **269**, 207 (1995).
81. G. D. Billing, *J. Chem. Phys.* **107**, 4286 (1997); **110**, 5526 (1999).
82. L. Schnieder, K. Seekamp-Rahn, E. Wrede, and K. H. Welge, *J. Chem. Phys.* **107**, 6175 (1997).
83. E. Wrede and L. Schnieder, *J. Chem. Phys.* **107**, 786 (1997).
84. E. Wrede, L. Schnieder, K. H. Welge, F. J. Aoiz, L. Banares, and V. J. Herrero, *Chem. Phys. Lett.* **265**, 129 (1997).
85. E. Wrede, L. Schnieder, K. H. Welge, F. J. Aoiz, L. Banares, V. J. Herrero, B. Martinez-Haya, and V. Saez Rabanos, *J. Chem. Phys.* **106**, 7862 (1997).
86. L. Banares, F. J. Aoiz, V. J. Herrero, M. J. D'Mello, B. Niederjohann, K. Seekamp-Rahn, E. Wrede, and L. Schnieder, *J. Chem. Phys.* **108**, 6160 (1998).
87. E. Wrede, L. Schnieder, K. H. Welge, F. J. Aoiz, L. Banares, J. F. Castillo, B. Martinez-Haya, and V. J. Herrero, *J. Chem. Phys.* **110**, 9971 (1999).
88. S. Adhikari and G. D. Billing, *Chem. Phys.* **259**, 149 (2000).
89. G. D. Billing, *Chem. Phys. Lett.* **321**, 197 (2000).
90. G. D. Billing, *J. Chem. Phys.* **114**, 6641 (2001).
91. G. D. Billing, *Quantum-Classical Methods* in *Reaction and Molecular Dynamics*, Lecture Notes in Chemistry, A. Lagana and A. Riganelli, eds., Springer-Verlag, Berlin, 2000.

COMPLEX STATES OF SIMPLE MOLECULAR SYSTEMS

R. ENGLMAN

Department of Physics and Applied Mathematics, Soreq NRC, Yavne, Israel;
College of Judea and Samaria, Ariel, Israel

A. YAHALOM

College of Judea and Samaria, Ariel, Israel

CONTENTS

The Role of Degenerate States in Chemistry: A Special Volume of Advances in Chemical Physics, Volume 124, Edited by Michael Baer and Gert Due Billing. Series Editors I. Prigogine and Stuart A. Rice.
ISBN 0-471-43817-0.

I. INTRODUCTION AND PREVIEW

In quantum theory, physical systems move in vector spaces that are, unlike those in classical physics, essentially complex. This difference has had considerable impact on the status, interpretation, and mathematics of the theory. These aspects will be discussed in this chapter within the general context of simple molecular systems, while concentrating at the same time on instances in which the electronic states of the molecule are exactly or nearly degenerate. It is hoped that as the chapter progresses, the reader will obtain a clearer view of the relevance of the complex description of the state to the presence of a degeneracy.

The difficulties that arose from the complex nature of the wave function during the development of quantum theory are recorded by historians of science [1–3]. For some time during the early stages of the new quantum theory the existence of a complex state defied acceptance ([1], p. 266). Thus, both de Broglie and Schrödinger believed that material waves (or "matter" or "de Broglie" waves, as they were also called) are real (i.e., not complex) quantities, just as electromagnetic waves are [3]. The decisive step for the acceptance of the complex wave came with the probabilistic interpretation of the theory, also known as Born's probability postulate, which placed the modulus of the wave function in the position of a (and, possibly, unique) connection between theory and experience. This development took place in the year 1926 and it is remarkable that already in the same year Dirac embraced the modulus-based interpretation wholeheartedly [4]. Oddly, it was Schrödinger who appears to have, in 1927, demurred at accepting the probabilistic interpretation ([2], p. 561, footnote 350). Thus, the complex wave function was at last legitimated, but the modulus was and has remained for a considerable time the focal point of the formalism.

A somewhat different viewpoint motivates this chapter, which stresses the added meaning that the complex nature of the wave function lends to our understanding. Though it is only recently that this aspect has come to the forefront, the essential point was affirmed already in 1972 by Wigner [5] in his famous essay on the role of mathematics in physics. We quote from this here at some length:

"The enormous usefulness of mathematics in the natural sciences is something bordering on the mysterious and there is no rational explanation for... this uncanny usefulness of mathematical concepts...

The complex numbers provide a particularly striking example of the foregoing. Certainly, nothing in our experience suggests the introducing of these quantities... Let us not forget that the Hilbert space of quantum mechanics is the complex Hilbert space with a Hermitian scalar product. Surely to the unpreoccupied mind, complex numbers... cannot be suggested by physical observations. Furthermore, the use of complex numbers is not a calculational trick of applied mathematics, but comes close to being a necessity in the formulation of the laws of quantum mechanics. Finally, it now (1972) begins to appear that not only complex numbers but analytic functions are destined to play a decisive role in the formulation of quantum theory. I am referring to the rapidly developing theory of dispersion relations. It is difficult to avoid the impression that a miracle confronts us here [i.e., in the agreement between the properties of the hypernumber $\sqrt{(-1)}$ and those of the natural world]."

A shorter and more recent formulation is "The concept of analyticity turns out to be astonishingly applicable" ([6], p. 37).

What is addressed by these sources is the ontology of quantal description. Wave functions (and other related quantities, like Green functions or density matrices), far from being mere compendia or short-hand listings of observational data, obtained in the domain of real numbers, possess an actuality of their own. From a knowledge of the wave functions for real values of the variables and by relying on their analytical behavior for complex values, new properties come to the open, in a way that one can perhaps view, echoing the quotations above, as "miraculous."

A term that is nearly synonymous with complex numbers or functions is their "phase." The rising preoccupation with the wave function phase in the last few decades is beyond doubt, to the extent that the importance of phases has of late become comparable to that of the moduli. (We use Dirac's terminology [7], which writes a wave function by a set of coefficients, the "amplitudes," each expressible in terms of its absolute value, its "modulus," and its "phase.") There is a related growth of literature on interference effects, associated with Aharonov–Bohm and Berry phases [8–14]. In parallel, one has witnessed in recent years a trend to construct selectively and to manipulate wave functions. The necessary techniques to achieve these are also anchored in the phases of the wave function components. This trend is manifest in such diverse areas as coherent or squeezed states [15,16], electron transport in mesoscopic systems [17], sculpting of Rydberg-atom wavepackets [18,19], repeated and nondemolition quantum measurements [20], wavepacket collapse [21], and quantum computations [22,23]. Experimentally, the determination of phases frequently utilizes measurement of Ramsey fringes [24] or similar methods [25].

The *status* of the phase in quantum mechanics has been the subject of debate. Insomuch as classical mechanics has successfully formulated and solved problems using action-angle variables [26], one would have expected to see in the phase of the wave function a fully "observable" quantity, equivalent to and having a status similar to the modulus, or to the equivalent concept of the "number variable". This is not the case and, in fact, no exact, well-behaved Hermitean phase operator conjugate to the number is known to exist. (An article by Nieto [27] describes the early history of the phase operator question, and gives a feeling of the problematics of the field. An alternative discussion, primarily related to phases in the electromagnetic field, is available in [28]). In Section II, a brief review is provided of the various ways that phase is linked to molecular properties.

Section III presents results that the analytic properties of the wave function as a function of time t imply and summarizes previous publications of the authors and of their collaborators [29–38]. While the earlier quote from Wigner has prepared us to expect some general insight from the analytic behavior of the wave function, the equations in this section yield the specific result that, due to the analytic properties of the *logarithm* of wave function amplitudes, certain forms of phase changes lead immediately to the logical necessity of enlarging the electronic set or, in other words, to the presence of an (otherwise) unsuspected state.

In the same section, we also see that the source of the appropriate analytic behavior of the wave function is outside its defining equation (the Schrödinger equation), and is in general the consequence of either some very basic consideration or of the way that experiments are conducted. The analytic behavior in question can be in the frequency or in the time domain and leads in either case to a Kramers–Kronig type of reciprocal relations. We propose that behind these relations there may be an "equation of restriction," but while in the former case (where the variable is the frequency) the equation of restriction expresses causality (no effect before cause), for the latter case (when the variable is the time), the restriction is in several instances the basic requirement of lower boundedness of energies in (no-relativistic) spectra [39,40]. In a previous work, it has been shown that analyticity plays further roles in these reciprocal relations, in that it ensures that time causality is not violated in the conjugate relations and that (ordinary) gauge invariance is observed [40].

As already mentioned, the results in Section III are based on dispersions relations in the complex time domain. A complex time is not a new concept. It features in wave optics [28] for "complex analytic signals" (which is an electromagnetic field with only positive frequencies) and in nondemolition measurements performed on photons [41]. For transitions between adiabatic states (which is also discussed in this chapter), it was previously introduced in several works [42–45].

Interestingly, the need for a multiple electronic set, which we connect with the reciprocal relations, was also a keynote of a recent review ([46] and previous publications cited there and in [47]). Though the considerations relevant to this effect are not linked to the complex nature of the states (but rather to the *stability* of the adiabatic states in the real domain), we have included in Section III a mention of, and some elaboration on, this topic.

In further detail, Section III stakes out the following claims: For time-dependent wave functions, rigorous conjugate relations are derived between analytic decompositions (in the complex t plane) of phases and of log *moduli*. This entails a reciprocity, taking the form of Kramers–Kronig integral relations (but in the time domain), holding between observable phases and moduli in several physically important cases. These cases include the nearly adiabatic (slowly varying) case, a class of cyclic wave functions, wavepackets, and noncyclic states in an "expanding potential." The results define a unique phase through its analyticity properties and exhibit the interdependence of geometric phases and related decay probabilities. It turns out that the reciprocity property obtained in this section holds for several textbook quantum mechanical applications (like the minimum width wavepacket).

The multiple nature of the electronic set becomes especially important when the potential energy surfaces of two (or more) electronic states come close, namely, near a "conical intersection" (ci). This is also the point in the space of nuclear configurations at which the phase of wave function components becomes anomalous. The basics of this situation have been extensively studied and have been reviewed in various sources [48–50]. Recent works [51–57] have focused attention on a new contingency: when there may be several ci's between two adiabatic surfaces, their combined presence needs to be taken into account for calculations of the non-adiabatic corrections of the states and can have tangible consequences in chemical reactions. Section IV presents an analytic modeling of the multiple ci model, based on the superlinear terms in the coupling between electronic and nuclear motion. This section describes in detail a tracing method that keeps track of the phases, even when these possess singular behavior (viz., at points where the moduli vanish or become singular). The continuous tracing method is applicable to real states (including stationary ones). In these, the phases are either zero or π. At this point, it might be objected that in so far as numerous properties of molecular systems (e.g., those relating to questions of stability and, in general, to static situations and not involving a magnetic field) are well described in terms of *real* wave functions, the *complex* form of the wave function need, after all, not be regarded as a fundamental property. However, it will be shown in Section IV that wave functions that are real but are subject to a sign change, can be best treated as limiting cases in complex variable theory. In fact, the "phase tracing" method is logically connected to the time-dependent wave functions (and represents a case of mathematical "embedding").

A specific result in Section IV is the construction of highly nonlinear vibronic couplings near a ci. The construction shows, inter alia, that the connection between the Berry (or "topological," or "geometrical") phase, acquired during cycling in a parameter space, and the number of ci's circled depends on the details of the case that is studied and can vary from one situation to another. Though the subject of Berry phase is reviewed in Chapter 12 in this volume [58], we note here some recent extensions in the subject [59–61]. In these works, the phase changes were calculated for *two-electron* wave functions that are subject to interelectronic forces . An added complication was also considered, for the case in which the two electrons are acted upon by different fields. This can occur when the two electrons are placed in different environments, as in asymmetric dimers. By and large, intuitively understandable results are found for the combined phase factor but, under conditions of accidental degeneracies, surprising jumps (named "switching") are noted. Some applications to quantum computations seem to be possible [61].

The theory of Born–Oppenheimer (BO) [62,63] has been hailed (in an authoritative but unfortunately unidentified source) as one of the greatest advances in theoretical physics. Its power is in disentangling the problem of two kinds of interacting particles into two separate problems, ordered according to some property of the two kinds. In its most frequently encountered form, it is the nuclei and electrons that interact (in a molecule or in a solid) and the ordering of the treatment is based on the large difference between their masses. However, other particle pairs can be similarly handled, like hadronic mesons and baryons, except that a relativistic or field theoretical version of the BO theory is not known. The price that is paid for the strength of the method is that the remaining coupling between the two kinds of particles is dynamic. This coupling is expressed by the so-called non-adiabatic coupling terms (NACTs), which involve derivatives of (the electronic) states rather than the states themselves. "Correction terms" of this form are difficult to handle by conventional perturbation theory. For atomic collisions the method of "perturbed stationary states" was designed to overcome this difficulty [64,65], but this is accurate only under restrictive conditions. On the other hand, the circumstance that this coupling is independent of the potential, indicates that a general procedure can be used to take care of the NACTs [66]. Such general procedure was developed by Yang and Mills in 1954 [66] and has led to far reaching consequences in the theory of weak and strong interactions between elementary particles.

The interesting history of the Yang–Mills field belongs essentially to particle physics [67–70]. The reason for mentioning it here in a chemical physics setting, is to note that an apparently entirely different procedure was proposed for the equivalent problem arising in the molecular context, namely, for the elimination of the derivative terms (the NACTs) from the nuclear part of the BO Schrödinger equation through an adiabatic–diabatic transformation (ADT) matrix [71,72]. It turns out that the quantity known as the tensorial field [or covariant, or Yang–Mills (YM) field, with some other names also in use] enters also into the ADT description, though from a completely different viewpoint, namely, through ensuring the validity of the ADT matrix method by satisfaction of what is known as the "curl condition." Formally, when the "curl condition" holds, the (classical) YM field is zero and this is also the requirement for the strict validity of the ADT method. [A review of the ADT and alternative methods is available in, e.g., [48,49], the latter of which also discusses the YM field in the context of the BO treatment.] However, it has recently been shown by a formal proof, that an *approximate* construction of the ADT matrix (using only a finite, and in practice small, number of BO, adiabatic states) is possible even though the "curl condition" may be *formally* invalid [36]. An example for such an approximate construction in a systematic way was provided in a model that uses Mathieu functions for the BO electronic states [73].

As noted some time ago, the NACTs, can be incorporated in the nuclear part of the Schrödinger equation as a vector potential [74,75]. The question of a possible magnetic field, associated with this vector potential has also been considered [76–83]. For an electron occupying an admixture of two or more states (a case that is commonly designated as noncommutative, "non-Abelian"), the fields of physical interest are not only the magnetic field, being the curl of the "vector potential," but also tensorial (YM) fields. The latter is the sum of the curl field and of a vector-product term of the NACTs. Physically, these fields represent the reaction of the electron on the nuclear motion via the NACTs.

In a situation characteristic of molecular systems, a conical intersection ci arises from the degeneracy point of adiabatic potential energy surfaces in a plane of nuclear displacement coordinates. There are also a number of orthogonal directions, each representing a so-called "seam" direction. In this setting, it emerges that both kinds of fields are aligned with the seam direction of the ci and are zero everywhere outside the seam, but they differ as regards the flux that they produce. Already in a two-state situation, the fields are representation dependent and the values of the fluxes depend on the states the electron occupies. (This evidently differs from conventional electro-magnetism, in which the magnetic field and the flux are unchanged under a gauge transformation.)

Another subject in which there are implications of phase is the time evolution of atomic or molecular wavepackets. In some recently studied cases, these might be a superposition of a good 10 or so energy eigenstates. Thanks to the availability of short, femtosecond laser pulses both the control of reactions by coherent light [16,84–94] and the probing of phases in a wavepacket are now experimental possibilities [19,95–97]. With short duration excitations the initial form of the wavepacket is a *real* "doorway state" [98–100] and this develops phases for each of its component amplitudes as the wavepacket evolves. It has recently been shown that the phases of these components are signposts of a time arrow [101,102] and of the irreversibility; both of these are inherent in the quantum mechanical process of preparation and evolution [34]. It was further shown in [34] (for systems that are invariant under time reversal, e.g., in the absence of a magnetic field) that the preparation of an initially *complex* wavepacket requires finite times for its construction (and cannot be achieved instantaneously).

The quantum phase factor is the exponential of an imaginary quantity (i times the phase), which multiplies into a wave function. Historically, a natural extension of this was proposed in the form of a gauge transformation, which both multiplies into and admixes different components of a multicomponent wave function [103]. The resulting "gauge theories" have become an essential tool of quantum field theories and provide (as already noted in the discussion of the YM field) the modern rationale of basic forces between elementary particles [67–70]. It has already been noted that gauge theories have also made notable impact on molecular properties, especially under conditions that the electronic state basis in the molecule consists of more than one component. This situation also characterizes the conical intersections between potential surfaces, as already mentioned. In Section V, we show how an important theorem, originally due to Baer [72], and subsequently used in several equivalent forms, gives some new insight to the nature and source of these YM fields in a molecular (and perhaps also in a particle field) context. What the above theorem shows is that it is the *truncation* of the BO set that leads to the YM fields, whereas for a complete BO set the field is inoperative for molecular vector potentials.

Section VI shows the power of the modulus-phase formalism and is included in this chapter partly for methodological purposes. In this formalism, the equations of continuity and the Hamilton–Jacobi equations can be naturally derived in both the nonrelativistic and the relativistic (Dirac) theories of the electron. It is shown that in the four-component (spinor) theory of electrons, the two extra components in the spinor wave function will have only a minor effect on the topological phase, provided certain conditions are met (*nearly* nonrelativistic velocities and external fields that are not excessively large).

So as to make the individual sections self-contained, we have found it advisable to give some definitions and statements more than once.

II. ASPECTS OF PHASE IN MOLECULES

This section attempts a brief review of several areas of research on the significance of phases, mainly for quantum phenomena in molecular systems. Evidently, due to limitation of space, one cannot do justice to the breadth of the subject and numerous important works will go unmentioned. It is hoped that the several cited papers (some of which have been chosen from quite recent publications) will lead the reader to other, related and earlier, publications. It is essential to state at the outset that the *overall* phase of the wave function is arbitrary and only the relative phases of its components are observable in any meaningful sense. Throughout, we concentrate on the relative phases of the components. (In a coordinate representation of the state function, the "phases of the components" are none other than the coordinate-dependent parts of the phase, so it is also true that *this* part is susceptible to measurement. Similar statements can be made in momentum, energy, etc., representations.)

A further preliminary statement to this section would be that, somewhat analogously to classical physics or mechanics where positions and momenta (or velocities) are the two conjugate variables that determine the motion, moduli and phases play similar roles. But the analogy is not perfect. Indeed, early on it was questioned, apparently first by Pauli [104], whether a wave function can be constructed from the knowledge of a set of moduli alone. It was then argued by Lamb [105] that from a set of values of wave function moduli and of their rates of change, the wave function, including its phase, is uniquely found. Counterexamples were then given [106,107] and it now appears that the knowledge of the moduli and some information on the analytic properties of the wave function are both required for the construction of a state. (The following section contains a formal treatment, based partly on [30–32] and [108,109].) In a recent research effort, states with definite phases were generated for either stationary or traveling type of fields [110].

Recalling for a start phases in *classical* waves, these have already been the subject of consideration by Lord Rayleigh [111], who noted that through interference between the probed and a probing wave the magnitude and phase of acoustic waves can be separately determined, for example, by finding surfaces of minimum and zero magnitudes. A recent review on classical waves is given by Klyshko [112]. The work of Pancharatnam on polarized light beams [113,114] is regarded as the precursor of later studies of topological phases in quantum systems [9]. This work contained a formal expression for the relative phase between beams in different elliptic polarizations of light, as well as a construction (employing the so-called "Poincare sphere") that related the phase difference to a geometrical, area concept. (For experimental realizations with polarized light beams we quote [115,116]; the issue of any arbitrariness in experimentally pinning down the topological part of the phase was raised in [117].) Regarding the interesting question of any common ground between classical and quantal phases, the relation between the adiabatic (Hannay's) angle in mechanics and the phase in wave functions was the subject of [118]. The difference in two-particle interference patterns of electromagnetic and matter waves was noted, rather more recently, in [119]. The two phases, belonging to light and to the particle wave function, are expected to enter on an equal footing when the material system is in strong interaction with an electromagnetic field (as in the Jaynes–Cummings model). An example of this case was provided in a study of a two-level atom, which was placed in a cavity containing an electromagnetic field. Using one or two photon excitations, it was found possible to obtain from the Pancharatnam phase an indication of the statistics of the quantized field [120].

Several essential basic properties of phases in optics are contained in [28,41,121]. It was noted in [28], with reference to the "complex analytic signal" (an electromagnetic field with positive frequency components), that the position of zeros (from which the phase can be determined) and the intensity represent two sets of information that are intetwined by the analytic property of the wave. In Section III, we shall again encounter this finding, in the context of complex matter (Schrödinger) waves. Experimentally, observations in wave guide structures of the positions of amplitude zeros (which are just the "phase singularities") were made in [122]. An alternative way for the determination of phase is from location of maxima in interference fringes ([28], Section VII.C.2).

Interference in optical waves is clearly a phase phenomenon; in classical systems it arises from the signed superposition of positive and negative *real* wave amplitudes. To single out some special results in the extremely broad field of interference, we point to recent observations using two-photon pulse transition [94] in which a differentiation was achieved between interferences due to temporal overlap (with finite pulse width) and quantum interference caused by delay. The (component-specific) topological phase in wave functions has been measured, following the proposal of Berry in [9], by neutron interferometry in a number of works, for example, [123,124] with continual improvements in the technique. The difficulties in the use of coherent neutron beams and the possibility of using conventional neutron sources for phase-sensitive neutron radiography have been noted in a recent review [125].

Phase interference in optical or material systems can be utilized to achieve a type of quantum measurement, known as nondemolition measurements ([41], Chapter 19). The general objective is to make a measurement that does not change some property of the system at the expense of some other property(s) that is (are) changed. In optics, it is the phase that may act as a probe for determining the intensity (or photon number). The phase can change in the course of the measurement, while the photon number does not [126].

In an intriguing and potentially important proposal (apparently not further followed up), a filtering method was suggested for image reconstruction (including phases) from the modulus of the correlation function [127]. [In mathematical terms this amounts to deriving the behavior of a function in the full complex (frequency) plane from the knowledge of the absolute value of the function on the real axis, utilizing some physically realizable kernel function.] A different spectral filtering method was discussed in [128].

Before concluding this sketch of optical phases and passing on to our next topic, the status of the "phase" in the representation of observables as quantum mechanical operators, we wish to call attention to the theoretical demonstration, provided in [129], that any (discrete, finite dimensional) operator can be constructed through use of optical devices only.

The appropriate quantum mechanical *operator* form of the phase has been the subject of numerous efforts. At present, one can only speak of the best approximate operator, and this also is the subject of debate. A personal historical account by Nieto of various operator definitions for the phase (and of its probability distribution) is in [27] and in companion articles, for example, [130–132] and others, that have appeared in Volume **48** of *Physica Scripta T* (1993), which is devoted to this subject. (For an introduction to the unitarity requirements placed on a phase operator, one can refer to [133]). In 1927, Dirac proposed a quantum mechanical operator $\hat{\phi}$, defined in terms of the creation and destruction operators [134], but London [135] showed that this is not Hermitean. (A further source is [136].) Another candidate, $e^{i\hat{\phi}}$ is not unitary, as was demonstrated, for example, in [28], Section 10.7. Following that, Susskind and Glogower proposed a pair of operators $\hat{\cos}$ and $\hat{\sin}$ [137], but it was found that these do not commute with the number operator $\hat{n}$. In 1988, Pegg and Barnett introduced a Hermitean phase operator through a limiting procedure based on the state with a definite phase in a truncated Hilbert space [138]. Some time ago a comparison was made between different phase operators when used on squeezed states [139]. Unfortunately, there is as yet no consensus on the status of the Pegg–Barnett operators [121,140–142]. Maybe at least part of the difficulties are rooted in problems that arise from the coupling between the quantum system and the measuring device. However, this difficulty is a moot point in quantum mechanical measurement theory, in general.

(For the special case of a two-state systems, a Hermitean phase operator was proposed, [143], which was said to provide a quantitative measure for "phase information.")

A related issue is the experimental *accessibility* of phases: It is now widely accepted that there are essentially two experimental ways to observe phases [9,124,144]: (1) through a two-Hamiltonian, one-state method, interferometrically (viz., by sending two identically prepared rays across two regions having different fields), (2) a one-Hamiltonian, two-state method (meaning, a difference in the preparation of the rays), for example, [89,92]. (One recalls that already several years ago it was noted that there are the two ways for measuring the phase of a four-component state, a spinor [145].) One can also note a further distinction proposed more recently, namely, that between "observabilities" of bosonic and fermionic phases [146]: Boson phases are observable both *locally* (at one point) and nonlocally (at extended distances, which the wave reaches as it progresses). They can lead to phase values that are incompatible with the Bell inequalities, while fermion phases are only nonlocally observable (i.e., by interference) and do not violate Bell's inequalities. The difference resides in that only the former type of particles gives rise to a coherent state with arbitrarily large occupation number n, whereas for the latter the exclusion principle allows only $n = 0$ or 1.

The question of determination of the phase of a field (classical or quantal, as of a wave function) from the modulus (absolute value) of the field along a *real* parameter (for which alone experimental determination is possible) is known as "*the phase problem*" [28]. (True also in crystallography.) The reciprocal relations derived in Section III represent a formal scheme for the determination of phase given the modulus, and vice versa. The physical basis of these singular integral relations was described in [147] and in several companion articles in that volume; a more recent account can be found in [148]. Thus, the reciprocal relations in the time domain provide, under certain conditions of analyticity, solutions to the phase problem. For electromagnetic fields, these were derived in [120,149,150] and reviewed in [28,148]. Matter or Schrödinger waves were considered in a general manner in [39]. The more complete treatment, presented in Section III applies the results to several situations in molecular and solid-state physics. It is likely that the full scope and meaning of the modulus-phase relationship await further and deeper going analyses.

In 1984, Berry made his striking discovery of time scale independent phase changes in many-component states [9] (now variously known as Berry or topological or *geometric phase*) . This followed a line of important developments regarding the role of phases and phase factors in quantum mechanics. The starting point of these may be taken with Aharonov and Bohm's discovery of the topologically acquired phase [8], named after them. (As a curiosity, it is recorded that Bohm himself referred to the "ESAB effect" [151,152].) The achievement, stressed by the authors of [8], was to have been able to show that when an electron traverses a closed path along which the magnetic field is zero, it acquires an observable phase change, which is proportional to the "vector potential." The "topological" aspect, namely, that the path is inside a multiply connected portion of space (or that, in physical terms, the closed path cannot be shrunk without encountering an infinite barrier), has subsequently turned out to be also of considerable importance [153,154], especially through later extensions and applications of the Aharonov–Bohm phase change [155] (cf. the paper by Wu and Yang [156] that showed the importance of the phase *factor* in quantum mechanics, which has, in turn, led to several developments in many domains of physics).

In molecular physics, the "topological" aspect has met its analogue in the Jahn–Teller effect [47,157] and, indeed, in any situation where a degeneracy of electronic states is encountered. The phase change was discussed from various viewpoints in [144,158–161] and [163].

For the Berry phase, we shall quote a definition given in [164]: "*The phase that can be acquired by a state moving adiabatically (slowly) around a closed path in the parameter space of the system.*" There is a further, somewhat more general phase, that appears in any cyclic motion, not necessarily slow in the Hilbert space, which is the Aharonov–Anandan phase [10]. Other developments and applications are abundant. An interim summary was published in 1990 [78]. A further, more up-to-date summary, especially on progress in experimental developments, is much needed. (In Section IV we list some publications that report on the experimental determinations of the Berry phase.) Regarding theoretical advances, we note (in a somewhat subjective and selective mode) some clarifications regarding parallel transport, e.g., [165]. This paper discusses the "projective Hilbert space" and its metric (the Fubini-Study metric). The projective Hilbert space arises from the Hilbert space of the electronic manifold by the removal of the overall phase and is therefore a central geometrical concept in any treatment of the component phases, such as this chapter.

The term "Open-path phase" was coined for a non-fully cyclic evolution [11,14]. This, unlike the Berry-phase proper, is not gauge invariant, but is, nevertheless (partially) accessible by experiments ([30–32]). The Berry phase for nonstationary states was given in [13], the interchange between dynamic and geometric phases is treated in [117]. A geometrical interpretation is provided in [166] and a simple proof for Berry's area formula in [167]. The phases in off-diagonal terms form the basis of generalizations of the Berry phase in [168,169]; an experimental detection by neutron interferometry was recently accomplished [170]. The treatment by Garrison and Wright of complex topological phases for non-Hermitean Hamiltonians [171] was extended in [172–174]. Further advances on Berry phases are corrections due to non-adiabatic effects (resulting, mainly, in a *decrease* from the value of the phase in the adiabatic, infinitely slow limit) [30,175,176]. In [177], the complementarity between local and nonlocal effects is studied by means of some examples. For more general time-dependent Hamiltonians than the cyclic one, the method of the Lewis and Riesenfeld invariant spectral operator is in use. This is discussed in [178].

Note that the Berry phase and the open-path phase designate changes in the phases of the state components, rather than the total phase change of the wave function, which belongs to the so-called "Dynamic phase" [9,10]. The existence of more than one component in the state function is a topological effect. This assertion is based on a theorem by Longuet-Higgins ([158], "Topological test for intersections"), which states that, if the wave function of a given electronic state changes sign when transported around a loop in nuclear configuration space, then the state must become degenerate with another at some point within the loop.

From this theorem it follows that, close to the point of intersection and slightly away from it, the corresponding adiabatic or BO electronic wave functions will be given (to a good approximation) by a superposition of the two degenerate states, with coefficients that are functions of the nuclear coordinates. (For a formal proof of this statement, one has to assume, as is done in [158], that the state is continuous function of the nuclear coordinates.) Moreover, the coefficients of the two states have to differ from each other, otherwise they can be made to disappear from the normalized electronic state. Necessarily, there is also a second "superposition state," with coefficients such that it is orthogonal to the first at all points in the configuration space. (If more than two states happen to be codegenerate at a point, then the adiabatic states are mutually orthogonal superpositions of all these states, again with coefficients that are functions of the nuclear coordinates.)

If now the nuclear coordinates are regarded as dynamical variables, rather than parameters, then in the vicinity of the intersection point, the energy eigenfunction, which is a combined *electronic–nuclear* wave function, will contain a superposition of the two adiabatic, superposition states, with nuclear wave functions as cofactors. We thus see that the topological phase change leads, first, to the adiabatic electronic state being a multicomponent superposition (of diabatic states) and, second, to the full solution being a multicomponent superposition (of adiabatic states), in each case with nuclear-coordinate-dependent coefficients.

The design and *control of molecular processes* has of late become possible thanks to advances in laser technology, at first through the appearance of femtosecond laser pulses and of pump–probe techniques [179] and, more recently, through the realization of more advanced ideas, including feedback and automated control [180–183]. In a typical procedure, the pump pulse prepares a coherent superposition of energy eigenstates, and a second delayed pulse probes the time-dependent transition between an excited and a lower potential energy surface. When the desired outcome is a particular reaction product, this can be promoted by the control of the *relative phases* of two fast pulses emanating from the same coherent laser source. One of the earliest works to achieve this is [184]. A recent study focuses on several basic questions, for example, those regarding pulsed preparation of an excited state [92]. In between the two, numerous works have seen light in this fast expanding and technologically interesting field. The purpose of mentioning them here is to single out this field as an application of phases in atomic [25,95,96] and molecular [84–90] spectroscopies. In spite of the achievements in photochemistry, summarized, for example, in [185], one hardly expects phases to play a role in ordinary (i.e., not state-selective or photon-induced) chemical reactions. Still, interference (of the kind seen in double-slit experiments) has been observed between different pathways during the dissociation of water [186,187]. Moreover, several theoretical ideas have also been put forward to produce favored reaction products through the involvement of phase effects [188–194]. Calculations for the scattering cross-sections in the four-atom reaction $OH + H_2 \rightarrow H_2O + H$ showed a few percent change due to the effect of phase [195].

Wavepacket reconstruction, or imaging from observed data, requires the derivation of a *complex* function from a set of real quantities. Again, this is essentially the "phase problem," well known also from crystallography and noted above in a different context than the present one [28]. An experimental study yielded the Wigner position-momentum distribution function [88]. This approach was named a "tomographic" method, since a single beam scans the whole phase space and is distinct from another approach, in which two different laser pulses create two wavepackets: an object and a reference. When the two states are superimposed, as in a conventional holographic arrangement, the cross-term in the modulus squared retains the phase information [16,90,196]. Computer simulations have shown the theoretical proposal to be feasible. In a different work, the preparation of a long-lived atomic electron wavepacket in a Rydberg state, with principal quantum numbers around $n = 30$, was achieved [197].

Rydberg states, as well as others, can provide an illustration for another, spectacular phenomenon: wavepacket revivals [15]. In this, a superposition of $\sim$10 energy states first spreads out in phase space (due to phase decoherence), only to return to its original shape after a time that is of the order of the deviation of the spacing of the energy levels from a uniform one [198,199]. Not only is the theory firmly based, and simulations convincing, but even an application, based on this phenomenon and aimed at separation of isotopes, has been proposed [200]. Elsewhere, it was shown that the effect of slow cycling on the evolving wavepacket is to leave the revival period unchanged, but to cause a shift in the position of the revived wavepacket [201].

Coherent states and diverse *semiclassical approximations* to molecular wavepackets are essentially dependent on the relative phases between the wave components. Due to the need to keep this chapter to a reasonable size, we can mention here only a sample of original works (e.g., [202–205]) and some summaries [206–208]. In these, the reader will come across the *Maslov index* [209], which we pause to mention here, since it links up in a natural way to the modulus-phase relations described in Section III and with the phase-tracing method in Section IV. The Maslov index relates to the phase acquired when the semiclassical wave function traverses a zero (or a singularity, if there be one) and it (and, particularly, its sign) is the consequence of the analytic behavior of the wave function in the complex time plane.

The subject of *time* connects with the complex nature of the wave function in a straightforward way, through the definition in quantum mechanics of the Wigner time-reversal operator [210,211]. In a rough way, the definition implies that the conjugate of the complex wave function describes (in several instances) the behavior of the system with the time running backward. Given, on one hand, "the time-reversal invariant" structure of accepted physical theories and, on the other hand, the experience of passing time and the successes of nonequilibrium statistical mechanics and thermodynamics, the question that is being asked is: When and where does a physical theory pick out a preferred direction of time (or a "time arrow")? From the numerous sources that discuss this subject, we call attention to some early controversies [212–214] and to more recent accounts [101,215–217], as well as to a volume with philosophical orientation [102]. Several attempts have been made recently to change the original formalism of quantum mechanics by adding non-Hermitean terms [218–220], or by extending (rigging) the Hilbert space of admissible wave functions [221,222]. The last two papers emphasize the preparation process as part of the wave evolution. By an extension of this idea, it has recently been shown that the *relative* phases in a wavepacket, brought to life by fast laser pulses, constitute a unidirectional clock for the passage of time (at least for the initial stages of the wavepacket) [34]. Thus, developing phases in real life are hallmarks of both a time arrow and of irreversibility. It also emerged that, in a setting that is invariant under time reversal, the preparation of an "initially" complex wavepacket needs finite times to accomplish, that is, it is not instanteneous [34,92].

Time shifts or delays in scattering processes are present in areas as diverse as particle, molecular, and solid-state phenomena, all of which are due to the complex nature of the wave function. For a considerable time, it was thought that the instance of formation of a particle or of an excited state is restricted only by the time-energy uncertainty relation. The time *delay* τ was first recognized by Bohm [223] and by Eisenbud and Wigner [224], and was then given by Smith [67] a unifying expression in terms of the frequency (ω) derivative of the scattering (or S) matrix, as

$$\tau = \mathrm{Re}\frac{\partial \ln S}{i\partial\omega} \tag{1}$$

The Re presymbol signifies that essentially it is the phase part of the scattering matrix that is involved. A conjugate quantity, in which the imaginary part is taken, was later identifed as the particle formation time [225–228]. Real and imaginary parts of derivatives were associated with the delay time in tunneling processes across a potential barrier in the Buttiker–Landauer approach (a review is in [229].) Experimentally, an example of time delay in reflection was found recently [230]. The question of time reversal invariance, or of its default, is naturally a matter of great and continued interest for theories of interaction between the fundamental constituents of matter. A summary that provides an updating, good to its time of printing, is found in [231].

Another type of invariance, namely, with respect to unitary or gauge transformation of the wave functions (without change of norm) is a cornerstone of modern physical theories [66]. Such transformations can be global (i.e., coordinate independent) or local (coordinate dependent). Some of the observational aspects arising from gauge transformation have caused some controversy; for example, what is the effect of a gauge transformation on an observable [232,233]. The justification for gauge invariance goes back to an argument due to Dirac [134], reformulated more recently in [234], which is based on the observability of the moduli of overlaps between different wave function, which then leads to a definite phase difference between any two coordinate values, the same for all wave functions. From this, Dirac goes on to deduce the invariance of Abelian systems under an arbitrary local phase change, but the same argument holds true also for the local gauge invariance of non-Abelian, multicomponent cases [70].

We end this section of phase effects in complex states by reflecting on how, in the first place, we have arrived at a complex description of phenomena that take place in a real world. There are actually two ways to come by this situation.

First, the time-dependent wave function is necessarily complex and is due to the form of the time-dependent Schrödinger equation for *real* times, which contains i. This equation will be the starting point of Section III, where we derive some consequence arising from the analytic properties of the complex wave function. But, second, there are also defining equations that do not contain i (like the time-independent Schrödinger equation). Here, also, the wave function can be made complex through making some or other of the variables take complex values. The advantage lies frequently in removing possible ambiguities that arise in the solution at a singular point, which may be an infinity. Complex times have been considered in several theoretical works (e.g., [42,43]). It is possible to associate a purely imaginary time with *temperature*. Then, recognizing that negative temperatures are unphysical in an unrestricted Hilbert space, we immediately see that the upper and lower halves of the complex t plane are nonequivalent. Specifically, regions of nonanalytic behavior are expected to be found in the upper half, which is the one that corresponds to negative temperatures, and analytic behavior is expected in the lower half plane that corresponds to positive temperatures. The formal extension of the nuclear coordinate space onto a complex plane, as is done in [44,45], is an *essentially* equivalent procedure, since in the semiclassical formalism of these works the particle coordinates depend parametrically on time. Complex topological phases are considered in, e.g., [171,172], which can arise from a non-Hermitean Hamiltonian. The so-called Regge poles are located in the complex region of momentum space. (A brief review well suited for molecular physicists is in [235]). The plane of complex-valued interactions is the subject of [236].

In addition, it can occasionally be useful to regard some physical parameter appearing in the theory as a complex quantity and the wave function to possess analytic properties with regard to them. This formal procedure might even include fundamental constants like e, h, and so on.

III. ANALYTIC THEORY OF COMPLEX COMPONENT AMPLITUDES

A. Modulus and Phase

With the time-dependent Schrödinger equation written as

$$i\frac{\partial\Psi(x,t)}{\partial t} = H(x,\, t)\Psi(x,\, t) \tag{2}$$

[in which t is time, x denotes all particle coordinates, $H(x, t)$ is a real Hamiltonian, and $\hbar = 1$], the presence of i in the equation causes the solution

$\Psi(x, t)$ to be complex valued. Writing $\Psi(x, t)$ in a logarithmic form and separating as

$$\ln \Psi(x, t) = \ln (|\Psi(x, t)|) + i\arg(\Psi(x, t)) \tag{3}$$

we have in the first term the modulus $|\Psi(x, t)|$ and in the second term arg, the "phase." It is the latter that expresses the signed or complex valued nature of the wave function. In this section, we shall investigate what, if any, interrelations exist between moduli and phases? Are they independent quantities or, more likely since they derive from a single equation (2), are they interconnected? The result will be of the form of "reciprocal" relations, shown in Eqs. (9) and (10). Some approximate and heuristic connections between phases and moduli have been known before ([2] Vol. 5, Part 2, Section IV.5); [237–241]; we shall return to these in Section III.C.3.

B. Origin of Reciprocal Relations

Contrary to what appears at a first sight, the integral relations in Eqs. (9) and (10) are not based on causality. However, they can be related to another principle [39]. This approach of expressing a general principle by mathematical formulas can be traced to von Neumann [242] and leads in the present instance to an "equation of restriction," to be derived below. According to von Neumann complete description of physical systems must contain:

1. A set of quantitative characterizations (energy, positions, velocities, charges, etc.).
2. A set of "properties of states" (causality, restrictions on the spectra of self-energies, existence or absence of certain isolated energy bands, etc.).

As has been shown previously [243], both sets can be described by eigenvalue equations, but for the set 2 it is more direct to work with projectors Pr taking the values 1 or 0. Let us consider a class of functions $f(x)$, describing the state of the system or a process, such that (for reasons rooted in physics) $f(x)$ should vanish for $x \notin D$ (i.e., for supp $f(x) = D$, where D can be an arbitrary domain and x represents a set of variables). If $\mathrm{Pr}_D(x)$ is the projector onto the domain D, which equals 1 for $x \in D$ and 0 for $x \notin D$, then all functions having this state property obey an "equation of restriction" [244]:

$$f(x) = \mathrm{Pr}_D(x) f(x) \tag{4}$$

The "equation of restriction" can embody causality, lower boundedness of energies in the spectrum, positive wavenumber in the outgoing wave (all these in nonrelativistic physics) and interactions inside the light cone only, conditions of mass spectrality, and so on in relativistic physics. In the case of interest in this chapter, the "equation of restriction" arises from the lower boundedness of energies (E), or the requirement that (in nonrelativistic physics) one must have $E > 0$ (where we have arbitrarily chosen the energy lower bound as equal to zero).

Applying to Eq. (4) an integral transform (usually, a Fourier transform) F_k, one derives by (integral) convolution, symbolized by $\otimes_k$, the expression

$$f(k) = F_k[\mathrm{Pr}_D(x)] \bigotimes_k f(k) = \int F_{k-k'}[\mathrm{Pr}_D(x)] f(k') dk' \tag{5}$$

For functions of a single variable (e.g., energy, momentum or time) the projector $\mathrm{Pr}_D(x)$ is simply $\Theta(x)$, the Heaviside step function, or a combination thereof. When also replacing x, k by the variables E, t, the Fourier transform in Eq. (5) is given by

$$F_t[\Theta(E)] = \delta_+(t) \equiv \frac{1}{2}\left[\delta(t) - \frac{i}{\pi} P\left(\frac{1}{t}\right)\right] \tag{6}$$

where P designates the principal part of an integral. Upon substitution into Eq. (5) (with k replaced by t) one obtains after a slight simplification

$$f(t) = \frac{i}{\pi} P \int_{-\infty}^{\infty} \frac{1}{t' - t} f(t') dt' \tag{7}$$

Real and imaginary parts of this yield the basic equations for the functions appearing in Eqs. (9) and (10). (The choice of the upper sign in these equations will be justified in a later subsection for the ground-state component in several physical situations. In some other circumstances, such as for excited states in certain systems, the lower sign can be appropriate.)

1. *A General Wavepacket*

We can state the form of the conjugate relationship in a setting more general than $\Psi(x, t)$, which is just a particular, the coordinate representation of the evolving state. For this purpose, we write the state function in a more general way, through

$$|\Psi(t)\rangle = \sum_n \phi_n(t)|n\rangle \tag{8}$$

where $|n\rangle$ represent some time-independent orthonormal set and $\phi_n(t)$ are the corresponding amplitudes. We shall write generically $\phi(t)$ for any of the "component amplitudes" $\phi_n(t)$ and derive from it, in Eq. (15), a new function $\chi_{\pm}(t)$ that retains all the *fine-structured* time variation in $\ln \Psi(t)$ and is free of the large-scale variation in the latter. We then derive in several physically important cases, but not in all, reciprocal relations between the modulus and phase of $\chi(t)$ taking the form

$$\frac{1}{\pi} P \int_{-\infty}^{\infty} dt' [\ln |\chi(t')|]/(t' - t) = \pm \arg \chi(t) \tag{9}$$

and

$$\frac{1}{\pi} P \int_{-\infty}^{\infty} dt' [\arg \chi(t')]/(t' - t) = \mp \ln |\chi(t)| \tag{10}$$

The sign alternatives depend on the location of the zeros (or singularities) of $\chi(t)$. The above conjugate, or reciprocal, relations are the main results in this section. When Eqs. (9) and (10) hold, $\ln |\chi(t)|$ and $\arg \chi(t)$ are "Hilbert transforms" [245,246].

Later in this section, we shall specify the analytic properties of the functions involved and obtain exact formulas similar to Eqs. (9) and (10), but less simple and harder to apply to observational data of, say, moduli.

In Section III.C.5, we give conditions under which Eqs. (9) and (10) are exactly or approximately valid. Noteworthy among these is the nearly adiabatic (slowly evolving) case, which relates to the Berry phase [9].

C. Other Phase-Modulus Relations

As a prelude to the derivation of our results, we note here some of the relations between phases and moduli that have been known previously. The following is a list (presumably not exhaustive) of these relations. Some of them are standard textbook material.

1. The Equation of Continuity

This was first found by Schrödinger in 1926 starting with Eq. (2), which he called the "eigentliche Wellengleichung." (Paradoxically, this got translated to "real wave equation" [2].) In the form

$$2m \frac{\partial \ln |\Psi(x, t)|}{\partial t} + 2\nabla \ln |\Psi(x, t)|.\nabla \arg[\Psi(x, t)] + \nabla \cdot \nabla \arg[\Psi(x, t)] = 0 \tag{11}$$

(where m is the particle mass), it is clearly a differential relation between the modulus and the phase. As such, it does not show up any discontinuity in the phase [125], whereas Eqs. (9) and (10) do that. We further note that the above form depends on the Hamiltonian and looks completely different for, for example, a Dirac electron.

2. *The WKB Formula*

In the classical region of space, where the potential is less than the energy, the standard formula leads to an approximate relation between phase and modulus in the form of the following path integral ([237], Section 28)

$$\arg \Psi(x) = \pm C \int_0^{x(t)} |\Psi(x)|^{-2} dx \tag{12}$$

where C is a normalization constant. This and the following example do not arise from the time-dependent Schrödinger equation; nevertheless, time enters naturally in a semiclassical interpretation [205].

3. *Extended Systems*

By extending some previous heuristic proposal [238,239], the phase in the polarized state of a 1D solid of macroscopic length L was expressed in [240] as

$$\arg \Psi(x) = \text{Im ln} \int_0^L e^{2\pi i x/L} |\Psi(x)|^2 dx \tag{13}$$

Note [240] that the phase in Eq. (13) is gauge independent. Based on the above mentioned heuristic conjecture (but fully justified, to our mind, in the light of our rigorous results), Resta noted that "Within a finite system two alternative descriptions [in terms of the squared modulus of the wave function, or in terms of its phase] are equivalent" [247].

4. *Loss of Phase in a Quantum Measurement*

In a self-consistent analysis of the interaction between an observed system and the apparatus (or environment), Stern et al. [241] proposed both a phase-modulus relationship ([241], Eq. (3.10)) and a deep lying interpretation. According to the latter, the decay of correlation between states in a superposition can be seen, equivalently, as the effect of either the environment upon the system or the back-reaction of the system on its environment. The reciprocal relations refer to the wave function of the (microscopic) system and not to its surroundings, thus there is only a change of correlation not a decay. Still it seems legitimate to speculate that the dual representation of the change that we have found (viz., through the phase or through the modulus) might be an expression of the reciprocal effect of the coupling between the system (represented by its states) and its environment (acting through the potential).

D. The Cauchy-Integral Method for the Amplitudes

Since the amplitude $\phi(t)$ arises from integration of Eq. (2), it can be assumed to be uniquely given. We can further assume that $\phi(t)$ has no zeros on the real t axis, except at those special points, where this is demanded by symmetry. The reason for this is that, in general, $\phi(t) = 0$ requires the solution of two equations, for the real and the imaginary parts of $\phi(t)$ and this cannot be achieved with a single variable: a real t. (Arguing from a more physical angle, if there is a zero somewhere on a the real t axis, then a small change in some parameter in the Hamiltonian, will shift this zero to a complex t. However, this small change cannot change the physical content of the problem and thus we can just as well start with the case where the zeros are away from the real axis.) We can therefore perform the decomposition of $\ln \phi(t)$, following [248,249]:

$$\ln \phi(t) = \ln \phi_+(t) + \ln \phi_-(t) \tag{14}$$

where $\ln \phi_+(t)$ is analytic in a portion of the complex t plane that includes the real axis (or, as stipulated in [248], "including a strip of finite width about the real axis") and a large semicircular region above it and $\ln \phi_-(t)$ is analytic in the corresponding portion below and including the real axis. By defining new functions $\chi_\pm(t)$, we separate off those parts of $\ln \phi_\pm(t)$ that do not vanish on the respective semicircles, in the form:

$$\ln \phi_\pm(t) = P_\pm(t) + \ln \chi_\pm(t) \tag{15}$$

where $\ln \chi_+(t)$ and $\ln \chi_-(t)$ are, respectively, analytic in the upper and lower half of the complex t plane and vanish in their respective half-planes for large $|t|$. The choices for suitable $P_\pm(t)$ are not unique, and only the end result for $\ln \phi_\pm(t)$ is. In the interim stage, we apply to the functions $\ln \chi_\pm(t)$ Cauchy's theorem with a contour C that consists of an infinite semicircle in the upper $(+)$, or lower $(-)$ half of the complex t' plane traversed anticlockwise $(+)$ or clockwise $(-)$ and a line along the real t' axis from $-\infty$ to $+\infty$ in which the point $t' = t$ is avoided with a small semicircle. We obtain

$$\oint_C \frac{\ln \chi_\pm(t')}{(t'-t)} dt' = \pm 2\pi i \ln \chi_\pm(t) \qquad \text{or zero} \tag{16}$$

depending on whether the small semicircle is outside or inside the half-plane of analyticity and the sign $\pm$ is taken to be consistently throughout. Further, writing the logarithms as

$$\ln \chi_\pm(t) = \ln |\chi_\pm(t)| + i \arg \chi_\pm(t) \tag{17}$$

and separating real and imaginary parts of the functions in Eq. (16) we derive the following relations between the amplitude moduli and phases in the wave function:

$$\left(\frac{1}{\pi}\right)P\int_{-\infty}^{\infty}\frac{[\log|\chi_-(t')|-\log|\chi_+(t')|]}{(t'-t)}dt' = \arg\chi_+(t)+\arg\chi_-(t) = \arg\chi(t) \tag{18}$$

and

$$\left(\frac{1}{\pi}\right)P\int_{-\infty}^{\infty}\frac{[\arg\chi_+(t')-\arg\chi_-(t')]}{(t'-t)}dt' = \log|\chi_-(t)|+\log|\chi_+(t)| = \log|\chi(t)| \tag{19}$$

E. Simplified Cases

We shall now concentrate on several cases where relations equations (18) and (19) simplify. The most favorable case is where $\ln\phi(t)$ is analytic in one half-plane, (say) in the lower half, so that $\ln\phi_+(t)=0$. Then one obtains reciprocal relations between observable amplitude moduli and phases as in Eqs. (9) and (10), with the upper sign holding. Solutions of the Schrödinger equation are expected to be regular in the lower half of the complex t plane (which corresponds to positive temperatures), but singularities of $\ln\phi(t)$ can still arise from zeros of $\phi(t)$. We turn now to the location of these zeros.

1. *The Near-Adiabatic Limit*

We wish to prove that as the adiabatic limit is approached, the zeros of the component amplitude for the "time-dependent ground state" (TDGS, to be presently explained) are such that for an overwhelming number of zeros t_r, $\operatorname{Im} t_r > 0$ and for a fewer number of other zeros $|\operatorname{Im} t_s| \ll 1/\Delta E \ll 2\pi/\omega$, where ΔE is the characteristic spacing of the eigenenergies of the Hamiltonian, and $2\pi/\omega$ is the timescale (e.g., period) for the temporal variation of the Hamiltonian. The TDGS is that solution of the Schrödinger equation (2) that is initially in the ground state of $H(x, 0)$, the Hamiltonian at $t=0$. It is known that in the extreme adiabatic (infinitesimally slow) limit a system not crossing degeneracies stays in the ground state (the adiabatic principle). We shall work in the nearly adiabatic limit, where the principle is approximately, but not precisely, true.

By expanding $\Psi(x, t)$ in the eigenstates $|n\rangle$ of $H(x, 0)$, we have

$$\Psi(x, t) = \sum_n C_n(t)\langle x|n\rangle \tag{20}$$

and we assume (for simplicity's sake) that the expansion can be halted after a finite number (say, $N + 1$) of terms, or that the coefficients decrease in a sufficiently fast manner (which will not be discussed here). Expressing the matrix of the Hamiltonian H as $Gh_{nm}(t)$, where $h_{nm}(t)$ is of the order of unity and G positive, we obtain (with the dot denoting time differentiation)

$$\dot{C}_n(t) = -iG \sum_m h_{nm}(t) C_m(t) \tag{21}$$

The adiabatic limit is characterized by

$$|\dot{h}_{nm}(t)| \ll |G| \tag{22}$$

We shall find that in the TDGS [i.e., $\Psi_g(x, t)$], the coefficient $C_g(t)$ of $\langle x|g\rangle$ has the form

$$C_g(t) = B_{gg}(t)e^{-iG\varphi_g} + \sum_m B_{gm}(t)e^{-iG\varphi_m} \tag{23}$$

Here, $\varphi_m = \varphi_m(t)$ are time integrals of the eigenvalues $e_m(t)$ of the matrix $h_{nm}(t)$

$$\varphi_m(t) = \int_0^t e_m(t')dt' \tag{24}$$

In the sum, the value $m = g$ is excluded and (as will soon be apparent) B_{gm}/B_{gg} is small of the order of

$$\frac{|\dot{h}_{nm}(t)|}{G} \tag{25}$$

To find the roots of $C_g(t) = 0$ we divide Eq. (23) by the first term shown and transfer the unity to the left-hand side to obtain an equation of the form

$$1 = c_1(t)e^{-iG\delta e_1 t} + c_2(t)e^{-iG\delta e_2 t} + \cdots \qquad \text{to } N \text{ terms} \tag{26}$$

where $\delta e_1 t$, and so on represent the differences $\varphi_m - \varphi_g$ and are necessarily positive and increasing with t, for noncrossing eigenvalues of $h_{nm}(t)$. (They are written in the form shown to make clear their monotonically increasing character and are exact, by the mean value theorem, with δe_1, etc., being some positive function of t.) The parameters $c_1(t)$, and so on, are small near the adiabatic limit, where G is large. It is clear that Eq. (26) has solutions only at points where $\operatorname{Im} t > 0$. That the number of (complex) roots of Eq. (26) is very large in the adiabatic limit, even if Eq. (26) has only a few number of terms, can be seen upon writing $e^{it|h_{nm}|} = z$ and regarding Eq. (26) as a polynomial equation in z^{-1}. Then the number of solutions increases with G. Moreover, these solutions can be expected to recur periodically provided the δe values approach to being commensurate.

It remains to investigate the zeros of $C_g(t)$ arising from having divided out by $B_{gg}(t)e^{-iG\varphi_g}$. The position and number of these zeros depend only weakly on G, but depends markedly on the form that the time-dependent Hamiltonian $H(x, t)$ has. It can be shown that (again due to the smallness of $c_1, c_2, \ldots$) these zeros are near the real axis. If the Hamiltonian can be represented by a small number of sinusoidal terms, then the number of fundamental roots will be small. However, in the t plane these will recur with a period characteristic of the periodicity of the Hamiltonian. These are relatively long periods compared to the recurrence period of the roots of the previous kind, which is characteristically shorter by a factor of

$$\frac{|\dot{h}_{nm}(t)|}{G} \tag{27}$$

This establishes our assertion that the former roots are overwhelmingly more numerous than those of the latter kind. Before embarking on a formal proof, let us illustrate the theorem with respect to a representative, though specific example. We consider the time development of a doublet subject to a Schrödinger equation whose Hamiltonian in a doublet representation is [13,29]

$$H(t) = G/2\begin{pmatrix} -\cos(\omega t) & \sin(\omega t) \\ \sin(\omega t) & \cos(\omega t) \end{pmatrix} \tag{28}$$

Here, ω is the angular frequency of an external disturbance. The eigenvalues of Eq. (28) are $G/2$ and $-G/2$. If $G > 0$, then in the ground state the amplitude of $|g\rangle$ [=the vector $\binom{1}{0}$ in Eq. (28)] is

$$\begin{aligned} C_g = {} & \cos(Kt)\cos(\omega t/2) + (\omega/2K)\sin(Kt)\sin(\omega t/2) \\ & + i(G/2K)\sin(Kt)\cos(\omega t/2) \end{aligned} \tag{29}$$

with

$$K = 0.5\sqrt{G^2 + \omega^2} \approx G/2 \qquad \text{since} \qquad G/\omega \gg 1 \tag{30}$$

Thus the amplitude in Eq. (29) becomes

$$\begin{aligned} C_g(t) & \approx \exp(iKt)\cos(\omega t/2) + (\omega/2K)\sin(Kt)\sin(\omega t/2) \\ & \approx \exp(iGt/2)[\cos(\omega t/2) - i(\omega/2G)\sin(\omega t/2)] \\ & \quad + i\exp(-iGt/2)(\omega/2G)\sin(\omega t/2) \end{aligned} \tag{31}$$

This is precisely of the form Eq. (23), with the second term being smaller than the first by the small factor shown in Eq. (25). Equating (31) to zero and dividing by the first term, we recover the form in Eq. (26), whose right-hand side consists now of just one term. For an integer value of $G/\omega = M$ (say), which is large and $\exp(-i\omega t) = Z$, the resulting equation in Z has $\sim M$ roots with $|Z| > 1$ (or, what is the same, Im $t > 0$). As noted above, further roots of $C_g(t)$ will arise from the neighborhood of $\cos(\omega t/2) = 0$, or $Z = -1$. [The upper state of the doublet states has the opposite properties, viz., $\sim M$ roots with $\text{Im}\, t < 0$. We have treated this case (in collaboration with Baer) in a previous work [29].]

A formal derivation of the location of the zeros of $C_g(t)$ for a general adiabatic Hamiltonian can be given, following proofs of the adiabatic principle (e.g., [250–252]). The last source, [252] derives an evolution operator U, which is written there, with some slight notational change, in the form

$$U(t) = A(t)\Phi(t)W(t) \tag{32}$$

(Eq. XVII.86 in the reference source [252]). Here $A(t)$ is a unitary transformation (Eq. XVII.70 in [252]) that "takes any set of basis vectors of $H(x, 0)$ into a set of basis vectors of $H(x, t)$ in a continuous manner" and is independent of G. In the previously worked example, its components are of the form $\cos(\omega t/2)$ and $\sin(\omega t/2)$ [252]. The next factor in Eq. (32) is diagonal [252] Eq. (XVII.68) and consists of terms of the form:

$$\Phi(t) = \exp(-iG\varphi_m)\delta_{nm} \tag{33}$$

Finally, the unitary transformation $W(t)$ was shown to have a near-diagonal form ([252], Eq. XVII.97)

$$W(t) = \delta_{nm} + \left(\frac{|\dot{h}(t)|}{G}\right)\delta W_{nm} \tag{34}$$

The gg component of the evolution operator U is just C_g and is, upon collecting the foregoing,

$$C_g(t) = \sum_m A_{gm}(t)\exp(-iG\varphi_m)[\delta_{mg} + \left(\frac{|\dot{h}(t)|}{G}\right)\delta W_{mg}] \tag{35}$$

This can be rewritten as

$$\begin{aligned} C_g(t) = A_{gg}(t)\left[1 + \left(\frac{|\dot{h}(t)|}{G}\right)\delta W_{gg}\right]\exp(-iG\varphi_g) \\ + \left(\frac{|\dot{h}(t)|}{G}\right)\sum_m{}' A_{gm}(t)\exp(-iG\varphi_m)\delta W_{mg} \end{aligned} \tag{36}$$

with the summation excluding g. This is again of the form of Eq. (23), establishing the generality of the location of the eigenvalues for the nearly adiabatic case.

2. *Cyclic Wave Functions*

This case is particularly interesting for two reasons. First, time-periodic potentials such that arise from external periodic forces, frequently give rise to cyclically varying states. (According to the authors of [253]: "The universal existence of the cyclic evolution is guaranteed for any quantum system.") The second reason is that the Fourier expansion of the cyclic state spares us the consideration of the convergence of the infinite-range integrals in Eqs. (9) and (10); instead, we need to consider the convergence of the (discrete) coefficients of the expansion. In this section, we show that in a broad class of cyclic functions one-half of the complex t plane is either free of amplitude zeros, or has zeros whose contributions can be approximately neglected. As already noted, in such cases, the reciprocal relations connect observable phases and moduli (exactly or approximately). The essential step is that a function $\phi(t)$ cyclic in time with period 2π can be written as a sine–cosine series. We assume that the series terminates at the Nth trigonometric function, with N finite. We can write the series as a polynomial in z, where $z = \exp(it)$, in the form

$$\phi(t) = \sum_{m=0}^{2N} c_m z^{m-N} \tag{37}$$

$$= z^{-N} c_0 \chi(t)$$

$$= z^{-N} c_0 \sum_{m=0}^{2N} \frac{c_m}{c_0} z^m \tag{38}$$

If $\phi(t)$ is a wave function amplitude arising from a Hamiltonian that is time-inversion invariant, then we can choose $\phi(-t) = \phi^*(t)$ for real t, where the star denotes the complex conjugate. Then, the coefficients c_m are all real. Next, factorize in products as

$$\chi(t) = \prod_{k=1}^{2N} (1 - z/z_k) \tag{39}$$

where z_k are the (complex) zeros of $\chi(t)$ or $\phi(t)$, $2N$ in number. Then the decomposition shown in Eq. (15), namely, $\ln\chi(t) = \ln\chi_+(t) + \ln\chi_-(t)$, will be achieved with

$$\ln\chi_+(t) = \sum_{k=1}^{R} \ln(1 - z/z_{k+}) \qquad |z_{k+}| \geq 1 \tag{40}$$

$$\ln\chi_-(t) = \sum_{k=R+1}^{2N} \ln(1 - z/z_{k-}) \qquad |z_{k-}| < 1 \tag{41}$$

provided that R of the roots are on or outside the unit circle in the z plane and $2N - R$ roots are inside the unit circle. The results in Eqs. (18) and (19) for the phases and amplitudes can now be applied directly. But it is more enlightening to obtain the coefficients in the complex Fourier series for the phases and amplitudes. This is easily done for Eq. (40), since for each term in the sum

$$|z/z_{k+}| = |\exp(it)/z_{k+}| < 1 \tag{42}$$

and the series expansion of each logarithm converges. (When, in Eq. (42) equality reigns, which is the case when the roots are upon the unit circle, the convergence is "in the mean" [254].) Then the nth Fourier coefficient is simply the coefficient of the term $\exp(int)$ in the expansion, namely, $-(1/n)(1/z_{k+})^n$.

The corresponding series expansion of $\ln \chi_-(t)$ in Eq. (41) is not legitimate, since now in every term

$$|z/z_{k-}| = |\exp(it)/z_{k-}| > 1 \tag{43}$$

Therefore, we rewrite

$$\ln \chi_-(t) = -\sum_{k=R+1}^{2N} \ln(-z_{k-}) + (2N-R)it + \sum_{k=R+1}^{2N} \ln(1 - z_{k-}/z) \tag{44}$$

Each logarithm in the last term can now be expanded and the $(-n)$th Fourier coefficient arising from each logarithm is $-(1/n)(z_{k-})^n$. To this must be added the $n = 0$ Fourier coefficient coming from the first, t-independent term and that arising from the expansion of second term as a periodic function, namely,

$$it = -2i\sum_n (-1)^n \sin(nt)/n \tag{45}$$

For the Fourier coefficients of the modulus and the phase we note that, because of the time-inversion invariance of the amplitude, the former is even in t and the latter is odd. Therefore the former is representable as a cosine series and the latter as a sine series. Formally,

$$\ln(\chi) = \ln|(\chi)| + i\arg(\chi) = \sum_n A_n \cos(nt) + i\sum_n B_n \sin(nt) \tag{46}$$

When expressed in terms of the zeros of χ, the sin–cos coefficients of the log modulus and of the phase are, respectively,

$$A_0 = -\sum_{k=R+1}^{2N} \ln|z_{k-}| \tag{47}$$

[This is written in terms of $|z_{k-}|$, the moduli of the roots z_{k-}, since the roots are either real or come in mutually complex conjugate pairs. In any case, this constant term can be absorbed in the polynomial $P(t)$ in Eq. (15).]

$$A_n = \left[\sum_{k=1}^{R} 1/(z_{k+})^n + \sum_{k=R+1}^{2N} (z_{k-})^n \right] \Big/ n \tag{48}$$

$$B_n = \left[\sum_{k=1}^{R} 1/(z_{k+})^n - \sum_{k=R+1}^{2N} [(z_{k-})^n - 2(-1)^n] \right] \Big/ n \tag{49}$$

Equations (47)–(49) are the central results of this section. Though somewhat complicated, they are easy to interpret, especially in the limiting cases (a–d), to follow. In the general case, the equations show that the Fourier coefficients are given in terms of the amplitude zeros. (a) When there are no amplitude zeros in one of the half-planes, then only one of the sums in Eq. (48) or (49) is nonzero (R is either 0 or $2N$). Consequently, the Fourier coefficients of the log modulus and of the phase are the same (up to a sign) and the two quantities are logically interconnected as functions of time. The connection is identical with that exhibited in Eqs. (9) and (10). In the two-state problem formulated by Eq. (28), the solution (29) is cyclic provided K/ω is an integer. A "Mathematica" output of the zeros of Eq. (29) for $K/\omega = 8$ gives the following results: None of the zeros is located in the lower half-plane, seven pairs and an odd one are in the upper half-plane proper, a pair of zeros is on the real t axis. The reciprocal integral relations in Eqs. (9) and (10) are verified numerically, as seen in Figure 1. (The equality between the Fourier coefficients A_n and B_n was verified independently.) (b) It is a characteristic of the above two-state problem (with general values of K/ω), and of other problems of similar type that there is one or more roots at or near $z_{k+} = -1(t = \pm\pi$; the generality of the occurrence of these roots goes back to a classic paper on conical intersection [255].) By inspection of the second sum in Eq. (49), we find that, if all the roots located in the upper half-plane are of this type, then $A_n = B_n$ up to small quantities of the order of $(z_{k+} + 1)$. Then again Eqs. (9) and (10) can be employed. (c) As a corollary to the previous observation (and an important one in view of the stipulation in Section III.C.4, that the wave function has no real zeros) a small shift in the location of a zero originally at $t = \pm\pi$ into the complex plane either just above or just below this location, will only have a small consequence on the Fourier coefficients. Therefore, zeros of this type do not violate the assumptions of the theory. (d) If either $|z_{k+}| \gg 1$ or $|z_{k-}| \ll 1$, it is clear from Eqs. (48) and (49) that the contribution of such roots is small. This circumstance is important for the following reason: Suppose that the model is changed slightly by adding to the potential a small term, for example, adding $\epsilon \cos 2\omega t$ to a diagonal matrix element

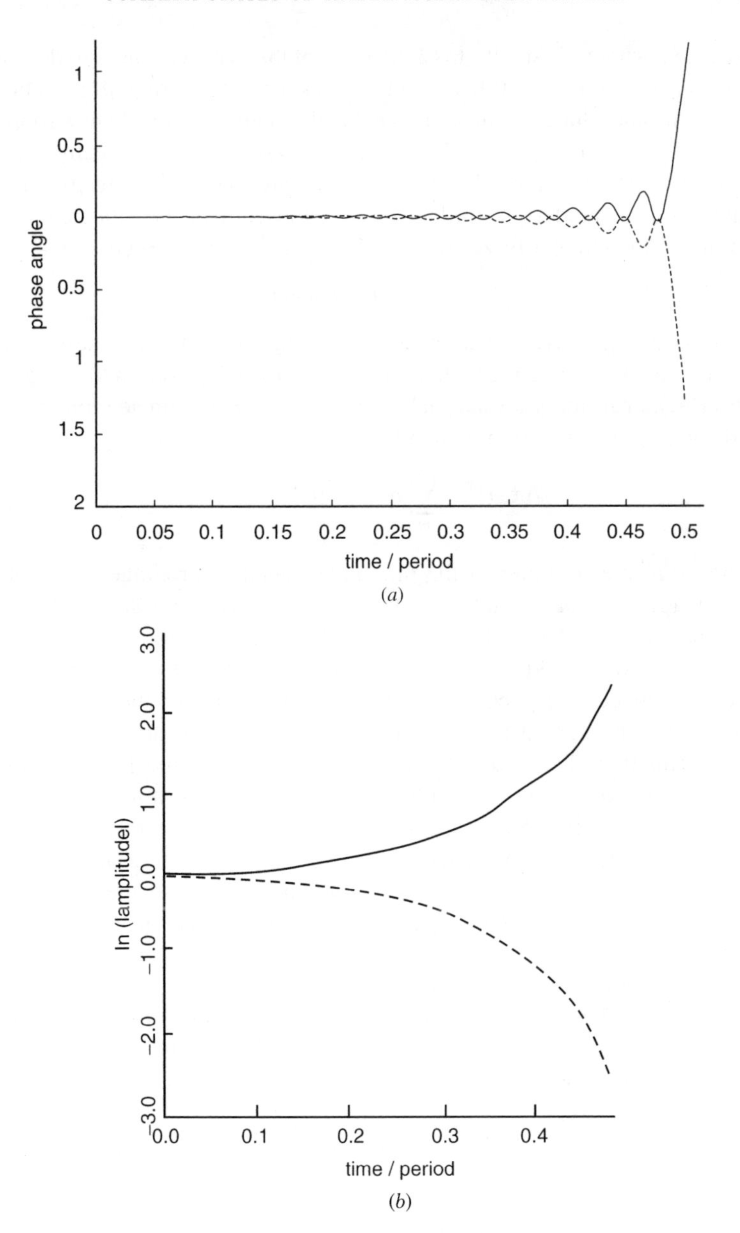

Figure 1. Numerical test of the reciprocal relations in Eqs. (9) and (10) for C_g shown in Eq. (29). The values computed directly from Eq. (29) are plotted upward and the values from the integral downward (by broken lines) for $K/\omega = 16$. The two curves are clearly identical. (*a*) $\ln|C_g(t)|$ against (t/period). The modulus is an even function of t. (*b*) $\arg C_g(t)$ against (t/period). The phase is odd in t.

in Eq. (28), where ϵ is small. (In [256] terms of this type were used to describe the nonlinear part of a Jahn–Teller effect.) Necessarily, this term will introduce new zeros in the amplitude. It can be shown that this addition will add new roots of the order $|z_{k+}| \approx 1/\epsilon$ or $|z_{k-}| \approx \epsilon$. The effects of these are asymptotically negligible. In other words, the formula (48) and (49) are stable with respect to small variations in the model. [A similar result is known as Rouche's theorem about the stability of the number of zeros in a finite domain ([257], Section 3.42).]

3. *Wave Packets*

A time-varying wave function is also obtained with a time-independent Hamiltonian by placing the system initially into a superposition of energy eigenstates ($|n\rangle$), or forming a wavepacket. Frequently, a coordinate representation is used for the wave function, which then may be written as

$$\Psi(x, t) = \sum_m a_m(t) \exp(-iE_m t)\langle x|m\rangle \tag{50}$$

where $\langle x|m\rangle$ are solutions of the time-independent Schrödinger equation, with eigenenergies E_m that are taken as nondegenerate and increasing with m. In this coordinate representation, the "component amplitudes" in the introduction are just fancy words for $\Psi(x, t)$ at fixed x [so that the discrete state label n that we have used in Eq. (8) is equivalent to the continuous variable x] and $\phi_n(t)$ is simply $\langle x(t)|\Psi(x, t)\rangle$. The results in the earlier section are applicable to the present situation. Thus, to test Eq. (9) or (10), one would look for any fixed position x in space at the moduli (or state populations) as a function of time, as with repeated state-probing set ups. In turn, by some repeated interference experiments at the same point x, one would establish the phase and then compare the results with those predicted by the equations. (Of course, the same equations can also be used to predict one quantity, provided the time history of the second is known.)

As in previous sections, the zeros of $\Psi(x, t)$ in the complex t plane at fixed x are of interest. This appears a hopeless task, but the situation is not that bleak. Thus, let us consider a wavepacket initially localized in the ground state in the sense that in Eq. (50), for some given x,

$$\sum_{m>0} |a_m\langle x|m\rangle|^2 < |a_0\langle x|0\rangle|^2 \tag{51}$$

Then, we expect that for this value(s) of the coordinate x, the t zeros of the wavepacket will be located in the upper t half-plane only. The reason for this is similar to the reasoning that led to the theorem about the location of zeros in the near-adiabatic case. (Section III.E.1). Actually, empirical investigation of wavepackets appearing in the literature indicates that the expectation holds in a broader range of cases, even when the condition (51) is not satisfied. It should be mentioned that much of the wavepacket work is numerical and it is not easy to theorize about it. (A review describing certain aspects of wavepackets is found in [258].)

We now present some examples of studied wavepackets for which the reciprocal relations hold (exactly or approximately), but have not been noted.

1. *Free Particle in 1D.* The Hamiltonian consists only of the kinetic energy of the particle having mass m ([237] Section 28, [259]). The (unnormalized) energy eigenstates labeled by the momentum index k are

$$\psi_k(x) = \exp(ikx) \tag{52}$$

with corresponding energies $E_k = k^2/2m$. Initially, the wave packet is centered on $x = 0$ and has mean momentum K. As shown in [259], the coefficients a_k appearing in Eq. (50) are

$$a_k = \exp[-(k-K)^2\Delta^2] \tag{53}$$

where $\Delta(> 0)$ is the root-mean-square width in the initial wave packet. The expanding wave packet can be written as

$$\ln \Psi(x, t) = -1/2\ln[\Delta + (it/2m\Delta)] - \frac{[x^2 - 4i\Delta^2 K(x - Kt/2m)]}{[4\Delta^2 + 2it/m]} + \text{constant} \tag{54}$$

which is clearly analytic in the lower half t plane. (The singularity arises because free electron wave functions are not normalizable.) We can therefore identify this function with $\ln\phi(t) = \ln\phi_-(t)$ in Eq. (14), and put $\ln\phi_+(t) = 0$. As a numerical test we have inserted Eq. (54) in Eqs. (9) and (10), integrated numerically and found (for $K = 0$) precise agreement.

2. "*Frozen Gaussian Approximation.*" Semianalytical and semiclassical wavepackets suitable for calculating evolution on an excited state multidimensional potential energy surface were proposed in pioneering studies by Heller [260]. In this method (called the frozen Gaussian approximation), the last two factors in the summand of Eq. (50) were replaced by time-dependent Gaussians. The time dependence arose through having time varying average energies, momenta, and positions. Specifically, each coefficient a_m in Eq. (50) was followed by a function $g(x, t)$ of the form

$$\begin{aligned}\ln g(x, t) = {} & -m\omega(x - \langle x\rangle_t)^2/2 + i\langle p\rangle_t(x - \langle x\rangle_t) \\ & + i\int_0^t (\langle p\rangle_t^2/m - \langle g|H|g\rangle_t)dt\end{aligned} \tag{55}$$

where ω is an energy characteristic of the upper potential surface, the angular brackets are the average position and momenta of the classical trajectory and the Dirac bracket of the Hamiltonian H is to be evaluated for each component g separately.

For a set of Gaussians, it is rather difficult to establish the analytic behavior of Eqs. (55), or of (50), in the t plane. However, with a single Gaussian (in one spatial dimension) and a harmonic potential surface one classically has

$$\langle x \rangle_t = x_0 \cos \omega t \tag{56}$$

$$\langle p \rangle_t = (m)\frac{d\langle x \rangle_t}{dt} = -(\omega x_0 m)\sin \omega t \tag{57}$$

$$\langle g|H|g \rangle_t = \omega/2 \tag{58}$$

By substituting these expressions into Eq. (55), one can see after some algebra that $\ln g(x, t)$ can be identified with $\ln \chi_-(t) + P(t)$ shown in Section III.C.4. Moreover, $\ln \chi_+(t) = 0$. It can be verified, numerically or algebraically, that the log-modulus and phase of $\ln \chi_-(t)$ obey the reciprocal relations (9) and (10). In more realistic cases (i.e., with several Gaussians), Eq. (56–58) do not hold. It still may be true that the analytical properties of the wavepacket remain valid and so do relations (9) and (10). If so, then these can be thought of as providing numerical checks on the accuracy of approximate wavepackets.

3. *Expanding Waves.* As a further application we turn to the expanding potential problem [261–263], where we shall work from the amplitude modulus to the phase. The time-dependent potential is of the form

$$V(x, t) = \zeta^{-2}(t)V(x/\zeta(t)) \tag{59}$$

Here, $\zeta^2(t) = c + t^2$, which differs from the more general case considered in [261–263], by putting their timescale factor $a = 1$ and making the potential real and regular for real t, as well as time-inversion invariant. Then c is positive and, in [261,262] $b = 0$. The Hamiltonian is singular at $t = \pm i\sqrt{c}$, away from the real axis. As first shown in [261], the generic form of the solution of the time-dependent Schrödinger equation is the same for a wide range of potentials. We shall consider the ground state for a harmonic potential $V(x) = 1/2m\omega_o^2 x^2$. The log (amplitude-modulus) of the ground-state wave function (in the coordinate representation) is according to [261] for real t

$$\ln|\phi(x, t)| = -(1/4)\ln(c + t^2) - 1/2[m\omega x^2/(c + t^2)] \tag{60}$$

where $\omega^2 = \omega_o^2 + c$. Processing the expression in Eq. (60) as in Eq. (14), we can arbitrarily decompose $\phi(x, t)$ into factors that are analytic above and below the real t axis. Thus, let us suppose that in Eq. (60) a fraction f_1 of the first term and a fraction f_2 of the second term is analytic in the upper half and, correspondingly, fractions $(1 - f_1)$ and $(1 - f_2)$ are analytic in the lower half. Explicitly, for complex t

$$\begin{aligned} \ln|\phi(x,t)| = \mathrm{Re}\{&-1/2[f_1 \ln(\sqrt{c} - it) + (1 - f_1)\log(\sqrt{c} + it)] \\ &- 1/2(m\omega x^2/\sqrt{c})[f_2/(\sqrt{c} - it) + (1 - f_2)/(\sqrt{c} + it)]\} \end{aligned} \tag{61}$$

Next, for the log term (which normalizes the wave function), we have to choose, as in Eq. (15), suitable functions $P_{\pm}(t)$ that will "correct" the behavior of that term along the large semicircles. Among the multiplicity of choices, the following are the most rewarding (since they completely cancel the log term):

$$\begin{aligned} P_+(t) &= -f_1\left(\frac{1}{2}\right)\ln(\sqrt{c} - it) \\ &= -(f_1/4)[\ln(t^2 + c) - 2i\arctan(t/\sqrt{c})] \\ P_-(t) &= -(1 - f_1)\left(\frac{1}{2}\right)\log(\sqrt{c} + it) \\ &= -\left(\frac{1}{4}\right)(1 - f_1)[\ln(t^2 + c) + 2i\arctan(t/\sqrt{c})] \end{aligned}$$

The right-hand side of Eq. (18) comes from the second term of Eq. (60) alone and is

$$\arg\chi(x,t) = (1 - 2f_2)[m\omega x^2/(t^2 + c)](t/4\sqrt{c}) \tag{62}$$

To complete the phase of the wave function, $\arg\phi(x,t)$, we have to reinstate the terms $P_{\pm}(t)$ that were removed in Eq. (15) so as to get $\chi_{\pm}(x,t)$. The result is

$$\arg\phi(x,t) = -(1 - 2f_1)\left(\frac{1}{2}\right)\arctan(t/\sqrt{c}) + (1 - 2f_2)[m\omega x^2/(t^2 + c)](t/4\sqrt{c}) \tag{63}$$

This establishes the functional form of the phase for real (physical) times. The phase of the solution given in [261,262] indeed has this functional form. The fractions f_1 and f_2 cannot be determined from our Eqs. (17) and (18). However, by comparing with the wave functions in [261,262], we get the following values for them:

$$1 - 2f_1 = \omega/\sqrt{c} \qquad 1 - 2f_2 = 4\sqrt{c}/\omega \tag{64}$$

In the excited states for the same potential, the log modulus contains higher order terms in x (x^3, x^4, etc.) with coefficients that depend on time. Each term can again be decomposed (arbitrarily) into parts analytic in the t half-planes, but from elementary inspection of the solutions in [261,262] it turns out that every term except the lowest [shown in Eq. (59)] splits up equally (i.e., the f's are just $1/2$) and there is no contribution to the phases from these terms. Potentials other than the harmonic can be treated in essentially identical ways.

F. Consequences

The following theoretical consequences of the reciprocal relations can be noted:

1. They unfold a connection between parts of time-dependent wave functions that arises from the structure of the defining equation (2) and some simple properties of the Hamiltonian.
2. The connection holds separately for the coefficient of each state component in the wave function and is not a property of the total wave function (as is, e.g., the "dynamical" phase [9]).
3. The relations pertain to the fine, small-scale time variations in the phase and the log modulus, not to their large-scale changes.
4. One can define a phase that is given as an integral over the log of the amplitude modulus and is therefore an observable and is gauge invariant. This phase [which is unique, at least in the cases for which Eq. (9) holds] differs from other phases, those that are, for example, a constant, the dynamic phase or a gauge-transformation induced phase, by its satisfying the analyticity requirements laid out in Section I.C.3.
5. Experimentally, phases can be obtained by measurements of occupation probabilities of states using Eq. (9). (We have calculationally verified this for the case treated in [264].)
6. Conversely, the implication of Eq. (10) is that a geometrical phase appearing on the left-hand side entails a corresponding geometric probability change, as shown on the right-hand side. Geometrical decay probabilities have been predicted in [162] and experimentally tested in [265].
7. An important ingredient in the analysis has been the positions of zeros of $\Psi(x, t)$ in the complex t plane for a fixed x. Within quantum mechanics the zeros have not been given much attention, but they have been studied in a mathematical context [257] and in some classical wave phenomena ([266] and references cited therein). Their relevance to our study is evident since at its zeros the phase of $\Psi(x, t)$ lacks definition. Future theoretical work shall focus on a systematic description of the location of zeros. Further, practically oriented work will seek out computed or

experimentally acquired time dependent wave functions for tests or application of the present results.

8. Finally, and probably most importantly, the relations show that changes (of a nontrivial type) in the phase imply necessarily a change in the occupation number of the state components and vice versa. This means that for time-reversal-invariant situations, there is (at least) one partner state with which the phase-varying state communicates.

IV. NON-LINEARITIES THAT LEAD TO MULTIPLE DEGENERACIES

In previous sections, we treated molecules and other localized systems in which a linear electron-nuclear coupling resulted in a single degeneracy, or ci of the electronic potential energy surfaces. A notable, symmetry-caused example of this is the linear $E \otimes \epsilon$ Jahn–Teller effect (a pair of degenerate electronic states, that can happen under trigonal or higher symmetry, which is coupled to two energetically degenerate displacement modes) [47,157,267]. Still, some time ago nonlinear coupling was also considered within the $E \otimes \epsilon$ case in [268,269] and subsequently in [256]. Such coupling can result in a more complex situation, in which there is a quadruplet of ci's, such that one ci is situated at the origin of the mode coordinates (as before) and three further ci's are located farther outside in the plane, at points that possess trigonal symmetry.

As of late, nonlinear coupling has become of increased interest, partly through evidence for a weak linear coupling in the metallic cluster Na_3 [52,270] (computations of vibrational levels in a related molecule Li_3 were performed in [271,272]), and partly by attempts to computationally locate ci's in the potential energy landscape with a view to estimate their effect on intersurface, non-adiabatic transitions [273]. The method used in the last reference was based on the acquisition of the geometric phase by the total function as a ci is circled [9,158,159,274]. Independently, the authors of [54] theoretically found a causal connection between the number of ci effectively circled (one or four) and the important question of the nature of the ground state. They showed that, contrary to what had been widely thought before, the ground state may be either a vibronic doublet or a singlet, depending on the distance (which is a function of the parameters in the vibronic Hamiltonian) of these trigonal ci's from the centre. (A similar instance of "quantum phase transition" was noted for a threefold degenerate system in [275] and, earlier, for an icosahedral system, in [276]).

In a different field, location, and characteristics of ci's on diabatic potential surfaces have been recognized as essential for the evaluation of dynamic parameters, like non-adiabatic coupling terms, needed for the dynamic and static properties of some molecules ([193,277–280]). More recently, pairs of ci's have been studied [281,282] in greater detail. These studies arose originally in connection with a ci between the $1^2A'$ and $2^2A'$ states found earlier in computed potential energy surfaces for C_2H in C_s symmetry [278]. Similar ci's appear between the potential surfaces of the two lowest excited states $^1A_2$ and $^1B_2$ in H_2S or of $^2B_2$ and $^2A_1$ in Al—H_2 within C_{2v} symmetry [283]. A further, closely spaced pair of ci's has also been found between the $3^2A'$ and $4^2A'$ states of the molecule C_2H. Here the separation between the twins varies with the assumed C—C separation, and they can be brought into coincidence at some separation [282].

In this section we investigate the phase changes that characterize the double and trigonal (or cubic) ci's. We shall find that the Berry phases upon circling around *all* the ci's can take the values of 0 or $2N\pi$ (where N is an integer). It can be shown that the different values of N can be made experimentally observable (through probing the state populations after inducing changes in the amplitudes of the components), in a way that is not marred by the fast oscillating dynamic phase. Apart from the results regarding the integer N in the Berry phase, the difference between our approach to the phase changes and those in some previous works, especially in [273,283], needs to be noted. While these consider the topological phase belonging to the total wave packet, we continue in the spirit of the previous Section III and treat the open phase belonging to a single component of the wave packet. (For the topological, full-cycle phase the two are equivalent, but not for the open phase, that is present at interim stages.) Explicitly, we write the (in general) time (t)-dependent molecular wave function $\Psi(t)$ as a superposition of (diabatic) electronic states χ_k as

$$\Psi(t) = \sum_k a_k(t)\chi_k \tag{65}$$

where the amplitudes a_k are functions of the nuclear coordinates. In Section III, we developed and used the reciprocal relations between the phases (arg a_k) and the (observable) moduli ($|a_k|$).

We also describe a "tracing" method to obtain the phases after a full cycling. We shall further consider wave functions whose phases at the completion of cycling differ by integer multiples of 2π (a situation that will be written, for brevity, as "$2N\pi$"). Some time ago, these wave functions were shown to be completely equivalent, since only the phase factor (viz., $e^{i\text{Phase}}$) is observable [156]; however, this is true only for a set of measurements that are all made at instances where the phase difference is $2N\pi$. We point out simple, necessary connections between having a certain $2N\pi$ situation and observations made prior to the achievement of that situation. The phase that is of interest in this chapter is the Berry phase of the wave function [9], not its total phase, though this distinction will not be restated.

A. Conical Intersection Pairs

We treat this case first, since it is simpler than the trigonal case. The molecular displacements are denoted by x and y (with suitable choice of their origins and of scaling). Then, without loss of generality we can denote the positions of the ci pairs in Cartesian coordinates by

$$x = \pm 1 \qquad y = 0 \tag{66}$$

or, in polar coordinates, where $x = q\cos\phi$, $y = q\sin\phi$, by

$$q = 1 \qquad \phi = 0, \pi \tag{67}$$

To obtain potential surfaces for two electronic states that will be degenerate at these points, we write a Hamiltonian as a 2×2 matrix in a diabatic representation in the following form:

$$\begin{aligned} H(x,y) &= K\begin{pmatrix} -(x^2-1) & yf(x) \\ yf(x) & (x^2-1) \end{pmatrix} \\ &= K\begin{pmatrix} -(q^2\cos^2\phi - 1) & q\sin\phi f(q\cos\phi) \\ q\sin\phi f(q\cos\phi) & (q^2\cos^2\phi - 1) \end{pmatrix} = H(q,\phi) \end{aligned} \tag{68}$$

whose two eigenvalues are

$$\begin{aligned} E_{\pm}(x,y) &= \pm K\sqrt{(x^2-1)^2 + [yf(x)]^2} \\ &= \pm K\sqrt{(q^2\cos^2\phi - 1)^2 + [q\sin\phi f(q\cos\phi)]^2} \\ &= E_{\pm}(q,\phi) \end{aligned} \tag{69}$$

For K, a (positive) constant and $f(x)$ a function that is nonzero at $x = \pm 1$, the Hamiltonian in Eq. (68) can be taken as a model that yields the postulated ci pairs, since the two eigenvalues coincide just at the points given by Eq. (66) or (67). There may be more general models that give the same two ci's. (Note, however, that if $f(x)$ had a zero at $x = \pm 1$, the degeneracy of energies would not be conical.) We now make the above model more specialized and show that different values of the Berry phase can be obtained for different choices of $f(x)$. For definiteness we consider specific molecular situations, but these are just instances of wider categories. (The notation of Herzberg [284] is used.)

1. $1A_1$ and $2A_2$ States in C_{2v} Symmetry (Exemplified by ${}^1A_1^{(1)}$ and ${}^1A_1^{(2)}$ in Bent HCH)

If the x coordinate represents a mode displacement that transforms as B_1 (e.g., an asymmetric stretch of CH) and y transforms as A_1 (a flapping motion of the

H_2, this coordinate being the same as y in Fig. 169 of [284]), then $f(x)$ in Eq. (68) can be taken as a constant. Without loss of generality we put for this case $f(x) = 1$ and find that cycling adiabatically counterclockwise around the ci that is at $(-1, 0)$ induces (in the component that is unity at $\phi = 0$) a topological phase of π, and that around $(1, 0)$ yields $-\pi$. Cycling either fully inside or outside $q = 1$ (the latter case encircling both ci's), gives zero phase. We now describe a "continuous (phase-) tracing method" that obtains in an unambiguous way the phase of a *real* wave function. The alternative, "adiabatic cycling" method Section III gave the same phase change in terms of the evolution of the *complex* solution of the time-dependent Schrödinger equation in the extremely slow (adiabatic) limit. Other methods will be briefly referred to.

B. Continuous Tracing of the Component Phase

In this method, one notes that real-valued solutions of the time-independent Hamiltonian of a 2×2 matrix form can be written in terms of an $\theta(\phi, q)$, which is twice the "mixing angle," such that the electronic component which is "initially" 1 is $\cos[\theta(\phi, q)/2]$, while that which is initially 0 is $\sin[\theta(\phi, q)/2]$. For the second matrix form in Eq. (68) (in which, for simplicity $f(x) = 1$), we get

$$\theta(\phi, q) = \arctan \frac{q \sin \phi}{q^2 \cos^2 \phi - 1} \tag{70}$$

One can trace the continuous evolution of θ (or of $\theta/2$) as ϕ describes the circle $q =$ constant. This will yield the topological phase (as well as intermediate, open-path phase during the circling). We illustrate this in the next two figures for the case $q > 1$ (encircling the ci's).

In Figure 2*a* several important stages in the circling are labeled with Arabic numerals. In the adjacent Figure 2*b* the values of $\theta(\phi, q)$ are plotted as ϕ increases continuously. The labeled points in the two Figures correspond to each other. (The notation is that points that represent zeros of $\tan \theta$ are marked with numbers surrounded by small circles, those that represent poles are marked by numbers placed inside squares, other points of interest that are neither zero nor poles are labeled by free numbers.) The zero value of the topological phase $(\theta/2)$ arises from the fact that at the point 3 (at which $\phi = \pi/2$), θ retraces its values, rather than goes on to decrease.

1. A_1 and B_2 States in C_{2v} Symmetry (Exemplified by $^2A_1$ and $^2B_2$ in AlH_2 [285])

Symmetry considerations forbid any nonzero off-diagonal matrix elements in Eq. (68) when $f(x)$ is even in x, but they can be nonzero if $f(x)$ is odd, for example, $f(x) = x$. (Note that x itself transforms as B_2 [284].) Figure 3 shows the outcome for the phase by the continuous phase tracing method for cycling

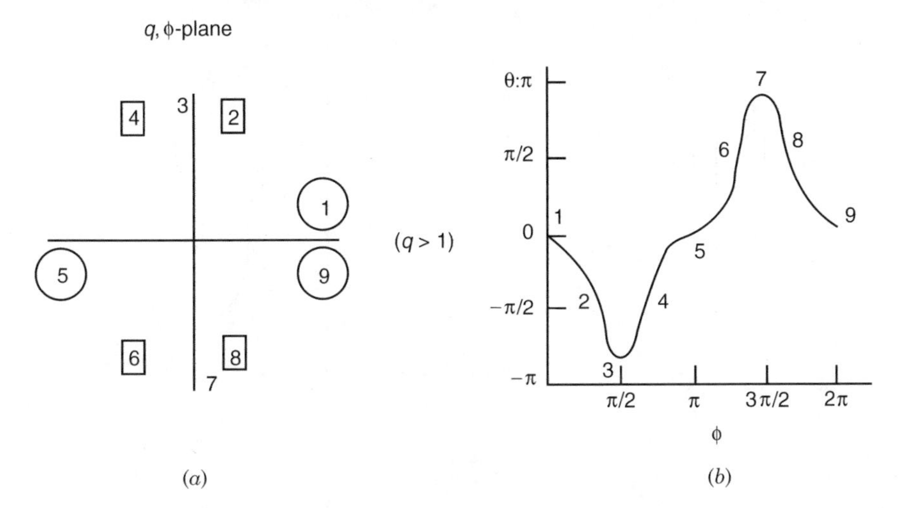

Figure 2. Phase tracing for the case of $1A_1$ and $2A_2$ states in C_{2v} symmetry: (*a*) The left-hand side shows the labels for the significant stages during the circling the (q, ϕ) plane. In this and the following figures, numbers in circles represent the positions of zeros in the argument of the arctan in the expression of the angle [Eq. (70)], numbers in squares are poles and free numbers are other significant stages in the circling. (*b*) The angle θ in Eq. (70) as a function of the circling angle. The numbers correspond to those on the adjacent part (*a*) of the figure. (*Note*: The angle θ is defined as twice the transformation or mixing angle.) The circling is with $q > 1$, namely, outside the ci pair.

outside the ci's $(q > 1)$. The difference between the present case and the previous one [in which $f(x)$ was an even function of x] is that now, in the second half of circling in the q, ϕ plane, the wave function component angle θ does not retrace its path, but goes on decreasing. [It is interesting to remark here on an analogy between the present results and the well-known results of contour integration in the complex z plane. An integration of $(z^2 - 1)^{-1}$ over a path that encircles the two poles of the function gives zero result, but the same path integration of $z(z^2 - 1)^{-1}$, gives $2\pi i$. However, the analogy does not work fully. Thus, a simple multiplication of the integrand by a positive constant alters the residues, but not the phase.]

However, the resulting Berry phase of -2π depends on (1) having reached the adiabatic limit and (2) circling well away from the ci's; that is, it is necessary that the circling shall be done with a value of q that is either $\ll$ or $\gg$ 1. A contrary case not satisfying these conditions, for example, when either $q < 3$ or $K < 60$, would give a Berry phase of $\sim 4\pi, 6\pi, \ldots$, or a number $N \approx 2, 3, \ldots$ rather than 1, as might have been expected. What is perhaps remarkable is that even in the not quite adiabatic or not very large q cases, N (though plainly different from 1) is still close to being an integer. More study may be needed on this result, especially in view of the possibility of observable consequences of

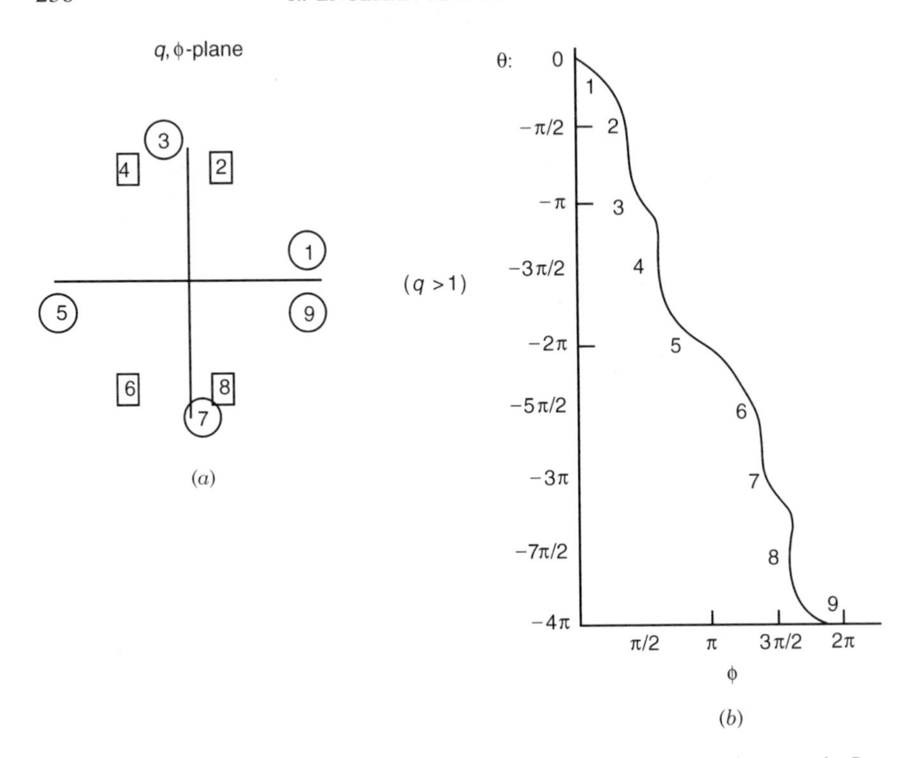

Figure 3. Phase tracing for circling outside the ci pair for the model in A_1 and B_2 states in C_{2v} symmetry. The Berry phase (half the angle shown at the extremity of the figure) is here -2π.

the value of N. The cases of "$1A_1$ and $2A_2$ states in C_{2v} symmetry" and of "A_1 and B_2 states in C_{2v} symmetry" are, of course, inequivalent, since they arise from different Hamiltonians. Their nonequivalence results not only in different topological phases (zero and 2π), but in different state occupation probabilities. These are defined as the probabilities of the systems being in one of the states χ_k, of which the superposition in Eq. (65) is made up. In Figure 4, we show these probabilities as functions of time for systems that differ by their having different functions $f(x)$ in the off-diagonal positions of the Hamiltonian. The differences in the probabilities are apparent.

2. *Trigonal Degeneracies*

The simplest way to write down the 2×2 Hamiltonian for two states such that its eigenvalues coincide at trigonally symmetric points in (x, y) or (q, ϕ), plane is to consider the matrices of vibrational–electronic coupling of the $E \otimes \epsilon$ Jahn–Teller problem in a diabatic electronic state representation. These have been constructed by Halperin, and listed in Appendix IV of [157], up to the third

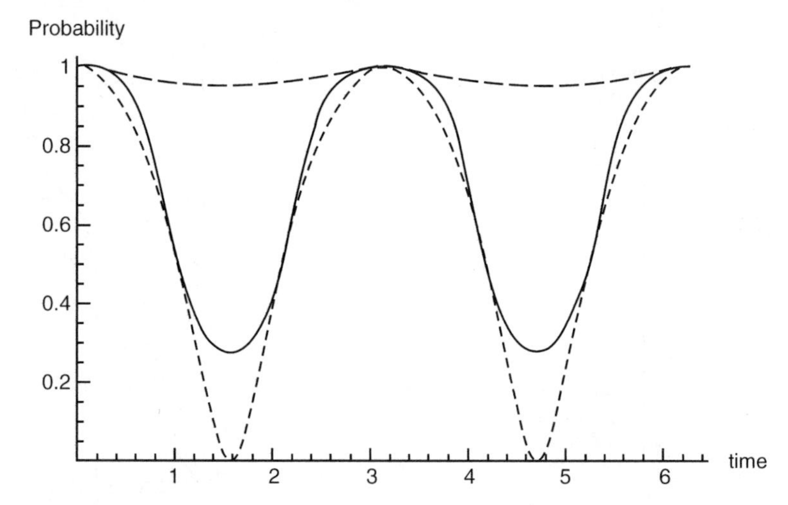

Figure 4. Probabilities in different models during adiabatic circling around ci's. The square moduli of component amplitudes as function of time are seen to be different for different models. Long-dashed lines: the model in $1A_1$ and $2A_2$ states in C_{2v} symmetry for circling inside the ci's. Full lines: the model in $1A_1$ and $2A_2$ states in C_{2v} symmetry for circling outside the ci's. Broken lines: the model in A_1 and B_2 states in C_{2v} symmetry (with the "xy" off-diagonal matrix element) for circling outside the ci's. In the latter model, circling inside the ci's gives probabilities that would be indistinguishable from unity on the figure (and are not shown).

order in q. The first order or linear coupling in the displacement coordinates is of the well-known form (shown by the first term in the Hamiltonian presented below) and yields the familiar ci at the origin, $q = 0$. When one adds to this the quadratic coupling, designated $I(E)$ in Section IV.3 (A) of [157] and quoted below, one obtains three further, trigonally situated ci, namely, at either $\phi = 0, \pm 2\pi/3$, or $\phi = \pi, \pm 4\pi/3$, depending on whether the signs of the linear and quadratic couplings are the same or opposite. The distance of the trigonal ci's from the origin varies with the relative magnitudes of the couplings: The higher the strength of the quadratic term, the nearer the trigonal ci are to the center. This was, of course, the physical basis of [54], in which a ground vibronic *singlet* state for strong quadratic coupling was found. The resulting Hamiltonian is of the form (to be compared with the two matrices in Eq. (68)):

$$\begin{aligned} H(x,y) &= K\begin{pmatrix} -(x - 2\kappa(x^2 - y^2)) & y + 4\kappa xy \\ y + 4\kappa xy & (x - 2\kappa(x^2 - y^2)) \end{pmatrix} \\ &= K\begin{pmatrix} -(q\cos\phi - 2\kappa q^2 \cos 2\phi) & q\sin\phi + 2\kappa q^2 \sin 2\phi \\ q\sin\phi + 2\kappa q^2 \sin 2\phi & (q\cos\phi - 2\kappa q^2\cos 2\phi) \end{pmatrix} \\ &= H(q,\phi) \end{aligned} \tag{71}$$

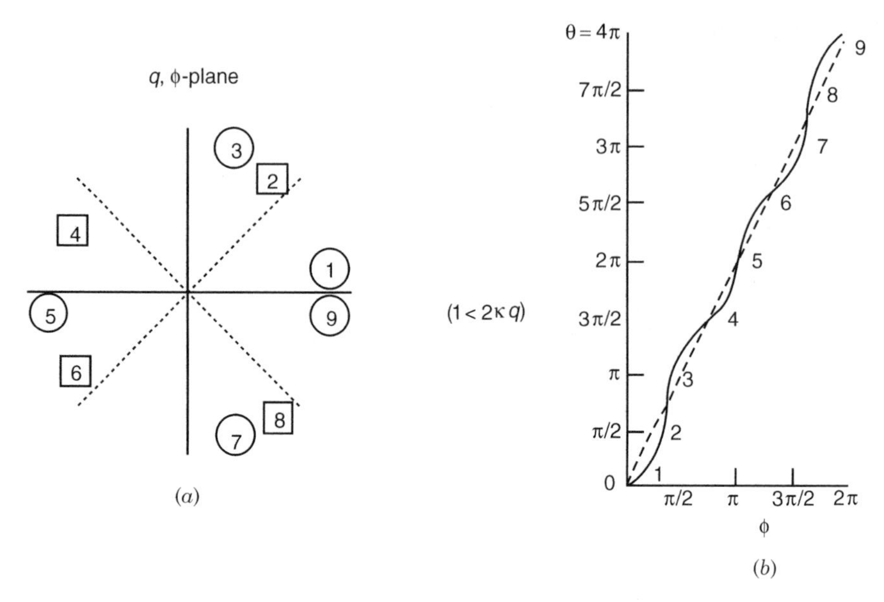

Figure 5. Phase tracing for the case of trigonal degeneracies when the circle encompasses all four ci's and the Berry phase is 2π.

where κ represents the ratio of the strength of the quadratic coupling to the linear one. The trigonal ci's are at a distance $q = (2\kappa)^{-1}$, with angular positions as described above. Now by employing the continuous phase-tracing method introduced in Section I.D.1, one again obtains the graphs for the mixing angle. There are now three cases to consider, namely (1) for cycling that encloses all four ci's ($q > (2\kappa)^{-1}$) the resulting phase acquired being now 2π (shown in Fig. 5). This is an even multiple of π, as expected for four ci's [274], but differs from 4π (or from zero). Then (2) for intermediate radius cycling ($(2\kappa)^{-1} > q > (4\kappa)^{-1}$) (which is shown in Fig. 6) that terminates with a Berry phase of $-\pi$. Lastly, (3) for small radius cycling ($q < (2\kappa)^{-1}$). The last case has also the Berry phase of π, but differs from the intermediate case (2), in that the initial increase is absent.

It might be asked what happens when one adds further couplings beyond the quadratic one? In the next higher order one finds a scalar cubic term of the form:

$$q^3 \cos 3\pi \mathbf{I} \tag{72}$$

where $\mathbf{I}$ is the unit matrix. This gives rise to three trigonally aligned degeneracies ([157], Appendix IV). However, these are parabolic (touching) degeneracies, not conical intersections, and do not cause changes of sign in the wave function upon circling round them. Higher order terms (not listed in that appendix) can give rise

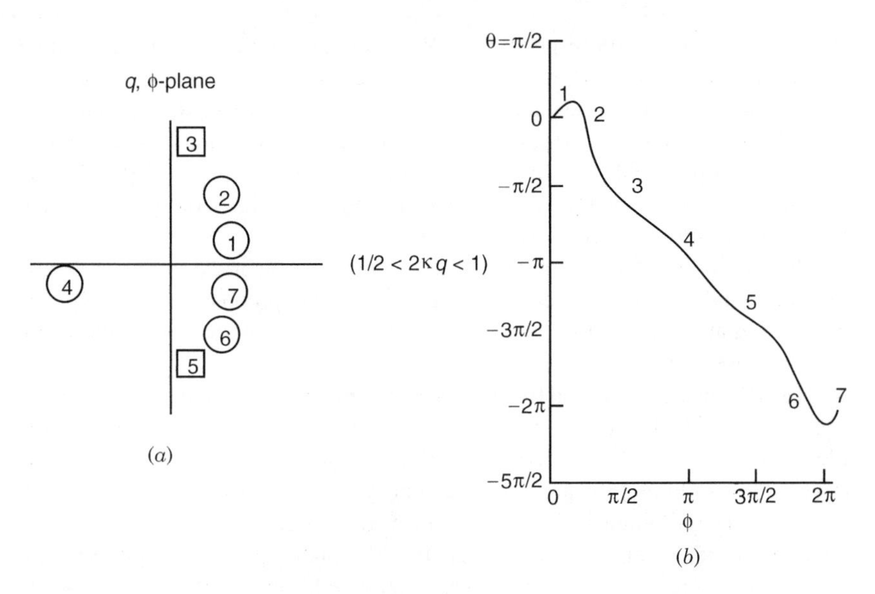

Figure 6. Phase tracing for the trigonal degeneracies. The drawings (which are explained in the caption to Fig. 3) are for intermediate radius (q) circling.

to additional ci's of trigonal symmetry, but the strength of these terms is expected to be less and therefore the resulting ci's will be farther outside, where they are without importance for low-lying states. Still, their presence is of interest for revealing the connection between the Berry angle and the number of ci's circled and we shall presently obtain nonlinear coupling terms to an arbitrarily high power of q.

C. The Adiabatic-to-Diabatic Transformation (ADT)

Several years ago Baer proposed the use of a matrix A, that transforms the adiabatic electronic set to a diabatic one [72]. (For a special twofold set this was discussed in [286,287].) Computations performed with the diabatic set are much simpler than those with the adiabatic set. Subject to certain conditions, A is the solution of a set of first order partial differential equations. A is unitary and has the form of a "path-ordered" phase factor, in which the phase can be *formally* written as

$$\int_{\mathbf{R}_0}^{\mathbf{R}} f^{IJ}(\mathbf{R}) \cdot \mathbf{dR} \tag{73}$$

Here, the integrand is the off-diagonal gradient matrix element between adiabatic electronic states,

$$f^{IJ}(\mathbf{R}) = \langle I|\boldsymbol{\nabla}|J\rangle \tag{74}$$

(∇ is the derivative with respect to **R**.) We stress that in this formalism, I and J denote the complete adiabatic electronic state, and not a component thereof. Both $|I\rangle$ and $|J\rangle$ contain the nuclear coordinates, designated by **R**, as parameters. The above line integral was used and elaborated in calculations of nuclear dynamics on potential surfaces by several authors [273,283,288–301]. (For an extended discussion of this and related matters the reviews of Sidis [48] and Pacher et al. [49] are especially informative.)

The possibility of a nongradient component in the integrand introduces some difficulty and an alternative formulation has been proposed [273,283]. [At positions that are close to a ci, the alternative approximates well to the angle shown above in Eq. (73).]

The ADT phase, computed for ci pairs in [56] and denoted (in their Figs. 1–3) by $\gamma(\phi \mid q)$, is related to the "open-path phase" defined below in Eq. (75), but is identical with it (and with Berry's angle) only at $\phi = 2\pi$. At this value, the computed results of [56] are in agreement with those that were derived with the model Hamiltonian in Eq. (68). However, in some of the cases, when the coupling terms became zero, the sign that the phase $\gamma(\phi \mid q)$ acquires might become ambiguous (e.g., whether it is even or odd under reflection about the line $\phi = \pi$). In the above analytic models the signs are given unambiguously.

Since to date summaries about the practical implementation of the line integral have been given recently (in [108,282]; as also in the chapter by Baer in the present volume), and the method was applied also to a pair of ci's [282], we do not elaborate here on the form of the phase associated with one or more ci's, as obtained through this method.

D. Direct Integration

The open-path phase [11,14] associated with a component amplitude can be obtained as the imaginary part of an integral

$$\gamma_k(t) = \text{Im} \int_0^t dt' \frac{\partial a_k(t')}{\partial t'} \Big/ a_k(t') \tag{75}$$

where, as before at several places in this chapter, $a_k(t)$ is the amplitude of the k component in the solution of the time-dependent Schrödinger equation in the near-adiabatic limit. The (complex) amplitude in the integrand is (in general) non-vanishing (unlike the real wave function amplitude in the strictly adiabatic solution) and thus the integral is nondivergent. However, in practice, even fairly close to the adiabatic limit, the convergence is very slow, due to oscillations in the amplitude, noted in Section IV.C and in [29–32]. For this reason, the formula in Eq. (75) is not a convenient one to use. Still, using the formula for increasing values of the adiabaticity parameter [i.e., increasing K in Eq. (68) to $K > 10^2$], we have evaluated the topological phase for the case with trigonal symmetry and have found it to converge close to the value $2N\pi$, with N = 1 (and not 0). Because of the difficulties in its practical implementation, we shall not further consider the direct integration method.

E. Higher Order Coupling in Some Jahn–Teller and Renner–Teller Effects

A systematic derivation of forms of coupling that is superlinear in the nuclear motion amplitude was given, partly based on Racah coefficients, in [157], Appendix IV, but these went up only as far as the third order in the amplitudes q. As will shortly be made apparent, there is some theoretical need to obtain higher order terms. For the $E \otimes \epsilon$ Jahn–Teller case, the form of coupling to arbitrary powers was given in [302]. Here we give a different and arguably simpler derivation using the vector-coupling formalism of Appendix IV in [157], the complex representation form given in [303], and a mathematical induction type of argument.

1. Complex Representation

The mode coordinates, transforming, respectively, as the -1 and 1 components of the E (doubly degenerate) modes, have the form:

$$(g_{-1}, g_1) = -\left(\frac{i}{\sqrt{2}}\right)(g_\theta, g_\epsilon)\begin{pmatrix} -1 & 1 \\ i & i \end{pmatrix} = \left(\frac{i}{\sqrt{2}}\right)(qe^{-i\phi}, -qe^{i\phi}) \tag{76}$$

where the extreme right-hand member recalls the "modulus" q and the "phase" ϕ used in the real representation. The vector coupling coefficients for for example, octahedral symmetry, can be obtained from Table A.20 in Appendix 2 of [304], upon performing the transformation shown (76) on the real representation. In the following Table we show the Clebsch–Gordan coefficients defined in equation (79) below.

TABLE I
The Coupling–Coefficients $U(ABC|abc)$ for the Complex Form of a Doubly Degenerate Representation in the Octahedral Group, Following G. F. Koster et al., *Properties of the Thirty-Two Point Groups*, MIT Press, MA, 1963, pp. 8, 52.

A:	B:	C:	A_1	A_2	E	
E	E					
a:	b:	c:	a_1	a_2	-1	-1
		U:				
1	1		0	0	1	0
1	-1		$1/\sqrt{2}$	$-i/\sqrt{2}$	0	0
-1	1		$1/\sqrt{2}$	$i/\sqrt{2}$	0	0
-1	-1		0	0	0	1

To obtain nonlinear coupling terms, we consider two linearly independent, not identical E modes, namely,

$$(g_{-1}, g_1) \tag{77}$$

$$(G_{-1}, G_1) \tag{78}$$

and construct bilinear expressions from these. The combination that transforms as components of an E mode is given by

$$\chi(E|x) = \sum_{a,b} U(EEE|abx) g_a G_b \tag{79}$$

The above table for U immediately shows that there are only two bilinear combinations of g and G, namely, those $a = 1$, $b = 1$ and $a = -1$, $b = -1$. These lead to the quadratic terms belonging to the components -1 and 1: $\chi(E|-1) = g_1 G_1$ and $\chi(E|1) = g_{-1} G_{-1}$.

Equation A IV.4 of [157] tells us which ket–bra operator $|d\rangle\langle e|$ is multiplied by the above combinations or, equivalently, where in the 2×2 electronic-nuclear coupling matrix each of these terms sit. Here again we adopt for the electronic kets a complex representation, analogous to that shown in Eq. (76). To use the vector coupling coefficients for these, we recall that in the complex representation the bra's transform as the corresponding "*minus-label* ket's" [cf. Eq. (2.34) in [303]). By using again the vector coupling coefficients, we see that $\chi(E|-1)$ is the factor that multiplies $|1\rangle\langle -1|$ (and $\chi(E|1)$ is the factor that multiplies $|-1\rangle\langle 1|$). In the usual matrix notation [in which rows and columns are taken in the order $(-1, 1)$] this means that in the upper right corner one has (for linear coupling) g_{-1} and G_{-1}, and (for quadratic coupling), $g_1 G_1$, and similarly in the lower left corner g_1, G_1 , as well as $g_{-1} G_{-1}$. Both linear and quadratic terms will be multiplied by different constants, whose values depend on the physical situation and cannot be given by symmetry considerations, except that the electronic–nuclear interaction must be Hermitean and invariant under the symmetry operations of the group. The same construction can be employed to derive bilinear terms on the diagonal part of the coupling matrix. By again using the U coefficients in the above table, one obtains the forms (not normalized)

$$(g_1 G_{-1} + g_{-1} G_1)(|1\rangle\langle 1| + |-1\rangle\langle -1|) \tag{80}$$

where each factor belongs to the A_1 representation and

$$(g_1 G_{-1} - g_{-1} G_1)(|1\rangle\langle 1| - |-1\rangle\langle -1|) \tag{81}$$

where each factor belongs to the A_2 representation.

2. *Squaring of Off-Diagonal Elements*

The method shown affords easy generalization to higher order coupling in the important case where a single mode is engaged, that is, $G_{\pm 1} = g_{\pm 1} = \pm(1/i\sqrt{2})qe^{\pm i\phi}$. Then the two off-diagonal terms derived above are, after physics-based constant coefficients have been affixed, in the upper right corner

$$(qe^{-i\phi} - 2\kappa q^2 e^{2i\phi})|1\rangle\langle -1| \tag{82}$$

with another Hermitean conjugate expression on the other (lower left) off-diagonal position. These were previously given in a similar form in, for example, [157,256]. The A_1 term in Eq. (80) only renormalizes the vibrational frequency. The A_2 term vanishes (for terms up to second-order in q^2). Proceeding in the same way to get further terms by cross-multiplying the second-order expression in Eq. (82), and continuing the procedure, we obtain the following terms in the upper right corner correct up to the fourth order in q

$$q^3 e^{-i\phi}, q^4 e^{2i\phi}, q^4 e^{-4i\phi} \tag{83}$$

The first and second terms contain phase factors identical to those previously met in Eq. (82). The last term has the "new" phase factor $e^{-4i\phi}$. [Though the power of q in the second term is different from that in Eq. (82), this term enters with a physics-based coefficient that is independent of κ in Eq. (82), and can be taken for the present illustration as zero. The full expression is shown in Eq. (86) and the implications of higher powers of q are discussed thereafter.] Then a new off-diagonal matrix element enlarged with the third term only, multiplied by a (new) coefficient λ, is

$$(qe^{-i\phi} - 2\kappa q^2 e^{2i\phi} + \lambda q^4 e^{-4i\phi}) = qe^{-i\phi}(1 - 2\kappa q e^{3i\phi} + \lambda q^3 e^{-3i\phi}) \tag{84}$$

There is going to be an A_1 (scalar) term of the form that is well known in the literature (e.g., [157]), $q^3 \cos 3\phi$, and an A_2 (pseudoscalar) term of the form $q^3 \sin 3\phi$. We may once again suppose that the coefficients of all these terms are independent (i.e., their physical origins are different) and that we may discuss terms in diagonal and off-diagonal positions separately. Let us consider the off-diagonal term, as given on the right-hand member of Eq. (84). The vanishing of the first factor gives the traditional ci at the origin. The zeros of the second factor give additional ci's. These are all trigonally positioned, due to the phase factors $e^{\pm 3i\phi}$, which induce trigonal symmetry. The maximum number of trigonal ci's (to this, fourth-order approximation in q) is clearly $3 \times 3 = 9$. Thus, to give a numerical example in which $\kappa = 0.15$, $\lambda = 0.003$, we obtain the following nine trigonal roots of Eq. (84)

$$\begin{aligned} q &= 3.95 \qquad \phi = 0, \frac{2\pi}{3}, \frac{4\pi}{3} \\ q &= 7.42 \qquad \phi = 0, \frac{2\pi}{3}, \frac{4\pi}{3} \\ q &= 11.37 \qquad \phi = \pi, \frac{\pi}{3}, \frac{5\pi}{3} \end{aligned} \tag{85}$$

(Clearly, the pseudoscalar term vanishes at these points; so the ci character at the roots is maintained, no matter whether there are or are not A_2 terms. Also, the vanishing of A_2 terms will not lead to new ci's.) On the other hand, by circling over a large radius path $q \rightarrow \infty$, so that all ci's are enclosed, the dominant term in Eq. (84) is the last one and the acquired Berry phase is $-4(2\pi)/2 = -4\pi$.

To see that this phase has no relation to the number of ci's encircled (if this statement is not already obvious), we note that this last result is true no matter what the values of the coefficients κ , λ, and so on are provided only that the latter is nonzero. In contrast, the number of ci's depends on their values; for example, for some values of the parameters the vanishing of the off-diagonal matrix elements occurs for complex values of q, and these do not represent physical ci's. The model used in [270] represents a special case, in which it was possible to derive a relation between the number of ci's and the Berry phase acquired upon circling about them. We are concerned with more general situations. For these it is not warranted, for example, to count up the total number of ci's by circling with a large radius.

3. *General Off-Diagonal Coupling*

The construction given above to obtain off-diagonal nonlinear couplings up to order q^4 can be generalized to arbitrary order. Only the final result is given. This gives for the off-diagonal term in the upper right corner

$$Kqe^{-i\phi}[1 + q^{-2} \sum_{m=1,\dots} q^{3m} Q_{m+} e^{3mi\phi} + \sum_{m=1,\dots} q^{3m} Q_{m-} e^{-3mi\phi}] \tag{86}$$

where Q_{m+} and Q_{m-} are polynomials in q^2 with coefficients that depend on the physical system and whose leading terms will be q^0. When transformed back to the real representation, by applying the inverse of the transformation in Eq. (76), one regains the expressions of [302]. Normally, for stable physical systems, it is expected that, with increasing m , Q_{m+} and Q_{m-} will numerically decrease and so will, in each polynomial, the coefficients of successively higher powers q^2. If we assume only a finite number of summands in the above sums and that the highest power of q in Eq. (86) has the phase factor $e^{3Mi\phi}$ (where M is a positive or negative integer), then the path along a very large circle will add a topological phase of $(3M - 1)\pi$. In general, $3|M|$ is different (either smaller or larger) than the number of ci's enclosed by the large contour, though it equals the number of ci's for the case $M = +1$ treated in [270]. When there are two or more different phase factors with the same highest power of q, then the amount of topological phase is not simply given, but can be determined, using the continuous phase-tracing method described in Section IV.B.

4. *Nonlinear Diagonal Elements*

Their forms are

$$A_1: \sum_{m=0,\ldots} q^{3m} D_{1,m}(q) \cos 3m\phi \tag{87}$$

$$A_2: \sum_{m=0,\ldots} q^{3m} D_{2,m}(q) \sin 3m\phi \tag{88}$$

where $D_{1,m}$ and $D_{2,m}$ are polynomials in q^2 with coefficients that again depend on the physical system and whose leading terms is q^0. The scalar term evidently does not produce a ci. The zeros of the A_2 term (which is applicable for systems not invariant under time reversal) by themselves do not add to or subtract from the ci's.

5. *Generalized Renner–Teller Coupling*

The foregoing formulas in Eqs. (86 and 88) can be applied immediately to two physically interesting situations (not treated in [302], but very recently considered for a special model in [305]). The first is the vibronic interaction in a system having inversion symmetry between a doubly degenerate electronic state and an *odd* vibrational mode. The second situation is the more common one of Renner–Teller coupling (e.g., a linear molecule whose doubly degenerate orbital is coupled to a bending-type distortion) [47], formally identical to the previous. To write out the coupling to any order, one simply removes in the previous formulas all terms having odd powers of q. In the *real* representation, the coupling matrix correct to the fourth harmonics in the angular coordinate has the following form

$$\begin{pmatrix} R_1 q^2 \cos 2\phi + R_2 q^4 \cos 4\phi + \cdots & R_1 q^2 \sin 2\phi - R_2 q^4 \sin 4\phi + \cdots \\ R_1 q^2 \sin 2\phi - R_2 q^4 \sin 4\phi + \cdots & -(R_1 q^2 \cos 2\phi + R_2 q^4 \cos 4\phi + \cdots) \end{pmatrix} \tag{89}$$

where, as in the instances of Eqs. (86 and 88) above, R_1 and R_2 are polynomials in q^2 with coefficients that again depend on the physical system and whose leading terms are of order q^0.

6. *Interpretation*

The key of constructing vibronic coupling terms for doubly degenerate states and modes to an arbitrary order is the use of a complex representation. The formal essence of the method is that in the complex representation $U(EEE|xyz)$ is nonzero only for a single z. (In Table I there is only one entry in a row. Figuratively speaking: All coupled "coaches" travel to a unique "train station" and all trains in that station consist of coupled coaches. Moreover, this goes also for the coupling of coupled trains, and so on. From our result, we conclude that the Berry phase around more than one conical intersection is not uniquely given by the number of conical intersections enclosed, but is model dependent. This has consequences for experimental tracing of the phase, as well as for computations of line integrals with the purpose of obtaining non-adiabatic surface jumping in chemical rearrangement processes (e.g., in [186–195,300,301]) and as discussed in Section II.

F. Experimental Phase Probing

Experimental observation of topological phases is difficult, for one reason (among others) that the dynamic-phase part (which we have subtracted off in our formalism, but is present in any real situation) in general oscillates much faster than the topological phase and tends to dominate the amplitude behavior [306–312]. Several researches have addressed this difficulty, in particular, by neutron-interferometric methods, which also can yield the open-path phase [123], though only under restricted conditions [313].

The continuous tracing method and other methods for cycling reviewed in this section can be used in several very different areas. An example is a mesoscopic system composed of quantum dots that is connected to several capacitors. For this, a network of singularities was described in the parameter space of the gate voltages [314]. It has been suggested that the outcome of circling around these singularities, through a phased alteration of the charges on the capacitors, is formally similar to that of circling around ci's [211]. Although the physical effects are different (i.e., the acquisition of a π phase by the wave function has the effect of transferring a single electronic charge), the results of circling obtained in this section can be associated with quantized charges passing between quantum dots. Some related topics, for which the results of this section can be used or extended are phase behavior in a different type of multiple ci's, located in a single point but common to several states. This was studied in [169] and for an electronic quartet state in [36,59,61]. A further future extension of the theory is to try to correlate the topological phase with a general (representation-independent) property of the system (or of the Hamiltonian).

The phases studied in the present work are those of material, Schrödinger waves, rather than of electromagnetic, light waves. Recently, it has been shown that it is possible to freeze coherent information (= phases) from light into material degrees of freedom and vice versa [315,316]. This development extends the relevance of this section to light, also. Among fields of application not directly addressed in their recent work, let us quote from the authors of [315] quantum information transfer [317] and Bose–Einstein condensates.

V. MOLECULAR YANG–MILLS FIELDS

A. A Nuclear Lagrangean

One starts with the Hamiltonian for a molecule $H(\mathbf{r}, \mathbf{R})$ written out in terms of the electronic coordinates ($\mathbf{r}$) and the nuclear displacement coordinates ($\mathbf{R}$, this being a vector whose dimensionality is three times the number of nuclei) and containing the interaction potential $V(\mathbf{r}, \mathbf{R})$. Then, following the BO scheme, one can write the combined wave function $\Psi(\mathbf{r}, \mathbf{R})$ as a sum of an infinite number of terms

$$\Psi(\mathbf{r}, \mathbf{R}) = \sum_k \zeta_k(\mathbf{r}, \mathbf{R})\chi_k(\mathbf{R}) \tag{90}$$

Here, the first factor $\zeta_k(\mathbf{r}, \mathbf{R})$ in the sum is one of the solutions of the electronic BO equation and its partner in the sum, $\chi_k(\mathbf{R})$ is the solution of the following equation for the nuclear motion, with total eigenvalue E_k

$$\left\{-\frac{1}{2M}\partial_b\partial^b\delta_m^k + V_m^k(\mathbf{R}) - \frac{1}{M}\tau_{bm}^k(\mathbf{R})\partial^b + \frac{1}{2M}\tau_{bn}^k(\mathbf{R})\tau_m^{bn}(\mathbf{R})\right\}\chi^m(\mathbf{R}) = E_k\chi^k(\mathbf{R}) \tag{91}$$

The symbol M represents the masses of the nuclei in the molecule, which for simplicity are taken to be equal. The symbol δ_m^k is the Kronecker delta. The tensor notation is used in this section and the summation convention is assumed for all repeated indexes not placed in parentheses. In Eq. (91) the NACT τ_{bm}^k appears (this being a matrix in the electronic Hilbert space, whose components are denoted by labels k, m, and a "vector" with respect to the b component of the nuclear coordinate $\mathbf{R}$). It is given by an integral over the electron coordinates

$$\tau_{bm}^k(\mathbf{R}) = \int d\mathbf{r}\zeta_k(\mathbf{r}, \mathbf{R})\partial_b\zeta_m(\mathbf{r}, \mathbf{R}) := \langle k|\partial_b|m\rangle = -\langle m|\partial_b|k\rangle \tag{92}$$

The effective potential matrix for nuclear motion, which is a diagonal matrix for the adiabatic electronic set, is given by

$$V_m^k(\mathbf{R}) = \langle k|V(\mathbf{r}, \mathbf{R})|m\rangle \tag{93}$$

In the algebraic, group-theoretical treatments of non-Abelian systems [66,67–70,77–80] the NACT is usually written in a decomposed form as

$$\tau^k_{bm}(\mathbf{R}) = d^{(b)}(\mathbf{R})(t_b)^k_m \tag{94}$$

where t_b is one of the set of constant (noncommuting) matrices (the "generators") that define the Lie group of the system. So far, with the summation in Eq. (90) over k running over the full electronic Hilbert space spanned by $\zeta_k(\mathbf{r}, \mathbf{R})$, the Hamiltonian treatment is exact. Shortly, we shall see that the truncation of the summation in Eq. (90), which in practice is almost inevitable, has far-reaching effects in the YM theory. Before that, we turn to an equivalent description, standardly used in field theories but that has not been in use for the BO treatment of molecules, namely, to write down a "nuclear" Lagrangean density $\mathcal{L}_M$ for the vector $\psi(\mathbf{R})$ whose (transposed) row vector form is

$$\psi^T = (\chi_1, \chi_2, \chi_3, \ldots, \chi_N)* \tag{95}$$

(The mixed, $\psi - \chi$ notation here has historic causes.) The Schrödinger equation is obtained from the nuclear Lagrangean by functionally deriving the latter with respect to ψ. To get the *exact* form of the Schrödinger equation, we must let N in Eq. (95) to be equal to the dimension of the electronic Hilbert space (viz., ∞), but we shall soon come to study approximations in which N is finite and even small (e.g., 2 or 3). The appropriate nuclear Lagrangean density is for an arbitrary electronic states

$$\begin{aligned}\mathcal{L}_M(\psi, \partial_a\psi) = {} & (2M)^{-1}(\partial^a\psi)^k(\partial_a\psi)_k - M^{-1}\psi^k\tau^m_{ka}(\partial^a\psi)_m \\ & - (2M)^{-1}(\psi)^k\tau_{kb}{}^m\tau^{bn}_m(\psi)_n - \psi^k V^m_k \psi_m\end{aligned} \tag{96}$$

The non-Abelian nature of the formalism is apparent from the presence of nondiagonal matrices τ and V. The parameter V can be diagonalized, leading to adiabatic energy surfaces and states, but not simultaneously with the $(\tau\partial)$ term. Requiring now only *global* gauge invariance of the Lagrangean, we obtain the usual phase-gauge theories [76,163], incorporating a vector potential. However, by requiring invariance under a *local* gauge transformation, we obtain the extension of the vector potential to a YM field [66,67]. [Actually, the local gauge invariance is not a "luxury" because, if the Lagrangean is invariant under global (constant) transformation, then it is also invariant under a gauge transformation with general position dependent parameters (Section 15.2 in [70]). [A remark on nomenclature: "field" and "fields" are used interchangeably.] Before obtaining the equation for the field, we return for a moment to the (simpler) Abelian case.

B. Pure versus Tensorial Gauge Fields

To start, it is useful to put the previous result in a more elementary setting, familiar in the context of electromagnetic force between charged particles, say electrons. Thus, we recapitulate as follows.

In an Abelian theory [for which $\Psi(\mathbf{r}, \mathbf{R})$ in Eq. (90) is a scalar rather than a vector function, $N = 1$], the introduction of a gauge field $g(\mathbf{R})$ means premultiplication of the wave function $\chi(\mathbf{R})$ by $\exp(ig\mathbf{R})$, where $g(\mathbf{R})$ is a scalar. This allows the definition of a "gauge"-vector potential, in natural units

$$A_a = \partial_a g \tag{97}$$

and if we define a field intensity tensor, as in electromagnetism, by

$$F_{bc} = \partial_b A_c - \partial_c A_b \tag{98}$$

we find that F_{bc} is 0, excluding singularities of A_a. Therefore, a vector potential arising from a gauge transformation g does not give a true field (since it can be transformed away by another gauge $-g$). Conversely, a vector potential A_a for which F_{bc} in Eq. (98) is not zero, gives a true field and cannot be transformed away by a choice of gauge.

In a non-Abelian theory (where the Hamiltonian contains noncommuting matrices and the solutions are vector or spinor functions , with N in Eq. (90) >1) we also start with a vector potential A_b. [In the manner of Eq. (94), this can be decomposed into components A_b^a, in which the superscript ^{a} labels the matrices in the theory). Next, we define the field intensity tensor through a "covariant curl" by

$$F_{bc}^a = \partial_b A_c^a - \partial_c A_b^a + C_{de}^a A_b^d A_c^e \tag{99}$$

Here, C_{bc}^a are the structure constants for the Lie group defined by the set of the noncommuting matrices t_a appearing in Eq. (94) and which also appear both in the Lagrangean and in the Schrödinger equation. We further define the "covariant derivative" by

$$(D_a\psi)_k = (\partial_a\psi)_k - iA_a^b(t_b)_k^m\psi_m \tag{100}$$

and write the field equations for A and F as

$$\partial_a F_b^{ac} = F_d^{cf} C_{be}^d A_f^e + i\frac{\delta\mathcal{L}_M(\psi, D\psi)}{\delta D_c\psi_k}(t_b)_n^m\psi_k \tag{101}$$

If the vector potential components A_b^a have the property that the derived field intensity, the YM field in Eq. (99) is nonzero, then the vector potential cannot be transformed away by a gauge phase $g(\mathbf{R})$ through premultiplication of the wave function $\chi^m(\mathbf{R})$ by the (unitary) factor $\exp(ig(\mathbf{R}))$. There is no $g(\mathbf{R})$ that will do this. Conversely, if there is a $g(\mathbf{R})$, one obtains a vector potential-matrix A_a whose km components satisfy

$$A^k_{am} = (\exp(g(\mathbf{R}))^{-1})^k_n \partial_a [\exp(ig(\mathbf{R}))]^n_m \tag{102}$$

Thus, the existence of a (matrix-type) phase g represents the "pure-gauge case" and the nonexistence of g represents the nonpure YM field case, which cannot be transformed away by a gauge.

C. The "Curl Condition"

We now return to the nuclear BO Eq. (91) in the molecular context. Consider the derivative coupling term in it, having the form

$$M^{-1}\tau^k_{bm}(\mathbf{R})\partial^b \chi^m(\mathbf{R}) \tag{103}$$

Suppose that we want this to be transformed away by a pure gauge factor having the form

$$[\exp(ig(\mathbf{R}))]^k_m = [G(\mathbf{R})]^k_m \tag{104}$$

where $\mathbf{g}$ and $\mathbf{G}$ are matrices. That is, we require

$$\tau^k_{bm}(\mathbf{R}) = [G(\mathbf{R})^{-1}]^k_s \partial_b [G(\mathbf{R})]^s_m \tag{105}$$

for all b, or

$$[G(\mathbf{R})]^s_k \tau^k_{bm}(\mathbf{R}) = \partial_b [G(\mathbf{R})]^s_m \tag{106}$$

The consistency condition for this set of equations to possess a (unique) solution is that the field intensity tensor defined in Eq. (99) is zero [72], which is also known as the "curl condition" and is written in an abbreviated form as

$$\text{curl}\,\tau = -\tau \times \tau \tag{107}$$

Under circumstances that this condition holds an ADT matrix, A exists such that the adiabatic electronic set can be transformed to a diabatic one. Working with this diabatic set, at least in some part of the nuclear coordinate space, was the objective aimed at in [72].

Starting from a completely different angle, namely, the nuclear Lagrangean and the requirement of local gauge invariance, we have shown in Section IV.B that if the very same curl condition is satisfied, there is a pure gauge field. If it is not satisfied, then the field is not a gauge field, but something more complicated, namely, the YM field. The set of equations that give the pure gauge g is identical to that which yields the ADT matrix A, which was introduced in [72]. The equivalence between a (pure) gauge phase factor and the ADT matrix does not seem to have been made in the literature before, though the conditionality of a pure gauge on the satisfaction of the "curl relation" was common knowledge. (Indeed, they are regarded as tautologously the same.) The reason for the omission may have been that, possibly, the ADT matrix was not thought to have the respectability of the pure gauge. (From a naive, superficial angle it is not evident, why one and the same condition should guarantee the elimination of the cross-term in the molecular Schrödinger equation, which is a nonrelativistic, second-order differential equation, and the possibility of a pure gauge for a hadron field, which obeys entirely different equations, for example, relativistic, first-order ones.)

D. The Untruncated Hilbert Space

Now, we recall the remarkable result of [72] that if the adiabatic electronic set in Eq. (90) is complete ($N = \infty$), then the curl condition is satisfied and the YM field is zero, except at points of singularity of the vector potential. (An algebraic proof can be found in Appendix 1 in [72]. An alternative derivation, as well as an extension, is given below.) Suppose now that we have a (pure) gauge $g(\mathbf{R})$, that satisfies the following two conditions:

1. The electronic set (represented in the following by Greek indexes) is *complete*.
2. The vector potential-matrix A present in the Hamiltonian (or in the Lagrangean) arises from a dynamic coupling: meaning, that it has the form

$$A^{\alpha}_{a\beta} \propto \langle\alpha|\partial_a|\beta\rangle \tag{108}$$

Then, two things (that are actually interdependent) happen: (1) The field intensity $F = 0$, (2) There exists a unique gauge $g(\mathbf{R})$ and, since $F = 0$, any apparent field in the Hamiltonian can be transformed away by introducing a new gauge. If, however, condition (1) does not hold, that is, the electronic Hilbert space is truncated, then F is in general not zero within the truncated set, In this event, the fields A and F cannot be nullified by a new gauge and the resulting YM field is true and irremovable.

Attention is directed to a previous discussion of what happens when the electronic basis is extended to the complete Hilbert space, [79] p. 60; especially Eqs. (2.17)–(2.18). It is shown there that in that event the full symmetry of the invariance group is regained (in effect, through the cancellation of the transformation matrix operating on the electronic and on the nuclear functions spaces). From this result, it is only a short step to conclude that the YM field coming from electron–nuclear coupling must be zero for a full set. However, this conclusion is not drawn in the article, nor is the vanishing of the YM field shown explicitly.

As was already noted in [9], the primary effect of the YM field is to induce transitions ($\zeta_m \rightarrow \zeta_k$) between the nuclear states (and, perhaps, to cause finite lifetimes). As already remarked, it is not easy to calculate the probabilities of transitions due to the derivative coupling between the zero-order nuclear states (if for no other reason, then because these are not all mutually orthogonal). Efforts made in this direction are successful only under special circumstances, for example, the perturbed stationary state method [64,65] for slow atomic collisions. This difficulty is avoided when one follows Yang and Mills to derive a mediating tensorial force that provide an alternative form of the interaction between the zero-order states and, also, if one introduces the ADT matrix to eliminate the derivative couplings.

E. An Alternative Derivation

The vanishing of the YM field intensity tensor can be shown to follow from the gauge transformation properties of the potential and the field. It is well known (e.g., Section II in [67]) that under a unitary transformation described by the matrix

$$U = U(\mathbf{R}) \tag{109}$$

(which induces a rotation in the nuclear function space) the vector potential transforms as

$$A_a = A_a(\mathbf{R}) \rightarrow U^{-1}A_aU + U^{-1}\partial_aU \tag{110}$$

whereas the field intensity transforms covariantly, homogeneously as

$$F_{ab} \rightarrow U^{-1}F_{ab}U \tag{111}$$

If now there exists a representation in which A_a is zero, then in this representation F_{ab} is also zero [by Eq. (99)]. Now, in a U-transformed representation (which can be chosen to be completely general), one finds that

$$A_a \rightarrow U^{-1}\partial_aU \tag{112}$$

since the first term in Eq. (110) is zero, but not the second. Thus A_a is not zero. However, the transformed F_{ab} has no such inhomogeneous term [see Eq. (111)]

and therefore, in the transformed representation $F_{ab} = 0$ and this is true in all representations, that is, generally true. The crucial assumption was that there is a representation in which A_a are all zero, and this holds in any diabatic representation (where the electronic functions $\zeta_k(\mathbf{r}, \mathbf{R})$ are independent of $\mathbf{R}$). Then, also, derivative matrices, defined in Eq. (92), are zero and so are the potentials A_a depending linearly on the derivative matrices. On the other hand, the possibility of a diabatic set is rigorously true only for a full electronic set. The existence of such a set is thus a (sufficient) condition for the vanishing of the YM field intensity tensor F_{ab}.

F. General Implications

The foregoing indicate that there are three alternative ways to represent the combined field in the degrees of freedom written as $\mathbf{r}, \mathbf{R}$.

1. By starting with a Lagrangean having the full symmetry, including that under local gauge transformations, and solving for $\Psi(\mathbf{r}, \mathbf{R})$ (this being a solution of the corresponding Schrödinger equation in the variables $\mathbf{r}, \mathbf{R}$). The solutions can then be expanded as in Eq. (90), utilizing the full electronic set [the first factor on the sum Eq. (90)], or, for that matter, employing any other full electronic set.
2. Projecting the nuclear solutions $\chi_k(\mathbf{R})$ on the Hilbert space of the electronic states $\zeta_k(\mathbf{r}, \mathbf{R})$ and working in the projected Hilbert space of the nuclear coordinates $\mathbf{R}$. The equation of motion (the nuclear Schrödinger equation) is shown in Eq. (91) and the Lagrangean in Eq. (96). In either expression, the terms with τ^k_{bm} represent couplings between the nuclear wave functions $\chi_k(\mathbf{R})$ and $\chi_m(\mathbf{R})$, that is, (virtual) transitions (or admixtures) between the nuclear states. (These may represent transitions also for the electronic states, which would get expressed in finite electronic lifetimes.) The expression for the transition matrix is not elementary, since the coupling terms are of a derivative type.

   Now the Lagrangean associated with the nuclear motion is not invariant under a local gauge transformation. For this to be the case, the Lagrangean needs to include also an "interaction field." This field can be represented either as a vector field (actually a four-vector, familiar from electromagnetism), or as a tensorial, YM type field. Whatever the form of the field, there are always two parts to it. First, the field induced by the nuclear motion itself and second, an "externally induced field," actually produced by some other particles $\mathbf{r}', \mathbf{R}'$, which are not part of the original formalism. (At our convenience, we could include these and then these would be part of the extended coordinates $\mathbf{r}, \mathbf{R}$. The procedure would then result in the appearance of a potential interaction, but not having the "field.") At a first glance, the field (whether induced internally or externally) is expected to be a YM type tensorial field since the system is non-Abelian, but here we meet a surprise. When the couplings τ^k_{bm} are of the derivative form shown in Eq. (92) *and* when a complete set is taken for the electronic states $\zeta_k(\mathbf{r}, \mathbf{R})$, then the YM field intensity tensor F^{bc}_a induced by the $\mathbf{r}, \mathbf{R}$ system vanishes and the induced field is a "pure gauge field." Just as the induced four-vector potential in the Abelian case can be transformed away by a choice of gauge, so also can the τ^k_{bm} interaction terms. (See our previous proof, which shows that the vanishing of the YM tensor is the condition for the possibility to transform away the interaction term.) This serves as a reminder that with choice of a full electronic set, the solutions $\Psi(\mathbf{r}, \mathbf{R})$ are exact and there is no residual interaction between different $\Psi(\mathbf{r}, \mathbf{R})$'s. Such interaction can, of course, be externally induced by an "external" YM field intensity tensor F^{bc}_a, which is rooted, as before, in $\mathbf{r}', \mathbf{R}'$, and it could be got rid off by including these in the Hamiltonian.

3. Finally, there is the case that the electronic set $\zeta_k(\mathbf{r}, \mathbf{R})$ is not a complete set. Then, neither $\Psi(\mathbf{r}, \mathbf{R})$ in Eq. (90), nor the nuclear equation (91) is exact. Moreover, the truncated Lagrangean in Eq. (96) is not exact either and this shows up by its not possessing a full symmetry (viz., lacking invariance under local gauge transformation). We can (and should) remedy this by introducing a YM field, which is not now a pure gauge field. This means that the internally induced YM field cannot be transformed away by a (local) gauge transformation and that it brings in (through the back door, so to speak) the effect of the excluded electronic states on the nuclear states, these being now dynamically coupled between themselves.

At this stage, it would be too ambitious to extrapolate the implications of the above molecular theory for to elementary particles and forces but, by analogy with the fully worked out molecular model and disregarding any complications due to the fully relativistic covariance, one might argue that particle states are also eigenstates of some operators (Hamiltonians) and constitute full sets. Interactions between different particles (leptons, muons, etc.) exist and when these interactions (in their minimal form) are incorporated in the formalism, one gets exact eigenstates (and at this stage, as yet, *no* interaction fields). It is only when one truncates the particle state manifolds to finite subsets, which may have some internal symmetry [as the $SU(2)$ multiplets: "neutron, proton," or "electron, neutrino"], that one finds that one has to pay some price for the approximation involved in this truncation. Namely, the Lagrangean loses its original gauge invariance, which is the formal reflection of the fact that the original interaction field is not fully accounted for in the truncated representation. To remedy both the formal deficiency and the neglect of part of the interaction, one has to introduce some new forces (electromagnetic, or YM types and possibly others). These do both jobs.

Moreover, if the molecular analogy is further extended, these residual forces play a further role, in addition to the two already mentioned (viz., restoring formal invariance and reinstating the missed interaction). They bring in extra degrees of freedoms (e.g., photons), which act on the particles (but, supposedly, not between themselves). (In a vernacular locution, the tail that was wagged by the dog, can also wag the dog.) In the consistent scheme that we describe here, these extra degrees of freedom are illusory in that the residual forces are only convenient expressions of the presence of some other particles, and would be eliminated by including these other particles in a broader scheme. Evidently, the above description steers clear of field theory and is not relativistic (covariant). These, as well as other shortcomings that need to be supplied, require us to stop our speculations at this stage.

G. An Extended (Sufficiency) Criterion for the Vanishing of the Tensorial Field

We define the field intensity tensor F_{bc} as a function of a so far undetermined vector operator $\mathbf{X} = \mathbf{X}_b$ and of the partial derivatives ∂_b

$$F_{bc\ mn}(\mathbf{X}) = \partial_b \mathbf{X}_{c\ mn} - \partial_c \mathbf{X}_{b\ mn} - [\mathbf{X}_{b\ mk}\mathbf{X}_{c\ kn} - \mathbf{X}_{c\ mk}\mathbf{X}_{b\ kn}] \tag{113}$$

(The summation convention for double indices, for example, k in Eq. (113), is assumed, as before. However, we no longer make distinction between covariant and contravariant sets.) We set ourselves the task to find anti-Hermitean operators $\mathbf{X}_b$ such that

$$F_{bc\ mn}(\mathbf{X}) = 0 \tag{114}$$

The matrix elements are given by

$$\mathbf{X}_{b\ km} := \langle m|\mathbf{X}_b|n\rangle := \int d\mathbf{r}\zeta_m(\mathbf{r},\mathbf{R})\mathbf{X}_b\zeta_n(\mathbf{r},\mathbf{R}) \tag{115}$$

that is, the brackets represent integration over the electron coordinate $\mathbf{r}$. The $\zeta_m(\mathbf{r},\mathbf{R})$ are a real orthonormal set for any fixed $\mathbf{R}$. By anti-Hermiticity of the derivative operator ∂_b, we have already noted that

$$\langle m|\partial_b|n\rangle = -\langle n|\partial_b|m\rangle \tag{116}$$

Now (with ∂_b designating a differential that operates to the right until it encounters a closing bracket symbol) one finds that

$$\partial_b\langle m|\mathbf{X}_c|n\rangle = \langle(\partial_b m)|\mathbf{X}_c|n\rangle + \langle m|\partial_b(\mathbf{X}_c|n)\rangle \tag{117}$$

and, further, that

$$\partial_b \mathbf{X}_{c\ mn} - \partial_c \mathbf{X}_{b\ mn} = \langle(\partial_b m)|\mathbf{X}_c|n\rangle - \langle(\partial_c m)|\mathbf{X}_b|n\rangle + \langle m|\mathbf{X}_c(\partial_b n)\rangle$$
$$- \langle n|\mathbf{X}_b(\partial_c m)\rangle + \text{Commut} \tag{118}$$

where we designate

$$\text{Commut} \equiv \langle m|[\partial_b \mathbf{X}_c - \partial_c \mathbf{X}_b]|n\rangle \tag{119}$$

By subtracting the derivatives in the first four terms from the $\mathbf{X}$'s and adding to compensate, we have for Eq. (118)

$$= \langle(\partial_b m)|(\mathbf{X}_c - \partial_c)|n\rangle - \langle(\partial_c m)|(\mathbf{X}_b - \partial_b)|n\rangle + \langle m|(\mathbf{X}_c - \partial_c)|(\partial_b n)\rangle$$
$$- \langle n|(\mathbf{X}_b - \partial_b)|(\partial_c m)\rangle + \langle(\partial_b m)|(\partial_c n)\rangle - \langle(\partial_c m)|(\partial_b n)\rangle + \text{Commut} \tag{120}$$

We have ignored a term $\langle m|(\partial_c\partial_b - \partial_b\partial_c)|n\rangle$, which is zero by the commutativity of derivatives. The crucial step is now, as in [72] and in other later derivations, the evaluation of the fifth and sixth terms by insertion of $|k\rangle\langle k|$ (which is the unity operator, when k is summed over a complete set)

$$\langle(\partial_b m)|(\partial_c n)\rangle - \langle(\partial_c m)|(\partial_b n)\rangle = \langle(\partial_b m)|k\rangle\langle k|\partial_c|n\rangle - \langle(\partial_c m)|k\rangle\langle k|\partial_b|n\rangle$$
$$= -\langle m|\partial_b|k\rangle\langle k|\partial_c|n\rangle + \langle m|\partial_c|k\rangle\langle k|\partial_b|n\rangle \tag{121}$$

where Eq. (116) has been used. We replace any derivative ∂ by $\partial - \mathbf{X}$ and compensate, so as to get for Eq. (121) the expression

$$= \langle m|\partial_b - \mathbf{X}_c|k\rangle\langle k|\partial_c - \mathbf{X}_b|n\rangle - \langle m|\partial_c - \mathbf{X}_c|k\rangle\langle k|\partial_b - \mathbf{X}_b|n\rangle$$
$$+ \langle m|\mathbf{X}_b|k\rangle\langle k|\partial_c - \mathbf{X}_c|n\rangle - \langle m|\mathbf{X}_c|k\rangle\langle k|\partial_b - \mathbf{X}_b|n\rangle$$
$$+ \langle m|\partial_b - \mathbf{X}_b|k\rangle\langle k|\mathbf{X}_c|n\rangle - \langle m|\partial_c - \mathbf{X}_c|k\rangle\langle k|\mathbf{X}_b|n\rangle$$
$$+ \langle m|\mathbf{X}_b|k\rangle\langle k|\mathbf{X}_c|n\rangle - \langle m|\mathbf{X}_c|k\rangle\langle k|\mathbf{X}_b|n\rangle \tag{122}$$

We now put

$$\mathbf{X}_b = \partial_b + f_b(\mathbf{R}) \tag{123}$$

where the function $f_b(\mathbf{R})$ is a c number (not an operator) and can be taken outside brackets (where the integration variable is $\mathbf{r}$). Then we find that the first three lines in Eq. (122) cancel, and so do the four matrix elements in Eq. (120) (involving $\partial_c - \mathbf{X}_b$). The surviving contributions to the right-hand side of Eq. (118) are, first, the last line of Eq. (122), which is nothing else than the square brackets in the expression Eq. 113 for the field intensity tensor and, second, the term in Eq. (118), defined in the line following Eq. (118). For this term to vanish for all values of n, m, we require

$$\partial_b \mathbf{X}_c - \partial_c \mathbf{X}_b = 0 \tag{124}$$

or, in view of Eq. (123), that the function $f_b(\mathbf{R})$ be the gradient of a scalar $G(\mathbf{R})$

$$f_b(\mathbf{R}) = \partial_b G(\mathbf{R}) \tag{125}$$

In conclusion, we have shown that the non-Abelian gauge-field intensity tensor $F_{bc}(\mathbf{X})$ shown in Eq. (113) vanishes when

1. The electronic set is complete.
2. The $\mathbf{X}$ operator has the form $\mathbf{X}_b = \partial_b + \partial_b G(\mathbf{R})$.

It will be recognized that this generalizes the result proved by Baer in [72]. Though that work did establish the validity of the curl condition for the derivative operator as long as some 25 years ago and the validity is nearly trivial for the second term taken separately, the same result is not self-evident for the combination of the two terms, due to the nonlinearity of $F(\mathbf{X})$. An important special case is when $G(\mathbf{R}) = \mathbf{R}^2/2$. Then

$$\mathbf{X}_b = \partial_b + \mathbf{R}_b \tag{126}$$

and the last expression is recognized as a multiple of the creation operator a_b^+. This result paves the way for second-quantized or field theoretic treatments. An additional extension is to the time derivative operator, appropriate when the electronic states are time dependent. This extension is elementary (though this has not been noted before), since the key relation that leads to the vanishing of the field intensity, $F_{bc} = 0$, is Eq. (116), and this also holds when the subscript b stands for the time variable. What makes this result of special interest is the way that it provides an extension of the results to relativistic theories, in particular to a combination of Hamiltonians that (for the electron) is the Dirac Hamiltonian and (for the nuclear coordinates) is the Schrödinger Hamiltonian.

H. Observability of Molecular States in a Hamiltonian Formalism

We now describe the relation between a purely formal calculational device, like a gauge transformation that merely admixes the basis states, and observable effects.

Let us start, for simplicity, with a Hamiltonian $H(\mathbf{r}, \mathbf{R})$ for two types of particles. The particles can have similar or very different masses, but for clarity of exposition we continue to refer to the two types of particle as electrons ($\mathbf{r}$) and nuclei ($\mathbf{R}$). As before, we posit solutions of the time independent

Schrödinger equation that have the form shown in Eq. (90) but, for completeness, we attach an energy label e to each solution

$$\Psi^e(\mathbf{r},\mathbf{R}) = \sum_k \zeta_k(\mathbf{r},\mathbf{R})\chi_k^e(\mathbf{R}) \qquad e = 0, 1, \ldots \tag{127}$$

The electronic factor in the sum $\zeta_k(\mathbf{r},\mathbf{R})$ arises from the familiar BO electronic Hamiltonian defined for a fixed $\mathbf{R}$. Since this Hamiltonian is independent of the nuclear set $\chi_k^e(\mathbf{R})$, it does not carry the e label. As is well known, with each k there is associated a potential surface $V_k(\mathbf{R})$ (the eigenenergies of the electronic Hamiltonian). Therefore, by holding the nuclear positions fixed for a sufficiently long time and choosing an excitation wavelength appropriate to $V_k(\mathbf{R})$, it is possible to excite into any of the mutually orthogonal electronic states, $\zeta_k(\mathbf{r},\mathbf{R})$. The dependence of these functions on both of their variables can therefore be experimentally obtained. Turning now to the nuclear equation, Eq. (91), when the derivative terms are excluded, this equation yields the nuclear set $\chi_k^e(\mathbf{R})$ with a set of (constant) eigenenergies E_k^e for any given diagonal V_k. The set $\chi_k^e(\mathbf{R})$ is orthogonal for different e's and the same k, but not orthogonal for different k's and the same e (say, the lowest energy $e = 0$) or different e's. By returning to Eq. (127), it becomes clear that we can select any stationary eigenstate $\Psi^e(\mathbf{r},\mathbf{R})$ of the combined system by exciting with the proper wavelength for a sufficiently long time (in this case, of course, without constraint on $\mathbf{R}$). Thus, the dependence of any of these superpositions on the two variables $\mathbf{r},\mathbf{R}$ can also be ascertained and $\Psi^e(\mathbf{r},\mathbf{R})$ thereby operationally obtained. By computing the projections

$$\langle\zeta_k(\mathbf{r},\mathbf{R})|\Psi^e(\mathbf{r},\mathbf{R})\rangle \tag{128}$$

(in which both factors have been experimentally determined) we obtain the nuclear cofactors $\chi_k^e(\mathbf{R})$. [See again Eq. (127).] Actually, one could have written, instead of Eq. (127), a different superposition, sometimes called the "crude BO" wave function

$$\Psi^e(\mathbf{r},\mathbf{R}) = \sum_k \zeta_k(\mathbf{r},\mathbf{R}_0)\chi_k^e(\mathbf{R}|\mathbf{R}_0) \qquad e = 0, 1, \ldots \tag{129}$$

in which the electronic state refers to a fixed nuclear position $\mathbf{R}_0$ rather than to all values of the nuclear coordinate. This electronic state can be operationally obtained in a manner similar to, but actually more simple, than that which has already been proposed to obtain $\zeta_k(\mathbf{r},\mathbf{R})$ in Eq. (127), namely, by exciting at a wavelength corresponding to $V_k(\mathbf{R}_0)$ and probing the $\mathbf{r}$ dependence of $\zeta_k(\mathbf{r},\mathbf{R}_0)$. Determining $\Psi^e(\mathbf{r},\mathbf{R})$ as before and forming the projection $\langle\zeta_k(\mathbf{r},\mathbf{R}_0)|\Psi^e(\mathbf{r},\mathbf{R})\rangle$ we again obtain (gedanken experimentally) the nuclear factors $\chi_k^e(\mathbf{R}|\mathbf{R}_0)$. While this procedure is legitimate (and even simpler than the previous), it suffers from the more problematic convergence of the superposition (129) in comparison to (127). One could next try Eq. (127) with a truncated superposition, say involving only N summand terms (in practice $N = 2$ or 3 are common), rather than an infinite number of terms. The electronic functions $\zeta_k(\mathbf{r}, \mathbf{R})$ $(k = 1, \ldots, N)$ can be determined as before, and so can be the associated nuclear factors $\chi_k^e(\mathbf{R})$, but here one risks to come upon inconsistencies, when from the observationally obtained full wave function $\Psi^e(\mathbf{r}, \mathbf{R})$ one computes the overlaps $\langle \zeta_k(\mathbf{r}, \mathbf{R}) | \Psi^e(\mathbf{r}, \mathbf{R}) \rangle$ for any k above N. Then, the truncated sum on the right-hand side vanishes, while the computed overlap on the left-hand side will in general be nonzero. In a sense, it may be said that it is this inconsistency that the introduction of the YM field tries to resolve. The resulting eigen state $\Psi^e(\mathbf{r}, \mathbf{R})$ is an "entangled state," in the terminology of measurement theory [242]. While there appears to be no problem in principle to extract by experiment any $\zeta_k(\mathbf{r}, \mathbf{R})$ (as already indicated), the question arises whether one can put the nuclear part into any particular k state $\chi_k^e(\mathbf{R})$. This does not appear possible for the form in Eq. (127) and the source of the difficulty may again be the presence of derivatives in the nuclear equation. Can one select some observable nuclear set? It turns out that the set $\phi_h^e(\mathbf{R})$ in the transformed eigenstate

$$\Psi^e(\mathbf{r}, \mathbf{R}) = \sum_{kh} \zeta_k(\mathbf{r}, \mathbf{R})[G(\mathbf{R})^{-1}]_{kh} \phi_h^e(\mathbf{R}) \qquad e = 0, 1, \ldots \tag{130}$$

is observable. The matrix $G(\mathbf{R})$ is the gauge factor introduced in Eq. (104). The product $\zeta_k(\mathbf{r}, \mathbf{R}) G(\mathbf{R})^{-1}$ is independent of $\mathbf{R}$. [Recall that $G(\mathbf{R})$ is identical with the ADT matrix A]. Then $\phi_h^e(\mathbf{R})$ can be selected by exciting an e state such as in Eq. (128) and then selecting one of the $\mathbf{r}$ states. The coefficient of the selection will be (apart from a phase factor) the nuclear state $\phi_h^e(\mathbf{R})$.

However, this procedure depends on the existence of the matrix $G(\mathbf{R})$ (or of any pure gauge) that predicates the expansion in Eq. (90) for a full electronic set. Operationally, this means the preselection of a full electronic set in Eq. (129). When the preselection is only to a partial, truncated electronic set, then the relaxation to the truncated nuclear set in Eq. (128) will not be complete. Instead, the now truncated set in Eq. (128) will be subject to a YM force F. It is not our concern to fully describe the dynamics of the truncated set under a YM field, except to say (as we have already done above) that it is the expression of the residual interaction of the electronic system on the nuclear motion.

I. An Interpretation

As shown in Eq. (92), the gauge field A_c^b is simply related to the non-adiabatic coupling elements τ_{bm}^k. For an infinite set of electronic adiabatic states [$N = \infty$ in Eq. (90)], $F_{bc} = 0$. This important results seems to have been first established by [72] and was later rederived by others. [In the original formulation of [72] only the contracted form of the field A_c^b (appearing in the definition of F)

$$A^k{}_{cm} = A_c^b (t_b)_m^k \tag{131}$$

enters. This has the form

$$A^k{}_{cm} = \tau^k{}_{cm}(\mathbf{R}) \tag{132}$$

If the intermediate summations are over a complete set, then

$$(F_{bc})_m^k = F^a{}_{bc}(t_a)_m^k = 0] \tag{133}$$

This result extends the original theorem [72] and is true due to the linear independence of the t-matrices [67]. The meaning of the vanishing of F is that, if $\chi_k(\mathbf{R})$ is the partner of the electronic states spanning the whole Hilbert space, there is no indirect coupling (via a gauge field) between the nuclear states; the only physical coupling being that between the electronic and nuclear coordinates, which is given by the potential energy part of $H(\mathbf{r}, \mathbf{R})$. When the electronic N set is only part of the Hilbert space (e.g., N is finite), then the underlying electron–nuclei coupling gets expressed by an additional, residual coupling between the nuclear states. Then $F^a{}_{bc} \neq 0$ and the Lagrangean has to be enlarged to incorporate these new forces.

We further make the following tentative conjecture (probably valid only under restricted circumstances, e.g., minimal coupling between degrees of freedom): In quantum field theories, too, the YM residual fields, A and F, arise because the particle states are truncated (e.g., the proton-neutron multiplet is an isotopic doublet, without consideration of excited states). Then, it is within the truncated set that the residual fields reinstate the neglected part of the interaction. If all states were considered, then eigenstates of the form shown in Eq. (90) would be exact and there would be no need for the residual interaction negotiated by A and F.

VI. LAGRANGEANS IN PHASE-MODULUS FORMALISM

A. Background to the Nonrelativistic and Relativistic Cases

The aim of this section is to show how the modulus-phase formulation, which is the keytone of our chapter, leads very directly to the equation of continuity and to the Hamilton–Jacobi equation. These equations have formed the basic building blocks in Bohm's formulation of non-relativistic quantum mechanics [318]. We begin with the nonrelativistic case, for which the simplicity of the derivation has mainly pedagogical merits, but then we go over to the relativistic case that involves new results, especially regarding the topological phase. Our conclusions (presented in VI.H) are that, for a broad range of commonly encountered situations, the relativistic treatment will not affect the presence or absence of the Berry phase that arises from the Schrödinger equation.

The earliest appearance of the nonrelativistic continuity equation is due to Schrödinger himself [2,319], obtained from his time-dependent wave equation. A relativistic continuity equation (appropriate to a scalar field and formulated in terms of the field amplitudes) was found by Gordon [320]. The continuity equation for an electron in the relativistic Dirac theory [134,321] has the well-known form [322]:

$$\partial_\nu J^\nu = 0 \tag{134}$$

where the four-current J^ν is given by

$$J^\nu = \bar{\psi}\gamma^\nu\psi \tag{135}$$

(The symbols in this equation are defined below). It was shown by Gordon [323], and further discussed by Pauli [104] that, by a handsome trick on the four current, this can be broken up into two parts $J^\nu = J^\nu_{(0)} + J^\nu_{(1)}$ (each divergence-free), representing, respectively, a conductivity current (Leitungsstrom):

$$J^\nu_{(0)} = -\frac{i}{2mc}\left\{\left[\left(\hbar\partial^\nu - i\frac{e}{c}A^\nu\right)\bar{\psi}\right]\psi - \bar{\psi}\left[\left(\hbar\partial^\nu + i\frac{e}{c}A^\nu\right)\psi\right]\right\} \tag{136}$$

and a polarization current [324]

$$J^\nu_{(1)} = -\frac{i\hbar}{2mc}\partial_\mu(\bar{\psi}\gamma^\mu\gamma^\nu\psi) \qquad \nu \neq \mu \tag{137}$$

Again, the summation convention is used, unless we state otherwise. As will appear below, the same strategy can be used upon the Dirac Lagrangean density to obtain the continuity equation and Hamilton–Jacobi equation in the modulus-phase representation.

Throughout, the space coordinates and other vectorial quantities are written either in vector form $\mathbf{x}$, or with Latin indices x_k $(k = 1, 2, 3)$; the time (t) coordinate is $x_0 = ct$. A four vector will have Greek lettered indices, such as x_ν $(\nu = 0, 1, 2, 3)$ or the partial derivatives ∂_ν. m is the electronic mass, and e the charge.

B. Nonrelativistic Electron

The phase-modulus formalism for nonrelativistic electrons was discussed at length by Holland [324], as follows.

The Lagrangean density $\mathcal{L}$ for the nonrelativistic electron is written as

$$\mathcal{L} = -\frac{\hbar^2}{2m}\nabla\psi^* \cdot \nabla\psi - eV\psi^*\psi - \frac{ie\hbar}{2cm}\mathbf{A}\cdot(\psi^*\nabla\psi - \nabla\psi^*\psi) + \frac{1}{2}i\hbar(\psi^*\dot{\psi} - \dot{\psi}^*\psi) \tag{138}$$

Here dots over symbols designate time derivatives. If now the modulus a and phase ϕ are introduced through

$$\psi = ae^{i\phi} \tag{139}$$

the Lagrangean density takes the form

$$\mathcal{L} = -\frac{\hbar^2}{2m}[(\nabla a)^2 + a^2(\nabla\phi)^2] - ea^2V + \frac{e\hbar}{cm}a^2\nabla\phi\cdot\mathbf{A} - \hbar a^2\frac{\partial\phi}{\partial t} \tag{140}$$

The variational derivative of this with respect to ϕ yields the continuity equation

$$\frac{\delta\mathcal{L}}{\delta\phi} = 0 \rightarrow \frac{\partial\rho}{\partial t} + \nabla\cdot(\rho\mathbf{v}) = 0 \tag{141}$$

in which the charge density is defined as: $\rho = ea^2$ and the velocity is

$$v = \frac{1}{m}\left(\hbar\nabla\phi - \frac{e}{c}\mathbf{A}\right)$$

Variationally deriving with respect to a leads to the Hamilton–Jacobi equation

$$\frac{\delta\mathcal{L}}{\delta a} = 0 \rightarrow \frac{\partial S}{\partial t} + \frac{1}{2m}\left(\nabla S - \frac{e}{c}\mathbf{A}\right)^2 + eV = \frac{\hbar^2\nabla^2 a}{2ma} + \frac{e^2\mathbf{A}^2}{2mc^2} \tag{142}$$

in which the action is defined as: $S = \hbar\phi$. The right-hand side of Eq. (142) contains the "quantum correction" and the electromagnetic correction. These results are elementary, but their derivation illustrates the advantages of using the two variables, phase and modulus, to obtain equations of motion that are substantially different from the familiar Schrödinger equation and have straightforward physical interpretations [318]. The interpretation is, of course, connected to the modulus being a physical observable (by Born's interpretational postulate) and to the phase having a similar though somewhat more problematic status. (The "observability" of the phase has been discussed in the literature by various sources, e.g., in [28] and, in connection with a recent development, in [31,33]. Some of its aspects have been reviewed in Section II.)

Another possibility to represent the quantum mechanical Lagrangian density is using the logarithm of the amplitude $\lambda = \ln a, \quad a = e^{\lambda}$. In that particular representation, the Lagrangean density takes the following symmetrical form

$$\mathcal{L} = e^{2\lambda}\left\{-\frac{\hbar^2}{2m}[(\nabla\lambda)^2 + (\nabla\phi)^2] - \hbar\frac{\partial\phi}{\partial t} - eV + \frac{e\hbar}{cm}\nabla\phi\cdot\mathbf{A}\right\} \tag{143}$$

C. Similarities Between Potential Fluid Dynamics and Quantum Mechanics

In writing the Lagrangean density of quantum mechanics in the modulus-phase representation, Eq. (140), one notices a striking similarity between this Lagrangean density and that of potential fluid dynamics (fluid dynamics without vorticity) as represented in the work of Seliger and Whitham [325]. We recall briefly some parts of their work that are relevant, and then discuss the connections with quantum mechanics. The connection between fluid dynamics and quantum mechanics of an electron was already discussed by Madelung [326] and in Holland's book [324]. However, the discussion by Madelung refers to the equations only and does not address the variational formalism which we discuss here.

If a flow satisfies the condition of zero vorticity, that is, the velocity field $\mathbf{v}$ is such that $\nabla \times \mathbf{v} = 0$, then there exists a function ν such that $\mathbf{v} = \nabla\nu$. In that case, one can describe the fluid mechanical system with the following Lagrangean density

$$\mathcal{L} = \left[-\frac{\partial\nu}{\partial t} - \frac{1}{2}(\nabla\nu)^2 - \varepsilon(\rho) - \Phi\right]\rho \tag{144}$$

in which ρ is the mass density, ε is the specific internal energy and Φ is some arbitrary function representing the potential of an external force acting on the fluid. By taking the variational derivative with respect to ν and ρ, one obtains the following equations

$$\frac{\partial\rho}{\partial t} + \nabla\cdot(\rho\nabla\nu) = 0 \tag{145}$$

$$\frac{\partial\nu}{\partial t} = -\frac{1}{2}(\nabla\nu)^2 - h - \Phi \tag{146}$$

in which $h = \partial(\rho\varepsilon)/\partial\rho$ is the specific enthalpy. The first of those equations is the continuity equation, while the second is Bernoulli's equation.

Going back to the quantum mechanical system described by Eq. (140), we introduce the following variables $\hat{\nu} = \hbar\phi/m, \ \hat{\rho} = ma^2$. In terms of these new variables the Lagrangean density in Eq. (140) will take the form

$$\mathcal{L} = \left[-\frac{\partial \hat{\nu}}{\partial t} - \frac{1}{2}(\nabla \hat{\nu})^2 - \frac{\hbar^2}{2m^2} \frac{(\nabla\sqrt{\hat{\rho}})^2}{\hat{\rho}} - \frac{e}{m} V \right] \hat{\rho} \tag{147}$$

in which we assumed that no magnetic fields are present and thus $\mathbf{A} = 0$. When compared with Eq. (144) the following correspondence is noted

$$\hat{\nu} \Leftrightarrow \nu \qquad \hat{\rho} \Leftrightarrow \rho \qquad \frac{\hbar^2}{2m^2} \frac{(\nabla\sqrt{\hat{\rho}})^2}{\hat{\rho}} \Leftrightarrow \varepsilon \qquad \frac{e}{m} V \Leftrightarrow \Phi \tag{148}$$

The quantum "internal energy" $(\hbar^2/2m^2)(\nabla\sqrt{\hat{\rho}})^2/\hat{\rho}$ depends also on the derivative of the density, unlike in the fluid case, in which internal energy is a function of the mass density only. However, in both cases the internal energy is a positive quantity.

Unlike classical systems in which the Lagrangean is quadratic in the time derivatives of the degrees of freedom, the Lagrangeans of both quantum and fluid dynamics are linear in the time derivatives of the degrees of freedom.

D. Electrons in the Dirac Theory

(Henceforth, for simplicity, the units $c = 1, \hbar = 1$ will be used, except at the end, when the results are discussed.) The Lagrangean density for the particle is in the presence of external forces

$$\mathcal{L} = \frac{i}{2}[\bar{\psi}\gamma^{\mu}(\partial_{\mu} + ieA_{\mu})\psi - (\partial_{\mu} - ieA_{\mu})\bar{\psi}\gamma^{\mu}\psi] - m\bar{\psi}\psi \tag{149}$$

Here, ψ is a four-component spinor, A_{μ} is a four potential, and the 4×4 matrices γ^{μ} are given by

$$\gamma^0 = \begin{pmatrix} I & 0 \\ 0 & -I \end{pmatrix} \qquad \gamma^k = \begin{pmatrix} 0 & \sigma^k \\ -\sigma^k & 0 \end{pmatrix} \qquad (k = 1, 2, 3) \tag{150}$$

where we have the 2×2 matrices

$$I = \begin{pmatrix} 1 & 0 \\ 0 & 1 \end{pmatrix} \qquad \sigma^1 = \begin{pmatrix} 0 & 1 \\ 1 & 0 \end{pmatrix}$$
$$\sigma^2 = \begin{pmatrix} 0 & -i \\ i & 0 \end{pmatrix} \qquad \sigma^3 = \begin{pmatrix} 1 & 0 \\ 0 & -1 \end{pmatrix} \tag{151}$$

By following [323], we substitute in the Lagrangean density, Eq. (149), from the Dirac equations [322], namely, from

$$\psi = \frac{i}{m}\gamma^{\nu}(\partial_{\nu} + ieA_{\nu})\psi \qquad \bar{\psi} = -\frac{i}{m}(\partial_{\nu} - ieA_{\nu})\bar{\psi}\gamma^{\nu} \tag{152}$$

and obtain

$$\mathcal{L} = \frac{1}{m}(\partial_{\nu} - ieA_{\nu})\bar{\psi}\gamma^{\nu}\gamma^{\mu}(\partial_{\mu} + ieA_{\mu})\psi - m\bar{\psi}\psi \tag{153}$$

We thus obtain a Lagrangean density, which is equivalent to Eq. (149) for all solutions of the Dirac equation, and has the structure of the nonrelativistic Lagrangian density, Eq. (140). Its variational derivations with respect to ψ and $\bar{\psi}$ lead to the solutions shown in Eq. (152), as well as to other solutions.

The Lagrangean density can be separated into two terms

$$\mathcal{L} = \mathcal{L}^0 + \mathcal{L}^1 \tag{154}$$

according to whether the summation symbols ν and μ in (149) are equal or different. The form of $\mathcal{L}^0$ is

$$\mathcal{L}^0 = \frac{1}{m}(\partial^{\mu} - ieA^{\mu})\bar{\psi}(\partial_{\mu} + ieA_{\mu})\psi + m\bar{\psi}\psi \tag{155}$$

Contravariant V^{μ} and covariant V_{ν} four vectors are connected through the metric $g^{\mu\nu} = \text{diag}\ (1, -1, -1, -1)$ by

$$V^{\mu} = g^{\mu\nu}V_{\nu} \tag{156}$$

The second term in Eq. (154), $\mathcal{L}^1$ will be shown to be smaller than the first in the near nonrelativistic limit.

Introducing the moduli a_i and phases ϕ_i for the four spinor components ψ_i $(i = 1, 2, 3, 4)$, we note the following relations (in which no summations over i are implied):

$$\begin{aligned} \psi_i &= a_i e^{i\phi_i} \\ \bar{\psi}_i &= \gamma^0_{ii} a_i e^{-i\phi_i} \\ \bar{\psi}_i\psi_i &= a_i^2 \gamma^0_{ii} \end{aligned} \tag{157}$$

The Lagrangean density eqaution (153) rewritten in terms of the phases and moduli takes a form that is much simpler (and shorter), than that which one would obtain by substituting from Eq. (139) into Eq. (149). It is given by

$$\mathcal{L}^0 = \frac{1}{m}\sum_i \gamma_{ii}^0[\partial_\nu a_i \partial^\nu a_i + a_i^2((\partial_\nu \phi_i + eA_\nu)(\partial^\nu \phi_i + eA^\nu) - m^2)] \tag{158}$$

When one takes its variational derivative with respect to the phases ϕ_i, one obtains the continuity equation in the form

$$\frac{\delta \mathcal{L}^0}{\delta \phi_i} = -\frac{\delta \mathcal{L}^1}{\delta \phi_i} \tag{159}$$

The right-hand side will be treated in a following section VI.E, where we shall see that it is small in the nearly nonrelativistic limit and that it vanishes in the absence of an electromagnetic field. The left-hand side can be evaluated to give

$$\frac{\delta \mathcal{L}^0}{\delta \phi_i} = -\frac{2}{m}\partial_\nu[a_i^2(\partial^\nu \phi_i + eA^\nu)] \equiv 2\partial_\nu J_i^\nu \qquad \text{(no summation over } i\text{)} \tag{160}$$

The above defined currents are related to the conductivity current by the relation

$$J_{(0)}^\nu = \sum_i \gamma_{ii}^0 J_i^\nu \tag{161}$$

Although the conservation of J_i^ν separately is a stronger result than the result obtained in [104], one should bear in mind that the present result is only approximate.

The variational derivatives of $\mathcal{L}^0$ with respect to the moduli a_i give the following equations:

$$\frac{\delta \mathcal{L}^0}{\delta a_i} = -\frac{2}{m}[\partial_\nu \partial^\nu a_i - a_i((\partial_\nu \phi_i + eA_\nu)(\partial^\nu \phi_i + eA^\nu) - m^2)] \tag{162}$$

The result of interest in the expressions shown in Eqs. (160) and (162) is that, although one has obtained expressions that include corrections to the nonrelativistic case, given in Eqs. (141) and (142), still both the continuity equations and the Hamilton–Jacobi equations involve each spinor component separately. To the present approximation, there is no mixing between the components.

E. The Nearly Nonrelativistic Limit

In order to write the previously obtained equations in the nearly nonrelativistic limit, we introduce phase differences s_i that remain finite in the limit $c \to \infty$. Then

$$\phi_i = \gamma_{ii}^0(-mx_0 + s_i) \qquad \partial_0 \phi_i = \gamma_{ii}^0(-m + \partial_0 s_i) \qquad \mathbf{\nabla}\phi_i = \gamma_{ii}^0 \mathbf{\nabla} s_i \tag{163}$$

We reinstate the velocity of light c in this and in Section VI.F in order to appreciate the order of magnitude of the various terms. When contributions from $\mathcal{L}^1$ are neglected, the expression in Eq. (162) equated to zero gives the following equations, in which the large $(i = 1, 2)$ and small $(i = 3, 4)$ components are separated.

$$\partial_t s_i + \frac{1}{2m}(\nabla s_i - \frac{e}{c}\mathbf{A})^2 + eA_0 = \frac{\nabla^2 a_i}{2ma_i} + \frac{e^2}{2mc^2}\mathbf{A}^2 + \frac{1}{2mc^2}\left[-\frac{\partial_t^2 a_i}{a_i} + (\partial_t s_i)^2 + 2eA_0\partial_t s_i + e^2A_0^2 - e^2\mathbf{A}^2\right] \quad (i = 1, 2) \tag{164}$$

$$\partial_t s_i + \frac{1}{2m}(\nabla s_i - \frac{(-e)}{c}\mathbf{A})^2 + (-e)A_0 = \frac{\nabla^2 a_i}{2ma_i} + \frac{e^2}{2mc^2}\mathbf{A}^2 + \frac{1}{2mc^2}\left[-\frac{\partial_t^2 a_i}{a_i} + (\partial_t s_i)^2 + 2(-e)A_0\partial_t s_i + e^2A_0^2 - e^2\mathbf{A}^2\right] \quad (i = 3, 4) \tag{165}$$

In the same manner, we obtain the following equations from Eq. (160)

$$\partial_t \rho_i + \nabla \cdot (\rho_i \mathbf{v}_i) = \frac{1}{c^2}\partial_t\left[\rho_i\left(\frac{\partial_t s_i + eA_0}{m}\right)\right] \quad (i = 1, 2) \quad \rho_i = ma_i^2 \quad \mathbf{v}_i = \frac{\nabla s_i - \frac{e}{c}\mathbf{A}}{m} \tag{166}$$

$$\partial_t \rho_i + \nabla \cdot (\rho_i \mathbf{v}_i) = \frac{1}{c^2}\partial_t\left[\rho_i\left(\frac{\partial_t s_i + (-e)A_0}{m}\right)\right] \quad (i = 3, 4) \quad \rho_i = ma_i^2 \quad \mathbf{v}_i = \frac{\nabla s_i - \frac{(-e)}{c}\mathbf{A}}{m} \tag{167}$$

The terms before the square brackets give the nonrelativistic part of the Hamilton–Jacobi equation and the continuity equation shown in Eqs. (142) and (141), while the term with the square brackets contribute relativistic corrections. All terms from $\mathcal{L}^0$ are of the nonmixing type between components. There are further relativistic terms, to which we now turn.

F. The Lagrangean-Density Correction Term

As noted above, $\mathcal{L}^1$ in Eq. (154) arises from terms in which $\mu \neq \nu$. The corresponding contribution to the four current was evaluated in [104,323] and was shown to yield the polarization current. Our result is written in terms of the magnetic field $\mathbf{H}$ and the electric field $\mathbf{E}$, as well as the spinor four-vector ψ and the vectorial 2×2 sigma matrices given in Eq. (151).

$$\mathcal{L}^1 = -\frac{e}{mc}\bar{\psi}(\mathbf{H} \cdot \boldsymbol{\sigma})\begin{pmatrix} I & 0 \\ 0 & I \end{pmatrix}\psi + \frac{ie}{mc}\bar{\psi}(\mathbf{E} \cdot \boldsymbol{\sigma})\begin{pmatrix} 0 & I \\ I & 0 \end{pmatrix}\psi \tag{168}$$

These terms are analogous to those on p. 265 of [7]. It will be noted that the symbol c has been reinstated as in Section VI.F, so as to facilitate the order of magnitude estimation in the nearly nonrelativistic limit. We now proceed based on Eq. (168) as it stands, since the transformation of Eq. (168) to modulus and phase variables and functional derivation gives rather involved expressions and will not be set out here.

To compare $\mathcal{L}^1$ with $\mathcal{L}^0$ we rewrite the latter in terms of the phase variables introduced in Eq. (163)

$$\mathcal{L}^0 = 2\sum_i \gamma_{ii}^0 \left\{ -\frac{1}{2m}[(\nabla a_i)^2 + a_i^2(\nabla s_i)^2] - (\gamma_{ii}^0 e)a_i^2 A^0 - a_i^2 \frac{\partial s_i}{\partial t} \right\}$$
$$+ \frac{2e}{mc}\sum_i a_i^2 \nabla s_i \cdot \mathbf{A} + O\left(\frac{1}{c^2}\right) \tag{169}$$

which contains terms independent of c as well as terms of the order $O(1/c)$ and $O(1/c^2)$.

In Eq. (168), the first, magnetic-field term admixes different components of the spinors both in the continuity equation and in the Hamilton–Jacobi equation. However, with the z axis chosen as the direction of $\mathbf{H}$, the magnetic-field term does not contain phases and does not mix component amplitudes. Therefore, there is no contribution from this term in the continuity equations and no amplitude mixing in the Hamilton–Jacobi equations . The second, electric-field term is nondiagonal between the large and small spinor components, which fact reduces its magnitude by a further small factor of $O(\textit{particle velocity}/c)$. This term is therefore of the same small order $O(1/c^2)$, as those terms in the second line in Eqs. (164) and (166) that refer to the upper components.

We conclude that in the presence of electromagnetic fields the components remain unmixed, correct to the order $O(1/c)$.

G. Topological Phase for Dirac Electrons

The topological (or Berry) phase [9,11,78] has been discussed in previous sections. The physical picture for it is that when a periodic force, slowly (adiabatically) varying in time, is applied to the system then, upon a full periodic evolution, the phase of the wave function may have a part that is independent of the amplitude of the force. This part exists in addition to that part of the phase that depends on the amplitude of the force and that contributes to the usual, "dynamic" phase. We shall now discuss whether a relativistic electron can have a Berry phase when this is absent in the framework of the Schrödinger equation, and vice versa. (We restrict the present discussion to the nearly nonrelativistic limit, when particle velocities are much smaller than c.)

The following lemma is needed for our result. Consider a matrix Hamiltonian h coupling two states, whose energy difference is $2m$

$$h = \begin{pmatrix} m + E_1 \cos(\omega t + \alpha) & E_2 \sin(\omega t) \\ E_2 \sin(\omega t) & -m - E_1 \cos(\omega t + \alpha) \end{pmatrix} \tag{170}$$

The Hamiltonian contains two fields, periodically varying in time, whose intensities E_1 and E_2 are nonzero. The parameter ω is their angular frequency and is (in appropriate energy units) assumed to be much smaller than the field strengths. This ensures the validity of the adiabatic approximation [33]. The parameter α is an arbitrary angle. It is assumed that initially, at $t = 0$, only the component with the positive eigenenergy is present. Then after a full revolution the initially excited component acquires or does not acquire a Berry phase (i.e., returns to its initial value with a changed or unchanged sign) depending on whether $|E_1|$ is greater or less than m (= half the energy difference).

Proof: When the time-dependent Schrödinger equation is solved under adiabatic conditions, the upper, positive energy component has the coefficient: the dynamic phase factor $\times C$, where

$$C = \cos\left[\frac{1}{2}\arctan\left(\frac{E_2 \sin(\omega t)}{m + E_1 \cos(\omega t + \alpha)}\right)\right] \tag{171}$$

Tracing the arctan over a full revolution by the method described in Section IV and noting the factor $1/2$ in Eq. (171) establishes our result. (The case that $|E_1| = m$ needs more careful consideration, since it leads to a breakdown of the adiabatic theorem. However, this case will be of no consequence for the results.)

We can now return to the Dirac equations, in which the time varying forces enter through the four-potentials $(A_0, \mathbf{A})$. [The "two states" in Eq. (28) refer now to a large and to a small (positive and negative energy) component in the solution of the Dirac equation in the near nonrelativistic limit.] In the expressions (164 and 165) obtained for the phases s_i and arising from the Lagrangean $\mathcal{L}^0$, there is no coupling between different components, and therefore the small relativistic correction terms will clearly not introduce or eliminate a Berry phase. However, terms in this section supply the diagonal matrix elements in Eq. (28). Turning now to the two terms in Eq. (168), the first, magnetic field term again does not admix the large and small components, with the result that for either of these components previous treatments based on the Schrödinger or the Pauli equations [321,324] should suffice. Indeed, this term was already discussed by Berry [9]. We thus need to consider only the second, electric-field term that admixes the two types of components. These are the source of the off-diagonal matrix elements in Eq. (28). However, we have just shown that in order to introduce a new topological phase, one needs field strengths matching the electronic rest energies, namely electric fields of the order of 10^{14}V/cm. (For comparison, we note that the electric field that binds an electron in a hydrogen atom is four orders of magnitudes smaller than this. Higher fields can also be produced in the laboratory, but, in general, are not of the type that can be used to guide the motion of a charged particle during a revolution.) As long as we exclude from our considerations such enormous fields, we need not contemplate relativistically induced topological phases. Possibly, there may be cases (e.g., many electron systems or magnetic field effects) that are not fully covered by the model represented in Eq. (28). Still, the latter model should serve as an indicator for relativistic effects on the topological phase.

H. What Have We Learned About Spinor Phases?

This part of our chapter has shown that the use of the two variables, moduli and phases, leads in a direct way to the derivation of the continuity and Hamilton–Jacobi equations for both scalar and spinor wave functions. For the latter case, we show that the differential equations for each spinor component are (in the nearly nonrelativistic limit) approximately decoupled. Because of this decoupling (mutual independence) it appears that the reciprocal relations between phases and moduli derived in Section III hold to a good approximation for each spinor component separately, too. For velocities and electromagnetic field strengths that are normally below the relativistic scale, the Berry phase obtained from the Schrödinger equation (for scalar fields) will not be altered by consideration of the Dirac equation.

VII. CONCLUSION

This chapter has treated a number of properties that arise from the presence of degeneracy in the electronic part of the molecular wave function. The existence of more than one electronic state in the superposition that describes the molecular state demands attention to the phase relations between the different electronic component amplitudes. Looked at from a different angle, the phase relations are the consequence of the complex form of the molecular wave functions, which is grounded in the time dependent Schrödinger equation. Beside reviewing numerous theoretical and experimental works relating to the phase properties of complex wave functions, the following general points have received emphasis in this chapter: (1) Relative phases of components that make up, by the superposition principle, the wave function are observable. (2) The analytic behavior of the wave function in a complex parameter plane is in several instances traceable to a physics-based "equation of restriction." (3) Phases and moduli in the superposition are connected through reciprocal integral relations. (4) Systematic treatment of zeros and singularities of component amplitudes are feasible by a phase tracing method. (5) The molecular

Yang–Mills field is conditioned by the finiteness of the basic Born–Oppenheimer set. Detailed topics are noted in the introductory Section I.

Acknowledgments

The authors are indebted to Professor Michael Baer for many years of exciting collaboration, to Dr. B. Halperin for advice, to Professor Mark Pere'lman for discussion and permission to quote from his preprint *Temporal Magnitudes and Functions of Scattering Theory*, to Professor Shmuel Elitzur for suggesting the approach leading to "Alternative Derivation" in Section V and to Professor Igal Talmi for an inspiration [327].

References

1. M. Jammer, *The Conceptual Development of Quantum Mechanics*, McGraw-Hill, New York, 1966.
2. J. Mehra and H. Rechenberg, *The Historical Development of Quantum Theory*, Springer-Verlag, New York, 1987, Vol. 5, Part 2.
3. T. Y. Cao, *Conceptual Development of 20th Century Field Theory*, University Press, Cambridge, UK, 1997.
4. P. A. M. Dirac, *Proc. R. Soc. London Ser. A* **112**, 661 (1926).
5. E. P. Wigner, in *The Place of Consciousness in Modern Physics*, C. Muses and A. M. Young, eds., *Consciousness and Reality*, Outerbridge and Lazzard, New York, 1972; reprinted in E. P. Wigner, *Philosophical Reflections and Syntheses*, Springer-Verlag, Berlin, 1997.
6. M. Steiner, *The Applicability of Mathematics as a Philosophical Problem*, Harvard University Press, Cambridge, MA, 1998.
7. P. A. M. Dirac, *The Principles of Quantum Mechanics*, Clarendon Press, Oxford, UK, 1958.
8. Y. Aharonov and D. Bohm, *Phys. Rev.* **115**, 485 (1959).
9. M. V. Berry, *Proc. R. Soc. London Ser. A* **392**, 45 (1984).
10. Y. Aharonov and J. Anandan, *Phys. Rev. Lett.* **58**, 1593 (1987).
11. S. R. Jain and A. K. Pati, *Phys. Rev. Lett.* **80**, 650 (1980).
12. C. M. Cheng and P. C. W. Fung, *J. Phys. A* **22**, 3493 (1989).
13. D. J. Moore and G. E. Stedman, *J. Phys. A* **23**, 2049 (1990).
14. A. K. Pati, *Phys. Rev. A* **52**, 2576 (1995).
15. I. Sh. Averbukh and N. F. Perel'man, *Sov. Phys. Uspekhi* **34**, 572 (1991).
16. I. Sh. Averbukh, M. Shapiro, C. Leichtle, and W. P. Schleich, *Phys. Rev. A* **59**, 2163 (1999).
17. R. Schuster, E. Buks, M. Heiblum, D. Mahalu, V. Umansky, and H. Shtrikman, *Nature (London)* **385**, 417 (1997).
18. C. Leichtle, W. P. Schleich, I. Averbukh, and M. Shapiro, *Phys. Rev. Lett.* **80**, 1418 (1998).
19. T. C. Weinacht, J. Ahn and P. H. Buchsbaum, *Phys. Rev. Lett.* **80**, 5508 (1998).
20. G. Nogues, A. Rauschenbeutel, S. Osnaghi, M. Brune, J. M. Raimond, and S. Haroche, *Nature (London)* **400**, 239 (1999).
21. G. C. Ghirardi, A. Rimini, and T. Weber, *Phys. Rev. D* **34**, 470 (1986).
22. D. V. Averin, *Nature (London)* **398**, 748 (1999).
23. L. Grover, *Sciences* **39**, 27 (1999, No. 2).
24. N. F. Ramsey, *Molecular Beams*, Oxford University Press, New York, 1985.
25. M. W. Noel and C. R. Stroud, Jr., *Phys. Rev. Lett.* **75**, 1252 (1995).

26. H. Goldstein, *Classical Mechanics*, Addison-Wesley, Reading MA, 1959.
27. M. M. Nieto, *Phys. Scr. T* **48**, 5 (1993).
28. L. Mandel and E. Wolf, *Optical Coherence and Quantum Optics*, University Press, Cambridge, 1995, Section 3.1.
29. R. Englman, A. Yahalom, and M. Baer, *J. Chem. Phys.* **109**, 6550 (1998).
30. R. Englman, A. Yahalom, and M. Baer, *Phys. Lett. A* **251**, 223 (1999).
31. R. Englman and A. Yahalom, *Phys. Rev. A* **60**, 1802 (1999).
32. R. Englman, A. Yahalom, and M. Baer, *Eur. Phys. J. D* **8**, 1 (2000).
33. R. Englman and A. Yahalom, *Phys. Rev. B* **61**, 2716 (2000).
34. R. Englman and A. Yahalom, *Found. Phys. Lett.* **13**, 329 (2000).
35. R. Englman and A. Yahalom, in *The Jahn Teller Effect: A Permanent Presence in the Frontiers of Science*, M. D. Kaplan and G. Zimmerman, eds., *Proceedings of the NATO Advanced Research Workshop, Boston, Sept. 2000*, Kluwer, Dordrecht, 2001.
36. M. Baer and R. Englman, *Chem. Phys. Lett.* **335**, 85 (2001).
37. A. Mebel, M. Baer, R. Englman, and A. Yahalom, *J. Chem. Phys.* **115**, 3673 (2001).
38. R. Englman, A. Yahalom, and M. Baer, *Int. J. Quant. Chem.* (in press).
39. L. A. Khalfin, *Sov. Phys. JETP* **8**, 1053 (1958) [*J. Exp. Teor. Fyz.* (*USSR*) **33**, 1371 (1957)].
40. M. E. Perel'man and R. Englman, *Modern Phys. Lett. B* **14**, 907 (2000).
41. M. N. Scully and M. S. Zubairy, *Quantum Optics*, University Press, Cambridge, UK, 1997.
42. W. H. Miller, *J. Chem. Phys.* **55**, 3146 (1971).
43. J.-T. Hwang and P. Pechukas, *J. Chem. Phys.* **67**, 4640 (1977).
44. E. E. Nikitin, *J. Chem. Phys.* **102**, 6768 (1997).
45. C. Zhu, E. E. Nikitin, and H. Nakamura, *J. Chem. Phys.* **104**, 7059 (1998).
46. I. B. Bersuker, *Chem. Rev.* **101**, 1067 (2001).
47. I. B. Bersuker and V. Z. Polinger, *Vibronic Interactions in Molecules and Crystals*, Springer-Verlag, New York, 1989.
48. V. Sidis, in *State Selected and State-to-State Ion-Molecule Reaction Dynamics, Part 2, Theory* M. Baer and C. Y. Ng eds., John Wiley & Sons, inc., New York, 1992; *Adv. Chem. Phys.* **82**, 73 (2000).
49. T. Pacher, L. S. Cederbaum, and H. Koppel, *Adv. Chem. Phys.* **84**, 293 (1993).
50. D. R. Yarkony, *Rev. Mod. Phys.* **68**, 985 (1996); *Adv. At. Mol. Phys.* **31**, 511 (1998).
51. H. Thummel, M. Peric, S. D. Peyerimhoff, and R. J. Buenker, *Z. Phys. D* **13**, 307 (1989).
52. H. Koppel and R. Meiswinkel, *Z. Phys. D* **32**, 153 (1994).
53. D. R. Yarkony, *Acc. Chem. Res.* **31**, 511 (1998).
54. H. Koizumi and I. B. Bersuker, *Phys. Rev. Lett.* **83**, 3009 (1999).
55. A. M. Mebel, M. Baer, and S. H. Lin, *J. Chem. Phys.* **112**, 10703 (2000).
56. A. M. Mebel, M. Baer, and S. H. Lin, *J. Chem. Phys.* **114**, 5109 (2001).
57. J. Avery, M. Baer, and G. D. Billing (2001, to be published).
58. M. S. Child, *Adv. Chem. Phys.* (Chapter 1 in this volume).
59. M. Baer, R. Englman, and A. J. C. Varandas, *Mol. Phys.* **97**, 1185 (1999).
60. M. Baer, A. J. C. Varandas, and R. Englman, *J. Chem. Phys.* **111**, 9493 (1999).
61. A. Yahalom and R. Englman, *Phys. Lett. A* **272**, 166 (2000).
62. M. Born and R. J. Oppenheimer, *Ann. Physik (Leipzig)* **89**, 457 (1927).

63. M. Born and K. Huang, *Dynamical Theory of Crystal Lattices*, Clarendon Press, Oxford, UK, 1951, Appendix VII.
64. D. R. Bates and R. McCarroll, *Proc. R. Soc. London Ser. A* **245**, 175 (1958).
65. N. F. Mott and H. S. W. Massey, *The Theory of Atomic Collisions*, Oxford University Press, London, 1965.
66. C. N. Yang and R. Mills, *Phys. Rev.* **96**, 191 (1954).
67. R. Jackiw, *Rev. Mod. Phys.* **52**, 661 (1980).
68. C. Itzykson and J.-B. Zuber, *Quantum Field Theories*, McGraw-Hill, New York, 1980.
69. M. E. Peskin and D. V. Schroeder, *An Introduction to Quantum Field Theory*, Perseus Books, Reading MA, 1995, Chap. 14.
70. S. Weinberg, *The Quantum Theory of Fields*, University Press, Cambridge, 1996, Vol. 2, Chap. 15.
71. F. T. Smith, *Phys. Rev.* **115**, 349 (1960); *ibid.* **179**, 111 (1969).
72. M. Baer, *Chem. Phys. Lett.* **35**, 112 (1975).
73. R. Englman, A. Yahalom, and M. Baer, *Intern. J. Quantum Chem.* (2002, to appear.)
74. C. A. Mead, *Phys. Rev. Lett.* **59**, 161 (1987).
75. B. Zygelman, *Phys. Lett. A* **125**, 476 (1987).
76. C. A. Mead and D. G. Truhlar, *J. Chem. Phys.* **70**, 2284 (1979).
77. J. Moody, A. Shapere, and F. Wilczek, *Phys. Rev. Lett.* **56**, 893 (1986).
78. A. Shapere and F. Wilczek, eds., *Geometrical Phases in Physics*, World Scientific, Singapore, 1990.
79. J. Moody, A. Shapere, and F. Wilczek in *Geometrical Phases in Physics*, A. Shapere and F. Wilczek, eds., World Scientific, Singapore, 1989, p. 160.
80. B. Zygelman, *Phys. Rev. Lett.* **64**, 256 (1990).
81. Y. Aharonov, E. Ben-Reuven, S. Popescu, and D. Rohrlich, *Nucl. Phys. B* **350**, 818 (1991).
82. M. V. Berry in *Geometrical Phases in Physics*, A. Shapere, and F. Wilczek, eds., World Scientific, Singapore, 1989, p. 7.
83. M. Baer, *Chem. Phys. Lett.* **349**, 84 (2001).
84. P. Brumer and M. Shapiro, *Chem. Phys. Lett.* **12**, 541 (1986).
85. M. Shapiro and P. Brumer, *Faraday Discuss. Chem. Soc.* **82**, 177 (1987).
86. M. Shapiro and P. Brumer, *J. Chem. Phys.* **90**, 6179 (1989).
87. M. Shapiro, *J. Phys. Chem.* **97**, 7396 (1993).
88. T. J. Dunn, I. A. Walmsley, and S. Mukamel, *Phys. Rev. Lett.* **74**, 884 (1995).
89. R. Uberna, M. Khalil, R. M. Williams, J. M. Papanikolas, and S. George, *J. Chem. Phys.* **108**, 9259 (1998).
90. C. Leichtle, W. P. Schleich, I. Sh. Averbukh, and M. Shapiro, *Phys. Rev. Lett.* **80**, 1418 (1998).
91. M. Shapiro, *J. Phys. Chem.* **102**, 9570 (1998).
92. A. Zucchetti, W. Vogel, D.-G. Welsch, and I. A. Walmsley, *Phys. Rev. A* **60**, 2716 (1999).
93. M. Shapiro, E. Frishman, and P. Brumer, *Phys. Rev. Lett.* **84**, 1669 (2000).
94. V. Blanchet, C. Nicole, M.-A. Bouchene, and B. Girard, *Phys. Rev. Lett.* **78**, 2716 (1997).
95. T. C. Weinacht, J. Ahn, and P. H. Buchsbaum, *Nature (London)* **397**, 233 (1999).
96. L. E. E. de Araujo, I. A. Walmsley, and C. R. Stroud, *Phys. Rev. Lett.* **81**, 955 (1998).
97. W. P. Schleich, *Nature (London)* **397**, 207 (1999).

98. H. Feshbach, K. Kerman, and R. H. Lemmer, *Ann. Phys. (NY)* **41**, 230 (1967).
99. J. Jortner and S. Mukamel, in *International Review Science & Theoretical Chemistry, Physical Chemistry*, A. D. Buckingham and C. A. Coulson, eds., Butterworth, London, 1975, Series 2, Vol. 1, p. 327.
100. R. Englman, *Non-Radiative Decay of Ions and Molecules in Solids*, North-Holland, Amsterdam, 1979, p. 155.
101. H. D. Zeh, *The Physical Basis of the Direction of Time*, Springer-Verlag, Berlin, 1992.
102. S. F. Savitt, ed., *Time's Arrow Today, Recent Physical and Philosophical Work on the Direction of Time*, University Press, Cambridge, 1995.
103. H. Weyl, *Space, Time, Matter*, Dover Books, New York, 1950; *The Theory of Groups and Quantum Mechanics*, Dover Books, New York, 1950.
104. W. Pauli, *General Principles of Quantum Mechanics*, Springer-Verlag, Berlin, 1980.
105. W. E. Lamb, *Phy. Today* **22**, 23 (April 1969).
106. W. Gale, E. Guth, and G. T. Trammel, *Phys. Rev.* **165**, 1434 (1968).
107. A. Royer, *Found. Phy.* **19**, 3 (1989) 3.
108. M. Baer and R. Englman, *Chem. Phys. Lett.* **265**, 105 (1997).
109. R. Englman and M. Baer, *J. Phys.:Condensed Matter* **11**, 1059 (1999).
110. Y. Guimare, B. Baseia, C. J. Villas-Boas, and M. H. Y. Moussa, *Phys. Lett. A* **268**, 260 (2000).
111. J. W. S. Rayleigh, *The Theory of Sound* Volume II, Dover, New York, 1945, section 282.
112. D. N. Klyshko, *Phys.-Usp.* **36**, 1005 (1993).
113. S. Pancharatnam, *Proc. Ind. Acad. Sci. A* **44**, 247 (1956).
114. G. N. Ramachandran and S. Ramaseshan, in *Handbuch de Physik*, S. Flugge, ed., Springer, Berlin, 1961, Vol. XXV.1, p. 1.
115. H. Schmitzer, S. Klein, and W. Dultz, *Physica B* **175**, 148 (1991).
116. H. Schmitzer, S. Klein, and W. Dultz, *Phys. Rev. Lett.* **71**, 1530 (1993).
117. G. Giavarini, E. Gozzi, D. Rohrlich, and W. D. Thacker, *J. Phys. A: Math. Gen.* **22**, 3513 (1989).
118. M. V. Berry, *J. Phys. A: Math. Gen.* **18**, 15 (1985).
119. C. Brukner and A. Zeilinger, *Phys. Rev. Lett.* **79**, 2599 (1997).
120. Q. V. Lawande, S. V. Lawande, and A. Joshi, *Phys. Lett. A* **251**, 164 (1999).
121. J. H. Shapiro and S. R. Shepard, *Phys. Rev. A* **43**, 3795 (1991).
122. M. L. M. Balisteri, J. P. Korterik, L. Kuipers and N. F. van Hulst, *Phys. Rev. Lett.* **85**, 294 (2000).
123. A. G. Wagh, V. C. Rakhecha, P. Fischer, and A. Ioffe, *Phys. Rev. Lett.* **81**, 1992 (1998).
124. A. G. Wagh, G. Badurek, V. C. Rakhecha, R. J. Buchelt, and A. Schricker, *Phys. Lett. A* **268**, 209 (2000).
125. K. A. Nugent, A. Paganin, and T. E. Gureyev, *Phy. Today* **54**(8), 27 (2001).
126. P. Grangier, J. A. Levenson, and J.-P. Poizat, *Nature* (*London*) **396**, 537 (1998).
127. D. Kohler and L. Mandel, in *Coherence and Quantum Optics*, L. Mandel and E. Wolf, eds., Plenum, New York, 1973, p. 387.
128. A. Vijay, R. E. Wyatt, and G. D. Billing, *J. Chem. Phys.* **111**(10), 794 (1999).
129. M. Reck, A. Zeilinger, H. J. Bernstein, and P. Bertani, *Phys. Rev. Lett.* **73**, 58 (1994).
130. J. W. Noh, A. Fougeres, and L. Mandel, *Phys. Scr. T* **48**, 29 (1993).
131. S. Stenholm, *Phys. Scr. T* **48**, 77 (1993).
132. A Vourdas, *Phys. Scr. T* **48**, 84 (1993).

133. R. Loudon, *The Quantum Theory of Light*, Clarendon Press, Oxford, UK, 1983, p. 140.
134. P. A. M. Dirac, *Proc. R. Soc. London Ser. A* **114**, 243 (1927).
135. F. London, *Z. Phys.* **37**, 915 (1926); **40**, 193 (1927).
136. W. H. Louisell, *Quantum Statistical Properties of Radiation*, Wiley, London, 1973.
137. L. Susskind and J. Glogower, *Physics* **1**, 49 (1964).
138. D. T. Pegg and S. M. Barnett, *Europhys. Lett.* **8**, 463 (1988).
139. N. Gronbech-Jensen, P. L. Christiansen, and P. S. Ramanujam, *J. Opt. Soc. Am.* **6**, 2423 (1989).
140. Yu. I. Vorontsov and Yu. A. Rembovsky, *Phys. Lett. A* **254**, 7 (1999).
141. J. A. Vaccaro, D. T. Pegg, and S. M. Barnett, *Phys. Lett. A* **262**, 483 (2000).
142. Yu. I. Vorontsov and Yu. A. Rembovsky, *Phys. Lett. A* **262**, 486 (2000).
143. A. Muller, *Phys. Rev. A* **57**, 731 (1998).
144. M. V. Berry, *Phys. Today* **43**, 34 (1990).
145. Y. Aharonov and L. Susskind, *Phys. Rev.* **158**, 1237 (1967).
146. Y. Aharonov and L. Vaidman, *Phys. Rev. A* **61**, 052108 (2000).
147. M. Floissart, in *Dispersion Relations and their Connection with Causality*, E. P. Wigner, ed., Academic Press, New York, 1964, p. 1.
148. K. E. Peiponen, E. M. Vertiainen, and T. Asakura, *Dispersion, Complex Analysis and Optical Spectroscopy. (Classical theory)*, Springer-Verlag, Berlin, 1999.
149. J. S. Toll, *Phys. Rev.* **104**, 1760 (1956).
150. R. E. Burge, M. A. Fiddy, A. H. Greenaway, and G. Ross, *Proc. R. Soc. London Ser. A* **350**, 191 (1976).
151. F. D. Peat, *Infinite Potential, The Life and Times of David Bohm*, Addison-Wesley, Reading, MA, 1997, p. 192.
152. W. Ehrenburg and R. E. Siday, *Proc. R. Soc. London Ser. B* **62**, 8 (1949).
153. M. Peshkin, I. Talmi, and L. J. Tassie, *Ann. Phys. (NY)* **12**, 426 (1960).
154. B. Simon, *Phys. Rev. Lett.* **51**, 2167 (1983).
155. M. Peshkin and A. Tonomura, *The Aharonov Bohm Effect*, Springer-Verlag, Berlin, 1989.
156. T. T. Wu and C. N. Yang, *Phys. Rev. D* **12**, 3845 (1975).
157. R. Englman, *The Jahn–Teller Effect in Molecules and Crystals*, Wiley, Chichester, UK, 1972.
158. H. C. Longuet-Higgins, *Proc. R. Soc. London Ser. A* **344**, 147 (1975).
159. A. J. Stone, *Proc. R. Soc. London Ser. A* **351**, 141 (1976).
160. C. A. Mead and D. G. Truhlar, *J. Chem. Phys.* **70**, 2284 (1979).
161. C. A. Mead, *Chem. Phys.* **49**, 23, 33 (1980).
162. F. S. Ham, *Phys. Rev. Lett.* **58**, 725 (1987).
163. C. A. Mead, *Rev. Mod. Phys.* **64**, 51 (1992).
164. C. C. Chancey and M. C. M. O'Brien, *The Jahn-Teller Effect in C_{60} and other Icosahedral Complexes*, Princeton University Press, Princeton, NJ, 1997.
165. J. Anandan, *Phys. Lett. A* **147**, 3 (1990).
166. B. Kohler, *Phys. Lett. A* **237**, 195 (1998).
167. L. Yang and F. Yan, *Phys. Lett. A* **265**, 326 (2000).
168. N. Manini and F. Pistolesi, *Phys. Rev. Lett.* **85**, 3067 (2000).
169. D. E. Manolopoulos and M. S. Child, *Phys. Rev. Lett.* **82**, 2223 (1999).

170. Y. Hasegawa, R. Loidl, M. Baron, G. Badurek, and M. Rauch, *Phys. Rev. Lett.* **87**, 070401 (2001).

171. J. C. Garrison and E. M. Wright, *Phys. Lett. A* **128**, 17781 (1988).

172. Ch. Miniatura, C. Sire, J. Baudon, and J. Belissard, *Europhys. Lett.* **13**, 199 (1990).

173. Y. C. Ge and M. S. Child, *Phys. Rev. Lett.* **78**, 2507 (1997).

174. Y. C. Ge and M. S. Child, *Phys. Rev. A* **58**, 872 (1998).

175. Z.-M. Bai, G.-Z. Chen, and M.-L. Ge, *Phys. Lett. A* **262**, 137 (1999).

176. R. S. Whitney and Y. Gefen (preprint, 2001).

177. Y. Aharonov and B. Reznik, *Phys. Rev. Lett.* **84**, 490 (2000).

178. J. M. Cervero, *Int. J. Theor. Phys.* **38**, 2095 (1999).

179. P. M. Felker and A. H. Zewail, *Adv. Chem. Phys.* **70**, 265 (1988).

180. P. Gaspard and I. Burghardt, eds., *Chemical Reactions and their Control on the Femtosecond Time Scale*, in Advances in Chemical Physics Vol. 101, John Wiley & Sons, New York, 1997.

181. D. Meshulach and Y. Silberberg, *Nature (London)* **396**, 239 (1998); *Phys. Rev. A* **60**, 1287 (1999).

182. H. Rabitz, R. de Vivie-Riedle, M. Motzkus, and K.-L. Kompa, *Science* **259**, 1581 (1993).

183. D. Zeidler, S. Frey, K.-L. Kompa, and M. Motzkus, *Phys. Rev. A* **64**, 023420 (2001).

184. F. Scherer, A. J. Ruggiero, M. Du, and G. R. Fleming, *J. Chem. Phys.* **93**, 856 (1990).

185. D. J. Tannor, *Design of Femtosecond Optical Pulse Sequences to Control Photochemical Products*, in A. D. Bandrark, ed., *Molecules in Laser Fields*, Dekker, New York, 1994.

186. R. N. Dixon, D. W. Hwang, X. F. Yang, S. Harich, J. J. Lin, and X. Yang, *Science* **285**, 1249 (1999).

187. D. C. Clary, *Science* **285**, 1218 (199).

188. J. E. Avron and J. Berger, *Chem. Phys. Lett.* **294**, 13 (1998).

189. S. Adhikari and G. Billing, *J. Chem. Phys.* **107**, 6213 (1997).

190. S. Adhikari and G. Billing, *Chem. Phys. Lett.* **284**, 31 (1998).

191. S. Adhikari and G. Billing, *J. Chem. Phys.* **111**, 40 (1999).

192. S. Zilberg and Y. Haas, *J. Phys. Chem. A* **103**, 2364 (1999).

193. M. Baer, S.-H. Lin, A. Alijah, S. Adhikari, and G. D. Billing, *Phys. Rev. A* **62**, 032506 (2000).

194. S. Adhikari, G. D. Billing, A. Alijah, S.-H. Lin, and M. Baer, *Phys. Rev. A* **62**, 032507 (2000).

195. G. D. Billing and A. Kuppermann, *Chem. Phys. Lett.* **294**, 26 (1998).

196. M. Shapiro, *J. Chem. Phys.* **103**, 1748 (1995); *J. Phys. Chem.* **100**, 7859 (1996).

197. J. Bromage and C. R. Stroud, Jr., *Phys. Rev. Lett.* **83**, 4963 (1999).

198. J. H. Eberly, J. J. Sanchez-Mondragon, and N. B. Narozhny, *Phys. Rev. Lett.* **44**, 1323 (1980).

199. I. Sh. Averbukh and N. F. Perel'man, *Phys. Lett. A* **139**, 449 (1989).

200. I. Sh. Averbukh, M. J. J. Vrakking, D. M. Villeneuve, and A. Stolow, *Phys. Rev. Lett.* **77**, 3518 (1996).

201. J. Jarzynski, *Phys. Rev. Lett.* **74**, 1264 (1995).

202. E. J. Heller, *J. Chem. Phys.* **62**, 1544 (1975).

203. M. F. Herman and E. Kluk, *Chem. Phys.* **91**, 27 (1984).

204. D. Huber and E. J. Heller, *J. Chem. Phys.* **87**, 5302 (1987).

205. K. G. Kay, *Phys. Rev. Lett.* **83**, 5190 (1999).

206. I. C. Percival, *Semiclassical Theory of Bound States*, in *Advances in Chemical Physics*, Vol. 36, p. 1, 1977.
207. W. M. Zhang, D. H. Feng, and R. Gilmore, *Rev. Mod. Phys.* **62**, 867 (1990).
208. M. S. Child, *Semiclassical Mechanics with Molecular Applications*, Clarendon Press, Oxford, UK, 1991.
209. V. P. Maslov and M. V. Fedoriuk, *Semi-classical Approximations in Quantum Mechanics*, Reidel, Boston, 1981.
210. A. S. Davydov, *Quantum Mechanics*, Pergamon Press, Oxford, UK, 1965, Section 108.
211. R. D. Levine, *Quantum Mechanics of Molecular Rate Processes*, Clarendon Press, Oxford, UK, 1969, Section 2.8.1.
212. K. R. Popper, *Nature (London)* **177**, 538 (1956) ; *ibid.* **178**, 382 (1956); *ibid.* **179**, 1293 (1957).
213. E. L. Hill and A. Grunbaum, *Nature (London)* **179**, 1292 (1957).
214. O. Costa de Beauregard, *La Second Principe de la Science du Temps*, De Seuil, Paris, 1963.
215. J. L. Lebowitz, *Phys. Today* **Sep. 1993**, 32 (1993). [Various responses in Physics Today **Nov. 1994**, 11 (1994)].
216. J. J. Halliwell, J. Perez-Mercader, and W. H. Zurek, eds., *Physical Origin of Time Asymmetry*, University Press, Cambridge, 1994.
217. L. S. Schulman, *Time Arrows and Quantum Measurement*, University Press, Cambridge, 1997.
218. T. Banks, L. Susskind, and M. E. Peskin, *Nucl. Phys. B* **244**, 125 (1984).
219. G. P. Beretta, E. P. Gyftopoulos, J. L. Park, and G. N. Hatsopoulos. *Nuovo. Cim.* **82B**, 169 (1984).
220. S. Gheorghiu-Svirshevski, *Phys. Rev. A* **63**, 022105 (2001); **63**, 054102 (2001).
221. A. Bohm, I. Antoniou, and P. Kielanowski, *Phys. Lett. A* **189**, 442 (1994).
222. A. Bohm, *Phys. Rev.* **A60**, 861 (1999).
223. D. Bohm, *Quantum Theory*, Prentice-Hall, New York, 1952.
224. E. P. Wigner, *Phys. Rev.* **98**, 145 (1955).
225. E. Pollak and W. H. Miller, *Phys. Rev. Lett.* **53**, 115 (1984).
226. E. Pollak. *J. Chem. Phys.* **83** , 1111 (1985).
227. M. E. Perel'man, *Kinetical Quantum Theory of Optical Dispersion*, Meznereba, Tblisi, 1989 (in Russian).
228. M. E. Perel'man, *Sov. Phys. JETP* **23**, 407 (1966); [*J. Eksp. Teoret. Fyz.* **50**, 613 (1966)]; *Sov. Phys. Dok.* **14**, 772 (1970) [DAN SSSR **187**, 781–783 (1969)].
229. R. Landauer and Th. Martin, *Rev. Mod. Phys.* **66**, 217 (1994).
230. D. Chauvat, O. Emile, F. Bretenaker, and A. Le Floch, *Phys. Rev. Lett.* **84**, 71 (1999).
231. P. Debu, *Europhys. News* **31**, 5 (May/June 2000).
232. Y. Aharonov and C. K. Au, *Phys. Lett. A* **86**, 269 (1981).
233. T. E. Feuchtwang, E. Kazes, H. Grotch, and P. H. Cutler, *Phys. Lett. A* **93**, 4 (1982).
234. F. A. Kaempffer, *Concepts in Quantum Mechanics*, Academic Press, New York, 1965, p. 169.
235. P. A. Ozimba and A. S. Msezane, *Chem. Phys.* **246**, 87 (1999).
236. W. D. Heuss, *Eur. Phys. J.* D **7**, 1 (1999).
237. L. I. Schiff, *Quantum Mechanics*, McGraw-Hill, New York, 1955, Section 12.
238. A. Selloni, P. Carnevali, R. Car, and M. Parinello, *Phys. Rev. Lett.* **59**, 823 (1987).
239. F. Ancilotti and F. Toigo, *Phys. Rev. A* **45**, 4015 (1992).
240. R. Resta, *Phys. Rev. Lett.* **80**, 1800 (1998).

241. A. Stern, Y. Aharonov, and Y. Imry, *Phys. Rev. A* **41**, 3436 (1990).
242. J. von Neumann, *The Mathematical Foundations of Quantum Mechanics*, Dover, New York, 1959.
243. M. E. Perel'man, *Sov. Phys. Dok.* **14**, 772 (1969)[DAN SSR **187**, 781 (1969)].
244. M. E. Perel'man, *Sov. Phys. JETP* **23**, 407 (1966) [*Zh. Eksp. Teor. Fiz.* **50**, 613 (1966)].
245. E. C. Titchmarsh, *Introduction to the Theory of Fourier Integrals*, Clarendon Press, Oxford, UK, 1948, Chap. V.
246. C. Caratheodory, *Theory of Functions*, Chelsea, New York, 1958, Vol. I, Chap. 3.
247. R. Resta, *Rev. Mod. Phys.* **66**, 899 (1994).
248. R. A. E. C. Paley and N. Wiener, *Fourier Transforms in the Complex Domain*, American Physical Society, New York, 1934.
249. B. Davison, *Neutron Transport Theory*, Clarendon Press, Oxford, UK, 1957, Chap. VI , Section 1.
250. M. Born and V. Fock, *Z. Phys.* **51**, 165 (1928).
251. T. Kato, *Prog. Theor. Phys.* **5**, 435 (1950).
252. A. Messiah, *Quantum Mechanics*, Vol. II North-Holland, Amsterdam, 1961, Chap. XVII.
253. J. Liu, B. Hu, and B. Li, *Phys. Rev. Lett.* **81**, 1749 (1998).
254. A. Zigmund, *Trigonometrical Series*, Dover, New York, 1955, Chap. III.
255. J. von Neumann and E. P. Wigner, *Phys. Z.* **30**, 467 (1929).
256. J. W. Zwanziger and E. R. Grant, *J. Chem. Phys.* **87**, 2954 (1987).
257. E. C. Titchmarsh, *The Theory of Functions*, Clarendon Press, Oxford, UK, 1932, Sections 3.42, 3.45, 7.8, 8.1.
258. R. Kosloff, *Ann. Rev. Phys. Chem.* **45**, 145 (1994).
259. S.-I. Tomonaga, *Quantum Mechanics*, Vol. II, North Holland, Amsterdam, 1966, Sections 41, 61.
260. E. J. Heller, *J. Chem. Phys.* **75**, 2923 (1981).
261. M. V. Berry and J. Klein, *J. Phys. A: Math. General* **17**, 1805 (1984).
262. V. V. Dodonov, V. I. Manko, and D. E. Nikonov, *Phys. Lett. A* **162**, 359 (1992).
263. C. Grosche, *Phys. Lett. A* **182**, 28 (1993).
264. Y. Aharonov, T. Kaufherr, S. Popescu, and B. Reznik, *Phys. Rev. Lett.* **80**, 2023 (1998).
265. J. W. Zwanziger, S. P. Rucker, and G. C. Chingas, *Phys. Rev. A* **43**, 3232 (1991).
266. N. Shvartsman and I. Freund, *Phys. Rev. Lett.* **72**, 1008 (1994).
267. A. H. Jahn, and E. Teller, *Proc. R. Soc. London Ser. A* **161**, 220 (1937).
268. M. C. M. O'Brien, *Proc. R. Soc. London Ser. A* **281**, 323 (1964).
269. F. S. Ham, *Jahn-Teller Effects in Electron Paramagnetic Spectra*, Plenum Press, New York, 1971.
270. D. Yarkony, *J. Chem. Phys.* **111**, 4906 (1999).
271. D. J. A. Varandas, H. G. Yu, and Z. R. Xu, *Mol. Phys.* **96**, 1193 (1999).
272. D. J. A. Varandas and Z. R. Xu, *Int. J. Quant. Chem.* **75**, 89 (1999).
273. D. Yarkony, *J. Chem. Phys.* **110**, 701 (1999).
274. G. Herzberg and H. C. Longuet-Higgins, *Discuss. Faraday Soc.* **35**, 77 (1963).
275. H. Koizumi, I. B. Bersuker, J. Boggs, and V. Z. Polinger, *J. Chem. Phys.* **112**, 8470 (2000).
276. M. C. P. Moate, M. C. O. O'Brien, J. I. Dunn, C. A. Bates, Y. M. Liu, and V. Z. Polinger, *Phys. Rev. Lett.* **77**, 4362 (1996).

277. P. Saxe, B. H. Lengsfield, and D. H. Yarkony, *Chem. Phys. Lett.* **113**, 19 (1985).
278. H. Thummel, M. Peric, S. D. Peyerimhoff, and R. J. Buenker, *Z. Phys. D* **13**, 307 (1989).
279. I. N. Radazos, M. A. Robb, M. A. Bernardi, and M. Olivucci, *Chem. Phys. Lett.* **192**, 217 (1992).
280. G. C. G. Waschewsky, P. W. Kash, T. L. Myers, D. C. Kitchen, and L. J. Butler, *J. Chem. Soc. Faraday Trans.* **90**, 1581 (1994).
281. A. M. Mebel, M Baer, and S. H. Lin, *J. Chem. Phys.* **112** 10,703 (2000).
282. A. M. Mebel, M. Baer and S. H. Lin, *J. Chem. Phys.* **114**, 5109 (2001).
283. D. Yarkony, *Acc. Chem. Res.* **31**, 511 (1998).
284. G. Herzberg, *Molecular Spectra and Molecular Structure*, Vol. 3, Van Nostrand: Princeton, NJ, 1966.
285. G. Chaban, M. S. Gordon, and D. Yarkony, *J. Phys. Chem.* **43**, 7953 (1997).
286. W. D. Hobey and A. D. McLachlan, *J. Chem. Phys.* **33**, 1695 (1961).
287. A. D. McLachlan, *Mol. Phys.* **4**, 417 (1961).
288. B. Lepetit and A. Kuppermann, *Chem. Phys. Lett.* **166**, 581 (1990).
289. Y.-S. M. Wu and A. Kuppermann, *Chem. Phys. Lett.* **201**, 178 (1993); *ibid.* **235**, 105 (1995).
290. M. Baer, *Chem. Phys.* **15**, 49 (1976).
291. M. Baer, *Mol. Phys.* **40**, 1011 (1980).
292. M. Baer, in *Theory of Chemical Reaction Dynamics*, M. Baer, ed., CRC Press, Boca Raton, FL, 1985, Vol. II, Chap. 4.
293. M. Baer, *J. Chem. Phys.* **107**, 2694, 10662 (1997).
294. X. Chapuisat, A. Nauts, and D. Hehaureg-Dao, *Chem. Phys. Lett.* **95**, 139 (1983).
295. D. Hehaureg, X. Chapuisat, J. C. Lorquet, G. Galloy, and G. Raseev, *J. Chem. Phys.* **78**, 1246 (1983).
296. L. S. Cederbaum, H. Koppel, and W. Domcke, *Int. J. Quant. Chem. Symp.* **15**, 251 (1981).
297. T. Pacher, L. S. Cederbaum, and H. Koppel, *J. Chem. Phys.* **84**, 293 (1993).
298. M. Baer and A. Alijah, *Chem. Phys. Lett.* **319**, 489 (2000).
299. M. Baer, *J. Phys. Chem. A* **104**, 3181 (2000).
300. R. Baer, D. Charutz, R. Kosloff, and M. Baer, *J. Chem. Phys.* **105**, 9141 (1996).
301. D. Charutz, R. Baer, and M. Baer, *Chem. Phys. Lett.* **265**, 629 (1997).
302. T. C. Thompson and C. A. Mead, *J. Chem. Phys.* **82**, 2408 (1985).
303. J. S. Griffith, *The Irreducible Tensor Method for Molecular Symmetry Groups*, Prentice-Hall, Englewood Cliffs, NJ, 1962, p. 20.
304. J. S. Griffith, *The Theory of Transition Metal Ions*, University Press, Cambridge, 1964.
305. G. Bevilacqua, L. Martinelli, and G. P. Parravicini (preprint 2001).
306. T. Bitter and D. Dubbens, *Phys. Rev. Lett.* **59**, 251 (1988).
307. D. Suter, K. T. Mueller, and A. Pines, *Phys. Rev. Lett.* **60**, 1218 (1988).
308. R. Simon, H. J. Kimble, and E. C. G. Sudarshan, *Phys. Rev. Lett.* **61**, 19 (1988).
309. A. F. Morpurgo, J. P. Heida, T. M. Klapwijk, B. J. Van Wees, and G. Borghs, *Phys. Rev. Lett.* **80**, 1050 (1998).
310. H. Von Busch, V. Dev, H.-E. Eckel, S. Kasahara, J. Wang, W. Demtroder, P. Sebald, and W. Meyer, *Phys. Rev. Lett.* **81**, 4584 (1998).
311. D. Loss, H. Schoeller, and P. M. Goldbart, *Phys. Rev. B* **59**, 13, 328 (1999).
312. I. Fuentes-Guri, S. Bose, and V. Vedral, *Phys. Rev. Lett.* **85**, 5018 (2000).

313. E. Sjöquist, *Phys. Lett. A* **286**, 4 (2001).

314. H. Pothier, P. Lafarge, C. Urbina, D. Esteve, and M. H. Devoret, *Europhys. Lett.* **17**, 1183 (1992).

315. C. Liu, Z. Dutton, C. Behroozi, and L. V. Hau, *Nature (London)* **409**, 490 (2001).

316. D. F. Phillips, A. Fleischhauer, A. Mair, R. L. Walsworth, and M. D. Lukin, *Phys. Rev. Lett.* **86**, 783 (2001).

317. D. DiVicenzo and B. Terhal, *Phys. World*, **March 1998**, 53 (1998).

318. D. Bohm, *Quantum Theory*, Prentice-Hall, New York, 1966, Section 12.6.

319. E. Schrödinger, *Ann. Phys.* **81**, 109 (1926). English translation appears in *Collected Papers in Wave Mechanics*, E. Schrödinger, ed., Blackie and Sons, London, 1928, p. 102.

320. W. Gordon, *Z. Phys.* **41**, 117 (1926).

321. W. Greiner, *Relativistic Quantum Mechanics: Wave Equations*, Springer-Verlag, Berlin, 1997.

322. N. N. Bogoliubov and D. V. Shirkov, *Introduction to the Theory of Quantized Fields*, Interscience, New York, 1959.

323. W. Gordon, *Z. Phys.* **50**, 630 (1928).

324. P. R. Holland, *The Quantum Theory of Motion*, Cambridge University Press, Cambridge, UK, 1993.

325. R. L. Seliger and G. B. Whitham, *Proc. R. Soc. London Ser. A* **305**, 1 (1968).

326. E. Madelung, *Z. Phys.* **40**, 322 (1926).

327. I. Talmi, *Simple Models of Complex Nuclei*, Harcourt, Academic, Chur, 1993.

QUANTUM REACTION DYNAMICS FOR MULTIPLE ELECTRONIC STATES

ARON KUPPERMANN and RAVINDER ABROL

Arthur Amos Noyes Laboratory of Chemical Physics, Division of Chemistry and Chemical Engineering, California Institute of Technology, Pasadena, CA

CONTENTS

I. INTRODUCTION

Electronic transitions (excitations or deexcitations) can take place during the course of a chemical reaction and have important consequences for its dynamics. The motion of electrons and nuclei were first analyzed in a quantum mechanical framework by Born and Oppenheimer [1], who separated the

The Role of Degenerate States in Chemistry: A Special Volume of Advances in Chemical Physics, Volume 124, Edited by Michael Baer and Gert Due Billing. Series Editors I. Prigogine and Stuart A. Rice. ISBN 0-471-43817-0.

motion of the light electrons from that of the heavy nuclei and assumed that the nuclei moved on a single adiabatic electronic state or potential energy surface (PES). This Born–Oppenheimer (BO) approximation can break down due to the presence of strong nonadiabatic couplings between degenerate electronic states (due to conical or glancing intersections between those states) or between the near-degenerate ones (due to avoided crossings). These couplings allow for the motion of nuclei on coupled multiple adiabatic electronic states, with the BO approximation replaced by the Born–Huang expansion [2,3] in which an arbitrary number of electronic states can be included.

These nonadiabatic couplings that give rise to electronic transitions can be classified into two categories: (1) Radial couplings, which have been treated by Zener [4], Landau [5], and others [6–11], arise due to translational, vibrational, and angular motions of the atomic or molecular species involved in the chemical process. These couplings allow for transitions to occur between electronic states of the same symmetry. (2) Rotational couplings, which have been studied by Kronig [12] and others [13–19], arise as a result of a transformation of molecular coordinates from a space-fixed (SF) frame to a body-fixed (BF) one due to the conservation of total electron plus nuclear angular momentum. These couplings allow for transitions between electronic states of the same as well as of different symmetries.

An important consequence of the presence of degenerate electronic states is the geometric phase effect. For a polyatomic system involving N atoms, where $N \geq 3$, any two adjacent adiabatic electronic states can be degenerate for a set of nuclear geometries even if those electronic states have the same symmetry and spin multiplicity [20]. These intersections, occur more frequently in such polyatomic systems than was previously believed. The reason is that these systems possess three or more internal nuclear motion degrees of freedom, and only two independent relations between three electronic Hamiltonian matrix elements (in a simple two electronic state picture) are sufficient for the existence of doubly degenerate electronic energy eigenvalues. As a result, these relations can easily be satisfied explaining thereby the frequent occurrence of intersections. If the lowest order terms in the expansion of these elements in displacements away from the intersection geometry are linear (as is usually the case), these intersections are conical, the most common type of intersection. Assuming the adiabatic electronic wave functions of the two intersecting states to be real and as continuous as possible in nuclear coordinate space, if the polyatomic system is transported around a closed loop in that space (a so-called pseudorotation) that encircles one conical intersection geometry, these electronic wave functions must change sign [20,21]. This change of sign requires the adiabatic nuclear wave functions to undergo a compensatory change of sign, known as the geometric phase (GP) effect [22–26], to keep the total wave function single valued. This sign change of the nuclear wave function, which is a special case of Berry's geometric phase [25], is also referred to as the molecular Aharonov–Bohm effect [27] and has important consequences for the structure and dynamics of the polyatomic system being considered, as it greatly affects the nature of the solutions of the corresponding nuclear motion Schrödinger equation [26].

The dynamics of chemical reactions on a single ground adiabatic electronic PES has been studied extensively over the last few decades using accurate quantum mechanical time-dependent and time-independent methods. These studies have been successfully applied to triatomic [28–30] and tetraatomic [31,32] reactions in the absence of conical intersections. In the last few years, these studies have been extended to triatomic reactions on a single adiabatic PES including the geometric phase effect [33–37] and to include one or more excited adiabatic electronic PESs [38–42]. These latter studies have been made possible by the availability of ab initio non-adiabatic couplings, the calculation of which has been reviewed previously by Lengsfield and Yarkony [43]. The singular nature of these couplings at the conical intersections of two electronic states, introduces numerical difficulties in the solution of the corresponding coupled adiabatic nuclear motion Schrödinger equations. These difficulties are circumvented by transforming the electronically adiabatic representation into a quasidiabatic one [44–55], in which couplings still exist but do not display the singular behavior of the adiabatic representation.

In this chapter, we present a rigorous quantum formalism for studying the dynamics of a polyatomic system (comprising of N atoms) on n electronically adiabatic states, in the absence of spin–orbit interactions. These can be introduced subsequently as perturbative corrections, if they are not too large. In Section II, we present the adiabatic n-electronic-state coupled nuclear motion Schrödinger equations and discuss the properties of first- and second-derivative non-adiabatic couplings in this adiabatic representation. Section III deals with the adiabatic-to-diabatic transformation that produces an optimal diabatic representation, in which the nonremovable couplings are minimized. The application of this transformation to the lowest two adiabatic electronic states of H_3 [55] is also presented. In Section IV, we introduce the full three-dimensional (3D) quantum reactive scattering formalism for a triatomic system on two adiabatic electronic PESs, capable of providing state-to-state differential and integral cross-sections. This formalism is an extension of the time-independent hyperspherical formalism of Kuppermann and co-workers [33] for a triatomic reaction on a single adiabatic electronic PES, that has been used to perform accurate quantum mechanical reactive scattering calculations (with and without the GP effect included) on the $H + H_2$ system and its isotopic variants ($D + H_2$ and $H + D_2$) [33–37] to obtain differential and integral cross-sections. The cross-sections obtained with the GP effect included were in much better agreement with the experimental results [56–59] than those obtained with the GP effect excluded. The two-electronic-state reactive scattering formalism and the associated nuclear motion hyperspherical coordinate coupled equations presented in Section IV should provide cross-sections that can be compared with those obtained from a one-electronic-state formalism and yield the energy range of validity of the one-electronic-state BO approximation.

II. *n*-ELECTRONIC STATE ADIABATIC REPRESENTATION

A. Born–Huang Expansion

Consider a polyatomic system consisting of N_{nu} nuclei (where $N_{\mathrm{nu}} \geq 3$) and N_{el} electrons. In the absence of any external fields, we can rigorously separate the motion of the center of mass $\bar{G}$ of the whole system as its potential energy function V is independent of the position vector of $\bar{G}$ ($\mathbf{r}_{\bar{G}}$) in a laboratory-fixed frame with origin O. This separation introduces, besides $\mathbf{r}_{\bar{G}}$, the Jacobi vectors $\mathbf{R}'_{\lambda} \equiv (\mathbf{R}'_{\lambda_1}, \mathbf{R}'_{\lambda_2}, \ldots, \mathbf{R}'_{\lambda_{N_{\mathrm{nu}}-1}})$ and $\mathbf{r}' \equiv (\mathbf{r}'_1, \mathbf{r}'_2, \ldots, \mathbf{r}'_{N_{\mathrm{el}}})$ for nuclei and electrons, respectively [26]. These Jacobi vectors are simply related to the position vectors of those nuclei and electrons in the laboratory-fixed frame. The parameter λ refers to an arbitrary clustering scheme for the N_{nu} nuclei [60,61] and helps specify different product arrangement channels during a chemical reaction.

We will omit the kinetic energy operator $\hat{T}_{\bar{G}}$ of the center of mass $\bar{G}$, since no external fields act on the system and consider only its internal kinetic energy operator $\hat{T}^{\mathrm{int}}$ given by [26]

$$\hat{T}^{\mathrm{int}} = \hat{T}^{\mathrm{int}}_{\mathrm{nu}} + \hat{T}_{\mathrm{el}} \tag{1}$$

where, $\hat{T}^{\mathrm{int}}_{\mathrm{nu}}$ and $\hat{T}_{\mathrm{el}}$ are, respectively, internal nuclear and electronic kinetic energy operators in the Jacobi vectors mentioned above. If these Jacobi vectors $\mathbf{R}'_{\lambda_i} (i = 1, 2, \ldots, N_{\mathrm{nu}} - 1)$ and $\mathbf{r}'_j$ $(j = 1, 2, \ldots, N_{\mathrm{el}})$ are transformed to their mass-scaled counterparts [61] $\mathbf{R}_{\lambda_i}$ and $\mathbf{r}_j$, the kinetic energy operators have relatively simple expressions given by

$$\hat{T}^{\mathrm{int}}_{\mathrm{nu}} = -\frac{\hbar^2}{2\mu} \nabla^2_{\mathbf{R}_\lambda} \qquad \text{and} \qquad \hat{T}_{\mathrm{el}} = -\frac{\hbar^2}{2\nu} \nabla^2_{\mathbf{r}} \tag{2}$$

where

$$\nabla^2_{\mathbf{R}_\lambda} \equiv \sum_{i=1}^{N_{\mathrm{nu}}-1} \nabla^2_{\mathbf{R}_{\lambda_i}} \qquad \text{and} \qquad \nabla^2_{\mathbf{r}} \equiv \sum_{j=1}^{N_{\mathrm{el}}} \nabla^2_{\mathbf{r}_j} \tag{3}$$

with the Laplacians on the left of these equivalence relations being independent of the choice of the clustering scheme λ. The transformation of Jacobi vectors to the mass-scaled ones is defined by

$$\mathbf{R}_{\lambda_i} = \left(\frac{\mu_{\lambda_i}}{\mu}\right)^{1/2} \mathbf{R}'_{\lambda_i} \qquad \text{and} \qquad \mathbf{r}_j = \left(\frac{\nu_j}{\nu}\right)^{1/2} \mathbf{r}'_j \tag{4}$$

where

$$\mu = \left(\frac{1}{M}\prod_{i=1}^{N_{\text{nu}}} M_i\right)^{1/(N_{\text{nu}}-1)} \qquad \text{and} \qquad \nu = m_{\text{el}}\left(\frac{M}{M + N_{\text{el}}m_{\text{el}}}\right)^{1/N_{\text{el}}} \tag{5}$$

are the effective reduced masses of the nuclei and electrons, respectively, with M_i being the mass of the ith nucleus. μ_{λ_i} and ν_j in Eq. (4) are the effective masses [26] associated with the corresponding vectors $\mathbf{R}'_{\lambda_i}$ and $\mathbf{r}'_j$, with

$$\nu_j = \frac{[M + (j-1)m_{\text{el}}]m_{\text{el}}}{M + jm_{\text{el}}} \tag{6}$$

In Eqs. (5) and (6), M is the total mass of the nuclei and m_{el} is the mass of one electron. By using Eq. (2), the system's internal kinetic energy operator is given in terms of the mass-scaled Jacobi vectors by

$$\hat{T}^{\text{int}} = -\frac{\hbar^2}{2\mu}\nabla^2_{\mathbf{R}_\lambda} - \frac{\hbar^2}{2\nu}\nabla^2_{\mathbf{r}} \tag{7}$$

If V is the total Coulombic potential between all the nuclei and electrons in the system, then, in the absence of any spin-dependent terms, the electronic Hamiltonian $\hat{H}^{\text{el}}$ is given by

$$\hat{H}^{\text{el}}(\mathbf{r}; \mathbf{q}_\lambda) = -\frac{\hbar^2}{2\nu}\nabla^2_{\mathbf{r}} + V(\mathbf{r}; \mathbf{q}_\lambda) \tag{8}$$

where, $\mathbf{q}_\lambda$ is a set of $3(N_{\text{nu}} - 2)$ internal nuclear coordinates obtained by removing from the set $\mathbf{R}_\lambda$ three Euler angles that orient a nuclear body-fixed frame with respect to the laboratory-fixed (or space-fixed) frame. Due to the small ratio of the electron mass to the total mass of the nuclei, $\nu \approx m_{\text{el}}$. This approximation is used in the ab initio electronic structure calculations that use the electronic Hamiltonian given in Eq. (8) but with the ν replaced by m_{el}. Figure 1 illustrates for a three-nuclei, four-electron system, the corresponding nonmass-scaled Jacobi vectors. The nuclear center of mass G is distinct from the overall system's center of mass $\bar{G}$. This distinction of the centers of mass and the difference between ν and m_{el} is responsible for the so-called mass polarization effect in the electronic spectra of these systems that produces

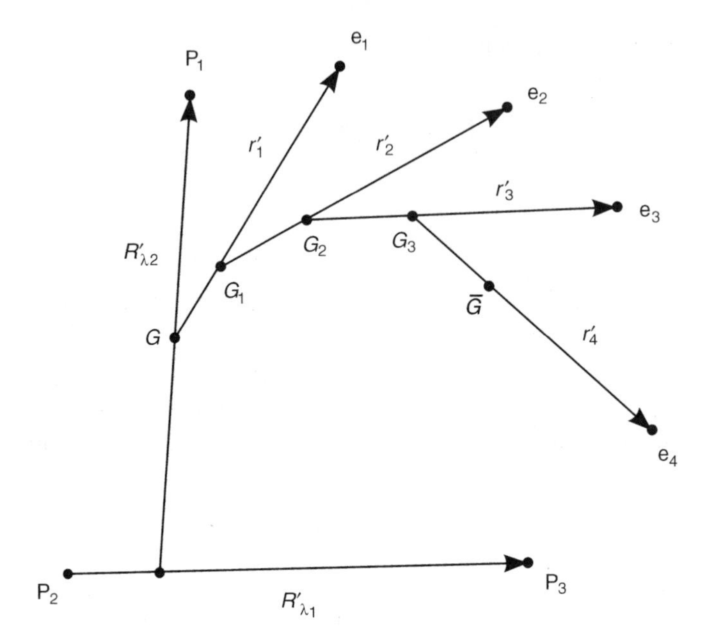

Figure 1. Jacobi vectors for a three-nuclei, four-electron system. The nuclei are P_1, P_2, P_3, and the electrons are e_1, e_2, e_3, e_4.

relative shifts in the energy levels of 10^{-4} or less. In actual scattering calculations, these differences are normally ignored as they introduce relative changes in the cross-sections of the order of 10^{-4} or less [26].

The electronically adiabatic wave functions $\psi_i^{\text{el,ad}}(\mathbf{r};\mathbf{q}_\lambda)$ are defined as eigenfunctions of the electronic Hamiltonian $\hat{H}^{\text{el}}$ with electronically adiabatic potential energies $\varepsilon_i^{\text{ad}}(\mathbf{q}_\lambda)$ as their eigenvalues:

$$\hat{H}^{\text{el}}(\mathbf{r};\mathbf{q}_\lambda)\psi_i^{\text{el,ad}}(\mathbf{r};\mathbf{q}_\lambda) = \varepsilon_i^{\text{ad}}(\mathbf{q}_\lambda)\psi_i^{\text{el,ad}}(\mathbf{r};\mathbf{q}_\lambda) \tag{9}$$

The electronic Hamiltonian and the corresponding eigenfunctions and eigenvalues are independent of the orientation of the nuclear body-fixed frame with respect to the space-fixed one, and hence depend only on $\mathbf{q}_\lambda$. The index i in Eq. (9) can span both discrete and continuous values. The $\psi_i^{\text{el,ad}}(\mathbf{r};\mathbf{q}_\lambda)$ form a complete orthonormal basis set and satisfy the orthonormality relations

$$\langle\psi_i^{\text{el,ad}}(\mathbf{r};\mathbf{q}_\lambda)|\psi_{i'}^{\text{el,ad}}(\mathbf{r};\mathbf{q}_\lambda)\rangle_{\mathbf{r}} = \begin{cases} \delta_{i,i'} & \text{for } i \text{ and } i' \text{ discrete} \\ \delta(i-i') & \text{for } i \text{ and } i' \text{ continuous} \\ 0 & \text{for } i \text{ discrete and } i' \text{ continuous or vice versa} \end{cases} \tag{10}$$

The total orbital wave function for this system is given by an electronically adiabatic n-state Born–Huang expansion [2,3] in terms of this electronic basis set $\psi_i^{\mathrm{el,ad}}(\mathbf{r};\mathbf{q}_\lambda)$ as

$$\Psi^{\mathrm{O}}(\mathbf{r},\mathbf{R}_\lambda) = \sum\!\!\!\!\!\!\int_i \chi_i^{\mathrm{ad}}(\mathbf{R}_\lambda)\psi_i^{\mathrm{el,ad}}(\mathbf{r};\mathbf{q}_\lambda) \tag{11}$$

where, $\sum\!\!\!\!\!\!\int_i$ is a sum over the discrete and an integral over the continuous values of i. The $\chi_i^{\mathrm{ad}}(\mathbf{R}_\lambda)$, which are the coefficients in this expansion, are the adiabatic nuclear motion wave functions. The number of electronic states used in the Born–Huang expansion of Eq. (11) can, in most cases of interest, be restricted to a small number n of discrete states, and Eq. (11) replaced by

$$\Psi^{\mathrm{O}}(\mathbf{r},\mathbf{R}_\lambda) \approx \sum_{i=1}^{n} \chi_i^{\mathrm{ad}}(\mathbf{R}_\lambda)\psi_i^{\mathrm{el,ad}}(\mathbf{r};\mathbf{q}_\lambda) \tag{12}$$

where n is a small number. This corresponds to restricting the motion of nuclei to only those n electronic states. In particular, if those n states constitute a sub-Hilbert space that interacts very weakly with higher states [62], this would be a very good approximation. The orbital wave function Ψ^{O} satisfies the Schrödinger equation

$$\hat{H}^{\mathrm{int}}(\mathbf{r},\mathbf{R}_\lambda)\Psi^{\mathrm{O}}(\mathbf{r},\mathbf{R}_\lambda) = E\Psi^{\mathrm{O}}(\mathbf{r},\mathbf{R}_\lambda) \tag{13}$$

where

$$\hat{H}^{\mathrm{int}}(\mathbf{r},\mathbf{R}_\lambda) = \hat{T}^{\mathrm{int}}(\mathbf{r},\mathbf{R}_\lambda) + V(\mathbf{r};\mathbf{q}_\lambda) \tag{14}$$

is the internal Hamiltonian of the system that excludes the motion of its center of mass and any spin-dependent terms and E is the corresponding system's total energy.

B. Adiabatic Nuclear Motion Schrödinger Equation

Let us define $\boldsymbol{\chi}^{\mathrm{ad}}(\mathbf{R}_\lambda)$ as an n-dimensional nuclear motion column vector, whose components are $\chi_1^{\mathrm{ad}}(\mathbf{R}_\lambda)$ through $\chi_n^{\mathrm{ad}}(\mathbf{R}_\lambda)$. The n-electronic-state nuclear motion Schrödinger equation satisfied by $\boldsymbol{\chi}^{\mathrm{ad}}(\mathbf{R}_\lambda)$ can be obtained by inserting Eqs. (12) and (14) into Eq. (13) and using Eqs. (7)–(10). The resulting Schrödinger equation can be expressed in compact matrix form as [26]

$$\left[-\frac{\hbar^2}{2\mu}\{\mathbf{I}\nabla^2_{\mathbf{R}_\lambda} + 2\mathbf{W}^{(1)\mathrm{ad}}(\mathbf{R}_\lambda)\cdot\nabla_{\mathbf{R}_\lambda} + \mathbf{W}^{(2)\mathrm{ad}}(\mathbf{R}_\lambda)\} + \{\boldsymbol{\varepsilon}^{\mathrm{ad}}(\mathbf{q}_\lambda) - E\mathbf{I}\}\right]\boldsymbol{\chi}^{\mathrm{ad}}(\mathbf{R}_\lambda) = \mathbf{0} \tag{15}$$

where $\mathbf{I}$, $\mathbf{W}^{(1)\text{ad}}$, $\mathbf{W}^{(2)\text{ad}}$, and $\boldsymbol{\varepsilon}^{\text{ad}}$ are $n \times n$ matrices and $\boldsymbol{\nabla}_{\mathbf{R}_\lambda}$ is the column vector gradient operator in the $3(N_{\text{nu}} - 1)$-dimensional space-fixed nuclear configuration space. The parameter $\mathbf{I}$ is the identity matrix and $\boldsymbol{\varepsilon}^{\text{ad}}(\mathbf{q}_\lambda)$ is the diagonal matrix whose diagonal elements are the n electronically adiabatic PESs $\varepsilon_i^{\text{ad}}(\mathbf{q}_\lambda)$ $(i = 1, \ldots, n)$ being considered. All matrices appearing in this n-electronic state nuclear motion Schrödinger equation (15) are n-dimensional diagonal except for $\mathbf{W}^{(1)\text{ad}}$ and $\mathbf{W}^{(2)\text{ad}}$, which are, respectively, the first- and second-derivative [26,43,63–68] nonadiabatic coupling matrices discussed below. These coupling matrices allow the nuclei to sample more than one adiabatic electronic state during a chemical reaction, and hence alter its dynamics in an electronically nonadiabatic fashion. It should be stressed that the effect of the geometric phase on Eqs. (15) must be added by either appropriate boundary conditions [26,33] or the introduction of an appropriate vector potential [23,26,69].

C. First-Derivative Coupling Matrix

The matrix $\mathbf{W}^{(1)\text{ad}}(\mathbf{R}_\lambda)$ in Eq. (15) is an $n \times n$ adiabatic first-derivative coupling matrix whose elements are defined by

$$\mathbf{w}_{i,j}^{(1)\text{ad}}(\mathbf{R}_\lambda) = \langle \psi_i^{\text{el,ad}}(\mathbf{r}; \mathbf{q}_\lambda) | \boldsymbol{\nabla}_{\mathbf{R}_\lambda} \psi_j^{\text{el,ad}}(\mathbf{r}; \mathbf{q}_\lambda) \rangle_{\mathbf{r}} \qquad i,j = 1, \ldots, n \tag{16}$$

These coupling elements are $3(N_{\text{nu}} - 1)$-dimensional vectors. If the Cartesian components of $\mathbf{R}_\lambda$ in $3(N_{\text{nu}} - 1)$ space-fixed nuclear congifuration space are $X_{\lambda 1}, X_{\lambda 2}, \ldots, X_{\lambda 3(N_{\text{nu}}-1)}$, the corresponding Cartesian components of $\mathbf{w}_{i,j}^{(1)\text{ad}}(\mathbf{R}_\lambda)$ are

$$\left[\mathbf{w}_{i,j}^{(1)\text{ad}}(\mathbf{R}_\lambda)\right]_l = \left\langle \psi_i^{\text{el,ad}}(\mathbf{r}; q_\lambda) \middle| \frac{\partial}{\partial X_{\lambda l}} \psi_i^{\text{el,ad}}(\mathbf{r}; q_\lambda) \right\rangle_{\mathbf{r}} \qquad l = 1, 2, \ldots, 3(N_{\text{nu}} - 1) \tag{17}$$

The matrix $\mathbf{W}^{(1)\text{ad}}$ is in general skew-Hermitian due to Eq. (10), and hence its diagonal elements $\mathbf{w}_{i,i}^{(1)\text{ad}}(\mathbf{R}_\lambda)$ are pure imaginary quantities. If we require that the $\psi_i^{\text{el,ad}}$ be real, then the matrix $\mathbf{W}^{(1)\text{ad}}$ becomes real and skew-symmetric with the diagonal elements equal to zero and the off-diagonal elements satisfying the relation

$$\mathbf{w}_{i,j}^{(1)\text{ad}}(\mathbf{R}_\lambda) = -\mathbf{w}_{j,i}^{(1)\text{ad}}(\mathbf{R}_\lambda) \qquad i \neq j \tag{18}$$

As with any vector, the above nonzero coupling vectors $(\mathbf{w}_{i,j}^{(1)\text{ad}}(\mathbf{R}_\lambda),\ i \neq j)$ can be decomposed, due to an extension beyond three dimensions [26] of the

Helmholtz theorem [70], into a longitudinal part $\mathbf{w}_{i,j,\mathrm{lon}}^{(1)\mathrm{ad}}(\mathbf{R}_\lambda)$ and a transverse one $\mathbf{w}_{i,j,\mathrm{tra}}^{(1)\mathrm{ad}}(\mathbf{R}_\lambda)$ according to

$$\mathbf{w}_{i,j}^{(1)\mathrm{ad}}(\mathbf{R}_\lambda) = \mathbf{w}_{i,j,\mathrm{lon}}^{(1)\mathrm{ad}}(\mathbf{R}_\lambda) + \mathbf{w}_{i,j,\mathrm{tra}}^{(1)\mathrm{ad}}(\mathbf{R}_\lambda) \tag{19}$$

where, by definition, the curl of $\mathbf{w}_{i,j,\mathrm{lon}}^{(1)\mathrm{ad}}(\mathbf{R}_\lambda)$ and the divergence of $\mathbf{w}_{i,j,\mathrm{tra}}^{(1)\mathrm{ad}}(\mathbf{R}_\lambda)$ vanish

$$\mathrm{curl}\ \mathbf{w}_{i,j,\mathrm{lon}}^{(1)\mathrm{ad}}(\mathbf{R}_\lambda) = \mathbf{0} \tag{20}$$

$$\nabla_{\mathbf{R}_\lambda} \cdot \mathbf{w}_{i,j,\mathrm{tra}}^{(1)\mathrm{ad}}(\mathbf{R}_\lambda) = 0 \tag{21}$$

The curl in Eq. (20) is the skew-symmetric tensor of rank 2, whose k, l element is given by [26,71]

$$\left[\mathrm{curl}\ \mathbf{w}_{i,j,\mathrm{lon}}^{(1)\mathrm{ad}}(\mathbf{R}_\lambda)\right]_{k,l} = \frac{\partial}{\partial X_{\lambda l}}\left[\mathbf{w}_{i,j,\mathrm{lon}}^{(1)\mathrm{ad}}(\mathbf{R}_\lambda)\right]_k - \frac{\partial}{\partial X_{\lambda k}}\left[\mathbf{w}_{i,j,\mathrm{lon}}^{(1)\mathrm{ad}}(\mathbf{R}_\lambda)\right]_l \qquad k,l = 1,2,\ldots,3(N_{\mathrm{nu}}-1) \tag{22}$$

As a result of Eq. (20), a scalar potential $\alpha_{i,j}(\mathbf{R}_\lambda)$ exists for which

$$\mathbf{w}_{i,j,\mathrm{lon}}^{(1)\mathrm{ad}}(\mathbf{R}_\lambda) = \nabla_{\mathbf{R}_\lambda}\alpha_{i,j}(\mathbf{R}_\lambda) \tag{23}$$

At conical intersection geometries, $\mathbf{w}_{i,j,\mathrm{lon}}^{(1)\mathrm{ad}}(\mathbf{R}_\lambda)$ is singular because of the $\mathbf{q}_\lambda$-dependence of $\psi_i^{\mathrm{el,ad}}(\mathbf{r};\mathbf{q}_\lambda)$ and $\psi_j^{\mathrm{el,ad}}(\mathbf{r};\mathbf{q}_\lambda)$ in the vicinity of those geometries and therefore so is the $\mathbf{W}^{(1)\mathrm{ad}}(\mathbf{R}_\lambda)\cdot\nabla_{\mathbf{R}_\lambda}$ term in Eq. (15). As a result of Eq. (19), $\mathbf{W}^{(1)\mathrm{ad}}$ can be written as a sum of the corresponding skew-symmetric matrices $\mathbf{W}_{\mathrm{lon}}^{(1)\mathrm{ad}}$ and $\mathbf{W}_{\mathrm{tra}}^{(1)\mathrm{ad}}$, that is,

$$\mathbf{W}^{(1)\mathrm{ad}}(\mathbf{R}_\lambda) = \mathbf{W}_{\mathrm{lon}}^{(1)\mathrm{ad}}(\mathbf{R}_\lambda) + \mathbf{W}_{\mathrm{tra}}^{(1)\mathrm{ad}}(\mathbf{R}_\lambda) \tag{24}$$

This decomposition into a longitudinal and a transverse part, as will be discussed in Section III, plays a crucial role in going to a diabatic representation in which this singularity is completely removed. In addition, the presence of the first derivative gradient term $\mathbf{W}^{(1)\mathrm{ad}}(\mathbf{R}_\lambda)\cdot\nabla_{\mathbf{R}_\lambda}\chi^{\mathrm{ad}}(\mathbf{R}_\lambda)$ in Eq. (15), even for a nonsingular $\mathbf{W}^{(1)\mathrm{ad}}(\mathbf{R}_\lambda)$ (e.g., for avoided intersections), introduces numerical inefficiencies in the solution of that equation.

D. Second-Derivative Coupling Matrix

The matrix $\mathbf{W}^{(2)\mathrm{ad}}(\mathbf{R}_\lambda)$ in Eq. (15) is an $n \times n$ adiabatic second-derivative coupling matrix whose elements are defined by

$$w_{i,j}^{(2)\mathrm{ad}}(\mathbf{R}_\lambda) = \langle\psi_i^{\mathrm{el,ad}}(\mathbf{r};\mathbf{q}_\lambda)|\nabla_{\mathbf{R}_\lambda}^2\psi_j^{\mathrm{el,ad}}(\mathbf{r};\mathbf{q}_\lambda)\rangle_{\mathbf{r}} \qquad i,j = 1,\ldots,n \tag{25}$$

These coupling matrix elements are scalars due to the presence of the scalar Laplacian $\nabla^2_{\mathbf{R}_\lambda}$ in Eq. (25). These elements are, in general, complex but if we require the $\psi_i^{\text{el,ad}}$ to be real they become real. The matrix $\mathbf{W}^{(2)\text{ad}}(\mathbf{R}_\lambda)$, unlike its first-derivative counterpart, is neither skew-Hermitian nor skew-symmetric.

The $w_{i,j}^{(2)\text{ad}}(\mathbf{R}_\lambda)$ are also singular at conical intersection geometries. The decomposition of the first-derivative coupling vector, discussed in Section II.C, also facilitates the removal of this singularity from the second-derivative couplings. Being scalars, the second-derivative couplings can be easily included in the scattering calculations without any additional computational effort. It is interesting to note that in a one-electronic-state BO approximation, the first-derivative coupling element $\mathbf{w}_{1,1}^{(1)\text{ad}}(\mathbf{R}_\lambda)$ is rigorously zero (assuming real adiabatic electronic wave functions), but $w_{1,1}^{(2)\text{ad}}(\mathbf{R}_\lambda)$ is not and might be important to predict sensitive quantum phenomena like resonances that can be experimentally verified.

III. ADIABATIC-TO-DIABATIC TRANSFORMATION

A. Electronically Diabatic Representation

As mentioned at the end of Section II.C, the presence of the $\mathbf{W}^{(1)\text{ad}}(\mathbf{R}_\lambda)\cdot\nabla_{\mathbf{R}_\lambda}\chi^{\text{ad}}(\mathbf{R}_\lambda)$ term in the n-adiabatic-electronic-state Schrödinger equation (15) introduces numerical inefficiencies in its solution, even if none of the elements of the $\mathbf{W}^{(1)\text{ad}}(\mathbf{R}_\lambda)$ matrix is singular.

This makes it desirable to define other representations in addition to the electronically adiabatic one [Eqs. (9)–(12)], in which the adiabatic electronic wave function basis set used in the Born–Huang expansion (12) is replaced by another basis set of functions of the electronic coordinates. Such a different electronic basis set can be chosen so as to minimize the above mentioned gradient term. This term can initially be neglected in the solution of the n-electronic-state nuclear motion Schrödinger equation and reintroduced later using perturbative or other methods, if desired. This new basis set of electronic wave functions can also be made to depend parametrically, like their adiabatic counterparts, on the internal nuclear coordinates $\mathbf{q}_\lambda$ that were defined after Eq. (8). This new electronic basis set is henceforth referred to as "diabatic" and, as is obvious, leads to an electronically diabatic representation that is not unique unlike the adiabatic one, which is unique by definition.

Let $\psi_n^{\text{el},d}(\mathbf{r};\mathbf{q}_\lambda)$ refer to that alternate basis set. Assuming that it is complete in $\mathbf{r}$ and orthonormal in a manner similar to Eq. (10), we can use it to expand the total orbital wave function of Eq. (11) in the diabatic version of Born–Huang expansion as

$$\Psi^{\text{O}}(\mathbf{r},\mathbf{R}_\lambda) = \sum\!\!\!\!\!\!\int_i \chi_i^d(\mathbf{R}_\lambda)\psi_i^{\text{el},d}(\mathbf{r};\mathbf{q}_\lambda) \tag{26}$$

where, the $\psi_i^{\mathrm{el},d}(\mathbf{r};\mathbf{q}_\lambda)$ form a complete orthonormal basis set in the electronic coordinates and the expansion coeffecients $\chi_i^d(\mathbf{R}_\lambda)$ are the diabatic nuclear wave functions.

As in Eq. (12), we also usually replace Eq. (26) by a truncated n-term version

$$\Psi^{\mathrm{O}}(\mathbf{r},\mathbf{R}_\lambda) \approx \sum_{i=1}^{n} \chi_i^d(\mathbf{R}_\lambda)\psi_i^{\mathrm{el},d}(\mathbf{r};\mathbf{q}_\lambda) \tag{27}$$

In the light of Eqs. (12) and (27), the diabatic electronic wave function column vector $\boldsymbol{\psi}^{\mathrm{el},d}(\mathbf{r};\mathbf{q}_\lambda)$ (with elements $\psi_i^{\mathrm{el},d}(\mathbf{r};\mathbf{q}_\lambda)$, $i = 1,\ldots,n$) is related to the adiabatic one $\boldsymbol{\psi}^{\mathrm{el,ad}}(\mathbf{r};\mathbf{q}_\lambda)$ (with elements $\psi_i^{\mathrm{el,ad}}(\mathbf{r};\mathbf{q}_\lambda)$, $i = 1,\ldots,n$) by an n-dimensional unitary transformation

$$\boldsymbol{\psi}^{\mathrm{el},d}(\mathbf{r};\mathbf{q}_\lambda) = \tilde{\mathbf{U}}(\mathbf{q}_\lambda)\boldsymbol{\psi}^{\mathrm{el,ad}}(\mathbf{r};\mathbf{q}_\lambda) \tag{28}$$

where

$$\mathbf{U}^\dagger(\mathbf{q}_\lambda)\mathbf{U}(\mathbf{q}_\lambda) = \mathbf{I} \tag{29}$$

$\mathbf{U}(\mathbf{q}_\lambda)$ is referred to as an adiabatic-to-diabatic transformation (ADT) matrix. Its mathematical structure is discussed in detail in Section III.C. If the electronic wave functions in the adiabatic and diabatic representations are chosen to be real, as is normally the case, $\mathbf{U}(\mathbf{q}_\lambda)$ is orthogonal and therefore has $n(n-1)/2$ independent elements (or degrees of freedom). This transformation matrix $\mathbf{U}(\mathbf{q}_\lambda)$ can be chosen so as to yield a diabatic electronic basis set with desired properties, which can then be used to derive the diabatic nuclear motion Schrödinger equation. By using Eqs. (27) and (28) and the orthonormality of the diabatic and adiabatic electronic basis sets, we can relate the adiabatic and diabatic nuclear wave functions through the same n-dimensional unitary transformation matrix $\mathbf{U}(\mathbf{q}_\lambda)$ according to

$$\boldsymbol{\chi}^d(\mathbf{R}_\lambda) = \tilde{\mathbf{U}}(\mathbf{q}_\lambda)\boldsymbol{\chi}^{\mathrm{ad}}(\mathbf{R}_\lambda) \tag{30}$$

In Eq. (30), $\boldsymbol{\chi}^{\mathrm{ad}}(\mathbf{R}_\lambda)$ and $\boldsymbol{\chi}^d(\mathbf{R}_\lambda)$ are the column vectors with elements $\chi_i^{\mathrm{ad}}(\mathbf{R}_\lambda)$ and $\chi_i^d(\mathbf{R}_\lambda)$, respectively, where $i = 1,\ldots,n$.

B. Diabatic Nuclear Motion Schrödinger Equation

We will assume for the moment that we know the ADT matrix of Eqs. (28) and (30) $\mathbf{U}(\mathbf{q}_\lambda)$, and hence have a completely determined electronically diabatic basis set $\boldsymbol{\psi}^{\mathrm{el},d}(\mathbf{r};\mathbf{q}_\lambda)$. By replacing Eq. (27) into Eq. (13) and using Eqs. (7) and (8) along with the orthonormality property of $\boldsymbol{\psi}^{\mathrm{el},d}(\mathbf{r};\mathbf{q}_\lambda)$, we obtain for $\boldsymbol{\chi}^d(\mathbf{R}_\lambda)$

the n-electronic-state diabatic nuclear motion Schrödinger equation

$$\left[-\frac{\hbar^2}{2\mu}\{\mathbf{I}\nabla^2_{\mathbf{R}_\lambda} + 2\mathbf{W}^{(1)d}(\mathbf{R}_\lambda)\cdot\nabla_{\mathbf{R}_\lambda} + \mathbf{W}^{(2)d}(\mathbf{R}_\lambda)\} + \{\boldsymbol{\varepsilon}^d(\mathbf{q}_\lambda) - E\mathbf{I}\}\right]\boldsymbol{\chi}^d(\mathbf{R}_\lambda) = \mathbf{0} \tag{31}$$

which is the diabatic counterpart of Eq. (15). The parameter $\boldsymbol{\varepsilon}^d(\mathbf{q}_\lambda)$ is an $n \times n$ diabatic electronic energy matrix that in general is nondiagonal (unlike its adiabatic counterpart) and has elements defined by

$$\varepsilon^d_{i,j}(\mathbf{q}_\lambda) = \langle\psi_i^{\mathrm{el},d}(\mathbf{r};\mathbf{q}_\lambda)|\hat{H}^{\mathrm{el}}(\mathbf{r};\mathbf{q}_\lambda)|\psi_j^{\mathrm{el},d}(\mathbf{r};\mathbf{q}_\lambda)\rangle_{\mathbf{r}} \qquad i,j = 1,\ldots,n \tag{32}$$

$\mathbf{W}^{(1)d}(\mathbf{R}_\lambda)$ is an $n \times n$ diabatic first-derivative coupling matrix with elements defined using the diabatic electronic basis set as

$$\mathbf{w}^{(1)d}_{i,j}(\mathbf{R}_\lambda) = \langle\psi_i^{\mathrm{el},d}(\mathbf{r};\mathbf{q}_\lambda)|\nabla_{\mathbf{R}_\lambda}\psi_j^{\mathrm{el},d}(\mathbf{r};\mathbf{q}_\lambda)\rangle_{\mathbf{r}} \qquad i,j = 1,\ldots,n \tag{33}$$

Requiring $\psi_i^{\mathrm{el},d}(\mathbf{r};\mathbf{q}_\lambda)$ to be real, the matrix $\mathbf{W}^{(1)d}(\mathbf{R}_\lambda)$ becomes real and skew-symmetric (just like its adiabatic counterpart) with diagonal elements equal to zero. Similarly, $\mathbf{W}^{(2)d}(\mathbf{R}_\lambda)$ is an $n \times n$ diabatic second-derivative coupling matrix with elements defined by

$$w^{(2)d}_{i,j}(\mathbf{R}_\lambda) = \langle\psi_i^{\mathrm{el},d}(\mathbf{r};\mathbf{q}_\lambda)|\nabla^2_{\mathbf{R}_\lambda}\psi_j^{\mathrm{el},d}(\mathbf{r};\mathbf{q}_\lambda)\rangle_{\mathbf{r}} \qquad i,j = 1,\ldots,n \tag{34}$$

An equivalent form of Eq. (31) can be obtained by inserting Eq. (30) into Eq. (15). Comparison of the result with Eq. (31) furnishes the following relations between the adiabatic and diabatic coupling matrices

$$\mathbf{W}^{(1)d}(\mathbf{R}_\lambda) = \tilde{\mathbf{U}}(\mathbf{q}_\lambda)[\nabla_{\mathbf{R}_\lambda}\mathbf{U}(\mathbf{q}_\lambda) + \mathbf{W}^{(1)\mathrm{ad}}(\mathbf{R}_\lambda)\mathbf{U}(\mathbf{q}_\lambda)] \tag{35}$$

$$\mathbf{W}^{(2)d}(\mathbf{R}_\lambda) = \tilde{\mathbf{U}}(\mathbf{q}_\lambda)[\nabla^2_{\mathbf{R}_\lambda}\mathbf{U}(\mathbf{q}_\lambda) + 2\mathbf{W}^{(1)\mathrm{ad}}(\mathbf{R}_\lambda)\cdot\nabla_{\mathbf{R}_\lambda}\mathbf{U}(\mathbf{q}_\lambda) + \mathbf{W}^{(2)\mathrm{ad}}(\mathbf{R}_\lambda)\mathbf{U}(\mathbf{q}_\lambda)] \tag{36}$$

It also furnishes the following relation between the diagonal adiabatic energy matrix and the nondiagonal diabatic energy one

$$\boldsymbol{\varepsilon}^d(\mathbf{q}_\lambda) = \tilde{\mathbf{U}}(\mathbf{q}_\lambda)\boldsymbol{\varepsilon}^{\mathrm{ad}}(\mathbf{q}_\lambda)\mathbf{U}(\mathbf{q}_\lambda) \tag{37}$$

It needs mentioning that the diabatic Schrödinger equation (31) also contains a gradient term $\mathbf{W}^{(1)d}(\mathbf{R}_\lambda)\cdot\nabla_{\mathbf{R}_\lambda}\chi(\mathbf{R}_\lambda)$ like its adiabatic counterpart [Eq. (15)].

The presence of this term can also introduce numerical inefficiency problems in the solution of Eq. (31). Since the ADT matrix $\mathbf{U}(\mathbf{q}_\lambda)$ is arbitrary, it can be chosen to make Eq. (31) have desirable properties that Eq. (15) does not possess. The parameter $\mathbf{U}(\mathbf{q}_\lambda)$ can, for example, be chosen so as to automatically minimize $\mathbf{W}^{(1)d}(\mathbf{R}_\lambda)$ relative to $\mathbf{W}^{(1)\text{ad}}(\mathbf{R}_\lambda)$ everywhere in internal nuclear configuration space and incorporate the effect of the geometric phase. Next, we will consider the structure of this ADT matrix for an n-electronic-state problem and a general evaluation scheme that minimizes the magnitude of $\mathbf{W}^{(1)d}(\mathbf{R}_\lambda)$.

C. Diabatization Matrix

In the n-electronic-state adiabatic representation involving real electronic wave functions, the skew-symmetric first-derivative coupling vector matrix $\mathbf{W}^{(1)\text{ad}}(\mathbf{R}_\lambda)$ has $n(n-1)/2$ independent nonzero coupling vector elements $\mathbf{w}_{i,j}^{(1)\text{ad}}(\mathbf{R}_\lambda)$, $(i \neq j)$. The ones having the largest magnitudes are those that couple adjacent adiabatic PESs, and therefore the dominant $\mathbf{w}_{i,j}^{(1)\text{ad}}(\mathbf{R}_\lambda)$ are those for which $j = i \pm 1$, that is, lying along the two off-diagonal lines adjacent to the main diagonal of zeros. Each one of the $\mathbf{w}_{i,j}^{(1)\text{ad}}(\mathbf{R}_\lambda)$ elements is associated with a scalar potential $\alpha_{i,j}(\mathbf{R}_\lambda)$ through their longitudinal component [see Eqs. (19) and (23)]. A convenient and general way of parametrizing the $n \times n$ orthogonal ADT matrix $\mathbf{U}(\mathbf{q}_\lambda)$ of Eqs. (28) and (30) is as follows. Since the coupling vector element $\mathbf{w}_{i,j}^{(1)\text{ad}}(\mathbf{R}_\lambda)$ couples the electronic states i and j, let us define an $n \times n$ orthogonal i,j-diabatization matrix [$\mathbf{u}_{i,j}(\mathbf{q}_\lambda)$, with $j > i$] whose row k and column l element $(k, l = 1, 2, \ldots, n)$ is designated by $u_{i,j}^{k,l}(\mathbf{q}_\lambda)$ and is defined in terms of a set of diabatization angles $\beta_{i,j}(\mathbf{q}_\lambda)$ by the relations

$$\begin{aligned}
u_{i,j}^{k,l}(\mathbf{q}_\lambda) &= \cos\beta_{i,j}(\mathbf{q}_\lambda) && \text{for } k = i \text{ and } l = i \\
&= \cos\beta_{i,j}(\mathbf{q}_\lambda) && \text{for } k = j \text{ and } l = j \\
&= -\sin\beta_{i,j}(\mathbf{q}_\lambda) && \text{for } k = i \text{ and } l = j \\
&= \sin\beta_{i,j}(\mathbf{q}_\lambda) && \text{for } k = j \text{ and } l = i \\
&= 1 && \text{for } k = l \neq i \text{ or } j \\
&= 0 && \text{for the remaining } k \text{ and } l
\end{aligned} \tag{38}$$

This choice of elements for the $\mathbf{u}_{i,j}(\mathbf{q}_\lambda)$ matrix will diabatize the adiabatic electronic states i and j while leaving the remaining states unaltered.

As an example, in a four-electronic-state problem $(n = 4)$ consider the electronic states $i = 2$ and $j = 4$ along with the first-derivative coupling vector element $\mathbf{w}_{2,4}^{(1)\text{ad}}(\mathbf{R}_\lambda)$ that couples those two states. The ADT matrix $\mathbf{u}_{2,4}(\mathbf{q}_\lambda)$ can then be expressed in terms of the corresponding diabatization angle $\beta_{2,4}(\mathbf{q}_\lambda)$ as

$$\mathbf{u}_{2,4}(\mathbf{q}_\lambda) = \begin{pmatrix} 1 & 0 & 0 & 0 \\ 0 & \cos\beta_{2,4}(\mathbf{q}_\lambda) & 0 & -\sin\beta_{2,4}(\mathbf{q}_\lambda) \\ 0 & 0 & 1 & 0 \\ 0 & \sin\beta_{2,4}(\mathbf{q}_\lambda) & 0 & \cos\beta_{2,4}(\mathbf{q}_\lambda) \end{pmatrix} \tag{39}$$

This diabatization matrix only mixes the adiabatic states 2 and 4 leaving the states 1 and 3 unchanged.

In the n-electronic-state case, $n(n-1)/2$ such matrices $\mathbf{u}_{i,j}(\mathbf{q}_\lambda)$ $(j > i$ with $i = 1, 2, \ldots, n-1$ and $j = 2, \ldots, n)$ can be defined using Eq. (38). The full ADT matrix $\mathbf{U}(\mathbf{q}_\lambda)$ is then defined as a product of these $n(n-1)/2$ matrices $\mathbf{u}_{i,j}(\mathbf{q}_\lambda)$ $(j > i)$ as

$$\mathbf{U}(\mathbf{q}_\lambda) = \prod_{i=1}^{n-1} \prod_{j=i+1}^{n} \mathbf{u}_{i,j}(\mathbf{q}_\lambda) \tag{40}$$

which is the n-electronic-state version of the expression that has appeared earlier [72,73] for three electronic states. This $\mathbf{U}(\mathbf{q}_\lambda)$ is orthogonal, as it is the product of orthogonal matrices. The matrices $\mathbf{u}_{i,j}(\mathbf{q}_\lambda)$ in Eq. (40) can be multiplied in any order without loss of generality. A different multiplication order leads to a different set of solutions for the diabatization angles $\beta_{i,j}(\mathbf{q}_\lambda)$. However, since the matrix $\mathbf{U}(\mathbf{q}_\lambda)$ is a solution of a set of Poisson-type equations with fixed boundary conditions, as will be discussed next, it is uniquely determined and therefore independent of this choice of the order of multiplication, that is, all of these sets of $\beta_{i,j}(\mathbf{q}_\lambda)$ give the same $\mathbf{U}(\mathbf{q}_\lambda)$ [73]. Remembered, however, that these are purely formal considerations, since the existence of solutions of Eq. (44) presented next, requires the set of adiabatic electronic states to be complete; a truncated set no longer satisfies the conditions of Eq. (43) for the existence of solutions of Eq. (44). These formal considerations are nevertheless useful for the consideration of truncated Born–Huang expansion which follows Eq. (46).

We want to choose the ADT matrix $\mathbf{U}(\mathbf{q}_\lambda)$ that either makes the diabatic first-derivative coupling vector matrix $\mathbf{W}^{(1)d}(\mathbf{R}_\lambda)$ zero if possible or that minimizes its magnitude in such a way that the gradient term $\mathbf{W}^{(1)d}(\mathbf{R}_\lambda) \cdot \nabla_{\mathbf{R}_\lambda} \chi^d(\mathbf{R}_\lambda)$ in Eq. (31) can be neglected. By rewriting the relation between $\mathbf{W}^{(1)d}(\mathbf{R}_\lambda)$ and $\mathbf{W}^{(1)\mathrm{ad}}(\mathbf{R}_\lambda)$ of Eq. (35) as

$$\mathbf{W}^{(1)d}(\mathbf{R}_\lambda) = \tilde{\mathbf{U}}(\mathbf{q}_\lambda)[\nabla_{\mathbf{R}_\lambda} \mathbf{U}(\mathbf{q}_\lambda) + \mathbf{W}^{(1)\mathrm{ad}}(\mathbf{R}_\lambda)\mathbf{U}(\mathbf{q}_\lambda)] \tag{41}$$

we see that all elements of the diabatic matrix $\mathbf{W}^{(1)d}(\mathbf{R}_\lambda)$ will vanish if and only if all elements of the matrix inside the square brackets in the right-hand side of this equation are zero, that is,

$$\nabla_{\mathbf{R}_\lambda}\mathbf{U}(\mathbf{q}_\lambda) + \mathbf{W}^{(1)\text{ad}}(\mathbf{R}_\lambda)\mathbf{U}(\mathbf{q}_\lambda) = \mathbf{0} \tag{42}$$

The structure of $\mathbf{W}^{(1)\text{ad}}(\mathbf{R}_\lambda)$ discussed at the beginning of this section, will reflect itself in some interrelations between the $\beta_{i,j}(\mathbf{q}_\lambda)$ obtained by solving this equation. More importantly, this equation has a solution if and only if the elements of the matrix $\mathbf{W}^{(1)\text{ad}}(\mathbf{R}_\lambda)$ satisfy the following curl-condition [26,47,74–76] for all values of $\mathbf{R}_\lambda$:

$$\text{curl } \mathbf{w}_{i,j}^{(1)\text{ad}}(\mathbf{R}_\lambda)]_{k,l} = -[\mathbf{w}_k^{(1)\text{ad}}(\mathbf{R}_\lambda), \mathbf{w}_l^{(1)\text{ad}}(\mathbf{R}_\lambda)]_{i,j} \qquad k,l = 1,2,\ldots,3(N_{\text{nu}}-1) \tag{43}$$

In this equation, $\mathbf{w}_p^{(1)\text{ad}}(\mathbf{R}_\lambda)$ (with $p = k, l$) is the $n \times n$ matrix whose row i and column j element is the p Cartesian component of the $\mathbf{w}_{i,j}^{(1)\text{ad}}(\mathbf{R}_\lambda)$ vector, that is, $[\mathbf{w}_{i,j}^{(1)\text{ad}}(\mathbf{R}_\lambda)]_p$, and the square bracket on its right-hand side is the commutator of the two matrices within. This condition is satisfied for an $n \times n$ matrix $\mathbf{W}^{(1)\text{ad}}(\mathbf{R}_\lambda)$ when n samples the complete infinite set of adiabatic electronic states. In that case, we can rewrite Eq. (42) using the unitarity property [Eq. (29)] of $\mathbf{U}(\mathbf{q}_\lambda)$ as

$$[\nabla_{\mathbf{R}_\lambda}\mathbf{U}(\mathbf{q}_\lambda)]\tilde{\mathbf{U}}(\mathbf{q}_\lambda) = -\mathbf{W}^{(1)\text{ad}}(\mathbf{R}_\lambda) \tag{44}$$

This matrix equation can be expressed in terms of individual matrix elements on both sides as

$$\sum_k (\nabla_{\mathbf{R}_\lambda} f_{i,k}[\boldsymbol{\beta}(\mathbf{q}_\lambda)]) f_{j,k}[\boldsymbol{\beta}(\mathbf{q}_\lambda)] = -\mathbf{w}_{i,j}^{(1)\text{ad}}(\mathbf{R}_\lambda) \tag{45}$$

where $\boldsymbol{\beta}(\mathbf{q}_\lambda) \equiv (\beta_{1,2}(\mathbf{q}_\lambda), \ldots, \beta_{1,n}(\mathbf{q}_\lambda), \beta_{2,3}(\mathbf{q}_\lambda), \ldots, \beta_{2,n}(\mathbf{q}_\lambda), \ldots, \beta_{n-1,n}(\mathbf{q}_\lambda))$ is a set of all unknown diabatization angles and $f_{p,q}[\boldsymbol{\beta}(\mathbf{q}_\lambda)]$ with $p, q = i, j, k$ are matrix elements of the ADT matrix $\mathbf{U}(\mathbf{q}_\lambda)$, which are known trignometric functions of the unknown $\boldsymbol{\beta}(\mathbf{q}_\lambda)$ due to Eqs. (38) and (40). Equation (45) are a set of coupled first-order partial differential equations in the unknown diabatization angles $\beta_{i,j}(\mathbf{q}_\lambda)$ in terms of the known first-derivative coupling vector elements $\mathbf{w}_{i,j}^{(1)\text{ad}}(\mathbf{R}_\lambda)$ obtained from ab initio electronic structure calculations [43]. This set of coupled differential equations can be solved in principle with some appropriate choice of boundary conditions for the angles $\beta_{i,j}(\mathbf{q}_\lambda)$.

The ADT matrix $\mathbf{U}(\mathbf{q}_\lambda)$ obtained in this way makes the diabatic first-derivative coupling matrix $\mathbf{W}^{(1)d}(\mathbf{R}_\lambda)$ that appears in the diabatic Schrödinger equation (31) rigorously zero. It also leads to a diabatic electronic basis set that is independent of $\mathbf{q}_\lambda$ [76], which, in agreement with the present formal considerations, can only be a correct basis set if it is complete, that is, infinite. It can be proved using Eqs. (35), (36), and (42) that this choice of the ADT matrix also makes the diabatic second-derivative coupling matrix $\mathbf{W}^{(2)d}(\mathbf{R}_\lambda)$ appearing in Eq. (31) equal to zero. As a result, when n samples the complete set of adiabatic electronic states, the corresponding diabatic nuclear motion Schrödinger equation (31) reduces to the simple form

$$\left[-\frac{\hbar^2}{2\mu}\mathbf{I}\nabla^2_{\mathbf{R}_\lambda} + \{\boldsymbol{\varepsilon}^d(\mathbf{q}_\lambda) - E\mathbf{I}\}\right]\boldsymbol{\chi}^d(\mathbf{R}_\lambda) = \mathbf{0} \tag{46}$$

where the only term that couples the diabatic nuclear wave functions $\boldsymbol{\chi}^d(\mathbf{R}_\lambda)$ is the diabatic energy matrix $\boldsymbol{\varepsilon}^d(\mathbf{q}_\lambda)$.

The curl condition given by Eq. (43) is in general not satisfied by the $n \times n$ matrix $\mathbf{W}^{(1)\mathrm{ad}}(\mathbf{R}_\lambda)$, if n does not span the full infinite basis set of adiabatic electronic states and is truncated to include only a finite small number of these states. This truncation is extremely convenient from a physical as well as computational point of view. In this case, since Eq. (42) does not have a solution, let us consider instead the equation obtained from it by replacing $\mathbf{W}^{(1)\mathrm{ad}}(\mathbf{R}_\lambda)$ by its longitudinal part

$$\nabla_{\mathbf{R}_\lambda}\mathbf{U}(\mathbf{q}_\lambda) + \mathbf{W}^{(1)\mathrm{ad}}_{\mathrm{lon}}(\mathbf{R}_\lambda)\mathbf{U}(\mathbf{q}_\lambda) = \mathbf{0} \tag{47}$$

This equation does have a solution, because in view of Eq. (20) the curl condition of Eq. (43) is satisfied when $\mathbf{W}^{(1)\mathrm{ad}}(\mathbf{R}_\lambda)$ is replaced by $\mathbf{W}^{(1)\mathrm{ad}}_{\mathrm{lon}}(\mathbf{R}_\lambda)$.

We can now rewrite Eq. (47) using the orthogonality of $\mathbf{U}(\mathbf{q}_\lambda)$ as

$$[\nabla_{\mathbf{R}_\lambda}\mathbf{U}(\mathbf{q}_\lambda)]\tilde{\mathbf{U}}(\mathbf{q}_\lambda) = -\mathbf{W}^{(1)\mathrm{ad}}_{\mathrm{lon}}(\mathbf{R}_\lambda) \tag{48}$$

The quantity on the right-hand side of this equation is not completely specified since the decomposition of $\mathbf{W}^{(1)\mathrm{ad}}(\mathbf{R}_\lambda)$ into its longitudinal and transverse parts given by Eq. (24) is not unique. By using that decomposition and the property of the transverse part $\mathbf{W}^{(1)\mathrm{ad}}_{\mathrm{tra}}(\mathbf{R}_\lambda)$ given by Eq. (21), we see that

$$\nabla_{\mathbf{R}_\lambda}\cdot\mathbf{W}^{(1)\mathrm{ad}}_{\mathrm{lon}}(\mathbf{R}_\lambda) = \nabla_{\mathbf{R}_\lambda}\cdot\mathbf{W}^{(1)\mathrm{ad}}(\mathbf{R}_\lambda) \tag{49}$$

and since $\mathbf{W}^{(1)\mathrm{ad}}(\mathbf{R}_\lambda)$ is assumed to have been previously calculated, $\nabla_{\mathbf{R}_\lambda}\cdot\mathbf{W}^{(1)\mathrm{ad}}_{\mathrm{lon}}(\mathbf{R}_\lambda)$ is known. If we take the divergence of both sides of

Eq. (48), we obtain [using Eq. (49)]

$$[\nabla^2_{\mathbf{R}_\lambda}\mathbf{U}(\mathbf{q}_\lambda)]\tilde{\mathbf{U}}(\mathbf{q}_\lambda) + [\nabla_{\mathbf{R}_\lambda}\mathbf{U}(\mathbf{q}_\lambda)]\cdot[\nabla_{\mathbf{R}_\lambda}\tilde{\mathbf{U}}(\mathbf{q}_\lambda)] = -\nabla_{\mathbf{R}_\lambda}\cdot\mathbf{W}^{(1)\mathrm{ad}}(\mathbf{R}_\lambda) \tag{50}$$

By using the parametrization of $\mathbf{U}(\mathbf{q}_\lambda)$ given by Eqs. (38) and (40) for a finite n, this matrix equation can be expressed in terms of the matrix elements on both sides as

$$\sum_k [(\nabla^2_{\mathbf{R}_\lambda} f_{i,k}[\boldsymbol{\beta}(\mathbf{q}_\lambda)])f_{j,k}[\boldsymbol{\beta}(\mathbf{q}_\lambda)] + (\nabla_{\mathbf{R}_\lambda} f_{i,k}[\boldsymbol{\beta}(\mathbf{q}_\lambda)])\cdot(\nabla_{\mathbf{R}_\lambda} f_{j,k}[\boldsymbol{\beta}(\mathbf{q}_\lambda)])]$$
$$= -\nabla_{\mathbf{R}_\lambda}\cdot\mathbf{w}^{(1)\mathrm{ad}}_{i,j}(\mathbf{R}_\lambda) \tag{51}$$

where $f_{p,q}$ are the same as defined after Eq. (45). Equations (51) are a set of coupled Poisson-type equations in the unknown angles $\beta_{i,j}(\mathbf{q}_\lambda)$. For $n = 2$, this becomes Eq. (68), as shown in Section III.D. The structure of this set of equations is again dependent on the order of multiplication of matrices $\mathbf{u}_{i,j}(\mathbf{q}_\lambda)$ in Eq. (40). Each choice of the order of multiplication will give a different set of $\beta_{i,j}(\mathbf{q}_\lambda)$ as before but the same ADT matrix $\mathbf{U}(\mathbf{q}_\lambda)$ after they are solved using the same set of boundary conditions.

By using the fact that for a finite number of adiabatic electronic states n, we choose a $\mathbf{U}(\mathbf{q}_\lambda)$ that satisfies Eq. (47) [rather than Eq. (42) that has no solution], Eq. (35) now reduces to

$$\mathbf{W}^{(1)d}(\mathbf{R}_\lambda) = \tilde{\mathbf{U}}(\mathbf{q}_\lambda)\mathbf{W}^{(1)\mathrm{ad}}_{\mathrm{tra}}(\mathbf{R}_\lambda)\mathbf{U}(\mathbf{q}_\lambda) \tag{52}$$

This can be used to rewrite the diabatic nuclear motion Schrödinger equation for an incomplete set of n electronic states as

$$\left[-\frac{\hbar^2}{2\mu}\{\mathbf{I}\nabla^2_{\mathbf{R}_\lambda} + 2\tilde{\mathbf{U}}(\mathbf{q}_\lambda)\mathbf{W}^{(1)\mathrm{ad}}_{\mathrm{tra}}(\mathbf{R}_\lambda)\mathbf{U}(\mathbf{q}_\lambda)\cdot\nabla_{\mathbf{R}_\lambda} + \mathbf{W}^{(2)d}(\mathbf{R}_\lambda)\}\right.$$
$$\left. + \{\boldsymbol{\varepsilon}^d(\mathbf{q}_\lambda) - E\mathbf{I}\}\right]\boldsymbol{\chi}^d(\mathbf{R}_\lambda) = \mathbf{0} \tag{53}$$

In this equation, the gradient term $\tilde{\mathbf{U}}(\mathbf{q}_\lambda)\mathbf{W}^{(1)\mathrm{ad}}_{\mathrm{tra}}(\mathbf{R}_\lambda)\mathbf{U}(\mathbf{q}_\lambda)\cdot\nabla_{\mathbf{R}_\lambda}\boldsymbol{\chi}^d(\mathbf{R}_\lambda) = \mathbf{W}^{(1)d}(\mathbf{R}_\lambda)\cdot\nabla_{\mathbf{R}_\lambda}\boldsymbol{\chi}^d(\mathbf{R}_\lambda)$ still appears and, as mentioned before, introduces numerical inefficiencies in its solution. Even though a truncated Born–Huang expansion was used to obtain Eq. (53), $\mathbf{W}^{(1)\mathrm{ad}}_{\mathrm{tra}}(\mathbf{R}_\lambda)$, although no longer zero, has no poles at conical intersection geometries [as opposed to the full $\mathbf{W}^{(1)\mathrm{ad}}(\mathbf{R}_\lambda)$ matrix].

The set of coupled Poisson equations (50) can, in principle, be solved with any appropriate choice of boundary conditions for $\beta_{i,j}(\mathbf{q}_\lambda)$. There is one choice, however, for which the magnitude of $\mathbf{W}_{\text{tra}}^{(1)\text{ad}}(\mathbf{R}_\lambda)$ is minimized. If at the boundary surfaces $\mathbf{R}_\lambda^{\text{b}}$ of the nuclear configuration space spanned by $\mathbf{R}_\lambda$ (and the corresponding subset of boundary surfaces $\mathbf{q}_\lambda^{\text{b}}$ in the internal configuration space spanned by $\mathbf{q}_\lambda$), one imposes the following mixed Dirichlet–Neumann condition [based on Eq. (48)],

$$[\nabla_{\mathbf{R}_\lambda^{\text{b}}}\mathbf{U}(\mathbf{q}_\lambda^{\text{b}})]\tilde{\mathbf{U}}(\mathbf{q}_\lambda^{\text{b}}) = -\mathbf{W}^{(1)\text{ad}}(\mathbf{R}_\lambda^{\text{b}}) \tag{54}$$

it minimizes the average magnitude of the vector elements of the transverse coupling matrix $\mathbf{W}_{\text{tra}}^{(1)\text{ad}}(\mathbf{R}_\lambda)$ over the entire internal nuclear configuration space as shown for the $n = 2$ case [55] and hence the magnitude of the gradient term $\mathbf{W}^{(1)d}(\mathbf{R}_\lambda)\cdot\nabla_{\mathbf{R}_\lambda}\boldsymbol{\chi}^d(\mathbf{R}_\lambda)$. To a first very good approximation, this term can be neglected in the diabatic Schrödinger Eq. (53) resulting in a simpler equation

$$\left[-\frac{\hbar^2}{2\mu}\{\mathbf{I}\nabla_{\mathbf{R}_\lambda}^2 + \mathbf{W}^{(2)d}(\mathbf{R}_\lambda)\} + \{\boldsymbol{\varepsilon}^d(\mathbf{q}_\lambda) - E\mathbf{I}\}\right]\boldsymbol{\chi}^d(\mathbf{R}_\lambda) = \mathbf{0} \tag{55}$$

In this diabatic Schrödinger equation, the only terms that couple the nuclear wave functions $\chi_i^d(\mathbf{R}_\lambda)$ are the elements of the $\mathbf{W}^{(2)d}(\mathbf{R}_\lambda)$ and $\boldsymbol{\varepsilon}^d(\mathbf{q}_\lambda)$ matrices. The $-(\hbar^2/2\mu)\mathbf{W}^{(2)d}(\mathbf{R}_\lambda)$ matrix does not have poles at conical intersection geometries [as opposed to $\mathbf{W}^{(2)\text{ad}}(\mathbf{R}_\lambda)$] and furthermore it only appears as an additive term to the diabatic energy matrix $\boldsymbol{\varepsilon}^d(\mathbf{q}_\lambda)$ and does not increase the computational effort for the solution of Eq. (55). Since the neglected gradient term is expected to be small, it can be reintroduced as a first-order perturbation afterward, if desired.

In this section, it was shown how an optimal ADT matrix for an n-electronic-state problem can be obtained. In Section III.D, an application of the method outlined above to a two-state problem for the H_3 system is described.

D. Application to Two Electronic States

In the two-electronic-state case (with real electronic wave functions as before), Eqs. (12) and (27) become

$$\Psi^{\text{O}}(\mathbf{r},\mathbf{R}_\lambda) = \chi_1^{\text{ad}}(\mathbf{R}_\lambda)\psi_1^{\text{el,ad}}(\mathbf{r};\mathbf{q}_\lambda) + \chi_2^{\text{ad}}(\mathbf{R}_\lambda)\psi_2^{\text{el,ad}}(\mathbf{r};\mathbf{q}_\lambda) \tag{56}$$

$$= \chi_1^{d}(\mathbf{R}_\lambda)\psi_1^{\text{el},d}(\mathbf{r};\mathbf{q}_\lambda) + \chi_2^{d}(\mathbf{R}_\lambda)\psi_2^{\text{el},d}(\mathbf{r};\mathbf{q}_\lambda) \tag{57}$$

Equations (28) and (30) are unchanged, with $\boldsymbol{\psi}^{\text{el},d}(\mathbf{r};\mathbf{q}_\lambda)$, $\boldsymbol{\psi}^{\text{el,ad}}(\mathbf{r};\mathbf{q}_\lambda)$, $\boldsymbol{\chi}^d(\mathbf{R}_\lambda)$ and $\boldsymbol{\chi}^{\text{ad}}(\mathbf{R}_\lambda)$ now being two-dimensional (2D) column vectors, and Eq. (40) having the much simpler form

$$\mathbf{U}[\beta(\mathbf{q}_\lambda)] = \begin{pmatrix} \cos\beta(\mathbf{q}_\lambda) & -\sin\beta(\mathbf{q}_\lambda) \\ \sin\beta(\mathbf{q}_\lambda) & \cos\beta(\mathbf{q}_\lambda) \end{pmatrix} \tag{58}$$

involving the single real diabatization or mixing angle $\beta(\mathbf{q}_\lambda)$.

Equations (31) and (32) are unchanged, with $\mathbf{W}^{(1)d}(\mathbf{R}_\lambda)$, $\mathbf{W}^{(2)d}(\mathbf{R}_\lambda)$, and $\boldsymbol{\varepsilon}^d(\mathbf{q}_\lambda)$ now being 2×2 matrices. The adiabatic-to-diabatic transformation, as for the n-state case, eliminates any poles in both the first- and second-derivative coupling matrices at conical intersection geometries but in this case Eq. (52) yields

$$\mathbf{W}^{(1)d}(\mathbf{R}_\lambda) = \mathbf{W}_{\text{tra}}^{(1)\text{ad}}(\mathbf{R}_\lambda) \tag{59}$$

Elements of the matrix $-(\hbar^2/2\mu)\mathbf{W}^{(2)d}$ are usually small in the vicinity of a conical intersection and can be added to $\boldsymbol{\varepsilon}^d$ to give a corrected diabatic energy matrix. As can be seen, whereas in Eq. (15) $\mathbf{W}^{(1)\text{ad}}$ contains both the singular matrix $\mathbf{W}_{\text{lon}}^{(1)\text{ad}}$ and the nonsingular one $\mathbf{W}_{\text{tra}}^{(1)\text{ad}}$, Eq. (31) contains only the latter. Nevertheless, the residual first-derivative coupling term $\mathbf{W}_{\text{tra}}^{(1)\text{ad}} \cdot \nabla_{\mathbf{R}_\lambda}$ does not vanish.

A "perfect" diabatic basis would be one for which the first-derivative coupling $\mathbf{W}^{(1)d}(\mathbf{R}_\lambda)$ in Eq. (31) vanishes [10]. From the above mentioned considerations, we conclude, as is well known, that a "perfect" diabatic basis cannot exist for a polyatomic system (except when the complete infinite set of electronic adiabatic functions is included [26,47,74,76]), which means that $\mathbf{W}^{(1)\text{ad}}(\mathbf{R}_\lambda)$ cannot be "transformed away" to zero. Consequently, the longitudinal and transverse parts of the first-derivative coupling vector are referred to as *removable* and *nonremovable* parts, respectively. As mentioned in the introduction, a number of formulations of approximate or quasidiabatic (or 'locally rigorous') diabatic states [44,45,47–54] have been considered. Only very recently [77–82] have there been attempts to use high quality ab initio wave functions to evaluate the *nonremovable* part of the first-derivative coupling vector. In one such attempt [81], a quasidiabatic basis was reported for the triatomic HeH_2 system by solving a 2D Poisson equation on the plane in 3D configuration space passing through the conical intersection configuration of smallest energy. It seems that no attempt has been made to get an optimal diabatization over the entire configuration space even for triatomic systems until now [55], aimed at facilitating accurate two-electronic-state scattering dynamics calculations for such systems. Conical intersections being omnipresent, such scattering calculations will permit a test of the validity of the one-electronic-state BO approximation as a function of energy in the presence of conical intersections, by comparing the results of these two kinds of calculations.

The ADT matrix for the lowest two electronic states of H_3 has recently been obtained [55]. These states display a conical intersection at equilateral triangle geometries, but the GP effect can be easily built into the treatment of the reactive scattering equations. Since, for two electronic states, there is only one nonzero first-derivative coupling vector, $\mathbf{w}_{1,2}^{(1)\text{ad}}(\mathbf{R}_\lambda)$, we will refer to it in the rest of this section as $\mathbf{w}^{(1)\text{ad}}(\mathbf{R}_\lambda)$. For a triatomic system, this vector is six dimensional (6D).

As discussed in Section II.A, the adiabatic electronic wave functions $\psi_i^{\text{el,ad}}$ and $\psi_j^{\text{el,ad}}$ depend on the nuclear coordinates $\mathbf{R}_\lambda$ only through the subset $\mathbf{q}_\lambda$ (which in the triatomic case consists of a nuclear coordinate hyperradius ρ and a set of two internal hyperangles ξ_λ), this permits one to relate the 6D vector $\mathbf{w}^{(1)\text{ad}}(\mathbf{R}_\lambda)$ to another one $\mathsf{w}^{(1)\text{ad}}(\mathbf{q}_\lambda)$ that is 3D. For a triatomic system, let $\mathbf{a}^{I\lambda} \equiv (a^{I\lambda}, b^{I\lambda}, c^{I\lambda})$ be the Euler angles that rotate the space-fixed Cartesian frame into the body-fixed principal axis of inertia frame $I\lambda$, and let $\nabla_{\mathbf{R}_\lambda}^{I\lambda}$ be the 6D gradient vector in this rotated frame. The relation between the space-fixed $\nabla_{\mathbf{R}_\lambda}$ and $\nabla_{\mathbf{R}_\lambda}^{I\lambda}$ is given by

$$\nabla_{\mathbf{R}_\lambda} = \tilde{\mathcal{R}}(\mathbf{a}^{I\lambda})\ \nabla_{\mathbf{R}_\lambda}^{I\lambda} \tag{60}$$

where $\mathcal{R}(\mathbf{a}^{I\lambda})$ is a 6×6 block-diagonal matrix whose two diagonal blocks are both equal to the 3×3 rotational matrix $\mathsf{R}(\mathbf{a}^{I\lambda})$. The $\nabla_{\mathbf{R}_\lambda}^{I\lambda}$ operator can be written as [83]

$$\nabla_{\mathbf{R}_\lambda}^{I\lambda} = \mathbf{G}^{I\lambda}(\xi_\lambda)\hat{\mathbf{p}}^{I\lambda}(\mathbf{q}_\lambda) + \mathbf{H}^{I\lambda}(\xi_\lambda)\hat{\mathbf{J}}^{I\lambda}(\mathbf{a}^{I\lambda}) \tag{61}$$

In this expression, $\mathbf{G}^{I\lambda}$ and $\mathbf{H}^{I\lambda}$ are both 6×3 rectangular matrices whose elements are known functions of the internal hyperangles ξ_λ. $\hat{\mathbf{p}}^{I\lambda}$ is a 3×1 column vector operator whose elements contain first derivatives with respect to the three $\mathbf{q}_\lambda$ coordinates and $\hat{\mathbf{J}}^{I\lambda}$ is the 3×1 column vector operator whose elements are the components $\hat{J}_x^{I\lambda}$, $\hat{J}_y^{I\lambda}$, and $\hat{J}_z^{I\lambda}$ of the system's nuclear motion angular momentum operator $\hat{\mathbf{J}}$ in the $I\lambda$ frame. From these properties, it can be shown that

$$\mathbf{w}^{(1)\text{ad}}(\mathbf{R}_\lambda) = \mathcal{R}(\mathbf{a}^{I\lambda})\mathbf{G}^{I\lambda}(\xi_\lambda)\mathsf{w}^{(1)\text{ad}}(\mathbf{q}_\lambda) \tag{62}$$

and that

$$\mathbf{W}^{(1)\text{ad}}(\mathbf{R}_\lambda) \cdot \nabla_{\mathbf{R}_\lambda}\chi^{\text{ad}}(\mathbf{R}_\lambda) = \mathbf{G}^{I\lambda}(\xi_\lambda)\mathsf{W}^{(1)\text{ad}}(\mathbf{q}_\lambda) \cdot \nabla_{\mathbf{R}_\lambda}^{I\lambda}\chi^{\text{ad}}(\mathbf{R}_\lambda) \tag{63}$$

where

$$\mathsf{w}^{(1)\text{ad}}(\mathbf{q}_\lambda) = \langle \psi_1^{\text{el,ad}}(\mathbf{r}; \mathbf{q}_\lambda) | \hat{\mathbf{p}}^{I\lambda}(\mathbf{q}_\lambda) \psi_2^{\text{el,ad}}(\mathbf{r}; \mathbf{q}_\lambda) \rangle_{\mathbf{r}} \tag{64}$$

is a 3D column vector and $\mathsf{W}^{(1)\text{ad}}(\mathbf{q}_\lambda)$ is a 2×2 skew-symmetric matrix whose only nonzero element is the $\mathsf{w}^{(1)\text{ad}}(\mathbf{q}_\lambda)$ vector. By using the symmetrized hyperspherical coordinates defined in Section IV.A for a triatomic system, the elements of $\hat{\mathbf{p}}^{I\lambda}$ are the spherical polar components of the 3D gradient associated with the polar coordinates $\rho, \theta, \phi_\lambda$ [84]:

$$\hat{\mathbf{p}}^{I\lambda} = \begin{pmatrix} \frac{\partial}{\partial \rho} \\ \frac{1}{\rho}\frac{\partial}{\partial \theta} \\ \frac{1}{\rho \sin\theta}\frac{\partial}{\partial \phi_\lambda} \end{pmatrix} \tag{65}$$

The corresponding cartesian gradient $\mathbf{\nabla}_{\mathbf{q}_\lambda}$ is given by

$$\mathbf{\nabla}_{\mathbf{q}_\lambda} = \begin{pmatrix} \sin\theta\cos\phi_\lambda & \cos\theta\cos\phi_\lambda & -\sin\phi_\lambda \\ \sin\theta\sin\phi_\lambda & \cos\theta\sin\phi_\lambda & \cos\phi_\lambda \\ \cos\theta & -\sin\theta & 0 \end{pmatrix} \hat{\mathbf{p}}^{I\lambda} \tag{66}$$

in a space whose polar coordinates are $\rho, \theta, \phi_\lambda$.

The $\mathsf{w}^{(1)\text{ad}}(\mathbf{q}_\lambda)$ vector can also be decomposed into a longitudinal and a transverse part

$$\mathsf{w}^{(1)\text{ad}}(\mathbf{q}_\lambda) = \mathbf{\nabla}_{\mathbf{q}_\lambda}\alpha(\mathbf{q}_\lambda) + \mathsf{w}^{(1)\text{ad}}_{\text{tra}}(\mathbf{q}_\lambda) \tag{67}$$

where, $\alpha(\mathbf{q}_\lambda)$ is a scalar potential. It can be shown using Eq. (58) and the two-electronic-state counterpart of Eq. (47) that $\beta(\mathbf{q}_\lambda) = \alpha(\mathbf{q}_\lambda)$. The diabatization angle $\beta(\mathbf{q}_\lambda)$ can be obtained by taking the divergence of Eq. (67) and solving for the resulting Poisson equation

$$\nabla^2_{\mathbf{q}_\lambda}\beta(\mathbf{q}_\lambda) = \sigma(\mathbf{q}_\lambda) \tag{68}$$

where

$$\sigma(\mathbf{q}_\lambda) = \nabla_{\mathbf{q}_\lambda} \cdot \mathsf{w}^{(1)\text{ad}}(\mathbf{q}_\lambda) \tag{69}$$

is known because $\mathsf{w}^{(1)\text{ad}}(\mathbf{q}_\lambda)$ has been accurately calculated and fitted over the entire $\mathbf{q}_\lambda$ space of interest [84]. The nuclear–electronic rotational couplings associated with the rotation of the H_3 molecular plane relative to a space-fixed frame vanish identically if the mass-scaled nuclear and electronic coordinates of Eqs. (4) and (5) are used and the electronically adiabatic PESs are calculated accordingly. This is not, however, done in standard electronic structure calculations, as mentioned after Eq. (8), and as a result such couplings do not vanish. We have, however, in our electronic wave function calculations [84], found them to be at least two orders of magnitude smaller than the $|\mathsf{w}^{(1)\text{ad}}(\mathbf{q}_\lambda)|$.

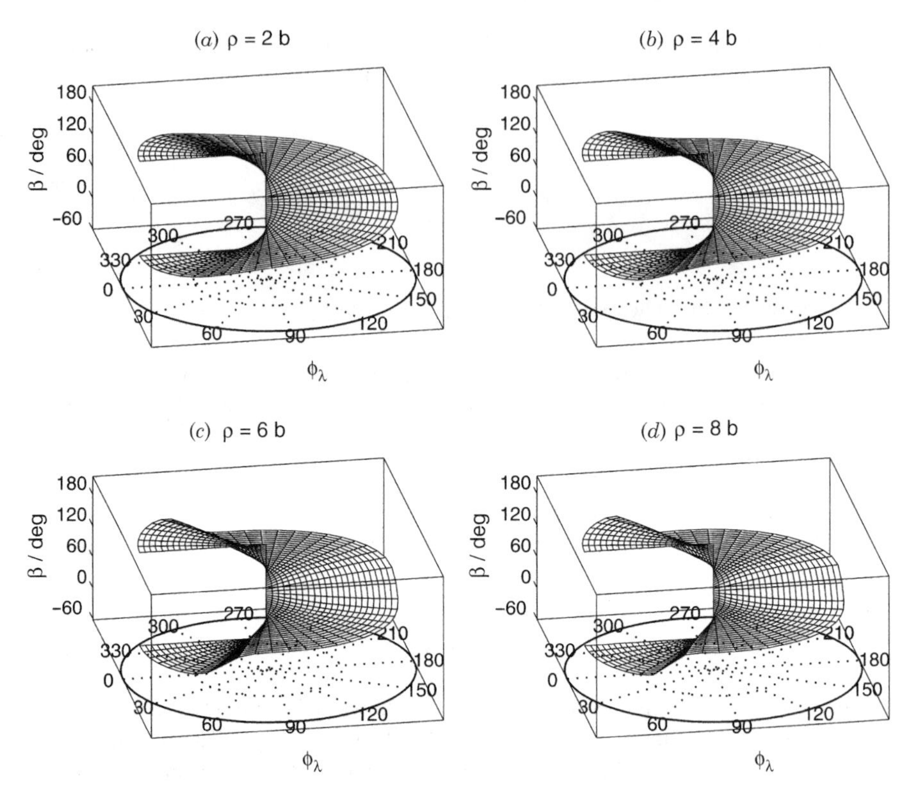

Figure 2. The diabatization angle $\beta(\rho, \theta, \phi_\lambda)$, in degrees, for the H_3 system at (*a*) $\rho = 2$ b, (*b*) $\rho = 4$ b, (*c*) $\rho = 6$ b, and (*d*) $\rho = 8$ b. The equatorial view of β contours is also given at (*e*) $\rho = 2$ b, (*f*) $\rho = 4$ b, (*g*) $\rho = 6$ b, and (*h*) $\rho = 8$ b.

This justifies the use of the simpler $\mathbf{q}_\lambda$ language over the $\mathbf{R}_\lambda$ one. The solution of the Poisson equation and the boundary conditions used are explained in detail elsewhere [55]. Here, we will present some selected results.

The internal coordinates used in the calculation are the hyperradius ρ, and the hyperangles θ and ϕ_λ, described in Section IV.A. The equation $\theta = 0^\circ$ corresponds to conical intersection geometries and $\theta = 90^\circ$ to collinear ones. For a fixed ρ and θ, as ϕ_λ is varied from 0 to 2π, the system executes a loop in internal configuration space around the corresponding conical intersection geometry. In Figure 2, the diabatization angle β is displayed for several values of ρ as a function of θ and ϕ_λ. Use of this β and Eq. (67), furnishes the transverse part $\mathsf{w}_{\text{tra}}^{(1)\text{ad}}(\mathbf{q}_\lambda)$ over the entire dynamically important region of internal configuration space. This is displayed in Figure 3. These sets of β and $\mathsf{w}_{\text{tra}}^{(1)\text{ad}}(\mathbf{q}_\lambda)$ were obtained using an optimal mixture of Dirichlet and Neumann

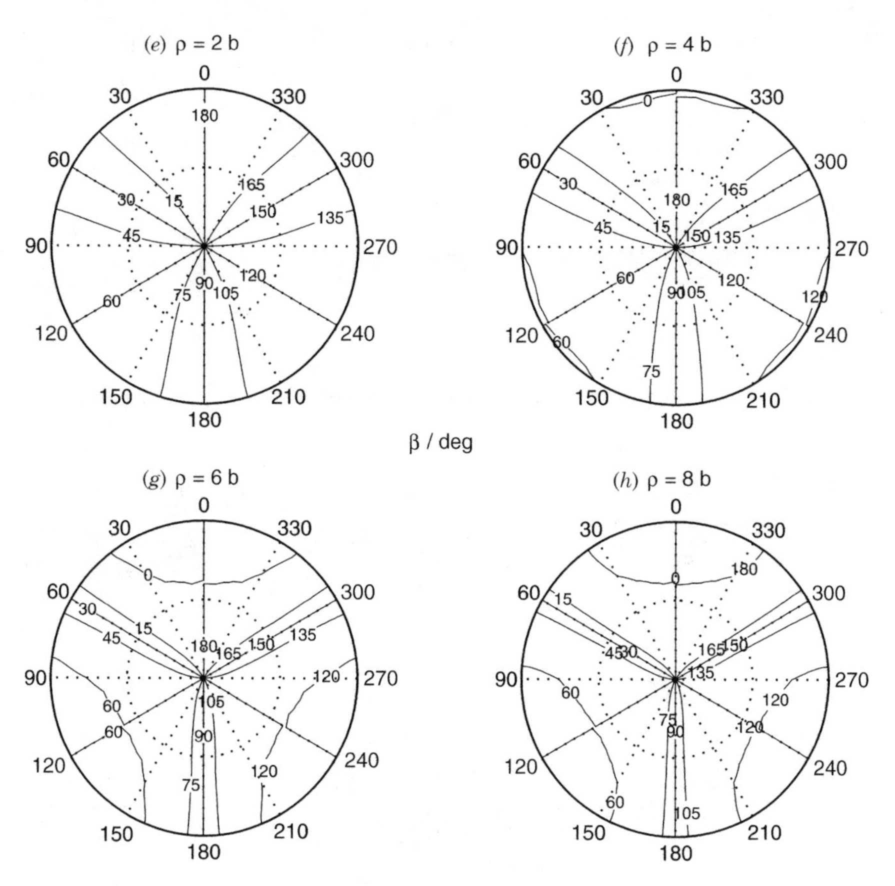

Figure 2 (*Continued*)

conditions for the solution of the above-mentioned Poisson equation. Using pure Dirichlet conditions instead gives a different transverse part, displayed in Figure 4. Comparison with Figure 3 clearly shows that the optimal boundary conditions significantly reduce the magnitude of the transverse part as compared to all-Dirichlet condition. Comparison of the average magnitude of the transverse vector over the entire internal configuration space for both the optimal and the all-Dirichlet boundary conditions, shows that the optimal condition average was $\sim$4.7 times smaller than the all-Dirichlet one. This result indicates that use of the optimal mixed set of Neumann and Dirichlet boundary conditions for solving the Poisson Eq. (68) does indeed significantly reduce the average magnitude of the transverse part of the first-derivative coupling vector.

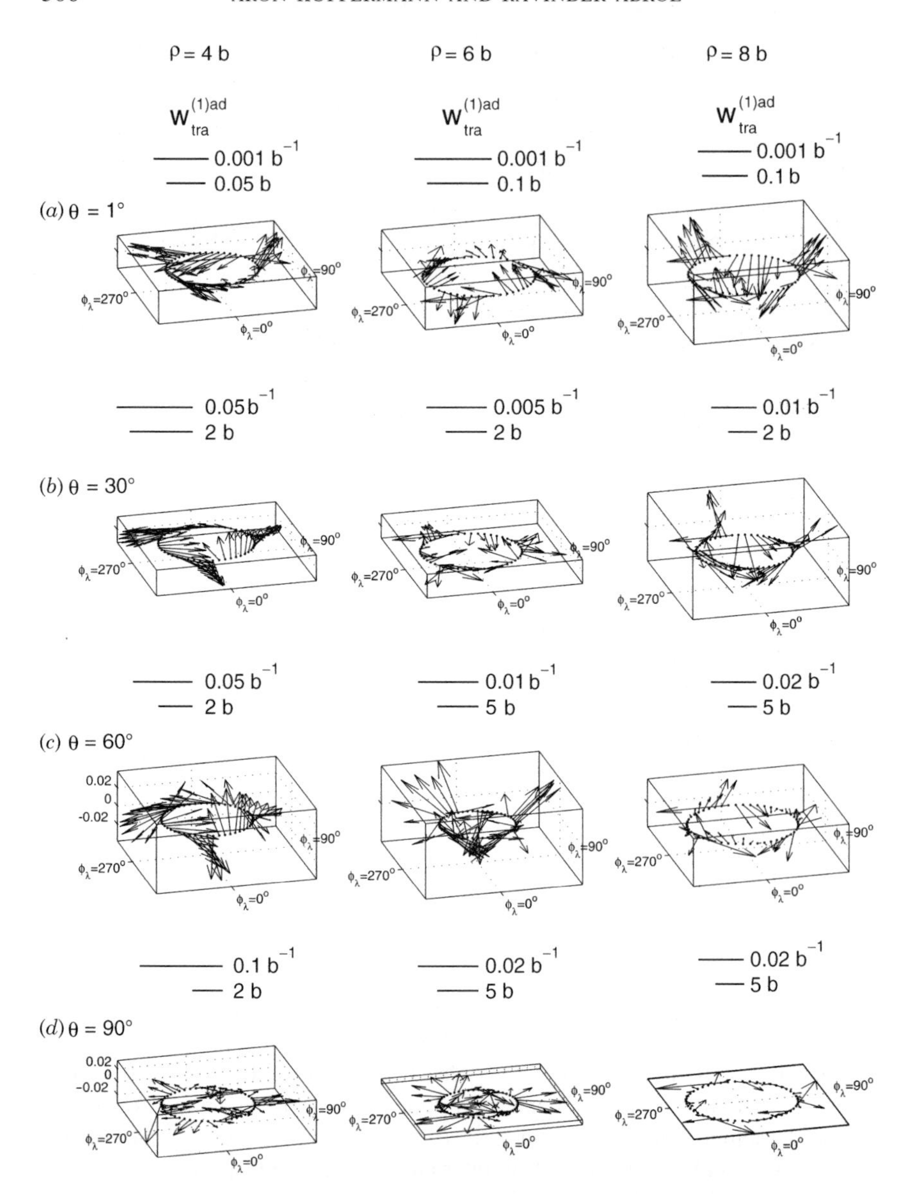

Figure 3. Transverse (nonremovable) part of the ab initio first-derivative coupling vector, $\mathbf{w}_{\text{tra}}^{(1)\text{ad}}(\rho, \theta, \phi_\lambda)$ as a function of ϕ_λ for $\rho = 4$, 6, and 8 b and (*a*) $\theta = 1°$ (near-conical intersection geometries), (*b*) $\theta = 30°$, (*c*) $\theta = 60°$, and (*d*) $\theta = 90°$ (collinear geometries).

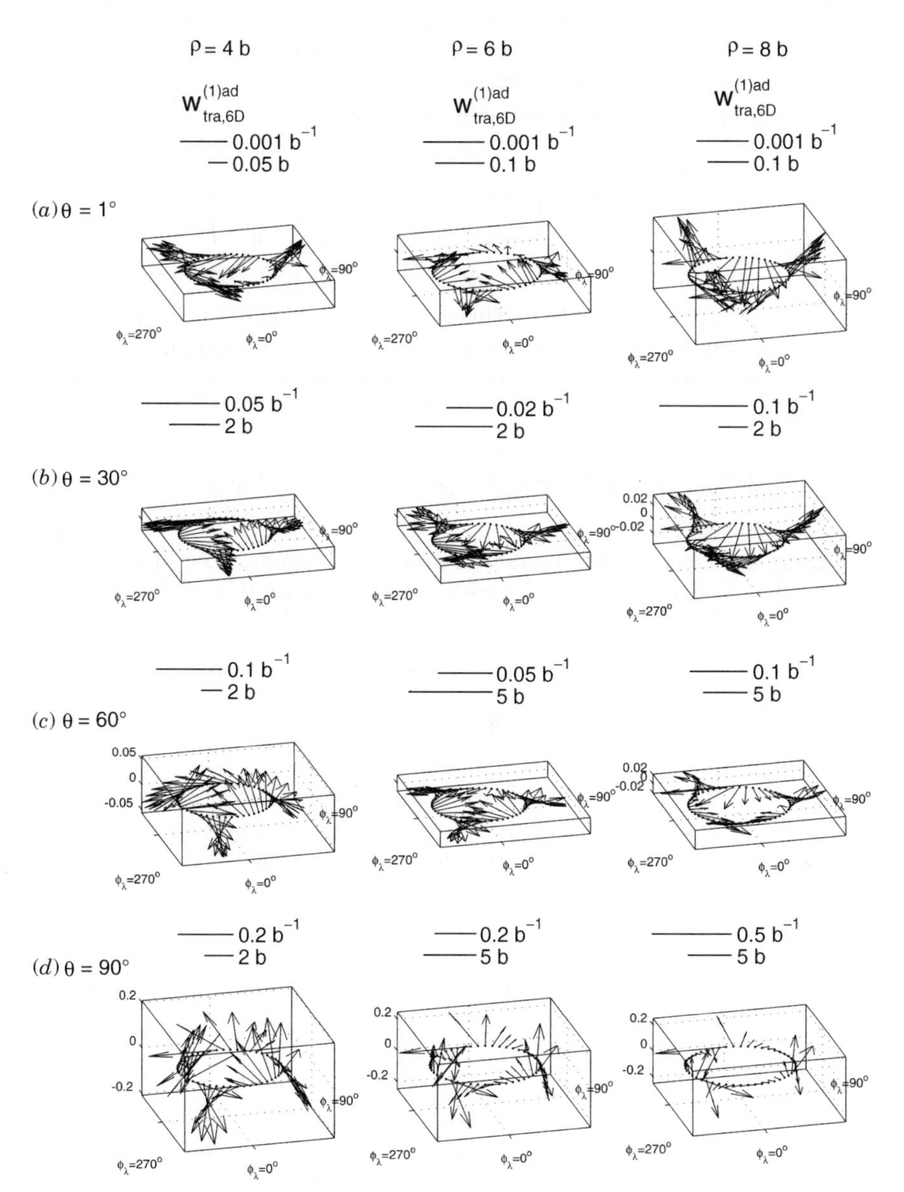

Figure 4. Same as Figure 3 for transverse (nonremovable) part of the ab initio first-derivative coupling vector $\mathbf{w}^{(1)\mathrm{ad}}_{\mathrm{tra.6D}}(\rho, \theta, \phi_\lambda)$, obtained using the all-Dirichlet boundary conditions.

The vector $\mathbf{w}^{(1)\mathrm{ad}}(\mathbf{q}_\lambda)$ [or $\mathbf{w}_{1,2}^{(1)\mathrm{ad}}(\mathbf{q}_\lambda)$] can also provide a good first approximation to the second-derivative coupling matrix $\mathbf{W}^{(2)\mathrm{ad}}(\mathbf{q}_\lambda)$, which in a two-electronic-state approximation is given by

$$\mathbf{W}^{(2)\mathrm{ad}}(\mathbf{q}_\lambda) = \begin{pmatrix} w_{1,1}^{(2)\mathrm{ad}}(\mathbf{q}_\lambda) & w_{1,2}^{(2)\mathrm{ad}}(\mathbf{q}_\lambda) \\ w_{2,1}^{(2)\mathrm{ad}}(\mathbf{q}_\lambda) & w_{2,2}^{(2)\mathrm{ad}}(\mathbf{q}_\lambda) \end{pmatrix} \tag{70}$$

In the two-electronic-state Born–Huang expansion, the full-Hilbert space of adiabatic electronic states is approximated by the lowest two states and furnishes for the corresponding electronic wave functions the approximate closure relation

$$|\psi_1^{\mathrm{el,ad}}(\mathbf{r};\mathbf{q}_\lambda)\rangle\langle\psi_1^{\mathrm{el,ad}}(\mathbf{r};\mathbf{q}_\lambda)| + |\psi_2^{\mathrm{el,ad}}(\mathbf{r};\mathbf{q}_\lambda)\rangle\langle\psi_2^{\mathrm{el,ad}}(\mathbf{r};\mathbf{q}_\lambda)| \approx 1 \tag{71}$$

By using this equation and the fact that for real electronic wave functions the diagonal elements of $\mathbf{W}^{(1)\mathrm{ad}}(\mathbf{q}_\lambda)$ vanish, it can be shown that

$$\begin{aligned} w_{1,1}^{(2)\mathrm{ad}}(\mathbf{q}_\lambda) &= w_{2,2}^{(2)\mathrm{ad}}(\mathbf{q}_\lambda) = -\mathbf{w}_{1,2}^{(1)\mathrm{ad}}(\mathbf{q}_\lambda)\cdot\mathbf{w}_{1,2}^{(1)\mathrm{ad}}(\mathbf{q}_\lambda) \\ w_{1,2}^{(2)\mathrm{ad}}(\mathbf{q}_\lambda) &= w_{2,1}^{(2)\mathrm{ad}}(\mathbf{q}_\lambda) = 0 \end{aligned} \tag{72}$$

For the H_3 system, since $\mathbf{w}_{1,2}^{(1)\mathrm{ad}}(\mathbf{q}_\lambda)$ is known over the entire $\mathbf{q}_\lambda$ space [84], Eq. (72) can be used to obtain the two equal nonzero diagonal elements of the $\mathbf{W}^{(2)\mathrm{ad}}(\mathbf{q}_\lambda)$ matrix. Since this matrix appears with a multiplicative factor of $(-\hbar^2/2\mu)$ in the adiabatic nuclear motion Schrödinger equation giving it the units of energy, both $(-\hbar^2/2\mu)\, w_{1,1}^{(2)\mathrm{ad}}(\mathbf{q}_\lambda)$ and $(-\hbar^2/2\mu)\, w_{2,2}^{(2)\mathrm{ad}}(\mathbf{q}_\lambda)$ can be labeled as $\varepsilon^{(2)\mathrm{ad}}$. In Figure 5, this quantity is displayed in units of kilocalories per mole (kcal/mol) for several values of ρ as a function of θ and ϕ_λ. It shows the singular behavior of the diagonal elements of $\mathbf{W}^{(2)\mathrm{ad}}(\mathbf{q}_\lambda)$ at conical intersection geometries ($\theta = 0°$). Being a repulsive correction to the adiabatic energies, this singular behavior prevents any hopping of the nuclei from one electronic state to another in the close vicinity of the conical intersection.

By using the diabatic version of the closure relation (71), and Eq. (59), the elements of the diabatic second-derivative coupling matrix $\mathbf{W}^{(2)d}(\mathbf{q}_\lambda)$ of Eq. (36) can be expressed as

$$\begin{aligned} w_{1,1}^{(2)d}(\mathbf{q}_\lambda) &= w_{2,2}^{(2)d}(\mathbf{q}_\lambda) = -\mathbf{w}_{1,2,\mathrm{tra}}^{(1)\mathrm{ad}}(\mathbf{q}_\lambda)\cdot\mathbf{w}_{1,2,\mathrm{tra}}^{(1)\mathrm{ad}}(\mathbf{q}_\lambda) \\ w_{1,2}^{(2)d}(\mathbf{q}_\lambda) &= w_{2,1}^{(2)d}(\mathbf{q}_\lambda) = 0 \end{aligned} \tag{73}$$

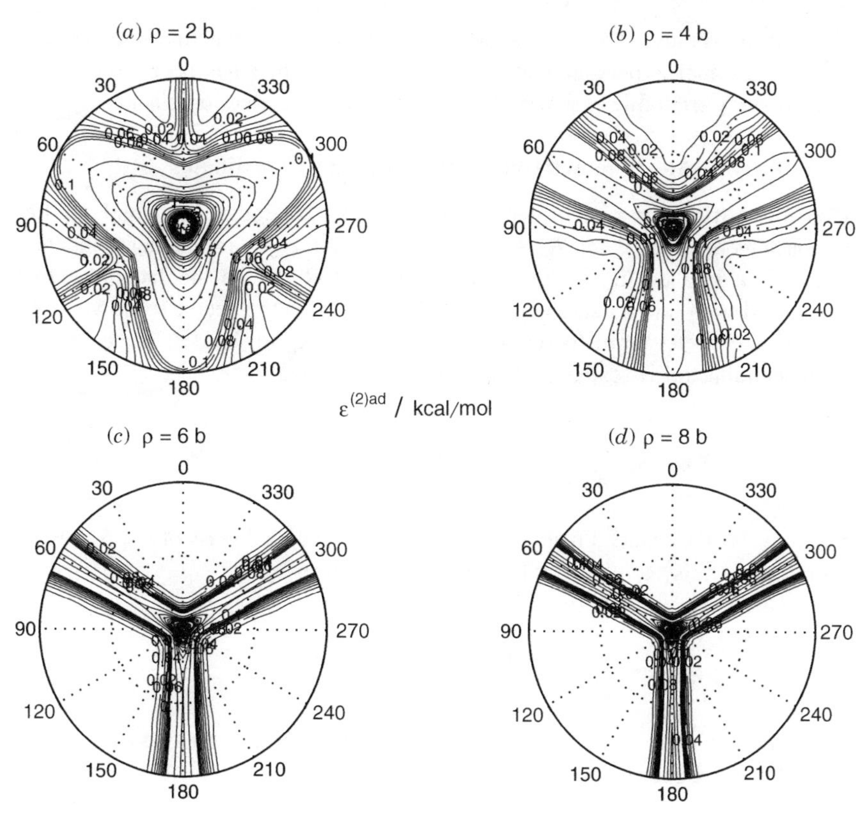

Figure 5. Second-derivative coupling term $\varepsilon^{(2)\mathrm{ad}}(\mathbf{q}_\lambda)$ defined at the end of Section III.D for the H_3 system at (*a*) $\rho = 2$ b, (*b*) $\rho = 4$ b, (*c*) $\rho = 6$ b, and (*d*) $\rho = 8$ b. The following contours are displayed 0–0.1 kcal/mol every 0.01 kcal/mol, 0.1–1.0 kcal/mol every 0.1 kcal/mol and 1.0–10.0 kcal/mol every 0.5 kcal/mol.

where, both $(-\hbar^2/2\mu)\, \mathrm{w}_{1,1}^{(2)d}(\mathbf{q}_\lambda)$ and $(-\hbar^2/2\mu)\, \mathrm{w}_{2,2}^{(2)d}(\mathbf{q}_\lambda)$ can be labeled as $\varepsilon^{(2)d}(\mathbf{q}_\lambda)$. The values of this (approximate) $\varepsilon^{(2)d}(\mathbf{q}_\lambda)$ calculated from this equation are smaller than 0.08 kcal/mol over the entire nuclear configuration space involved, and to a very good approximation can be neglected.

IV. TWO-ELECTRONIC-STATE QUANTUM REACTION DYNAMICS FORMALISM FOR TRIATOMIC REACTIONS

In a two-lowest-electronic-state Born–Huang description for a chemical reaction, the nuclei can move on both of two corresponding PESs during the reaction, due to the electronically non-adiabatic couplings between those states. A reactive scattering formalism for such a reaction involving a triatomic system is presented below. This formalism is an extension of the time-independent coupled-channel hyperspherical method [26,33–37] that has been used in the past to study triatomic reactions on a single adiabatic electronic state.

A. Symmetrized Hyperspherical Coordinates

Consider a triatomic system with the three nuclei labeled A_α, A_β, and A_γ. Let the arrangement channel $A_\lambda + A_\nu A_\kappa$ be called the λ arrangement channel, where $\lambda\nu\kappa$ is a cyclic permutation of $\alpha\beta\gamma$. Let $\mathbf{R}'_\lambda, \mathbf{r}'_\lambda$ be the Jacobi vectors associated with this arrangement channel, where $\mathbf{r}'_\lambda$ is the vector from A_ν to A_κ and $\mathbf{R}'_\lambda$ the vector from the center of mass of $A_\nu A_\kappa$ to A_λ. Let $\mathbf{R}_\lambda, \mathbf{r}_\lambda$ be the corresponding mass-scaled Jacobi coordinates defined by

$$\mathbf{R}_\lambda = \left(\frac{\mu_{\lambda,\nu\kappa}}{\mu}\right)^{1/2} \mathbf{R}'_\lambda \qquad \text{and} \qquad \mathbf{r}_\lambda = \left(\frac{\mu_{\nu\kappa}}{\mu}\right)^{1/2} \mathbf{r}'_\lambda \tag{74}$$

where $\mu_{\nu\kappa}$ is the reduced mass of $A_\nu A_\kappa$, $\mu_{\lambda,\nu\kappa}$ the reduced mass of the $A_\lambda, A_\nu A_\kappa$ pair, and μ the system's overall reduced mass given by

$$\mu = \left(\frac{m_\alpha m_\beta m_\gamma}{m_\alpha + m_\beta + m_\gamma}\right)^{1/2}$$

m_λ being the mass of atom A_λ ($\lambda = \alpha, \beta, \gamma$). We define a set of symmetrized hyperspherical coordinates $\rho, \omega_\lambda, \gamma_\lambda$ [85,86] by

$$\rho = \left(R_\lambda^2 + r_\lambda^2\right)^{1/2} \tag{75}$$

and

$$R_\lambda = \rho\cos(\omega_\lambda/2) \qquad r_\lambda = \rho\sin(\omega_\lambda/2) \qquad 0 \le \omega_\lambda \le \pi \tag{76}$$

where ρ is independent of the arrangement channel [60,61]. The corresponding internal configuration space Cartesian coordinates are defined by

$$\begin{aligned} X_\lambda &= \rho\sin\omega_\lambda\cos\gamma_\lambda \\ Y &= \rho\sin\omega_\lambda\sin\gamma_\lambda \\ Z_\lambda &= \rho\cos\omega_\lambda \end{aligned} \tag{77}$$

where γ_λ is the angle between $\mathbf{R}_\lambda$ and $\mathbf{r}_\lambda$ (or $\mathbf{R}'_\lambda$ and $\mathbf{r}'_\lambda$) in the 0 to π range and $\omega_\lambda, \gamma_\lambda$ are the polar angles of a point in this space. The alternate internal configuration space symmetrized hyperspherical coordinates θ, ϕ_λ are defined as the polar angles associated with the interchanged axes $O\bar{X}_\lambda = OZ_\lambda$, $O\bar{Y}_\lambda = OX_\lambda$, and $O\bar{Z}_\lambda = OY_\lambda$ for which

$$\begin{aligned} \bar{X}_\lambda &= Z_\lambda = \rho \sin\theta \cos\phi_\lambda \\ \bar{Y}_\lambda &= X_\lambda = \rho \sin\theta \sin\phi_\lambda \\ \bar{Z} &= Y = \rho \cos\theta \end{aligned} \tag{78}$$

The coordinates ρ, θ and ϕ_λ are limited to the ranges

$$0 \leq \rho < \infty \qquad 0 \leq \theta \leq \pi/2 \qquad 0 \leq \phi_\lambda < 2\pi \tag{79}$$

The relation between θ, ϕ_λ and $\omega_\lambda, \gamma_\lambda$ is [using Eqs. (77) and (78)]

$$\begin{aligned} \sin\theta \cos\phi_\lambda &= \cos\omega_\lambda \\ \sin\theta \sin\phi_\lambda &= \sin\omega_\lambda \cos\gamma_\lambda \\ \cos\theta &= \sin\omega_\lambda \sin\gamma_\lambda \end{aligned} \tag{80}$$

Let $Gx^{I\lambda}y^{I}z^{I\lambda}$ be a body-fixed frame $I\lambda$, whose axes are the principal axes of inertia of the three nuclei and whose Euler angles with respect to the space-fixed frame $Gx^{\text{sf}}y^{\text{sf}}z^{\text{sf}}$ are $a_\lambda, b_\lambda, c_\lambda$ with G being the center of mass of the three nuclei. The senses of these axes are chosen to result in a 1:1 correspondence between $\rho, \theta, \phi_\lambda, a_\lambda, b_\lambda, c_\lambda$ coordinates and the space-fixed Cartesian coordinates of $\mathbf{R}_\lambda$ and $\mathbf{r}_\lambda$. In addition, the $I\lambda$ axes are labeled so as to order the corresponding principal moments of inertia according to

$$I_z^\lambda \leq I_x^\lambda \leq I_y \tag{81}$$

Furthermore, let Υ_λ refer collectively to the five hyperangles $(\theta, \phi_\lambda, a_\lambda, b_\lambda, c_\lambda)$, $\mathbf{q}_\lambda$ to the three internal coordinates $(\rho, \theta, \phi_\lambda)$ and $\mathbf{R}_\lambda$ to all six hyperspherical coordinates.

The coordinates ρ, Υ_λ are called the principal axes of inertia symmetrized hyperspherical coordinates. The nuclear kinetic energy operator in these coordinates is given by

$$\hat{T}_{\text{nu}}(\mathbf{R}_\lambda) = -\frac{\hbar^2}{2\mu}\nabla^2_{\mathbf{R}_\lambda} = \hat{T}_\rho(\rho) + \frac{\hat{\Lambda}^2(\Upsilon_\lambda)}{2\mu\rho^2} \tag{82}$$

where, $\hat{T}_\rho(\rho)$ is the hyperradial kinetic energy operator

$$\hat{T}_\rho(\rho) = -\frac{\hbar^2}{2\mu}\frac{1}{\rho^5}\frac{\partial}{\partial\rho}\rho^5\frac{\partial}{\partial\rho} \tag{83}$$

and $\hat{\Lambda}^2(\Upsilon_\lambda)$ is the grand canonical angular momentum operator

$$\begin{aligned}\hat{\Lambda}^2(\Upsilon_\lambda) = {} & \hat{\Lambda}_o^2(\theta, \phi_\lambda) + \frac{4\hat{J}_z^{I\lambda^2}}{\cos^2\theta} \\ & + \frac{2}{1+\sin\theta}\left[\frac{\hat{J}^2 - \hat{J}_z^{I\lambda^2}}{2} + \frac{\hat{J}_+^{I\lambda^2} + \hat{J}_-^{I\lambda^2}}{4} - \hat{J}_z^{I\lambda^2}\right] \\ & + \frac{1}{\sin^2\theta}\left[\frac{\hat{J}^2 - \hat{J}_z^{I\lambda^2}}{2} - \frac{\hat{J}_+^{I\lambda^2} + \hat{J}_-^{I\lambda^2}}{4}\right] \\ & - 2\hbar\frac{\cos\theta}{\sin^2\theta}(\hat{J}_+^{I\lambda} - \hat{J}_-^{I\lambda})\frac{\partial}{\partial\phi_\lambda}\end{aligned} \tag{84}$$

where

$$\hat{\Lambda}_o^2(\theta, \phi_\lambda) = -4\hbar^2\left(\frac{1}{\sin 2\theta}\frac{\partial}{\partial\theta}\sin 2\theta\frac{\partial}{\partial\theta} + \frac{1}{\sin^2\theta}\frac{\partial^2}{\partial\phi_\lambda^2}\right) \tag{85}$$

and

$$\hat{J}_\pm^{I\lambda} = \hat{J}_x^{I\lambda} \pm i\hat{J}_y^{I\lambda} \tag{86}$$

$\hat{J}_x^{I\lambda}$, $\hat{J}_y^{I\lambda}$, and $\hat{J}_z^{I\lambda}$ are the components of the total orbital angular momentum $\hat{\mathbf{J}}$ of the nuclei in the $I\lambda$ frame. The Euler angles $a_\lambda, b_\lambda, c_\lambda$ appear only in the $\hat{J}^2$, $\hat{J}_z^{I\lambda}$, and $\hat{J}_\pm^{I\lambda}$ angular momentum operators. Since the results of their operation on Wigner rotation functions are known, we do not need their explicit expressions in terms of the partial derivatives of those Euler angles.

B. Partial Wave Expansion

In the two-adiabatic-electronic-state Born–Huang description of the total orbital wave function, we wish to solve the corresponding nuclear motion Schrödinger equation in the diabatic representation

$$\left[-\frac{\hbar^2}{2\mu}\{\mathbf{I}\nabla_{\mathbf{R}_\lambda}^2 + \mathbf{W}^{(2)d}(\mathbf{q}_\lambda)\} + \{\boldsymbol{\varepsilon}^d(\mathbf{q}_\lambda) - E\mathbf{I}\}\right]\boldsymbol{\chi}^d(\mathbf{R}_\lambda) = \mathbf{0} \tag{87}$$

for the diabatic orbital nuclear wave function column vector $\boldsymbol{\chi}^d(\mathbf{R}_\lambda)$

$$\boldsymbol{\chi}^d(\mathbf{R}_\lambda) = \begin{pmatrix}\chi_1^d(\mathbf{R}_\lambda) \\ \chi_2^d(\mathbf{R}_\lambda)\end{pmatrix} \tag{88}$$

In Eq. (87), the gradient term containing the transverse coupling has been dropped because its inclusion in this formalism leads to numerical inefficiencies in the very efficient logarithmic derivative propagator [87,88] used in solving Eq. (103). In the process of obtaining the ADT matrix, the magnitude of the transverse coupling vector is minimized over the entire internal nuclear configuration space following the procedure described in Section III. This makes dropping the gradient term a very good approximation. After the diabatization, since we know the transverse coupling vector, the effect of the gradient term on the scattering results obtained without it can be assessed using perturbative or other methods. In Eq. (87), $\boldsymbol{\varepsilon}^d(\mathbf{q}_\lambda)$ is a 2×2 diabatic energy matrix

$$\boldsymbol{\varepsilon}^d(\mathbf{q}_\lambda) = \begin{pmatrix} \varepsilon^d_{11}(\mathbf{q}_\lambda) & \varepsilon^d_{12}(\mathbf{q}_\lambda) \\ \varepsilon^d_{12}(\mathbf{q}_\lambda) & \varepsilon^d_{22}(\mathbf{q}_\lambda) \end{pmatrix} \tag{89}$$

and $\mathbf{W}^{(2)d}(\mathbf{q}_\lambda)$ is a 2×2 second-derivative diabatic coupling matrix

$$\mathbf{W}^{(2)d}(\mathbf{q}_\lambda) = \begin{pmatrix} \mathrm{w}^d_{1,1}(\mathbf{q}_\lambda) & \mathrm{w}^d_{1,2}(\mathbf{q}_\lambda) \\ \mathrm{w}^d_{2,1}(\mathbf{q}_\lambda) & \mathrm{w}^d_{2,2}(\mathbf{q}_\lambda) \end{pmatrix} \tag{90}$$

The $\mathbf{W}^{(2)d}$ and $\mathrm{w}^{(2)d}_{i,j}$ $(i, j = 1, 2)$ now depend on $\mathbf{q}_\lambda$ only rather than on the full $\mathbf{R}_\lambda$. The reason is as follows. The $\nabla^2_{\mathbf{R}_\lambda}$ appearing in the three body and two-electronic-state version of Eq. (36) contains terms in $\rho, \theta, \phi_\lambda$ as well as in the $\mathbf{a}^{I\lambda}$, the latter through the angular momentum operators $\hat{J}_x^{I\lambda^2}$, $\hat{J}_y^{I\lambda^2}$, $\hat{J}_z^{I\lambda^2}$, which are the squares of the components of the total angular momentum vector $\hat{\mathbf{J}}$ in the principal axes of inertia frame that also appeared in Eq. (61). Since, as discussed in Section II.A, $\psi_i^{\mathrm{el,ad}}$ and therefore $\psi_i^{\mathrm{el},d}$ $(i = 1, 2)$ depend only on $\mathbf{q}_\lambda$ (rather on the full $\mathbf{R}_\lambda$), the result of the application of those angular momentum operators on these diabatic electronic wave functions is zero. Therefore, the only contributions to $\nabla^2_{\mathbf{R}_\lambda}\psi_i^{\mathrm{el},d}(\mathbf{r}; \mathbf{q}_\lambda)$ come from the terms in $\nabla^2_{\mathbf{R}_\lambda}$ that contain $\mathbf{q}_\lambda$ only, which is different from the first-derivative $\boldsymbol{\nabla}_{\mathbf{R}_\lambda}$ coupling elements for which the $\tilde{\mathcal{R}}(\mathbf{a}^{I\lambda})$ factor in the right-hand side of Eq. (60) results in a dependence of $\mathbf{W}^{(1)d}$ on $\mathbf{a}^{I\lambda}$ when using Eq. (61).

Since the second-derivative coupling matrix $\mathbf{W}^{(2)d}$ is only an additive term in Eq. (87), we can merge it with the diabatic energy matrix and define a 2×2 diabatic matrix

$$\bar{\boldsymbol{\varepsilon}}^d(\mathbf{q}_\lambda) = \boldsymbol{\varepsilon}^d(\mathbf{q}_\lambda) - \frac{\hbar^2}{2\mu}\mathbf{W}^{(2)d}(\mathbf{q}_\lambda) \tag{91}$$

By using Eq. (91), we can rewrite Eq. (87) as

$$\left[-\frac{\hbar^2}{2\mu}\mathbf{I}\nabla^2_{\mathbf{R}_\lambda} + \{\bar{\boldsymbol{\varepsilon}}^d(\mathbf{q}_\lambda) - E\mathbf{I}\}\right]\boldsymbol{\chi}^d(\mathbf{R}_\lambda) = \mathbf{0} \tag{92}$$

The two diabatic nuclear wave functions χ_1^d and χ_2^d can be expressed as linear combinations of auxiliary nuclear wave functions $\chi_1^{d,JM\Pi\Gamma}$ and $\chi_2^{d,JM\Pi\Gamma}$, respectively (the linear combinations referred to as partial wave expansions and the individual $\chi_1^{d,JM\Pi\Gamma}$ and $\chi_2^{d,JM\Pi\Gamma}$ referred to as partial waves), such that if we define another nuclear wave function column vector

$$\boldsymbol{\chi}^{d,JM\Pi\Gamma}(\mathbf{R}_\lambda) = \begin{pmatrix} \chi_1^{d,JM\Pi\Gamma}(\mathbf{R}_\lambda) \\ \chi_2^{d,JM\Pi\Gamma}(\mathbf{R}_\lambda) \end{pmatrix} \tag{93}$$

then $\boldsymbol{\chi}^{d,JM\Pi\Gamma}$ is a simultaneous eigenfunction of the diabatic matrix $\hat{\mathbf{H}}^{\mathrm{nu}}$ (given by the expression inside the square brackets in the left-hand side of Eq. (87) with the $E\mathbf{I}$ term omitted), of the square of the total nuclear orbital angular momentum $\hat{\mathbf{J}}$, of its space-fixed z-component $\hat{\mathbf{J}}_z$ and of the inversion operator $\hat{I}$ of the nuclei through their center of mass G according to the expressions

$$\begin{aligned} \hat{\mathbf{H}}^{nu}\boldsymbol{\chi}^{d,JM\Pi\Gamma} &= E\boldsymbol{\chi}^{d,JM\Pi\Gamma} \\ \hat{J}^2\boldsymbol{\chi}^{d,JM\Pi\Gamma} &= J(J+1)\hbar^2\boldsymbol{\chi}^{d,JM\Pi\Gamma} \\ \hat{J}_z\boldsymbol{\chi}^{d,JM\Pi\Gamma} &= M\hbar\boldsymbol{\chi}^{d,JM\Pi\Gamma} \\ \hat{I}\boldsymbol{\chi}^{d,JM\Pi\Gamma} &= (-1)^{\Pi}\boldsymbol{\chi}^{d,JM\Pi\Gamma} \end{aligned} \tag{94}$$

In these equations, J and M are quantum numbers associated with the angular momentum operators $\hat{J}^2$ and $\hat{J}_z$, respectively. The number $\Pi = 0, 1$ is a parity quantum number that specifies the symmetry or antisymmetry of the $\boldsymbol{\chi}^{d,JM\Pi\Gamma}$ column vector with respect to the inversion of the nuclei through G. Note that the same parity quantum number Π appears for $\chi_1^{d,JM\Pi\Gamma}$ and $\chi_2^{d,JM\Pi\Gamma}$. Also, the same irreducible representation symbol Γ in these two components of $\boldsymbol{\chi}_d$ appears, which does not mean that these diabatic nuclear wave functions transform according to the irreducible representation Γ. Its meaning instead is as follows. The electronuclear Hamiltonian of the system is invariant under the group of permutations of identical $A_\lambda A_\nu A_\kappa$ atoms. For A_3 it is the P_3 group, for A_2B it is the P_2 group and for three distinct atoms ABC it is the trivial identity group. As a result, the $\Psi^{\mathrm{O}}(\mathbf{r}, \mathbf{R}_\lambda)$ that appears in Eq. (56) must transform according to an irreducible representation Γ of the corresponding permutation group. The superscript Γ signifies that the transformation properties of $\boldsymbol{\chi}^{d,JM\Pi\Gamma}$ are such that when taken together with the transformation properties of $\boldsymbol{\psi}^{\mathrm{el},d}(\mathbf{r}, \mathbf{q}_\lambda)$, they make $\Psi^{\mathrm{O}}(\mathbf{r}, \mathbf{R}_\lambda)$ belong to Γ. The separate factors $\chi_i^{d,JM\Pi\Gamma}$ and $\psi_i^{\mathrm{el},d}(\mathbf{r}, \mathbf{q}_\lambda)$ do not individually belong to Γ but that their product does. In addition, it is important to stress that these diabatic $\boldsymbol{\chi}^{d,JM\Pi\Gamma}$ are single valued, that is, are unchanged under a pseudorotation [26]. This behavior is the opposite to that of the adiabatic $\chi^{ad,JM\Pi\Gamma}$, which must change sign under such pseudorotations, due to the geometric-phase effect.

Let us now expand the two nuclear motion partial waves $\chi_1^{d,JM\Pi\Gamma}$ and $\chi_2^{d,JM\Pi\Gamma}$ according to the following vector equation:

$$\begin{pmatrix} \chi_1^{d,JM\Pi\Gamma,\mathbf{n}'_\lambda\Omega'_\lambda}(\rho,\Upsilon_\lambda) \\ \chi_2^{d,JM\Pi\Gamma,\mathbf{n}'_\lambda\Omega'_\lambda}(\rho,\Upsilon_\lambda) \end{pmatrix}$$
$$= \rho^{-5/2}\sum_{\Omega_\lambda} D_{M\Omega_\lambda}^{J\Pi}(\Upsilon_\lambda^{(1)}) \begin{pmatrix} \sum_{n_{1_\lambda}} b_{1,n_{1_\lambda},\Omega_\lambda}^{d,J\Pi\Gamma\mathbf{n}'_\lambda\Omega'_\lambda}(\rho;\bar{\rho})\Phi_{1,n_{1_\lambda},\Omega_\lambda}^{d,\Pi\Gamma}(\Upsilon_\lambda^{(2)};\bar{\rho}) \\ \sum_{n_{2_\lambda}} b_{2,n_{2_\lambda},\Omega_\lambda}^{d,J\Pi\Gamma\mathbf{n}'_\lambda\Omega'_\lambda}(\rho;\bar{\rho})\Phi_{2,n_{2_\lambda},\Omega_\lambda}^{d,\Pi\Gamma}(\Upsilon_\lambda^{(2)};\bar{\rho}) \end{pmatrix} \tag{95}$$

where, $\Upsilon_\lambda^{(1)}$ refers to the set of three Euler angles $\mathbf{a}^{I\lambda}$, $\Upsilon_\lambda^{(2)}$ refers to the set of two hyperangles θ, ϕ_λ and $\Omega_\lambda \geq 0$ is the absolute magnitude of the quantum number for the projection of the total angular momentum onto the body-fixed $Gz^{I\lambda}$ axis. Furthermore, the $D_{M\Omega_\lambda}^{J\Pi}(\Upsilon_\lambda^{(1)})$ are the parity-symmetrized Wigner rotation functions defined as [33]

$$D_{M\Omega_\lambda}^{J\Pi}(\Upsilon_\lambda^{(1)}) = \left\{ \frac{2J+1}{16\pi^2[1+(-1)^{J+\Pi}\delta_{\Omega_\lambda,0}]} \right\}^{1/2}$$
$$\times \left[D_{M\Omega_\lambda}^{J}(\Upsilon_\lambda^{(1)}) + (-1)^{J+\Pi+\Omega_\lambda} D_{M,-\Omega_\lambda}^{J}(\Upsilon_\lambda^{(1)}) \right] \tag{96}$$

where $D_{M\Omega_\lambda}^{J}(\Upsilon_\lambda^{(1)})$ is a Wigner rotation function of the Euler angles $\Upsilon_\lambda^{(1)}$ [89]. The symmetrized Wigner functions have been orthonormalized according to

$$\int D_{M'\Omega'_\lambda}^{J'\Pi'} D_{M\Omega_\lambda}^{J\Pi}\, d\tau = \delta_{J\Pi M\Omega_\lambda}^{J'\Pi'M'\Omega'_\lambda} \tag{97}$$

where $d\tau$ is the volume element for the Euler angles.

In Eq. (95), $\Phi_{1,n_{1_\lambda},\Omega_\lambda}^{d,\Pi\Gamma}(\Upsilon_\lambda^{(2)};\bar{\rho})$ and $\Phi_{2,n_{2_\lambda},\Omega_\lambda}^{d,\Pi\Gamma}(\Upsilon_\lambda^{(2)};\bar{\rho})$ are the diabatic 2D (in θ,ϕ_λ) local hyperspherical surface functions (LHSFs) that depend parametrically on $\bar{\rho}$ and are defined as the eigenfunctions of a diabatic reference hamiltonian $\hat{\mathbf{h}}_d^{\Omega_\lambda}$. This Hamiltonian can be chosen to be block diagonal, that is,

$$\hat{\mathbf{h}}_d^{\Omega_\lambda}(\theta,\phi_\lambda;\bar{\rho}) = \frac{1}{2\mu\bar{\rho}^2}\left[\hat{\Lambda}_o^2(\theta,\phi_\lambda) + \frac{4\Omega_\lambda^2\hbar^2}{\cos^2\theta}\right]\begin{pmatrix} 1 & 0 \\ 0 & 1 \end{pmatrix}$$
$$+ \begin{pmatrix} \bar{\varepsilon}_{11}^d(\theta,\phi_\lambda;\bar{\rho}) & 0 \\ 0 & \bar{\varepsilon}_{22}^d(\theta,\phi_\lambda;\bar{\rho}) \end{pmatrix} \tag{98}$$

or have the off-diagonal diabatic couplings built in, that is,

$$\hat{\mathbf{h}}_d^{\Omega_\lambda}(\theta, \phi_\lambda; \bar{\rho}) = \frac{1}{2\mu\bar{\rho}^2}\left[\hat{\Lambda}_o^2(\theta, \phi_\lambda) + \frac{4\Omega_\lambda^2\hbar^2}{\cos^2\theta}\right]\begin{pmatrix} 1 & 0 \\ 0 & 1 \end{pmatrix} + \begin{pmatrix} \bar{\varepsilon}_{11}^d(\theta, \phi_\lambda; \bar{\rho}) & \bar{\varepsilon}_{12}^d(\theta, \phi_\lambda; \bar{\rho}) \\ \bar{\varepsilon}_{21}^d(\theta, \phi_\lambda; \bar{\rho}) & \bar{\varepsilon}_{22}^d(\theta, \phi_\lambda; \bar{\rho}) \end{pmatrix} \tag{99}$$

In the former case, $\Phi_{1,n_{1_\lambda},\Omega_\lambda}^{d,\Pi\Gamma}(\Upsilon_\lambda^{(2)}; \bar{\rho})$ and $\Phi_{2,n_{2_\lambda},\Omega_\lambda}^{d,\Pi\Gamma}(\Upsilon_\lambda^{(2)}; \bar{\rho})$ are solutions of uncoupled second-order partial differential equations, whereas in the latter case they are solutions of coupled differential equations and therefore their calculation requires a larger computational effort than to obtain the former. Since, however, the reference Hamiltonian $\hat{\mathbf{h}}_d^{\Omega_\lambda}$ is independent of the total energy E of the system, the LHSFs need to be evaluated only once whereas the resulting scattering equations given by Eq. (101) must be solved for many values of E. As the off-diagonal diabatic couplings are built into Eq. (99), a smaller number of the corresponding LHSFs will be needed for convergence of the solutions of the scattering equations, as opposed to the ones resulting from Eq. (98), which do not have this off-diagonal coupling built in. Given the fact that the computational effort for solving those scattering equations scales with the cube of the number of LHSFs used, it is desirable to use LHSFs obtained from Eq. (99) rather than Eq. (98).

With either of these diabatic reference Hamiltonians, the LHSFs satisfy the eigenvalue equation

$$\hat{\mathbf{h}}_d^{\Omega_\lambda}(\theta, \phi_\lambda; \bar{\rho})\begin{pmatrix} \Phi_{1,n_{1_\lambda},\Omega_\lambda}^{d,\Pi\Gamma}(\theta, \phi_\lambda; \bar{\rho}) \\ \Phi_{2,n_{2_\lambda},\Omega_\lambda}^{d,\Pi\Gamma}(\theta, \phi_\lambda; \bar{\rho}) \end{pmatrix} = \begin{pmatrix} \epsilon_{1,n_{1_\lambda},\Omega_\lambda}^{d,\Pi\Gamma}(\bar{\rho})\Phi_{1,n_{1_\lambda},\Omega_\lambda}^{d,\Pi\Gamma}(\theta, \phi_\lambda; \bar{\rho}) \\ \epsilon_{2,n_{2_\lambda},\Omega_\lambda}^{d,\Pi\Gamma}(\bar{\rho})\Phi_{2,n_{2_\lambda},\Omega_\lambda}^{d,\Pi\Gamma}(\theta, \phi_\lambda; \bar{\rho}) \end{pmatrix} \tag{100}$$

The diabatic LHSFs are not allowed to diverge anywhere on the half-sphere of fixed radius $\bar{\rho}$. This boundary condition furnishes the quantum numbers n_{1_λ} and n_{2_λ}, each of which is 2D since the reference Hamiltonian $\hat{\mathbf{h}}_d^{\Omega_\lambda}$ has two angular degrees of freedom. The superscripts $\mathbf{n}'_\lambda, \Omega'_\lambda$ in Eq. (95), with $\mathbf{n}'_\lambda$ refering to the union of n'_{1_λ} and n'_{2_λ}, indicate that the number of linearly independent solutions of Eqs. (94) is equal to the number of diabatic LHSFs used in the expansions of Eq. (95).

In the strong interaction region, the diabatic eigenfunctions $\Phi_{i,n_{i_\lambda},\Omega_\lambda}^{d,\Pi\Gamma}(\theta, \phi_\lambda; \bar{\rho})$, $i = 1, 2$ are themselves expanded in a direct product of two orthonormal basis sets [90], $f_{n_{i\theta_\lambda}}^{\Omega_\lambda}(\theta; \bar{\rho})$ and $g_{n_{i\phi_\lambda}}^{\Pi\Gamma\Omega_\lambda}(\phi_\lambda; \bar{\rho})$, where $n_{i_\lambda} \equiv (n_{i\theta_\lambda}, n_{i\phi_\lambda})$. Both $f_{n_{i\theta_\lambda}}^{\Omega_\lambda}$ and $g_{n_{i\phi_\lambda}}^{\Pi\Gamma\Omega_\lambda}$ are chosen to be simple linear combinations of trignometric functions [33] such that the resulting diabatic nuclear wave functions transform under the operations of the permutation symmetry group of identical atoms as described after Eqs. (94). Equations (100) are then transformed into an algebraic eigenvalue eigenvector equation involving the coefficients of these expansions, which is solved numerically by linear algebra methods. In the weak interaction region, where the coordinates $\rho, \omega_\lambda, \gamma_\lambda$ of Eq. (77) are used, the diabatic LHSFs are eigenfunctions of the appropriate reference hamiltonian expressed in those coordinates [33,90] and are labelled $\Phi^{d,\Pi\Gamma}_{i,n_{i_\lambda},\Omega_\lambda}(\omega_\lambda, \gamma_\lambda; \bar{\rho})$, $i = 1, 2$. These new LHSFs are expanded in the direct product of the associated Legendre functions of $\cos\gamma_\lambda$ and at a set of functions of ω_λ determined by the numerical solution of a one-dimensional (1D) eigenfunction equation in ω_λ [33,90]. Once the diabatic LHSFs are known, they provide the basis of functions in terms of which the expansion in Eq. (95) is defined. The diabatic nuclear wave function vector of that equation is then inserted into the first equation of Eqs. (94). Use of the orthonormality of the symmetrized Wigner functions (Eq. (97)) and integration over the 2D diabatic LHSFs, yields a set of coupled hyperradial second-order ordinary differential equations (also called coupled-channel equations) in the coefficients $b^{d,J\Pi\Gamma\mathbf{n}'_\lambda\Omega'_\lambda}_{1,n_{1_\lambda},\Omega_\lambda}(\rho; \bar{\rho})$ and $b^{d,J\Pi\Gamma\mathbf{n}'_\lambda\Omega'_\lambda}_{2,n_{2_\lambda},\Omega_\lambda}(\rho; \bar{\rho})$. Let us define the column vectors $\mathbf{b}^{d,J\Pi\Gamma\mathbf{n}'_\lambda\Omega'_\lambda}_i(\rho; \bar{\rho})$ ($i = 1, 2$) as the vectors whose elements are scanned by $n_{i_\lambda}, \Omega_\lambda$ considered as a single row index.

Let us also define a matrix $\mathbf{B}^{d,J\Pi\Gamma}(\rho; \bar{\rho})$ whose $\mathbf{n}'_\lambda, \Omega'_\lambda$ column vector is obtained by stacking the vector $\mathbf{b}^{d,J\Pi\Gamma\mathbf{n}'_\lambda\Omega'_\lambda}_2(\rho; \bar{\rho})$ under the vector $\mathbf{b}^{d,J\Pi\Gamma\mathbf{n}'_\lambda\Omega'_\lambda}_1(\rho; \bar{\rho})$. These vectors, for different $\mathbf{n}'_\lambda, \Omega'_\lambda$, are then placed side-by-side thereby generating a square matrix $\mathbf{B}^{d,J\Pi\Gamma}$ whose dimensions are the total number of LHSFs (channels) used. The coupled hyperradial equation satisfied by this matrix has the form

$$\left[-\frac{\hbar^2}{2\mu}\mathbf{I}\frac{d^2}{d\rho^2} + \mathbf{V}^{d,J\Pi\Gamma}(\rho; \bar{\rho})\right]\mathbf{B}^{d,J\Pi\Gamma}(\rho; \bar{\rho}) = E\mathbf{B}^{d,J\Pi\Gamma}(\rho; \bar{\rho}) \tag{101}$$

where $\mathbf{V}^{d,J\Pi\Gamma}(\rho; \bar{\rho})$ is the interaction potential matrix obtained by this derivation procedure and that encompasses $\bar{\boldsymbol{\varepsilon}}^d(\bar{\rho})$:

$$\mathbf{V}^{d,J\Pi\Gamma}(\rho; \bar{\rho}) = \begin{pmatrix} \mathbf{V}^{d,J\Pi\Gamma}_{11}(\rho; \bar{\rho}) & \mathbf{V}^{d,J\Pi\Gamma}_{12}(\rho; \bar{\rho}) \\ \mathbf{V}^{d,J\Pi\Gamma}_{21}(\rho; \bar{\rho}) & \mathbf{V}^{d,J\Pi\Gamma}_{22}(\rho; \bar{\rho}) \end{pmatrix} \tag{102}$$

Its dimensions are those of $\mathbf{B}^{d,J\Pi\Gamma}(\rho; \bar{\rho})$.

C. Propagation Scheme and Asymptotic Analysis

The strong and weak interaction regions of the internal configuration space is divided into a certain number of spherical hyperradial shells. The 2D diabatic

LHSFs are determined at the center $\bar{\rho}$ of each shell. These LHSFs are then used to obtain the coupling matrix $\mathbf{V}^{d,J\Pi\Gamma}(\rho;\bar{\rho})$ given in Eq. (102). The coupled hyperradial equations in Eq. (101) are transformed into the coupled first-order nonlinear Bessel–Ricatti logarithmic matrix differential equation

$$\frac{d\mathbf{F}^{d,J\Pi\Gamma}(\rho;\bar{\rho})}{d\rho} + [\mathbf{F}^{d,J\Pi\Gamma}(\rho;\bar{\rho})]^2 + \frac{2\mu}{\hbar^2}[E\mathbf{I} - \mathbf{V}^{d,J\Pi\Gamma}(\rho;\bar{\rho})] = \mathbf{0} \tag{103}$$

where

$$\mathbf{F}^{d,J\Pi\Gamma}(\rho;\bar{\rho}) = [(d/d\rho)\mathbf{B}^{d,J\Pi\Gamma}(\rho;\bar{\rho})][\mathbf{B}^{d,J\Pi\Gamma}(\rho;\bar{\rho})]^{-1} \tag{104}$$

is the logarithmic derivative matrix and associated with $\mathbf{B}^{d,J\Pi\Gamma}$. Equation (103) is integrated from the beginning of each sector to its end using a highly efficient fourth-order logarithmic-derivative method [87,88] , and matched smoothly from one shell to another.

By using this method, the $\mathbf{F}^{d,J\Pi\Gamma}$ matrix is propagated from a very small value of $\rho = \rho_o$, where a WKB solution is applicable, through a value ρ_s that separates the strong and weak interaction regions, to an asymptotic value $\rho = \rho_a$ where the interactions between different arrangement channels λ has become negligible. At this asymptotic ρ_a, the diabatic $\chi^{d,JM\Pi\Gamma}$ is transformed to its adiabatic representation using the ADT matrix and matched to the asymptotic atom–diatom wave functions. This asymptotic analysis furnishes the reactance matrix $\mathsf{R}^{J\Pi\Gamma}$ and from it the scattering matrix $\mathsf{S}^{J\Pi\Gamma}$ [91,92]. For total energies E at which no electronically excited states of the isolated atoms or diatomic molecules are open, the elements of the open parts of these matrices correspond to the ground electronic atom and diatom products only. This is done for all Γ and both parities ($\Pi = 0, 1$) and for a sufficiently large number of values of J (i.e., of partial waves) for the resulting differential and integral cross-sections to be converged. This numerical procedure for the current two-electronic-state case is closely related to that for a single-electronic-state described in [33].

V. SUMMARY AND CONCLUSIONS

A general treatment of quantum reaction dynamics for multiple interacting electronic states is considered for a polyatomic system. In the adiabatic representation, the n-electronic-state nuclear motion Schrödinger equation is presented along with the structure of the first- and second-derivative nonadiabatic coupling matrices. In this representation, the geometric phase must be introduced separately and the presence of a gradient term introduces numerical inefficiencies for the solution of that Schrödinger equation, even if the nonadiabatic couplings do not display any singular behavior at the intersections of adjacent electronic states. This makes it desirable to go to a diabatic representation that incorporates automatically the geometric phase effect. In addition, appropriate boundary conditions can be chosen so as to impart desired properties on the diabatic version of the n-electronic-state nuclear motion Schrödinger equation. One such property is the minimization of the magnitude of that gradient term. If a complete (infinite) set of adiabatic electronic wave functions is used in a Born–Huang expansion of the system's electronuclear wave function (which is not possible in practice), this term vanishes automatically. In practice, a finite number n of adiabatic states are included for the treatment of chemical reactions. For this case, the gradient term survives in the diabatic representation as a nonremovable derivative coupling term, which, however, does not diverge at conical intersection geometries. A general method is presented that minimizes this nonremovable coupling term over the entire internal nuclear configuration space, leading to an optimal diabatization. As a very good first approximation, this gradient term can be neglected in the diabatic nuclear motion Schrödinger equation. Since that nonremovable coupling is obtained as a part of the diabatization process, its effect on the scattering cross-sections can be studied subsequently by perturbative or other methods.

A reactive scattering formalism for a triatomic reaction on two interacting electronic states is also presented. This formalism is an extension of the time-independent hyperspherical method [26,33] for one adiabatic electronic state. The extended formalism involves obtaining diabatic local hyperspherical surface functions (LHSFs) for each hyperradial shell. The partial wave diabatic nuclear wave functions are expanded in terms of these diabatic surface functions and the coefficients of the expansion propagated to an asymptotic value of the hyperradius, where the diabatic nuclear wave function is transformed to its adiabatic counterpart. An asymptotic analysis of the adiabatic nuclear wave function gives the partial wave scattering matrices needed to obtain the desired differential and integral cross-sections. A comparison of the cross-sections obtained using this two-electronic-state formalism with those obtained using only the adiabatic ground electronic state with the geometric phase included, should provide an estimation of the energy range for which the one-electronic-state BO approximation is valid.

Acknowledgment

This work was supported in part by NSF Grant CHE-98-10050.

References

1. M. Born and J. R. Oppenheimer, *Ann. Phys. (Leipzig)* **84**, 457 (1927).
2. M. Born, *Nachr. Akad. Wiss. Gött. Math.-Phys. Kl.* Article No. 6, 1 (1951).

3. M. Born and K. Huang, *Dynamical Theory of Crystal Lattices*, Oxford University Press, Oxford, UK, 1954, pp. 166–177 and 402–407.
4. C. Zener, *Proc. R. Soc. London Ser. A* **137**, 696 (1932).
5. L. D. Landau, *Phys. Z. Sowjetunion* **2**, 46 (1932).
6. E. C. G. Stuckelberg, *Helv. Phys. Acta.* **5**, 369 (1932).
7. N. Rosen and C. Zener, *Phys. Rev.* **40**, 502 (1932).
8. W. Lichten, *Phys. Rev.* **131**, 229 (1963).
9. E. E. Nikitin, in *Chemische Elementarprozesse*, H. Hartmann, ed., Springer-Verlag, Berlin, 1968, p. 43.
10. F. T. Smith, *Phys. Rev.* **179**, 111 (1969).
11. M. S. Child, *Mol. Phys.* **20**, 171 (1971).
12. R. de L Kronig, *Band Spectra and Molecular Structure*, Cambridge University Press, New York, 1930, p. 6.
13. D. R. Bates, *Proc. R. Soc. London Ser. A* **240**, 437 (1957).
14. D. R. Bates, *Proc. R. Soc. London Ser. A* **257**, 22 (1960).
15. W. R. Thorson, *J. Chem. Phys.* **34**, 1744 (1961).
16. D. J. Kouri and C. F. Curtiss, *J. Chem. Phys.* **44**, 2120 (1966).
17. R. T. Pack and J. O. Hirschfelder, *J. Chem. Phys.* **49**, 4009 (1968).
18. C. Gaussorgues, C. Le Sech, F. Mosnow-Seeuws, R. McCarroll, and A. Riera, *J. Phys. B* **8**, 239 (1975).
19. C. Gaussorgues, C. Le Sech, F. Mosnow-Seeuws, R. McCarroll, and A. Riera, *J. Phys. B* **8**, 253 (1975).
20. G. Herzberg and H. C. Longuet-Higgins, *Discuss. Faraday Soc.* **35**, 77 (1963).
21. H. C. Longuet-Higgins, *Proc. R. Soc. London, Ser. A* **344**, 147 (1975).
22. H. C. Longuet-Higgins, *Adv. Spectrosc.* **2**, 429 (1961).
23. C. A. Mead and D. G. Truhlar, *J. Chem. Phys.* **70**, 2284 (1979).
24. C. A. Mead, *Chem. Phys.* **49**, 23 (1980).
25. M. V. Berry, *Proc. R. Soc. London, Ser. A* **392**, 45 (1984).
26. A. Kuppermann, in *Dynamics of Molecules and Chemical Reactions*, R. E. Wyatt and J. Z. H. Zhang, eds., Marcel Dekker, New York, 1996, pp. 411–472.
27. Y. Aharonov and D. Bohm, *Phys. Rev.* **115**, 485 (1969).
28. M. Baer, ed., *Theory of Chemical Reaction Dynamics* Vols. I and II, CRC Press, Boca Raton, FL, 1985.
29. R. E. Wyatt and J. Z. H. Zhang, eds., *Dynamics of Molecules and Chemical Reactions*, Marcel Dekker, New York, 1996.
30. G. Nyman and H.-G. Yu, *Rep. Prog. Phys.* **63**, 1001 (2000).
31. J. Z. H. Zhang, J. Q. Dai, and W. Zhu, *J. Phys. Chem. A* **101**, 2746 (1997).
32. S. K. Pogrebnya, J. Palma, D. C. Clary, and J. Echave, *Phys. Chem. Chem. Phys.* **2**, 693 (2000).
33. Y.-S. M. Wu, A. Kuppermann, and B. Lepetit, *Chem. Phys. Lett.* **186**, 319 (1991).
34. Y.-S. M. Wu and A. Kuppermann, *Chem. Phys. Lett.* **201**, 178 (1993).
35. A. Kuppermann and Y.-S. M. Wu, *Chem. Phys. Lett.* **205**, 577 (1993).
36. Y.-S. M. Wu and A. Kuppermann, *Chem. Phys. Lett.* **235**, 105 (1995).
37. A. Kuppermann and Y.-S. M. Wu, *Chem. Phys. Lett.* **241**, 229 (1995).

38. H. Flöthmann, C. Beck, R. Schinke, C. Woywod, and W. Domcke, *J. Chem. Phys.* **107**, 7296 (1997).
39. D. Simah, B. Hartke, and H.-J. Werner, *J. Chem. Phys.* **111**, 4523 (1999).
40. T. W. J. Whiteley, A. J. Dobbyn, J. N. L. Connor, and G. C. Schatz, *Phys. Chem. Chem. Phys.* **2**, 549 (2000).
41. M. H. Alexander, D. E. Manolopoulos, and H.-J. Werner, *J. Chem. Phys.* **113**, 11084 (2000).
42. S. Mahapatra, H. Köppel, and L. S. Cederbaum, *J. Phys. Chem. A* **105**, 2321 (2001).
43. B. H. Lengsfield and D. R. Yarkony, in *State-selected and State to State Ion-Molecule Reaction Dynamics: Part 2 Theory*, M. Baer and C.-Y. Ng, eds., John Wiley & Sons, Inc., New York, 1992, Vol. 82, pp. 1–71.
44. T. Pacher, L. S. Cederbaum, and H. Köppel, *J. Chem. Phys.* **89**, 7367 (1988).
45. T. Pacher, C. A. Mead, L. S. Cederbaum, and H. Köppel, *J. Chem. Phys.* **91**, 7057 (1989).
46. V. Sidis, in *State-selected and State to State Ion-Molecule Reaction Dynamics: Part 2 Theory*, M. Baer and C.-Y. Ng, eds., John Wiley & Sons, Inc., New York, 1992, Vol. 82, pp. 73–134.
47. M. Baer and R. Englman, *Mol. Phys.* **75**, 293 (1992).
48. K. Ruedenberg and G. J. Atchity, *J. Chem. Phys.* **99**, 3799 (1993).
49. M. Baer and R. Englman, *Chem. Phys. Lett.* **265**, 105 (1997).
50. M. Baer, *J. Chem. Phys.* **107**, 2694 (1997).
51. B. K. Kendrick, C. A. Mead, and D. G. Truhlar, *J. Chem. Phys.* **110**, 7594 (1999).
52. M. Baer, R. Englman, and A. J. C. Varandas, *Mol. Phys.* **97**, 1185 (1999).
53. A. Thiel and H. Köppel, *J. Chem. Phys.* **110**, 9371 (1999).
54. D. R. Yarkony, *J. Chem. Phys.* **112**, 2111 (2000).
55. R. Abrol and A. Kuppermann, *J. Chem. Phys.* **116**, 1035 (2002).
56. D. A. V. Kliner, K. D. Rinen, and R. N. Zare, *Chem. Phys. Lett.* **166**, 107 (1990).
57. D. A. V. Kliner, D. E. Adelman, and R. N. Zare, *J. Chem. Phys.* **95**, 1648 (1991).
58. D. Neuhauser, R. S. Judson, D. J. Kouri, D. E. Adelman, N. E. Shafer, D. A. V. Kliner, and R. N. Zare, *Science* **257**, 519 (1992).
59. D. E. Adelman, N. E. Shafer, D. A. V. Kliner, and R. N. Zare, *J. Chem. Phys.* **97**, 7323 (1992).
60. L. M. Delves, *Nucl. Phys.* **9**, 391 (1959).
61. L. M. Delves, *Nucl. Phys.* **20**, 275 (1960).
62. M. Baer, *Chem. Phys. Lett.* **322**, 520 (2000).
63. R. J. Buenker, G. Hirsch, S. D. Peyerimhoff, P. J. Bruna, J. Römelt, M. Bettendorff, and C. Petrongolo, *Current Aspects of Quantum Chemistry*, Elsevier, New York, 1981, pp. 81–97.
64. M. Desouter-Lecomte, C. Galloy, J. C. Lorquet, and M. V. Pires, *J. Chem. Phys.* **71**, 3661 (1979).
65. B. H. Lengsfield, P. Saxe, and D. R. Yarkony, *J. Chem. Phys.* **81**, 4549 (1984).
66. P. Saxe, B. H. Lengsfield, and D. R. Yarkony, *Chem. Phys. Lett.* **113**, 159 (1985).
67. B. H. Lengsfield and D. R. Yarkony, *J. Chem. Phys.* **84**, 348 (1986).
68. J. O. Jensen and D. R. Yarkony, *J. Chem. Phys.* **89**, 3853 (1988).
69. B. Kendrick and R. T. Pack, *J. Chem. Phys.* **104**, 7475 (1996).
70. P. M. Morse and H. Feshbach, *Methods of Theoretical Physics*, McGraw-Hill, New York, 1953, pp. 52–54, 1763.
71. H. Margenau and G. M. Murphy, *The Mathematics of Physics and Chemistry*, Van Nostrand, New York, 1943, p. 192.

72. M. Baer, ed., *Theory of Chemical Reaction Dynamics*, CRC Press, Boca Raton, FL, 1985, Vol. II, Chap. 4.
73. A. Alijah and M. Baer, *J. Phys. Chem. A* **104**, 389 (2000).
74. M. Baer, *Chem. Phys. Lett.* **35**, 112 (1975).
75. M. Baer, *Chem. Phys.* **15**, 49 (1976).
76. C. A. Mead and D. G. Truhlar, *J. Chem. Phys.* **77**, 6090 (1982).
77. D. R. Yarkony, *J. Chem. Phys.* **105**, 10456 (1996).
78. D. R. Yarkony, *J. Phys. Chem. A* **101**, 4263 (1997).
79. N. Matsunaga and D. R. Yarkony, *J. Chem. Phys.* **107**, 7825 (1997).
80. N. Matsunaga and D. R. Yarkony, *Mol. Phys.* **93**, 79 (1998).
81. R. G. Sadygov and D. R. Yarkony, *J. Chem. Phys.* **109**, 20 (1998).
82. D. R. Yarkony, *J. Chem. Phys.* **110**, 701 (1999).
83. A. Kuppermann, unpublished results.
84. R. Abrol, A. Shaw, A. Kuppermann, and D. R. Yarkony, *J. Chem. Phys.* **115**, 4640 (2001).
85. A. Kuppermann, *Chem. Phys. Lett.* **32**, 374 (1975).
86. A. Kuppermann, in *Advances in Molecular Vibrations and Collision Dynamics*, J. Bowman, ed., JAI Press, Greenwich, CT, 1994, Vol. 2B, pp. 117–186.
87. B. R. Johnson, *J. Comput. Phys.* **13**, 445 (1973).
88. D. E. Manolopoulos, *J. Chem. Phys.* **85**, 6425 (1986).
89. A. S. Davydov, *Quantum Mechanics*, 2nd ed., Pergamon Press, Oxford, UK, 1976, p. 151.
90. J. M. Launay and M. Le Dourneuf, *Chem. Phys. Lett.* **163**, 178 (1989).
91. P. G. Hipes and A. Kuppermann, *Chem. Phys. Lett.* **133**, 1 (1987).
92. S. A. Cuccaro, P. G. Hipes, and A. Kuppermann, *Chem. Phys. Lett.* **154**, 155 (1989); (E) **157**, 440 (1989).

ELECTRON NUCLEAR DYNAMICS

YNGVE ÖHRN and ERIK DEUMENS

University of Florida, Quantum Theory Project, Departments of Chemistry and Physics, Gainesville, FL

CONTENTS

I. INTRODUCTION

Reactive atomic and molecular encounters at collision energies ranging from thermal to several kiloelectron volts (keV) are, at the fundamental level, described by the dynamics of the participating electrons and nuclei moving under the influence of their mutual interactions. Solutions of the time-dependent Schrödinger equation describe the details of such dynamics. The representation of such solutions provide the pictures that aid our understanding of atomic and molecular processes.

Traditionally, for molecular systems, one proceeds by considering the electronic Hamiltonian H_{el}, which consists of the quantum mechanical operators for the kinetic energy of the electrons, their mutual Coulombic repulsions, and

The Role of Degenerate States in Chemistry: A Special Volume of Advances in Chemical Physics, Volume 124, Edited by Michael Baer and Gert Due Billing. Series Editors I. Prigogine and Stuart A. Rice. ISBN 0-471-43817-0.

their attractions to each of the atomic nuclei. Commonly, the nuclear–nuclear repulsion energy is included in H_{el}. The time-dependent Schrödinger equation with this Hamiltonian describes the electron dynamics in a field of stationary nuclei. Methods of solving the time-independent electronic Schrödinger equation, commonly referred to as electronic structure theory, have reached considerable refinement and accuracy over the past decades. The bulk of such work consists of development of approximate many-electron theory and its implementation in terms of sophisticated computer software for solution of

$$H_{\text{el}}|n\rangle = E_n(R)|n\rangle \tag{1}$$

that is, finding approximate stationary state solutions $|n\rangle$ with characteristic electronic energies $E_n(R)$ for one fixed nuclear geometry at a time.

While experiments by their very nature are carried out in a laboratory coordinate frame, theory commonly proceeds via the introduction of internal coordinates in terms of molecule fixed axes. Done properly, this means that the kinetic energy of the center-of-mass motion is first separated from the other degrees of freedom. The origin of the internal coordinates, of course, can be chosen in a number of ways. The center of mass of the nuclei is a convenient choice that does not introduce kinetic energy coupling terms between electronic and nuclear degrees of freedom. No matter what is the choice of origin of the internal system of coordinates the result is a set of modified kinetic energy operators with reduced particle masses and so-called mass polarization terms. The latter, which are sums of products of momenta of different particles, are as a rule small and usually neglected.

For some systems consisting of two-to-four atoms of light elements it is currently feasible to consider enough points for, say, the ground-state electronic energy $E_0(R)$ such that appropriate interpolation techniques can produce the energy for all nuclear geometries below some suitable energy cutoff. The resulting function $E_0(R)$ is the Born–Oppenheimer (BO) potential energy surface (PES) of the system. Traditional molecular reaction dynamics proceeds by considering such a PES to be the potential energy for the nuclear dynamics, which, of course, may be treated classically, semiclassically, or by employing quantum mechanical methods. The other energy eigenvalues $E_n(R)$ similarly yield potential energy surfaces for electronically excited states. Each PES usually exhibits considerable structure for a polyatomic system and can provide instructive pictures with reactant and product valleys, local minima identifying stable species, and transition states providing gateways for the system to travel from one local minimum to another. Avoided crossings or more generally conical intersections and potential surface crossings are regions of dramatic chemical change in the system. The PES in this way provides attractive pictures of dynamical processes, which since the very beginning of molecular reaction dynamics have dominated our ways of thinking about molecular processes.

Quantum mechanical methods using high-quality potential energy surfaces have produced results in excellent agreement with the best experiments for small systems of the lightest elements at low energies. However, high quality potential energy surfaces exist only for a few systems. The difficulty of their determination increases rapidly with the number of atoms in the system. The determination of a PES in $3n - 6$ dimensions for an n-atom system is not only costly, but a PES in six or more dimensions is very hard to visualize and thus less useful. One way to proceed for larger systems is to identify active modes and to freeze or discretize other degrees of freedom. Such procedures tend to be subjective, and may introduce artificial features into the dynamics. In this traditional approach to dynamics, each process is studied at a fixed total angular momentum.

Many molecular beam experiments are performed at collision energies from a fraction of an electron volt to tens of electron volts. In such cases two or more stationary molecular electronic states and their potential energy surfaces can provide an adequate description provided also the effects of the nonadiabatic coupling terms are taken into account. Even in cases where a single PES is sufficient to describe the relevant forces on the participating nuclei one should augment the Born–Oppenheimer PES with the diagonal kinetic energy correction to produce the so-called adiabatic approximation, something that is only rarely done in practice.

Ion–atom and ion–molecule collisions at energies in the kiloelectron volts range are common in studies of energy deposition and stopping of swift particles in various materials. Theoretical treatments of such processes often employ stationary electronic states and their potential energy surfaces. At such elevated energies the relevant state vector of the system is an evolving state, which may be expressed as a superposition of a number of such energy states. In fact, the system moves on an effective PES, which is the dynamical average of a number of adiabatic surfaces and should in principle also include effects of the nonadiabatic coupling terms.

Obviously, the BO or the adiabatic states only serve as a basis, albeit a useful basis if they are determined accurately, for such evolving states, and one may ask whether another, less costly, basis could be just as useful. The electron nuclear dynamics (END) theory [1–4] treats the simultaneous dynamics of electrons and nuclei and may be characterized as a time-dependent, fully nonadiabatic approach to direct dynamics. The END equations that approximate the time-dependent Schrödinger equation are derived by employing the time-dependent variational principle (TDVP).

II. STRUCTURE AND DYNAMICS

The most accurate information about quantum systems is obtained via spectroscopic measurements. Such measurements have, until quite recently, been capable of only rather long-time averages of molecular events. Such studies emphasize structure and the associated electronic structure theory can successfully calculate molecular spectra and properties by applying the time-independent Schrödinger equation and focusing on stationary electronic states. The methods and techniques of electronic structure theory have a long history, and coupled with the development of ever more powerful computers this area of study has reached a very high degree of sophistication.

Development of laser technology over the last decade or so has permitted spectroscopy to probe short-time events. Instead of having to resort to the study of reactants and products and their energetics and structures, one is now able to follow reactants as they travel toward products. Fast pulsed lasers provide snapshots of entire molecular processes [5] demanding similar capabilities of the theory. Thus, explicitly time-dependent methods become suitable theoretical tools.

The dominant theoretical approaches to study molecular processes break the problem down into separate parts, the first being the determination of one or more potential energy surfaces. This involves electronic structure calculations for a large number of nuclear geometries and interpolation techniques [6–8] to provide as accurate as possible a functional form of each PES. Electronic structure methods and algorithms have been developed into efficient codes such as Gaussian [9], and ACES II [10], which can be used with minimal knowledge of electronic structure theory. These and many other codes have made computational chemistry a working tool for the bench chemist on an equal footing with various spectroscopic methods.

Given a ground-state PES the dynamics methods that have dominated the field since the beginning proceed by treating the nuclear motion with quantum mechanics, semiclassical or quasiclassical techniques, or with classical trajectory methods. For processes where more than one electronic stationary state is involved one needs preconstructed PESs for all states and also nonadiabatic coupling terms in order to study the dynamics. Several workers have contributed significantly to the developments of a variety of methods for molecular dynamics with active electronic degrees of freedom [11–17]. The electronic basis employed for the evolution of approximate solutions to the time-dependent Schrödinger equation may consist of the electronic energy eigenstates and the nonadiabatic effects can be accounted for by calculated and interpolated coupling terms or simulated by phenomenological surface hopping.

A major drawback to the approaches that use preconstructed PESs is that there are many more interesting systems undergoing reactive dynamical processes than there are available PESs. It would be much better if the electronic structure part of the problem, which provides the forces for the nuclear dynamics, could be performed simultaneously with the dynamics part, thus making possible the treatment of systems for which preconstructed PESs do not exist. Such approaches are referred to as direct dynamics methods. The popularity of the Car–Parinello method [18] is ample evidence for the need of such theoretical dynamics treatments. Car–Parinello uses density functional theory for the electronic degrees of freedom and may be considered to be a direct dynamics method for processes that strictly follow the electronic ground state.

Potential energy surfaces, although purely theoretical constructs, are extremely useful and attractive tools in providing illustrative pictures of molecular processes and are essential for understanding the energetics. When the dynamics takes place on a single surface one can picture the dynamics moving from a reactant valley to a product valley passing through a transition state region. When several PESs are involved one tends to picture the dynamics as following one surface or another with the nonadiabatic coupling terms providing the means for transitions from one surface to the other. Strictly speaking, nonadiabatic dynamics takes place between PESs and one can very well use a different basis from that of electron energy eigenstates in describing the evolving system. The challenge is then to develop simple and illustrative alternative pictures of the molecular process to that provided by PESs.

Electron nuclear dynamics theory is a direct nonadiabatic dynamics approach to molecular processes and uses an electronic basis of atomic orbitals attached to dynamical centers, whose positions and momenta are dynamical variables. Although computationally intensive, this approach is general and has a systematic hierarchy of approximations when applied in an ab initio fashion. It can also be applied with semiempirical treatment of electronic degrees of freedom [4]. It is important to recognize that the reactants in this approach are not forced to follow a certain reaction path but for a given set of initial conditions the entire system evolves in time in a completely dynamical manner dictated by the interparticle interactions.

III. TDVP AND END

The TDVP employs the quantum mechanical action

$$A = \int_{t_1}^{t_2} L(\psi^*, \psi) dt \tag{2}$$

where the Lagrangian is ($\hbar = 1$)

$$L = \langle \psi | i \frac{\partial}{\partial t} - H | \psi \rangle / \langle \psi | \psi \rangle \tag{3}$$

with H the total Hamiltonian of the system and requires the action to be stationary under variations of the wave function, that is,

$$\delta A = \int_{t_1}^{t_2} \delta L(\psi^*, \psi) dt = 0 \tag{4}$$

The total molecular system wave function is subject to the boundary conditions

$$\delta|\psi\rangle = \delta\langle\psi| = 0 \tag{5}$$

at the endpoints t_1 and t_2.

When the wave function is completely general and permitted to vary in the entire Hilbert space the TDVP yields the time-dependent Schrödinger equation. However, when the possible wave function variations are in some way constrained, such as is the case for a wave function restricted to a particular functional form and represented in a finite basis, then the corresponding action generates a set of equations that approximate the time-dependent Schrödinger equation.

The time dependence of the molecular wave function is carried by the wave function parameters, which assume the role of dynamical variables [19,20]. Therefore the choice of parameterization of the wave functions for electronic and nuclear degrees of freedom becomes important. Parameter sets that exhibit continuity and nonredundancy are sought and in this connection the theory of generalized coherent states has proven useful [21]. Typical parameters include molecular orbital coefficients, expansion coefficients of a multiconfigurational wave function, and average nuclear positions and momenta. We write

$$|\psi\rangle \equiv |\psi(z)\rangle \equiv |\mathbf{z}\rangle \tag{6}$$

where $\mathbf{z}$ denotes an array of suitable, and in general complex, wave function parameters.

By using Eq. (5), we can write the Lagrangian in a more symmetric form as

$$L = \left[\frac{i}{2}(\langle \mathbf{z}|\dot{\mathbf{z}}\rangle - \langle \dot{\mathbf{z}}|\mathbf{z}\rangle) - \langle \mathbf{z}|H|\mathbf{z}\rangle\right] \Big/ \langle \mathbf{z}|\mathbf{z}\rangle \tag{7}$$

where the dot denotes differentiation with respect to the time parameter t. Variation of the Lagrangian, δL, with respect to all the parameters introduces $|\delta\dot{\mathbf{z}}\rangle$ and $\langle\delta\dot{\mathbf{z}}|$, which can be eliminated by the introduction of the total time derivatives

$$\frac{d}{dt}\frac{\langle \mathbf{z}|\delta\mathbf{z}\rangle}{\langle \mathbf{z}|\mathbf{z}\rangle} \tag{8}$$

and

$$\frac{d}{dt}\frac{\langle \delta \mathbf{z}|\mathbf{z}\rangle}{\langle \mathbf{z}|\mathbf{z}\rangle} \tag{9}$$

and the boundary conditions Eq. (5). This results in a set of equations

$$\begin{aligned} 0 = \delta A &= \int_{t_1}^{t_2} \delta L\, dt \\ &= \int_{t_1}^{t_2} \left(\sum_{\alpha} \left[\sum_{\beta} i \frac{\partial^2 \ln S}{\partial z_{\alpha}^* \partial z_{\beta}} \dot{z}_{\beta} - \frac{\partial E}{\partial z_{\alpha}^*} \right] \delta z_{\alpha}^* \right. \\ &\quad \left. + \sum_{\alpha} \left[\sum_{\beta} -i \frac{\partial^2 \ln S}{\partial z_{\alpha} \partial z_{\beta}^*} \dot{z}_{\beta}^* - \frac{\partial E}{\partial z_{\alpha}} \right] \delta z_{\alpha} \right) dt \end{aligned} \tag{10}$$

where $S = S(\mathbf{z}^*, \mathbf{z}) = \langle \mathbf{z}|\mathbf{z}\rangle$ and $E = E(\mathbf{z}^*, \mathbf{z}) = \langle \mathbf{z}|H|\mathbf{z}\rangle / \langle \mathbf{z}|\mathbf{z}\rangle$, and where the chain rule has been applied to the time differentiation.

Since δz_{α} and δz_{α}^* are independent variations it must follow that

$$i \sum_{\beta} C_{\alpha\beta} \dot{z}_{\beta} = \frac{\partial E}{\partial z_{\alpha}^*} \tag{11}$$

$$-i \sum_{\beta} C_{\alpha\beta}^* \dot{z}_{\beta}^* = \frac{\partial E}{\partial z_{\alpha}} \tag{12}$$

where $C_{\alpha\beta} = \partial^2 \ln S / \partial z_{\alpha}^* \partial z_{\beta}$. These equations govern the time evolution of the wave function parameters. The time evolution of the overall phase factor exp $i\gamma$ is controlled by the equation

$$\dot{\gamma} = \frac{i}{2} \sum_{\alpha} \left[\dot{z}_{\alpha} \frac{\partial}{\partial z_{\alpha}} - \dot{z}_{\alpha}^* \frac{\partial}{\partial z_{\alpha}^*} \right] \ln S - E \tag{13}$$

Note that for a stationary state all parameters satisfy $\dot{z} = 0$ and thus $\gamma = -Et$, yielding the phase-factor exp $- iEt$ as expected.

In matrix block form the equations that govern the time evolution of the parameters can be expressed as

$$\begin{pmatrix} i\mathbf{C} & 0 \\ 0 & -i\mathbf{C}^* \end{pmatrix} \begin{pmatrix} \dot{\mathbf{z}} \\ \dot{\mathbf{z}}^* \end{pmatrix} = \begin{pmatrix} \partial E / \partial \mathbf{z}^* \\ \partial E / \partial \mathbf{z} \end{pmatrix} \tag{14}$$

If the wave function parameters are chosen appropriately, then the Hermitian matrix $\mathbf{C} = [C_{\alpha\beta}]$ has an inverse and we can write

$$\begin{pmatrix} \dot{\mathbf{z}} \\ \dot{\mathbf{z}}^* \end{pmatrix} = \begin{pmatrix} -i\mathbf{C}^{-1} & \mathbf{0} \\ \mathbf{0} & -i\mathbf{C}^{*-1} \end{pmatrix} \begin{pmatrix} \partial E/\partial \mathbf{z}^* \\ \partial E/\partial \mathbf{z} \end{pmatrix} \tag{15}$$

It is possible to introduce a generalized Poisson bracket by considering two general differentiable functions $f(\mathbf{z}, \mathbf{z}^*)$ and $g(\mathbf{z}, \mathbf{z}^*)$ and write

$$\begin{aligned} \{f, g\} &= [\partial f/\partial \mathbf{z}^{\mathbf{T}} \quad \partial f/\partial \mathbf{z}^{\dagger}] \begin{pmatrix} -i\mathbf{C}^{-1} & \mathbf{0} \\ \mathbf{0} & -i\mathbf{C}^{*-1} \end{pmatrix} \begin{bmatrix} \partial g/\partial \mathbf{z}^* \\ \partial g/\partial \mathbf{z} \end{bmatrix} \\ &= -i \sum_{\alpha,\beta} \left[\frac{\partial f}{\partial z_\alpha} (\mathbf{C}^{-1})_{\alpha\beta} \frac{\partial g}{\partial z^*_\beta} - \frac{\partial g}{\partial z_\alpha} (\mathbf{C}^{-1})_{\alpha\beta} \frac{\partial f}{\partial z^*_\beta} \right] \end{aligned} \tag{16}$$

It follows straightforwardly that

$$\dot{\mathbf{z}} = \{\mathbf{z}, E\} \tag{17}$$

$$\dot{\mathbf{z}}^* = \{\mathbf{z}^*, E\} \tag{18}$$

which shows that the time evolution of the wave function parameters is governed by Hamilton-like equations. The time evolution of the molecular system can then be viewed as occurring on a phase space made up of the complex wave function parameters $\mathbf{z}$ and $\mathbf{z}^*$ acting as conjugate positions and momenta and $E(\mathbf{z}, \mathbf{z}^*)$ being the Hamiltonian or the generator of infinitesimal time translations. This is obviously not a flat phase space. Such coupled sets of first-order differential equations can be integrated by a great variety of methods (see [22]).

A. The Basic END Ansatz

The END theory can be implemented at various levels of approximation. The simplest approximation develops a Lagrangian for classical nuclei or distinguishable atomic nuclei represented by traveling Gaussian wavepackets in the narrow width limit, and for quantum electrons represented by a single determinant built from nonorthogonal, complex spin orbitals [23]. The principle of least action using this Lagrangian yields the dynamical equations of minimal END.

At this level of approximation, the molecular wave function can be expressed as

$$|\Psi(t)\rangle = |R(t), P(t)\rangle |z(t), R(t), P(t)\rangle \tag{19}$$

where

$$\langle X|R(t), P(t)\rangle = \prod_k \exp\left[-\frac{1}{2}\left(\frac{\mathbf{X}_k - \mathbf{R}_k}{b}\right)^2 + i\mathbf{P}_k \cdot (\mathbf{X}_k - \mathbf{R}_k)\right] \tag{20}$$

and

$$\langle x|z(t), R(t), P(t)\rangle = \det \chi_i(\mathbf{x}_j) \tag{21}$$

with the spin orbitals

$$\chi_i = u_i + \sum_{j=N+1}^{K} u_j z_{ji}(t) \tag{22}$$

expanded in terms of atomic spin orbitals

$$\{u_i\}_1^K \tag{23}$$

which in turn are expanded in a basis of traveling Gaussians,

$$(x - R_x)^l (y - R_y)^m (z - R_z)^n \exp\left[-a(\mathbf{x} - \mathbf{R})^2 - \frac{i}{\hbar M}\mathbf{P} \cdot (\mathbf{x} - \mathbf{R})\right] \tag{24}$$

centered on the average nuclear positions $\mathbf{R}$ and moving with velocity $\mathbf{P}/M$.

In the narrow wavepacket limit, $b \to \infty$, the Lagrangian may be expressed as

$$L = \sum_{i,j}\left\{\left[P_{jl} + \frac{i}{2}\left(\frac{\partial \ln S}{\partial R_{jl}} - \frac{\partial \ln S}{\partial R'_{jl}}\right)\right]\dot{R}_{jl} + \frac{i}{2}\left(\frac{\partial \ln S}{\partial P_{jl}} - \frac{\partial \ln S}{\partial P'_{jl}}\right)\dot{P}_{jl}\right\}$$
$$+ \frac{i}{2}\sum_{p,h}\left(\frac{\partial \ln S}{\partial z_{ph}}\dot{z}_{ph} - \frac{\partial \ln S}{\partial z^*{}_{ph}}\dot{z}^*_{ph}\right) - E \tag{25}$$

with $S = \langle z, R', P'|z, R, P\rangle$ and

$$E = \sum_{jl} P_{jl}^2/2M_l + \langle z, R', P'|H_{\text{el}}|z, R, P\rangle / \langle z, R', P'|z, R, P\rangle \tag{26}$$

Here, H_{el} is the electronic Hamiltonian including the nuclear–nuclear repulsion terms, P_{jl} is a Cartesian component of the momentum, and M_l the mass of nucleus l. One should note that the bra depends on z^* while the ket depends on z and that the primed R and P equal their unprimed counterparts and the prime simply denotes that they belong to the bra.

The Euler–Lagrange equations

$$\frac{d}{dt}\frac{\partial L}{\partial \dot{q}} = \frac{\partial L}{\partial q} \tag{27}$$

can now be formed for the dynamical variables

$$q = R_{jl}, P_{jl}, z_{ph}, z^*_{ph} \tag{28}$$

and collected into a matrix equation

$$\begin{bmatrix} i\mathbf{C} & \mathbf{0} & i\mathbf{C}_\mathbf{R} & i\mathbf{C}_\mathbf{P} \\ \mathbf{0} & -i\mathbf{C}^* & -i\mathbf{C}^*_\mathbf{R} & -i\mathbf{C}^*_\mathbf{P} \\ i\mathbf{C}^\dagger_\mathbf{R} & -i\mathbf{C}^\mathbf{T}_\mathbf{R} & \mathbf{C}_\mathbf{RR} & -\mathbf{I} + \mathbf{C}_\mathbf{RP} \\ i\mathbf{C}^\dagger_\mathbf{P} & -i\mathbf{C}^\mathbf{T}_\mathbf{P} & \mathbf{I} + \mathbf{C}_\mathbf{PR} & \mathbf{C}_\mathbf{PP} \end{bmatrix} \begin{bmatrix} \dot{\mathbf{z}} \\ \dot{\mathbf{z}}^* \\ \dot{\mathbf{R}} \\ \dot{\mathbf{P}} \end{bmatrix} = \begin{bmatrix} \partial E/\partial \mathbf{z}^* \\ \partial E/\partial \mathbf{z} \\ \partial E/\partial \mathbf{R} \\ \partial E/\partial \mathbf{P} \end{bmatrix} \tag{29}$$

where the dynamical metric contains the elements

$$(C_{XY})_{ik;jl} = -2Im \left.\frac{\partial^2 \ln S}{\partial X_{ik} \partial Y_{jl}}\right|_{R'=R,\ P'=P} \tag{30}$$

$$(C_{X_{ik}})_{ph} = (C_X)_{ph;ik} = \left.\frac{\partial^2 \ln S}{\partial z^*_{ik} \partial X_{ik}}\right|_{R'=R,\ P'=P} \tag{31}$$

which are the nonadiabatic coupling terms, and

$$C_{ph;qg} = \left.\frac{\partial^2 \ln S}{\partial z^*_{ph} \partial z_{qg}}\right|_{R'=R,\ P'=P} \tag{32}$$

In this minimal END approximation, the electronic basis functions are centered on the average nuclear positions, which are dynamical variables. In the limit of classical nuclei, these are conventional basis functions used in molecular electronic structure theory, and they follow the dynamically changing nuclear positions. As can be seen from the equations of motion discussed above the evolution of the nuclear positions and momenta is governed by Newton-like equations with Hellman–Feynman forces, while the electronic dynamical variables are complex molecular orbital coefficients that follow equations that look like those of the time-dependent Hartree–Fock (TDHF) approximation [24]. The coupling terms in the dynamical metric are the well-known nonadiabatic terms due to the fact that the basis moves with the dynamically changing nuclear positions.

The time evolution of molecular processes in the END formalism employs a Cartesian laboratory frame of coordinates. This means that in addition to the internal dynamics overall translation and rotation of the molecular system are treated. The six extra degrees of freedom add work, but become less of a burden as the complexity of the system grows. The advantage is that the kinetic energy terms are simple and no mass polarization terms need to be discarded. Furthermore, the complications of having to choose different internal coordinates for product channels exhibiting different fragmentations are not present. One can treat all product channels on an equal footing in the same laboratory frame. Since the fundamental invariance laws with respect to overall translation and rotation are satisfied within END [4] it is straightforward to extract the internal dynamics at any time in the evolution.

Better END approximations are defined by the introduction of more general molecular wave functions leading to larger and more involved parameter spaces.

B. Free Electrons

In this context, it is interesting to explore the possibilities of the END theory to describe molecular processes that involve free electrons either as reagents or as products. Electron-molecule scattering or ionization processes in molecular collisions are commonly treated separately from general molecular reaction dynamics. The principal idea in extending END to include free electron capabilities is to center electronic basis functions on independent positions in space. This means that such basis centers, so-called *free centers*, move on their own and are not associated with nuclear positions, however, the positions and conjugate momenta of these free centers are dynamical variables, which evolve according to the appropriate Euler–Lagrange equations.

The electronic basis for the free centers is similar to that in Eq. (24) and more precisely can be written as

$$(x-\rho_x)^l(y-\rho_y)^m(z-\rho_z)^n\exp\left[-c(\mathbf{x}-\rho)^2-\frac{i}{\hbar}\pi\cdot(\mathbf{x}-\rho)\right] \tag{33}$$

with $\mathbf{x}=(x,y,z)$ an electron coordinate, ρ the center coordinate, and π the average electronic momentum. We can add such electronic orbitals to the minimal END wave function considered in Section III.A. The electronic basis centered on the atomic nuclei are standard basis functions u_i suitable for the particular element, while on the free centers the basis is the union $w_i = u_i \cup v_i$, where v_i is a set of diffuse functions. In order to create an initial state for an ionizing atomic or molecular collision, one performs an self-consistent field (SCF) calculation in the bound state basis u_i to obtain the orbitals $\phi_i = \sum_k u_k c_{ki}$. The initial state component on a free center is then constructed using the projector $|w\rangle\langle w|w\rangle^{-1}\langle w|$ to obtain

$$\bar{\phi}_k = \sum_{l,k} w_l(\mathbf{\Delta}^{-1})_{lm}\langle w_m|\phi_k\rangle \tag{34}$$

In an electron scattering or recombination process, the free center of the incoming electron has the functions $w_i = u_i \cup v_i$ and the initial state of the free electron is some function v_i the width of which is chosen on the basis of the electron momentum and the time it takes the electron to arrive at the target. Such choice is important in order to avoid nonphysical behavior due to the natural spreading of the wavepacket.

In a completely general and flexible application of END one may choose to include some number, say N_{ion}, of nuclei described as in Eq. (20) completely void of electronic basis functions, and some number (N_A) of nuclei with electronic basis functions, as well as some number (N_F) of free centers.

C. General Electron Nuclear Dynamics

When constructing more general molecular wave functions there are several concepts that need to be defined. The concept of *geometry* is introduced to mean a (time-dependent) point in the generalized phase space for the total number of centers used to describe the END wave function. The notations R and P are used for the position and conjugate momenta vectors, such that

$$R = (\mathbf{R}_k, k = 1, \ldots, N_A + N_F + N_{\text{ion}}) \tag{35}$$

These notations are used for positions and momenta, when the nuclei are treated as classical particles and denote average positions and momenta when they are treated quantum mechanically.

Another concept is that of *electronic structure*, which is defined as an electronic wave function associated with a *geometry*. For the case that the electrons are described by a single determinantal wave function it would be meaningful to consider multiple different electronic structures associated with the same geometry. In general, it would also be meaningful to consider multiple geometries, each evolving with its own electronic structure. The reason for this particular definition of electronic structure is that it would not be meaningful to consider multiple geometries with a single electronic structure, since the BO approximation provides a very good description. In Table I, we list the three possible combinations of geometry and electronic structure.

The wave function for the electronic structure can in principle be any of the constructions employed in electronic structure theory. The preferred choice in this context is a wave functions that can be classified as single and multi-configurational, and for the latter type only complete active space (CAS) wave

TABLE I
The Three Meaningful Combinations of Electronic Structure[a]

	SG	MG
SES	$\Psi(X, x, R, P)$	Not meaningful
MES	$\sum_{\mu} c_{\mu} \Psi_{\mu}(X, x, R, P)$	$\sum_{\mu,\gamma} c_{\mu,\gamma} \Psi_{\mu}(X, x, R_{\gamma}, P_{\gamma})$

[a] single electronic structure (SES), multiple electronic structure (MES), single geometry (SG), and multiple geometry (MG).

The symbols X and x denote the quantum mechanical coordinates of the nuclei and electrons, respectively. The index μ runs over electronic structures and γ over geometries.

functions are really useful. The reason for this is that such constructions have a well-established coherent (or vector coherent) state description, so that the parameters define a well-behaved phase space for a dynamical Hamiltonian system. Because in the END formulation of molecular dynamics the wave function parameters are the dynamical variables it is essential that they are nonredundant and continuous.

The Thouless determinantal electronic wave function $|\mathbf{z}\rangle = \det \chi_i(x_j)$ in Eq. (21) is an example of such proper parametrization. The dynamical spin orbitals are expressed in terms of atomic spin orbitals centered on the various nuclei

$$\chi_i = u_i + \sum_j u_j z_{ji} \tag{36}$$

with time-dependent complex coefficients z_{ji} being the dynamical variables. This parameterization guarantees that all possible determinantal wave functions in terms of the atomic orbitals are accessible during imposed dynamical changes of the system. Numerical stability is ensured as long as the z coefficients are small in comparison to unity. This can be assured by the capability to switch from one local parameterization or chart that during the dynamics may have led to large parameter values and therefore numerically unstable equations, to another chart more suitable for that part of the dynamics. Such change of charts must be possible without introduction of any artificial discontinuities in trajectories and various calculated properties.

We consider the example of a particular trajectory of the $H^+ + H_2(0,0) \rightarrow H_2(v,j) + H$ at an energy of 1.2 eV in the center-of-mass frame. By using an atomic orbital basis and a representation of the electronic state of the system in terms of a Thouless determinant and the protons as classical particles, the leading term of the electronic state of the reactants is

$$|(1s_1 + 1s_2)\alpha(1s_1 + 1s_2)\beta 1s_3\alpha| \tag{37}$$

where 1 and 2 label the protons of the reactant molecule and 3 that of the projectile atom, and $1s_i$ is an atomic orbital centered on proton i. Let the reactive trajectory proceed by exchange of protons 2 and 3 making the leading term of the product electronic state

$$|(1s_1 + 1s_3)\alpha(1s_1 + 1s_3)\beta 1s_2\alpha| \tag{38}$$

The original chart or Thouless parameterization

$$\begin{aligned} &1s_1\alpha + 1s_2\alpha z_{12}^{\alpha} \\ &1s_1\beta + 1s_2\beta z_{12}^{\beta} + 1s_3\beta z_{13}^{\beta} \\ &1s_3\alpha + 1s_2\alpha z_{22}^{\alpha} \end{aligned} \tag{39}$$

represents the state in Eq. (37) with

$$\begin{aligned} z_{12}^{\alpha} &= 1 \\ z_{22}^{\alpha} &= 0 \\ z_{12}^{\beta} &= 1 \\ z_{13}^{\beta} &= 0 \end{aligned} \tag{40}$$

but cannot properly represent the state in Eq. (38), that is,

$$\begin{aligned} z_{12}^{\alpha} &= \text{undefined} \\ z_{22}^{\alpha} &= \infty \\ z_{12}^{\beta} &= 0 \\ z_{13}^{\beta} &= 1 \end{aligned} \tag{41}$$

Numerically, it shows up in z_{12}^{α} and z_{22}^{α} coefficients becoming very large in comparison to unity making the integration of the dynamical equations less accurate. The ENDyne code then automatically switches to a new chart with the coefficients more suitable to the product side, that is,

$$\begin{aligned} &1s_1\alpha + 1s_3\alpha z_{13}^{\alpha} \\ &1s_1\beta + 1s_2\beta z_{12}^{\beta} + 1s_3\beta z_{13}^{\beta} \\ &1s_2\alpha + 1s_3\alpha z_{23}^{\alpha} \end{aligned} \tag{42}$$

which represents the state Eq. (38) when

$$
\begin{aligned}
z_{13}^{\alpha} &= 1 \\
z_{23}^{\alpha} &= 0 \\
z_{12}^{\beta} &= 0 \\
z_{13}^{\beta} &= 1
\end{aligned}
\tag{43}
$$

Although the leading term of the electronic wave function of the system is thus changed, the total wave function has not and the calculated trajectory and properties exhibit no discontinuous behavior.

Some details of END using a multiconfigurational electronic wave function with a complete active space (CASMC) have been introduced in terms of an orthonormal basis and for a fixed nuclear framework [25], and were recently [26] discussed in some detail for a nonorthogonal basis with electron translation factors.

The full dynamical treatment of electrons and nuclei together in a laboratory system of coordinates is computationally intensive and difficult. However, the availability of multiprocessor computers and detailed attention to the development of efficient software, such as ENDyne, which can be maintained and debugged continually when new features are added, make END a viable alternative among methods for the study of molecular processes. Furthermore, when the application of END is compared to the *total* effort of accurate determination of relevant potential energy surfaces and nonadiabatic coupling terms, faithful analytical fitting and interpolation of the common pointwise representation of surfaces and coupling terms, and the solution of the coupled dynamical equations in a suitable internal coordinates, the computational effort of END is competitive.

IV. MOLECULAR PROCESSES

The END equations are integrated to yield the time evolution of the wave function parameters for reactive processes from an initial state of the system. The solution is propagated until such a time that the system has clearly reached the final products. Then, the evolved state vector may be projected against a number of different possible final product states to yield corresponding transition probability amplitudes. Details of the END dynamics can be depicted and cross-section cross-sections and rate coefficients calculated.

The approximations defining minimal END, that is, direct nonadiabatic dynamics with classical nuclei and quantum electrons described by a single complex determinantal wave function constructed from nonorthogonal spin orbitals with electron translation factors centered on the dynamically changing nuclear positions, yield results for hyperthermal atomic and molecular reactive collisions that are usually in agreement with available experimental data. It is interesting to ask to what extent this level of treatment applies to low energy processes. The experience gained from several applications is that some quantities that are not too sensitive to the detailed dynamics, such as integral cross-sections, can be described quite well, while other properties, notably differential cross-sections, are not. This is understandable from the fact that at thermal energies the dynamics follows closely the ground-state potential energy surface, which for minimal END is the ground-state SCF surface.

In order to make END better suited to the application of low energy events it is important to include an explicitly correlated description of the electron dynamics. Therefore multiconfigurational [25] augmentations of the minimal END are under development.

However, for molecular events involving more than one electronic state, even when they take place at low energies, minimal END direct dynamics appear to do well. Electron transfer is an example of such processes. Ion–atom collisions have been studied at a great variety of energies [27–29], ranging from a few tens of an electron volt to hundreds of kiloelectron volts, usually achieving agreement with available experimental data. Minimal END for $H_2^+ + H_2$ at 0.5–4.0 eV [30] yields integral cross-sections for formation of H_3^+ and for electron transfer in good agreement with experiment.

A. Reactive Collisions

Bimolecular reactive encounters, atom–molecule, ion–molecule, and ion–atom collisions at a great variety of energies and initial states can be studied with the END theory. If we use classical nuclei this means that in addition to the initial electronic state of the system the nuclear geometries or internal states of the participating molecular species must be chosen. Several END trajectories have to be calculated, which means that for, say, gas-phase processes a sufficient number of relative orientations of the reactants must be considered so that directional averages can be obtained. Also, a range of impact parameters must be employed ranging from zero for head on collisions to such values that produce nonreactive trajectories. This simply corresponds to studying the processes for a range of total angular momenta.

The general problem of molecular reactive scattering can be studied with the machinery of formal time-dependent (or time-independent) scattering theory. However, for the implementation of END theory with classical nuclei it is useful to remind ourselves of some of the concepts of classical potential scattering. The consideration of the scattering of two structureless particles interacting via a potential energy $U(R)$ can suffice for reminding the reader of some of the features of classical scattering. The collision energy is $E = \mu v^2/2$ with μ the reduced mass and v the relative speed. The angular momentum of the system is $J = \mu v b$ with b the impact parameter. The scattering angle θ in the laboratory frame is the absolute value of the deflection function $\Theta(b)$ as [31]

$$\theta = |\Theta(b)| = \left| \pi - 2b \int_{R_0}^{\infty} R^{-2} [1 - U(R)/E - b^2/R^2]^{-1/2} dR \right| \tag{44}$$

The classical scattering cross section for a given process is simply

$$\sigma(E) = 2\pi \int_0^{b_{\max}} P(E, b) b \, db \tag{45}$$

where $P(E, b)$ is the so-called opacity function, which can be directly obtained from the evolved END wave function and the appropriate final state in the same basis, giving the fraction of collisions leading to the considered reaction products for a given collision energy and impact parameter. The corresponding classical differential cross-section is

$$d\sigma(E, \theta) = P(E, b) \frac{b}{\sin\theta |d\theta/db|} \tag{46}$$

or when more than one impact parameter b_i produces the same scattering angle

$$d\sigma(E, \theta) = \sum_i P(E, b_i) \frac{b_i}{\sin\theta |d\theta/db_i|} \tag{47}$$

The well-known glory scattering or forward peak scattering for small θ and rainbow scattering at angles for which $d\theta/db = 0$ causes singularities in the classical differential cross-sections for which semiclassical corrections [32–34] usually work well. The particular considerations of semiclassical corrections in END theory have been thoroughly treated by Morales et al. [35]. A particularly elegant and useful semiclassical treatment of the scattering amplitude for small angle scattering at higher energies has been developed by Schiff [36]. He sums the infinite Born series for the scattering amplitude by approximating each term in the sum by the stationary phase method. This approach has been applied to minimal END [27] with great success for ion–atom, atom–atom, and ion–molecule collisions in the kiloelectron volt range. The scattering amplitude in the small angle Schiff (semiclassical) approximation is

$$f(\theta) = ik \int_0^{\infty} \{1 - \exp[-i\delta(b)]\} J_0(qb) b \, db \tag{48}$$

with J_0 a Bessel function of order zero, and where $q = |\mathbf{k}_i - \mathbf{k}_f|$ is the momentum transferred during the collision, θ is the angle between the initial wave vector of the projectile $\mathbf{k}_i$ and final wave vector in the direction of the detector $\mathbf{k}_f$. The semiclassical phase shift $\delta(b)$ is related to the deflection function through (see [31])

$$\Theta(b) = \frac{2}{k_i} \frac{d\delta(b)}{db} \tag{49}$$

The END trajectories for the simultaneous dynamics of classical nuclei and quantum electrons will yield deflection functions. For collision processes with nonspherical targets and projectiles, one obtains one deflection function per orientation, which in turn yields the semiclassical phase shift and thus the scattering amplitude and the semiclassical differential cross-section

$$\frac{d\sigma}{d\Omega} = \frac{k_f}{k_i} |f(\theta)|^2 \tag{50}$$

For a particular process, this expression should be multiplied with the probability for that process as determined by projection of the END evolved state $\psi(t)$ for the system on the appropriate final state ψ_f described within the same basis set and at the same level of approximation as the evolved state, that is, the amplitude $\langle \psi_f | \psi(t) \rangle$ at a sufficiently large time t.

It is interesting to note the similarity of the expression in Eq. (48) with the result obtained through a WKB or eikonal type of argument [37,38]. The eikonal approximation resorts to straight-line trajectories, while the END application of the Schiff approximation uses fully dynamical trajectories. Schiff [36] demonstrates that the scattering wave function obtained through his procedure of summing the Born series contains an additional term, which is essential for the correct treatment of the scattering and is not present in the eikonal or WKB approaches to the problem. This formula of the scattering amplitude [Eq. (48)] is also considered to be in principle valid for all scattering angles (see [38], p. 604).

Many experimental techniques now provide details of dynamical events on short timescales. Time-dependent theory, such as END, offer the capabilities to obtain information about the details of the transition from initial-to-final states in reactive processes. The assumptions of time-dependent perturbation theory coupled with Fermi's Golden Rule, namely, that there are well-defined (unperturbed) initial and final states and that these are occupied for times, which are long compared to the transition time, no longer necessarily apply. Therefore, truly dynamical methods become very appealing and the results from such theoretical methods can be shown as movies or time lapse photography.

We have found that display of nuclear trajectories and the simultaneous evolution of charge distributions to yield insightful details of complicated processes. Such descriptions also map more readily to the actual experimental conditions than do the more conventional time-independent scattering matrix descriptions.

As an illustration of how results from time-dependent treatments of reactive molecular collisions can be represented, we present some recent results [61] on the $D_2 + NH_3^+$ reaction at energies from 6 to 16 eV in the center-of-mass frame. Recent molecular beam experiments have been carried out on this system in the group of Zare [39–41] at energies from 1 to 10 eV in the center of mass. These studies seek to gain insight into the mechanisms of the reaction by considering several different initial conditions with varying amounts of energy in translational and vibrational degrees of freedom of the reactants. At these energies the two main mechanisms are the abstraction

$$NH_3^+ + D_2 \rightarrow NH_3D^+ + D \tag{51}$$

and the competing exchange reaction

$$NH_3^+ + D_2 \rightarrow NH_2D^+ + HD \tag{52}$$

In applying minimal END to processes such as these, one finds that different initial conditions lead to different product channels. In Figure 1, we show a somewhat truncated time lapse picture of a typical trajectory that leads to abstraction. In this rendering, one of the hydrogens of NH_3D^+ is hidden. As an example of properties whose evolution can be depicted we display interatomic distances and atomic electronic charges. Obviously, one can similarly study the time dependence of various other properties during the reactive encounter.

At low energies the abstraction process dominates and at higher energies the exchange mechanism becomes more important. The cross-sections for the two processes crossing at $\sim$10 eV. The END calculations yield absolute cross-sections that show the same trend as the experimentally determined relative cross-sections for the two processes. The theory predicts that a substantial fraction of the abstraction product NH_3D^+, which are excited above the dissociation threshold for an N—H bond actually dissociates to $NH_2D^+ + H$ or NH_3^+ during the almost 50-μs travel from the collision chamber to the detector, and thus affects the measured relative cross-sections of the two processes.

One can note some interesting features from these trajectories. For example, the Mulliken population on the participating atoms in Figure 1 show that the departing deuterium carries a full electron. Also, the deuterium transferred to the NH_3^+ undergoes an initial substantial bond stretch with the up spin and down spin populations separating so that the system temporarily looks like a biradical before it settles into a normal closed-shell behavior.

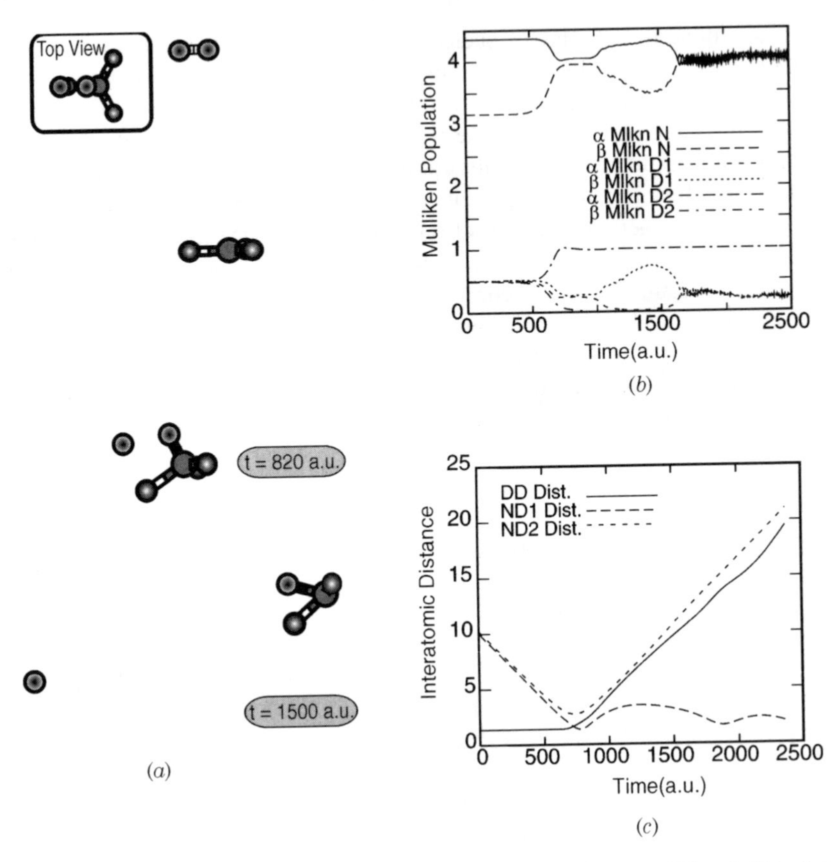

Figure 1. The trajectory of ground-state D_2 colliding with ground-state NH_3^+ at 8 eV leads to abstraction with the NH_3D^+ ion highly vibrationally excited. The time evolution of the interatomic distances (*c*) and of the atomic charges (*b*) show which product species are generated.

B. Final-State Analysis

The determinantal wave function in Eq. (21) is built [23] from complex dynamical spin orbitals χ_i. Even when the basis orbitals u_k in Eq. (22) are orthogonal these dynamical orbitals are nonorthogonal, and for a basis of nonorthogonal atomic orbitals based on Gaussians as those in Eq. (24) the metric of the basis becomes involved in all formulas and the END theory as implemented in the ENDyne code works directly in the atomic basis without invoking transformations to system orbitals.

The product analysis of the END system wave function is quite general, but for simplicity we consider the case of two product fragments, A and B. As these fragments separate the corresponding dynamical spin orbitals may be expressed as

$$\chi_i = \chi_i^{\mathrm{A}} + \chi_i^{\mathrm{B}} \tag{53}$$

with negligible overlap of the atomic orbitals centered on the nuclei of fragment A with those on fragment B. At any given point in time t after which the separation of products has taken place, a particular molecular product fragment has a particular nuclear geometry and its electronic wave function can be projected on an electronic eigenstate of that geometry determined in the same electronic basis set to obtain probabilities for state-to-state events. Specifically, a molecular orbital basis is obtained for each fragment by performing an SCF calculation at the geometry for a given final time t. Then Slater determinants are formed with these local fragment orbitals for the entire system exhibiting various degrees of intrafragment excitations. These Slater determinants are orthogonal and can be sorted according to charge and spin state depending on the number and spin of system electrons assigned to each fragment. Projection of the END evolved determinant against each of these determinants then yields the desired probabilities.

In many ion–atom and ion–molecule collisions, one is often only interested in the projections on various charge states, which can be given a very simple treatment. The Thouless determinant at separation of the two product fragments can be expressed as

$$\langle x|z(t), R(t), P(t)\rangle = \{|\chi_1^{\mathrm{A}}\chi_2^{\mathrm{A}}\cdots\chi_N^{\mathrm{A}}|\} + \{|\chi_1^{\mathrm{B}}\chi_2^{\mathrm{A}}\cdots\chi_N^{\mathrm{A}}|\} + \cdots \{|\chi_1^{\mathrm{B}}\chi_2^{\mathrm{B}}\cdots\chi_N^{\mathrm{B}}|\} \tag{54}$$

where each curly bracket contains all $\binom{N}{M}$ determinants with $M = 0, 1, \ldots, N$ fragment B orbitals, respectively. A canonical orthonormalization of the atomic orbitals of each fragment, that is,

$$\phi_i = \sum_j u_j(\mathbf{U}\mathbf{s}^{-1/2})_{ji} \tag{55}$$

with the atomic orbital metric $\boldsymbol{\Delta}$ being diagonalized by a unitary transformation $\mathbf{U}$, such that

$$\mathbf{s} = \mathbf{U}^\dagger \boldsymbol{\Delta} \mathbf{U} \tag{56}$$

makes it possible to write

$$\chi_i^{\mathrm{C}} = \sum_l u_l c_{li} = \sum_{k,l} \phi_k^{\mathrm{C}} (\mathbf{s}^{1/2}\mathbf{U}^{\mathrm{C}\dagger})_{kl} c_{li} = \sum_k \phi_k^{\mathrm{C}} d_{ki} \tag{57}$$

For each of the fragment determinants in Eq. (54), the following expansion or its analogues applies:

$$|\chi_1^A \chi_2^B \cdots \chi_N^B| = \sum_{i_1, i_2, \ldots, i_N} d_{i_1 1}^A d_{i_2 2}^B \cdots d_{i_N N}^B |\phi_{i_1}^A \phi_{i_2}^B \cdots \phi_{i_N}^B| \tag{58}$$

in terms of orthonormal determinants. The relevant transition probability to a particular charge state can then be obtained by squaring the coefficients $d_{i_1 1}^A d_{i_2 2}^B \cdots d_{i_N N}^B$, add them up, and divide by the total normalization of the Thouless determinant.

Rovibrational final-state analysis can also be achieved even for the case of classical nuclei. A product fragment with classical nuclei rotates and vibrates as a classical object. A classical quantum correspondence is adopted, such that this classical object is described by an evolving coherent state. For the case of a diatomic fragment when rotational excitations can be neglected or decoupled, the dynamics can be resolved into quantum states [42]. For low excitations with near equidistant splittings between consecutive vibrational energy levels the harmonic oscillator coherent state provides an excellent basis for obtaining vibrationally resolved cross-sections [43]. As a general approach valid for polyatomic molecular product fragment a multidimensional Prony [44] method has been developed [45], which can produce rovibrationally resolved cross-sections for the case of weak coupling between rotation and vibrational modes.

The mass weighted position of a single nucleus ν in the center-of-mass frame of a molecule with N atomic nuclei at time points t is obtained from an END trajectory and can be expressed as

$$\mathbf{R}_\nu[t] = O[t-1]\left[\mathbf{E}_\nu + \sum_{j=1}^{p} \mathbf{T}_{\nu,j} c_j e^{(2\pi i \Omega_j (t-1)\Delta t + \varphi_j)}\right] \tag{59}$$

where the interval (time step) between data points is Δt, p is the number of vibrational modes of the molecule, $O[t]$ is a rotation matrix ($O[0] = 1$), $\mathbf{E}_\nu$ is the equilibrium position of nucleus ν. The direction and magnitude of the displacement of nucleus ν in the jth normal mode is $\mathbf{T}_{\nu,j}$, the weight of this normal mode is c_j, and its frequency and phase is Ω_j and φ_j, respectively.

The generalized Prony analysis can extract a great variety of information from the ENDyne dynamics, such as the vibrational energy E_{vib} and the frequency for each normal mode. The classical quantum connection is then made via coherent states, such that, say, each normal vibrational mode is represented by an evolving state

$$|\alpha\rangle = \exp\left(-\frac{1}{2}|\alpha|^2\right) \sum_n \frac{\alpha^n}{\sqrt{n!}} |n\rangle \tag{60}$$

in terms of the harmonic oscillator eigenstates $|n\rangle$, and where α is a time-dependent complex parameter. Since the energy of such an evolving state above the ground state is $E_{\text{vib}} = \hbar\omega|\alpha|^2$ we find $|\alpha|^2 = E_{\text{vib}}/\hbar\omega$ and we can conclude that the probability of the fragment occupying a particular eigenstate $|n\rangle$ is

$$P_n = \frac{(E_{\text{vib}}/\hbar\omega)^n}{n!} \exp(-E_{\text{vib}}/\hbar\omega) \tag{61}$$

By using this approach, it is possible to calculate vibrational state-selected cross-sections from minimal END trajectories obtained with a classical description of the nuclei. We have studied vibrationally excited $H_2(\nu)$ molecules produced in collisions with 30-eV protons [42,43]. The relevant experiments were performed by Toennies et al. [46] with comparisons to theoretical studies using the trajectory surface hopping model [11,47] (TSHM). This system has also stimulated a quantum mechanical study [48] using diatomics-in-molecule (DIM) surfaces [49] and invoking the infinite-order sudden approximation (IOSA).

In Figure 2, we show the total differential cross-section for product molecules in the vibrational ground state (no charge transfer) of the hydrogen molecule in collision with 30-eV protons in the laboratory frame. The experimental results that are in arbitrary units have been normalized to the END

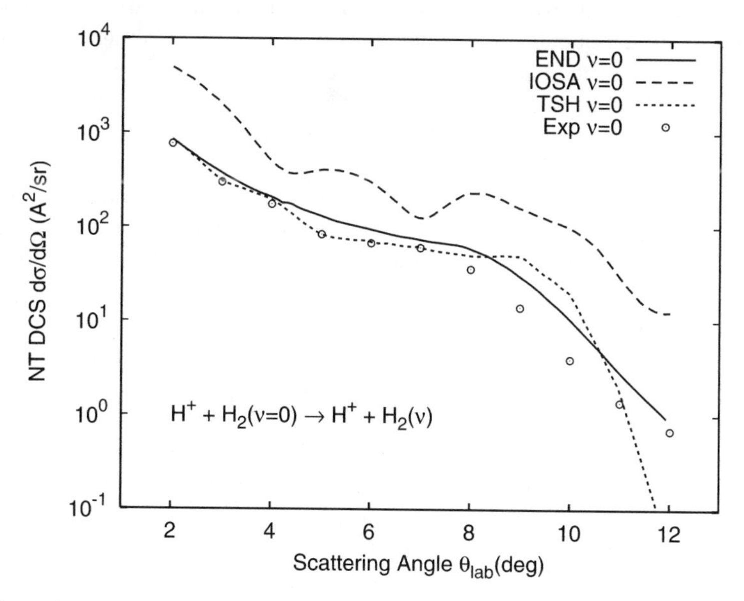

Figure 2. Total differential cross-section versus laboratory scattering angle for vibrational ground state of hydrogen molecules in single collisioins with 30-eV protons.

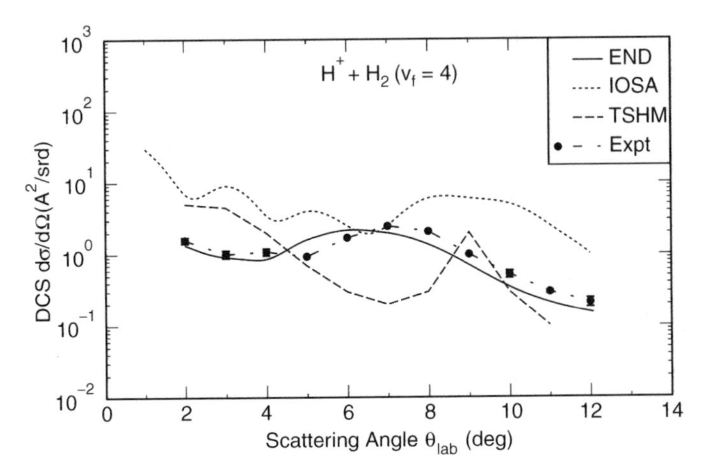

Figure 3. State resolved differential cross-section versus laboratory scattering angle for vibrational excitation of hydrogen molecules into state $\nu = 4$ in single collisions with 30-eV protons.

results at the rainbow angle. The experimental estimate of the rainbow angle and the END calculated one are very close, $\sim 7^\circ$. The theoretical results using THSM and IOSA are shown for comparison. In Figure 3, the vibrational state resolved differential cross-section is shown for the fourth excited state ($\nu = 4$). Results of similar quality are obtained for products in other vibrational states as long as the use of the harmonic oscillator coherent state can be justified.

State resolved differential cross-sections for H_2O in collisions with 46-eV protons in the center of mass were deduced [50] from time-of-flight energy loss spectra. The vibrational states are labeled $[\nu_1, \nu_2, \nu_3]$, where ν_1 denotes the number of quanta in the symmetric stretch mode, and ν_2 and ν_3, similarly denote the bending and asymmetric stretch, respectively. The experimental analysis assumes that the progressions $[\nu_1, 0, 0]$ and $[\nu_1, 1, 0]$ are the principal final states of the water molecules. This assumption is corroborated by our calculations. We show in Figure 4 the total differential cross-section for vibrational excitation (NT) and the state-resolved differential cross-sections for $[0, 0, 0]$, $[0, 1, 0]$, and $[1, 0, 0]$. The experimental energy resolution is such that it is not possible to distinguish between the symmetric and asymmetric stretching modes, so only one stretching mode is considered and denoted by ν_1.

The generalized Prony analysis of END trajectories for this system yield total and state resolved differential cross-sections. In Figure 5, we show the results. The theoretical analysis, which has no problem distinguishing between the symmetric and asymmetric stretch, shows that the asymmetric mode is only excited to a minor extent. The corresponding state resolved cross-section is about two orders of magnitude less than that of the symmetric stretch.

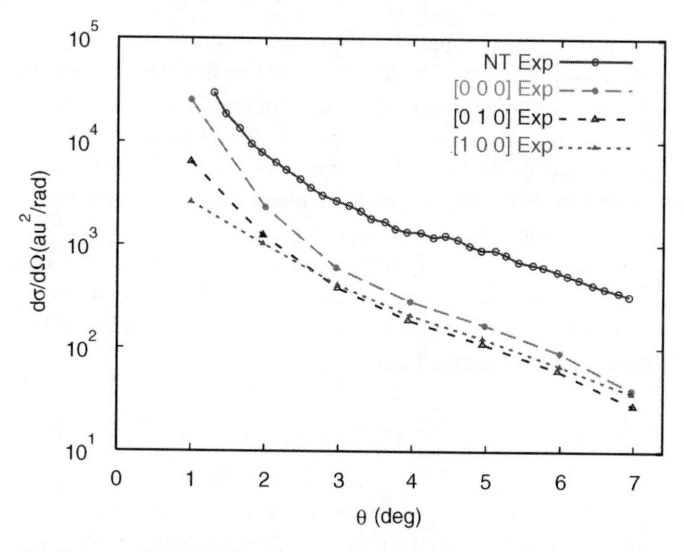

Figure 4. The experimental [50] total and three-state resolved differential cross-sections of vibrational excitations of the water molecule in collisions with 46-eV protons.

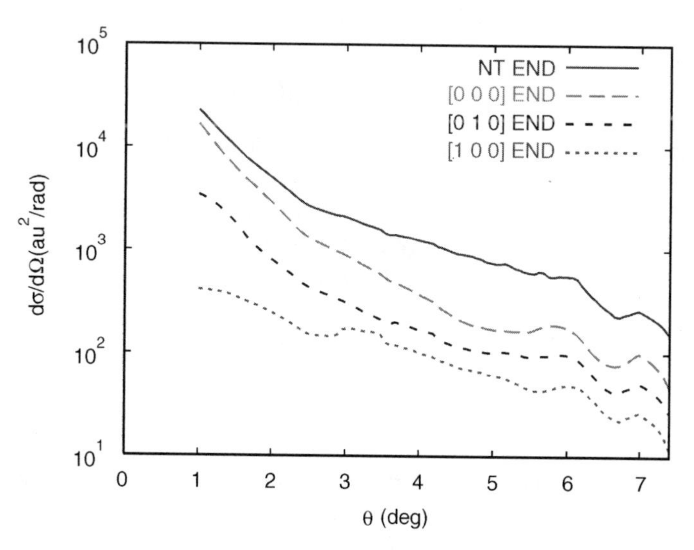

Figure 5. The theoretical total and state resolved differential cross-sections of vibrational excitations of water molecules in collisions with 46-eV protons.

One reason that the symmetric stretch is favored over the asymmetric one might be the overall process, which is electron transfer. This means that most of the END trajectories show a nonvanishing probability for electron transfer and as a result the dominant forces try to open the bond angle during the collision toward a linear structure of H_2O^+. In this way, the totally symmetric bending mode is dynamically promoted, which couples to the symmetric stretch, but not to the asymmetric one.

Also, rotational state resolution of cross-sections can be obtained by employing a coherent state analysis [51] for the situation of weak coupling between rotational and vibrational degrees of freedom. A suitable rotational coherent state can be expressed as

$$|\alpha, \beta, \gamma, \delta, \epsilon, \zeta\rangle = e^{-\zeta/2} \sum_{IMK} D^I_{MI}(\alpha, \beta, 0) D^I_{KI}(-\gamma, \delta, \epsilon) \frac{\zeta^{1/2}}{\sqrt{I!}} |IMK\rangle \tag{62}$$

where $D^I_{MK}(\alpha, \beta, \gamma) = e^{-i\alpha M} d^I_{MK} e^{-i\gamma K}$ are rotational matrices [52], and where the angle parameters are related to the average values of the body fixed L and space-fixed J angular momentum components calculated with this coherent state, that is,

$$\begin{aligned} \langle L_x \rangle &= \zeta \cos\gamma \sin\delta & \langle J_x \rangle &= \zeta \cos\alpha \sin\beta \\ \langle L_y \rangle &= \zeta \sin\gamma \sin\delta & \langle J_y \rangle &= \zeta \sin\alpha \sin\beta \\ \langle L_z \rangle &= \zeta \cos\delta & \langle J_z \rangle &= \zeta \cos\beta \end{aligned} \tag{63}$$

and

$$\langle L^2 \rangle = \langle J^2 \rangle = \zeta(\zeta + 2) \tag{64}$$

From these relations it follows that ζ is related to the angular momentum modulus, and that the pairs of angle α, β and γ, δ are the azimuthal, and the polar angle of the $\langle \mathbf{J}^2 \rangle$ and the $\langle \mathbf{L}^2 \rangle$ vector, respectively. The angle ϵ is associated with the relative orientation of the body-fixed and space-fixed coordinate frames. The probability to find the particular rotational state $|IMK\rangle$ in the coherent state is

$$P_{IMK}(\zeta, \beta, \delta) = [d^I_{MI}(\beta)]^2 [d^I_{KI}]^2 \frac{\zeta^I}{I!} e^{-\zeta} \tag{65}$$

The use of the rotational coherent state is then analogous to the use of the vibrational coherent state and can be used to study rotational state resolved properties. We note that the resolution of the identity applies here as well, that is,

$$\sum_{I=0}^{\infty} \sum_{M=-I}^{I} \sum_{K=-I}^{I} P_{IMK}(\zeta, \beta, \delta) = 1 \tag{66}$$

Final state analysis is where dynamical methods of evolving states meet the concepts of stationary states. By their definition, final states are relatively long lived. Therefore experiment often selects a single stationary state or a statistical mixture of stationary states. Since END evolution includes the possibility of electronic excitations, we analyze reaction products in terms of rovibronic states.

C. Intramolecular Electron Transfer

Minimal END has also been applied to a model system for intramolecular electron transfer. The small triatomic system LiHLi is bent C_{2v} structure. But the linear structure presents an unrestricted Hartree–Fock (UHF) broken symmetry solution with the two charge localized structures

$$\text{Li–H–Li}^{+} \rightleftharpoons \text{Li}^{+}\text{–H–Li} \tag{67}$$

These charge-transfer structures have been studied [4] in terms a very limited number of END trajectories to model vibrational induced electron transfer. An electronic 3-21G+ basis for Li [53] and 3-21G for H [54] was used. The equilibrium structure has the geometry with a long Li(2)—H bond (3.45561 a.u.) and a short Li(1)—H bond (3.09017 a.u.). It was first established that only the Li—H bond stretching modes will promote electron transfer, and then initial conditions were chosen such that the long bond was stretched and the short bond compressed by the same (%) amount. The small ensemble of six trajectories with 5.6, 10, 13, 15, 18, and 20% initial change in equilibrium bond lengths are sufficient to illustrate the approach.

The END approach to electron-transfer processes is quite different from the current paradigm of Marcus theory, which due to its conceptual simplicity has guided much theoretical and experimental development. Introduced in the late 1950s [55], this theory has been extensively reviewed, revised, and extended [56–59]. This approach is characterized by the assumptions that there is a reaction coordinate that the reactants travel to the products and that there is a coupling H_{12} between the donor and acceptor states. Figure 6 shows a typical picture of participating adiabatic and diabatic states along a reaction coordinate for normal electron transfer according to the Marcus theory. END by its very nature constructs dynamical trajectories in wave function phase space, including the electronic degrees of freedom, from which transition probabilities are obtained. In this approach, there is no need to break the transfer process into two separate steps, that is geometry change and electronic transition. Instead END describes the full evolution and the coupling of these two aspects of the process. Initiation of electron transfer is accomplished by simply distorting the molecule and letting the system evolve in time.

A simple measure of the electron density distribution over the participating atoms is the Mulliken population [60]. For linear Li–H–Li the alpha spin is

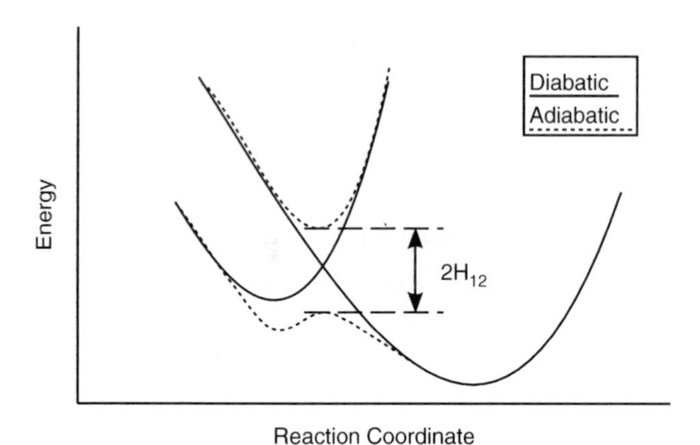

Figure 6. Diabatic and corresponding adiabatic potential energy along a relevant reaction coordinate for normal electron transfer.

arbitrarily chosen in excess in the single determinantal electronic state. In Figure 7, the alpha Mulliken populations are shown for the six END trajectories over 10,000 a.u. of time.

A transfer rate constant can be obtained by applying a Boltzmann distribution, and by writing the concentration of reactant present as

$$X(t) = \sum_n e^{E_n/kT} P_n(t) \tag{68}$$

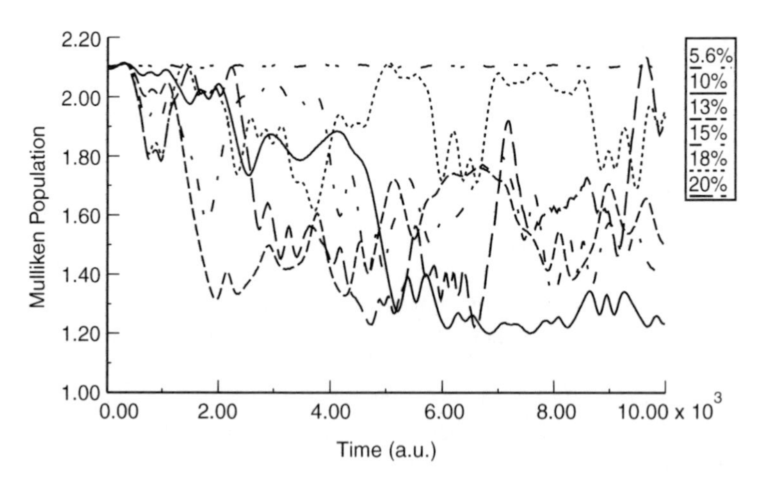

Figure 7. Alpha Mulliken population on Li(2) as functions of time for different initial conditions.

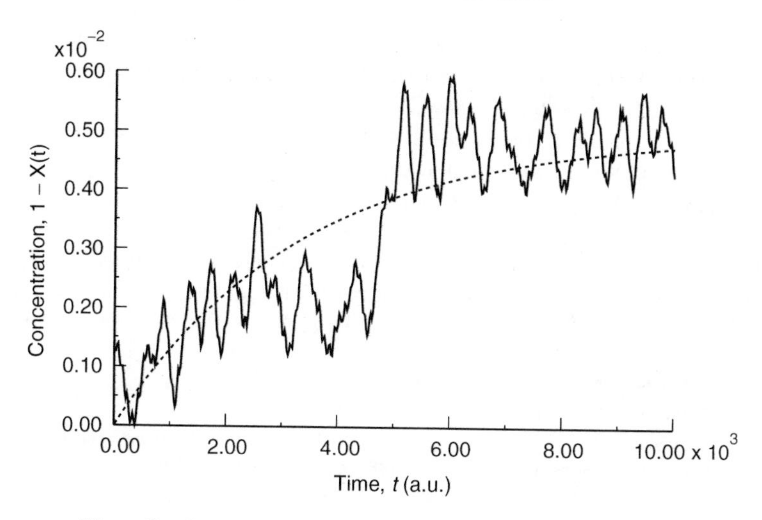

Figure 8. Concentration of the product species as a function of time.

where E_n is the energy above the ground state and P_n the probability of electron transfer for initial state n. The electron transfer in this case is effectively a one-electron process and since for such a case the transfer probability is directly related to the Mulliken population one may write

$$P_n(t) = 2 - 2M_n(t)/M_n^{\max} \tag{69}$$

where M_n is the alpha Mulliken population on Li(2) for initial state n, and $M_n^{\max}$ is the maximum value of this population. In this case, $M_N^{\max} = 2$ and P_n becomes a number between 0 and 1 yielding the probability that an electron will move from Li(2) to Li(1).

The small statistical sample leaves strong fluctuations on the timescale of the nuclear vibrations, which is a behavior typical of any detailed microscopic dynamics used as data for a statistical treatment to obtain macroscopic quantities.

However, as can be seen from Figure 8 a simple exponential expected from first-order kinetics can be fitted to the data yielding a limiting concentration of 0.005, and a rate constant of 0.0003 a.u., which translates to $1.25 \times 10^{13}\ \mathrm{s}^{-1}$ at 300 K.

References

1. Y. Öhrn et al., in *Time-Dependent Quantum Molecular Dynamics*, J. Broeckhove and L. Lathouwers, eds., Plenum, New York, 1992, pp. 279–292.
2. E. Deumens, A. Diz, H. Taylor, and Y. Öhrn, *J. Chem. Phys.* **96**, 6820 (1992).

3. R. Longo, E. Deumens, and Y. Öhrn, *J. Chem. Phys.* **99**, 4554 (1993).
4. E. Deumens, A. Diz, R. Longo, and Y. Öhrn, *Rev. Mod. Phys.* **66**, 917 (1994).
5. A. H. Zewail, *Science* **242**, 1645 (1989).
6. A. Gonzalez-Lafont, T. N. Truong, and D. S. Truhlar, *J. Chem. Phys.* **95**, 8875 (1991).
7. G. S. Wu, G. C. Schatz, and G. Lendvay, *J. Chem. Phys.* **113**, 7712 (2000).
8. T. Hollebeek, T.-S. Ho, and H. Rabitz, *J. Chem. Phys.* **114**, 3940 (2001).
9. M. J. Frisch et al., *Gaussian 92*, Gaussian Inc., Carnegie-Mellon University, Pittsburg, PA, 1992.
10. J. F. Stanton et al., *ACES II*, Quantum Theory Project, University of Florida, Gainesville, FL 32611-8435, 2001, integral packages included are VMOL, J. Almlöf and P. Taylor; VPROPS (P. R. Taylor); A modified version of ABACUS (T. U. Helgaker, H. J. Aa. Jensen, J. Olsen, and P. Jørgensen.
11. J. C. Tully and R. K. Preston, *J. Chem. Phys.* **55**, 562 (1971).
12. H.-D. Meyer and W. H. Miller, *J. Chem. Phys.* **70**, 3214 (1979).
13. R. D. Coalson, in *Laser, Molecules and Methods*, J. Hirshfelder, R. Wyatt, and R. Coalson, eds., John Wiley & Sons, Inc., New York, 1989, Chap. 13, pp. 605–636.
14. U. Manthe, H. Köppel, and L. S. Cederbaum, *J. Chem. Phys.* **95**, 1708 (1991).
15. J. Theilhaber, *Phys. Rev. B* **46**, 12990 (1992).
16. M. Ben-Nun and T. J. Martinez, *J. Chem. Phys.* **108**, 7244 (1998).
17. G. D. Billing, *Chem. Phys. Lett.* **342**, 65 (2001).
18. R. Car and M. Parrinello, *Phys. Rev. Lett.* **55**, 2471 (1985).
19. P. Kramer and M. Saraceno, *Geometry of the Time-Dependent Variational Principle in Quantum Mechanics*, Springer, New York, 1981.
20. E. Deumens and Y. Öhrn, *J. Chem. Soc., Faraday Trans.* **93**, 919 (1997).
21. J. R. Klauder and B.-S. Skagerstam, *Coherent States, Applications in Physics and Mathematical Physics*, World Scientific, Singapore, 1985.
22. W. H. Press, B. P. Flannery, S. A. Teutolsky, and W. T. Vetterling, *Numerical Recipes: The Art of Scientific Computing*, Cambridge University, New York, 1986.
23. D. J. Thouless, *Nucl. Phys.* **21**, 225 (1960).
24. E. Deumens and Y. Öhrn, *Int. J. Quant. Chem.: Quant. Chem. Symp.* **23**, 31 (1989).
25. E. Deumens, Y. Öhrn, and B. Weiner, *J. Math. Phys.* **32**, 1166 (1991).
26. E. Deumens and Y. Öhrn, *J. Phys. Chem.* **105**, 2660 (2001).
27. R. Cabrera-Trujillo, J. R. Sabin, Y. Öhrn, and E. Deumens, *Phys. Rev. A* **61**, 032719 1 (2000).
28. R. Cabrera-Trujillo, J. R. Sabin, Y. Öhrn, and E. Deumens, *Phys. Rev. Lett.* **84**, 5300 (2000).
29. R. Cabrera-Trujillo, Y. Öhrn, E. Deumens, and J. R. Sabin, *Phys. Rev. A* **62**, 052714 1 (2000).
30. Y. Öhrn, J. Oreiro, and E. Deumens, *Int. J. Quantum Chem.* **58**, 583 (1996).
31. N. F. Mott and H. S. W. Massey, *The Theory of Atomic Collisions*, Oxford University, Oxford, UK, 1965.
32. K. W. Ford and J. A. Wheeler, *Ann. Phys.* **7**, 259 (1959).
33. M. V. Berry, *Proc. Phys. Soc. London* **89**, 479 (1966).
34. J. N. L. Connor and P. R. Curtis, *J. Phys. A: Math. Gen.* **15**, 1179 (1982).
35. J. A. Morales, Ph.D. Dissertation, University of Florida, Gainesville, FL (unpublished).
36. L. I. Schiff, *Phys. Rev.* **103**, 443 (1956).
37. R. J. Glauber, *Phys. Rev.* **91**, 459 (1953).

38. R. G. Newton, *Scattering Theory of Waves and Particles*, Springer-Verlag, New York, 1982.
39. R. J. S. Morrison, W. E. Conaway, and R. N. Zare, *Chem. Phys. Lett.* **113**, 435 (1985).
40. R. J. S. Morrison, W. E. Conway, T. Ebata, and R. N. Z. re, *J. Chem. Phys.* **84**, 5527 (1986).
41. J. C. Poutsma et al., *Chem. Phys. Lett.* **305**, 343 (1999).
42. J. A. Morales, A. C. Diz, E. Deumens, and Y. Öhrn, *Chem. Phys. Lett.* **233**, 392 (1995).
43. J. A. Morales, A. C. Diz, E. Deumens, and Y. Öhrn, *J. Chem. Phys.* **103**, 9968 (1995).
44. B. de Prony, *J. E. Polytech.* **1**, 24 (1795).
45. A. Blass, E. Deumens, and Y. Öhrn, *J. Chem. Phys.* **115**, 8366 (2001).
46. G. Niedner, M. Noll, J. Toennies, and C. Schlier, *J. Chem. Phys.* **87**, 2686 (1987).
47. A. Bjerre and E. E. Nikitin, *Chem. Phys. Lett.* **1**, 179 (1967).
48. M. Baer, G. Niedner-Schatteburg, and J. P. Toennies, *J. Chem. Phys.* **91**, 4169 (1989).
49. F. O. Ellison, *J. Am. Chem. Soc.* **85**, 3540 (1963).
50. B. Friedrich, G. Niedner, M. Noll, and J. P. Toennies, *J. Chem. Phys.* **87**, 5256 (1987).
51. J. A. Morales, E. Deumens, and Y. Öhrn, *J. Math. Phys.* **40**, 766 (1999).
52. R. Zare, *Angular Momentum*, 1st ed., John Wiley & Sons, Inc., New York, 1988.
53. T. Clark, J. Chandrasekhar, G. W. Spitznagel, and P. v. R. Schleyer, *J. Comput. Chem.* **4**, 294 (1983).
54. J. S. Binkley, J. A. Pople, and W. J. Hehre, *J. Am. Chem. Soc.* **102**, 939 (1980).
55. R. A. Marcus, *J. Chem. Phys.* **24**, 966 (1956).
56. R. A. Marcus, *J. Chem. Phys.* **46**, 679 (1965).
57. R. A. Marcus and N. Sutin, *Biochim. Biophys. Acta* **811**, 265 (1985).
58. A. Broo and S. Larsson, *Chem. Phys.* **148**, 103 (1990).
59. M. D. Newton, *Chem. Rev.* **91**, 767 (1991).
60. R. S. Mulliken, C. A. Rieke, D. Orloff, and H. Orloff, *J. Chem. Phys.* **17**, 1248 (1949).
61. M. Coutinho-Neto, E. Deumens, and Y. Öhrn, *J. Chem. Phys.* **116**, 2794 (2002).

APPLYING DIRECT MOLECULAR DYNAMICS TO NON-ADIABATIC SYSTEMS

G. A. WORTH and M. A. ROBB

Department of Chemistry, King's College London, The Strand, London, U.K.

CONTENTS

The Role of Degenerate States in Chemistry: A Special Volume of Advances in Chemical Physics, Volume 124, Edited by Michael Baer and Gert Due Billing. Series Editors I. Prigogine and Stuart A. Rice.
ISBN 0-471-43817-0.

I. INTRODUCTION

Knowledge of the underlying nuclear dynamics is essential for the classification and description of photochemical processes. For the study of complicated systems, molecular dynamics (MD) simulations are an essential tool, providing information on the channels open for decay or relaxation, the relative populations of these channels, and the timescales of system evolution. Simulations are particularly important in cases where the Born–Oppenheimer (BO) approximation breaks down, and a system is able to evolve non-adiabatically, that is, in more than one electronic state.

In this chapter, we look at the techniques known as direct, or on-the-fly, molecular dynamics and their application to non-adiabatic processes in photochemistry. In contrast to standard techniques that require a predefined potential energy surface (PES) over which the nuclei move, the PES is provided here by explicit evaluation of the electronic wave function for the states of interest. This makes the method very general and powerful, particularly for the study of polyatomic systems where the calculation of a multidimensional potential function is an impossible task. For a recent review of standard non-adiabatic dynamics methods using analytical PES functions see [1].

Direct dynamics methods are, however, still in their infancy, and have a number of difficulties that need to be solved. One is the sheer size of the problem—all nuclear and electronic degrees of freedom are treated explicitely. A second is the restriction placed on the form of the nuclear wave function as a local, trajectory-based, representation is required. This introduces the problem of including quantum effects into methods that are often based on classical mechanics. For non-adiabatic processes, there is the additional complication of the treatment of the non-adiabatic coupling. In this chapter, we will show how progress has been made in this new and exciting field, highlighting the different problems and how they are being solved. Complimentary reviews on applying direct dynamics to adiabatic problems are given in [2,3].

Interaction with light changes the quantum state a molecule is in, and in photochemistry this is an electronic excitation. As a result, the system will no longer be in an eigenstate of the Hamiltonian and this nonstationary state evolves, governed by the time-dependent Schrödinger equation

$$i\hbar \frac{\partial}{\partial t} \Psi(\boldsymbol{R}, \boldsymbol{r}, t) = \hat{H}(\boldsymbol{R}, \boldsymbol{r}) \Psi(\boldsymbol{R}, \boldsymbol{r}, t) \tag{1}$$

Central to the description of this dynamics is the BO approximation. This separates the nuclear and electronic motion, and allows the system evolution to be described by a function of the nuclei, known as a wavepacket, moving over a PES provided by the (adiabatic) motion of the electrons.

Coupling between the electronic and nuclear motion can, however, result in the breakdown of the BO approximation, which leads to an effective coupling between the adiabatic states of the system, providing pathways for fast, radiationless, electronic transitions. The wavepacket in non-adiabatic systems, as these are known, must therefore be described as evolving over a manifold of coupled PES. Non-adiabatic coupling is particularly important in regions where the PES are degenerate, or near-degenerate, and it can lead to an interesting topology of the surfaces. Typical features are avoided crossings, where the surfaces seem to repel one another, or conical intersections, where the surfaces meet at a point or seam. While avoided crossings are well established in chemical ideas through the noncrossing rule, it is only in recent years that the importance of conical intersections in photochemistry has emerged [4–8]. The idea of conical intersections has a long history [9–14]. Their general acceptance was delayed by the difficulties in conclusively demonstrating their existence in large molecules, due to the problems in calculating wave functions for excited states.

If the PES are known, the time-dependent Schrödinger equation, Eq. (1), can in principle be solved directly using what are termed wavepacket dynamics [15–18]. Here, a time-independent basis set expansion is used to represent the wavepacket and the Hamiltonian. The evolution is then carried by the expansion coefficients. While providing a complete description of the system dynamics, these methods are restricted to the study of typically 3–6 degrees of freedom. Even the highly efficient multiconfiguration time-dependent Hartree (MCTDH) method [19,20], which uses a time-dependent basis set expansion, can handle no more than 30 degrees of freedom.

For larger systems, various approximate schemes have been developed, called mixed methods as they treat parts of the system using different levels of theory. Of interest to us here are quantum-semiclassical methods, which use full quantum mechanics to treat the electrons, but use approximations based on trajectories in a classical phase space to describe the nuclear motion. The prefix quantum may be dropped, and we will talk of semiclassical methods. There are a number of different approaches, but here we shall concentrate on the few that are suitable for direct dynamics molecular simulations. An overview of other methods is given in the introduction of [21].

As mentioned above, the correct description of the nuclei in a molecular system is a delocalized quantum wavepacket that evolves according to the Schrödinger equation. In the classical limit of the single surface (adiabatic) case, when effectively $\hbar \rightarrow 0$, the evolution of the wavepacket density (amplitude squared) can be simulated by a "swarm" of trajectories, each driven by classical (e.g., Newtonian) mechanics. Note that this does not mean that the nuclei are being treated as classical particles, each is being represented by a set of classical pseudoparticles that together simulate the behavior of the nucleus. Methods based on this approximation are sometimes termed quasiclassical.

A different approach is to represent the wavepacket by one or more Gaussian functions. When using a local harmonic approximation to the true PES, that is, expanding the PES to second-order around the center of the function, the parameters for the Gaussians are found to evolve using classical equations of motion [22–26]. Detailed reviews of Gaussian wavepacket methods are found in [27–29].

To add non-adiabatic effects to semiclassical methods, it is necessary to allow the trajectories to sample the different surfaces in a way that simulates the population transfer between electronic states. This sampling is most commonly done by using surface hopping techniques or Ehrenfest dynamics. Recent reviews of these methods are found in [30–32]. Gaussian wavepacket methods have also been extended to include non-adiabatic effects [33,34]. Of particular interest here is the spawning method of Martínez, Ben-Nun, and Levine [35,36], which has been used already in a number of direct dynamics studies.

In traditional dynamics calculations, the first step is to find a representation of the PES. For accurate calculations, this involves fitting a function to ab initio data, maybe with final adjustments using experimental data. A major hurdle to the calculation of information about the excited state PES of molecules, required for a description of photochemistry, is the development of appropriate quantum chemical methods. Probably the most general method is the complete active space self-consistent field (CASSCF) method [37]. This is a multi-configuration self-consistent field (MCSCF) method that uses a full configuration interaction (CI) within an active space of the important molecular orbitals. As it is an MCSCF method, both the orbitals and the CI coefficients are optimised. Unlike other, maybe more powerful methods, calculation of analytic gradients is relatively straightforward using CASSCF, which makes it suitable for direct dynamics. Although care is needed in its application, accurate results are possible, particularly when combined with perturbation theory to correct for the missing so-called dynamic electron correlation [38–41].

Techniques have been developed within the CASSCF method to characterize the critical points on the excited-state PES. Analytic first and second derivatives mean that minima and saddle points can be located using traditional energy optimization procedures. More importantly, intersections can also be located using constrained minimization [42,43]. Of particular interest for the mechanism of a reaction is the minimum energy path (MEP), defined as the line followed by a classical particle with zero kinetic energy [44–46]. Such paths can be calculated using intrinsic reaction coordinate (IRC) techniques

[47,48]. For systems in which conical intersections play a role, however, the concept of an IRC must be extended. Due to the topology of the region more than one path may be accessible after crossing from the upper electronic state to the lower one (i.e., the wavepacket may bifurcate). For this situation, the initial relaxation direction (IRD) method has been developed [49,50] to identify the open channels on the ground-state PES moving away from the minimum energy intersection point. The MEP can then be used to explore each channel to the products. For more information on the study of PES critical points using quantum chemistry techniques, see the recent reviews [51,52].

An alternative method that can be used to characterize the topology of PES is the line integral technique developed by Baer [53,54], which uses properties of the non-adiabatic coupling between states to identify and locate different types of intersections. The method has been applied to study the complex PES topologies in a number of small molecules such as H_3 [55,56] and C_2H [57].

Information about critical points on the PES is useful in building up a picture of what is important in a particular reaction. In some cases, usually thermally activated processes, it may even be enough to describe the mechanism behind a reaction. However, for many real systems dynamical effects will be important, and the MEP may be misleading. This is particularly true in non-adiabatic systems, where quantum mechanical effects play a large role. For example, the spread of energies in an excited wavepacket may mean that the system finds an intersection away from the minimum energy point, and crosses there. It is for this reason that molecular dynamics is also required for a full characterization of the system of interest.

Calculating points on a set of PES, and fitting analytic functions to them is a time-consuming process, and must be done for each new system of interest. It is also an impossible task if more than a few (typically 4) degrees of freedom are involved, simply as a consequence of the exponential growth in number of ab initio data points needed to cover the coordinate space.

For this reason, there has been much work on empirical potentials suitable for use on a wide range of systems. These take a sensible functional form with parameters fitted to reproduce available data. Many different potentials, known as molecular mechanics (MM) potentials, have been developed for ground-state organic and biochemical systems [58–60]. They have the advantages of simplicity, and are transferable between systems, but do suffer from inaccuracies and rigidity—no reactions are possible. Schemes have been developed to correct for these deficiencies. The empirical valence bond (EVB) method of Warshel [61,62], and the molecular mechanics–valence bond (MMVB) of Bernardi et al. [63,64] try to extend MM to include excited-state effects and reactions. The MMVB Hamiltonian is parameterized against CASSCF calculations, and is thus particularly suited to photochemistry.

A further model Hamiltonian that is tailored for the treatment of non-adiabatic systems is the vibronic coupling (VC) model of Köppel et al. [65]. This provides an analytic expression for PES coupled by non-adiabatic effects, which can be fitted to ab initio calculations using only a few data points. As a result, it is a useful tool in the description of photochemical systems. It is also very useful in the development of dynamics methods, as it provides realistic global surfaces that can be used both for exact quantum wavepacket dynamics and more approximate methods.

Direct dynamics attempts to break this bottleneck in the study of MD, retaining the accuracy of the full electronic PES without the need for an analytic fit of data. The first studies in this field used semiclassical methods with semiempirical [66,67] or simple Hartree–Fock [68] wave functions to treat the electrons. These first studies used what is called BO dynamics, evaluating the PES at each step from the electronic wave function obtained by solution of the electronic structure problem. An alternative, the Ehrenfest dynamics method, is to propagate the electronic wave function at the same time as the nuclei. Although early direct dynamics studies using this method [69–71] restricted themselves to adiabatic problems, the method can incorporate non-adiabatic effects directly in the electronic wave function.

Major impetus in the field was given by the introduction of the Car–Parrinello method [72–74]. Related to the Ehrenfest dynamics method, this is a very efficient algorithm that propagates the electronic wave function using a fictitious mass to produce classical equations of motion for the expansion coefficients. For full efficiency, however, it requires a plane-wave basis set, which is inefficient for the description of isolated molecules. Recent work using Gaussian functions points the way to the solution of this problem [75]. The method is usually restricted to adiabatic dynamics, although the method has been applied to excited states using a very simple wave function [76]. We shall ignore Car–Parrinello methods in the following.

An important step forward in the study of molecular systems was afforded by the introduction of an efficient propagation algorithm by Helgaker et al. [77] and further improved by Chen et al. [78]. With the large step-size made possible by this method it became feasible to simply reevaluate the electronic wave function at each step, thus opening up all the power of electronic structure calculations for direct BO dynamics. By combining the Helgaker–Chen algorithm with a surface hopping method, a number of dynamics studies of photochemical systems have been made using the MMVB empirical Hamiltonian [79–85]. These studies have allowed us to gain much experience in the behavior of trajectories over coupled PES. The method has then been applied to direct dynamics study using CASSCF wave functions [86,87].

The Gaussian wavepacket based spawning method, mentioned above, has also been used in direct dynamics where it is called ab initio multiple spawning

(AIMS) [88]. The inclusion of quantum effects directly in the nuclear motion may be a significant step, as the motion near a conical intersection is known to be very quantum mechanical.

The present state of the art is not able to use direct dynamics to calculate accurate dynamical properties: For this many trajectories are required, and it is simply too expensive. Even so, as we shall show, mechanistic information can be gained directly from the calculations, extending the minimum energy path picture to include a dynamical term, which is certainly important in the study of excited molecules. A further use, still to be explored fully, is to use the information from direct dynamics trajectories to efficiently generate the PES for more accurate calculations. The ground work for this has been laid by the work of Collins and co-workers [89–93], who developed a scheme to generate a PE function by interpolating information on the surface (the energy, and its first and second derivatives) at a set of points. These points could be generated by direct dynamics, thus sampling only the areas of configuration space important for the system dynamics. The accuracy of the method has been shown recently in state-of-the-art four-dimensional (4D) quantum scattering calculations [94].

By its nature, the application of direct dynamics requires a detailed knowledge of both molecular dynamics and quantum chemistry. This chapter is aimed more at the quantum chemist who would like to use dynamical methods to expand the tools at their disposal for the study of photochemistry, rather than at the dynamicist who would like to learn some quantum chemistry. It tries therefore to introduce the concepts and problems of dynamics simulations, stressing that one cannot strictly think of a molecule moving along a trajectory even though this is what is being calculated.

To demonstrate the basic ideas of molecular dynamics calculations, we shall first examine its application to adiabatic systems. The theory of vibronic coupling and non-adiabatic effects will then be discussed to define the sorts of processes in which we are interested. The complications added to dynamics calculations by these effects will then be considered. Some details of the mathematical formalism are included in appendices. Finally, examples will be given of direct dynamics studies that show how well the systems of interest can at present be treated.

Throughout, unless otherwise stated, R and r will be used to represent the nuclear and electronic coordinates, respectively. Boldface is used for vectors and matrices, thus $\mathbf{R}$ is the vector of nuclear coordinates with components R_α. The vector operator $\mathbf{\nabla}$, with components

$$\nabla_\alpha = \frac{\partial}{\partial R_\alpha} \tag{2}$$

forms the derivative vector when applied to a function, for example,

$$\nabla V = \left(\frac{\partial V}{\partial R_1}, \frac{\partial V}{\partial R_2}, \ldots \right) \tag{3}$$

If the nuclear coordinates are mass-scaled Cartesian coordinates,

$$R_\alpha = \sqrt{M_\alpha}\, x_\alpha \tag{4}$$

where M_α is the mass associated with the coordinate, then the kinetic energy operator can be written

$$\hat{T} = \sum_{\alpha=1}^{3N} -\frac{\hbar^2}{2M_\alpha} \frac{\partial^2}{\partial^2 x_\alpha} = -\frac{\hbar^2}{2} \nabla^2 \tag{5}$$

The full system Hamiltonian is partitioned so as to define an electronic Hamiltonian, $\hat{H}_{\mathrm{el}}$

$$\hat{H}(\boldsymbol{R}, \boldsymbol{r}) = \hat{T}_{\mathrm{n}}(\boldsymbol{R}) + \hat{H}_{\mathrm{el}}(\boldsymbol{R}, \boldsymbol{r}) \tag{6}$$

Here, $\hat{T}_{\mathrm{n}}$ is the nuclear kinetic energy operator, and so all terms describing the electronic kinetic energy, electron–electron and electron–nuclear interactions, as well as the nuclear–nuclear interaction potential function, are collected together. This sum of terms is often called the clamped nuclei Hamiltonian as it describes the electrons moving around the nuclei at a particular configuration $\boldsymbol{R}$.

II. ADIABATIC MOLECULAR DYNAMICS

In this section, the basic theory of molecular dynamics is presented. Starting from the BO approximation to the nuclear Schrödinger equation, the picture of nuclear dynamics is that of an evolving wavepacket. As this picture may be unusual to readers used to thinking about nuclei as classical particles, a few prototypical examples are shown.

In the full quantum mechanical picture, the evolving wavepackets are delocalized functions, representing the probability of finding the nuclei at a particular point in space. This representation is unsuitable for direct dynamics as it is necessary to know the potential surface over a region of space at each point in time. Fortunately, there are approximate formulations based on trajectories in phase space, which will be discussed below. These local representations, so-called as only a portion of the PES is examined at each point in time, have a classical flavor. The delocalized and nonlocal nature of the full solution of the Schrödinger equation should, however, be kept in mind.

In what is called BO MD, the nuclear wavepacket is simulated by a swarm of trajectories. We emphasize here that this does not necessarily mean that the nuclei are being treated classically. The difference is in the chosen initial conditions. A fully classical treatment takes the initial positions and momenta from a classical ensemble. The use of quantum mechanical distributions instead leads to a semiclassical simulation. The important topic of choosing initial conditions is the subject of Section II.C.

Finally, Gaussian wavepacket methods are described in which the nuclear wavepacket is described by one or more Gaussian functions. Again the equations of motion to be solved have the form of classical trajectories in phase space. Now, however, each trajectory has a quantum character due to its spread in coordinate space.

A. Quantum Wavepacket Propagation

Using the BO approximation, the Schrödinger equation describing the time evolution of the nuclear wave function, $\chi(\boldsymbol{R},t)$, can be written

$$i\hbar\frac{\partial}{\partial t}|\chi(\boldsymbol{R},t)\rangle = (\hat{T}_N + V(\boldsymbol{R}))|\chi(\boldsymbol{R},t)\rangle \tag{7}$$

In this picture, the nuclei are moving over a PES provided by the function $V(\boldsymbol{R})$, driven by the nuclear kinetic energy operator, $\hat{T}_N$. More details on the derivation of this equation and its validity are given in Appendix A. The potential function is provided by the solutions to the electronic Schrödinger equation,

$$H_{\text{el}}(\boldsymbol{r};\boldsymbol{R})|\psi(\boldsymbol{r};\boldsymbol{R})\rangle = V(\boldsymbol{R})|\psi(\boldsymbol{r};\boldsymbol{R})\rangle \tag{8}$$

where H_{el} is the electronic (clamped nucleus) Hamiltonian defined in Eq. (6). In this equation it must be remembered that $\boldsymbol{R}$ is a parameter defining the nuclear configuration, and $\psi(\boldsymbol{r};\boldsymbol{R})$ an electronic eigenfunction at this configuration. A PES is thus formed by following one of the roots of this equation (e.g., the second root for the first excited state) as the nuclear geometry changes. Approximate solutions to this equation are the results of the standard quantum chemistry computer packages, such as GAUSSIAN [95], GAMESS [96], MOLCAS [97], MOLPRO [98], and COLUMBUS [99].

To solve this equation, an appropriate basis set $\{\phi_\alpha(\boldsymbol{R})\}$ is required for the nuclear functions. These could be a set of harmonic oscillator functions if the motion to be described takes place in a potential well. For general problems, a discrete variable representation (DVR) [100,101] is more suited. These functions have mathematical properties that allow both the kinetic and potential energy operators to be easily represented. In coordinate space, they are effectively δ functions, and so the potential can be represented on a grid of points. The wave function is then expanded in this set

$$\chi(\boldsymbol{R}, t) = \sum_{\alpha} c_{\alpha}(t)\phi_{\alpha}(\boldsymbol{R}) \tag{9}$$

and Eq. (7) transformed to a matrix equation,

$$i\hbar\dot{\chi} = \boldsymbol{H}\chi \tag{10}$$

where the nuclear function is a vector in the space provided by the basis, that is, the components are the expansion coefficients c_{α}, and the Hamiltonian matrix elements are

$$H_{\alpha\beta} = \langle \phi_{\alpha} | \hat{T}_N + V(\boldsymbol{R})) | \phi_{\beta} \rangle \tag{11}$$

If $V(\boldsymbol{R})$ is known and the matrix elements $H_{\alpha\beta}$ are evaluated, then solution of Eq. (10) for a given initial wavepacket is the numerically exact solution to the Schrödinger equation.

Efficient techniques for the direct solution of Eq. (10) have been developed using either a DVR or FFT-based method [102] to generate a representation of the wavepacket and Hamiltonian on a grid in coordinate space [15,16,18,103]. In principle, the differential equation can be directly solved, using a standard integrator (predictor–corrector, Runge–Kutte, etc.) to propagate the vector χ forward in time using the time derivative, which is calculated using simple matrix–vector multiplication. Alternatively, for a time-independent Hamiltonian, Eq. (10) can be written in integral form

$$\chi(t) = \exp\left(-\frac{i}{\hbar}\boldsymbol{H}t\right)\chi(0) \tag{12}$$

The problem is then reduced to the representation of the time-evolution operator [104,105]. For example, the Lanczos algorithm could be used to generate the eigenvalues of $\boldsymbol{H}$, which can be used to set up the representation of the exponentiated operator. Again, the methods are based on matrix–vector operations, but now much larger steps are possible.

Unfortunately, the resources required for these numerically exact methods grow exponentially with the number of degrees of freedom in the system of interest. Without the use of clever algorithms to optimize the basis set used [106,107], this limits the range of systems treatable to 4–6 degrees of freedom (3–4 atoms). For larger systems, the MCTDH method [19,20,108] provides a flexible, yet accurate method. This method uses a time-dependent basis set, and has treated, for example, the dynamics of the pyrazine molecule after photoexcitation including all 24 vibrational modes and 2 coupled electronic states [109]. A time-dependent basis is efficient because it follows the evolving wavepacket, and does not waste effort in describing regions of empty space. In effect, the semiclassical methods described below are using a set of classical trajectories or Gaussian wavepackets as a time-dependent basis set. The connection between the MCTDH basis functions and trajectories has recently been explored [110], and it has been shown that a set of coupled trajectories can act as a basis set for full quantum dynamics calculations. The connection between a time-dependent basis set and the Gaussian wavepacket methods is more obvious.

Before progressing, it is useful to review the dynamics of typical molecular systems. We consider three types: scattering (chemical reaction), photodissociation, and bound-state photoabsorption (no reaction).

The $H + H_2 \rightarrow H_2 + H$ hydrogen atom exchange reaction is the simplest atom–molecule scattering system. Molecules and atoms colliding is a basic step in chemical reactivity, and much work has been made to understand this system in all its details [111,112]. As well as experimental work, extensive calculations have been made using both a time-independent framework [113] and wavepacket methods [114–116] to obtain fully state resolved cross-sections for the reaction. This system is best described by Jacobi coordinates, shown in Figure 1*a*, and the reaction is dominated by the colinear configuration. The PES for this configuration (i.e., a cut with $\theta = 0^\circ$) has a C-shaped minimum energy channel, with a saddle point as a transition region at the apex. This is shown in Figure 2.

The evolution of a wavepacket representing the $H + H_2$ scattering reaction for a particular set of initial conditions is plotted on Figure 2 as a series of snapshots. To display the three-dimensional (3D) wavepacket on a two-dimensional (2D) plot, the reduced density

$$\rho(R_d, R_v) = \int_0^{2\pi} d\theta\, \chi(R_d, R_v, \theta)\chi^*(R_d, R_v, \theta) \tag{13}$$

is plotted. The system stays close to the colinear configuration, and so integrating over the angular coordinate does not lead to significant loss of information. Note, however, that the results from a 2D calculation in which the angle is kept fixed would be different.

In the reactant channel leading up to the transition region, motion along R_d represents the H atom approaching the molecule, while motion along R_v is the vibrational motion of the atom. The initial wavepacket is chosen to represent the desired initial conditions. In Figure 2, the H_2 molecule is initially in the ground

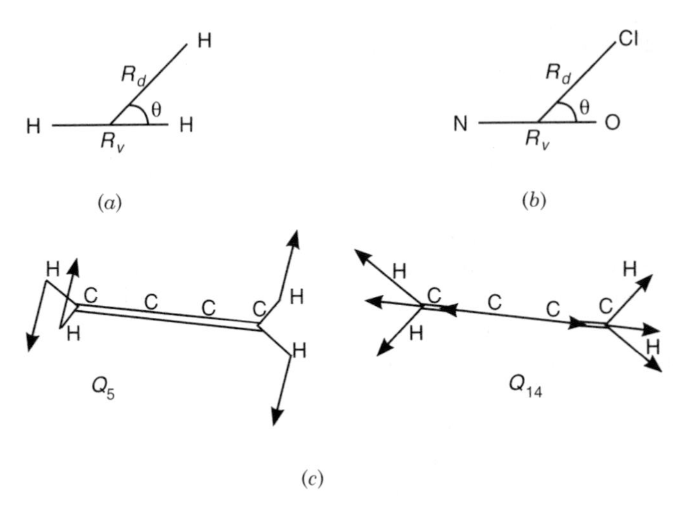

Figure 1. Coordinates used for describing the dynamics of (*a*) H + H_2 (*b*) NOCl, (*c*) butatriene. (*a*), (*b*) Are Jacobi coordinates, where R_d and R_v are the dissociative and vibrational coordinates, respectively. (*c*) Shows the two most important normal mode coordinates, Q_5 and Q_{14} which are the torsional and central C—C bond stretch, respectively.

vibrational state, and the atom is located relative to the molecule by a Gaussian distribution of positions, moving with an initial momentum toward the molecule. The initial packet is thus close to a product of Gaussian functions. The quantum mechanical nature of the system means that the wavepacket possesses a distribution of momenta, and therefore energies. The figure shows how the wavepacket moves along the reactant channel, and is split as it hits the energy barrier representing the molecule–atom collision. Part of the packet moves on into the product channel (hydrogen atom exchange), and part is reflected back to the reactants (no exchange). The wavepacket can then be analyzed to obtain information about the transfer of population from the initial state to the final states over the energies contained in the packet.

A different category of dynamics is found in photodissociation processes, in which a molecule breaks up after absorbing a photon. A simple example is found in the NOCl molecule after excitation to the first singlet, S_1, state [117]. The molecule is initially in the ground vibrational state on the ground electronic surface. After the photoexcitation, this nuclear wave function is moved vertically onto the excited state. The S_1 PES as calculated by Schinke et al. [118] is shown in Figure 3. This is again in Jacobi coordinates, which are shown in Figure 1*b*. For the plot the angular coordinate, which plays only a minor role in the process, is at the ground-state equilibrium value of 127°.

The evolution of the nuclear wavepacket is also traced by a number of snapshots of the absolute values of the wavepacket, again integrating over the

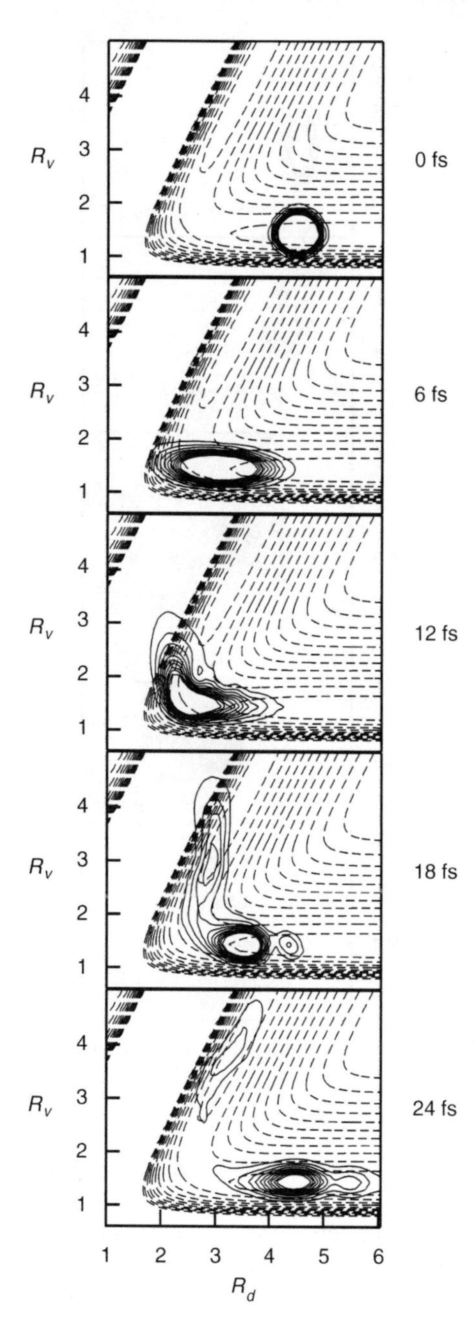

Figure 2. Wavepacket dynamics of the H + H $\rightarrow$ H_2 + H scattering reaction, shown as snapshots of the density (wave packet amplitude squard) at various times. The coordinates, in au, are described in Figure 1*a*, and the wavepacket is initially moving to describe the H atom approaching the H_2 molecule. The density has been integrated over the angular coordinate. The PES is plotted for the collinear interaction geometry, $\theta = 180°$.

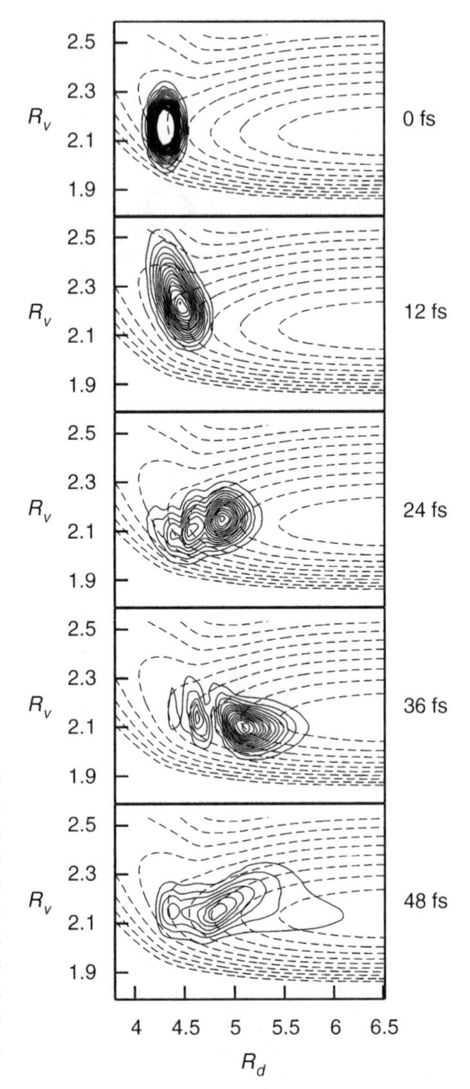

Figure 3. Wavepacket dynamics of the photodissociation of NOCl, shown as snapshots of the density (wavepacket amplitude squared) at various times. The coordinates, in au, are described in Figure 1*b*, and the wavepacket is initially the ground-state vibronic wave function vertically excited onto the S_1 state. Increasing R_d corresponds to chlorine dissociation. The density has been integrated over the angular coordinate. The S_1 PES is ploted for the geometry, $\theta = 127^\circ$, the ground-state equilibrium value.

angular coordinate. The wavepacket evolves away from its initial Gaussian-like form down the valley, which leads to direct dissociation. The structure formed in the wavepacket leads to structure in the absorption spectrum, which is absent if the angle θ is frozen, and is thus due to the flow of energy between the bend and stretch motions. Even more complicated behavior would be found if the PES contained a barrier to the dissociation, which would lead to a break up of the packet.

The third major category of processes are dynamics after bound–bound transitions, such as photoexcitation to a bound state. In Figure 4, the system dynamics of the butatriene radical cation are shown after excitation from the neutral molecule ground state to a simple model of the cationic first excited state, a process related to the first excited band in the photoelectron spectrum. The dynamics are dominated by two vibrational modes, the central C—C stretch, labeled Q_{14} and the torsion, Q_5. These coordinates are shown in Figure 1c. In this simple model, the PES is taken as a harmonic approximation around the minimum energy point, which is found to be shifted along the Q_{14} mode. Here, non-adiabatic effects have been ignored. As will be shown in Section III.D, there is in fact strong vibronic coupling to the cationic ground state via the torsional mode, and the true dynamics after excitation into this state is radically altered. This model is, however, a reasonable representation of a bound state in which vibronic coupling does not play a role. The systems dynamics in the space of the two normal modes shown is fairly simple. The initial Gaussian shaped wavepacket representing the neutral ground-state wave function moves back and forth across the well, driven by the initial force due to the shifted energy minimum.

For bound state systems, eigenfunctions of the nuclear Hamiltonian can be found by diagonalization of the Hamiltonian matrix in Eq. (11). These functions are the possible nuclear states of the system, that is, the vibrational states. If these states are used as a basis set, the wave function after excitation is a superposition of these vibrational states, with expansion coefficients given by the Frank–Condon overlaps. In this picture, the dynamics in Figure 4 can be described by the time evolution of these expansion coefficients, a simple phase factor. The periodic motion in coordinate space is thus related to a discrete spectrum in energy space.

B. Born–Oppenheimer Molecular Dynamics

In a classical limit of the Schrödinger equation, the evolution of the nuclear wave function can be rewritten as an ensemble of pseudoparticles evolving under Newton's equations of motion

$$M\ddot{\mathbf{R}} = -\nabla V \tag{14}$$

where $V(R)$ is the potential and $\ddot{\mathbf{R}}$ is the second-derivative with respect to time of the position, that is, the acceleration. They are referred to as pseudoparticle trajectories as, as explained above, the ensemble might be simulating the motion of a quantum wavepacket, in which case a single particle is being represented by a number of pseudoparticles.

This picture is often referred to as "swarms of trajectories," and details are given in Appendix B. The nuclear problem is thus reduced to solving Newton's equations of motion for a number of different starting conditions. To connect

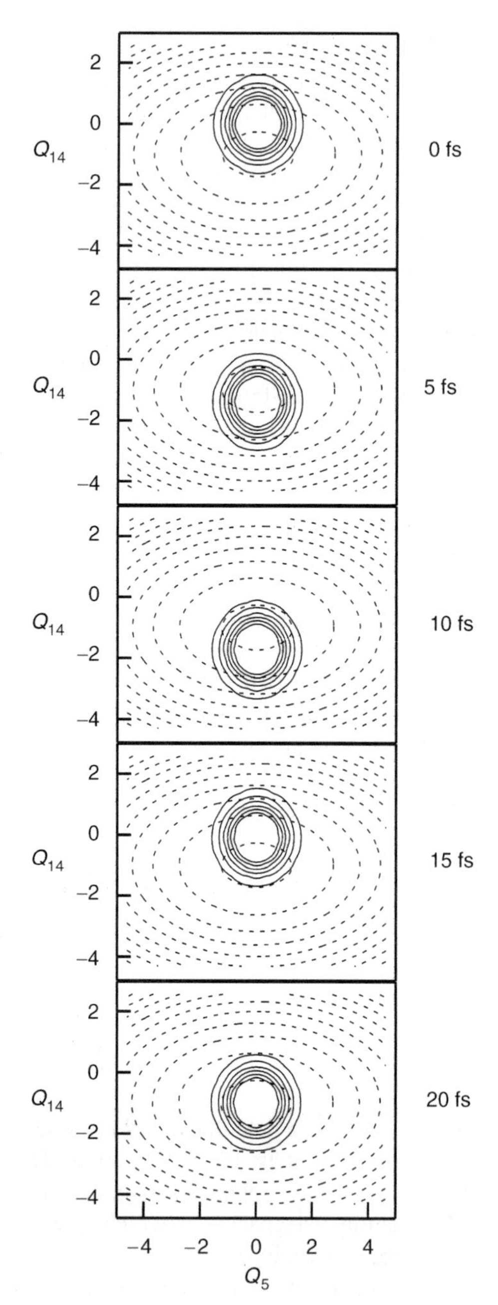

Figure 4. Wavepacket dynamics of photoexcitation, shown as snapshots of the density (wavepacket amplitude squared) at various times. The model is a 2D model based on a single, uncoupled, state of the butatriene redical cation. The initial structure represents the neutral ground-state vibronic wave function vertically excited onto the $\tilde{A}$ state of the radical cation.

this picture to the delocalized quantum mechanical one, the wavepacket is being represented by a time-dependent basis set of "functions" that are the points describing the trajectory of the classical pseudoparticle. The nuclear wavepacket is then being approximated by a vector, the elements of which are the populations of the trajectories in the initial ensemble.

The force experienced by a pseudoparticle is simply

$$F_\alpha(\boldsymbol{R}) = -\nabla_\alpha V(\boldsymbol{R}) \tag{15}$$

$$= -\nabla_\alpha \langle \psi(\boldsymbol{r};\boldsymbol{R}) | H_{\text{el}} | \psi(\boldsymbol{r};\boldsymbol{R}) \rangle \tag{16}$$

where ∇_α is a component of the derivative operator in coordinate space. The obvious approach is thus to calculate the electronic wave function at time t, and then directly calculate the required derivatives. The nuclei can then be propagated forward a step, and the process repeated. This algorithm is usually termed BO dynamics, and it was the method used in the first direct dynamics studies [66–68].

While it is conceptually simple, however, the method suffers from the expense of requiring the full electronic wave function at each step, for example, by solving the electronic structure problem using an SCF technique. For the method to be feasible, a large time-step is therefore required to minimise the number of these expensive evaluations that need to be made. Classical MD simulations typically use integration schemes based on either the Gear predictor-corrector [119] or the Verlet [120] algorithms (see [121] for overview of these methods, and [122] for other useful integrators). These give reasonable time-steps with low memory requirements for large systems, and require only first derivatives of the potential, the forces, at each step.

A different approach comes from the idea, first suggested by Helgaker et al. [77], of approximating the PES at each point by a harmonic model. Integration within an area where this model is appropriate, termed the trust radius, is then trivial. Normal coordinates, $\boldsymbol{Q}$, are defined by diagonalization of the mass-weighted Hessian (second-derivative) matrix, so if

$$\boldsymbol{\Delta} = \boldsymbol{R} - \boldsymbol{R}_0 \tag{17}$$

where $\boldsymbol{R}_0$ is the present position, then

$$\boldsymbol{Q} = \boldsymbol{L}\boldsymbol{m}^{\frac{1}{2}}\boldsymbol{\Delta} \tag{18}$$

$$\boldsymbol{g} = \boldsymbol{L}\boldsymbol{m}^{-\frac{1}{2}}\boldsymbol{G} \tag{19}$$

$$\boldsymbol{\omega}^2 = \boldsymbol{L}\boldsymbol{m}^{-\frac{1}{2}}\boldsymbol{H}\boldsymbol{m}^{-\frac{1}{2}}\boldsymbol{L}^\dagger \tag{20}$$

where $\boldsymbol{\omega}^2$ is the (diagonal) matrix of eigenvalues from transforming the mass-weighted Hessian, $\boldsymbol{H}$, using the unitary matrix $\boldsymbol{L}$, and $\boldsymbol{g}$ and $\boldsymbol{G}$ are the forces (first derivatives) in the two coordinate sets. The diagonal matrix $\boldsymbol{m}$ contains the masses associated with each coordinates. If $\boldsymbol{R}_0$ is a minimum on the PES, then $\boldsymbol{\omega}$ are the vibrational frequencies, and $\boldsymbol{Q}$ the vibrational modes of the molecule.

In this representation, Newton's equations of motion separate to $3N - 6$ equations

$$\ddot{Q}_\alpha = -g_\alpha - \omega_\alpha^2 Q_\alpha \tag{21}$$

that have different analytical solutions depending on whether the "frequency" ω is real, zero, or imaginary. These solutions are used to integrate the equations of motion from $\boldsymbol{R}_0$ to $\boldsymbol{R}_t$, where t is controlled by the trust radius. This radius changes, guided by the difference between the information about the PES calculated at $\boldsymbol{x}_t$, and that estimated from the harmonic model.

This algorithm was improved by Chen et al. [78] to take into account the surface anharmonicity. After taking a step from $\boldsymbol{R}_0$ to $\boldsymbol{R}'_t$ using the harmonic approximation, the true surface information at $\boldsymbol{R}'_t$ is then used to fit a (fifth-order) polynomial to form a better model of the surface. This polynomial model is then used in a corrector step to give the new $\boldsymbol{R}_t$.

The Helgaker–Chen algorithm results in very large steps being possible, and despite the extra cost of the required second derivatives, this is the method of choice for direct dynamics calculations. A number of systems have been treated, and a review of the method as applied to chemical reactions is given in [2].

The gradient of the PES (force) can in principle be calculated by finite difference methods. This is, however, extremely inefficient, requiring many evaluations of the wave function. Gradient methods in quantum chemistry are fortunately now very advanced, and analytic gradients are available for a wide variety of ab initio methods [123–127]. Note that if the wave function depends on a set of parameters $\{\lambda\}$, for example, the expansion coefficients of the basis functions used to build the orbitals in molecular orbital (MO) theory,

$$\psi \equiv \psi(\boldsymbol{R}; \boldsymbol{\lambda}) \tag{22}$$

then a component of the force, F_α is

$$F_\alpha = \frac{\partial V}{\partial R_\alpha} + \sum_i \frac{\partial V}{\partial \lambda_i} \frac{\partial \lambda_i}{\partial R_\alpha} \tag{23}$$

where V is defined in Eqs. (15) and (16). If the wave function is derived using a variational method, then $\partial V / \partial \lambda_i = 0$. Further, if the basis set is independent of $\boldsymbol{R}$, which is the case when it is complete, then Eq. (23) can be used to show that

$$\boldsymbol{\nabla} \langle \psi | \hat{H}_{\text{el}} | \psi \rangle = \langle \psi | \boldsymbol{\nabla} \hat{H}_{\text{el}} | \psi \rangle \tag{24}$$

This, the well-known Hellmann–Feynman theorem [128,129], can then be used for the calculation of the first derivatives. In normal situations, however, the use of an incomplete atom-centered (e.g., atomic orbital) basis set means that further terms, known as Pulay forces, must also be considered [130].

C. Choosing Initial Conditions

The time-dependent Schrödinger equation governs the evolution of a quantum mechanical system from an initial wavepacket. In the case of a semiclassical simulation, this wavepacket must be translated into a set of initial positions and momenta for the pseudoparticles. What the initial wavepacket is depends on the process being studied. This may either be a physically defined situation, such as a molecular beam experiment in which the particles are defined in particular quantum states moving relative to one another, or a theoretically defined situation suitable for a mechanistic study of the type "what would happen if . . . ."

In photochemistry, we are interested in the system dynamics after the interaction of a molecule with light. The absorption spectrum of a molecule is thus of primary interest which, as will be shown here, can be related to the nuclear motion after excitation by the capture of a photon. Experimentally, the spectrum is given by the Beer–Lambert law

$$I(z) = I_0 e^{-\sigma\rho z} \tag{25}$$

where $I(z)$ is the intensity of the light, propagated along the z axis, as it changes from I_0 due to absorption by molecules at a density of ρ. The molecular interaction with the light is contained here in the cross-section for the capture of a photon, σ, which describes the absorption properties of the sample.

The simplest theoretical description of the photon capture cross-section is given by Fermi's Golden Rule

$$\sigma(\omega) \sim \omega |\langle \chi_f(R) | \boldsymbol{\mu}_{fi}(R) | \chi_i(R) \rangle|^2 \delta(\omega_{fi} - \omega) \tag{26}$$

where

$$\boldsymbol{\mu}_{fi} = \langle \psi_f | \boldsymbol{e} \cdot \boldsymbol{d} | \psi_i \rangle \tag{27}$$

is the transition dipole moment, which connects the initial electronic state, ψ_i, to the final state, ψ_f, by the component of the molecular dipole moment operator, $\boldsymbol{d}$, along the electric field vector of the incident light, $\boldsymbol{e}$. The delta function ensures that spectral density is found only when the frequency of the incident light, ω, equals the frequency difference between the initial and final vibronic states, $\omega_{fi} = \omega_f - \omega_i$. This expression is valid for the usual light strengths used in spectroscopy, which act as a weak perturbation on the molecules. For more details on the derivation of this expression, and the time-dependent version below see [131].

Using the Condon approximation, the transition dipole moment is taken to be a constant with respect to the nuclear coordinates. Equation (26) then reduces to the familiar expression

$$\sigma(\omega) \sim \omega |\langle \chi_f(R) | \chi_i(R) \rangle|^2 \delta(\omega_{fi} - \omega) \tag{28}$$

where $\langle \chi_f(R) | \chi_i(R) \rangle$ is called the Frank–Condon factor. The spectral lines thus appear at a frequency of ω_{fi}, with an intensity related to the overlap between initial- and final-state functions.

To make a clearer connection to the molecular dynamics, this expression can be transformed to the time domain. In this picture, which was initially developed by Heller and co-workers [132,133], the absorption spectrum is given by the expression

$$\sigma(\omega) \sim \omega \int_{-\infty}^{\infty} dt\, e^{i\omega t} C(t) \tag{29}$$

which is the Fourier transform of the autocorrelation function

$$C(t) = \langle \chi(0) | \chi(t) \rangle \tag{30}$$

It is interesting to note that the use of correlation functions in spectroscopy is an old topic, and has been used to derive, for example, infrared (IR) spectra, from classical trajectories [134,135]. Stock and Miller have recently extended this approach, and derived expressions for obtaining electronic and femtosecond pump–probe spectra from classical trajectories [136].

Equation (29) directly incorporates our ideas about molecular dynamics after photoexcitation. The system is initially in a particular state at $t = 0$, for example, the ground vibrational state on the ground-state PES. On interacting with a photon, this state is vertically excited into the upper electronic state, that is, the electronic state changes while the nuclear function remains unchanged. Dynamics then takes place with this nuclear wavepacket, no longer an eigenfunction of the Schrödinger equation, driven by the appropriate Hamiltonian. The examples in Section II.B use this picture. Other functions have been derived for other spectra, for example, emission and Raman [133].

As it stands, the picture of dynamics from Eq. (29) is derived from the interaction of molecules with a continuous light source, that is, the system is at equilibrium with the oscillating light field. It is also valid if the light source is an infinitely short laser pulse, as here all frequencies are instantaneously excited.

Problems arise if a light pulse of finite duration is used. Here, different frequencies of the wave packet are excited at different times as the laser pulse passes, and thus begin to move on the upper surface at different times, with resulting interference. In such situations, for example, simulations of femtochemistry experiments, a realistic simulation must include the light field explicitely [1].

To return to the simple picture of vertical excitation, the question remains as to how a wavepacket can be simulated using classical trajectories? A classical ensemble can be specified by its distribution in phase space, $\rho_{\mathrm{cl}}(\boldsymbol{p}, \boldsymbol{q})$, which gives the probability of finding the system of particles with momentum $\boldsymbol{p}$ and position $\boldsymbol{q}$. In contrast, it is strictly impossible to assign simultaneously a position and momentum to a quantum particle.

A number of procedures have been proposed to map a wave function onto a function that has the form of a phase-space distribution. Of these, the oldest and best known is the Wigner function [137,138]. (See [139] for an exposition using Louiville space.) For a review of this, and other distributions, see [140]. The quantum mechanical density matrix is a matrix representation of the density operator

$$\hat{\rho} = |\Psi(\boldsymbol{x})\rangle\langle\Psi(\boldsymbol{x})| \tag{31}$$

where $\boldsymbol{x}$ is the variable being used here for the system particle coordinates. The density operator is used to link quantum mechanics to statistical mechanics, and effects of temperature are easily included via the concept of "mixed states" [141]. In coordinate space the matrix representation of the density operator is

$$\rho(\boldsymbol{x}, \boldsymbol{x}') = \langle \boldsymbol{x} | \hat{\rho} | \boldsymbol{x}' \rangle \tag{32}$$

which at $\boldsymbol{x} = \boldsymbol{x}'$ gives the probability of finding the particle at this point in space. The Wigner distribution uses the new coordinates $q = \frac{1}{2}(x + x')$ and $s = \frac{1}{2}(x - x')$ along with the momentum, p, conjugate to s to make the transformation, in one dimension (1D),

$$\rho_{\mathrm{w}}(p, q) = \int_{-\infty}^{\infty} ds\, e^{2ips} \langle q + s | \hat{\rho} | q - s \rangle \tag{33}$$

Extension to the multidimensional case is trivial. Wigner developed a complete mechanical system, equivalent to quantum mechanics, based on this distribution. He also showed that it satisfies many properties desired by a phase-space distribution, and in the high-temperature limit becomes the classical distribution.

Note that despite the form this cannot be interpreted as the probability of finding a particle at a point in phase space, and in fact the function can become negative. Obtaining ρ_{w} for a system is also not straightforward. For a harmonic oscillator, which can be taken as an approximation to the ground-state vibrational function, there is, however, an analytic expression

$$\rho_W(p, q; \beta) = \frac{1}{\pi\hbar}\tanh\left(\frac{1}{2}\beta\hbar\omega\right)\exp\left[-\frac{2}{\hbar\omega}\tanh\left(\frac{1}{2}\beta\hbar\omega\right)H_{HO}\right] \tag{34}$$

where $\beta = 1/kT$ is the thermodynamic temperature, and

$$H_{HO} = \frac{1}{2m}p^2 + \frac{m\omega}{2}q^2 \tag{35}$$

is the harmonic oscillator Hamiltonian. At zero temperature, when only the ground-vibrational state is occupied this expression becomes

$$\rho_W(p, q; \beta) = \frac{1}{2\pi\hbar}\exp\left[-\frac{1}{\hbar\omega}H_{HO}\right] \tag{36}$$

and the distribution is a product of Gaussian functions in p and q.

For many applications, it may be reasonable to assume that the system behaves classically, that is, the trajectories are real particle trajectories. It is then not necessary to use a quantum distribution, and the appropriate ensemble of classical thermodynamics can be taken. A typical approach is to use a microcanonical ensemble to distribute energy into the internal modes of the system. The normal-mode sampling algorithm [142–144], for example, assigns a desired energy to each normal mode, Q_α as a harmonic amplitude

$$A_\alpha = \frac{\sqrt{2E_\alpha}}{\omega_\alpha} \tag{37}$$

where ω_α is the harmonic frequency. The momentum and initial position are then sampled by adding a random phase, ξ_α

$$Q_\alpha = A_\alpha \cos(2\pi\xi_\alpha) \tag{38}$$

$$\dot{Q}_\alpha = -A_\alpha\,\omega_\alpha \sin(2\pi\xi_\alpha) \tag{39}$$

After transforming to Cartesian coordinates, the position and velocities must be corrected for anharmonicities in the potential surface so that the desired energy is obtained. This procedure can be used, for example, to include the effects of zero-point energy into a classical calculation.

One of the basic problems in molecular dynamics is how to sample infrequent events. Typically a reaction must pass over a barrier, and effort would be wasted if many trajectories are run that do not reach the reactant channel.

One way to overcome this problem is to start by setting up the ensemble of trajectories (or wavepacket) at the transition state. If these trajectories are then run back in time into the reactants region, they can be used to set up the distribution of initial conditions that reach the barrier. These can then be run forward to completion, that is, into the products, and by using transition state theory a reaction rate obtained [145]. These ideas have also been recently extended to non-adiabatic systems [146].

In a mechanistic study, the aim is not to quantitatively reproduce an experiment. As a result it is not necessary to use the methods outlined above. The question here is what drives a reaction in a particular direction, or what would happen if the molecule is driven in different ways. The initial conditions are then at the disposal of the investigator to be chosen in a way to answer the relevant question, using a suitable spread of positions and energies.

D. Gaussian Wavepacket Propagation

A different approach that also leads to a representation of the nuclear wave function suitable for direct dynamics is to follow the work of Heller on the time evolution of Gaussian wavepackets. The nuclear wave function in Eq. (7) is represented by one or more Gaussian functions. Equations of motion for the parameters defining these functions are then determined, which are found to have properties that can be related to classical mechanics. The underlying idea is the observation that a wavepacket with a Gaussian form retains this form when moving in a harmonic potential, and under these circumstances the method can be equivalent to full quantum mechanical wavepacket propagation [147]. In more complicated cases, a harmonic approximation to the true potential is used, and the method becomes a semiclassical one. The dynamics shown in Figures 3 and 4 support the idea, as the wavepacket retains a form that is approximately a distorted Gaussian at all times.

The fundamental method [22,24] represents a multidimensional nuclear wavepacket by a multivariate Gaussian with time-dependent width matrix, $\boldsymbol{A}_t$, center position vector, $\boldsymbol{R}_t$, momentum vector, $\boldsymbol{p}_t$, and phase, γ_t

$$G(\boldsymbol{R},t) = \exp\frac{i}{\hbar}\left[(\boldsymbol{R}-\boldsymbol{R}_t)^T\boldsymbol{A}_t(\boldsymbol{R}-\boldsymbol{R}_t) + \boldsymbol{p}_t^T(\boldsymbol{R}-\boldsymbol{R}_t) + \gamma_t\right] \tag{40}$$

where the superscript T denotes the transpose of a vector. Note that the width matrix allows the Gaussian to distort in any direction. The potential surface is represented by a harmonic expansion about the center point of the wavepacket,

$$H = -\sum_{\alpha=1}^{3N}\frac{\hbar^2}{2m_\alpha}\frac{\partial^2}{\partial R_\alpha^2} + V_t + (\boldsymbol{R}-\boldsymbol{R}_t)^T\boldsymbol{V}' + \frac{1}{2}(\boldsymbol{R}-\boldsymbol{R}_t)^T\boldsymbol{V}''(\boldsymbol{R}-\boldsymbol{R}_t) \tag{41}$$

where $V_t, \boldsymbol{V}'$, and $\boldsymbol{V}''$ are the value, first derivative vector, and second derivative matrix of the potential surface at $\boldsymbol{R}_t$. This is known as the local harmonic approximation (LHA). By using this approximate Hamiltonian, equations of motion for the parameters in Eq. (40) can be obtained using the time-dependent Schrödinger equation. These are

$$\dot{\boldsymbol{R}}_t = \boldsymbol{m}^{-1}\boldsymbol{p}_t \tag{42}$$

$$\dot{\boldsymbol{p}}_t = -\boldsymbol{V}' \tag{43}$$

$$\dot{\boldsymbol{A}}_t = -2\boldsymbol{A}_t\boldsymbol{m}^{-1}\boldsymbol{A}_t - \frac{1}{2}\boldsymbol{V}'' \tag{44}$$

$$\dot{\gamma}_t = i\hbar\,\mathrm{Tr}(\boldsymbol{m}^{-1}\cdot\boldsymbol{A}_t) + \boldsymbol{p}_t^T\dot{\boldsymbol{R}}_t - E \tag{45}$$

where $\boldsymbol{m}$ is the diagonal matrix of masses associated with each coordinate, Tr denotes the trace over the matrix product, and $E = \langle H(\boldsymbol{R}_t, \boldsymbol{p}_t)\rangle$ is the expectation value of the Hamiltonian at the center of the packet.

The center of the wavepacket thus evolves along the trajectory defined by classical mechanics. This is in fact a general result for wavepackets in a harmonic potential, and follows from the Ehrenfest theorem [147] [see Eqs. (154,155) in Appendix C]. The equations of motion are straightforward to integrate, with the exception of the width matrix, Eq. (44). This equation is numerically unstable, and has been found to cause problems in practical applications using Morse potentials [148]. As a result, Heller introduced the P–Z method as an alternative propagation method [24]. In this, the matrix $\boldsymbol{A}_t$ is rewritten as a product of matrices

$$\boldsymbol{A}_t = \frac{1}{2}\boldsymbol{P}_t\cdot\boldsymbol{Z}_t^{-1} \tag{46}$$

with the definition that

$$\dot{\boldsymbol{Z}}_t = \boldsymbol{m}^{-1}\cdot\boldsymbol{P}_t \tag{47}$$

From Eqs. (46), (47), and (44),

$$\dot{\boldsymbol{P}}_t = -\boldsymbol{V}''\cdot\boldsymbol{Z}_t \tag{48}$$

and so $\boldsymbol{P}_t, \boldsymbol{Z}_t$ have the form of equations of motion for a matrix harmonic oscillator. These new equations are stable and soluble.

The big advantage of the Gaussian wavepacket method over the swarm of trajectory approach is that a wave function is being used, which can be easily manipulated to obtain quantum mechanical information such as the spectrum, or reaction cross-sections. The initial Gaussian wave packet is chosen so that it describes the quantum mechanical function as well as possible, rather than selecting the initial momentum and position, $\boldsymbol{p}_0, \boldsymbol{R}_0$, from a phase-space distribution. A second advantage is the efficiency. Again looking at the dynamics of Figures 3 and 4 a qualitatively correct result would be expected propagating a single Gaussian function, a total of $N^2 + 2N + 1$ parameters, where N is the number of degrees of freedom. In contrast, hundreds of trajectories may be required in the swarm to attain reasonable results.

One drawback is that, as a result of the time-dependent potential due to the LHA, the energy is not conserved. Approaches to correct for this approximation, which is valid when the Gaussian wavepacket is narrow with respect to the width of the potential, include that of Coalson and Karplus [149], who use a variational principle to derive the equations of motion. This results in replacing the function values and derivatives at the central point, V_t, $\boldsymbol{V}'$, and $\boldsymbol{V}''$ in Eq. (41), by values averaged over the wavepacket.

The method will, however, fail badly if the Gaussian form is not a good approximation. For example, looking at the dynamics shown in Figure 2, a problem arises when a barrier causes the wavepacket to bifurcate. Under these circumstances it is necessary to use a superposition of functions. As will be seen later, this is always the case when non-adiabatic effects are present.

Sawada et al. [26] made a detailed study of the methodology and numerical properties of the method. They paid particular attention to the problem of using a superposition of Gaussian wavepackets

$$\chi(\boldsymbol{R}, t) = \sum_n G_n(\boldsymbol{R}, t) \tag{49}$$

In earlier work, the Gaussian functions were always taken to be independent of each other, the independent Gaussian approximation (IGA). Here the case was also studied for interacting Gaussians, and equations of motion worked out for the parameters. The minimum energy method (MEM) was used in the derivation, which like the variational methods used by Coalson and Karplus goes beyond the LHA approximation. The accuracy of both the IGA and the LHA were then tested, and found to be inadequate in a few cases. They also dealt with the problem of how to choose the initial Gaussians, as the flexibility in Eq. (49) allows many different ways in which the Gaussian parameters can be chosen. There is of course a balance between flexibility of the wave function (large numbers of functions) and efficiency (small number of functions). Furthermore, when using interacting Gaussians it was found that a large number of functions can lead to numerical instability if the overlap between the functions becomes too large.

The lack of generality and the numerical problems [150] seem to have effectively stopped this otherwise attractive and pictorial method. This line of investigation has, however, recently been reopened by Burghardt et al. [34], who have incorporated general Gaussian functions into the MCTDH wavepacket propagation method. How well this mixed scheme will perform is to be seen.

An alternative to using a superposition of Gaussian functions is to extend the basis set by using Hermite polynomials, that is, harmonic oscillator functions [24]. This provides an orthonormal, in principle complete, basis set along the trajectory, and the idea has been taken up by Billing [151,152]. The basic problem with this approach is the slow convergence of the basis set.

To deal with the problem of using a superposition of functions, Heller also tried using Gaussian wave packets with a fixed width as a time-dependent basis set for the representation of the evolving nuclear wave function [23]. Each "frozen Gaussian" function evolves under classical equations of motion, and the phase is provided by the classical action along the path

$$\dot{\gamma}_t = \boldsymbol{p}_t \cdot \dot{\boldsymbol{R}}_t - H(\boldsymbol{p}_t, \boldsymbol{R}_t) \tag{50}$$

Singly, these functions provide a worse description of the wave function than the "thawed" ones described above. Not requiring the propagation of the width matrix is, however, a significant simplification, and it was hoped that collectively the frozen Gaussian functions provide a good description of the changing shape of the wave function by their relative motions.

Coming from a different line of research, Herman and Klux [25] showed the relationship between the frozen Gaussian approximation and rigorous semiclassical mechanics. The initial wave function is represented by a superposition of an (overcomplete) set of Gaussian functions, which thus cover elements in phase space. Replacing the quantum mechanical propagator [shown in a matrix representation in Eq. (12)] by a semiclassical propagator

$$\exp\left(-\frac{i}{\hbar}\hat{H}t\right) \approx C(S)\exp\left(\frac{i}{\hbar}S\right) \tag{51}$$

where S is the classical action along a path in Eq. (50), and C is a preexponential factor depending on the action, then leads to a formula for the propagation of a wave packet in terms of the evolution of the fixed Gaussians. The initial conditions are taken from the classical phase space, typically using Monte Carlo integration to sample the space.

The Herman–Kluk method has been developed further [153–155], and used in a number of applications [156–159]. Despite the formal accuracy of the approach, it has difficulties, especially if chaotic regions of phase space are present. It also needs many trajectories to converge, and the initial integration is time consuming for large systems. Despite these problems, the frozen Gaussian approximation is the basis of the spawning method that has been applied to non-adiabatic systems with much success. This approach is described below in Section IV.C.

III. VIBRONIC COUPLING AND NON-ADIABATIC EFFECTS

The adiabatic picture developed above, based on the BO approximation, is basic to our understanding of much of chemistry and molecular physics. For example, in spectroscopy the adiabatic picture is one of well-defined spectral bands, one for each electronic state. The structure of each band is then due to the shape of the molecule and the nuclear motions allowed by the potential surface. This is in general what is seen in absorption and photoelectron spectroscopy. There are, however, occasions when the picture breaks down, and non-adiabatic effects must be included to give a faithful description of a molecular system [160–163].

Non-adiabatic coupling is also termed vibronic coupling as the resulting breakdown of the adiabatic picture is due to coupling between the nuclear and electronic motion. A well-known special case of vibronic coupling is the Jahn–Teller effect [14,164–168], in which a symmetrical molecule in a doubly degenerate electronic state will spontaneously distort so as to break the symmetry and remove the degeneracy.

The majority of photochemistry of course deals with nondegenerate states, and here vibronic coupling effects are also found. A classic example of non-Jahn–Teller vibronic coupling is found in the photoelectron spectrum of butatriene, formed by ejection of electrons from the electronic eigenfunctions (approximately the molecular orbitals). Bands due to the ground $\tilde{X}^2B_{2g}$ and first excited $\tilde{A}^2B_{2u}$ states of the radical cation are found at energies predicted by calculations. Between the bands, however, is a further band, which was termed the *mystery band* [169]. This band was then shown to be due to vibronic coupling between the states [170].

A different example of non-adiabatic effects is found in the absorption spectrum of pyrazine [171,172]. In this spectrum, the S_1 state is a weak structured band, whereas the S_2 state is an intense broad, fairly featureless band. Importantly, the fluorescence lifetime is seen to dramatically decrease in the energy region of the S_2 band. There is thus an efficient nonradiative relaxation path from this state, which results in the broad spectrum. Again, this is due to vibronic coupling between the two states [109,173,174].

Another example of the role played by a nonradiative relaxation pathway is found in the photochemistry of octatetraene. Here, the fluorescence lifetime is found to decrease dramatically with increasing temperature [175]. This can be assigned to the opening up of an efficient nonradiative pathway back to the ground state [6]. In recent years, nonradiative relaxation pathways have been frequently implicated in organic photochemistry, and a number of articles published on this subject [4–8].

In this section, the adiabatic picture will be extended to include the non-adiabatic terms that couple the states. After this has been done, a diabatic picture will be developed that enables the basic topology of the coupled surfaces to be investigated. Of particular interest are the intersection regions, which may form what are called conical intersections. These are a multimode phenomena, that is, they do not occur in 1D systems, and the name comes from their shape—in a special 2D space it has the form of a double cone. Finally, a model Hamiltonian will be introduced that can describe the coupled surfaces. This enables a global description of the surfaces, and gives both insight and predictive power to the formation of conical intersections. More detailed review on conical intersections and their properties can be found in [1,14,65,176–178].

A. The Complete Adiabatic Picture

In Section II, molecular dynamics within the BO approximation was introduced. As shown in Appendix A, the full nuclear Schrödinger equation is, however,

$$(\hat{T}_n + V_i)|\chi_i\rangle - \sum_j \hat{\Lambda}_{ij}|\chi_j\rangle = i\hbar\frac{\partial}{\partial t}|\chi_i\rangle \tag{52}$$

Comparison with Eq. (7) shows that the the non-adiabatic operator matrix, $\hat{\Lambda}$, has been added. This is responsible for mixing the nuclear functions associated with different BO PES.

The non-adiabatic operator matrix, $\hat{\mathbf{\Lambda}}$ can be written as a sum of two terms; a matrix of numbers, $\boldsymbol{G}$, and a derivative operator matrix

$$\hat{\Lambda}_{ij} = \frac{\hbar^2}{2}\left(G_{ij} + 2\boldsymbol{F}_{ij}\cdot\boldsymbol{\nabla}\right) \tag{53}$$

Both terms on the right are related to the rate of change of the adiabatic electronic functions with respect to the nuclear coordinates. The first term G_{ij} is given by

$$G_{ij} = \left\langle \psi_i^{\mathrm{ad}} \middle| \nabla^2 \psi_j^{\mathrm{ad}} \right\rangle \tag{54}$$

while the second term in Eq. (53) is the dot product of two vectors, the derivative operator with components

$$\nabla_\alpha = \frac{\partial}{\partial R_\alpha} \tag{55}$$

and the matrix elements with components

$$F_{ij}^{\alpha} = \left\langle \psi_i^{\mathrm{ad}} \middle| \nabla_\alpha \psi_j^{\mathrm{ad}} \right\rangle \tag{56}$$

Importantly, this term is a derivative (nonlocal) operator on the nuclear coordinate space.

The matrix $\boldsymbol{G}$ can in fact be expressed in terms of $\boldsymbol{F}$ (see Appendix A),

$$G_{ij} = \nabla \cdot \boldsymbol{F}_{ij} + \sum_k \boldsymbol{F}_{ik} \cdot \boldsymbol{F}_{kj} \tag{57}$$

And using this equation the nuclear Schrödinger equation can be written in matrix form [54,179]

$$\left[-\frac{\hbar^2}{2} (\nabla + \boldsymbol{F})^2 + \boldsymbol{V} \right] \chi = i\hbar \frac{\partial}{\partial t} \chi \tag{58}$$

The matrix of vectors $\boldsymbol{F}$ is thus the defining quantity, and is called the non-adiabatic coupling matrix. It gives the strength (and direction) of the coupling between the nuclear functions associated with the adiabatic electronic states.

The elements of these vectors can be evaluated using an off-diagonal form of the Hellmann–Feynman theorem, which in Cartesian coordinates, x_α, is

$$F_{ij}^{\alpha} = \frac{1}{\sqrt{M_\alpha}} \frac{1}{V_j - V_i} \langle \psi_i^{\text{ad}} | \frac{\partial \hat{H}_{\text{el}}}{\partial x_\alpha} | \psi_j^{\text{ad}} \rangle \tag{59}$$

Nonscaled coordinates are used here to explicitely include the mass to show that the coupling is modulated by two factors. The first is the mass associated with the coordinate, M_α (atomic mass in Cartesian coordinates, reduced mass in normal mode coordinates, etc.), and the larger the mass the smaller the coupling. This is the basis of the justification for the BO approximation: The mass of the electron is so much smaller than the mass of the nuclei that the motion of the electrons is effectively independent of the nuclear motion, and the electrons instantaneously adjust to the nuclear geometry. The second factor, however, depends inversely on the energy gap between the adiabatic surfaces. This will overwhelm the mass factor as the surfaces approach one another, until at a degenerate point the coupling is infinitely large.

As written, Eq. (52) depends on all the (infinite number of) adiabatic electronic states. Fortunately, the inverse dependence of the coupling strength on energy separation means that it is possible to separate the complete set of states into manifolds that effectively do not interact with one another. In particular, Baer has recently shown [54] that Eq. (57), and hence Eq. (58) also holds in the subset of mutually coupled states. This finding has important consequences for the use of diabatic states explored below.

Choosing a basis set for the nuclear functions $\{\phi_\alpha\}$ allows us to write Eq. (52) in a matrix form, similar to Eq. (10) for the single-surface case, now as a matrix of matrices

$$\begin{pmatrix} \boldsymbol{H}_{11} & \boldsymbol{H}_{12} & \cdots \\ \boldsymbol{H}_{21} & \boldsymbol{H}_{22} & \cdots \\ \vdots & \vdots & \ddots \end{pmatrix} \begin{pmatrix} \chi_1 \\ \chi_2 \\ \vdots \end{pmatrix} = i\hbar \begin{pmatrix} \dot{\chi}_1 \\ \dot{\chi}_2 \\ \vdots \end{pmatrix} \tag{60}$$

where the indexes of $\boldsymbol{H}_{ij}$ and χ_i, which relate to the adiabatic electronic states, run over the states included in the manifold.

The supermatrix notation emphasizes the structure of the problem. Each diagonal operator drives a wavepacket, just as in the adiabatic case of Eq. (10), but here the motion of the wavepackets in different adiabatic states is mixed by the off-diagonal non-adiabatic operators. In practice, a single matrix is built for the operator, and a single vector for the wavepacket. The operator matrix elements in the basis set $\{\phi_\alpha\}$ are

$$(\boldsymbol{H}_{ii})_{\alpha\beta} = \langle \phi_\alpha | \hat{T}_N + V_i - \hat{\Lambda}_{ii} | \phi_\beta \rangle \tag{61}$$

$$(\boldsymbol{H}_{ij})_{\alpha\beta} = \langle \phi_\alpha | - \hat{\Lambda}_{ij} | \phi_\beta \rangle \tag{62}$$

which are arranged in the blocks. All the methods mentioned in Section II.B for wavepacket dynamics can then be used.

B. The Diabatic Picture

The adiabatic picture is the standard one in quantum chemistry for the reason that, not only is it mathematically well defined, but it is also that used in ab initio calculations, which solve the electronic Hamiltonian at a particular nuclear geometry. To see the effects of vibronic coupling on the potential energy surfaces one must move to what is called a diabatic representation [1,65,180, 181].

In a diabatic representation, the electronic wave functions are no longer eigenfunctions of the electronic Hamiltonian. The aim is instead that the functions are so chosen that the (nonlocal) non-adiabatic coupling operator matrix, $\hat{\Lambda}$ in Eq. (52), vanishes, and the couplings are represented by (local) potential operators. The nuclear Schrödinger equation is then written

$$\hat{T}_n |\chi_i\rangle + \sum_j W_{ij} |\chi_j\rangle = i\hbar \frac{\partial}{\partial t} |\chi_i\rangle \tag{63}$$

where $\boldsymbol{W}$ is the new (nondiagonal) potential matrix, and the coupling between states is now achieved by the off-diagonal elements of this matrix. The adiabatic surfaces are the eigenfunctions of $\boldsymbol{W}$.

The diabatic electronic functions are related to the adiabatic functions by unitary transformations at each point in coordinate space

$$\phi(\boldsymbol{R}) = \boldsymbol{S}(\boldsymbol{R})\psi(\boldsymbol{R}) \tag{64}$$

From the time-dependent Schrödinger equation in the matrix form of Eq. (58), it can be shown [54] that if $\boldsymbol{S}(\boldsymbol{R})$ is chosen so that

$$\boldsymbol{S}^{\dagger}(\nabla + \boldsymbol{F})\boldsymbol{S} = 0 \tag{65}$$

the basis set transformation changes the operators in Eq. (58) to

$$-\frac{\hbar^2}{2}\boldsymbol{S}^{\dagger}(\nabla + \boldsymbol{F})^2\boldsymbol{S} = -\frac{\hbar^2}{2}\nabla^2 = \boldsymbol{T} \tag{66}$$

$$\boldsymbol{S}^{\dagger}\boldsymbol{V}\boldsymbol{S} = \boldsymbol{W} \tag{67}$$

and the diabatic representation is rigorously equivalent to the adiabatic representation in the subspace of the coupled states. Baer [53,54] has obtained solutions to this equation, and analyzed the validity of the transformation in regions where the non-adiabatic coupling becomes singular. Note that derivative operators in both terms act on $\boldsymbol{S}(\boldsymbol{R})$, and so this is not a local transformation. Note also that the diabatic basis is only defined up to a constant rotation. As a result, it is possible to select a point at which the diabatic and adiabatic functions are identical. This simplifies various mathematical manipulations.

Assuming that the diabatic space can be truncated to the same size as the adiabatic space, Eqs. (64) and (65) clearly define the relationship between the two representations, and methods have been developed to obtain the transformation matrices directly. These include the line integral method of Baer [53,54] and the block diagonalization method of Pacher et al. [179]. Failure of the truncation assumption, however, leads to possibly important nonremovable derivative couplings remaining in the diabatic basis [55,182].

Difficulties in obtaining the non-adiabatic coupling elements for polyatomic molecules have lead to the development of alternative approaches to provide the diabatic representation, typically using states that are smooth in a molecular property [183]. Although there is no formal justification for this approach, it seems to work well in practice. Properties used include the dipole moment [184], or retention of the configurational character from an MCSCF wave function [185], or maximization of the overlap between wave functions at neighbouring sites [186]. It has also been shown that the CASSCF method provides a good framework for the definition of diabatic states [187]. A simple scheme that removes the leading terms of the non-adiabatic coupling matrices using information from the adiabatic surfaces only has also been recently proposed [188]. For a fuller account of ways to construct diabatic states see [1].

Despite the difficulties in obtaining diabatic states they provide an extremely useful picture for many descriptive purposes. They are in many ways the natural choice for dynamics calculations as the kinetic energy operator is diagonal in this basis, and the singularities associated with the non-adiabatic operator where the adiabatic surfaces meet are not present. In principle, as the electronic basis set is only weakly dependent on $\boldsymbol{R}$, the electronic character of a state is preserved. The range of properties used to define diabatic states shows the sort of properties that are conserved within them. A typical example is in electron-transfer theory, which uses smooth diabatic states to define the donor–acceptor and charge-transfer states. A more important example in photochemistry is that photoexcitation in the Condon approximation should be modelled as taking place vertically to a single diabatic state, as in this picture the transition matrix element is relatively constant with respect to the nuclear geometry [1].

C. Conical Intersections

For a two-state system, the eigenfunctions of the diabatic potential matrix of Eq. (63) in terms of its elements are

$$V_{\pm} = \frac{1}{2}(W_{11} + W_{22}) \pm \sqrt{\Delta W^2 + W_{12}^2} \tag{68}$$

where $\Delta W = \frac{1}{2}(W_{22} - W_{11})$. The functions V_+ and V_- are the adiabatic PES, and they will meet when

$$\Delta W = 0 \tag{69}$$

$$W_{12} = 0 \tag{70}$$

In Section III.D, we shall investigate when this happens. For the moment, imagine that we are at a point of degeneracy. To find out the topology of the adiabatic PES around this point, the diabatic potential matrix elements can be expressed by a first order Taylor expansion.

Setting the diabatic basis equal to the adiabatic basis at the degenerate point, $\boldsymbol{R}_0$, the expansions can be written in vector notation as

$$\Delta W = \boldsymbol{x}_1 \cdot \boldsymbol{Q} \tag{71}$$

$$W_{12} = \boldsymbol{x}_2 \cdot \boldsymbol{Q} \tag{72}$$

where $\boldsymbol{Q}$ is the vector of nuclear displacements away from the intersection. Note that the constants in both expansions are zero due to the adiabatic–diabatic correspondence at $\boldsymbol{R}_0$. The vectors that form the first-order coefficients are

$$x_1^\alpha = \frac{\partial}{\partial Q_\alpha} \Delta W \tag{73}$$

$$x_2^\alpha = \frac{\partial}{\partial Q_\alpha} W_{12} \tag{74}$$

evaluated at $\boldsymbol{R}_0$. Also due to the adiabatic–diabatic correspondence at this point, these can be written as the gradient difference (GD) vector

$$x_1^\alpha = \frac{\partial}{\partial Q_\alpha} (V_+ - V_-) \tag{75}$$

and the derivative coupling (DC) vector

$$x_2^\alpha = \left\langle \psi_i^{\text{ad}} \middle| \frac{\partial \hat{H}_{\text{el}}}{\partial Q_\alpha} \middle| \psi_j^{\text{ad}} \right\rangle \tag{76}$$

This latter relationship is obtained by evaluating

$$\frac{\partial}{\partial Q_\alpha} \langle \psi_i | \hat{H}_{\text{el}} | \psi_j \rangle = \left\langle \psi_i \middle| \hat{H}_{\text{el}} \middle| \frac{\partial \psi_j}{\partial Q_\alpha} \right\rangle + \left\langle \frac{\partial \psi_i}{\partial Q_\alpha} \middle| \hat{H}_{\text{el}} \middle| \psi_j \right\rangle + \langle \psi_i | \frac{\partial \hat{H}_{\text{el}}}{\partial Q_\alpha} | \psi_j \rangle \tag{77}$$

at $\boldsymbol{R}_0$. In the diabatic basis, the first two terms on the right are zero. Due once again to the adiabatic–diabatic correspondence at $\boldsymbol{R}_0$ the third term on the right-hand side is

$$\langle \psi_i | \frac{\partial \hat{H}_{\text{el}}}{\partial Q_\alpha} | \psi_j \rangle = (V_j - V_i) \left\langle \psi_i^{\text{ad}} \middle| \frac{\partial \psi_j^{\text{ad}}}{\partial Q_\alpha} \right\rangle \tag{78}$$

[see Eqs. (130)–(132) in Appendix A]. And so the expansion coefficient vector lies in the same direction as the adiabatic coupling vector.

The major features of the PES around the degenerate point can now be easily analysed if we write the vector $\boldsymbol{Q}$ in the basis of $(\boldsymbol{x}_1, \boldsymbol{x}_2, \ldots)$ where the unspecified $n-2$ basis vectors are orthogonal to the $(\boldsymbol{x}_1, \boldsymbol{x}_2)$ plane, which is called the branching space. First, moving in the $n-2$-dimensional space orthogonal to the branching space the degeneracy is retained. Second, moving in the plane of the branching space, the degeneracy will be lifted. Ignoring the term $\frac{1}{2}(W_{11} + W_{22})$, which is the same for both surfaces, the adiabatic PES have the form

$$V_\pm \sim \pm \sqrt{x_1^2 + x_2^2} \tag{79}$$

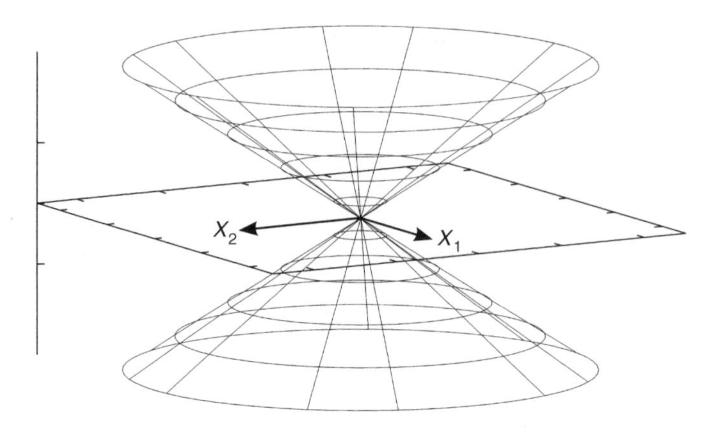

Figure 5. Sketch of a conical intersection. The vectors $\boldsymbol{x}_1$ and $\boldsymbol{x}_2$ are the GD and DC respectively, that lift the degeneracy of the two adiabatic surfaces. The plane containing these vectors is known as the branching space.

where x_1, x_2 are the components of $\boldsymbol{Q}$ along the respective vectors, and so the topology of the surfaces around the degenerate point is that of a double cone. Hence, this is called a conical intersection. A sketch of such a point is given in Figure 5.

Conical intersections can be broadly classified in two topological types: "peaked" and "sloped" [189]. These are sketched in Figure 6. The peaked case is the classical theoretical model from Jahn–Teller and other systems where the minima in the lower surface are either side of the intersection point. As indicated, the dynamics of a system through such an intersection would be expected to move fast from the upper to lower adiabatic surfaces, and not return. In contrast, the sloped form occurs when both states have minima that lie on the same side of the intersection. Here, after crossing from the upper to lower surfaces, recrossing is very likely before relaxation to the ground-state minimum can occur.

A final point to be made concerns the symmetry of the molecular system. The branching space vectors in Eqs. (75) and (76) can be obtained by evaluating the derivatives of matrix elements in the adiabatic basis

$$\frac{\partial}{\partial Q_\alpha} \langle \psi_i^{\text{ad}} | H_{\text{el}} | \psi_j^{\text{ad}} \rangle \tag{80}$$

with $i = j$ required for $\boldsymbol{x}_1$ and $i \neq j$ for $\boldsymbol{x}_2$. These elements are only nonzero if the product of symmetries of the adiabatic functions Γ_i, Γ_j, and the symmetry of the nuclear coordinate, Γ_{Q_α} contains the totally symmetric irrep

$$\Gamma_i \otimes \Gamma_{Q_\alpha} \otimes \Gamma_j \supset \text{A}_\text{g} \tag{81}$$

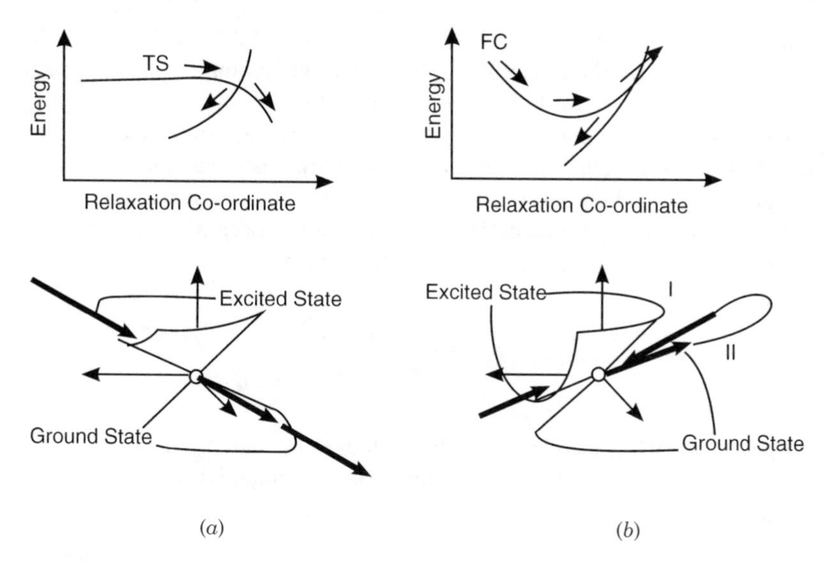

Figure 6. Two-dimensional (top) and 3D (bottom) representations of a peaked (*a*) and sloped (*b*) conical intersection topology. There are two directions that lift the degeneracy: the GD and the DC. The top figures have energy plotted against the DC while the bottom figures represent the energy plotted in the space of both the GD and DC vectors. At a peaked intersection, as shown at the bottom of (*a*), the probability of recrossing the conical intersection should be small whereas in the case of a sloped intersection [bottom of (*b*)], this possibility should be high. [Reproduced from [84] courtesy of Elsevier Publishers.]

If the symmetries of the two adiabatic functions are different at $\boldsymbol{R}_0$, then only a nuclear coordinate of appropriate symmetry can couple the PES, according to the point group of the nuclear configuration. Thus if $\boldsymbol{Q}$ are, for example, normal coordinates, $\boldsymbol{x}_1$ will only span the space of the totally symmetric nuclear coordinates, while $\boldsymbol{x}_2$ will have nonzero elements only for modes of the correct symmetry.

D. The Vibronic-Coupling Model Hamiltonian

A more general description of the effects of vibronic coupling can be made using the model Hamiltonian developed by Köppel, Domcke and Cederbaum [65]. The basic idea is the same as that used in Section III.C, that is to assume a quasidiabatic representation, and to develop a Hamiltonian in this picture. It is a useful model, providing a simple yet accurate analytical expression for the coupled PES manifold, and identifying the modes essential for the non-adiabatic effects. As a result it can be used for comparing how well different dynamics methods perform for non-adiabatic systems. It has, for example, been used to perform benchmark full-dimensional (24-mode) quantum dynamics calculations on the photoabsorption of pyrazine into the S_2/S_1 manifold [109]. Here it will be used to demonstrate the effect of a conical intersection on a systems dynamics using the butatriene radical cation as an example.

The Hamiltonian again has the basic form of Eq. (63). The system is described by the nuclear coordinates, $\boldsymbol{Q}$, which are relative to a suitable nuclear configuration $\boldsymbol{Q}_0$. In contrast to Section III.C, this may be any point in configuration space. As a diabatic representation has been assumed, the kinetic energy operator matrix, $\boldsymbol{T}$, is diagonal with elements

$$T_{ii} = \sum_{\alpha=1}^{f} -\frac{\hbar^2}{2}\frac{\partial^2}{\partial Q_\alpha^2} \tag{82}$$

The potential matrix elements are then obtained by making Taylor expansions around $\boldsymbol{Q}_0$, using suitable zero-order diabatic potential energy functions, $V_\alpha^{(0)}(\boldsymbol{Q})$.

$$W_{ij} - V_i^{(0)}\delta_{ij} = \langle\psi_i|H_{\text{el}}|\psi_j\rangle + \sum_{\alpha=1}^{f}\frac{\partial}{\partial Q_\alpha}\langle\psi_i|H_{\text{el}}|\psi_j\rangle Q_\alpha + \cdots \tag{83}$$

where the integrals and derivatives are evaluated at the point $\boldsymbol{Q}_0$. The diabatic functions are again taken to be equal to the adiabatic functions at $\boldsymbol{Q}_0$, and so

$$W_{ii} = V_i^{(0)} + E_i + \sum_{\alpha=1}^{f}\kappa_\alpha^{(i)} Q_\alpha + \cdots \tag{84}$$

$$W_{ij} = \sum_{\alpha=1}^{f}\lambda_\alpha^{(ij)} Q_\alpha + \cdots \tag{85}$$

The model is that the ground-state PES is first altered by the electronic excitations (on-diagonal coupling leads to a change in equilibrium geometry and frequency), and these smooth diabatic states are then further altered by vibronic (off-diagonal) coupling.

The eigenvalues of this matrix have the form of Eq. (68), but this time the matrix elements are given by Eqs. (84) and (85). The symmetry arguments used to determine which nuclear modes couple the states, Eq. (81), now play a crucial role in the model. Thus the linear expansion coefficients are only nonzero if the products of symmetries of the electronic states at $\boldsymbol{Q}_0$ and the relevant nuclear mode contain the totally symmetric irrep. As a result, on-diagonal matrix elements are only nonzero for totally symmetric nuclear coordinates and, if the electronic states have different symmetry, the off-diagonal elements will only be nonzero for a suitable nonsymmetric mode, as given by the product of the electronic-state symmetries.

For states of different symmetry, to first order the terms ΔW and W_{12} are independent. When they both go to zero, there is a conical intersection. To connect this to Section III.C, take $\boldsymbol{Q}_0$ to be at the conical intersection. The gradient difference vector in Eq. (75) is then a linear combination of the symmetric modes, while the non-adiabatic coupling vector in Eq. (76) is a linear combination of the appropriate nonsymmetric modes. States of the same symmetry may also form a conical intersection. In this case it is, however, not possible to say a priori which modes are responsible for the coupling. All totally symmetric modes may couple on- or off-diagonal, and the magnitudes of the coupling determine the topology.

A conical intersection needs at least two nuclear degrees of freedom to form. In a 1D system states of different symmetry will cross as $W_{ij} = 0$ for $i \neq j$ and so when $W_{ii} = 0$ the surfaces are degenerate. There is, however, no coupling between the states. States of the same symmetry in contrast cannot cross, as both W_{ij} and W_{ii} are nonzero and so the square root in Eq. (68) is always nonzero. This is the basis of the well-known non-crossing rule.

If the states are degenerate rather than of different symmetry, the model Hamiltonian becomes the Jahn–Teller model Hamiltonian. For example, in many point groups $E \otimes E \supset E$ and so a doubly degenerate electronic state can interact with a doubly degenerate vibrational mode. In this, the $E \times \epsilon$ Jahn–Teller effect the first-order Hamiltonian is then [65]

$$\boldsymbol{H} = (T + V_0)\mathbf{1} + \kappa \begin{pmatrix} Q_x & Q_y \\ Q_y & -Q_x \end{pmatrix} \tag{86}$$

where x, y denote the two components of the degenerate vibrational mode, $\mathbf{1}$ is the 2×2 unit matrix, and the zero-order Hamiltonian

$$T + V_0 = \sum_{i=x,y} \frac{\omega_i}{2} \left(-\frac{\partial^2}{\partial Q_i^2} + Q_i^2 \right) \tag{87}$$

is the unperturbed harmonic state (written here in mass-frequency scaled coordinates). This model results in the splitting of the degeneracy to form a symmetrical moat around a central conical intersection.

The Hamiltonian provides a suitable analytic form that can be fitted to the adiabatic surfaces obtained from quantum chemical calculations. As a simple example we take the butatriene molecule. In its neutral ground state it is a planar molecule with D_{2h} symmetry. The lowest two states of the radical cation, responsible for the first two bands in the photoelectron spectrum, are $\tilde{X}^2B_{2g}$ and $\tilde{A}^2B_{2u}$. The vibronic coupling model Hamiltonian is set up using the ground-state

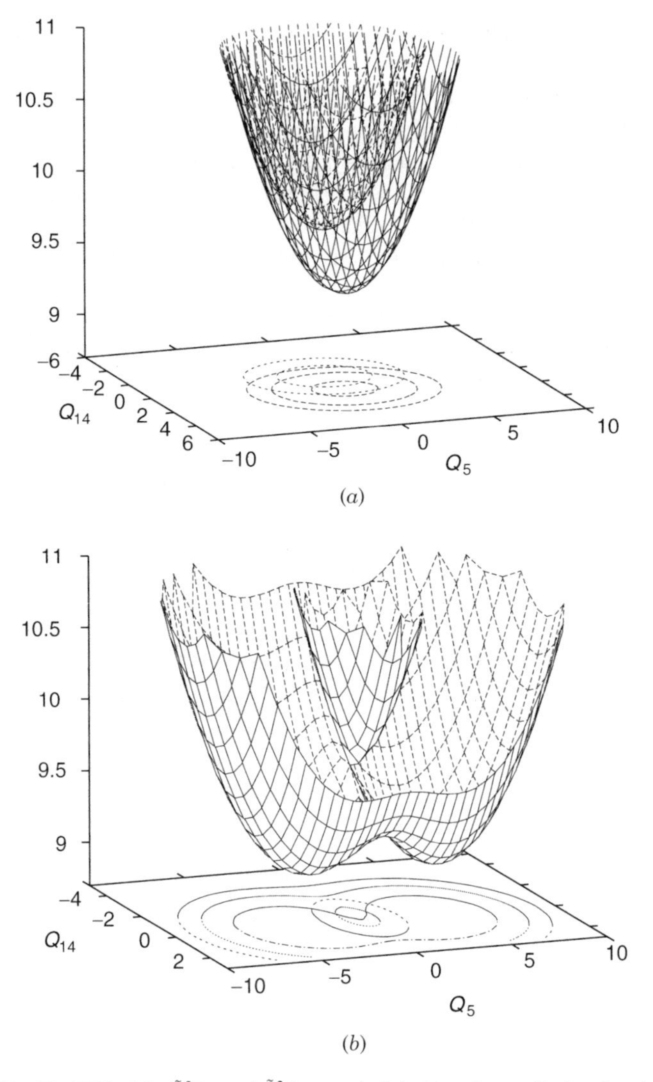

Figure 7. The PES of the $\tilde{X}^2B_{2g}$ and $\tilde{A}^2B_{2u}$ states of the butatriene radical cation. (*a*) Diabatic surfaces. (*b*) Adiabatic surfaces. The surfaces are obtained as eigenfucations of the vibronic coupling model Hamiltonain that fitted to reproduce quantum chemical calculations. The coordinates are shown in Figure 1*c*. See Section III. D for further details.

normal modes, of which there are 15, expanding around the neutral equilibrium geometry using the harmonic ground state as a zero-order surface. Taking symmetry into account, to first order these states can only be coupled by a nuclear degree of freedom with A_u symmetry, of which there is only one, the torsional motion labeled Q_5. Four A_g modes are present, which may have first-order expansion coefficients on the diagonal of the diabatic potential matrix. After the parameters for the Taylor expansions are fitted to quantum chemical calculations [170,190], it is found that only one symmetric mode, the central C—C stretch Q_{14}, has a significant linear coupling constant, κ. Thus this system can be well described considering only two modes, Q_5 and Q_{14}.

In Figure 7*a* the diabatic surfaces are plotted, that is, the on-diagonal functions from the potential matrix. These diabatic PES are interlocking harmonic wells, and they would be the adiabatic surfaces in the absence of non-adiabatic coupling. Compared to the neutral ground-state surface, the two minima have been shifted along the totally symmetric coordinate. Now, including the off-diagonal vibronic coupling term, the adiabatic surfaces change dramatically. They are plotted in Figure 7*b*, where the PES has been cut away to reveal the conical intersection between the two surfaces. Note also that the minima are now shifted significantly along the torsional, Q_5, axis. This deformation away from the D_{2h} symmetry is thus due to non-adiabatic effects.

In Section II.B, the molecular dynamics was examined after excitation to the $\tilde{\text{A}}$ state ignoring the coupling to the $\tilde{\text{X}}$ state, that is, the PES in Figure 4 is the higher energy diabatic well in Figure 7*a*. Figure 8 shows the same dynamics including the non-adiabatic coupling. Starting in the $\tilde{A}$ state, the wavepacket is seen to transfer very fast to the lower $\tilde{X}$ state, with the transfer taking place around the intersection point. Notice the complicated dynamics of the wavepacket on the lower surface that runs around the double well. After 40 fs, the wavepacket has returned to the intersection point, and a small recrossing is seen to the upper surface.

The vibronic coupling model has been applied to a number of molecular systems, and used to evaluate the behavior of wavepackets over coupled surfaces [191]. Recent examples are the radical cation of allene [192,193], and benzene [194] (for further examples see references cited therein). It has also been used to explain the lack of structure in the S_2 band of the pyrazine absorption spectrum [109,173,174,195], and recently to study the photoisomerization of retinal [196].

IV. NON-ADIABATIC MOLECULAR DYNAMICS

As shown above in Section III.A, the use of wavepacket dynamics to study non-adiabatic systems is a trivial extension of the methods described for adiabatic systems in Section II.B. The equations of motion have the same form, but now there is a wavepacket for each electronic state. The motions of these packets are then coupled by the non-adiabatic terms in the Hamiltonian operator matrix elements. In contrast, the methods in Section II that use trajectories in phase space to represent the time evolution of the nuclear wave function cannot be

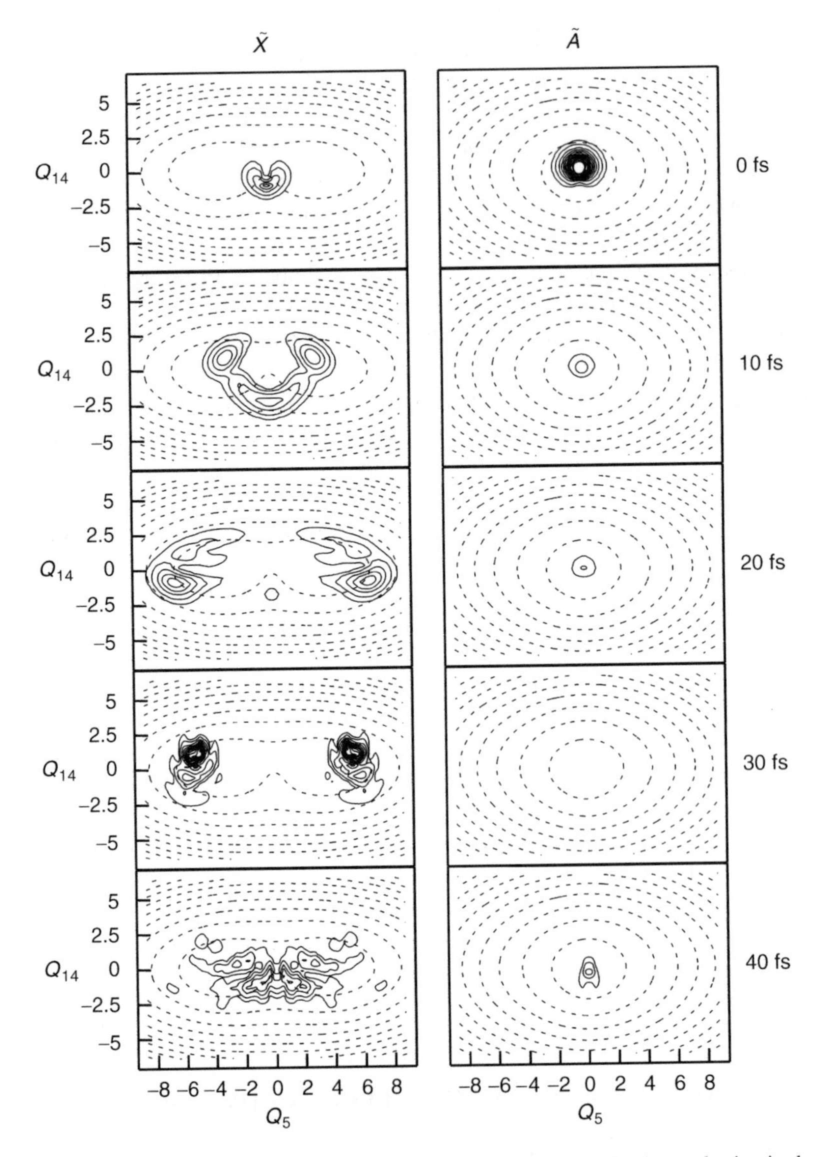

Figure 8. Wavepacket dynamics of the butatriene radical cation after its production in the $\tilde{A}$ state, shown as snapshots of the adiabatic density (wavepacket amplitude squared) at various times. The 2D model uses the coordinates in Figure 1*c*, and includes the coupled $\tilde{A}$ and $\tilde{X}$ states. The PES are plotted in the adiabatic picture (see Fig. 7*b*). The initial structure represents the neutral ground-state vibronic wave function vertically excited onto the diabatic $\tilde{A}$ state of the radical cation.

easily extended to systems that evolve in a manifold of coupled electronic states. For this, we need mixed methods that treat the electronic degrees of freedom using quantum mechanics, while using the classical or semiclassical methods for the nuclei.

The standard semiclassical methods are surface hopping and Ehrenfest dynamics (also known as the classical path (CP) method [197]), and they will be outlined below. More details and comparisons can be found in [30–32]. The multiple spawning method, based on Gaussian wavepacket propagation, is also outlined below. See [1] for further information on both quantum and semiclassical non-adiabatic dynamics methods.

A. Ehrenfest Dynamics

Both the BO dynamics and Gaussian wavepacket methods described above in Section II separate the nuclear and electronic motion at the outset, and use the concept of potential energy surfaces. In what is generally known as the Ehrenfest dynamics method, the picture is still of semiclassical nuclei and quantum mechanical electrons, but in a fundamentally different approach the electronic wave function is propagated at the same time as the pseudoparticles. These are driven by standard classical equations of motion, with the force provided by an instantaneous potential energy function

$$\dot{R}_\alpha = \frac{P_\alpha}{m_\alpha} \tag{88}$$

$$\dot{P}_\alpha = -\frac{\partial}{\partial R_\alpha}\langle\psi(\boldsymbol{r},t)|H_{\text{el}}(\boldsymbol{R})|\psi(\boldsymbol{r},t)\rangle \tag{89}$$

and a time-dependent Schrödinger-like equation for the electronic wave function

$$i\dot{\psi}(\boldsymbol{r},t) = H_{\text{el}}(\boldsymbol{R})\psi(\boldsymbol{r},t) \tag{90}$$

Note that the Hamiltonian is time dependent due to the time dependence of $\boldsymbol{R}$. There is also a phase corresponding to each trajectory

$$i\dot{A} = -\langle\psi(\boldsymbol{r},t)|H_{\text{el}}(\boldsymbol{R})|\psi(\boldsymbol{r},t)\rangle A \tag{91}$$

Details of the derivation of these equations are given in Appendix C.

The expression for the force on the nuclei, Eq. (89), has the same form as the BO force Eq. (16), but the wave function here is the time-dependent one. As can be shown by perturbation theory, in the limit that the nuclei move very slowly compared to the electrons, and if only one electronic state is involved, the two expressions for the wave function become equivalent. This can be shown by comparing the time-independent equation for the eigenfunction of H_{el} at time t

with the wave function at time t obtained from time-dependent perturbation theory expression using a suitable slow perturbation due to the nuclear motion. Apart from a phase factor the two functions are the same (see e.g., [198]). Away from this limit, however, $\psi(t)$ does not remain an eigenfunction of the electronic Hamiltonian, and this provides correlation between the electronic and nuclear motion as the electrons do not instantaneously follow the nuclei.

If more than one electronic state is involved, then the electronic wave function is free to contain components from all states. For example, for non-adiabatic systems the electronic wave function can be expanded in the adiabatic basis set at the nuclear geometry $\boldsymbol{R}(t)$

$$\psi(\boldsymbol{r},t) = \sum_j c_j(t)\psi_j^{\mathrm{ad}}(\boldsymbol{r};\boldsymbol{R}(t)) \tag{92}$$

Setting this into Eq. (90) and multiplying from the left by $\langle\psi_i^{\mathrm{ad}}(\boldsymbol{r};\boldsymbol{R}(t))|$ then gives

$$i\hbar\dot{c}_i = \sum_j \langle\psi_i^{\mathrm{ad}}|H_{\mathrm{el}}(R)|\psi_j^{\mathrm{ad}}\rangle c_j - i\hbar\sum_\alpha \dot{R}_\alpha\left\langle\psi_i^{\mathrm{ad}}\left|\frac{\partial\psi_j^{\mathrm{ad}}}{\partial R_\alpha}\right.\right\rangle \tag{93}$$

where the chain rule for the time-derivative operator has been used to take care of the implicit time dependence (through $\boldsymbol{R}$) of the adiabatic functions in the expansion. This can be rearranged to

$$i\hbar\dot{c}_i = c_i V_i - i\hbar\sum_j \dot{\boldsymbol{R}}\cdot\boldsymbol{F}_{ij}c_j \tag{94}$$

where $\boldsymbol{F}_{ij}$ are the derivative coupling vector matrices defined in Eq. (56). This expression for the state amplitudes provides a simple measure for the population (amplitudes squared) of the different adiabatic electronic states at $\boldsymbol{R}(t)$ as time progresses.

This method thus leads to the concept of a mixed-state trajectory. A trajectory starting on one surface starts to evolve driven by this PES. As the non-adiabatic coupling increases (the surfaces approach one another), population will be transferred from the initial state to the other. In a region where the non-adiabatic coupling is negligible, however, there is no population transfer. Thus if a trajectory comes out of a region of strong non-adiabatic coupling with appreciable populations in both states, the trajectory will continue in both states at the same time.

The mixed-state character of a trajectory outside a non-adiabatic region is a serious weakness of the method. As the time-dependent wave function does not depend on $\boldsymbol{R}$, the Ehrenfest force in Eq. (89) can be evaluated using the wave function Eq. (92) by

$$\langle \psi(t) | \boldsymbol{\nabla} H_{\text{el}} | \psi(t) \rangle = \sum_{ij} c_i^* c_j \langle \psi_i^{\text{ad}} | \boldsymbol{\nabla} H_{\text{el}} | \psi_j^{\text{ad}} \rangle \tag{95}$$

Note that the exact adiabatic functions are used on the right-hand side, which in practical calculations must be evaluated by the full derivative on the left of Eq. (24) rather than the Hellmann–Feynman forces. This form has the advantage that the $\boldsymbol{R}$ dependence of the coefficients, c_i, does not have to be considered. Using the relationship Eq. (78) for the off-diagonal matrix elements of the right-hand side then leads directly to

$$\dot{\boldsymbol{P}} = -\sum_i |c_i|^2 \boldsymbol{\nabla} V_i - \sum_{i \neq j} c_i^* c_j (V_j - V_i) \boldsymbol{d}_{ij} \tag{96}$$

The first term on the right of this equation is the average force from the adiabatic potential energy surfaces. The second term is a force due to the non-adiabatic coupling. This mean-field potential is inherent in the method. That it leads to practical problems can be seen by considering the case of a bound state coupled to a dissociative state. Non-adiabatic forces will cause the dissociative state to be populated. The mean-field force, however, gives a bound component to the experienced potential, which may prevent the trajectory from reaching the dissociative region. A discussion of this incorrect behavior is found in [199].

B. Trajectory Surface Hopping

The simplest way to add a non-adiabatic correction to the classical BO dynamics method outlined above in Section II.B is to use what is known as surface hopping. First introduced on an intuitive basis by Bjerre and Nikitin [200] and Tully and Preston [201], a number of variations have been developed [202–205], and are reviewed in [28,206]. Reference [204] also includes technical details of practical algorithms. These methods all use standard classical trajectories that use the hopping procedure to sample the different states, and so add non-adiabatic effects. A different scheme was introduced by Miller and George [207] which, although based on the same ideas, uses complex coordinates and momenta.

The motivation comes from the early work of Landau [208], Zener [209], and Stueckelberg [210]. The Landau–Zener model is for a classical particle moving on two coupled 1D PES. If the diabatic states cross so that the energy gap is linear with time, and the velocity of the particle is constant through the non-adiabatic region, then the probability of changing adiabatic states is

$$P_{2 \to 1} = \exp\left(\frac{-2\pi H_{12}^2}{\hbar v |F_1 - F_2|} \right) \tag{97}$$

where v is the velocity of the particle, $F_i = -dH_{ii}/dR$ is the force on the ith diabatic surface, and H_{ij} is the Hamiltonian matrix elements in the diabatic basis. All quantities are evaluated at the crossing point. The result is that for high velocity, or small diabatic coupling, the probability of staying on a diabatic surface (changing adiabatic state) approaches 1. The opposite happens for low velocities and strong couplings.

Stueckelberg derived a similar formula, but assumed that the energy gap is quadratic. As a result, electronic coherence effects enter the picture, and the transition probability oscillates (known as Stueckelberg oscillations) as the particle passes through the non-adiabatic region (see [204] for details).

The basic idea is that non-adiabatic interactions occur in localized regions of configuration space, where the adiabatic surfaces are close together, and away from these regions the BO description is a useful one. In the interaction regions, the non-adiabatic interactions are such that they cause population transfer from one state to the other. This can be simulated by the trajectory "hopping" from one surface to the other with a certain probability. The ensemble of trajectories on each state thus simulates the relevant wavepackets, with the population transfer made by the hopping. The trajectories are driven only by a single potential surface, which means that they are able to behave suitably in the asymptotic limit.

In principle, the Landau–Zener formula could be used to calculate a hop probability for a trajectory, but this is often not practical as it requires knowledge about the position of the crossing point. Studies [32,211] indicate instead that the best method for accuracy and simplicity is the fewest switches algorithm [203]. The aim is that the percentage of trajectories in each state equals the state populations with a minimum number of transitions occurring to maintain this. The state populations are provided by integrating the equation for state amplitudes Eq. (94). Changes in the populations over a time step then mean that for a two-state system the probability of a trajectory changing out of state 2 into state 1 is

$$P_{2\to 1} = -\frac{d}{dt}\log|c_2|^2 \tag{98}$$

This expression being set to zero if the right-hand side is negative. The switching probability is then

$$g_{2\to 1} = P_{2\to 1}\Delta t \tag{99}$$

which achieves the desired result. Notice in particular that no switches occur when the coupling is weak as then $P_{2\to 1} \sim 0$.

After a hop has been made, adjustments have to be made to conserve the energy of a trajectory. There is a variety of ways in which this can be done, but the most common way is to rescale the momentum in the direction of the derivative coupling. This has been justified by semiclassical arguments [205,212] and experience [213]. Other possibilities include the Miller–George expression, which is used in [214].

The use of the time-dependent Schrödinger equation to calculate the state populations means that coherence effects due to the electronic states are correctly accounted for, although this coherence is lost if many passes through a non-adiabatic region are made. One drawback of the method is that a double ensemble of trajectories is required for convergence: One is required for the initial conditions, and then each initial trajectory requires an ensemble of hops in the non-adiabatic region to generate good statistics. A second problem is that situations can arise where not enough energy is available to make a predicted hop. These aborted hops means that the state populations are not correctly reflected by the ensemble of trajectories. Despite these problems, the methods have often given good results.

Formulations have also been made that try to combine the best of the Ehrenfest and surface hopping methods. These effectively use the mixed-state approach through a non-adiabatic region, and then force the trajectories to exit the region on a single surface. This can be achieved, for example, by using a complex Hamiltonian to project the electronic wave function into a single adiabatic state after coming out of the non-adiabatic region [199]. Alternatively, a switching function may be used as in the recently proposed continuous surface switching algorithm [215], where the function is designed to preserve the electronic populations over the ensemble of trajectories.

C. Gaussian Wavepackets and Multiple Spawning

The first work on generalizing the Gaussian wavepacket methods to account for non-adiabatic effects was made by Sawada and Metiu [33]. They used a wave function described by a single Gaussian function for the nuclear wavepacket in each electronic state, and derived equations of motion for the Gaussian parameters that are similar to the Heller equations Eqs. (42)–(45), but include terms with the non-adiabatic coupling. This direct Gaussian wavepacket approach has been applied to model systems [216], but the inflexibility of the wave function form makes it unable to obtain more than qualitative information. Recently, the method has been extended to use a harmonic oscillator (Gauss–Hermite) basis set representing the packets on each surface [217], which may add enough flexibility for reasonable results.

A more comprehensive Gaussian wavepacket based method has been introduced by Martínez et al. [35,36,218]. Called the multiple spawning method, it has already been used in direct dynamics studies (see Section V.B), and shows much promise. It has also been applied to adiabatic problems in which tunneling plays a role [219], as well as the interaction of a molecule with an ultrashort laser pulse [220]. The method has two elements. The first part sets up equations of motion for the nuclear wavepacket using Gaussian wavepackets as a basis set. The second part is an algorithm to place basis functions when and where they are required to describe non-adiabatic (or tunneling) events.

In Section III.A, it was shown that the nuclear wavepacket can be represented by a packet associated with each electronic state, χ_i. Each of these packets can be expanded in a set of Gaussian functions, $\chi_\alpha^{(i)}$,

$$\chi_i(\boldsymbol{R}) = \sum_\alpha D_\alpha^{(i)} \chi_\alpha^{(i)}(\boldsymbol{R}) \tag{100}$$

where i labels the different electronic states. While the Gaussian functions evolve along classical trajectories using the Heller equations of motion, Eqs. (42), (43), (45), equations of motion for the expansion coefficients, $D_\alpha^{(i)}$, are obtained from a variational solution of the Schrödinger equation. For the expansion coefficients for the wavepacket on the ith state in vector notation these are

$$\dot{\boldsymbol{D}}^{(i)} = -i\boldsymbol{S}^{-1}[(\boldsymbol{H}^{(ii)} - i\dot{\boldsymbol{S}})\boldsymbol{D}^{(i)} + \boldsymbol{H}^{(ij)}\boldsymbol{D}^{(j)}] \tag{101}$$

$\boldsymbol{H}$ are the Hamiltonian matrices

$$H_{\alpha\beta}^{(ij)} = \langle \chi_\alpha^{(i)} | \hat{H}_{ij} | \chi_\beta^{(j)} \rangle \tag{102}$$

where the operators $\hat{H}_{ij}$ in the adiabatic picture are those in Eqs. (61) and (62). The matrix $\boldsymbol{S}$ is the overlap

$$S_{\alpha\beta} = \langle \chi_\alpha^{(i)} | \chi_\beta^{(i)} \rangle \tag{103}$$

and $\dot{\boldsymbol{S}}$ is related to the time evolution of the overlap of the Gaussian functions

$$\dot{S}_{\alpha\beta} = \langle \chi_\alpha^{(i)} | \frac{\partial}{\partial t} | \chi_\beta^{(i)} \rangle \tag{104}$$

The picture here is of uncoupled Gaussian functions roaming over the PES, driven by classical mechanics. The coefficients then add the quantum mechanics, building up the nuclear wavepacket from the Gaussian basis set. This makes the treatment of non-adiabatic effects simple, as the coefficients are driven by the Hamiltonian matrices, and these elements couple basis functions on different surfaces, allowing transfer of population between the states. As a variational principle was used to derive these equations, the coefficients describe the time dependence of the wavepacket as accurately as possible using the given basis, and if the basis is complete at all times the method will deliver the full quantum wavepacket.

For efficiency the number of Gaussian functions used must be kept as small as possible, otherwise time spent building and inverting the matrices will become prohibitive. The big question is where to put the Gaussian functions for the initially unoccupied state to ensure that they are present in regions of strong non-adiabatic coupling when required. The multiple spawning method does this by generating new functions in non-adiabatic regions when required, that is, when the wavepacket enters the region [218].

For simplicity, imagine that the wavepacket is initially described by a single Gaussian function, which evolves along a trajectory as in the simple Heller method. The first problem is to define when it enters a non-adiabatic region. For a calculation using an adiabatic electronic basis this is done using an effective non-adiabatic coupling [36]

$$H_{ij}^{\text{eff}}(\boldsymbol{R}) = \left|\dot{\boldsymbol{R}} \cdot \boldsymbol{F}_{ij}\right| \tag{105}$$

For diabatic calculations, the equivalent expression uses the diabatic potential matrix elements [218]. When the value of this coupling becomes greater than a pre-defined cutoff, the trajectory has entered a non-adiabatic region. The propagation is continued from this time, t_1, until the trajectory moves out of the region at time t_2.

The time spent in the non-adiabatic region, $t_2 - t_1$, is then divided into N_s equal intervals, where N_s is a predefined parameter. At each interval, a new basis function is "spawned" (generated) on the PES of the initially unoccupied state. In line with the practices of surface hopping, the function is placed at the same position as the parent function, adjusting the momentum along the non-adiabatic coupling vector to conserve energy. Other possible choices for the function placement are discussed in [218]. To avoid the linear dependence of spawned functions, the overlap between the new function and all other basis functions is calculated and the spawn attempt rejected if an overlap is large. The parameter N_s thus controls the number of spawned functions. If it is too small the basis set will be poor, if it is too large, effort will be wasted in generating rejected functions. Calculations should be converged with respect to this parameter, to ensure that the coupling is correctly treated.

The parent and spawned functions provide the basis set for the propagation in the non-adiabatic region, which now needs to be repeated as the evolution of the coefficients $\boldsymbol{D}$ have not yet been calculated including the new functions. The new and old functions are propagated back in time to t_1, and the equations of motion solved anew from this point including coupling between all of them. Any spawned functions that fail to become populated during the passage through the region of non-adiabatic events are subsequently removed.

While it is not essential to the method, frozen Gaussians have been used in all applications to date, that is, the width is kept fixed in the equation for the phase evolution. The widths of the Gaussian functions are then a further parameter to be chosen, although it appears that the method is relatively insensitive to the choice. One possibility is to use the width taken from the harmonic approximation to the ground-state potential surface [221].

As usual there is the question of the initial conditions. In general, more than one frozen Gaussian function will be required in the initial set. In keeping with the frozen Gaussian approximation, these basis functions can be chosen by selecting the Gaussian momenta and positions from a Wigner, or other appropriate phase space, distribution. The initial expansion coefficients are then defined by the equation

$$D_{\alpha}^{(i)} = \sum_{\beta} \boldsymbol{S}_{\alpha\beta}^{-1} \langle \chi_{\beta}^{(i)} | \chi_i(t=0) \rangle \tag{106}$$

where $\boldsymbol{S}$ is the overlap matrix for the Gaussian functions associated with the ith state and $\chi_i(t=0)$ is the initial wavepacket on the ith state.

A technical difference from other Gaussian wavepacket based methods is that the local harmonic approximation has not been used to evaluate any integrals, but instead Martínez et al. use what they term a saddle-point approximation. This uses the localization of the functions to evaluate the integrals by

$$\langle \chi_{\alpha}^{i} | f(\boldsymbol{R}) | \chi_{\beta}^{j} \rangle = \langle \chi_{\alpha}^{i} | \chi_{\beta}^{j} \rangle f(\bar{\boldsymbol{R}}) \tag{107}$$

where $\bar{\boldsymbol{R}} = \langle \chi_{\alpha}^{i} | \hat{R} | \chi_{\beta}^{j} \rangle$ is the center of the function overlap [36]. The quality of this approximation is difficult to ascertain. It does, however, result in significant simplification as only first derivatives are now required for the propagation scheme.

In addition to the full multiple spawning (FMS) described here, in which all basis functions—original and spawned—are coupled, it is also possible to use simplified versions [222]. One possibility is to ignore coupling between spawned functions from different initial starting points. A second possibility, more radical still, is to run trajectories from different starting points independently of one another. This method, which is closer to the other mixed methods discussed above that also use independent trajectories, is called the FMS–M (M for minimal) method [also called the multiple independent spawning (MIS) method [35]]. It should still produce qualitative correct results with significant savings of computational effort due to the smaller size of the matrices $\boldsymbol{H}$ and $\boldsymbol{S}$ involved in the propagation of the expansion coefficients, Eq. (101).

D. Validation of Mixed Methods

How well do these quantum-semiclassical methods work in describing the dynamics of non-adiabatic systems? There are two sources of errors, one due to the approximations in the methods themselves, and the other due to errors in their application, for example, lack of convergence. For example, an obvious source of error in surface hopping and Ehrenfest dynamics is that coherence effects due to the phases of the nuclear wavepackets on the different surfaces are not included. This information is important for the description of short-time (few femtoseconds) quantum mechanical effects. For longer timescales, however, this loss of information should be less of a problem as dephasing washes out this information. Note that surface hopping should be run in an adiabatic representation, whereas the other methods show no preference for diabatic or adiabatic.

A problem in the evaluation of their validity is the lack of exact quantum mechanical results for realistic systems. One-dimensional models covering a range of situations have been used to discuss the performance of the Ehrenfest and surface hopping methods [30,203,205,223]. The results were found to be generally of good accuracy compared to exact quantum mechanical calculations. As expected, Ehrenfest dynamics have problems when trajectories are in mixed states that have very different characteristics. In contrast, surface hopping suffers when trajectories have to recross a region of non-adiabatic coupling many times, due to loss of electronic phase coherence.

Truhlar and co-workers have also made studies of the performance of these two methods applying them to atom–molecule scattering reactions containing non-adiabatic effects [32,211,213,224]. The reaction studied is for the quenching of an excited atom by collision with a diatomic, and these references provide good sources of how to run and analyze semiclassical scattering calculations. Systems both with avoided crossings and conical intersections were examined. In these cases, qualitative agreement was found between the exact calculations and all methods tried. The errors in more detailed properties such as rearrangement channel probabilities are, however, quite large and system dependent. It seems that the continuous surface switching method [215] shows promise, being in general more robust and accurate than the other methods [32]. The same test systems have also been used to test the minimal model of the multiple spawning method, denoted FMS–M, whereby the initial trajectories are independent [222]. This method was found to perform at least as well as fewest switches surface hopping.

Other studies have also been made on the dynamics around a conical intersection in a model 2D system, both for dissociative [225] and bound-state [226] problems. Comparison between surface hopping and exact calculations show reasonable agreement when the coupling between the surfaces is weak, but larger errors are found in the strong coupling limit.

Müller and Stock [227] used the vibronic coupling model Hamiltonian, Section III.D, to compare surface hopping and Ehrenfest dynamics with exact calculations for a number of model cases. The results again show that the semiclassical methods are able to provide a qualitative, if not quantitative, description of the dynamics. A large-scale comparison of mixed method and quantum dynamics has been made in a study of the pyrazine absorption spectrum, including all 24 degrees of freedom [228]. Here a method related to Ehrenfest dynamics was used with reasonable success, showing that these methods are indeed able to reproduce the main features of the dynamics of non-adiabatic molecular systems.

V. DIRECT DYNAMICS OF NON-ADIABATIC SYSTEMS

In the preceeding sections, the dynamics theory required to study non-adiabatic systems has been outlined. Now, a review will be made of direct dynamics studies on such systems in the literature. The number of studies is small, but growing. A range of photochemical systems have been covered, mostly using MCSCF electronic wave functions, but semiempirical methods have also been used to study some large molecules. Studies using the MMVB empirical Hamiltonian are also included. Although no wave function is explicitely calculated, the Hamiltonian is a matrix for which the integrals are parametrised against CASSCF calculations, and the surfaces are calculated on-the-fly from this matrix rather than from an analytic function. These are thus direct dynamics studies in the sense that they simulate CASSCF direct dynamics calculations at a low cost, so enable valuable experience to be gained in this new field.

The aim here is not to give exhaustive descriptions, but to emphasize the questions being asked and the information obtained. With a few exceptions the studies are mechanistic in nature, and we will show the additional, sometimes critical, insight gained over traditional nondynamics studies.

A. Using CASSCF Methods

To use direct dynamics for the study of non-adiabatic systems it is necessary to be able to efficiently and accurately calculate electronic wave functions for excited states. In recent years, density functional theory (DFT) has been gaining ground over traditional Hartree–Fock based SCF calculations for the treatment of the ground state of large molecules. Recent advances mean that so-called time-dependent DFT methods are now also being applied to excited states. Even so, at present, the best general methods for the treatment of the photochemistry of polyatomic organic molecules are MCSCF methods, of which the CASSCF method is particularly powerful.

MCSCF methods describe a wave function by the linear combination of M configuration state functions (CSFs), Φ_K, with CI coefficients, C_K,

$$\Psi(\boldsymbol{r}) = \sum_{K=1}^{M} C_K \Phi_K \tag{108}$$

In practice, each CSF is a Slater determinant of molecular orbitals, which are divided into three types: inactive (doubly occupied), virtual (unoccupied), and active (variable occupancy). The active orbitals are used to build up the various CSFs, and so introduce flexibility into the wave function by including configurations that can describe different situations. Approximate electronic-state wave functions are then provided by the eigenfunctions of the electronic Hamiltonian in the CSF basis. This contrasts to standard HF theory in which only a single determinant is used, without active orbitals. The use of CSFs, gives the MCSCF wave function a structure that can be interpreted using chemical pictures of electronic configurations [229]. An interpretation in terms of valence bond structures has also been developed, which is very useful for description of a chemical process (see the appendix in [230] and references cited therein).

The MCSCF method then optimizes both the molecular orbitals, represented as usual in SCF calculations by linear combinations of atomic orbitals (LCAO), and the CI expansion coefficients to obtain the variational wave function for one state. The optimization of the orbitals for a particular state, however, will not converge if a degeneracy, or a near degeneracy, of states is present, as the wave function will have problems following a single state. To overcome this, state-averaged orbitals (SA–MCSCF) must be used [231,232]. Rather than optimizing a single eigenvalue of the Hamiltonian matrix, an averaged energy function is used so that the orbitals describe all the states of interest simultaneously to the same accuracy.

CASSCF is a version of MCSCF theory in which all possible configurations involving the active orbitals are included. This leads to a number of simplifications, and good convergence properties in the optimization steps. It does, however, lead to an explosion in the number of configurations being included, and calculations are usually limited to 14 electrons in 14 active orbitals.

A simple example would be in a study of a diatomic molecule that in a Hartree–Fock calculation has a bonded σ orbital as the highest occupied MO (HOMO) and a σ^* lowest unoccupied MO (LUMO). A CASSCF calculation would then use the two σ electrons and set up four CSFs with single and double excitations from the HOMO into the σ^* orbital. This allows the bond dissociation to be described correctly, with different amounts of the neutral atoms, ion pair, and bonded pair controlled by the CI coefficients, with the optimal shapes of the orbitals also being found. For more complicated systems the orbitals involved in a particular process must be selected and included in the captive space. This is both the strength and weakness of the method. Only the important orbitals are used, so accurate calculations can be made relatively cheaply. If the active space is, however, badly chosen, this may lead to qualitatively incorrect results due to imbalances in the basis set.

Importantly for direct dynamics calculations, analytic gradients for MCSCF methods [124–126] are available in many standard quantum chemistry packages. This is a big advantage as numerical gradients require many evaluations of the wave function. The evaluation of the non-Hellmann–Feynman forces is the major effort, and requires the solution of what are termed the coupled-perturbed MCSCF (CP–MCSCF) equations. The large memory requirements of these equations can be bypassed if a direct method is used [233]. Modern computer architectures and codes then make the evaluation of first and second derivatives relatively straightforward in this theoretical framework.

Using MCSCF methods it is also possible to obtain the non-adiabatic coupling terms using analytic procedures [232,234,235]. SA–MCSCF must again be used in the calculation of the non-adiabatic coupling elements, as the functions for the two states must be described to the same level of accuracy. One important point to note is that the derivative coupling matrix elements contain a relative phase between the functions of the coupled states that must remain continuous along a trajectory. It is possible that standard computer packages ignore this phase, as it is not important for static properties, resulting in a random phase being generated as the geometry is changed. This can be eliminated by comparison between orbitals at neighboring steps [236].

1. The MMVB Method

The present high cost of full CASSCF direct dynamics means that it is not possible to use such calculations to run large numbers of trajectories. As a result it cannot be used to build up experience of the types of effects to be found in dynamical studies of organic photochemistry, and in their interpretation. This problem can be remedied by performing calculations using the MMVB force field [63,64].

MMVB is a hybrid force field, which uses MM to treat the unreactive molecular framework, combined with a valence bond (VB) approach to treat the reactive part. The MM part uses the MM2 force field [58], which is well adapted for organic molecules. The VB part uses a parametrized Heisenberg spin Hamiltonian, which can be illustrated by considering a two orbital, two electron description of a sigma bond described by the VB determinants

$$|\phi_a(1)\bar{\phi}_b(2)| \qquad |\phi_b(1)\bar{\phi}_a(2)| \tag{109}$$

The Hamiltonian in the basis set of these configurations is

$$H_s = \begin{vmatrix} Q_{ab} & K_{ab} \\ K_{ab} & Q_{ab} \end{vmatrix} \tag{110}$$

where Q_{ab} and K_{ab} are the usual Coulomb and exchange integrals in the atomic basis. The eigenvalues and eigenfunctions of this determinant are the ground-state singlet and excited-state triplet functions. If the integrals are fitted as functions of bond length to the full CASSCF values, then this Hamiltonian can be used as a model of the PES. Adding many determinants, a model Hamiltonian for a complicated molecular system can be built up.

The method has been validated by comparison against full CASSCF calculations for a number of systems (see references below and the references cited therein). In general, the topology of the surface is faithfully reproduced, although the energetics may sometimes differ. Dynamics calculations have been made using this force field on a number of systems. In most cases, a simple surface hopping model, based on the fewest switches method described above, was used. A trajectory is propagated on the initial (upper) surface until the state population, calculated by solving Eq. (94), approaches a value of 0.5, when an unconditional hop is made. No return hop was then considered. The initial conditions were chosen by adding random energy, up to a given threshold, to the normal modes. This gives a wavepacket character to the set of trajectories.

For the mechanistic studies made, this protocol is able to give information about how dynamical properties affect the evolution of a photochemical reaction, but is not accurate enough for quantitative results. The information obtained relates to aspects of the surface such as the relative steepness of regions on the lower slopes of the conical intersection, and the relative width of alternative channels.

The first study was made on the benzene molecule [79]. The S_0/S_1 photochemistry of benzene involves a conical intersection, as the fluorescence vanishes if the molecule is excited with an excess of 3000 cm^{-1} of energy over the excitation energy, indicating that a pathway is opened with efficient nonradiative decay to the ground state. After irradiation, most of the molecules return to benzene. A low yield of benzvalene, which can lead further to fulvene, is, however, also obtained.

Calculations indicate that the S_1 surface does have a conical intersection leading to the ground state [237], as well as a minimum. The starting point for the trajectories was then taken as halfway between the minimum and the lowest energy conical intersection point, and the vector connecting these points taken as the reaction coordinate. Various trajectories were started with different random sampling of the normal modes orthogonal to the reaction path, and adding different excess energy in the form of momentum along this path.

Trajectories with low excess energies do not reach the conical intersection, but are trapped in the S_1 minimum and lead to fluorescence. Increasing the excess energy leads to nonradiative transfer to the ground state, seen by hops on the trajectory, which indicates quenching of the fluorescence. Interestingly, higher excess energies lead to a higher proportion of benzvalene being formed after crossing between the states. Increasing the width of a packet, however, leads to a decrease of benzvalene production. This is related to the fine details of the intersection topology, the benzvalene structure is on a narrow plateau, and by spreading the wavepacket or reducing the excess energy the wavepacket "falls off" the plateau back into the benzene minimum.

A second study [80] looked at the anomalous fluorescence of azulene (from S_2 rather than S_1), which has been known about for many years. Despite a paper from Beer and Longuet-Higgins [238] suggesting fast $S_1 \rightarrow S_0$ internal conversion via an intersection, this system has a long history of measurements trying to ascertain the mechanism. These conclusively show that the lifetime of the S_1 state is under 1 ps. The MMVB dynamics calculations support these findings by showing that, not only is there a conical intersection between the surfaces, but also that a nuclear wave packet would find the intersection within a single vibrational period. This results in extremely efficient internal conversion.

Trajectories were run from around the Franck–Condon point. Even starting with no excess energy, that is, at the Franck–Condon point with an initial momentum of zero, the energy in running down into the S_1 minimum is enough to reach the conical intersection and cross to the ground state. Increasing the excess energy by sampling the normal modes does not change the general picture as they all find the crossing within a vibrational period. It is, however, found that at higher energies trajectories cross with increasingly large S_1–S_0 energy gap. This can be simply understood from the effective increase in non-adiabatic coupling due to the higher momentum [see Eq. (94)]. These trajectories are thus crossing near to, but not at, the conical intersection.

The dynamics after excitation of fulvene similarly shows that high-energy starting points can cross away from the minimum energy conical intersection point [81]. The ground-state equilibrium structure is planar. The S_1 surface has a double minima on the crossing where the methyl group is rotated perpendicular to the ring in either direction. Crossing via these minima could thus lead to cis–trans isomerization. Increased kinetic energy leads, however, to crossing where the structure remains planar, and so isomerization is not likely to take place.

A more demanding dynamical study aimed to rationalize the product distribution in photochemical cycloaddition, looking at butadiene–butadiene [82]. A large number of products are possible, with two routes on the excited S_1 state leading back to channels on the ground state. The results are promising, as the MMVB dynamics find the major products found experimentally. They also indicate that the distributions depend on the initial conditions, and thus altering the experimental parameters could lead to different product ratios. As the exact decay path is unknown, it is assumed that the reaction must go through the two transitions states on S_1, and so the initial conditions sampled around these points. No sampling was made along the transition vector, so the trajectory could find its own (excess energy dependent) path across the PES to the products.

One transition state involves an intermolecular interaction between the two butadiene molecules. With low excess energy, half of the trajectories run into the minimum on the S_1 surface (resulting in fluorescence), while the other half cross to the ground state—most of which ends as 1,3-divinylcyclobutane, with some unreacted butadiene. Increasing the excess energy leads to a lower rate of crossing, indicating that the channel from the transition state to the S_1 minimum is wider than that for crossing to S_0. The other transition state involves an intramolecular (bonded) interaction between the butadiene molecules. Over 90% of trajectories now run to the ground state, and 40% end up as the major photoproducts. These results are fairly independent of initial conditions, due to the steepness of the PES at the transition state that produces a large kinetic energy at the crossing point.

The photochemistry of polyenes is another complicated process. The MMVB dynamics with surface hopping has also been used to study what happens after photoexcitation in the alternant hydrocarbons C_6H_8, C_8H_{10}, and $C_{12}H_{14}$ [83]. A conical intersection has been identified between S_1 and S_0 that involves an out-of-plane $-(CH)_3-$ kink, with four unpaired electrons spread over the three methyl groups. Two paths have also been identified from this intersection back to the ground state. One is direct relaxation, with reformation of the ground-state double bonds. The second is more interesting, and has a plateau with a π-diradical structure, with a π bond sandwiched between two radical structures, for example, $(C_1{=}C_2{-}C_3)^{\bullet}{-}C_4{=}C_5{-}(C_6)^{\bullet}$. Such structures under the name of neutral soliton pairs have been used to explain the absorption spectrum in polyacetylene.

Trajectories starting from structures sampled around the conical intersection are found to decay by three different mechanisms. The first is direct decay, while the second and third involve the π-diradical structures. Interestingly, even though there is no minimum on the PES for these structures, a trajectory can become locked into this region of configuration space for significant amounts of time—and in the case of $C_{10}H_{12}$ up to 1 ps. This stability is seen to be due to motion of the $-(CH)_3-$ kink along the chain. "Locked" and "direct" trajectories have also been found in ground-state dynamics simulations, where they have been related to statistical and nonstatistical distributions of products, respectively [239].

The MMVB force field has also been used with Ehrenfest dynamics to propagate trajectories using mixed-state forces [84]. The motivation for this is that surface hopping may not cope too well with situations where trajectories pass more than once through a non-adiabatic region. Hexatriene and azulene provide two contrasting conical intersections. In the classification of Atchity et al. [189] (see Fig. 6), the former has a peaked, while the latter has a sloped topology. The sloped intersection results in multiple passes.

In both cases, about one-third of the trajectories decay directly to the ground-state. The remaining trajectories form mixed-states before decaying. For hexatriene, this decay is a steady process. Studying trajectories around the peaked conical intersection run separately on the two surfaces, the trajectories on the lower surface leave the non-adiabatic region immediately. On the upper surface, however, the trajectory stays near this region. As a result, the mixed-state trajectory is held near the intersection until decay has progressed far enough for the ground-state surface to dominate and the system moves away. In contrast, for azulene the population transfer takes place stepwise, each step corresponding to a recrossing of the non-adiabatic region. Such a stepwise transfer is compatible with time-resolved measurements [240]. Averaging over the trajectories produces a biexponential decay, again a behavior observed experimentally. These calculations support the idea that Ehrenfest dynamics perform well for bound-state systems—recrossings ensure that the system is not trapped in a mixed state.

Ehrenfest dynamics with the MMVB method has also been applied to the study of intermolecular energy transfer in anthryl–naphthylalkanes [85]. These molecules have a naphthalene joined to a anthracene by a short alkyl $-(CH)_n-$ chain. After exciting the naphthalene moiety, if $n = 1$ emission is seen from both parts of the system, if $n = 3$ emission is exclusively from the anthracene. The mechanism of this energy exchange is still not clear. This system is at the limits of the MMVB method, and the number of configurations required means that only a small number of trajectories can be run. The method is also unable to model the zwitterionic states that may be involved. Even so, the calculations provide some mechanistic information, which supports a stepwise exchange of energy, rather than the conventional direct process.

2. *Direct Dynamics*

The first study in which a full CASSCF treatment was used for the non-adiabatic dynamics of a polyatomic system was a study on a model of the retinal chromophore [86]. The cis–trans photoisomerization of retinal is the primary event in vision, but despite much study the mechanism for this process is still unclear. The minimal model for retinal is 2-cis-$C_5H_6NH_2^+$, which had been studied in an earlier quantum chemistry study [230]. There, it had been established that a conical intersection exists between the S_1 and S_0 states with the cis–trans defining torsion angle at approximately $\alpha = 80°$ (cis is at 0°). Two paths run away from this intersection, leading to either trans (products) or cis (reactants) isomers.

Four trajectories were run, starting at the Franck–Condon point, varying the torsion angle α from 0–20°. In all cases, the same behavior was seen. Initial motion away from the Franck–Condon region involved stretching motion along the molecular backbone. After ~ 60 fs, the motion then changes, with energy being transfered to the torsional mode. This motion then takes the system to the intersection, and the resulting (diabatic) hop takes the system to the trans isomer. This dynamic behavior is consistent with calculations on retinal using semiempirical surfaces [241], and using adiabatic direct dynamics on the excited state [242]. It also supports the use of low-dimensional models that have been used in quantum mechanical calculations on retinal [196].

Model systems for cyanine dyes have also been studied [87]. In this case, it is important to understand the mechanism by which relaxation to the ground-state occurs so as to design efficient dye molecules, that is, without fast internal conversion. The simplest model is trans-NH_2—$(CH)_3$–NH_2^+. Although this molecule has a structural similarity to the retinal model investigated above, the dynamics after photoexcitation are quite different. A trajectory starting from near the Franck–Condon point is sketched in Figure 9. The initial motion is dominated by conrotatory torsional motion around the C—C bonds, which after 50 fs changes to disrotatory motion. This last only 20 fs until the molecule reaches the minima on the S_1 surface. Here, the torsion remains twisted at $\sim 104°$, and large amplitude motion involving skeletal stretching and pyramidalization of a terminal nitrogen atom. The system oscillates in the minima for ~ 50 fs, before crossing to the ground-state near the conical intersection. This crossing leads to the cis conformer, and so isomerization has taken place.

This behavior is consistent with experimental data. For high-frequency excitation, no fluorescence rise-time and a biexponential decay is seen. The lack of rise-time corresponds to a very fast internal conversion, which is seen in the trajectory calculation. The biexponential decay indicates two mechanisms, a fast component due to direct crossing (not seen in the trajectory calculation but would be the result for other starting conditions) and a slow component that samples the excited-state minima (as seen in the trajectory). Long wavelength excitation, in contrast, leads to an observable rise time and monoexponential decay. This corresponds to the dominance of the slow component, and more time spent on the upper surface.

B. Ab Initio Multiple Spawning

The multiple spawning method described in Section IV.C has been applied to a number of photochemical systems using analytic potential energy surfaces. As well as small scattering systems [36,218], the large retinal molecule has been treated [243,244]. It has also been applied as a direct dynamics method,

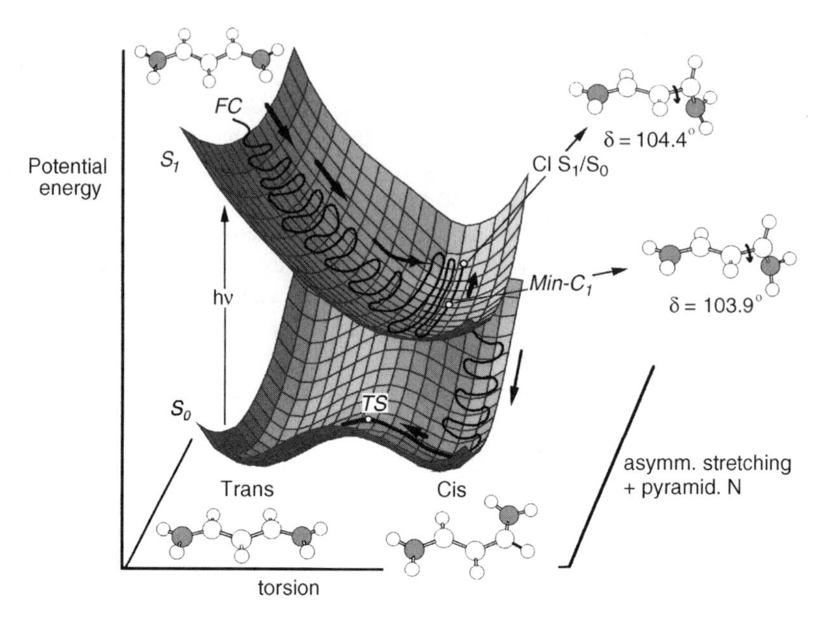

Figure 9. Schematic reprsentation of a classical trajectory moving on the S_1 and S_0 energy surfaces of the NH_2–$(CH)_3$–NH_2^+ trans→cis photoisomerization, starting near the planar Franck–Condon geometry. The geometric coordinates are (*a*) torsion of the C_2–C_3 and C_3–C_4 bonds and (*b*) asymmetric stretching coupled with pyramidalization. Both S_1 and S_0 intersect at a conical intersection (**S$_1$/S$_0$ CI**) located near the minimum of the S_1 surface (**Min-C$_1$**) where the $C_2C_3C_4N_5$ torsion angle is 104°. [Reproduced with permission from [87]. Copyright © 2000 Amercian Chemical Society].

combining the basic algorithm with quantum chemical input for the potential surfaces and non-adiabatic coupling elements, when it is called AIMS.

In contrast to both the surface hopping and Ehrenfest methods, for the spawning method it is necessary to calculate the full non-adiabatic operator, Eq. (53), as the equation of motion for the expansion coefficients, Eq. (101) involves the full Hamiltonian in the matrix elements of $\boldsymbol{H}$. To simplify the calculations, the orbital contribution to the derivative coupling was ignored in the calculation of the non-adiabatic coupling matrix, $\boldsymbol{F}_{ij}$. The second-order derivative terms, G_{ij}, were also ignored. In fact, in all studies second-derivative (Hessian) information was not used. A standard Burlisch–Stoer integrator was used in place of the Helgaker–Chen algorithm, with step sizes ~ 0.25 fs. Small steps of 0.025 fs were required in non-adiabatic regions. The saddle-point approximation for the evaluation of the matrix elements in the Gaussian basis set also means that the Hessian is not required for the description of the potential surface.

The first direct dynamics application of the spawning method was on the collision dynamics of sodium iodide [245,246]. This is a classic diatomic system in which the lowest two adiabatic electronic states change character on dissociation. Thus the neutral atoms approach each other on the ground-state, and when close enough together an electron is exchanged to form the ionic species. As a 1D nuclear problem this system does not contain a conical intersection, but the non-adiabatic coupling is still strong where the surfaces come close together. This is the crossing point, where the electron transfer takes place.

This system is small enough that the full multiple spawning method, with coupled trajectories could be applied. The number of spawns, however, was restricted to one per pass of the intersection. In the system studied, the atoms cross the non-adiabatic region on their approach. As there is no third body to remove the energy, after formation of the ionic collision complex the atoms bounce off the repulsive wall, pass back through the non-adiabatic region, and separate. Some population is now in the excited state, and so a proportion separate as ions. The biggest challenge here was to develop a method to describe the long-range harpoon mechanism that is involved in the electron exchange. For this a method termed the localized molecular orbital/generalized valence bond (LMO/GVB) method has been developed, which combines the pictorial nature of VB theory with the computational efficiency of an MO method.

The second system studied was the quenching of excited Li($2p$) by collision with H_2 [236], which is a simple system for the study of energy transfer. The electronic wave function was treated using configurations based on Hartree–Fock molecular orbitals. The orbitals were "occupation averaged", which means that the lithium valence electron was split between the ground-state HOMO and LUMO [$\sim$ the Li ($2s$) and Li ($2p$)]. This is a simplified form of the state averaging used in the SA–CASSCF methods mentioned above, used to prevent bias of the basis toward the ground state. The singly occupied HOMO and LUMO then provide the reference configurations. A basis set with all single excitations from the references were then used as a basis set for the wave function. As the reference orbitals were not reoptimized this is termed a CAS–CI rather than a full SA–CASSCF calculation.

There is a sloped conical intersection between the adiabatic states. Thus as the reactants approach on the upper surface, they are seen to cross to the ground state, followed shortly afterward by a recrossing to the upper state. The system is thus only partially quenched, parting still in the excited state. Up to 25 initial functions were included, and up to 15 functions per state were spawned, resulting in a large nuclear basis set for the description of the problem. Interestingly, even though a low-impact energy was used, some trajectories were found to describe reactions of $Li + H_2 \rightarrow LiH + H$.

After this, Martínez and Ben-Nun applied the method to the photoexcitation of ethylene [88,247]. The lowest energy excitation is the HOMO–LUMO $\pi \rightarrow \pi^*$ transition. These states are labeled $N^1\Sigma_g^+$ and $V^1\Sigma_u^+$. Close in energy to the excited state is also the doubly excited Z^1A_1. The electronic wave function was again treated using the occupation averaged orbital reference configurations for the three states to build a set of single excitation configurations. The lowest two eigenfunctions of this CI space were then taken to form the surfaces of interest in the AIMS studies, ignoring any non-adiabatic coupling between V and Z states.

A number of independent trajectories, with up to 10 spawns each, were run to study the dynamics after excitation, with the initial conditions taken from the Wigner distribution. The results shows that initial motion is along the torsional motion to form the D_{2d} twisted conformation. After a slight lag of 50–250 fs, this structure starts to distort by pyramidalization of one of the ethylene groups. Crucially for the system dynamics, this leads to a conical intersection between S_1 and S_0. At this point, the system relaxes to the ground-state, but with an efficiency much less than 100% per pass of the intersection region. Interestingly, the character of the wave function at this point indicates that in fact the molecule is in the Z state, which in the distorted structure lies lower than the V. A study of the ethylene PES using more sophisticated quantum chemical methods [248] supports the observations from the dynamics that the relaxation mechanism for this system is not from the twisted structure as conventionally thought.

In an ambitious study, the AIMS method was used to calculate the absorption and resonance Raman spectra of ethylene [221]. In this, sets starting with 10 functions were calculated. To cope with the huge resources required for these calculations the code was parallelized. The spectra, obtained from the auto-correlation function, compare well with the experimental ones. It was also found that the non-adiabatic processes described above do not influence the spectra, as their profiles are formed in the time before the packet reaches the intersection, that is, the observed dynamic is dominated by the torsional motion. Calculations using the Condon approximation were also compared to calculations implicitly including the transition dipole, and little difference was seen.

C. Other Studies

Jones et al. [144,214] used direct dynamics with semiempirical electronic wave functions to study electron transfer in cyclic polyene radical cations. Semiempirical methods have the advantage that they are cheap, and so a number of trajectories can be run for up to 50 atoms. Accuracy is of course sacrificed in comparison to CASSCF techniques, but for many organic molecules semiempirical methods are known to perform adequately.

The AM1 Hamiltonian with a 2×2 CAS-CI (two electrons in the space of the HOMO and LUMO) was used to describe the surfaces and coupling elements. The electron-transfer process studied takes place on the ground-state, with the upper state providing "diabatic effects," that is, passage to this surface can delay, or even hinder, the transfer process. A surface hopping approach was used for the dynamics with a Landau–Zener hopping probability and using the Miller–George correction for the momentum after a hop. The charge distribution was used to describe the positions along the reaction coordinate with charge localization on the left and right corresponding to reactant and product, and the symmetric delocalized charge denoting the non-adiabatic region. The studies used trajectories taken from thermalized ensembles to provide detailed dynamic information for the transfer processes, and the relationship between energy gap, electronic coupling between states and rates of transfer.

A final study that must be mentioned is a study by Hartmann et al. [249] on the ultrafast spectroscopy of the Na_3F_2 cluster. They derived an expression for the calculation of a pump–probe signal using a Wigner-type density matrix approach, which requires a time-dependent ensemble to be calculated after the initial excitation. This ensemble was obtained using fewest switches surface hopping, with trajectories initially sampled from the thermalized vibronic Wigner function vertically excited onto the upper surface.

The process of interest is the photoisomerization taking place via a conical intersection, which is reached after the breaking of two bonds. The electronic structure problem was solved using a simplified restricted open-shell Hartree–Fock (ROHF) procedure, which seems to produce reasonable results for this system at a low cost, and for which analytic gradients and non-adiabatic coupling elements are possible. As a result, a connection between the pump–probe signal and the underlying dynamics could be made. For example, timescales for the breaking of the two bonds, and for reaching the conical intersection could be ascertained.

VI. SUMMARY AND CONCLUSIONS

For the understanding of photochemical systems it is necessary to look carefully at non-adiabatic effects, as these may provide unexpected pathways for efficient transitions between electronic states. In cases where non-adiabatic coupling is strong, which is seen as conical intersections or avoided crossings between adiabatic PES, dynamic effects also need to be considered. This is because where a system undergoes interstate crossing depends not only on the PES topology, but on the initial conditions, that is, the spread of momenta and positions in the wavepacket after the interaction with the light field.

This dependency is seen in the Landau–Zener expression for the probability of a classical particle changing states while moving through a non-adiabatic region, Eq. (97), and in the Ehrenfest dynamics expression for the state amplitudes, Eq. (94), which both depend on the particle velocity. It is also seen in the dynamics calculations reviewed in Section V.A, where higher initial kinetic energy often leads to crossing away from the lowest energy point on an intersection seam. Where the crossing occurs is important, as this determines the outcome of a photochemical process—returning to the ground state at different points may lead to different products.

Full quantum wavepacket studies on large molecules are impossible. This is not only due to the scaling of the method (exponential with the number of degrees of freedom), but also due to the difficulties of obtaining accurate functions of the coupled PES, which are required as analytic functions. Direct dynamics studies of photochemical systems bypass this latter problem by calculating the PES on-the-fly as it is required, and only where it is required. This is an exciting new field, which requires a synthesis of two existing branches of theoretical chemistry—electronic structure theory (quantum chemistry) and mixed nuclear dynamics methods (quantum-semiclassical).

Quantum chemical methods, exemplified by CASSCF and other MCSCF methods, have now evolved to an extent where it is possible to routinely treat accurately the excited electronic states of molecules containing a number of atoms. Mixed nuclear dynamics, such as swarm of trajectory based surface hopping or Ehrenfest dynamics, or the Gaussian wavepacket based multiple spawning method, use an approximate representation of the nuclear wavepacket based on classical trajectories. They are thus able to use the information from quantum chemistry calculations required for the propagation of the nuclei in the form of forces. These methods seem able to reproduce, at least qualitatively, the dynamics of non-adiabatic systems. Test calculations have now been run using direct dynamics, and these show that even a small number of trajectories is able to produce useful mechanistic information about the photochemistry of a system. In some cases it is even possible to extract some quantitative information.

Many problems still remain, in particular the question of the best dynamics method is still not clear. Surface hopping is the simplest method, but suffers from its ad hoc nature—it is not possible to say when it will fail, although the fewest switched method seems reasonably reliable as long as a non-adiabatic region is not recrossed. It may also suffer from slow convergence. Ehrenfest dynamics has some intuitive appeal, and correctly treats electronic coherences through the non-adiabatic region. The problem of getting stuck in a mixed state can, however, be serious, although this does not seem to be the case in bound-state photochemical systems. The continuous surface switching method may solve this problem. Using Gaussian wavepackets to include some quantum effects into the nuclear dynamics using multiple spawning also has some advantages, although the overhead seems to be large.

Being able to run direct dynamics calculations will add an extra, important, tool to help chemists understand photochemical systems. This chapter has outlined the present standpoint of the theory and practice of such calculations showing that, although much work remains to be done, they are already bringing new insight to mechanistic studies of photochemistry.

APPENDIX A: THE NUCLEAR SCHRÖDINGER EQUATION

The starting point for the theory of molecular dynamics, and indeed the basis for most of theoretical chemistry, is the separation of the nuclear and electronic motion. In the standard, adiabatic, picture this leads to the concept of nuclei moving over PES corresponding to the electronic states of a system.

In its Cartesian form, the Hamiltonian can be written

$$\hat{H}(\boldsymbol{R},\boldsymbol{r}) = \hat{T}_{\mathrm{n}}(\boldsymbol{R}) + \hat{V}_{\mathrm{nn}}(\boldsymbol{R}) + \hat{T}_{\mathrm{e}}(\boldsymbol{r}) + \hat{V}_{\mathrm{ee}}(\boldsymbol{r}) + \hat{V}_{\mathrm{en}}(\boldsymbol{R},\boldsymbol{r}) \tag{A.1}$$

$$= \hat{T}_{\mathrm{n}}(\boldsymbol{R}) + \hat{H}_{\mathrm{el}}(\boldsymbol{R},\boldsymbol{r}) \tag{A.2}$$

where $\hat{T}$ is the kinetic energy operator, and $\hat{V}$ the potential energy operator with subscripts n and e relating to the nuclei and electrons, respectively. The second line sums together the last four terms of Eq. (A.1) to define the clamped nucleus electronic Hamiltonian, $\hat{H}_{\mathrm{el}}$, which depends on both the electronic and nuclear coordinates.

The separation of nuclear and electronic motion may be accomplished by expanding the total wave function in functions of the electron coordinates, $\boldsymbol{r}$, parametrically dependent on the nuclear coordinates

$$\Psi(\boldsymbol{R},\boldsymbol{r},t) = \sum_i \chi_i(\boldsymbol{R},t)\psi_i^{\mathrm{ad}}(\boldsymbol{r};\boldsymbol{R}) \tag{A.3}$$

Further, the time-independent electronic basis functions are taken to be the eigenfunctions of the electronic Hamiltonian,

$$\hat{H}_{\mathrm{el}}(\boldsymbol{r};\boldsymbol{R})\psi_i^{\mathrm{ad}}(\boldsymbol{r};\boldsymbol{R}) = V_i\psi_i^{\mathrm{ad}}(\boldsymbol{r};\boldsymbol{R}) \tag{A.4}$$

and there is one set of eigenfunctions for each value of $\boldsymbol{R}$. This is known as the Born representation [250]. The superscript ad denotes the functions as "adiabatic."

Substituting the expansion Eq. (A.3) in the time-dependent Schrödinger equation, Eq. (1), and multiplying from the left by the bra $\langle\psi_i^{\mathrm{ad}}|$ leads to

$$\sum_j \langle\psi_i^{\mathrm{ad}}|\hat{T}_n|\psi_j^{\mathrm{ad}}\chi_j\rangle + V_i|\chi_i\rangle = i\hbar\frac{\partial}{\partial t}|\chi_i\rangle \tag{A.5}$$

Introducing the non-adiabatic operators

$$\hat{\Lambda}_{ij}(R) = \hat{T}_n - \langle \psi_i^{\text{ad}} | \hat{T}_n | \psi_j^{\text{ad}} \rangle \tag{A.6}$$

$$= \langle \psi_i^{\text{ad}} | [\hat{T}_n, | \psi_j^{\text{ad}} \rangle] \tag{A.7}$$

the time-dependent Schrödinger equation can finally be written as

$$(\hat{T}_n + V_i)|\chi_i\rangle - \sum_j \hat{\Lambda}_{ij}|\chi_j\rangle = i\hbar \frac{\partial}{\partial t}|\chi_i\rangle \tag{A.8}$$

The familiar BO approximation is obtained by ignoring the operators $\hat{\boldsymbol{\Lambda}}$ completely. This results in the picture of the nuclei moving over the PES provided by the electrons, which are moving so as to instantaneously follow the nuclear motion. Another common level of approximation is to exclude the off-diagonal elements of this operator matrix. This is known as the Born–Huang, or simply the adiabatic, approximation (see [250] for further details of the possible approximations and nomenclature associated with the nuclear Schrödinger equation).

Finally, we shall look briefly at the form of the non-adiabatic operators. Taking the kinetic energy operator in Cartesian form, and using mass-scaled coordinates $R_\alpha = \sqrt{M_\alpha} x_\alpha$, where M_α is the nuclear mass associated with the αth nuclear coordinate,

$$\hat{T} = \sum_{\alpha=1}^{3N} -\frac{\hbar^2}{2} \frac{\partial^2}{\partial R_\alpha^2} = -\frac{\hbar^2}{2} \nabla^2 \tag{A.9}$$

the non-adiabatic operators can be written as

$$\hat{\Lambda}_{ij} = -\frac{\hbar^2}{2} \left(\nabla^2 - \langle \psi_i^{\text{ad}} | \nabla^2 | \psi_j^{\text{ad}} \rangle \right) \tag{A.10}$$

$$= \frac{\hbar^2}{2} \left(G_{ij} + 2\boldsymbol{F}_{ij} \cdot \boldsymbol{\nabla} \right) \tag{A.11}$$

the expression in Eq. (53) in Section III.A. Both G_{ij} and $\boldsymbol{F}_{ij}$ involve the first and second derivatives of the adiabatic electronic functions with the nuclear coordinates

$$G_{ij} = \langle \psi_i^{\text{ad}} | \nabla^2 \psi_j^{\text{ad}} \rangle \tag{A.12}$$

$$F_{ij}^\alpha = \langle \psi_i^{\text{ad}} | \nabla_\alpha \psi_j^{\text{ad}} \rangle \tag{A.13}$$

Note that these matrix elements are numbers, in comparison with the term on the right-hand side of Eq. (A.10), which involves matrix elements of the second derivative operator. Notice also that the product of $\boldsymbol{F}_{ij}$ and $\boldsymbol{\nabla}$ results in a nonlocal operator on the nuclear coordinate space.

The elements of the matrix $\boldsymbol{G}$ can be written in terms of $\boldsymbol{F}$, which is called the non-adiabatic coupling matrix. For a particular coordinate, α, and dropping the subscript for clarity,

$$\frac{\partial}{\partial R}\left\langle \psi_i^{\text{ad}} \middle| \frac{\partial \psi_j^{\text{ad}}}{\partial R} \right\rangle = \left\langle \frac{\partial \psi_i^{\text{ad}}}{\partial R} \middle| \frac{\partial \psi_j^{\text{ad}}}{\partial R} \right\rangle + \left\langle \psi_i^{\text{ad}} \middle| \frac{\partial^2 \psi_j^{\text{ad}}}{\partial^2 R} \right\rangle \tag{A.14}$$

As the eigenfunctions form a complete set

$$\left\langle \frac{\partial \psi_i^{\text{ad}}}{\partial R} \middle| \frac{\partial \psi_j^{\text{ad}}}{\partial R} \right\rangle = \sum_k \left\langle \frac{\partial \psi_i^{\text{ad}}}{\partial R} \middle| \psi_k^{\text{ad}} \right\rangle \left\langle \psi_k^{\text{ad}} \middle| \frac{\partial \psi_j^{\text{ad}}}{\partial R} \right\rangle \tag{A.15}$$

and as the derivative operator is anti-Hermitian,

$$\left\langle \psi_i^{\text{ad}} \middle| \frac{\partial \psi_j^{\text{ad}}}{\partial R} \right\rangle = -\left\langle \frac{\partial \psi_i^{\text{ad}}}{\partial R} \middle| \psi_j^{\text{ad}} \right\rangle \tag{A.16}$$

we obtain

$$G_{ij} = \boldsymbol{\nabla} \cdot \boldsymbol{F}_{ij} + \sum_k \boldsymbol{F}_{ik} \cdot \boldsymbol{F}_{kj} \tag{A.17}$$

While this derivation uses a complete set of adiabatic states, it has been shown [54] that this equation is also valid in a subset of mutually coupled states that do not interact with the other states.

By using this expression for $\boldsymbol{G}$, it is possible to write the nuclear Schrödinger equation (A.8) in matrix form [54,179] as

$$\left[-\frac{\hbar^2}{2}(\boldsymbol{\nabla} + \boldsymbol{F})^2 + \boldsymbol{V}\right]\chi = i\hbar\frac{\partial}{\partial t}\chi \tag{A.18}$$

where $\boldsymbol{V}$ is the diagonal potential operator matrix, and χ is the vector of nuclear functions. The first term stands for the product

$$(\boldsymbol{\nabla} + \boldsymbol{F})^2 = \nabla^2 \mathbf{1} + \boldsymbol{\nabla} \cdot \boldsymbol{F} + \boldsymbol{F} \cdot \boldsymbol{\nabla} + \boldsymbol{F} \cdot \boldsymbol{F} \tag{A.19}$$

where $\mathbf{1}$ is the unit matrix.

The adiabatic coupling matrix elements, $\boldsymbol{F}_{ij}^{\alpha}$, can be evaluated using an off-diagonal form of the Hellmann–Feynman theorem

$$\frac{\partial}{\partial R}\langle\psi_i^{\mathrm{ad}}|\hat{H}_{\mathrm{el}}|\psi_j^{\mathrm{ad}}\rangle = \left\langle\psi_i^{\mathrm{ad}}\middle|\hat{H}_{\mathrm{el}}\middle|\frac{\partial\psi_j^{\mathrm{ad}}}{\partial R}\right\rangle + \left\langle\frac{\partial\psi_i^{\mathrm{ad}}}{\partial R}\middle|\hat{H}_{\mathrm{el}}\middle|\psi_j^{\mathrm{ad}}\right\rangle + \langle\psi_i^{\mathrm{ad}}|\frac{\partial\hat{H}_{\mathrm{el}}}{\partial R}|\psi_j^{\mathrm{ad}}\rangle \tag{A.20}$$

As ψ_i^{ad} and ψ_j^{ad} are eigenvalues of $\hat{H}_{\mathrm{el}}$ at all values of $\boldsymbol{R}$, this expression at a particular set of nuclear coordinates reduces to

$$0 = V_i\left\langle\psi_i^{\mathrm{ad}}\middle|\frac{\partial\psi_j^{\mathrm{ad}}}{\partial R}\right\rangle + \left\langle\frac{\partial\psi_i^{\mathrm{ad}}}{\partial R}\middle|\psi_j^{\mathrm{ad}}\right\rangle V_j + \langle\psi_i^{\mathrm{ad}}|\frac{\partial\hat{H}_{\mathrm{el}}}{\partial R}|\psi_j^{\mathrm{ad}}\rangle \tag{A.21}$$

Finally, making use of the anti-Hermitian properties of the derivative operator,

$$\left\langle\psi_i^{\mathrm{ad}}\middle|\frac{\partial\psi_j^{\mathrm{ad}}}{\partial R}\right\rangle = \frac{1}{V_j - V_i}\langle\psi_i^{\mathrm{ad}}|\frac{\partial\hat{H}_{\mathrm{el}}}{\partial R}|\psi_j^{\mathrm{ad}}\rangle \tag{A.22}$$

Thus, as the adiabatic PES become degenerate the adiabatic coupling matrix elements become singular.

APPENDIX B: SWARMS OF TRAJECTORIES

As described above in Appendix A, within the BO approximation the nuclear Schrödinger equation is

$$(\hat{T}_n + V)|\chi\rangle = i\hbar\frac{\partial}{\partial t}|\chi\rangle \tag{B.1}$$

with the nuclear kinetic energy operator given by Eq. (A.9) and the potential provided by the eigenvalues of the electronic Hamiltonian. Subscripts labeling the state have been dropped for clarity. Following Bohm [251] and Messiah [147], it is possible to take a classical limit to this equation in which the nuclear wave function can be represented by a "swarm" of trajectories.

Making use of the polar representation of a complex number, the nuclear wave function can be written as a product of a real amplitude, A, and a real phase, S,

$$\chi(\boldsymbol{R}) = A(\boldsymbol{R})\exp\left(\frac{i}{\hbar}S(\boldsymbol{R})\right) \tag{B.2}$$

Inserting this completely general wave function into Eq. (B.1), multiplying by $\exp(-\frac{i}{\hbar}S(\boldsymbol{R}))$, and separating the real and imaginary parts leads to

$$\dot{S} + \frac{(\nabla S)^2}{2m} + V = \frac{\hbar^2}{2m}\frac{\nabla^2 A}{A} \tag{B.3}$$

$$\dot{A} + \frac{1}{2m}\left(2\nabla S \nabla A + A\nabla^2 S\right) = 0 \tag{B.4}$$

In the classical limit, $\hbar \rightarrow 0$, and so the right-hand side of Eq. (B.3) can be ignored. Multiplying Eq. (B.4) by $2A$ and rearranging, the classical equations of motion are

$$\dot{S} + \frac{(\nabla S)^2}{2m} + V = 0 \tag{B.5}$$

$$(\dot{A^2}) + \frac{1}{m}\nabla \cdot (A^2 \nabla S) = 0 \tag{B.6}$$

The hydrodynamical analogy now follows by comparing Eq. (B.6) to the conservation law for a classical fluid

$$\dot{P} + \nabla \cdot \boldsymbol{J} = 0 \tag{B.7}$$

that is, the rate of change of the density, P, and the divergence of the current vector, $\boldsymbol{J}$, is conserved. The quantum fluid "density" is thus defined as

$$P = A^2 \tag{B.8}$$

and the quantum "current" as

$$\boldsymbol{J} = \frac{1}{m} P \nabla S \tag{B.9}$$

By using the relationship between the fluid current and its velocity field, $\boldsymbol{J} = Pv$, a quantum fluid velocity field of

$$v = \frac{\nabla S}{m} \tag{B.10}$$

Substituting Eq. (B.10) into Eq. (B.5),

$$\dot{S} + \frac{mv^2}{2} + V = 0 \tag{B.11}$$

and substituting Eq. (B.10) into the gradient in space of this equation, we obtain

$$m\dot{v} = -\nabla V \tag{B.12}$$

This proves that the pseudoparticles in the quantum fluid obey classical mechanics in the classical limit.

APPENDIX C: PROPAGATING THE ELECTRONIC WAVE FUNCTION

In the classical picture developed above, the wavepacket is modeled by pseudoparticles moving along uncorrelated Newtonian trajectories, taking the electrons with them in the form of the potential along the trajectory. In this spirit, a classical wavepacket can be defined as an incoherent (i.e., noninteracting) superposition of configurations, $\chi_i(\boldsymbol{R}, t)\psi_i(\boldsymbol{r}, t)$

$$\Psi(\boldsymbol{R}, \boldsymbol{r}, t) = \sum_i A_i(t)\chi_i(\boldsymbol{R}, t)\psi_i(\boldsymbol{r}, t) \tag{C.1}$$

Note that, although there is a resemblance, this ansatz is quite different from the Born representation of Eq. (A.3) due to the time dependence of the electronic functions. By taking a single configuration,

$$\Psi(\boldsymbol{R}, \boldsymbol{r}, t) = A(t)\chi(\boldsymbol{R}, t)\psi(\boldsymbol{r}, t) \tag{C.2}$$

and inserting this form into the time-dependent Schrödinger equation leads to equations of motion for the coefficient and the functions.

$$i\dot{A} = -\langle\chi\psi|H_{\text{el}}|\chi\psi\rangle A \tag{C.3}$$

$$i\dot{\psi} = \langle\chi|H_{\text{el}}|\chi\rangle\psi \tag{C.4}$$

$$i\dot{\chi} = (T_n + \langle\psi|H_{\text{el}}|\psi\rangle)\chi \tag{C.5}$$

If we assume that χ are localized in space, these reduce to

$$i\dot{A} = -\langle\psi|H_{\text{el}}(R)|\psi\rangle A \tag{C.6}$$

$$i\dot{\psi} = H_{\text{el}}(R)\psi \tag{C.7}$$

$$i\dot{\chi} = (T_n + V(R))\chi \tag{C.8}$$

In a final step, we follow the ideas of Ehrenfest [252], who first looked for classical structures in the equations of quantum mechanics, and look at the time evolution of the expectation values of the nuclear position and momentum operators. For a general operator, $\hat{O}$

$$\frac{\partial}{\partial t}\langle\hat{O}\rangle = -i\langle[\hat{O},H]\rangle \tag{C.9}$$

where $\langle\hat{O}\rangle = \langle\chi|\hat{O}|\chi\rangle$ and $[\hat{O},H]$ is the commutator of the operator with the Hamiltonian. Evaluating the commutators $[\hat{R},H]$ and $[\hat{P},H]$ leads to the Ehrenfest theorem

$$\frac{\partial}{\partial t}\langle\hat{R}\rangle = \frac{1}{m}\langle\hat{P}\rangle \tag{C.10}$$

$$\frac{\partial}{\partial t}\langle\hat{P}\rangle = -\left\langle\frac{\partial V}{\partial R}\right\rangle \tag{C.11}$$

The localized nature of the nuclear functions means that these reduce to classical equations of motion

$$\dot{R}_i = \frac{P_i}{m} \tag{C.12}$$

$$\dot{P}_i = -\left.\frac{\partial V}{\partial R}\right|_{R=R_i} \tag{C.13}$$

Solving the Eqs. (C.6–C.8,C.12,C.13) comprise what is known as the Ehrenfest dynamics method. This method has appeared under a number of names and derivations in the literature such as the classical path method, eikonal approximation, and hemiquantal dynamics. It has also been put to a number of different applications, often using an analytic PES for the electronic degrees of freedom, but splitting the nuclear degrees of freedom into quantum and classical parts.

In the derivation used here, it is clear that two approximations have been made—the configurations are incoherent, and the nuclear functions remain localized. Without these approximations, the wave function form Eq. (C.1) could be an exact solution of the Schrödinger equation, as it is in 2D MCTDH form (in fact is in what is termed a natural orbital form as only "diagonal" configurations are included [20]).

Acknowledgments

Thanks are due to Luis Blancafort, Mike Bearpark, Irene Burghardt, and Adelaida Sanchez-Galvez for reading the manuscript, and for helpful hints in the presentation of this material.

References

1. W. Domcke and G. Stock, *Adv. Chem. Phys.* **100**, 1 (1997).
2. K. Bolton, W. Hase, and G. Peslherbe, in *Modern methods for multi-dimensional dynamics computations in chemistry*, D. Thompson, ed., World Scientific, Singapore, 1998, pp. 143–189.

3. D. Truhlar, in *The reaction path in chemistry*, D. Heidrich, ed., Kluwer Academic Publishers, Dordrecht, 1995, pp. 229–255.
4. M. Klessinger and J. Michl, *Excited states and photochemistry of organic molecules*, VCH New York, 1994.
5. J. Michl and Bonacic-Koutecky, *Electronic aspects of organic photochemistry*, John Wiley & Sons, Inc., New York, 1990.
6. F. Bernardi, M. Olivucci, and M. Robb, *Chem. Soc. Rev.*, **25**, 321 (1996).
7. M. Robb, F. Bernardi, and M. Olivucci, *Pure Appl. Chem.* **67**, 783 (1995).
8. M. Klessinger, *Angew. Chem.* **107**, 597 (1995).
9. E. Teller, *J. Phys. Chem.* **41**, 109, (1937).
10. G. Herzberg and H. C. Longuet-Higgins, *Discuss. Farad. Soc.* **35**, 77 (1963).
11. E. Teller, *Isr. J. Chem.* **7**, 227 (1969).
12. H. Zimmerman, *J. Am. Chem. Soc.* **88**, 1566 (1966).
13. J. Michl, *Mol. Photochem.* **4**, 243 (1972).
14. R. Englman, *The Jahn-Teller effect in molecules and crystals*, John Wiley & Sons, Inc., New York, 1972.
15. R. Kosloff, *J. Phys. Chem.* **92**, 2087 (1988).
16. G. G. Balint-Kurti, R. N. Dixon, and C. C. Marston, *Int. Rev. Phys. Chem.* **11**, 317 (1992).
17. R. Kosloff, in *Dynamics of Molecules and Chemical Reactions*, R. E. Wyatt and J. Z. H. Zhang, eds., Marcel Dekker, New York, 1996, pp. 185–230.
18. N. Balakrishnan, C. Kalyanaraman, and N. Sathyamurthy, *Phys. Rep.* **280**, 79 (1997).
19. H. -D. Meyer, U. Manthe, and L. S. Cederbaum, *Chem. Phys. Lett.* **165**, 73 (1990).
20. M. H. Beck, A. Jäckle, G. A. Worth, and H. -D. Meyer, *Phys. Rep.* **324**, 1 (2000).
21. A. Donoso, D. Kohen, and C. Martens, *J. Chem. Phys.* **112**, 7345 (2000).
22. E. J. Heller, *J. Chem. Phys.* **62**, 1544 (1975).
23. E. J. Heller, *J. Chem. Phys.* **75**, 2923 (1981).
24. S.-Y. Lee and E. Heller, *J. Chem. Phys.* **76**, 3035 (1982).
25. M. F. Herman and E. Kluk, *Chem. Phys.* **91**, 27 (1984).
26. S. Sawada, R. Heather, B. Jackson, and H. Methiu, *J. Chem. Phys.* **83**, 3009 (1985).
27. R. Littlejohn, *Phys. Rep.* **138**, 193 (1986).
28. M. F. Hermann, *Ann. Rev. Phys. Chem.* **45**, 83 (1994).
29. M. A. Sepulveda and F. Grossmann, *Adv. Chem. Phys.* **96**, 191 (1996).
30. J. Tully, *Farad. Discuss.* **110**, 407 (1998).
31. J. Tully, in *Modern methods for multi-dimensional dynamics computations in chemistry*, D. Thompson, ed., World Scientific, Singapore, 1998, pp. 34–72.
32. M. D. Hack and D. G. Truhlar, *J. Phys. Chem. A* **104**, 7917 (2000).
33. S. Sawada and H. Methiu, *J. Chem. Phys.* **84**, 227 (1986).
34. I. Burghardt, H.-D. Meyer, and L. S. Cederbaum, *J. Chem. Phys.* **111**, 2927 (1999).
35. T. J. Martínez, M. Ben-Nun, and R. D. Levine, *J. Phys. Chem.* **100**, 7884 (1996).
36. T. J. Martínez, M. Ben-Nun, and R. D. Levine, *J. Phys. Chem. A* **101**, 6389 (1997).
37. B. Roos, *Adv. Chem. Phys.* **87**, 399 (1987).
38. J. McDouall, K. Peasley, and M. Robb, *Chem. Phys. Lett.* **148**, 183 (1988).
39. K. Andersson, P. Malmqvist, and B. Roos, *J. Chem. Phys.* **96**, 1218 (1992).

40. K. Andersson and B. Roos, in *Modern Electronic structure theory*, D. Yarkony, ed., World Scientific, Singapore, 1995, pp. 55–109.
41. B. Roos, *Acc. Chem. Res.* **32**, 137 (1999).
42. I. Ragazos, M. Robb, F. Bernardi, and M. Olivucci, *Chem. Phys. Lett.* **197**, 217 (1992).
43. M. Bearpark, M. Robb, and H. Schlegel, *Chem. Phys. Lett.* **223**, 269 (1994).
44. D. Truhlar, R. Steckler, and M. Gordon, *Chem. Rev.* **87**, 217 (1987).
45. D. Truhlar and M. Gordon, *Science* **249**, 491 (1990).
46. D. Heidrich, ed., *The reaction path in chemistry*, Kluwer Academic Publishers, Dordrecht, 1995.
47. C. Gonzalez and H. Schlegel, *J. Chem. Phys.* **90**, 2154 (1989).
48. C. Gonzalez and H. Schlegel, *J. Phys. Chem.* **94**, 5523 (1990).
49. P. Celani, M. Robb, M. Garavelli, F. Bernardi, and M. Olivucci, *Chem. Phys. Lett.* **243**, 1 (1995).
50. M. Garavelli, P. Celani, M. Fato, M. Bearpark, B. Smith, M. Olivucci, and M. Robb, *J. Phys. Chem.* **101**, 2023 (1997).
51. M. Robb, M. Garavelli, M. Olivucci, and F. Bernardi, in *Reviews in Computational Chemistry*, K. Lipkowitz and D. Boyd, eds., Vol. 15, John Wiley & Sons, New York, 2000, pp. 87–146.
52. M. Olivucci, M. Robb, and F. Bernardi, in *Conformational analysis of molecules in excited states*, Wiley-VCH, New York, 2000, pp. 297–366.
53. M. Baer, *Chem. Phys. Lett.* **35**, 112 (1975).
54. M. Baer, *Chem. Phys.* **259**, 123 (2000).
55. D. Yarkony, *J. Chem. Phys.* **105**, 10456 (1996).
56. Z. Xu, M. Baer, and A. Varandas, *J. Chem. Phys.* **112**, 2746 (2000).
57. A. Mebel, A. Yahalom, R. Englman, and M. Baer, *J, Chem. Phys.* **115**, 3673 (2001).
58. N. Allinger, *Adv. Phys. Org. Chem.* **13**, 1 (1976).
59. C. Brooks III, M. Karplus, and B. Pettitt, *Proteins: A theoretical perspective of dynamics, structure, and thermodynamics*, John Wiley & Sons, Inc., New York, 1988, also *Adv. Chem. Phys.* **LXXI**.
60. J. McCammon and S. Harvey, *Dynamics of proteins and nucleic acids*, Cambridge University Press, Cambridge, U.K., 1987.
61. A. Warshel and R. Weiss, *J. Am. Chem. Soc.* **102**, 6218 (1980).
62. A. Warshel, *Computer modeling of chemical reactions in enzymes and solutions*, John Wiley & Sons, Inc., New York, 1991.
63. F. Bernardi, M. Olivucci, and M. Robb, *J. Am. Chem. Soc.* **114**, 1606 (1992).
64. M. Bearpark, M. Robb, F. Bernardi, and M. Olivucci, *Chem. Phys. Lett.* **217**, 513 (1994).
65. H. Köppel, W. Domcke, and L. S. Cederbaum, *Adv. Chem. Phys.* **57**, 59 (1984).
66. I. Wang and M. Karplus, *J. Am. Chem. Soc.* **95**, 8160 (1973).
67. A. Warshel and M. Karplus, *Chem. Phys. Lett.* **32**, 11 (1975).
68. C. Leforestier, *J. Chem. Phys.* **68**, 4406 (1978).
69. R. Barnett, U. Laudman, and A. Nitzan, *J. Chem. Phys.* **89**, 2242 (1988).
70. A. Selloni, P. Carnevali, R. Car, and M. Parinello, *Phys. Rev. Lett.* **59**, 823 (1987).
71. E. Fois, A. Selloni, M. Parinello, and R. Car, *J. Phys. Chem.* **92**, 3268 (1988).
72. R. Car and M. Parrinello, *Phys. Rev. Lett.* **55**, 2471 (1985).
73. D. Remler and P. Madden, *Mol. Phys.* **70**, 921 (1990).

74. D. Marx and J. Hutter, in *Modern Methods and Algorithms of Quantum Chemistry*, J. Grotendorst, ed., John von Neumann Institute for Computing, Jülich, Germany, 2000, pp. 301–449. See http://www.fz-juelich.de/nic-series.
75. H. Schlegel, J. Millam, S. Iyengar, G. Voth, A. Daniels, G. Scuseria, and M. Frisch, *J. Chem. Phys.* **114**, 9758 (2001).
76. C. Molteni, I. Frank, and M. Parrinello, *J. Am. Chem. Soc.* **121**, 12177 (1999).
77. T. Helgaker, E. Uggerud, and H. Jensen, *Chem. Phys. Lett.* **173**, 145 (1990).
78. W. Chen, W. Hase, and H. Schlegel, *Chem. Phys. Lett.* **228**, 446 (1994).
79. B. Smith, M. Bearpark, M. Robb, F. Bernardi, and M. Olivucci, *Chem. Phys. Lett.* **242**, 27 (1995).
80. M. Bearpark, F. Bernardi, S. Clifford, M. Olivucci, M. Robb, B. Smith, and T. Vreven, *J. Am. Chem. Soc.* **118**, 169 (1996).
81. M. Bearpark, F. Bernardi, M. Olivucci, M. Robb, and B. Smith, *J. Am. Chem. Soc.* **118**, 5254 (1996).
82. M. Deumal, M. Bearpark, B. Smith, M. Olivucci, F. Bernardi, and M. Robb, *J. Org. Chem.* **63**, 4594 (1998).
83. M. Garavelli, B. Smith, M. Bearpark, F. Bernardi, M. Olivucci, and M. Robb, *J. Am. Chem. Soc.* **122**, 5568 (2000).
84. S. Klein, M. Bearpark, B. Smith, M. Robb, M. Olivucci, and F. Bernardi, *Chem. Phys. Lett.* **292**, 259 (1998).
85. F. Jolibois, M. Bearpark, S. Klein, M. Olivucci, and M. Robb, *J. Am. Chem. Soc.* **122**, 5801 (2000).
86. T. Vreven, F. Bernardi, M. Garavelli, M. Olivucci, M. Robb, and H. Schlegel, *J. Am. Chem. Soc.* **119**, 12687 (1997).
87. A. Sanchez-Galvez, P. Hunt, M. Robb, M. Olivucci, T. Vreven, and H. Schlegel, *J. Am. Chem. Soc.* **122**, 2911 (2000).
88. M. Ben-Nun, J. Quenneville, and T. Martínez, *J. Phys. Chem. A* **104**, 5162 (2000).
89. J. Ischtwan and M. Collins, *J. Chem. Phys.* **100**, 8080 (1994).
90. M. Collins, *Adv. Chem. Phys.* **93**, 389 (1996).
91. R. Bettens and M. Collins, *J. Chem. Phys.* **111**, 816 (1999).
92. K. Thompson, M. Jordan, and M. Collins, *J. Chem. Phys.* **108**, 8302 (1998).
93. K. Thompson, M. Jordan, and M. Collins, *J. Chem. Phys.* **108**, 564 (1998).
94. M. Collins and D. Zhang, *J. Chem. Phys.* **111**, 9924 (1999).
95. M. J. Frisch, G. W. Trucks, H. B. Schlegel, G. E. Scuseria, M. A. Robb, J. R. Cheeseman, V. G. Zakrzewski, J. A. Montgomery, Jr., R. E. Stratmann, J. C. Burant, S. Dapprich, J. M. Millam, A. D. Daniels, K. N. Kudin, M. C. Strain, O. Farkas, J. Tomasi, V. Barone, M. Cossi, R. Cammi, B. Mennucci, C. Pomelli, C. Adamo, S. Clifford, J. Ochterski, G. A. Petersson, P. Y. Ayala, Q. Cui, K. Morokuma, D. K. Malick, A. D. Rabuck, K. Raghavachari, J. B. Foresman, J. Cioslowski, J. V. Ortiz, B. B. Stefanov, G. Liu, A. Liashenko, P. Piskorz, I. Komaromi, R. Gomperts, R. L. Martin, D. J. Fox, T. Keith, M. A. Al-Laham, C. Y. Peng, A. Nanayakkara, C. Gonzalez, M. Challacombe, P. M. W. Gill, B. Johnson, W. Chen, M. W. Wond, J. L. Andres, C. Gonzalez, M. Head-Gordon, E. S. Replogle, and J. A. Pople, GAUSSIAN 98, Gaussian, Inc., Pittsburgh PA, 1998.
96. M. Schmidt, K. Baldridge, J. Boatz, S. Elbert, M. Gordon, J. Jensen, S. Koseki, N. Matsunaga, K. Nguyen, S. Su, T. Windus, M. Dupuis, and J. Montgomery, *J. Comp. Chem.* **14**, 1347 (1993).

97. K. Andersson, M. Blomberg, M. Fülscher, G. Karlström, R. Lindh, P.-Å. Malmqvist, P. Neogrády, J. Olsen, B. Roos, A. Sadlej, M. Schütz, L. Seijo, L. Serrano-Andrés, P. Siegbahn, and P.-O. Widmark, *MOLCA. S, Version 4*, Lund University, Sweden, 1997.
98. H. -J. Werner and P. Knowles, MOLPR. O, Version 96.3, University of Birmingham, U.K., 1996.
99. H. Dachsel, R. Shepard, J. Nieplocha, and R. Harrison, *J. Comp. Chem.* **18**, 430 (1997).
100. J. C. Light, I. P. Hamilton, and J. V. Lill, *J. Chem. Phys.* **82**, 1400 (1985).
101. J. C. Light, in *Time-Dependent Quantum Molecular Dynamics* J. Broeckhove and L. Lathouwers, eds., Plenum, New York, 1992, pp. 185–199.
102. D. Kosloff and R. Kosloff, *J. Comp. Phys.* **52**, 35 (1983).
103. C. Cerjan, ed., *Numerical Grid Methods and their Application to Schrödinger's Equation*, Kluwer Academic Publishers, Dordrecht, 1993.
104. R. Kosloff, *Ann. Rev. Phys. Chem.* **45**, 145 (1994).
105. C. Leforestier, R. H. Bisseling, C. Cerjan, M. D. Feit, R. Friesner, A. Guldenberg, A. Hammerich, G. Jolicard, W. Karrlein, H. -D. Meyer, N. Lipkin, O. Roncero, and R. Kosloff, *J. Comp. Phys.* **94**, 59 (1991).
106. R. E. Wyatt and C. Iung, *J. Chem. Phys.* **98**, 6758 (1993).
107. A. Maynard, R. E. Wyatt, and C. Iung, *J. Chem. Phys.* **106**, 9483 (1997).
108. U. Manthe, H.-D. Meyer, and L. S. Cederbaum, *J. Chem. Phys.* **97**, 3199 (1992).
109. A. Raab, G. Worth, H. -D. Meyer, and L. S. Cederbaum, *J. Chem. Phys.* **110**, 936 (1999).
110. G. Worth, *J. Chem. Phys.* **114**, 1524 (2001).
111. H. Buchenau, J. P. Toennies, J. Arnold, and J. Wolfrum, *Ber. Bunsenges. Phys. Chem.* **94**, 1231 (1990).
112. G. C. Schatz, *J. Phys. Chem.* **100**, 12839 (1996).
113. G. Schatz and A. Kuppermann, *J. Chem. Phys.* **65**, 4668 (1976).
114. D. Neuhauser and M. Baer, *J. Chem. Phys.* **91**, 4651 (1989).
115. D. Neuhauser, M. Baer, and D. Kouri, *J. Chem. Phys.* **93**, 2499 (1990).
116. A. Jäckle and H.-D. Meyer, *J. Chem. Phys.* **109**, 2614 (1998).
117. A. Untch, K. Weide, and R. Schinke, *J. Chem. Phys.* **95**, 6496 (1991).
118. R. Schinke, M. Nonella, H. U. Suter, and J. R. Huber, *J. Chem. Phys.* **93**, 1098 (1990).
119. C. W. Gear, *Numerical Initial Value Problems in Ordinary Differential Equations*, Prentice-Hall, Englewood Cliffs, N.J. (1971).
120. L. Verlet, *Phys. Rev.* **159**, 98 (1967).
121. M. Allen and D. Tildesley, *Computer Simulation of Liquids*, OUP, Oxford, U.K., (1987).
122. S. Gray, D. Noid, and B. Sumpter, *J. Chem. Phys.* **101**, 4062 (1994).
123. P. Pulay, *Adv. Chem. Phys.* **69**, 241 (1987).
124. T. Helgaker and P. Jørgensen, *Adv. Quant. Chem.* **19**, 183 (1988).
125. T. Helgaker and P. Jørgensen, in *Methods in computational molecular physics*, S. Wilson and G. Diercksen, eds., Plenum Press, New York, 1992, pp. 353–421.
126. R. Shepard, in *Modern Electronic structure theory*, D. Yarkony, ed. World Scientific, Singapore, 1995, pp. 345–458.
127. P. Pulay, in *Modern Electronic Structure Theory*, D. Yarkony, ed., World Scientific, Singapore, 1995, pp. 1191–1240.
128. H. Hellmann, *Einführung in die Quantenchemie*, Franz Deutiche, Leipzig, 1937.
129. R. Feynman, *Phys. Rev.* **56**, 340, (1939).

130. T. Helgaker, in *The Encyclopedia of Computational Chemistry*, eds. P. v. R. Schleyer, N. L. Allinger, T. Clark, J. Gasteiger, P. A. Kollman, H. F. Schaefer III, and P. R. Schreiner, eds., Vol. 2, John Wiley & Sons, Inc., Chichester, 1998, pp. 1157–1169.
131. R. Schinke, *Photodissociation Dynamics*, Cambridge University Press, Cambridge, UK, 1993.
132. K. C. Kulander and E. J. Heller, *J. Chem. Phys.* **69**, 2439 (1978).
133. E. Heller, *Acc. Chem. Res.* **14**, 368 (1981).
134. D. Noid, M. Koszykowski, and R. Marcus, *J. Chem. Phys.* **67**, 404 (1977).
135. P. Berens and K. Wilson, *J. Chem. Phys.* **74**, 4872 (1981).
136. G. Stock and W. Miller, *J. Chem. Phys.* **99**, 1545 (1993).
137. E. Wigner, *Phys. Rev.* **40**, 749 (1932).
138. E. Heller, *J. Chem. Phys.* **65**, 1289 (1976).
139. S. Mukamel, *Principles of nonlinear optical spectroscopy*, Oxford University Press, Oxford, U. K., 1995.
140. M. Hillery, R. O'Connell, M. Scully, and E. Wigner, *Phys. Rep.* **106**, 122 (1984).
141. K. Blum, *Density matrix theory and applications*, Plenum Press, New York, 1981.
142. S. Chapman and Bunker, *J. Chem. Phys.* **62**, 2890 (1975).
143. C. Sloane and W. Hase, *J. Chem. Phys.* **66**, 1523 (1977).
144. G. Jones, B. Carpenter, and M. Paddon-Row, *J. Am. Chem. Soc.* **121**, 11171 (1999).
145. D. Chandler, *J. Chem. Phys.* **68**, 2959 (1978).
146. S. Hammes-Schiffer and J. Tully, *J. Chem. Phys.* **103**, 8528 (1995).
147. A. Messiah, *Quantum Mechanics*, Vol. 1, John Wiley & Sons, Inc., New York, 1962.
148. R. Skodje and D. Truhlar, *J. Chem. Phys.* **80**, 3123 (1984).
149. R. Coalson and M. Karplus, *J. Chem. Phys.* **93**, 3919 (1990).
150. F. Hansen, N. Henriksen, and G. Billing, *J. Chem. Phys.* **90**, 3060 (1989).
151. G. Billing, *J. Chem. Phys.* **107**, 4286 (1997).
152. G. Billing, *J. Chem. Phys.* **110**, 5526 (1999).
153. K. Kay, *J. Chem. Phys.* **100**, 4377 (1994).
154. Y. Elran and K. Kay, *J. Chem. Phys.* **110**, 3653 (1999).
155. Y. Elran and K. Kay, *J. Chem. Phys.* **110**, 8912 (1999).
156. K. Kay, *J. Chem. Phys.* **100**, 4432 (1994).
157. A. Walton and D. Manolopoulos, *Chem. Phys. Lett.* **244**, 448 (1995).
158. A. Walton and D. Manolopoulos, *Mol. Phys.* **87**, 961 (1996).
159. D. McCormack, *J. Chem. Phys.* **112**, 992 (2000).
160. A. Baede, *Adv. Chem. Phys.* **30**, 463 (1975).
161. W. Domcke, H. Köppel, and L. Cederbaum, *Mol. Phys.* **43**, 851 (1981).
162. R. Whetten, G. Ezra, and E. Grant, *Ann. Rev. Phys. Chem.* **36**, 277 (1985).
163. L. Butler, *Ann. Rev. Phys. Chem.* **49**, 125 (1998).
164. H. Jahn and E. Teller, *Proc. R. Soc. London Sec. A* **161**, 220 (1937).
165. I. Bersuker, *The Jahn-Teller effect and vibronic interactions in modern chemistry*, Plenum Press, New York, 1984.
166. T. Barckholtz and T. Miller, *Int. Rev. Phys. Chem.* **17**, 435 (1998).
167. T. Barckholtz and T. Miller, *J. Phys. Chem. A* **103**, 2321 (1999).

168. I. Bersuker, *Chem. Rev.* **101**, 1067 (2001).
169. F. Brogli, E. Heilbronner, E.Kloster-Jensen, A. Schmelzer, A. S. Manocha, J. A. Pople, and L. Radom, *Chem. Phys.* **4**, 107 (1974).
170. L. S. Cederbaum, W. Domcke, H. Köppel, and W. von Niessen, *Chem. Phys.* **26**, 169 (1977).
171. I. Yamazaki, T. Murao, T. Yamanaka, and K. Yoshihara, *Faraday Discuss. Chem. Soc.* **75**, 395 (1983).
172. K. K. Innes, I. G. Ross, and W. R. Moonaw, *J. Mol. Spec.* **132**, 492 (1988).
173. C. Woywod, W. Domcke, A. L. Sobolewski, and H.-J. Werner, *J. Chem. Phys.* **100**, 1400 (1994).
174. G. Stock, C. Woywod, W. Domcke, T. Swinney, and B. S. Hudson, *J. Chem. Phys.* **103**, 6851 (1995).
175. B. Kohler, *Chem. Rev.* **93**, 41 (1993).
176. D. Yarkony, *Rev. Mod. Phys.* **68**, 985 (1996).
177. D. Yarkony, *J. Phys. Chem.* **100**, 18612 (1996).
178. D. Yarkony, *Acc. Chem. Res.* **31**, 511 (1998).
179. T. Pacher, L. S. Cederbaum, and H. Köppel, *Adv. Chem. Phys.* **84**, 293 (1993).
180. W. Lichten, *Phys. Rev.* **164**, 131 (1967).
181. F. Smith, *Phys. Rev.* **179**, 111 (1969).
182. C. A. Mead and D. G. Truhlar, *J. Chem. Phys.* **77**, 6090 (1982).
183. A. Macias and A. Riera, *J. Phys. B* **11**, L489 (1978).
184. H.-J. Werner and W. Meyer, *J. Chem. Phys.* **74**, 5802 (1981).
185. R. Cimirglia, J.-P. Malrieu, M. Persico, and F. Spiegelmann, *J. Phys. B* **18**, 3073 (1985).
186. X. Gadéa and M. Péllisier, *J. Chem. Phys.* **93**, 545 (1990).
187. W. Domcke, C. Woywood, and M. Stengle, *Chem. Phys. Lett.* **226**, 257 (1996).
188. H. Köppel, J. Gronki, and S. Mahapatra, *J. Chem. Phys.* **115**, 2377 (2001).
189. G. Atchity, S. Xantheas, and K. Ruendenberg, *J. Chem. Phys.* **95**, 1862 (1991).
190. C. Cattarius, G. Worth, H.-D. Meyer, and L. Cederbaum, *J. Chem. Phys.* **115**, 2088 (2001).
191. U. Manthe and H. Köppel, *J. Chem. Phys.* **93**, 1658 (1990).
192. S. Mahapatra, G. Worth, H. -D. Meyer, L. Cederbaum, and H. Köppel, *J. Phys. Chem. A* **105**, 5567 (2001).
193. S. Mahapatra, L. Cederbaum, and H. Köppel, *J. Chem. Phys.* **111**, 10452 (1999).
194. H. Köppel, M. Doscher, and S. Mahapatra, *Int. J. Quant. Chem.* **80**, 942 (2000).
195. G. A. Worth, H.-D. Meyer, and L. S. Cederbaum, *J. Chem. Phys.* **109**, 3518 (1998).
196. H. S. and S. G, *J. Phys. Chem. B* **104**, 1146 (2000).
197. G. Billing, *Int. Rev. Phys. Chem.* **13**, 309 (1994).
198. P. Atkins, *Molecular Quantum Mechanics*, 2nd ed., OUP, Oxford, UK, 1983.
199. P. Kuntz, *J. Chem. Phys.* **95**, 141 (1991).
200. A. Bjerre and E. Nikitin, *Chem. Phys. Lett.* **1**, 179 (1967).
201. J. C. Tully and R. K. Preston, *J. Chem. Phys.* **55**, 562 (1971).
202. N. Blais and D. Truhlar, *J. Chem. Phys.* **79**, 1334 (1983).
203. J. C. Tully, *J. Chem. Phys.* **93**, 106 (1990).
204. D. F. Coker, in *Computer Simulation in Chemical Physics*, M. P. Allen and D. J. Tildesley, eds., Kluwer Academic, Dordrecht, 1993, pp. 315–377.
205. D. Coker and L. Xiao, *J. Chem. Phys.* **102**, 496 (1995).

206. S. Chapman, *Adv. Chem. Phys.* **82**(pt. II), 423 (1992).
207. W. Miller and T. George, *J. Chem. Phys.* **56**, 5637 (1972).
208. L. Landau, *Phys. Z. Sow.* **2**, 46 (1932).
209. C. Zener, *Proc. R. Soc. London, Ser. A* **137**, 596 (1932).
210. E. Stueckelberg, *Helv. Phys. Acta* **5**, 369 (1932).
211. M. Topaler, T. Allison, D. Schwenke, and D. Truhlar, *J. Chem. Phys.* **109**, 3321 (1998).
212. M. Herman, *J. Chem. Phys.* **81**, 754 (1984).
213. M. Hack, A. Jasper, Y. Volobuev, D. Schwenke, and D. Truhlar, *J. Phys. Chem. A* **103**, 6309 (1999).
214. G. Jones, B. Carpenter, and M. Paddon-Row, *J. Am. Chem. Soc.* **120**, 5499 (1998).
215. Y. Volobuev, M. Hack, M. Topaler, and D. Truhlar, *J. Chem. Phys.* **112**, 9716 (2000).
216. M. Krishna, *J. Chem. Phys.* **93**, 3258 (1990).
217. S. Adhikari and G. Billing, *J. Chem. Phys.* **111**, 48 (1999).
218. M. Ben-Nun and T. J. Martínez, *J. Chem. Phys.* **108**, 7244 (1998).
219. M. Ben-Nun and T. Martínez, *J. Chem. Phys.* **112**, 6113 (2000).
220. T. J. Martínez, M. Ben-Nun, and G. Ashkenazi, *J. Phys. Chem.* **100**, 2847 (1996).
221. M. Ben-Nun and T. Martínez, *J. Phys. Chem. A* **103**, 10517 (1999).
222. M. Hack, A. Wensmann, D. Truhlar, M. Ben-Nun, and T. Martinez, *J. Chem. Phys.* **115**, 1172 (2001).
223. D. Kohen, F. Stillinger, and J. Tully, *J. Chem. Phys.* **109**, 4713 (1998).
224. M. Topaler, M. Hack, T. Allison, Y.-P. Liu, S. Mielke, D. Schwenke, and D. Truhlar, *J. Chem. Phys.* **106**, 8699 (1997).
225. P. Cattaneo and M. Persico, *J. Phys. Chem. A* **101**, 3454 (1997).
226. A. Ferretti, G. Granucci, A. Lami, M. Persico, and G. Villani, *J. Chem. Phys.* **104**, 5517 (1996).
227. U. Müller and G. Stock, *J. Chem. Phys.* **107**, 6230 (1997).
228. M. Thoss, W. Miller, and G. Stock, *J. Chem. Phys.* **112**, 10282 (2000).
229. M. Schmidt and M. Gordon, *Ann. Rev. Phys. Chem.* **49**, 233 (1998).
230. M. Garavelli, P. Celani, F. Bernardi, M. Robb, and M. Olivucci, *J. Am. Chem. Soc.* **119**, 6891 (1997).
231. H.-J. Werner and W. Meyer, *J. Chem. Phys.* **74**, 5794 (1981).
232. B. Lengsfield and D. Yarkony, *Adv. Chem. Phys.* **82**(part 2), 1 (1992).
233. N. Yamamoto, T. Vreven, M. Robb, M. Frisch, and H. Schlegel, *Chem. Phys. Lett.* **250**, 373 (1996).
234. D. Yarkony, in *Modern Electronic structure theory*, D. Yarkony, ed., World Scientific, Singapore, 1995, pp. 643–721.
235. T. Neuhauser, N. Sukumar, and S. Peyerimhoff, *Chem. Phys.* **194**, 45 (1995).
236. T. Martínez, *Chem. Phys. Lett.* **272**, 139 (1997).
237. I. Palmer, I. Ragazos, F. Bernardi, M. Olivucci, and M. Robb, *J. Am. Chem. Soc.* **115**, 672 (1992).
238. M. Beer and H. Longuet–Higgins, *J. Chem. Phys.* **23**, 1390 (1955).
239. B. Carpenter, *Ang. Chem. Int. Ed. Engl.* **37**, 3340 (1998).
240. A. J. Wurzer, T. Wilhelm, J. Piel, and E. Riedle, *Chem. Phys. Lett.* **299**, 296 (1999).
241. A. Warshel, Z. Chu, and J.-T. Hwang, *Chem. Phys.* **158**, 303 (1991).
242. V. Buß, O, Weingart, and M. Sugihara, *Ang. Chem. Int. Ed. Engl.* **39**, 2784 (2000).

243. M. Ben-Nun, F. Molnar, L. H., J. Phillips, T. Martínez, and K. Schulten, Faraday Discuss. 447 (1998).

244. M. Ben-Nun and T. Martínez, *J. Phys. Chem. A* **102**, 9607 (1998).

245. T. Martínez and R. Levine, *Chem. Phys. Lett.* **259**, 252 (1996).

246. T. Martínez and R. Levine, *J. Chem. Phys.* **105**, 6334 (1996).

247. M. Ben-Nun and T. Martínez, *Chem. Phys. Lett.* **298**, 57 (1998).

248. M. Ben-Nun and T. Martínez, *Chem. Phys.* **259**, 237 (2000).

249. M. Hartmann, J. Pittner, and V. Bonačić-Koutecký, *J. Chem. Phys.* **114**, 2123 (2001).

250. C. Ballhausen and A. Hansen, *Ann. Rev. Phys. Chem.* **23**, 15 (1972).

251. D. Bohm, *Phys. Rev.* **85**, 166 (1952).

252. P. Ehrenfest, *Z. Phys.* **45**, 455 (1927).

CONICAL INTERSECTIONS IN MOLECULAR PHOTOCHEMISTRY: THE PHASE-CHANGE APPROACH

YEHUDA HAAS and SHMUEL ZILBERG

Department of Physical Chemistry and the Farkas Center for Light Induced Processes, Hebrew University of Jerusalem, Jerusalem, Israel

CONTENTS

The Role of Degenerate States in Chemistry: A Special Volume of Advances in Chemical Physics, Volume 124, Edited by Michael Baer and Gert Due Billing. Series Editors I. Prigogine and Stuart A. Rice.
ISBN 0-471-43817-0.

I. INTRODUCTION

Conical intersections, introduced over 60 years ago as possible efficient funnels connecting different electronically excited states [1], are now generally believed to be involved in many photochemical reactions. Direct laboratory observation of these subsurfaces on the potential surfaces of polyatomic molecules is difficult, since they are not stationary "points". The system is expected to pass through them very rapidly, as the transition from one electronic state to another at the conical intersection is very rapid. Their presence is surmised from the following data [2–5]:

Very rapid (subpicosecond) decay of electronically excited states.

Lack of fluorescence.

Rapid formation of ground-state products.

In recent years, computational testimonies for the existence of conical intersections in many polyatomic systems became abundant and compelling [6–11]. The current consensus concerning the ubiquitous presence of conical intersections in polyatomic molecules is due in large part to computational "experiments."

In this chapter, we present an analysis of conical intersections, based on chemical reaction concepts. It is argued that conical intersections leading to the ground state can be identified and characterized by considering properties of the ground-state surface only. The basis of the model is the Longuet-Higgins phase-change rule [12,13] (Section II), which provides a simple criterion for the existence of a degeneracy on the electronic ground state. Longuet-Higgins showed that a degeneracy necessarily exists within a region enclosed by a loop, if the total electronic wave function changes sign upon being transported around the loop. (For more details, see Section II). We propose to construct the loop discussed by Longuet-Higgins from reaction coordinates of elementary reactions converting the reactant to the desired product and other possible products. In this sense, our approach is "chemical" in nature. In order to properly search for the elementary reactions, the need for an agreed definition of common terms such as a molecule and a transition state arises. This reaction is carried out in this section, based on the concept of electron spin pairing [14,15]. The reacting system (reactant and product) are treated as a two-state system [16]. The spirit of this strategy is akin to the Evans–Dewar–Zimmerman approach [17–21], and is closely related to the concept of aromaticity and anti-aromaticity, which is dealt with in Section III.

The phase change of the total polyelectronic wave function in a chemical reaction [22–25], which is more extensively discussed in Section III, is central to the approach presented in this chapter. It is shown that some reactions may be classified as phase preserving (p) on the ground-state surface, while others are phase inverting (i). The distinction between the two can be made by checking the change in the spin pairing of the electrons that are exchanged in the reaction. A complete loop around a point in configuration space may be constructed using a number of consecutive elementary reactions, starting and ending with the reactant A. The smallest possible loop typically requires at least three reactions: two other molecules must be involved in order to complete a loop; they are the desired product B and another one C, so that the complete loop is $A \rightarrow B \rightarrow C \rightarrow A$. Two types of phase inverting loops may be constructed: those in which each reaction is phase inverting (an i^3 loop) and those in which one reaction is phase inverting, and the other two phase preserving (an ip^2 loop). At least one reaction must be phase inverting for the complete loop to be phase inverting and thus to encircle a conical intersection and lead to a photochemical reaction. It follows, that if a conical intersection is crossed during a photochemical reaction, in general at least two products are expected, B and C. A single product requires the existence of a two-component loop. This is possible if one of the molecules may be viewed as the out-of-phase combination of a two-state system. The allyl radical (Section IV, cf. Fig. 12) and the triplet state are examples of such systems. We restrict the discussion in this chapter to singlet states only.

In Section IV, the construction of phase inverting loops is described. A conical intersection is an example of an electronic degeneracy; A well-known case of electronic degeneracy in polyatomic molecules occurs in the Jahn–Teller effect. Systems of high symmetry tend to distort to lower symmetry if their electronic ground state is degenerate. We show (Section V) that the Longuet-Higgins loop treatment can be applied to these systems, making them part of the general conical intersections concept.

The method discussed in this chapter allows, in principle, the detection of all conical intersections connecting the ground with the excited state. Assuming that photochemical products are mainly formed through conical intersections, it therefore provides a means to design selection rules for photochemistry.

A. A Chemical Reaction as a Two-State System

The concept of two-state systems occupies a central role in quantum mechanics [16,26]. As discussed extensively by Feynmann et al. [16], benzene and ammonia are examples of simple two-state systems: Their properties are best described by assuming that the wave function that represents them is a combination of two base states. In the cases of ammonia and benzene, the two base states are equivalent. The two base states necessarily give rise to two independent states, which we named twin states [27,28]. One of them is the ground state, the other an excited states. The twin states are the ones observed experimentally.

The extra stabilization of benzene in the ground state, as compared to a single Kekulé structure, is assigned to a *resonance* between the two equivalent base states. In standard textbooks, the fact that the combination is in-phase (i.e., that the two Kekulé structures in the ground-state combination carry the same sign) is taken for granted. In Section III, it is shown that whether the ground state is representing by the in-phase or out-of-phase combination of the two states is determined by the permutational symmetry of the electronic wave function, and may be traced to Pauli's principle. Hückel's $4n + 2$ rule [29] arises from the fact that there is an odd number of electron pairs in this system.

Stabilizing resonances also occur in other systems. Some well-known ones are the allyl radical and square cyclobutadiene. It has been shown that in these cases, the ground-state wave function is constructed from the *out-of-phase* combination of the two components [24,30]. In Section III, it is shown that this is also a necessary result of Pauli's principle and the permutational symmetry of the polyelectronic wave function: When the number of electron pairs exchanged in a two-state system is even, the ground state is the out-of-phase combination [28]. Three electrons may be considered as two electron pairs, one of which is half-populated. When both electron pairs are fully populated, an antiaromatic system arises (Section III).

During a chemical reaction, a chemical system (or substance) A is converted to another, B. Viewed from a quantum chemical point of view, A and B together are a single system that evolves with time. It may be approximated by a combination of two states, A at time zero and B as time approaches infinity. The first is represented by the wave function $|\mathrm{A}\rangle$ and the second by $|\mathrm{B}\rangle$. At any time during the reaction, the system may be described by a combination of the two

$$|\mathrm{R}\rangle(t) = \mathrm{c_A}(t)|\mathrm{A}\rangle + \mathrm{c_B}(t)|\mathrm{B}\rangle \tag{1}$$

where $\mathrm{c_A}(t = 0) = 1, \mathrm{c_B}(t = 0) = 0, \mathrm{c_A}(t = \infty) = 0, \mathrm{c_B}(t = \infty) = 1$.

Within the Born–Oppenheimer (BO) approximation, $|\mathrm{A}\rangle$ and $|\mathrm{B}\rangle$ may be written as the product of an electronic wave function, $|\mathrm{M}\rangle_{\mathrm{el}}$ and a nuclear wave function $|\mathrm{M}\rangle_{\mathrm{n}}$.

$$|\mathrm{M}\rangle = |\mathrm{M}\rangle_{\mathrm{el}}|\mathrm{M}\rangle_{\mathrm{n}} \qquad (\mathrm{M} = \mathrm{A}, \mathrm{B}) \tag{2}$$

It is useful to represent the polyelectronic wave function of a compound by a valence bond (VB) structure that represents the bonding between the atoms. Frequently, a single VB structure suffices, sometimes it is necessary to use several. We assume for simplicity that a single VB structure provides a faithful representation. A common way to write down a VB structure is by the spin-paired determinant, that ensures the compliance with Pauli's principle: (It is assumed that there are $2n$ paired electrons in the system)

$$|\mathrm{A}\rangle_{\mathrm{el}} = \sum_{p} \epsilon_p \mathrm{P}1(1)2(2)\cdots 2\mathrm{n}(2n)[\alpha(1)\beta(2) - \beta(1)\alpha(2)][\alpha(3)\beta(4) - \beta(3)\alpha(4)]\cdots[\alpha(2n-1)\beta(2n) - \beta(2n-1)\alpha(2n)] \tag{3}$$

Where the summation is over all $2n!$ permutations P each with parity ϵ_p. We use a short-hand notation:

$$|\mathrm{A}\rangle_{\mathrm{el}} = (1\bar{2} - \bar{1}2)(3\bar{4} - \bar{3}4)\cdots(2n - 1\overline{2n} - \overline{2n-1}2n) \tag{4}$$

As the electronic and nuclear wave functions are separated in the BO approximation, a single electronic wave function may be associated with many different nuclear configurations. Furthermore, the electronic energy of the system depends parametrically on the nuclear configuration $\{\mathbf{Q}\}$. It is convenient to introduce a term for all systems having a specific spin-pairing scheme, independent of the nuclear configuration. We use the term *anchor* to represent this group of systems.

We may now distinguish two classes of reactions:

1. The system does not change the spin-pairing scheme during the process. In this case, $|\mathrm{A}\rangle_{\mathrm{el}}$ remains put throughout the reaction, and only the internuclear distances or angles change. Such transformations are called intraanchor reactions.
2. The spin-pairing scheme of the product, $|\mathrm{B}\rangle_{\mathrm{el}}$, is different from that of the reactant. This happens if at least two pairs of electrons have exchanged partners. In other words, at least three electrons need to be involved.

If the reaction is elementary, there is only a single transition state between A and B. At this point the derivative of the total electronic wave function with respect to the reaction coordinate $Q_{\mathrm{A}\rightarrow\mathrm{B}}$ vanishes:

$$\partial|\mathrm{R}\rangle_{\mathrm{el,TS}}/\partial Q_{\mathrm{A}\rightarrow\mathrm{B}} = 0 \tag{5}$$

In the transition state region, the spin-pairing change must take place. At this nuclear configuration, the electronic wave function may be written as

$$|\mathrm{R}\rangle_{\mathrm{el,TS}} = k_{\mathrm{A}}|A\rangle_{\mathrm{el}} + k_{\mathrm{B}}|\mathrm{B}\rangle_{\mathrm{el}} \tag{6}$$

If the sign of k_A is equal to that of k_B, the reaction is phase preserving, if the signs are different, the reaction is phase inverting.

We shall assume, for simplifying the notation, that the k values are positive. For a phase-inverting reaction, the wave function of the transition state is therefore written as

$$|R\rangle_{el,TS} = k_A|A\rangle_{el} - k_B|B\rangle_{el} \qquad \text{(phase inverting)} \tag{7}$$

It is important to recall, that the reaction takes place on the ground-state surface. Clearly, at the same nuclear configuration, the other combination

$$|R\rangle^*_{el,TS} = k^*_A|A\rangle_{el} + k^*_B|B\rangle_{el} \qquad \text{(phase preserving)} \tag{8}$$

lies on an excited state surface.

The distinction between an in-phase and an out-of-phase combination of the two base states is easy for degenerate two-state systems (such that the two components are equivalent). In these cases, the transition state has an additional symmetry element not present in either of the two base states. In other words, it belongs to a group of higher symmetry. The electronic wave function at the transition state nuclear configuration transforms as the totally symmetric representation of the new group if the transition state is the in-phase combination. If it is the out-of-phase combination, it transforms as one of the nontotally symmetric representations. In this case, the motion along the reaction coordinate is antisymmetric with respect to the new symmetry element [28]. For example, the ground state of the C_{2v} allyl radical transforms as B_1 (not A_2), and the ground state of square cyclobutadiene (D_{4h} symmetry) as B_{1g} (not A_{1g}). The symmetry properties of the transition states are more easily established using the VB approximation than the molecular orbital–configuration interaction (MO–CI) one. The character of the bonding before and after the reaction does not matter: The transfer of electrons from one atom to another to form a covalent, ionic, or coordinate bond is always accompanied by a change in spin pairing [31,32], and is clearly represented by the VB structures. The MO–CI method can also be used successfully, but several configurations are ordinarily required in the general case, as shown, for example, in [33].

By using the determinant form of the electronic wave functions, it is readily shown that a phase-inverting reaction is one in which an even number of electron pairs are exchanged, while in a phase-preserving reaction, an odd number of electron pairs are exchanged. This holds for Hückel-type reactions, and is demonstrated in Appendix A. For a definition of Hückel and Möbius-type reactions, see Section III.

B. Anchors

Intuitively, a molecule is defined as an assembly of atoms bound by chemical bonds, which lies in a local minimum on the potential surface. The molecule preserves its identity when the nuclei are transported from their minimum energy position, as long as the gradient of the electronic energy with respect to the displacement maintains its sign. As soon as this gradient changes its sign, the system undergoes a chemical change in an elementary reaction. The idea that a molecule is defined by the spin-pairing arrangement of the valence electrons seems to be at odds with this concept of a molecule. In particular, a definite structure is not assumed.

An anchor, as defined above, contains stable molecules, conformers, all pairs of radicals and biradicals formed by a simple bond fission in which no spin re-pairing took place, ionic species, and so on. Figure 1 shows some examples of species belonging to the same anchor. Thus, an anchor is a more general and convenient term used in the discussion of spin re-pairing.

C. Anchors, Molecules and Independent Quantum Species

At this point, it is instructive to discuss the distinction between molecules, anchors, and quantum mechanical wave functions that represent them. The topic is best introduced by using an example. Consider the H_4 system [34].

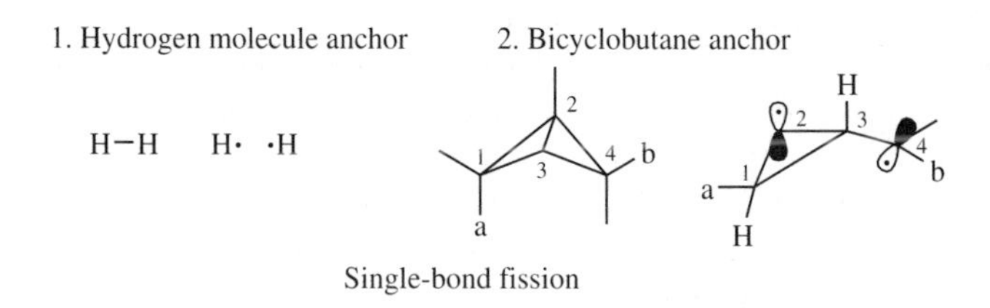

3. Butadiene anchor

4. Twisted ethylene anchor

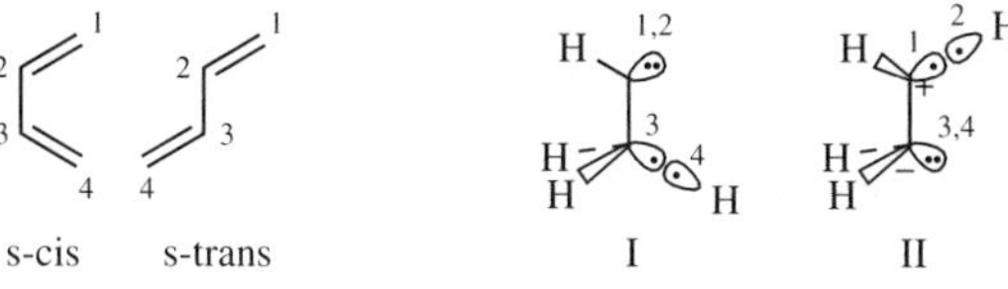

Two conformers

The conversion from I to II is by the transfer of a proton from one carbon atom to the other. No change in spin pairing.

Figure 1. Examples of species residing in the same anchor.

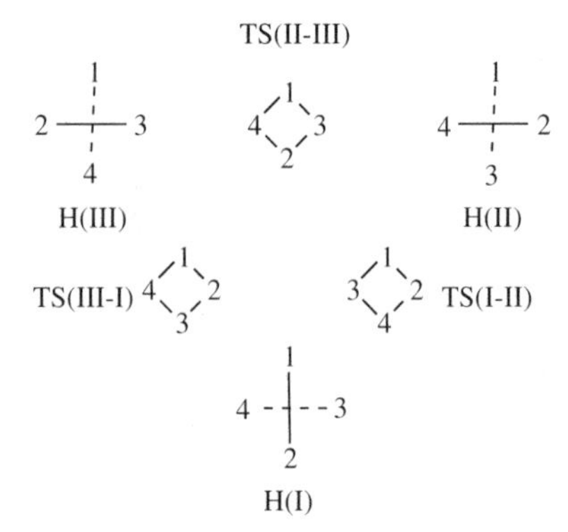

Figure 2. The H_4 system. TS are transition states.

The most stable nuclear configuration of this system is a pair of H_2 molecules. There are three possible spin coupling combinations for H_4 corresponding to three distinct stable product H_2 pairs: H1:H2 with H3:H4, H1:H3 with H2:H4, and H1:H4 with H2:H3. Each H atom contributes one electron, the dot diagrams indicate spin pairing. The three combinations are designated as H(I), H(II), and H(III), respectively. They may be interconverted via square transition states, Figure 2.

The electronic wave functions of the different spin-paired systems are not necessarily linearly independent. Writing out the VB wave function shows that one of them may be expressed as a linear combination of the other two. *Nevertheless, each of them is obviously a separate chemical entity, that can be clearly distinguished from the other two*. [This is readily checked by considering a hypothetical system containing four isotopic H atoms (H, D, T, and U). The anchors will be HD + TU, HT + DU, and HU + DT].

In short-hand notation, the electronic wave functions of the three spin-paired combinations may be written as

$$|\mathrm{H(I)}\rangle = (1\bar{2} - \bar{1}2)(3\bar{4} - \bar{3}4) = 1\bar{2}3\bar{4} - \bar{1}23\bar{4} - 1\overline{23}4 + \bar{1}2\bar{3}4 \tag{9a}$$

$$|\mathrm{H(II)}\rangle = (1\bar{3} - \bar{1}3)(2\bar{4} - \bar{2}4) = 1\bar{3}2\bar{4} - \bar{1}32\bar{4} - 1\overline{32}4 + \bar{1}3\bar{2}4 \tag{9b}$$

$$|\mathrm{H(III)}\rangle = (1\bar{4} - \bar{1}4)(2\bar{3} - \bar{2}3) = 1\bar{4}2\bar{3} - \bar{1}42\bar{3} - 1\overline{42}3 + \bar{1}4\bar{2}4 \tag{9c}$$

Since exchanging two columns in a determinant changes its sign, simple algebra shows that

$$-|\mathrm{H(III)}\rangle = |\mathrm{H(I)}\rangle + |\mathrm{H(II)}\rangle \tag{10}$$

Thus, the electronic wave function of H(III) is (to within a multiplication constant) equal to the in-phase combination of the electronic wave functions of

H(I) and H(II). This fact does not provide any information on the nuclear structure of this species at the energy minimum. By symmetry, it is clear that the system has three equivalent minima on the ground-state surface, which were designated as the three diatomic pairs. The nuclear geometry of each of these minima is quite different from that of the other two.

There are two nuclear configurations on the ground-state surface that are of special interest to the chemist: One is the energy minimum for the *in-phase* combination of $|\mathrm{H(I)}\rangle$ and $|\mathrm{H(II)}\rangle$, which is the equilibrium geometry of H(III). The second is also a stationary point on the ground-state surface, but for the *out-of-phase* combination of $|\mathrm{H(I)}\rangle$ and $|\mathrm{H(II)}\rangle$—it is the TS between H(I) and H(II). Clearly, the geometries (nuclear configuration) of these two species are quite different. Each of these structures is constructed from two base functions, and is therefore a two-state system. As for any two-state system, each has a twin state on the electronic excited surface. Thus, the *in-phase combination of the two electronic wave functions* $|\mathrm{H(I)}\rangle$ *and* $|\mathrm{H(II)}\rangle$ *at the nuclear configuration of the transition state* is found on the *excited-state* potential surface. Likewise, the *out-of-phase* combination at the nuclear geometry of the minimum energy of $|\mathrm{H(III)}\rangle$ also lies on the excited-state potential. Thus a given spin-paired scheme of the H_4 system is seen to support very different nuclear geometries on the each potential surfaces.

We can now proceed to discuss the phase-change rule and its use to locate conical intersections.

II. THE PHASE-CHANGE RULE AND THE CONSTRUCTION OF LOOPS

Herzberg and Longuet-Higgins noted the singular behavior of the electronic wave function around a degeneracy [12,13]. This observation is the basis of the present approach to molecular photochemistry. Let $\phi(\mathbf{r},\mathbf{R})$ be the total polyelectronic wave function of a polyatomic molecule, where $\mathbf{r}$ and $\mathbf{R}$ denote the electronic and nuclear coordinates, respectively. Within the BO approximation, this wave function is an explicit function of $\mathbf{r}$ for a given set of nuclear coordinates $\mathbf{R}_0$. It must be continuous everywhere in the electronic coordinates $\mathbf{r}$, but may change sign in an abrupt, seemingly discontinuous manner when the R's are slightly changed. If $\phi(\mathbf{r},\mathbf{R})$ is nondegenerate throughout a certain region of the nuclear configuration space, it will be a real continuous function of the $\mathbf{R}$'s as well as of the $\mathbf{r}$'s. If it changes abruptly at some point, there must be two electronic states with the same energy at this point, in other words the function is degenerate at that point.

Consider the function at a certain set of R's, $\mathbf{R}_0$, where $\phi(\mathbf{r},\mathbf{R})$ is nondegenerate. When the nuclei move away from that point, and approach it back via a different route, the wave function must return to its original value. However, in

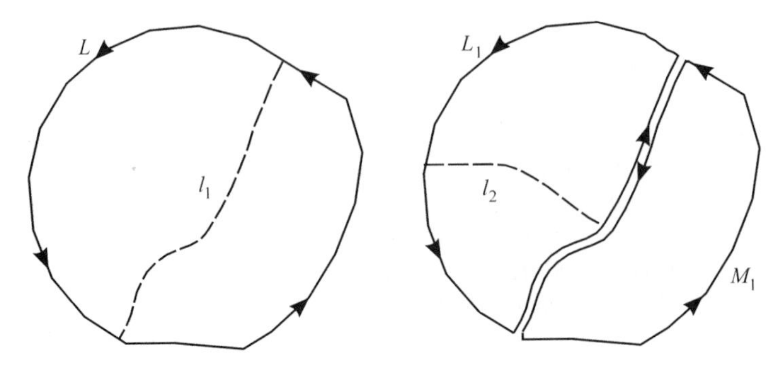

Figure 3. Longuet-Higgins' rule proof.

the process it may change sign, since if $\phi(\mathbf{r}, \mathbf{R}_0)$ is a solution of the electronic wave, so is $-\phi(\mathbf{r}, \mathbf{R}_0)$. Thus one can distinguish two kinds of paths leading by a closed loop from $\mathbf{R}_0$ back to itself: sign preserving and sign reversing.

Longuet-Higgins stated and proved the following theorem [13]:

Let S be any simply connected surface in nuclear configuration space, bounded by a closed-loop L. Then, if $\phi(\mathbf{r}, \mathbf{R})$ changes sign when transported adiabatically round L, there must be at least one point on S at which $\phi(\mathbf{r}, \mathbf{R})$ is discontinuous, implying that its potential energy surface intersects that of another electronic state.

The proof was by reduction ad absurdum.

Let l_1 be any line in S that bisects the area enclosed by L, and let L_1 and M_1 be the two loops created (Fig. 3). If L is sign reversing and if $\phi(\mathbf{r}, \mathbf{R})$ is continuous everywhere on S, than *either* L_1 or M_1 must be sign reversing. If L_1 and M_1 were both sign reversing or sign preserving, than L would also be sign preserving, in contradiction with the assumption. Let L_1 be the sign reversing loop. It encloses a simply connected surface S_1 which is smaller than S. We now bisect S_1 by a line l_2 and repeat the argument. In this fashion, a large number of successively smaller loops are created, all of them sign reversing. These loops converge to a point P on the surface, where $\phi(\mathbf{r}, \mathbf{R})$ is discontinuous in $\mathbf{R}$, because of the sign change. Thus, the function cannot be single valued—it must be degenerate. In other words, two potential surfaces cross at this point. Longuet-Higgins' proof assumed that the electronic wave function is real everywhere. Stone [35] showed that the theorem applies also for a general phase change. Thus, when the nuclei return to their original configuration $\mathbf{R}_0$, the wave function may undergo a change phase $Z = e^{i\Omega}$, since if $\phi(\mathbf{r}, \mathbf{R}_0)$ is a solution of the electronic wave equation, so is $e^{i\Omega}\phi(\mathbf{r}, \mathbf{R}_0)$.

The proof runs analogously to the original Longuet-Higgins one, and is not reproduced here.

A. Construction of Loops: Nature of the Coordinates

Herzberg and Longuet-Higgins [12] explicitly discussed the H_3 system. This is a three-electron problems, which has the same spin-pairing properties as the four-electron system H_4. The loop is constructed by considering the three possible spin-pairing options for these systems (Fig. 4), compare Figure 2. The transition states for the H_3 system are linear [36] and their wave functions are the out-of-phase combination of the two wave functions of the reactant and product systems. As mentioned above for H_4, Pauli's principle and the permutational symmetry of the polyelectronic wave function are the ultimate reason for the fact that the ground-state surface in this case is the out-of-phase combination, rather than the in-phase one.

Generalizing on [12], we construct a loop by using a sequence of three elementary reactions. It is emphasized that the reactions comprising the loop must be elementary ones: There should not be any other spin pairing combination that connects two anchors. This ensures that the loop in question is indeed the "smallest" possible one. Inspection of the loops depicted in Figure 4 shows that the H_3 and H_4 systems are entirely analogous. We include the H_3 system in order to introduce the coordinates spanning the plane in which the loop lies, and as a prototype of all three-electron systems.

There are two independent coordinates that define the plane of a loop. If the loop is phase inverting, one of these coordinates must be phase inverting, the other, phase preserving. Out of the infinite number of possible candidates, a convenient choice are reaction coordinates (Section I). Any one of the three reaction coordinates connecting two of the anchors can be used for the

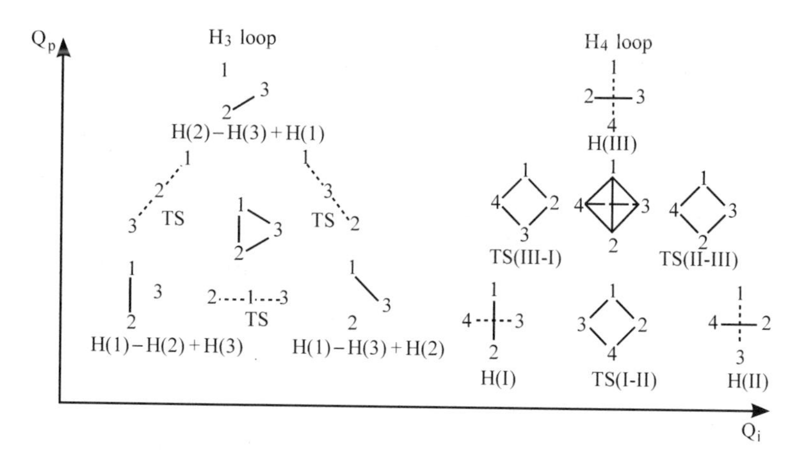

Figure 4. The H_3 and H_4 loops. At the center, the conical intersections are shown schematically: an equilateral triangle for H_3 and a perfect tetrahedron for H_4, Q_p, and Q_i are the phase-preserving and phase-inverting coordinates, respectively.

phase-inverting coordinates. Let us use the H_4 system, and choose the reaction $H_1H_2 + H_3H_4 \rightarrow H_1H_3 + H_2H_4$; this coordinate is designated as Q_i, Figure 4. The coordinate connecting the transition state of this reaction with the third anchor is phase preserving (Q_p in the figure), which may be shown as follows.

Consider the coordinate that transforms the nuclear configuration of H(III) at the minimum energy with the corresponding configuration of TS(I–II). In the former, atoms 1 and 4 are close together, as are atoms 2 and 3. The separation between the two pairs is large. In other words, if R_{ij} is the separation between atoms i and j, we have

$$R_{14} = R_{23} \ll R_{12}, R_{13}, R_{24}, R_{34} \qquad \text{(at H(III) minimum)}$$

In the latter,

$$R_{12} = R_{23} = R_{34} = R_{41} \qquad \text{(at the transition state between H(I) and H(II))}$$

Moving along this coordinate without changing the phase of the electronic wave function (solid line in Fig. 5), leads from the ground state to the excited state of the system. The dashed line shows the motion along the coordinate connecting the transition state with the configuration of H(III). Keeping the phase constant also leads from the ground to the excited state.

It is clear from Figure 5 that the phase of the electronic wave function of the *ground state* is constant when moving along the Q_p coordinate, until a certain

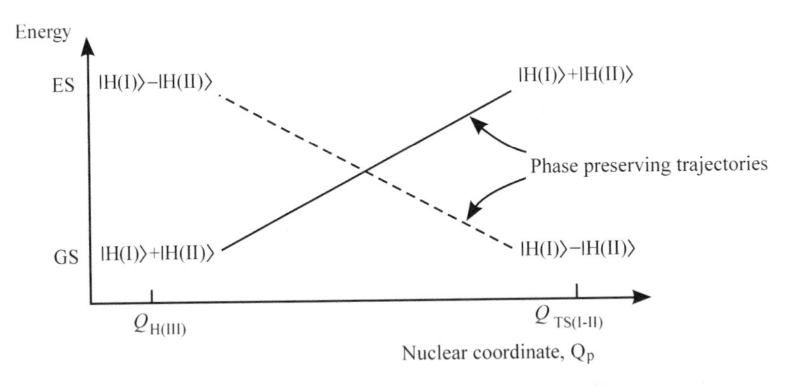

Figure 5. A cut across the ground state (GS) and the excited state (ES) potential surfaces of the H_4 system. The parameter Q_p is the phase preserving nuclear coordinate connecting the H(III) with the transition state between H(I) and H(II) (Fig. 4). Keeping the phase of the electronic wave function constant, this coordinate leads from the ground to the excited state. At a certain point, the two surfaces must touch. At the crossing point, the wave function is degenerate.

configuration (point in phase space) is reached, in which it changes sign. This "point" is the out-of-phase transition state. The same holds for the electronic wave function of the excited state. The two potential curves cross at some point, where the electronic wave function becomes degenerate. The crossing along this coordinate is permitted, since the two curves are of different symmetry.

This situation arises when the electronic wave function of the transition state is described by the out-of-phase combination of the two base functions. If the electronic wave function of the transition state is described by the in-phase combination, no curve crossing occurs.

In the vicinity of the crossing point of the two electronic states of the H_4 system, we can therefore define two coordinates along which the potential surface of the system is constructed. The phase-preserving coordinate Q_p connecting the H(I) minimum with TS(II–III), and the phase-inverting coordinate Q_i connecting the minima of H(II) and H(III). A plot of the energies of the two surfaces has the shape of double cone. Moreover, as nothing but spin pairing was assumed in the derivation, the situation is not unique to the H_4 system: It holds for any four-electron system. Figure 6 depicts the general case, in which the potential energy surface relevant to three chemical species A, B,

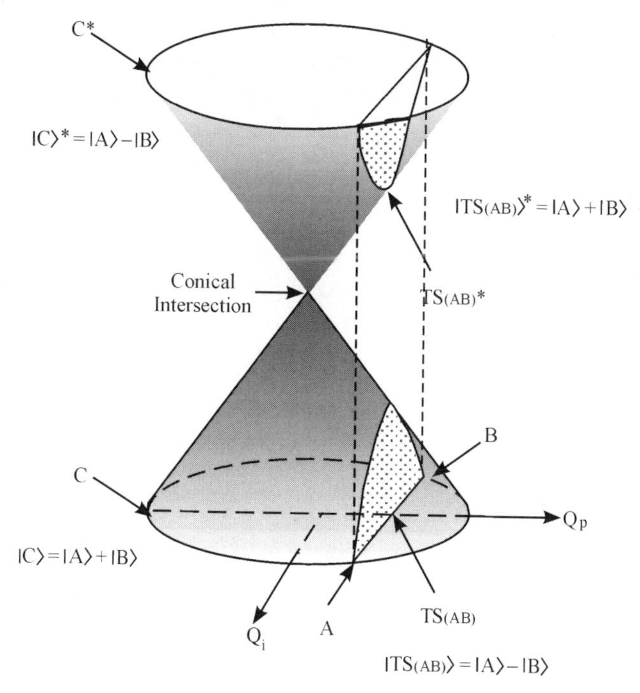

Figure 6. A schematic representation of a conical intersection. The bottom part of the cone belongs to the ground state, the upper, to the electronically excited state.

and C is shown. The species A, B, and C consist of the same atoms, and differ only in their spin pairing schemes. We assume that the electronic wave function of C, $|C\rangle$, may be expressed as the in-phase combination of $|A\rangle$ and $|B\rangle$, while the transition state between A and B TS(AB) is represented by the out-of-phase combination. Here C is a local minimum on the ground-state surface. At the geometry of the transition state, the in-phase combination of $|A\rangle$ and $|B\rangle$ lies on the excited potential surface.

The H_4 system is the prototype for many four-electron reactions [34]. The basic tetrahedral structure of the conical intersection is preserved in all four-electron systems. It arises from the fact that the four electrons are contributed by four different atoms. Obviously, the tetrahedron is in general not a perfect one. This result was found computationally for many systems (see, e.g., [37]). Robb and co-workers [38] showed that the structure shown (a tetraradicaloid conical intersection) was found for many different photochemical transformations. Having the form of a tetrahedron, the conical intersection can exist in two enantiomeric structures. However, this feature is important only when chiral reactions are discussed.

The two coordinates defined for H_4 apply also for the H_3 system, and the conical intersection in both is the most symmetric structure possible by the combination of the three equivalent structures: An equilateral triangle for H_3 and a perfect tetrahedron for H_4. These structures lie on the ground-state potential surface, at the point connecting it with the excited state. This result is generalized in the Section. IV.

III. PHASE CHANGE IN A CHEMICAL REACTION

The phase change occurring upon carrying out a complete loop around a point, as discussed by Longuet-Higgins and Herzberg, does not explicitly consider phase changes in chemical reactions. However, the proof concerning H_3 depends on the phase change taking place during the methathesis reaction $H_1H_2 + H_3 \rightarrow H_1 + H_2H_3$ [12]. In this section, we discuss the general case—a phase change taking place upon an arbitrary chemical reaction.

A chemical reaction takes place on a potential surface that is determined by the solution of the electronic Schrödinger equation. In Section, we defined an anchor by the spin-pairing scheme of the electrons in the system. In the discussion of conical intersections, the only important reactions are those that are accompanied by a change in the spin pairing, that is, interanchor reactions. We limit the following discussion to these class of reactions.

The concept of phase change in chemical reactions, was introduced in Section I, where it was shown that it is related to the number of electron pairs exchanged in the course of a reaction. In every chemical reaction, the fundamental law to be observed is the preservation permutational symmetry of the polyelectronic wave function: Pauli's principle must be obeyed. Phase changes are not generally regarded as important parameters in understanding the fundamentals of chemical reactions. Goddard tried to utilize the phase-change concept for a better understanding of chemical transformations [22,39]. Up to now, these efforts do not seem to have significantly affected the way chemists conceive reactions. We propose that the study of phase change in chemical reaction may become a substantial ingredient in photochemical theory, since conical intersections appear to be of major importance in this field. The phase-change rule is closely related to conical intersections on one hand, and to the phase change on the other. We suggest that a loop can always be constructed from several reactions, and the total phase change deduced from the combination of changes incurred in the individual steps [25,40,41]. If this hypothesis proves to be correct, phase changes taking place on the ground-state surface will play an important role in photochemistry.

In this section, we illustrate the applicability of the model to some important special cases, and summarize the relationship between aromaticity and chemical reactivity, expressed in the properties of transition states.

A. Pericyclic Reactions

The special case of pericyclic reactions is an appropriate means of introducing the subject: These reactions are very common, and were extensively studied experimentally and theoretically. They also provide a direct and straightforward connection with aromaticity and antiaromaticity, concepts that turn out to be quite useful in analyzing phase changes in chemical reactions.

Already in 1938, Evans and Warhurst [17] suggested that the Diels–Alder addition reaction of a diene with an olefin proceeds via a concerted mechanism. They pointed out the analogy between the delocalized electrons in the transition states for the reaction between butadiene and ethylene and the π electron system of benzene. They calculated the resonance stabilization of this transition state by the VB method earlier used by Pauling to calculate the resonance energy of benzene. They concluded that the extra "aromatic" stabilization of this transition state made the concerted route more favorable then a two-step process. In a subsequent paper [18], Evans used the Hückel MO theory to calculate the transition state energy of the same reaction and some others. These ideas essentially introduce a chemical reacting complex (reactants and products) as a two-state system. Dewar [42] later formulated a general principle for all pericyclic reactions (Evans' principle): Thermal pericyclic reactions take place preferentially via aromatic transition states. Aromaticity was defined by the amount of resonance stabilization. Evans' principle connects the problem of thermal pericyclic reactions with that of aromaticity: Any theory of aromaticity is also a theory of pericyclic reactions [43]. Evans' approach was more recently used to aid in finding conical intersections [44], (cf. Section VIII).

Several theories of aromaticity have been suggested, starting with Hückel's $4n + 2$ rule [29]. All of them consider the entities in question as two-state systems (Section I), and consider the relative phase relations of adjacent *p* orbitals as playing a major role. It has been pointed out by many authors [19,21,45–47] that in a conjugated systems there can be two cases. If all neighboring lobes of *p*- orbitals have a the same sign, the system is denoted as a Hückel one. If, on the other hand, there is a phase change at one point (termed phase dislocation by Craig [46]), the system is a Möbius one (one of the *p* orbitals is inverted by 180°, forming a Möbius strip). This definition may be generalized [19]: In a Hückel system, the number of dislocations is even, in a Möbius system, it is odd. Möbius systems are sometimes termed anti-Hückel.

A more general classification considers the phase of the total electronic wave function [13]. We have treated the case of cyclic polyenes in detail [28,48,49] and showed that for Hückel systems the ground state may be considered as the combination of two Kekulé structures. If the number of electron pairs in the system is odd, the ground state is the in-phase combination, and the system is aromatic. If the number of electron pairs is even (as in cyclobutadiene, pentalene, etc.), the ground state is the out-of-phase combination, and the system is antiaromatic. These ideas are in line with previous work on specific systems [40,50].

The results of the derivation (which is reproduced in Appendix A) are summarized in Figure 7. This figure applies to both reactive and resonance stabilized (such as benzene) systems. The compounds A and B are the reactant and product in a pericyclic reaction, or the two equivalent Kekulé structures in an aromatic system. The parameter ξ is the reaction coordinate in a pericyclic reaction or the coordinate interchanging two Kekulé structures in aromatic (and antiaromatic) systems. The avoided crossing model [26–28] predicts that the two eigenfunctions of the two-state system may be formed by in-phase and out-of-phase combinations of the noninteracting basic states $|A\rangle$ and $|B\rangle$. State $|A\rangle$ differs from $|B\rangle$ by the spin-pairing scheme.

The ground state is the in-phase combination $(|A\rangle + |B\rangle)$ for an odd number of electron pairs exchanged, while if the number is even, the out-of-phase $(|A\rangle - |B\rangle)$ combination is the ground state. The other combination is an electronically excited state. Classical VB theory predicts the curves shown as solid lines. The energy ordering of the in-phase and out-of-phase combinations, and the energy splitting between them is due primarily to the pairwise transposition permutations (Appendix A). The effect of the cyclic permutation term [32], shown by the dashed lines, is to modify the splitting: In systems for which an odd number of electron pairs are exchanged, the cyclic permutation term acts in harmony with the classical term, and increases the gap between the ground and excited state. In the even parity systems, it acts to decrease the gap. When the cyclic term is large enough, a single minimum in the ground state is obtained—this is the origin of the extra stability of benzene, for example. In

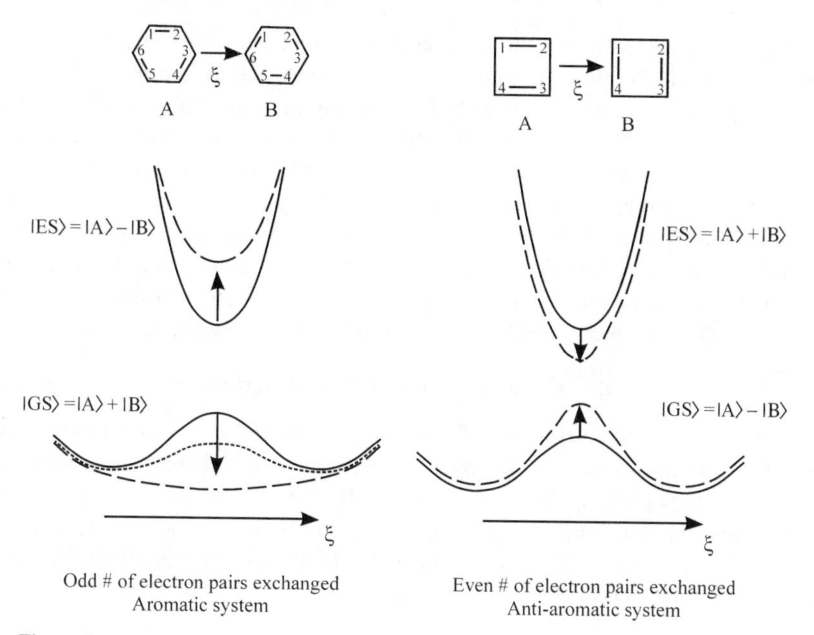

Figure 7. Aromatic and antiaromatic systems in the ground state (GS) and the twin excited state (ES). The parameter ξ is the coordinate that transforms A to B.

other systems, the barrier is reduced but not eliminated. The result is a stabilized aromatic transition state [51].

Adopting the view that any theory of aromaticity is also a theory of pericyclic reactions [19], we are now in a position to discuss pericyclic reactions in terms of phase change. Two reaction types are distinguished: those that preserve the phase of the total electronic wave-function – these are phase preserving reactions (p-type), and those in which the phase is inverted – these are phase inverting reactions (i-type). The former have an aromatic transition state, and the latter an antiaromatic one. The results of [28] may be applied to these systems. In distinction with the cyclic polyenes, the two basis wave functions need not be equivalent. The wave function of the reactants $|\mathrm{R}\rangle$ and the products $|\mathrm{P}\rangle$, respectively, can be used. The electronic wave function of the transition state may be represented by a linear combination of the electronic wave functions of the reactant and the product. Of the two possible combinations, the in-phase one [Eq. (11)] is phase preserving (p-type), while the out-of-phase one [Eq. (12)], is i-type (phase inverting), compare Eqs. (6) and (7). Normalization constants are assumed in both equations:

$$|\text{Aromatic TS}\rangle = |\mathrm{A}\rangle + |\mathrm{B}\rangle \qquad \text{Phase-preserving transition state} \tag{11}$$

$$|\text{Antiaromatic TS}\rangle = |\mathrm{A}\rangle - |\mathrm{B}\rangle \qquad \text{Phase-inverting transition state} \tag{12}$$

The Woodward–Hoffmann method [52], which assumes conservation of orbital symmetry, is another variant of the same idea. In it, the emphasis is put on the symmetries of molecular orbitals. Longuet-Higgins and Abramson [53] noted the necessity of state-to-state correlation, rather than the orbital correlation, which is not rigorously justified (see also, [30,44]). However, the orbital symmetry conservation rules appear to be very useful for most thermal reactions.

A symmetry that holds for any system is the permutational symmetry of the polyelectronic wave function. Electrons are fermions and indistinguishable, and therefore the exchange of any two pairs must invert the phase of the wave function. This symmetry holds, of course, not only to pericyclic reactions.

B. Generalization to Any Reactions

In this chapter, we restrict the discussion to elementary chemical reactions, which we define as reactions having a single energy barrier in both directions. As discussed in Section I, the wave function $|\mathrm{R}\rangle$ of any system undergoing an elementary reaction from a reactant A to a product B on the ground-state surface, is written as a linear combination of the wave functions of the reactant, $|\mathrm{A}\rangle$, and the product, $|\mathrm{B}\rangle$ [47,54]:

$$|\mathrm{R}\rangle = C_{\mathrm{A}}|\mathrm{A}\rangle + C_{\mathrm{B}}|\mathrm{B}\rangle \tag{13}$$

C_{A} and C_{B} may have the same or opposite signs.

Within the Born–Oppenheimer approximation, the electronic wave function $|\mathrm{R}\rangle_{\mathrm{el}}$, is well defined, throughout the reaction and may be written analogously [cf. Eq. (6)]

$$|\mathrm{R}(t)\rangle_{\mathrm{el}} = k_{\mathrm{A}}(t)|\mathrm{A}\rangle_{\mathrm{el}} + k_{\mathrm{B}}(t)|\mathrm{B}\rangle_{\mathrm{el}} \tag{14}$$

where k_{A} is unity in the beginning of the reaction ($t = 0$) and k_{B} is unity at the end ($t = \infty$) [26]. A phase change involves the introduction of a new node (or an odd number of nodes) along the reaction coordinate, which is equivalent to changing the total electronic angular momentum of the system along that coordinate. The role of nodes and nodal parity was discussed extensively for correlated molecular orbitals during a reaction [21]. A similar approach, using VB, was suggested by Mulder and Oosterhof for pericyclic reactions [32]. We emphasize the properties of the total wave function, a concept that is difficult to visualize in a graphic manner.

There are two mechanisms by which a phase change on the ground-state surface can take place. One, the *orbital overlap mechanism*, was extensively discussed by both MO [55] and VB [47] formulations, and involves the creation of a negative overlap between two adjacent atomic orbitals during the reaction (or an odd number of negative overlaps). This case was termed a phase dislocation by other workers [43,45,46]. A reaction in which this happens is termed Möbius- or anti-Hückel type. A well-known example is conrotatory ring closure in pericyclic reactions. Another is the antarafacial sigmatropic migration. A reaction in which all overlaps between adjacent atomic orbitals along the reaction coordinate are positive (or such that the number of negative overlaps is even) is termed Hückel type. Only Hückel-type reactions are possible when *s* orbitals are the sole ones involved in the reaction. With *p* orbitals, both Hückel- and Möbius-type reactions are possible. Figure 8 depicts two examples.

The second mechanism, due to the permutational properties of the electronic wave function is referred to as the *permutational mechanism*. It was introduced in Section I for the H_4 system, and above for pericyclic reactions and is closely related to the aromaticity of the reaction. Following Evans' principle, an aromatic transition state is defined in analogy with the hybrid of the two Kekulé structures of benzene. A cyclic transition state in pericyclic reactions is defined as aromatic or antiaromatic according to whether it is more stable or less stable than the open chain analogue, respectively. In [32], it was assumed that the in-phase combination in Eq. (14) lies *always* the on the ground state potential. As discussed above, it can be shown that the ground state of aromatic systems is always represented by the in-phase combination of Eq. (14), and antiaromatic ones—by the out-of-phase combination.

The concepts may be extended to describe transition states of *any* chemical reaction. Since the only assumption made was that the VB structures represent the molecules, and that an exchange of two columns in the determinant representing the VB structures changes the sign, the result is general (Appendix A).

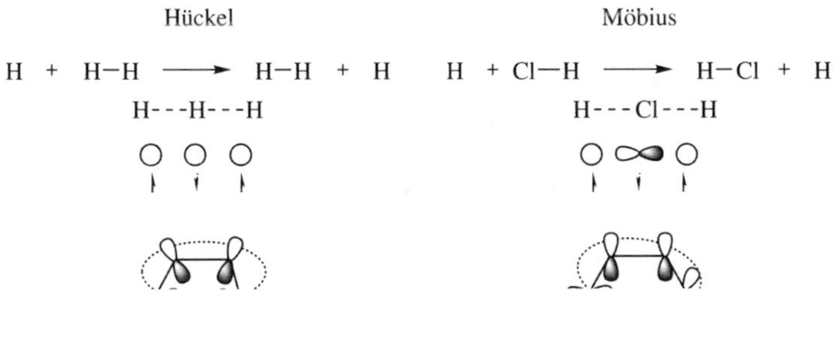

Figure 8. Hückel and Möbius reactions. Top left: the $H + H_2$ reaction. The second line shows the structure of the linear transition state, the circles in the third line denote *s*-type orbitals, the arrows designate the electron spin vectors. Top right: the H + ClH reaction. The second line shows the structure of the linear transition state, the symbols in the third line denote *s*- and *p*-type orbitals, the arrows designate the electron spin vectors. Bottom left: A disrotatory ring closure reaction of butadiene, showing the *p*-orbitals of the carbon atoms that exchange the spin-pairs. Bottom right: A conrotatory ring closure reaction of butadiene, showing one negative overlap between *p*-orbitals of the carbon atoms that exchange the spin-pairs.

We term the in-phase combination an aromatic transition state (ATS) and the out-of-phase combination an antiaromatic transition state (AATS). An ATS is obtained when an odd number of electron pairs are re-paired in the reaction, and an AATS, when an even number is re-paired. In the context of reactions, a system in which an odd number of electrons (3, 5, ...) are exchanged is treated in the same way—one of the electron pairs may contain a single electron. Thus, a three-electron system reacts as a four-electron one, a five-electron system as a six-electron one, and so on.

Finally, the distinction between Hückel and Möbius systems is considered. The above definitions are valid for Hückel-type reactions. For aromatic Möbius-type reations, the reverse holds: An ATS is formed when an even number of electron pairs is re-paired.

These general ideas will be demonstrated by considering a few examples.

1. Reactions Involving Sigma Bonds Only

The H_3 and H_4 systems were discussed above. Another type of sigma bonds involves a *p* orbital lying along the reaction coordinate, as, for example, in reaction (15) (Fig. 8).

$$H + ClH \rightarrow HCl + H \qquad (15)$$

This is an example of a Möbius reaction system—a node along the reaction coordinate is introduced by the placement of a phase inverting orbital. As in the $H + H_2$ system, a single spin-pair exchange takes place. Thus, the reaction is phase preserving. Möbius reaction systems are quite common when *p* orbitals (or hybrid orbitals containing *p* orbitals) participate in the reaction, as further discussed in Section III.B.2.

A phase change takes place when one enantiomer is converted to its optical isomer. As illustrated in Figure 9, when the chiral center is a tetra-substituted carbon atom, the conversion of one enantiomer to the other is equivalent to the exchange of two electron pairs. This transformation is therefore phase inverting.

2. Reactions Involving π Bonds

Hückel-type systems (such as Hückel pericyclic reactions and suprafacial sigmatropic shifts) obey the same rules as for sigma electron. The rationale for this observation is clear: If the overlap between adjacent *p*-electron orbitals is positive along the reaction coordinate, only the permutational mechanism can

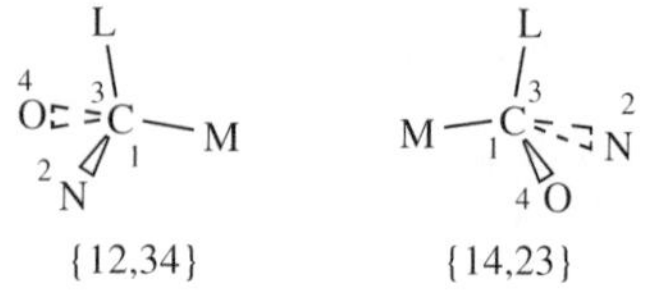

{12,34} {14,23}

Figure 9. The phase-inverting transformation of a chiral system with a tetra-substituted carbon atom.

lead to sign inversion. In Möbius-type systems (antarafacial reactions), the system will change phase only if an even number of electron spin pairs exchanges takes place.

Electrocyclic reactions are examples of cases where π-electron bonds transform to sigma ones [32,49,55]. A prototype is the cyclization of butadiene to cyclobutene (Fig. 8, lower panel). In this four electron system, phase inversion occurs if no new nodes are formed along the reaction coordinate. Therefore, when the ring closure is disrotatory, the system is Hückel type, and the reaction a phase-inverting one. If, however, the motion is conrotatory, a new node is formed along the reaction coordinate just as in the HCl + H system. The reaction is now Möbius type, and phase preserving. This result, which is in line with the Woodward–Hoffmann rules and with Zimmerman's Möbius–Hückel model [20], was obtained without consideration of nuclear symmetry. This conclusion was previously reached by Goddard [22,39].

IV. LOOP CONSTRUCTION FOR PHOTOCHEMICAL SYSTEMS

In this section, the systematic search for conical intersections based on the Longuet-Higgins phase-change rule is described. For conciseness sake, we limit the present discussion to Hückel-type systems only, unless specifically noted otherwise. The first step in the analysis is the determination of the LH loops containing a conical intersection for the reaction of interest.

In general, at least three anchors are required as the basis for the loop, since the motion around a point requires two independent coordinates. However, symmetry sometimes requires a greater number of anchors. A well-known case is the Jahn–Teller degeneracy of perfect pentagons, heptagons, and so on, which will be covered in Section V. Another special case arises when the electronic wave function of one of the anchors is an out-of-phase combination of two spin-paired structures. One of the vibrational modes of the stable molecule in this anchor serves as the out-of-phase coordinate, and the loop is constructed of only two anchors (see Fig. 12).

We have seen (Section I) that there are two types of loops that are phase inverting upon completing a round trip: an i^3 one and an ip^2 one. A schematic representation of these loops is shown in Figure 10. The other two options, p^3 and i^2p loops do not contain a conical intersection. Let us assume that A is the reactant, B the desired product, and C the third anchor. In an ip^2 loop, any one of the three reaction may be the phase-inverting one, including the B → C one. Thus, the A → B reaction may be phase preserving, and still B may be attainable by a photochemical reaction. This is in apparent contradiction with predictions based on the Woodward–Hoffmann rules (see Section VIII). The different options are summarized in Figure 11.

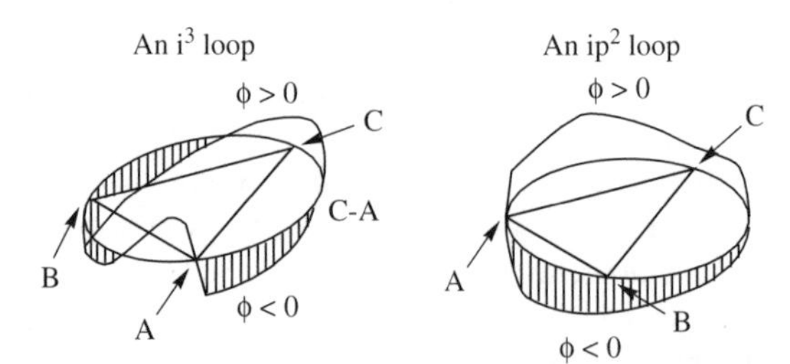

Figure 10. A cartoon showing the phase change in loops containing a conical intersection.

The two coordinates that define the "plane" in which the loop located were discussed in Section II. In loops that encircle a conical intersection, there is always at least one phase-inverting reaction—we can choose its coordinate as the phase-inverting one. Let us assume that this is the reaction connecting A and B. The phase changes near the transition state lying along this coordinate. It must therefore be positive close to that locality. The electronic wave function of C, the third anchor is obtained from the in-phase combination of $|A\rangle$ and $|B\rangle$, as shown in Section I. Therefore, there is always a phase-preserving coordinate connecting C and the vicinity of the TS between A and B. We shall make use of this property in the practical application of the method.

A given pair of anchors may be part of several loops, containing different conical intersections. A systematic search for the third anchor is conducted by considering the electrons that are to be re-paired (i.e., that form the chemical bonds that are created in the reaction). A pragmatic and systematic way of doing this is by considering first the re-pairing of the smallest possible number of

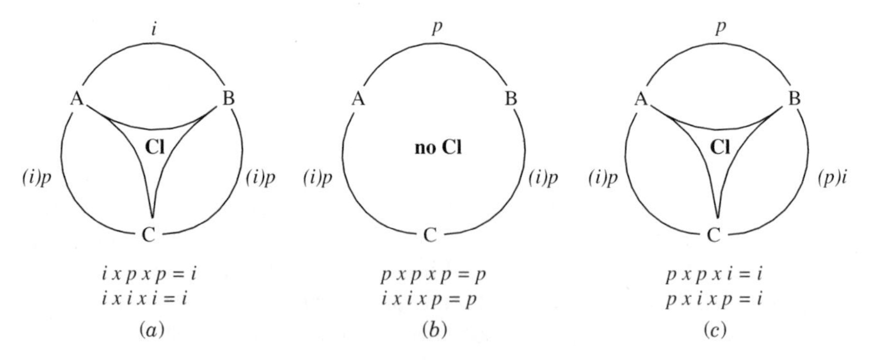

Figure 11. Three typical loops for the case where A is the reactant and B—the desired product. Loops in which a conical intersection may be found are (*a*) and (*c*). A loop that does not encircle a conical intersection is (*b*). In loop (*a*) the A → B reaction is phase inverting, and in loops (*b*) and (*c*) it is phase preserving.

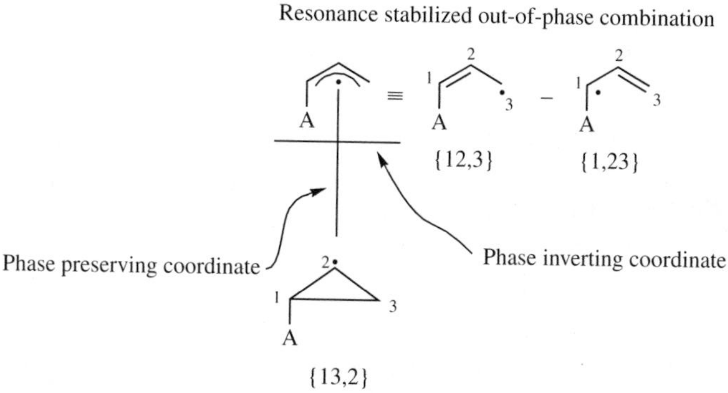

Figure 12. The allyl/cyclopropyl radical loop.

electrons that change their pairing under the energy constraints of the reaction. In closed shell systems, at least four electrons must be involved since at least one phase inverting reaction is required. Next, reactions involving six electrons are considered, and so on.

A. Three-Electron Systems

We begin by considering a three-atom system, the allyl radical. A two anchor loop applies in this case as illustrated in Figure 12: The phase change takes place at the allyl anchor, and the phase-inverting coordinate is the asymmetric stretch C_3 mode of the allyl radical. Quantum chemical calculations confirm this qualitative view [24,56]. In this particular case only one photochemical product is expected.

The allyl radical plays an important role in many photochemical transformations, as further discussed in Section IV.

B. Four-Electron Systems

Here the prototype is H_4—as only three spin-pairing arrangements are possible, this system is simple to analyze. It turns out to be very frequently encountered in practice, even in rather complex systems.

1. *Four π Electrons: Butadiene Ring Closure*

The classic example is the butadiene system, which can rearrange photochemically to either cyclobutene or bicyclobutane. The spin pairing diagrams are shown in Figure 13. The stereochemical properties of this reaction were discussed in Section III (see Fig. 8). A related reaction is the addition of two ethylene derivatives to form cyclobutanes. In this system, there are also three possible spin pairing options.

{12,34}
{14,23}
Phase preserving coordinate
Phase inverting coordinate
{13,24}

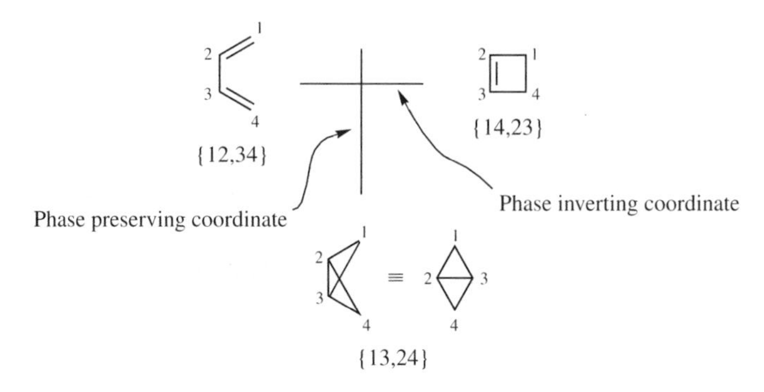

Figure 13. Anchors, coordinates, and loop for the butadiene system.

2. *cis–trans Isomerization: 2 π and 2 σ Electrons*

Although this reaction appears to involve only two electrons, it was shown by Mulder [57] that in fact two π and two σ electrons are required to account for this system. The three possible spin pairings become clear when it is realized that a pair of carbene radicals are formally involved, Figure 14. In practice, the conical intersection defined by the loop in Figure 14 is high-lying, so that often other conical intersections are more important in ethylene photochemistry. Hydrogen-atom shift products are observed [58]. This topic is further detailed in Section VI.

3. *Ammonia and Chiral Systems*

Ammonia is a two-state system [16], in which the two base states lie at a minimum energy. They are connected by the inversion reaction with a small barrier. The process proceeds upon the spin re-pairing of four electrons (Fig. 15) and has a very low barrier. The system is analogous to the tetrahedral carbon one

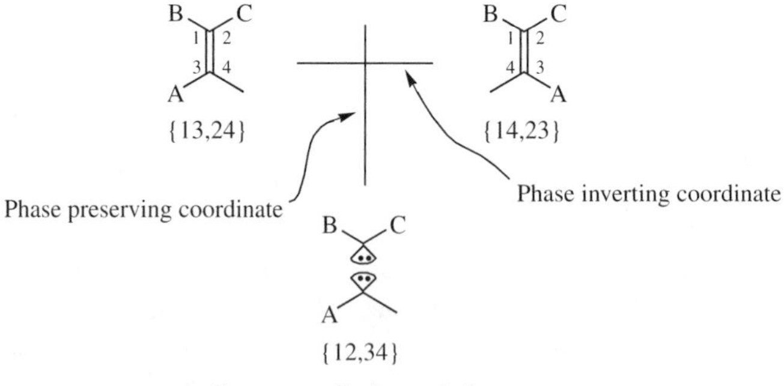

{13,24}
{14,23}
Phase preserving coordinate
Phase inverting coordinate
{12,34}
Two perpendicular methylenes

Figure 14. Same as Figure 13, for ethylene isomerization.

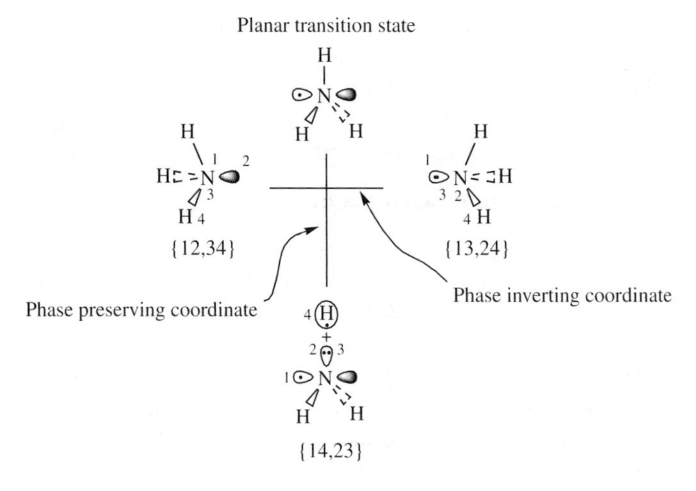

Figure 15. Three pairing schemes for the ammonia system.

(Fig. 9). Two loops based on these anchors are active in the photochemistry of ammonia, discussed in Section VI.

Another way to obtain the phase change taking place during this reaction is by assuming that the lone pair can tunnel through the barrier, while the spin-pairing of all NH bonds remains unchanged. This is then a two-electron Möbius-type reaction, which is phase inverting. Many examples of the equivalence of a four-electron Hückel system with a two-electron Möbius one are known.

A similar situation holds for a molecule containing a tetrahedral carbon is shown in (Figure 16). The reaction converting one enantiomer to another, is formally equivalent to the exchange of two sigma-bond electron pairs, and

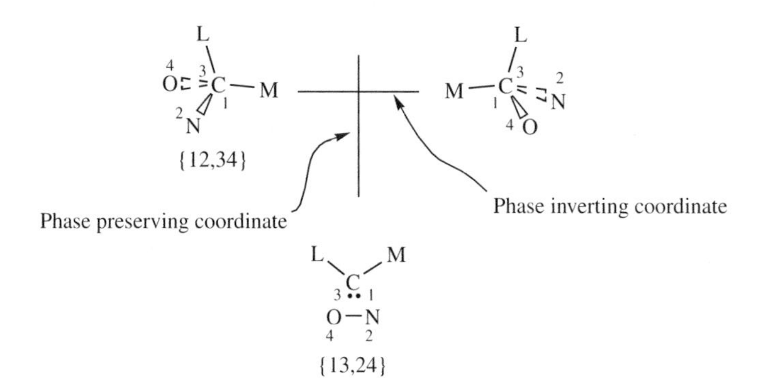

Figure 16. Chiral system anchors and coordinates.

therefore is a phase inverting reaction. A planar transition state may be imagined for this system [59]. The third anchor may be envisioned as a carbene plus a molecule, in which a new bond is formed between the two radicals created by the dissociation of two carbon–ligand bonds. The enantiomer conversion reaction does not take place thermally, possibly due to the very high barrier.

C. Four Electrons in Larger Systems

The main application of the loop method is to analyze complex systems, that can support several low-lying conical intersections. The idea is to provide a simple systematic, not intuition dependent, method for finding the accessible conical intersections.

The simplest loops would be i^3 loops in which all three reactions exchange two electron pairs (ip^2 loops require the re-pairing of at least three electron pairs). For a given system, valence electrons are considered (neglecting core electrons) in order of their increasing binding energy: π electrons first, then combination of π and σ electrons, and finally two pairs of σ electrons. Rydberg electrons need to be considered only in deep ultraviolet (UV) applications.

We illustrate the method for the relatively complex photochemistry of 1,4-cyclohexadiene (CHDN), a molecule that has been extensively studied [60–64]. There are four π electrons in this system. They may be paired in three different ways, leading to the anchors shown in Figure 17. The loop is phase inverting (type i^3), as every reaction is phase inverting), and therefore contains a conical intersection; Since the products are highly strained, the energy of this conical intersection is expected to be high. Indeed, neither of the two expected products was observed experimentally so far.

Next, we consider one pair of π electrons and one pair of σ electrons. The σ electrons may originate from a CH or from a CC bond. Let us consider the loop enclosed by the three anchors formed when the electron pair comes from a C–H bond. There are only three possible pairing options. The hydrogen-atom originally bonded to carbon atom 1, is shifted in one product to carbon atom 2,

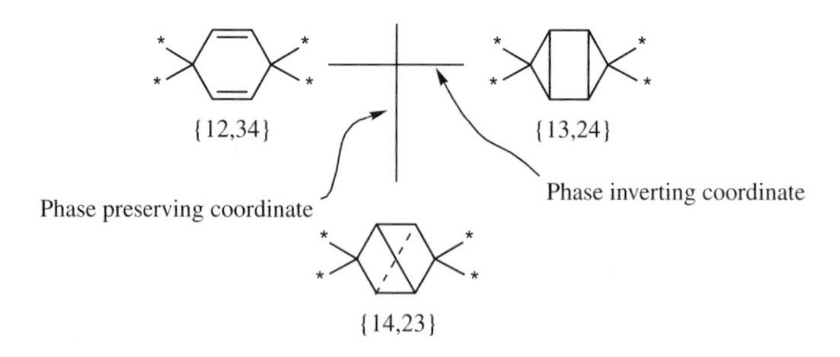

Figure 17. Possible spin-pairing schemes for CHDN, involving π electrons only.

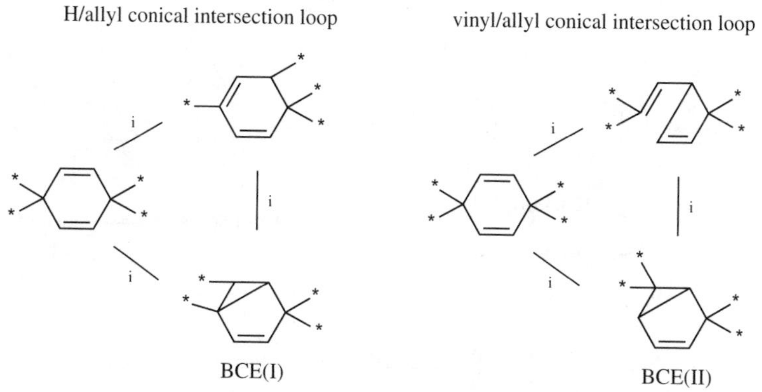

H/allyl conical intersection loop

vinyl/allyl conical intersection loop

BCE(I)

BCE(II)

Figure 18. H/allyl (left) and vinyl/allyl (right) loops for CHDN.

to form bicyclo-[3,1,0]hex-2-ene [BCE(I)], and in the other to carbon atom 3, to form 1,3-cyclohexadiene (1,3-CHDN). In case the two electrons originate from a C–C single bond, one product is again BCE, but using isotopic labeling, it is clear that it is different from the molecule formed upon hydrogen migration. It is therefore labeled as BCE(II). The second product is vinylcyclobutene. As seen from Figure 18, both loops are phase inverting (i^3 type), and enclose conical intersection. When a CH bond is cleaved, an H/allyl conical intersection is obtained and when a C–C bond is involved, a vinyl/allyl conical intersection. Both were reported in [65]. We designate the BCE isotopomer formed from the H/allyl loop as BCE(I). Both products of the loop encircling the vinyl/allyl CI were not observed experimentally. One of them, BCE(II), is the isotopomer expected from the di-π methane rearrangement [66].

The exchange of two pairs of σ electrons is expected to lead to a high-lying conical intersection that is not likely to be important in the UV photochemistry of CHDN. This winds up the possibilities of loops involving two-electron pair exchanges only.

D. More Than Four Electrons

The next simplest loop would contain at least one reaction in which three electron pairs are re-paired. Inspection of the possible combinations of two four-electron reactions and one six-electron reaction starting with CHDN reveals that they all lead to phase preserving i^2p loops that do not contain a conical intersection. It is therefore necessary to examine loops in which one leg results in a two electron-pair exchange, and the other two legs involve three electron-pair exchanges (ip^2 loops). As will be discussed in Section VI, all reported products (except the "helicopter-type" elimination of H_2) can be understood on the basis of four-electron loops. We therefore proceed to discuss the unique helicopter

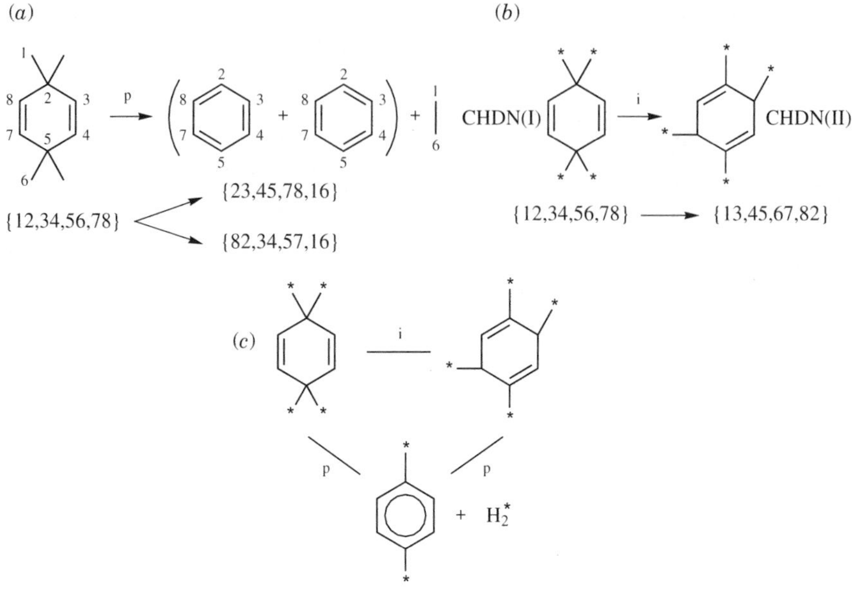

Figure 19. The proposed phase-inverting loop for the helicopter-type elimination of H_2 off CHDN. The asterisks denote the H atoms that were originally bonded in the 1,4 positions of CHDN. Parts (*a*) and are (*b*) the anchors and (*c*) is the loop.

reaction, in which the H_2 molecule departs from the carbon ring in a helicopter type motion [61,62].

The concerted CHDN → benzene + H_2 reaction (Fig. 19*a*) has an aromatic transition state [67,68] and is thermally allowed (phase preserving). Three electron pairs are re-paired in the reaction. In order to construct a conical intersection containing Longuet-Higgins loop that has this reaction as one of the legs, we must look for another reaction of CHDN (or benzene + H_2) that is phase inverting. The reaction must involve the two hydrogen atoms (that are eliminated in the benzene-forming reaction), so that all four electrons of the two CH bonds must participate in the reaction. Obviously, other bonds must also change, so that in order for the reaction to be phase inverting, at least *two* more electron pair exchanges are required. Thus the simplest loop that contains a CI and leads to benzene and H_2 in a concerted reaction is of ip^2 type, in which the phase-inverting leg involves *eight* electrons. A reaction that suggests itself is the isomerization CHDN(I) → CHDN(II), in which a shift of the two double bonds takes place, along with the associated transposition of hydrogen-atom bonds (Fig. 19*b*). Being a thermally "forbidden" reaction, it is likely to have a high barrier. The loop encircling the conical intersection that is defined by these three reactions is shown in Fig. 19*c*. The loops described in this section are the basis for the computation procedure detailed in Section VI.

V. LONGUET-HIGGINS LOOPS AND THE JAHN–TELLER THEOREM

Longuet-Higgins loops are closely related to the Jahn–Teller theorem [69,70]. In this section, we show that the Longuet-Higgins loop method renders the same results as the standard Jahn–Teller treatment. The H_3 system (a well-known Jahn–Teller case) was used as an example by Herzberg and Longuet-Higgins [12,13]. They showed that by symmetry, the electronic degeneracy occurs at the equilateral geometry. We shall extend the discussion to a more complicated case, and show that the correspondence holds for them. The case of several neighboring degeneracies will be covered. While the usual treatment of the Jahn–Teller problem emphasizes the degeneracy point, the Longuet-Higgins rule considers the neighborhood (loop) *around* the degeneracy (conical intersection).

The Jahn–Teller theorem [69] states that "the nuclear configuration of any nonlinear polyatomic system in a degenerate electronic state is unstable with respect to nuclear displacements that lower the symmetry and remove the degeneracy." A more rigorous formulation [71] is "If the potential energy surface of a nonlinear polyatomic system has two or more branches that intersect at one point, then at least one of them has no extremum at this point." An example (the $E \times \epsilon$ case, [70]) is shown in Figure 20. Since the nuclear displacement lowers the energy of the system, the point of degeneracy becomes

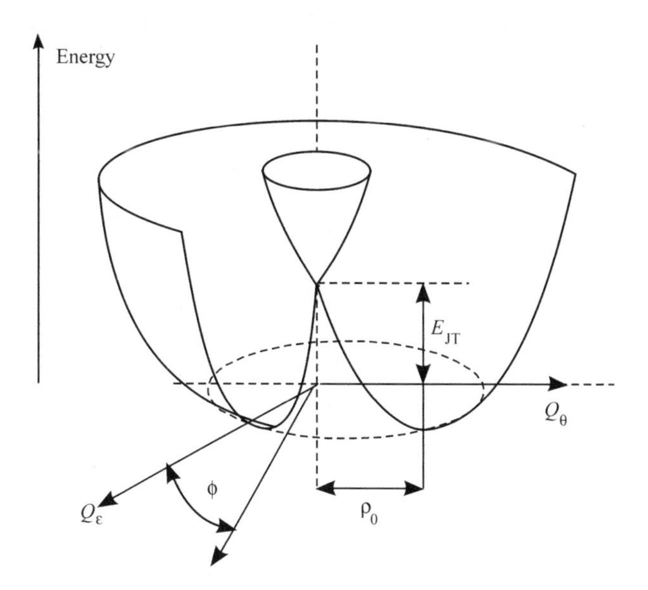

Figure 20. The potential surface near the degeneracy point of a degenerate E state that distorts along two coordinates Q_ϵ and Q_θ. The parameter E_{JT} is the stabilization energy of the ground state (the depth of the "moat"). [Adapted from [70]].

a crossing point between an excited and the ground state, that is, a conical intersection. In that sense, a Jahn–Teller system may be viewed as a special case of the more general problem of electronic degeneracy. In fact, by the Longuet–Higgins theorem, *any* degeneracy that connects the ground-state potential and the excited-state one, must be surrounded by a Longuet-Higgins loop, which can be formed by proper spin-paired combinations.

There is no analytic proof of the Jahn–Teller theorem. It was shown to be valid by considering all possible point groups one by one. The theorem is traditionally treated within perturbation theory: The Hamiltonian is divided into three parts

$$H = H(\mathbf{r}) + H(\mathbf{R}) + V(\mathbf{r}, \mathbf{R}) \tag{16}$$

where $H(\mathbf{r})$ is the pure electronic part, $H(\mathbf{R})$ is the nuclear kinetic energy, and $V(\mathbf{r}, \mathbf{R})$ is the electron–nuclear interaction. The parameter $V(\mathbf{r}, \mathbf{R})$ is expanded with respect to small nuclear displacement from the initial configuration $\mathbf{R}_0$:

$$V(\mathbf{r}, \mathbf{R}) = V(\mathbf{r}, \mathbf{R}_0) + \sum_{\alpha} (\partial V / \partial Q_\alpha) Q_\alpha + 1/2 \sum_{\alpha,\beta} (\partial^2 V / \partial Q_\alpha \partial Q_\beta) Q_\alpha \cdot Q_\beta + \cdots \tag{17}$$

The terms $\sum_\alpha (\partial V / \partial Q_\alpha) Q_\alpha$ and $1/2 \sum_{\alpha,\beta} (\partial^2 V / \partial Q_\alpha Q_\beta) Q_\alpha Q_\beta$ are the *linear and quadratic vibronic coupling terms, respectively.* For small Q_α values, they may be considered as a perturbation.

If the solution of the zero-order Schrödinger equation [i.e., all terms in (17) except $V(\mathbf{r}, \mathbf{R}_0)$ are neglected] yields an *f*-fold degenerate electronic term, the degeneracy may be removed by the vibronic coupling terms. If $|\Gamma\rangle$ and $|\Gamma'\rangle$ are the two degenerate wave functions, then the vibronic coupling constant

$$F_{Q\alpha}^{\Gamma\Gamma} = \langle \Gamma (\partial V / \partial Q_\alpha | \Gamma' \rangle \tag{18}$$

is nonzero for some coordinate Q_α (this is ensured by the lack of extremum at this point).

An example that is closely related to organic photochemistry is the $E \times \epsilon$ case [70]. A doubly degenerate E term is the ground or excited state of any polyatomic system that has at least one axis of symmetry of not less than third order. It may be shown [70] that if the quadratic term in Eq. (17) is neglected, the potential surface becomes a moat around the degeneracy, sometimes called "Mexican hat." The polar coordinates ρ and ϕ, shown in Figure 20, can be used to write an expression for the energy:

$$E_\pm(\rho, \phi) = 1/2 K_E \rho^2 \pm \rho [F_E^2 + G_E^2 \rho^2 + 2 F_E G_E \rho \cos 3\phi]^{1/2} \tag{19}$$

F_E and G_E are the linear and coupling quadratic terms, respectively.

If the quadratic coupling cannot be neglected, the potential surface acquires three minima at $\phi = 0, 2\pi/3$ and $4\pi/3$. The two wave functions corresponding to the two branches are

$$\Psi_{-} = \cos(\Omega/2)|\theta\rangle - \sin(\Omega/2)|\epsilon\rangle \tag{20}$$

$$\Psi_{+} = \sin(\Omega/2)|\theta\rangle + \cos(\Omega/2)|\epsilon\rangle \tag{21}$$

where $|\theta\rangle$ and $|\epsilon\rangle$ are the two electronic wave functions that are degenerate at $\rho = 0$. $\tan\theta = (F_E \sin\phi - |G_E|\rho \sin 2\phi)/F_E \cos\phi + |G_E|\rho\cos 2\phi)$.

When $G_E = 0$, it turns out that the two wave functions [Eqs. (20) and (21)] are not single valued: They change their sign when moving in a complete circle at the bottom of the moat! Since the *total* wave function must be single valued, this means that the *electronic* wave function must be multiplied by a phase factor $e^{im\phi}$, with half-integer values of m. The energy is a function of m^2, so that all levels are doubly degenerate, including the ground state.

It follows that the Jahn–Teller effect is a special case of the Longuet-Higgins rule, for systems of high nuclear symmetry. The degeneracy is removed as one moves away from the highly symmetric structure. The symmetry of the two electronic states that are formed for a given distortion may be determined from the symmetry of the problem, and was worked out for all point groups. The distortion in the Jahn–Teller problem are usually expressed in terms of the normal coordinates of the (fictitious) highly symmetric molecule that would have existed if a distortion did not take place.

In the more general case of nonsymmetric systems, we have shown that one can use reaction coordinates connecting two different spin-paired anchors. These two approaches should be equivalent; We shall show that this is indeed the case by discussing some examples.

Herzberg and Longuet-Higgins used the special case of the H_3 system to demonstrate the relation of the Jahn–Teller theorem to the Longuet-Higgins loop [12]. We repeated their arguments in Section II (Figs. 4 and 5). Longuet-Higgins went on to show, that the fact that three minima are obtained is not related to the C_{3v} symmetry of the problem—the rule works for an arbitrary ABC system [13]. According to VB theory, any three-atom system for which the wave function of the transition state on the ground state is an out-of-phase combination of the wave functions of the reactant and products behaves in the same way. As we have seen (Sections I and III), this arises from a more fundamental symmetry property of the system: the permutational symmetry of the polyelectronic wave function and Pauli's principle.

Accepting the Longuet-Higgins rule as the basis for the search of conical intersection, it is necessary to look for the appropriate loop. The E-type degeneracy of a Jahn–Teller system is removed by a nonsymmetric motion, leading to a splitting into two electronic states of A and B symmetry (in C_{2v}). *For a given geometry*, one of the states, say the A state, is the ground state and the other, an excited state. However, *for a different geometry*, their relative energy must switch, as we have seen for the H_3 and H_4 systems. We shall find that this is the case for other Jahn–Teller systems. In particular, if the minima transform as an A-type representation, the TS transform as B-type.

There is an odd number of equivalent spin-paired structures around the degeneracy: for a D_{3h} case, three, for D_{5h} case, five, and so on. These structures occupy separate minima around the conical intersection, or are located around an isoenergetic moat in the Mexican hat. The transition between them requires a change in the spin pairing, which, by the Longuet-Higgins rule, must be phase inverting. This means that the symmetry of electronic wave function at the transition state is in general *different* than that of the minima. Thus, the electronic wave function of different nuclear geometries on the ground-state surface may transform according to different symmetry species.

A. An Example: The Cyclopentadienyl Radical and Cation Systems

The cyclopentadienyl radical and the cyclopentadienyl cation are two well-known Jahn–Teller problems: The traditional Jahn–Teller treatment starts at the D_{5h} symmetry, and looks for the normal modes that reduce the symmetry by first- or second-order vibronic coupling. A Longuet-Higgins treatment will search for anchors that may be used to form the proper loop. The coordinates relevant to this approach are reaction coordinates.

1. *Cyclopentadienyl Radical (CPDR)*

This system was analyzed using ab initio (MO–CI) methods [72,73]. In the cyclopentadienyl radical, three electrons occupy a pair of e_1'' π MOs that are degenerate in D_{5h} symmetry. This gives rise to a degenerate $^2E_1''$ state, which by the Jahn–Teller theorem should distort away from D_{5h} symmetry along a degenerate e_2'-type vibration. The resulting states are of A_2 and B_1 symmetry. Five equivalent minima of C_{2v} symmetry are obtained, which may be connected by a motion that does not pass through the central D_{5h}-symmetric structure. It turns out [73,74] that the barriers between the five equivalent structures are small, so that the system can pseudorotate among them—a typical Mexican hat case.

A simple VB approach was used in [75] to describe the five structures. Only the lowest energy spin-pairing structures I (B_1 symmetry) of the type {12,34,5} were used (Fig. 21). We consider them as reactant–product pairs and note that the transformation of one structure (e.g., Ia) to another (e.g., Ib) is a three-electron phase-inverting reaction, with a type-II transition state. As shown in Figure 22, a type-II structure is constructed by an out-of-phase combination of

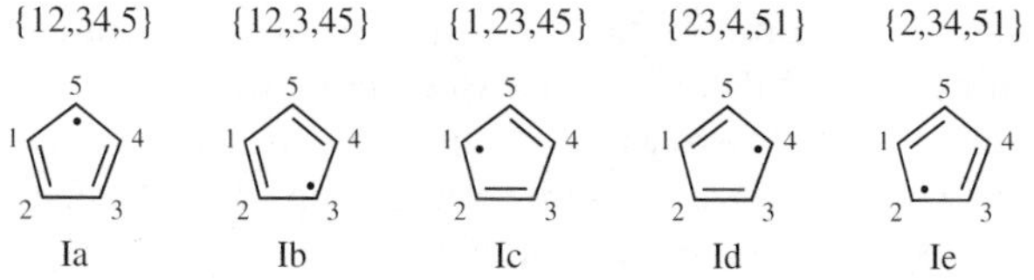

{12,34,5} {12,3,45} {1,23,45} {23,4,51} {2,34,51}

Ia Ib Ic Id Ie

Figure 21. The five equivalent spin-paired structures of CPDR.

the two type-I structures. The degenerate ${}^2E_1''$ state is the lowest state of D_{5h} symmetry: It lies on the ground-state surface and is constructed from a combination of the five. Type-II are A_2 symmetry structures, and they turn out to be isoenergetic with type-I. Their relatively strong stabilization is due to the allyl-type resonance. Their description as transition states is thus a matter of choice—type-I structures could be considered as transition states between two type-II structures.

The electronic spectrum of the radical has been recorded long before a satisfactory theoretical explanation could be provided. It was realized early on that the system should be Jahn–Teller distorted from the perfect pentagon symmetry (D_{5h} point group). Recently, an extensive experimental study of the high-resolution UV spectrum was reported [76], and analyzed using Jahn–Teller formalism [73].

It was shown by several workers that in this case the first-order Jahn–Teller distortion is due to an e_2' vibration, and that the second-order distortion vanishes. Therefore, in terms of simple Jahn–Teller theory, the "moat" around the symmetric point should be a Mexican hat type, without secondary minima. This expectation was borne out by high-level quantum chemical calculations, which showed that the energy difference between the two expected C_{2v} structures (${}^2A_2$ and ${}^2B_1$) were indeed very small [73].

The system provides an opportunity to test our method for finding the conical intersection and the stabilized ground-state structures that are formed by the distortion. Recall that we focus on the distinction between spin-paired structures, rather than true minima. A natural choice for anchors are the two C_{2v} structures having A_2 and B_1 symmetry shown in Figures 21 and 22: In principle, each set can serve as the anchors. The reaction converting one type-I structure to another is phase inverting, since it transforms one allyl structure to another (Fig. 12).

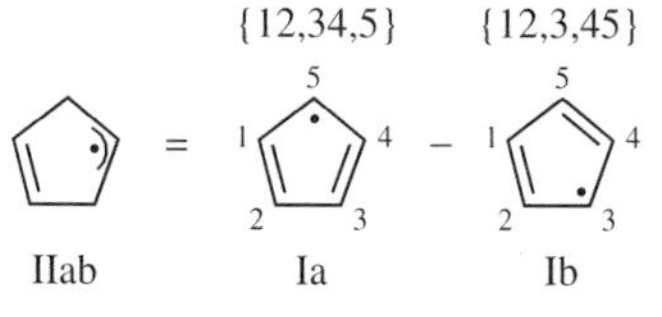

{12,34,5} {12,3,45}

IIab = Ia − Ib

Figure 22. An out-of-phase combination of two type-I (B_1 symmetry) structures yields a type-II structure (A_2 symmetry).

Type-II structures are formally the out-of-phase transition states between two type-I structures, even if there is no measurable barrier.

The complete loop is shown in Figure 23: It has five phase-inverting reactions, and therefore is a phase-inverting loop. The degeneracy that lies within the loop is the symmetric D_{5h} structure—at this symmetry all five type-I structures are degenerate. Alternatively, the five B_1 structures could serve as anchors, the transitions between any pair are also phase inverting, with the A_2 structures functioning as transition states. This example emphasizes the cardinal importance of spin pairing as the basis for choosing anchors—the conventional choice is a nuclear structure that lies in an energy minimum, but this is not an essential requirement.

This example may be used to address another issue concerning Longuet-Higgins loops: What is the minimum number of anchors needed to form a loop. Formally, one might choose three anchors (e.g., Ia, Ic, and Id), and use them as a loop. Inspection of Figure 23 shows that the conical intersection is formally encircled by a loop connecting these structures. It is also easily verified that the loop is an ip^2 phase inverting one: The Ic $\rightarrow$ Id reaction is phase inverting, and the Ia $\rightarrow$ Ic and the Ia $\rightarrow$ Id reactions are phase preserving. However, one of the conditions for a proper loop was that all reactions must be elementary, that is, there must not be an intermediate between any two anchors. This condition is not satisfied for the transformation Ia $\rightarrow$ Ic (or Ia $\rightarrow$ Id): motion along this trajectory leads up-hill (on the slope of the cone leading to the E_1'' degeneracy). Somewhere on the way, the minimum energy path will lead the system to Ib, which is therefore an intermediate. Thus, the smallest loop must pass through all five type-I structures.

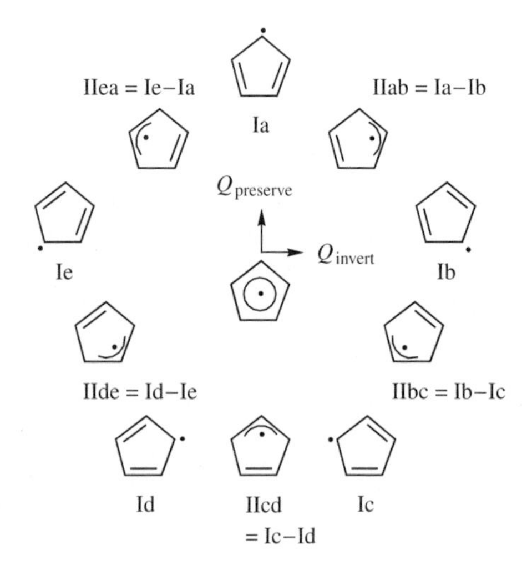

Figure 23. A Longuet-Higgins loop around the Jahn–Teller degeneracy of CPDR at D_{5h} symmetry. $Q_{preserve}$ and Q_{invert} are the phase-inverting and phase-preserving coordinates that define the loop.

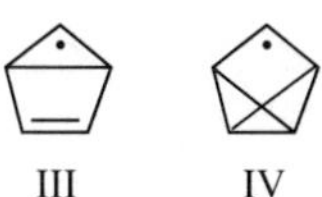

Figure 24. Other spin-paired structures of CPDR.

Other spin-pairing forms that may in principle be used to construct a loop are shown in Figure 24.

Structures III and IV that have different spin-pairing schemes are expected to be higher in energy than type-I because of the strain introduced by the cyclopropyl rings. They may be anchors for secondary conical intersections around the most symmetric one.

2. *Cyclopentadienyl Cation (CPDC)*

In the case of the cyclopentadienyl cation, there are only two electrons in the e_1'' π molecular orbitals that are degenerate in D_{5h} symmetry. The MO treatments [72,77] predict three low-lying electronic states: ${}^3A_2'$ (which is the ground state), ${}^1A_1'$, and a degenerate ${}^1E_2'$ state. As in the case of the radical, an e'-type vibrational mode is expected to lower the symmetry of the system and produce a lower energy singlet state. In this case, the E state splits to states of A_1 and B_2 symmetry. This system is thus analogous to the cyclopentadienyl radical one—five equivalent C_{2v} structures are expected to be formed upon distortion of the D_{5h} one. These structures are situated around the degeneracy in a Mexican hat arrangement. The MO calculations found two very close lying ${}^1A_1$ structures [72,77]. The B_1 state, expected from the Jahn–Teller treatment, was not discussed by [72].

Figure 25 shows the results of the C_{2v} distortion induced by a degenerate e_2' vibration that removes the D_{5h} degeneracy (compare Fig. 23). By symmetry, five

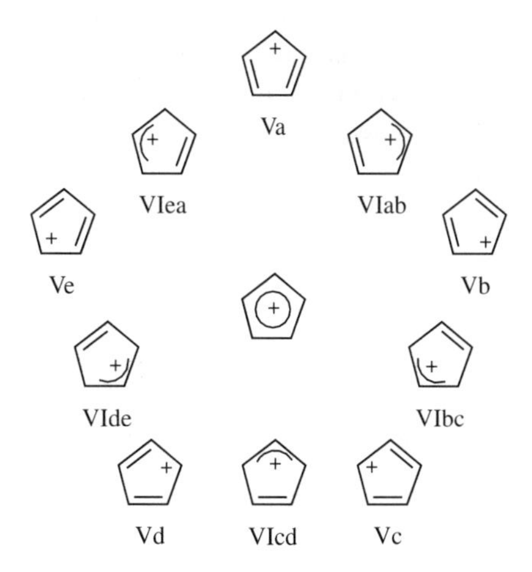

Figure 25. A suggested explanation for the pseudorotation motion around the degeneracy in the cyclopentadienyl cation. (Adapted from Ref. [72])

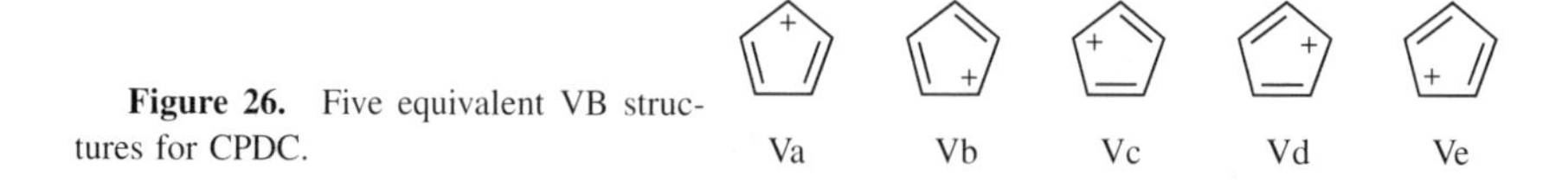

Figure 26. Five equivalent VB structures for CPDC.

equivalent type-V and five equivalent type-VI structures result (as shown). The system can pseudorotate around the degenerate D_{5h} structure, along the route delineated by the 10 C_{2v} structures. According to this scheme [72], the central symmetric structure is surrounded by totally symmetric structures (in C_{2v} symmetry). This is in apparent contradiction to the Jahn–Teller theorem, since a first order distortion must generate a nonsymmetric state.

Simple VB theory [75] uses for the basis set five low-lying structures that differ in their spin pairing characteristics, as shown in Figure 26. Similar to the case of the radical, the degenerate $^1E_2'$ state is the lowest singlet state of D_{5h} symmetry. It lies on the lowest singlet surface and is constructed by the combination of the five type-V structures (in fact, only four are linearly independent). These structures transform as A_1 in C_{2v}, and will be referred to as A_1(I) structures.

As shown in Figure 27, an in-phase combination of type-V structures leads to another A_1 symmetry structures (type-VI), which is expected to be stabilized by allyl cation-type resonance. However, calculation shows that the two structures are isoenergetic. The electronic wave function preserves its phase when transported through a complete loop around the degeneracy shown in Figure 25, so that no conical intersection (or an even number of conical intersections) should be enclosed in it. This is obviously in contrast with the Jahn–Teller theorem, that predicts splitting into A_1 and B_2 states.

The key to the correct answer is the fact that the conversion of one type-V (or VI) structures to another is a phase-inverting reaction, with a B_2 species transition state. This follows from the observation that the two type-V (or VI) structure differ by the spin pairing of four electrons. Inspection shows (Fig. 28), that the *out-of-phase* combination of two A_1 structures is in fact a B_2 one,

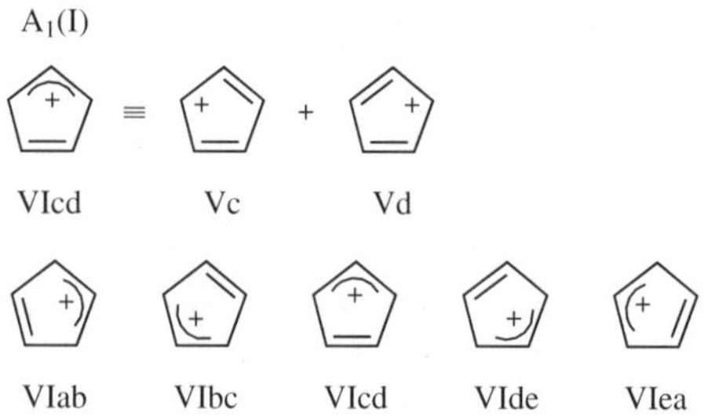

A_1(I)

VIcd ≡ Vc + Vd

VIab VIbc VIcd VIde VIea

Figure 27. Top: One of the allylic type-VI structures, formed by in-phase combination of type-V structures. Bottom: The five type-VI structures.

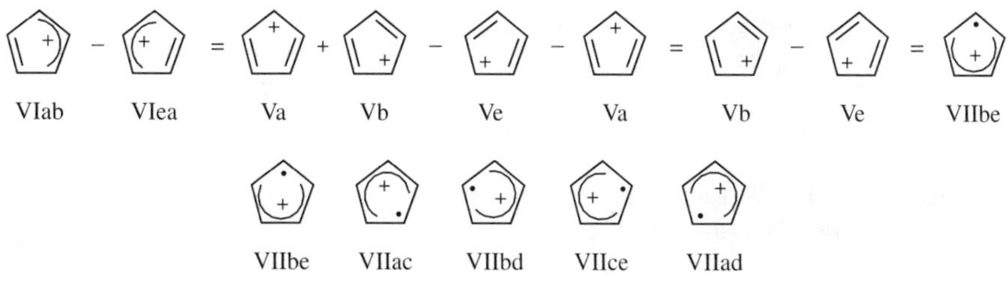

VIab VIea Va Vb Ve Va Vb Ve VIIbe

VIIbe VIIac VIIbd VIIce VIIad

Figure 28. Top: Construction of type-VII structure of B_2 symmetry. Bottom: the five type-VII structures.

type-VII. It may be conceived as a three-electron combination spread over four carbon atoms. A single electron resides on the fifth carbon atom.

Types-VI and -VII structures can be formed from the symmetric D_{5h} structure by the e_2' vibration as shown in Figure 29, in accord with the Jahn–Teller theorem.

The B_2 form of the cyclopentadienyl cation was not studied extensively. In [77], it was referred to as an electronically excited state. According to the Jahn–Teller theorem, at a certain geometry it should be part of the ground state, formed by the distortion of the degenerate structure. This requirement is fulfilled, of course, if it is indeed a TS between two ground-state species. We computationally verified this proposition as follows. The exact structure of the system at the degeneracy point was searched for under D_{5h} symmetry constraint. The C–C bond distance r_{C-C} for the symmetric D_{5h} cation at the conical intersection was found by calculating the energy of the two states point by point with different r_{C-C} values. A minimum energy was obtained at $r_{C-C} = 1.437$ Å. At this point, the D_{5h} symmetry was removed, and a search for an electronic state of B_2 symmetry was conducted. A structure with the geometry shown in Figure 30 was found. At this geometry, the B_2 state is lower in energy than any other, and therefore lies on the ground-state surface. Going either way to the type-VI structure, still on the ground-state surface, the energy decreased—the

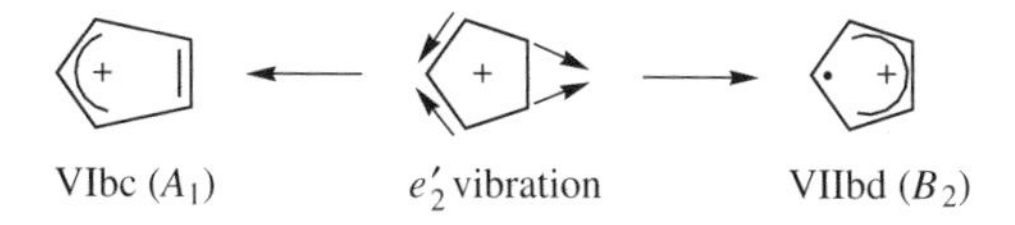

Figure 29. The effect of the phase-preserving component of the degenerate e_2' distorting mode. It may be regarded as a major component of the reaction coordinate that leads to the A_1 structure (going left, one phase of the mode). Going right, the other phase of the same vibration, the B_2 state is formed. (A type-V structure is also obtained along the same coordinate).

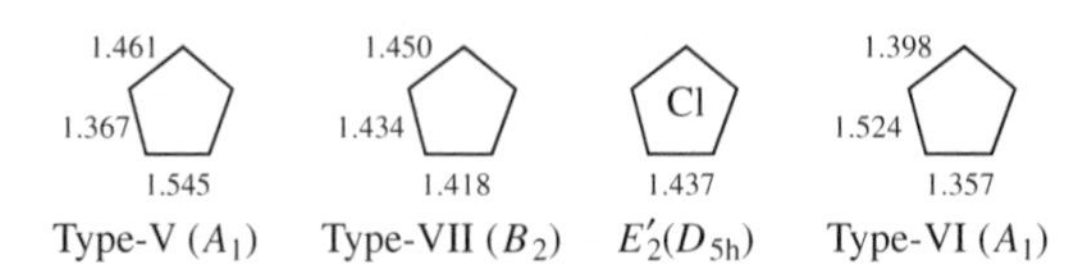

Figure 30. The calculated [CAS (4,5)/DZV] structures of the ground-state species V(A_1), VII(B_2), VI(A_1) and the conical intersection species.

type-VII species is indeed a transition state between two A_1 symmetry species. The ground-state equilibrium structures of the two A_1 species are also shown in Figure 30. Note that the type-VII structure is very slightly distorted from the perfect pentagon, in agreement with previous results [77].

In the B_1 structure, the positive charge is centered near one of the bases, rather than at a vertex, as in the A_1 structures.

Figure 31 shows the proposed Longuet-Higgins loop for the cyclopentadienyl cation. It uses the type-VI A_1 anchors, with the type-VII B_1 structures as transition states between them. This situation is completely analogous to that of the radical (Fig. 23). Since the loop is phase inverting, a conical intersection should be located at its center—as required by the Jahn–Teller theorem.

A final comment concerns the presence of other conical intersections near the central one. They are enclosed by loops consisting of two A_1 (type-VI) and one A_1 (type-V) species, as depicted in Figure 32. This is a phase-inverting ip^2 loop. Thus, the main Jahn–Teller degeneracy is surrounded by five further degeneracies, arranged in a symmetrical fashion.

This figure shows that there are many touching points between the lower and upper excited states. The shown structures are all on the ground-state surface. At

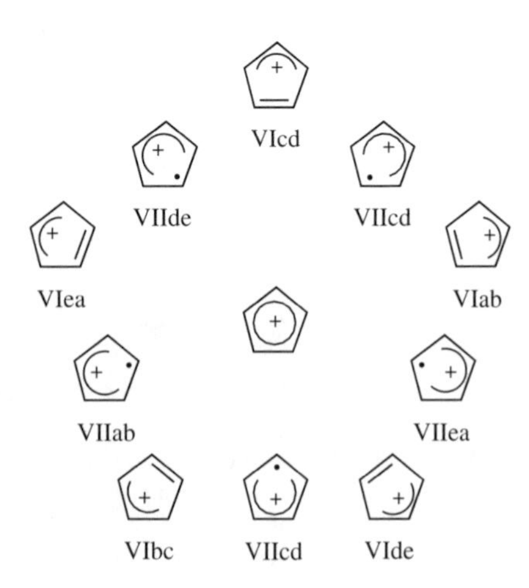

Figure 31. A phase-inverting loop accounting for the pseudorotation motion around the degeneracy in the cyclopentadienyl cation.

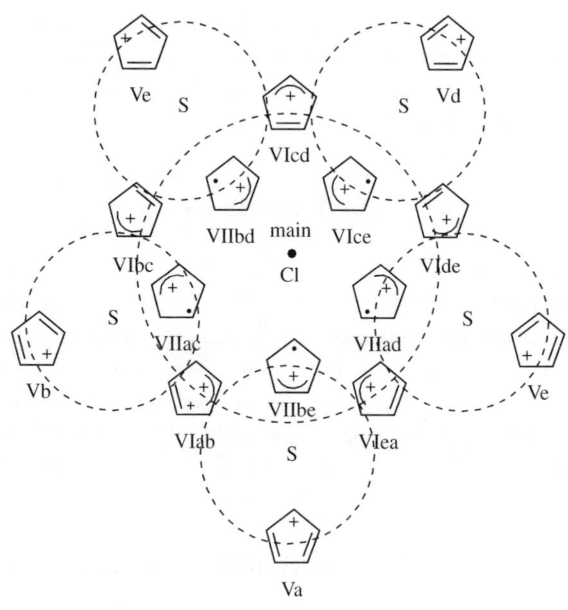

Figure 32. The main symmetric conical intersection of cyclopentadienyl cation, and five secondary conical intersections around it.

the geometry of the B_2 transition state, *the A_1 state is of higher energy*, and thus an excited states, even if by a small margin.

The case of CPDC can be used to address the issue of the loop's size and the number of conical intersections near a Jahn–Teller degeneracy. If we carry the system around a loop encircling all six degeneracies, the total electronic wave function will not change sign, and the phase-change theorem will appear to be violated. This point, relating to the "radius" of the loop, was raised by several authors [11,78]. Zwanziger and Grant [78] showed that in the case of the $2^2E'$ state of Na_3, three degeneracies are found nearby the main Jahn–Teller one. They analyzed the system using linear and quadratic coupling terms. Our discussion of the cyclopentadienyl radical shows how a problem of this nature is treated by the spin-pairing model. The extra five satellite conical intersections of CPDR arise from the fact that there are more spin-paired functions surrounding the main degeneracy point. In the case of the first degeneracies of the H_3 and Na_3 systems, the lowest $^2E'$ state ($1^2E'$) displays a single degeneracy. The Zwanziger–Grant effect arises in the *second* $^2E'$ state, since other spin pairing structure with similar energies are possible on the excited-state surface, for example, such that involve $3p$ electrons. Due to symmetry, only three such structures are possible, giving rise to three secondary loops, in analogy with the five loops shown in Figure 32.

VI. EXAMPLES

In this section, we apply the phase-change rule and the loop method to some representative photochemical systems. The discussion is illustrative, no comprehensive coverage is intended. It is hoped that the examples are sufficient to help others in applying the method to other systems. This section is divided into two parts: in the first, loops are constructed and a qualitative discussion of the photochemical consequences is presented. In the second, the method is used for an in-depth, quantitative analysis of one system—photolysis of 1,4-cyclohexadiene.

A. Loops in Molecular Photochemistry: Qualitative Discussion

Some prototype systems were presented in Sections I–IV; here, we offer a more extended discussion and application to realistic photochemical systems.

1. Four-Electron Problems

The prototype system for all four electron problem is the H_4 system, discussed in Section II.

Ethylene. A possible loop for ethylene photolysis was presented in Figure 14. Experimentally, irradiation into the first absorption band [populating the $B(1^1B_{1u})$ state] leads to cis–trans isomerization *as well as to a H atom shift.* The covalent A state lies at a very high energy in the planar form [79,80], but is the lowest excited singlet in the perpendicular one. This system was studied as early as 1985 by Ohmine [6], who reported a conical intersection that involves pyramidalization, and may lead to hydrogen-atom transfer. He calculated the conical intersection to be found along two coordinates—a phase inverting one (rotation around the double bond) and a phase preserving one, forming a methyl carbene (hydrogen-atom transfer) or a charge separated intermediate. Similar results were obtained more recently by the extensive studies of Ben-Nun and Martinez ([10] and references cited therein). Figure 33 shows a phase-inverting loop for the cis–trans isomerization, that leads also to a hydrogen-atom shift. The third anchor in this loop may appear as a carbene (CH_3CH:) or as a zwitterion ($CH_2^-CH_2^+$), see Figure 1(4). The conical intersection encircled by the loop is termed CI_H, for hydrogen-atom transfer.

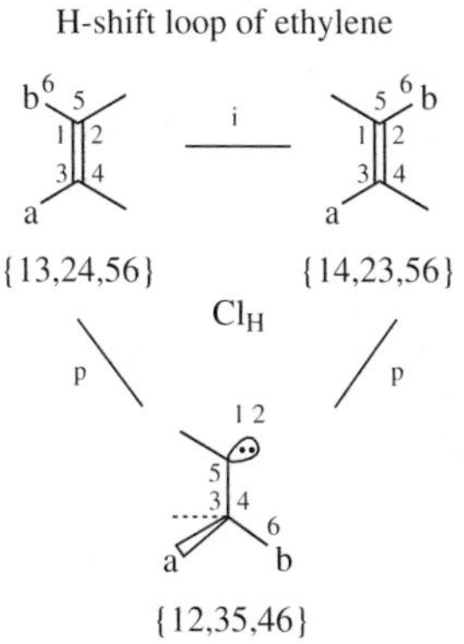

H-shift loop of ethylene

{13,24,56} {14,23,56}

CI_H

{12,35,46}

Figure 33. A loop containing a conical intersection for the cis–trans ethylene isomerization.

The transformation of ethylene to the carbene requires the re-pairing of three electron pairs. It is a phase-preserving reaction, so that the loop is an ip^2 one. The sp^3-hybridized carbon atom formed upon H transfer is a chiral center; consequently, there are two equivalent loops, and thus conical intersections, leading to two enantiomers.

Assuming that the cis isomer is the reactant, the trans isomer product is expected to be accompanied by others arising from secondary reactions of the biradical, as observed experimentally [58].

The carbene anchor includes the ionic pyramidal one, as established in Section I, since both have the same spin-pairing scheme [81]. The *geometry* of the ionic pyramidal structure is quite different from that of the carbene, raising the possibility of two minima for this anchor. Ionic structures like this were predicted by MO theory and since their dipole moment was found to depend steeply on geometry, this phenomenon was termed the sudden polarization effect [2, p. 212; 82]. It leads to reduction of the excited state's energy in the presence of polar solvents, which results in efficient crossing to the ground state. On the ground state, this structure is calculated to be a local minimum, but to our knowledge has not been observed experimentally. Since the structure is ionic, it may promote the coupling between the ionic B state and the otherwise largely covalent ground state.

The Cyclooctene Isomerization. A reaction that attracted some attention in recent years is the cis–trans isomerization of cyclooctene [84]. The cis isomer is much less strained than the trans, but the latter is readily formed upon direct photolysis and also upon photosensitization. In this case, two enantiomeric trans isomers are formed. The appropriate loop is a variation of that shown in Figure 14, as shown in Figure 34. This is a phase inverting i^3 -type loop, that

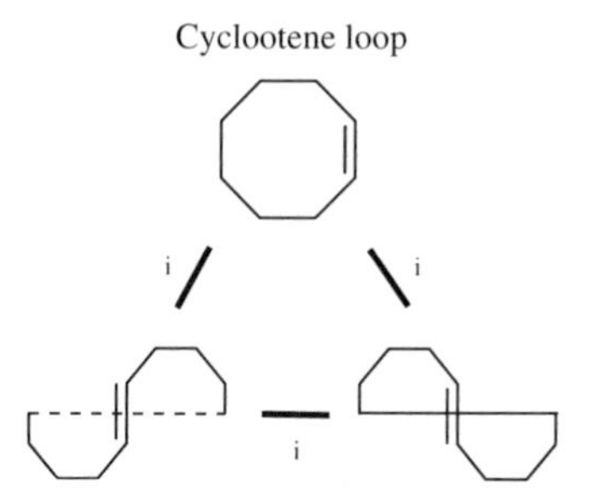

Figure 34. Anchors for the cis–trans isomerization of cyclooctene.

contains a conical intersection. In this case, the ring closure makes the two trans isomers distinguishable—being enantiomers, they belong to two different anchors. The proposed loop is consistent with the fact that the only reaction observed is cis–trans isomerization—the enclosed conical intersection is expected to be low lying, and to lead smoothly only to the trans isomers and back to the cis reactant.

Butadiene

UNSUBSTITUTED BUTADIENE. Butadiene anchors were presented in Figures 1(3) and 13. The basic tetrahedral character of the conical intersection (as for H_4) is expected to be maintained, when considering the re-pairing of four electrons. However, the situation is more complicated (and the photochemistry much richer), since here *p* electrons are involved rather than *s* electrons as in H_4. It is therefore necessary to consider the consequences of the *p*-orbital rotation, en route to a new sigma bond.

As shown in Figure 13, which is completely analogous to Figure 2, three independent spin pairing schemes exist: in the planar butadiene, the pairing is {12,34}; in cyclobutene, {14,23}. The third possibility {13,24} corresponds to bicyclobutane. Since the loop between these three structures is an i^3 one (for Hückel-type reactions), a conical intersection must be present inside the it. Irradiation of butadiene is thus expected to lead to both cyclobutene and bicyclobutane. The latter is not observed at room temperature, but is found when the reaction is carried out in cryogenic matrices, where highly strained structures may be observed due to rapid cooling [83,85]. A room temperature experiment that may reveal the bicyclobutane anchor, is the scrambling of carbon atoms expected to be found in recovered butadiene, using labeled ^{13}C isotopes. A given spin-pairing scheme (anchor) accommodates different conformers (Fig. 1), which may also be revealed in low-temperature

TABLE I
The Phase Change Upon Cyclization of Different s-cis Cyclobutadiene Isomers[a]

Product Reactant	XII	XIII	XIV	XV
VIII	i	i	p	p
IX	i	i	p	p
X	p	p	i	i
XI	p	p	i	i

[a] The p stands for phase-preserving reaction, the i for phase inverting.

experiments [83,85]. For example, photochemical s-cis/s-trans isomerization of butadiene is observed.

SUBSTITUTED BUTADIENES. The consequences of *p*-type orbitals rotations, become apparent when substituents are added. Many structural isomers of butadiene can be formed (Structures VIII–XI), and the electrocylic ring-closure reaction to form cyclobutene can be phase inverting or preserving if the motion is conrotatory or disrotatory, respectively. The four cyclobutene structures XII–XV of cyclobutene may be formed by cyclization. Table I shows the different possibilities for the cyclization of the four isomers VIII–XI. These structures are shown in Figure 35.

In a similar way Table II summarizes how the phase changes upon interconversion among the isomers. Inspection of the two tables shows that for any loop containing three of the possible isomers (open chain and cyclobutene ones), *the phase either does not change, or changes twice*. Thus, there cannot be a conical intersection inside any of these loops; in other words, *photochemical transformations between these species only cannot occur via a conical intersection, regardless of the nature of the excited state.*

TABLE II
The Phase Change Upon Interconversion Reactions Between Different s-cis Cyclobutadiene Isomers[a]

Product Reactant	VIII	IX	X	XI
VIII		p	i	i
IX	p		i	i
X	i	i		p
XI	i	i	p	

[a] The p stands for the phase-preserving reaction, the i for the phase inverting.

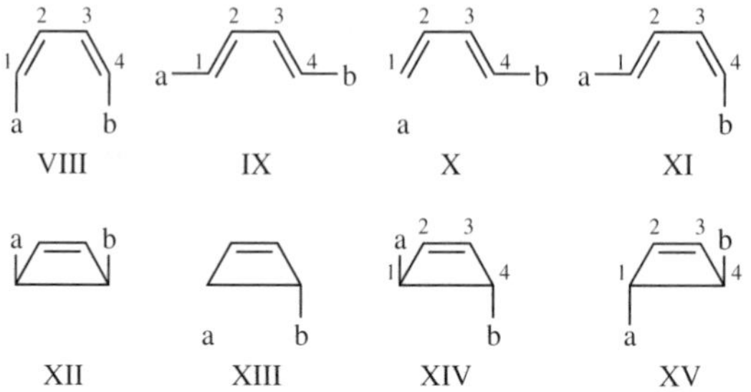

Figure 35. Structures VIII–XV.

With bicyclobutane as the third anchor, several different conical intersections may be found for certain reactant–product pairs, which in turn can be identified using Tables I and II. For example, structure IX may cyclize to XII or XIII, but not to XIV or XV, as shown in Figure 36. In a similar way, isomer VIII may convert to either X or XI but not to IV. In each of these cases, a second product that can be traced to either the cyclopropyl biradical or bicyclobutane (BCB) form of the third anchor (Fig. 1) is also formed. Although the relative yield of the products cannot be estimated from this analysis, these *selection rules* are strict for reactions in which conical intersections are involved, and they can predict which pairs of products are possible. In Figure 36*a* and 36*b*, the two possible modes of ring closure are shown: suprafacial and antarafacial—the BCBs (or cyclopropyl biradicals) formed are different isomers. In Figure 36*a*, we show the conical intersection as a tetra radicaloid, allowing re-pairing of all four electrons. The biradical form was chosen in order to explain more clearly the thermal formation of the two final products—cyclopropyl derivatives XVII and XVIII. This result may generalized as follows: Reactions in which all four π electrons participate, cannot result in cis–trans isomerization or cyclobutane formation *only*. These photochemical transformations involving a four-electron system must be accompanied by an additional, fairly strained one in which there is a cyclopropyl ring, or by a bicyclobutane product.

THE CYCLOBUTADIENE–TETRAHEDRANE SYSTEM. A related reaction is the photo-isomerization of cyclobutadiene (CBD). It was found that unsubstituted CBD does not react in an argon matrix upon irradiation, while the tri-butyl substituted derivative forms the corresponding tetrahedrane [86,87]. These results may be understood on the basis of a conical intersection enclosed by the loop shown in Figure 37. The analogy with the butadiene loop (Fig. 13) is obvious. The two CBDs and the biradical shown in the figure are the three anchors in this system. With small substituents, the two lobes containing the lone electrons can be far

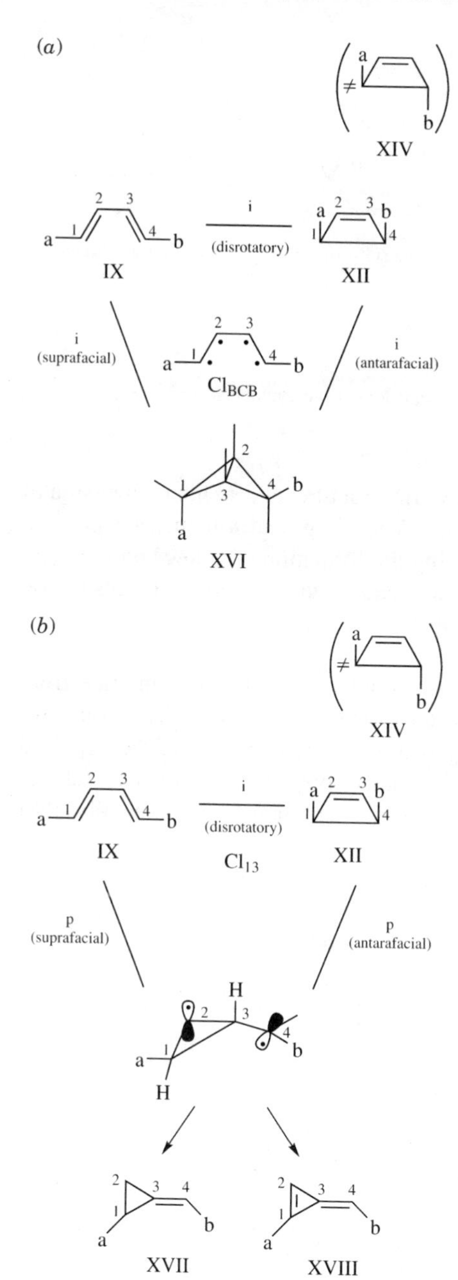

Figure 36. Possible butadiene photochemical reactions.

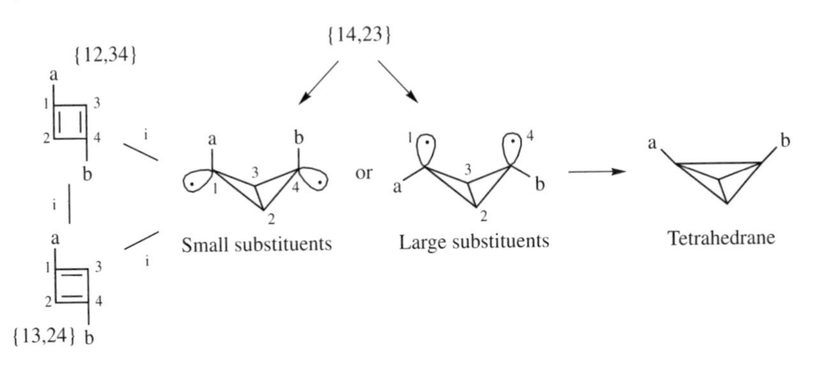

Figure 37. Formation of tetrahedrane from cyclobutadiene.

apart. The biradical reverts back to the CBD structure, and so interconversion of the two CBDs is the sole net reaction. When large substituents are present, the two lobes are pushed together, helping the formation of a new bond to yield a tetrahedrane. This expectation is in agreement with experimental results on CBD and tert-butyl substituted CBD [86].

1-BUTENE. As shown in Figure 38, a group attached to C-1 can migrate from position 1 to 3 (1,3 shift) to produce an isomer. If it is a methyl group, we recover a 1-butene. If it is a hydrogen atom, 2-butene is obtained. A third possible product is the cyclopropane derivative. The photochemical rearrangement of 1-butene was studied extensively both experimentally [88]

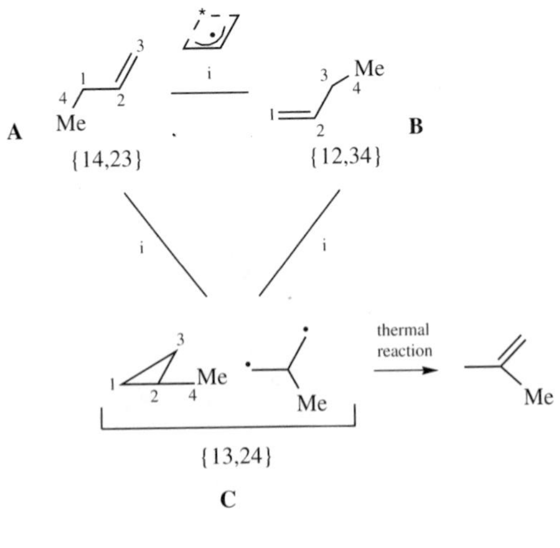

Figure 38. The photochemistry of 1-butene, explained with the methyl–allyl conical intersection.

and theoretically [89]. Orbital symmetry rules [52] predict that the major photochemical pathway be a [1,3] suprafacial sigmatropic shift that preserves the molecular stereochemistry. It is found experimentally that a [1,3] shift does indeed take place, but in addition, a cyclopropane derivative is obtained, and that the [1,3] shift reaction is indeed stereospecific—the methyl group migrates via a *supra* path, with retention of the configuration. No evidence for an *antara* path products was reported. This system provides an example of the H-allyl conical intersection: The transition between A and B is via an allyl type transition state. This is a four electron antiaromatic transition state, which is an out-of-phase combination of the two bond alternating structures of the reactant and product. Anchor C may include both the biradical and the cyclopropane derivative—they have the same spin-pairing scheme. All three reactions are phase inverting, the loop is an i^3 one and contains a conical intersection.

If A transforms to B by an *antara*-type process (a Möbius four electron reaction), the phase would be preserved in the reaction and in the complete loop (An i^2p loop), *and no conical intersection is possible for this case*. In that case, the only way to equalize the energies of the ground and excited states, is along a trajectory that *increases* the separation between atoms in the molecule. Indeed, the two are computed to meet only at infinite interatomic distances, that is, upon dissociation [89].

BENZENE–BENZVALENE ISOMERIZATION. The photochemical valence isomerization of benzene to form benzvalene [90, p. 357] is another example in which allyl radical structures (Fig. 12) play a central role. The system is analyzed in Figure 39. In order to follow the pattern of the previous examples, we show the in Figure 39*a* two benzvalene isomers as two anchors, and benzene as the third. This is an example of type C loop shown in Figure 11. The two benzvalene isomers, are connected via the shown allylic prebenzvalene TS structure, which is the *out-of-phase* combination of the two allyl structures shown at the top of the figure. The benzvalene → benzene transformation is thermally allowed (phase preserving) [91]. The coordinate connecting benzene with the allyl-type transition state is phase preserving. Thus, the phase-change rule predicts the existence of a conical intersection in the region enclosed by these anchors. High-level computations indicate that the prebenzvalene moiety is in fact an intermediate [92]. In that case, the two-anchor loop shown in Figure 39*b* applies, with the same results.

In a photochemical experiment, irradiation of benzene leads to S_1, which connects to the ground-state surface via the conical intersection shown. Benzene, the much more stable species, is expected to be recovered preferentially, but the prebenzvalene structure which transforms to benzvalene is also formed. Another possible route from the prebenzvalene, along a different coordinate, will lead to fulvene [90, p.357] after a hydrogen-atom transfer from

Allyl-type TS or intermediate

{13,26,45} {15,26,34}

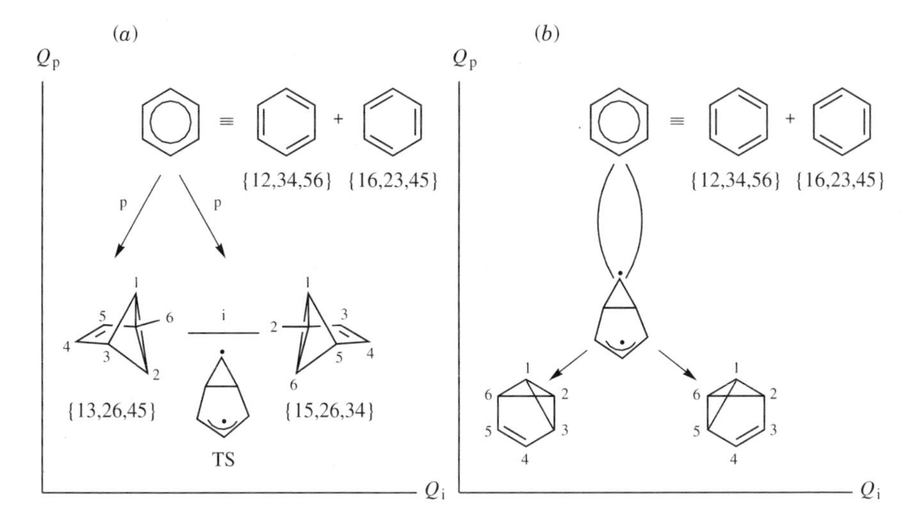

Figure 39. Benzene to benzvalene reaction. (*a*) Assuming that the prebenzvalene structure is a transition state. The two benzvalene isomers are anchors. (*b*) Assuming that prebenzvalene is an intermediate. A two-anchor loop results, compare Figure 12.

one of the carbon atoms to another. This scenario, which is observed experimentally [90], was corroborated computationally [92].

PHOTOLYSIS OF AMMONIA. Restricting the discussion to neutral species only (ionic ones require high energy, and are not important in the 170–220-nm UV range, where ammonia absorbs strongly), the two low-energy reaction channels to ground state products are

$$NH_3 \rightarrow NH_2(X^2B_1) + H(^2S) \qquad \text{Dissociation to ground-state atomic hydrogen}$$

$$NH_3 \rightarrow NH(a^1\Delta) + H_2 \qquad \text{Dissociation to molecular hydrogen}$$

The nonbonding electrons of the nitrogen atom are important in determining spin re-pairing, and thus the conical intersections. This is the physical origin of the topicity concept developed by Salem and co-workers [2,30]. Two different spin

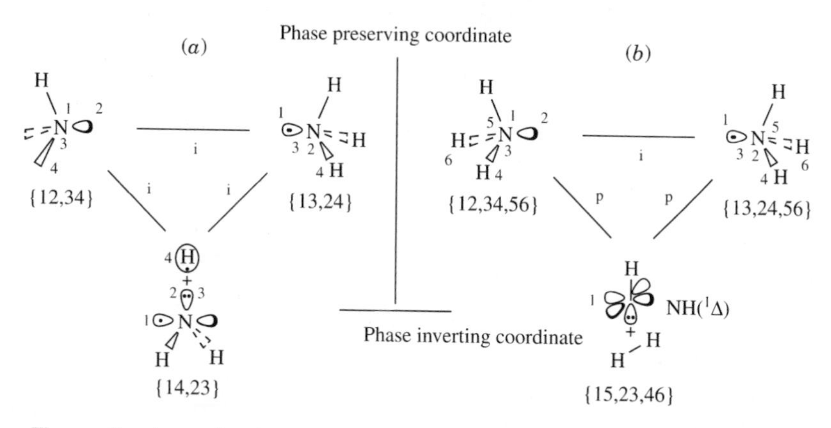

Figure 40. Ammonia photochemistry. (*a*) A loop for the $NH_3 \rightarrow NH_2(X^2B_1) + H(^2S)$ reaction. (*b*) A loop for the $NH_3 \rightarrow NH(a^1\Delta) + H_2$ reaction.

pairing schemes for ammonia were shown in Figure 15. They were based on the observation that the umbrella inversion is a reaction in which electrons are re-paired (a phase inverting reaction). In Figure 40 this reaction forms one leg of two possible Longuet-Higgins loops. In the left-hand one the third anchor is the $NH_2(^2B_1) + H(^2S)$, as shown in Figure 15 this is an i^3 loop. The right-hand loop has the pair $H_2(^1\Sigma_g^+) + NH(a^1\Delta)$ as the third anchor: the reactions connecting the two ammonia anchors with it re-pair six electrons—this is an ip^2 loop. Both loops therefore encircle conical intersection, and both channels are expected to be photochemically active on the basis of the phase-change rule. Note that the hydrogen-atom dissociation channel requires planarization en route to the conical intersection. This expectation was verified computationaly [93].

INORGANIC COMPLEXES. The cis–trans isomerization of a planar square form of a d^8 transition metal complex (e.g., of Pt^{2+}) is known to be photochemically allowed and thermally forbidden [94]. It was found experimentally [95] to be an intramolecular process, namely, to proceed without any bond-breaking step. Calculations show that the ground and the excited state touch along the reaction coordinate (see Fig. 12 in [96]). Although conical intersections were not mentioned in these papers, the present model appears to apply to these systems.

Consider a metal M bound to four ligands, L_1–L_4, lying at the corners of a square around the metal. Three anchors can be written for this system, as shown in Figure 41. They consist of all three possible geometrical permutations of pairs of ligands lying across the metal ion. The conical intersection in this case is a tetrahedron (cf. Fig. 2). There are two different (though energetically

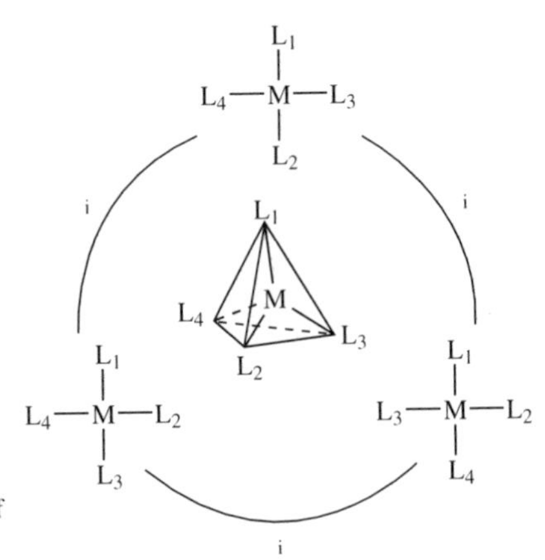

Figure 41. A loop for the photochemistry of a planar ML_4 complex.

equivalent) conical intersections, as the tetrahedral structure can exist in two enantiomeric forms. A possible way to distinguish between them is by using a chiral molecule as one of the ligands, so that the two conical intersections become diastereomers rather than enantiomers. These conical intersections allow the photoisomerization to proceed without breaking a single bond. In the ground state, a bond-breaking—bond-recombination mechanism is often energetically more favorable. An example for such a system is provided by the photoisomerization of $Pt(gly)_2Cl_2$ [95].

2. *More Than Four Electrons*

Cyclooctatetraene (COT) → Semibullvalene (SB) Photorearrangement. Irradiation of COT yields semibullvalene [97], in spite of the fact that this photochemical reaction is forbidden by orbital conservation rules. The Longuet-Higgins loop for this system actually predicts that this should happen, although the reaction is phase preserving. (Fig. 42). This is another example of type C loop (Fig. 11). Only six of the eight electrons re-pair as COT transforms to SB. The reaction is made possible by the fact that COT valence isomerization, a phase-inverting reaction (four electron-pair Hückel system), takes place simultaneously. One expects to produce in the reaction a COT isomer, that can be detected solely by proper substitution.

B. A Quantitative Example: The Photochemistry of 1,4-Cyclohexadiene (CHDN)

The construction of loops relevant to this system was described in Section IVD. In this section, the computational search for the conical intersection using the

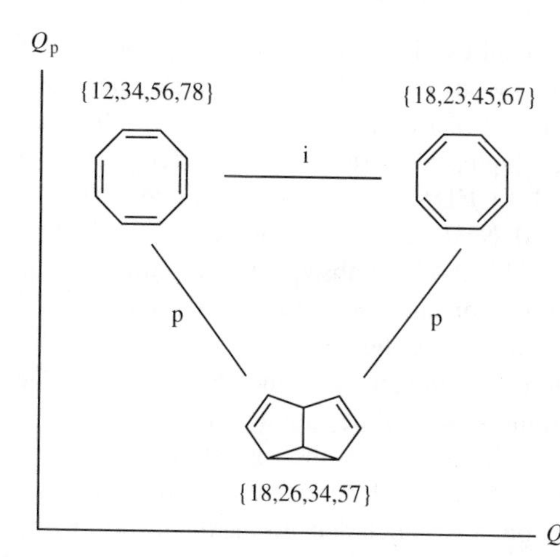

Figure 42. The cyclooctateteraene to semibullvalene reaction.

loop method is described in some detail [98]. The phase-change rule requires that in order for a conical intersection to be enclosed in a loop, at least one of the reaction coordinates forming it must be phase inverting. Motion along this coordinate leads to an antiaromatic transition state (AATS), discussed in Section III. The coordinate connecting this transition state and the third anchor is phase preserving, (cf. Fig. 5). Thus, the conical intersection lying within the region encircled by the three anchors, may be found by moving first along the phase-inverting reaction coordinate from CHDN to the AATS and then along the phase-preserving coordinate to the third anchor. In practice, the geometry of the AATS is calculated and the system is transported vertically to the first electronically excited-state surface. This state is the twin state of the AATS, and is low lying [28,49]. From this point, relaxation along the steepest gradient downward on the electronically excited state is performed, under two constraints: The system is kept at the same symmetry it had at the AATS geometry (viz., the migrating H atom(s) midway between the original and destination carbon atoms], and the molecule is kept along the phase-preserving reaction coordinate in the direction of the third anchor. Recall that the twin excited state and the third anchor have the same spin-pairing scheme (Section I.C). All other coordinates are optimized for minimum energy—this is a constrained minimum energy path leading to the product. The point at which the system reaches the ground-state potential lies on the CI. It is not necessarily the minimum energy point on the CI; rather, the locus reached by this process is obtained upon moving along the shortest path to the product from the excited-state surface (at the AATS geometry). The numerical data cited below are for these points on the CI.

We demonstrate this procedure by an explicit example—finding the H/allyl CI shown in Figure 18. In this case, any one of the three reaction channels may be used to begin the search, since all are phase inverting and therefore have an antiaromatic transition state. For example, the system is propagated along the reaction coordinate from CHDN to 1,3-CHDN, and the geometry and energy of the AATS of this reaction is computed. Next, vertical excitation leads to the first excited singlet state (S_1) lying above this AATS. Subsequently, motion along is initiated along the phase-preserving coordinate connecting this point on S_1 and the third anchor, BCE in this example, (which lies on S_0). This procedure did indeed result in locating the conical intersection, as confirmed by calculating the energy gap between the two electronic states, while moving along.

As a check on the performance of the procedure, the formation of the three possible products on the ground-state potential surface was validated after the search for the conical intersection was concluded. Immediately following the crossing of the conical intersection, the system was allowed to relax to an energy minimum on the S_0 surface. Removing all constraints led to one of the three anchors. The other two were sought by first "nudging" the atoms slightly in the direction of their geometry, and then letting the system find a minimum energy. The physical justification of the "nudging" is the ever present redistribution of energy on the ground-state surface (IVR). Recovery of the three anchors without encountering a barrier confirmed the location of the conical intersection in the loop and the validity of the process.

The energies of this CI and of the other ones calculated in this work are listed in Table III. The calculated CASSCF values of the energies of the two lowest electronically states are 9.0 eV (S_1, vertical) and 10.3 eV (S_2, vertical) [99]. They are considerably higher than the experimental ones, as noted for this method by other workers [65]. In all cases, the computed conical intersections lie at much lower energies than the excited state, and are easily accessible upon excitation to S_1. In the case of the H/allyl CI, the validity confirmation process recovered the CHDN and 1,3-CHDN anchors. An attempt to approach the third anchor [BCE(I)] resulted instead in a biradical, shown in Figure 43. The biradical may be regarded as a resonance hybrid of two allyl-type biradicals.

TABLE III
The CASSCF(8,8)/DZV Energies of Some Stable Molecules and Conical Intersections Relevant to 1,4-cyclohexadiene (CHDN) Photochemistry (kcal/mole with respect to CHDN)

Molecules	1,4-CHDN	1,3-CHDN	BCE	benzene + H_2	BCE/allyl biradical
Energy	0[1]	−8.1	18.9	13.7	39.6
Conical Intersections	H/Allyl		Helicopter-type		
Energy	103.2	148.1			

Note: In Hartree units, −231.84363

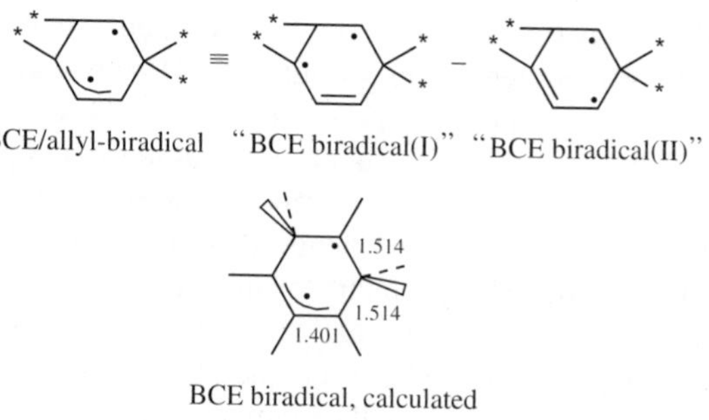

BCE/allyl-biradical "BCE biradical(I)" "BCE biradical(II)"

BCE biradical, calculated

Figure 43. The structure of the BCE biradical (CASSCF(8,8)/DZV).

The combination is in this case an out-of-phase one (Section I). This biradical was calculated to be at an energy of 39.6 kcal/mol above CHDN (Table III), and to lie in a real local minimum on the S_0 potential energy surface. A normal mode analysis showed that all frequencies were real. (Compare with the prebenzvalene intermediate, discussed above. The computational finding that these species are bound moieties is difficult to confirm experimentally, as they are highly reactive.)

The search for the conical intersection leading to the concerted ejection of H_2 (the helicopter-type reaction) was facilitated by the fact that the AATS between CHDN(I) and CHDN(II) has an added nuclear symmetry element with respect to the reactant: the molecule has a C_{2v} symmetry, which distorts to C_2 on the way to the transition state (compare Section I.A). When the system reaches the AATS, the symmetry becomes again C_{2v} (with different symmetry elements). The ground-state electronic wave function of this AATS (of A_2 symmetry) is formed by an *out-of-phase* combination of the electronic wave functions of the two VB structures having the same geometry but different spin pairing schemes. The electronic wave functions of such transition states transform as one of the nontotally symmetric irreducible representations of the group (Section I.A). In the case at hand, as the A_2 representation of C_{2v}. From Figure 19, it is clear, that the $(1^1A_1/1^1A_2)$ conical intersection, by symmetry, is to be found on the coordinate connecting the AATS between the two isomeric hexadienes and the third anchor—benzene and H_2 (A_1 symmetry in C_{2v}).

The potential surfaces of the ground and excited states in the vicinity of the conical intersection were calculated point by point, along the trajectory leading from the antiaromatic transition state to the benzene and H_2 products. In this calculation, the HH distance was varied, and all other coordinates were optimized to obtain the minimum energy of the system in the excited electronic state ($^1A_1$). The energy of the ground state was calculated at the geometry optimized for the excited state. In the calculation of the conical intersection locality, the system was constrained to C_{2v} symmetry, that is the rotational motion of the two hydrogen atoms forming the H_2 molecules was frozen. It was found that the two surfaces ($^1A_1$ and $^1A_2$, cf. Fig. 44) did cross at a certain geometry, representing a conical intersection, as expected from the phase-change rule. Some numerical results are reported in Figure 45. The approach to the conical intersection from the hexadiene side is much more gradual than from the benzene and H_2 side. The geometry of this conical intersection, shown in Figure 45, is found to be similar to that of the AATS. In both, the C–C bonds have very similar values. The HH distance is much larger in the AATS, while the H_2 center-of-mass distance to the carbon ring is larger in the conical intersection. The angle between the line connecting the two hydrogen atoms and the line connecting the two carbon atoms to which they were originally bonded changes due to the rotational motion of two hydrogen atoms with respect to the C_6H_6 fragment. It is 30° at both the AATS and at the conical intersection.

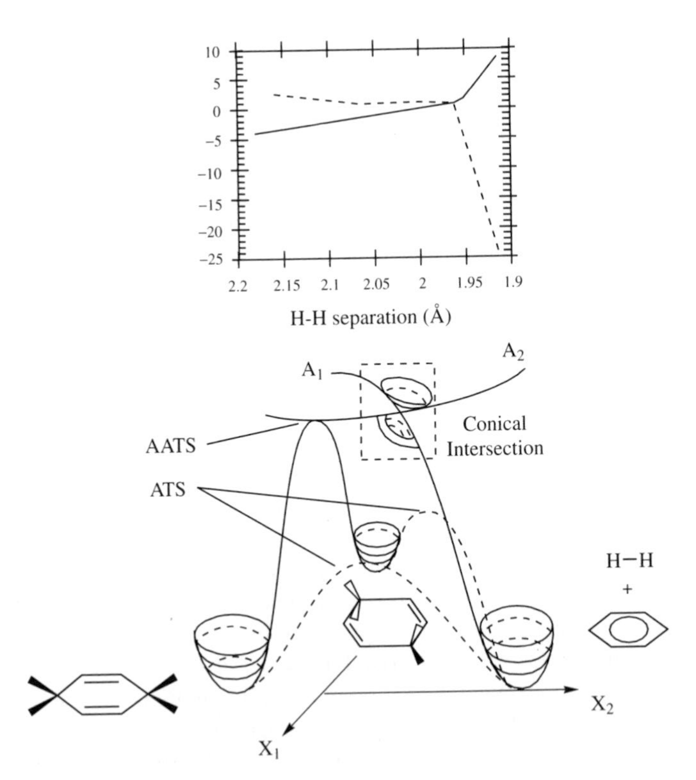

Figure 44. The helicopter-type conical intersection for CHDN. Bottom: A cartoon showing the anchors and the conical intersection. Top: The calculated energies (kcal/mole) of S_0 and S_1 near the conical intersection.

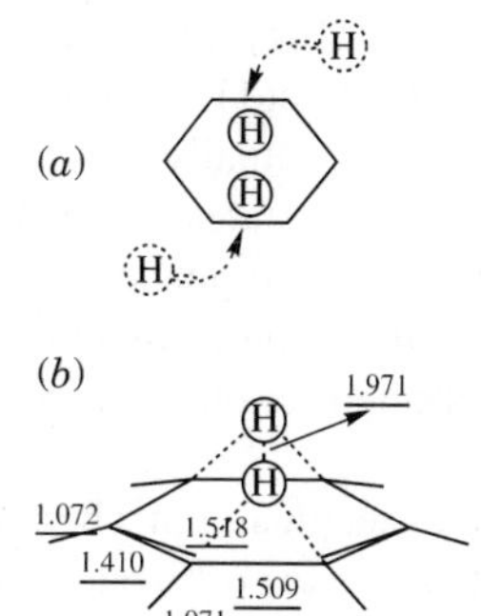

Figure 45. The structure of the helicopter-type conical intersection of CHDN.

The two departing hydrogen atoms perform a complex motion from their initial positions to the conical intersection: contracting of the distance between them, rotation that brings them to the position above the center of the ring and increasing the distance between the center of mass (CM) of the H_2 fragment and the carbon ring. The angular momentum created in this motion is the origin of the observed helicopter-type one. It may be traced to the participation of the isomerization reaction in the Longuet-Higgins loop encircling the conical intersection which induces the necessary angular momentum.

The two proposed conical intersections provide a model that is consistent with the experimental results on the CHDN system [60–64].

VII. COMPARISON WITH OTHER METHODS FOR LOCATING CONICAL INTERSECTIONS

The method presented in this chapter is aimed mainly at providing information on the presence of conical intersections in large molecules, and helps in the calculation of their energies and structures. In this section, we review briefly some other procedures used to characterize conical intersections, and compare them with the present method.

The practical implementation of all schemes requires high-level quantum mechanical calculations. The major advances achieved in this field in the last two decades appear to be a major factor for the revival of interest in the subject. Several efficient computational methods are currently available for finding conical intersections in polyatomic systems and calculating their properties. The field is very active at the moment, and the prospects for further progress are very promising. Yet, it is still true, as summarized by Worth and Cederbaum [100]: The complete evaluation of potential energy surfaces is an impossible task for systems comprising more than a few atoms. Approximations *have* to be made—the different methods are distinguished mostly by the nature of the approximations.

The phase-change rule, also known as the Berry phase [101], the geometric phase effect [102,103] or the molecular Aharonov–Bohm effect [104–106], was used by several authors to verify that two near-by surfaces actually cross, and are not repelled apart. This point is of particular relevance for states of the same symmetry. The total electronic wave function and the total nuclear wave function of both the upper and the lower states change their phases upon being transported in a closed loop around a point of conical intersection. Any one of them may be used in the search for degeneracies.

Ruedenberg and co-workers [33,107], found by exact quantum chemical calculations a crossing point between the two lowest $^1A_1$ states of ozone. They used the phase-change rule to verify that the electronic wave function changes its sign when transported in a closed loop around this point. This was done by considering the phase change of the dominant configurations of the ground-state wave functions. Initially, only C_{2v} symmetry was considered, later, a complete seam of conical intersections was calculated [107].

Yarkoni [108] developed a computational method based on a perturbative approach [109,110]. He showed that in the near vicinity of a conical intersection, the Hamiltonian operator may be written as the sum a nonperturbed Hamiltonian H_0 and a linear perturbative term. The expansion is made around a nuclear configuration $\mathbf{Q}_x$, at which an intersection between two electronic wave functions takes place. The task is to find out under what conditions there can be a crossing at a neighboring nuclear configuration $\mathbf{Q}_y$. The diagonal Hamiltonian matrix elements at $\mathbf{Q}_y$ may be written as

$$H_{\mathrm{II}}(\mathbf{Q}_y) = E_{\mathrm{I}}(\mathbf{Q}_x) + \mathbf{g}^{\mathrm{I}}(\mathbf{Q}_z)\delta\mathbf{Q} \tag{22}$$

where $\mathbf{g}^{\mathrm{I}}_{\alpha}(\mathbf{Q}_x) = \mathbf{c}^{\mathrm{I}}(\mathbf{Q}_x)^{\dagger}[\partial H(\mathbf{Q}_x)/\partial Q_{\alpha}]\mathbf{c}^{\mathrm{I}}(\mathbf{Q}_x) = [dE_{\mathrm{I}}(\mathbf{Q}_x)/\partial Q_{\alpha}.\ \mathbf{g}^{\mathrm{I}}_{\alpha}(\mathbf{Q}_x)$ is thus the gradient of the energy along a small displacement $\delta\mathbf{Q} = \mathrm{S}_{\alpha}[\partial Q_{\alpha}\mathbf{q}_{\alpha}]$, with a unit vector in the α direction.

The off-diagonal elements are written as

$$H_{\mathrm{II}}(\mathbf{Q}_y) = \mathbf{h}^{\mathrm{IJ}}(\mathbf{Q}_x)^{\dagger}\delta\mathbf{Q} \tag{23}$$

$$\mathbf{h}^{\mathrm{IJ}}_{\alpha}(\mathbf{Q}_x) = \mathbf{c}^{\mathrm{I}}(\mathbf{Q}_x)^{\dagger}[\partial H(\mathbf{Q}_x)/\partial \mathrm{Q}_{\alpha}]\mathbf{c}^{\mathrm{J}}(\mathbf{Q}_x) \tag{24}$$

In the special case of a triatomic system, $\mathbf{g}^{\mathrm{I}}_{\alpha}(\mathbf{Q}_x)$ and $\mathbf{h}^{\mathrm{IJ}}_{\alpha}(\mathbf{Q}_x)$ are one-dimensional vectors, spanning the *g–h* plane. A closed loop in the *g–h* plane will change the sign of the electronic wave function. The degeneracy is preserved through first order provided $\delta\mathbf{Q}$ is restricted to motion along the g–$h^{\dagger}$ plane, orthogonal to the *g–h* plane. The line connecting all degeneracies is the conical intersection *seam*. More generally, the conical intersection will be a $3N - 8$ dimensional surface.

The search for a conical intersection is based on the assumption that it is not feasible (or desirable) to characterize the entire surface of the conical intersection. Conical intersections of interest are those with energies accessible in chemical processes. Furthermore, it is desired to restrict the search to nuclear configurations of chemical interest. These constraints may be expressed mathematically (e.g., by Lagrange multiplier techniques). The Schrödinger equation is solved for the two states in question, and points of near crossing are checked by the phase change rule to verify whether the wave function indeed changes sign.

As shown by several authors, the sign ambiguity of the electronic wave function found by Longuet-Higgins near surface crossing in the adiabatic approximation (usual BO case) can be removed [102,103,111,112]. This is done by transforming to a *diabatic* framework, or representation, in which the Hamiltonian matrix is diagonalized near the conical intersection. The resulting diabatic states are coupled by potential coupling terms that can be calculated to a high degree of accuracy. It turns out that the number of important coupling terms is small, making the approximation practical. These coupling terms are closely related to the Jahn–Teller distortive modes, and in our model, to the coordinates connecting two anchors. Using this approach, Mebel et al. [113] showed that a conical intersection connecting the two low-lying states of C_2H actually cross. The change in phase angle upon a complete loop around the point was calculated, using the coupling elements that were obtained by quantum calculations. Only when the loop enclosed a single conical intersection, the phase changed. When the loop enclosed two (or no) conical intersection, the phase did not change.

Robb, Bernardi, and Olivucci (RBO) [37] developed a method based on the idea that a conical intersection can be found if one moves in a plane defined by two vectors: $\mathbf{x}_1$ and $\mathbf{x}_2$, defined in the adiabatic basis of the molecular Hamiltonian H. The direction of $\mathbf{x}_1$ corresponds to the gradient difference

$$\mathbf{x}_1 = \partial(E_1 - E_2)/\partial\mathbf{q} \tag{25}$$

where E_1 and E_2 are the energies of the two electronic states in the BO approximation, and $d\mathbf{q}$ is a vector of nuclear displacement.

The direction of $\mathbf{x}_2$ is parallel to the direction $\mathbf{g}$ of the diabatic coupling matrix mentioned above

$$\mathbf{g} = <\Psi_1 \cdot (\partial\Psi_2/\partial\mathbf{q}) \tag{26}$$

where Ψ_1 and Ψ_2 are the eigenfunctions of H. Note the formal similarity between the vectors $\mathbf{x}_1$ and $\mathbf{x}_2$ on one hand, and $\mathbf{g}_\alpha^{\mathrm{I}}(\mathbf{Q}_\mathbf{x})$ and $\mathbf{h}_\alpha^{\mathrm{IJ}}(\mathbf{Q}_\mathbf{x})$ [Eqs. (22–24)] on the other.

The system is propagated along the two vectors, until the separation between the two surfaces vanishes upon reaching the conical intersection geometry.

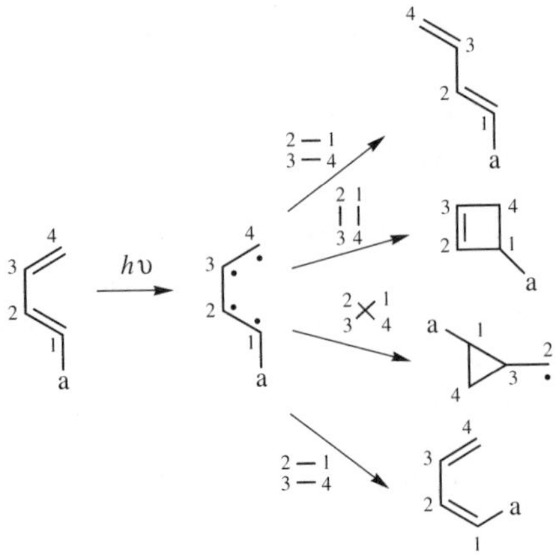

Figure 46. Butadiene conical intersection (Adapted from [117]).

The practical application of the RBO method is helped by selecting the required coordinates based on chemical intuition or calculations. The original papers made use of the MMVB method [114]. The advantage of the MMVB method is that it is many orders of magnitude less expensive computationally than the currently high-level CASSCF method. A disadvantage is that it can only describe covalent states [114]. Minima on the conical intersection hypersurface are optimized using an algorithm developed by Bearpark et al. [115].

For the purpose of finding an efficient funnel connecting an excited electronic state and the ground state, the scrupulous distinction between "real" crossing and "near" crossing may be of theoretical importance only. In practice, the system will jump from one to other with similar effectiveness. Therefore, the conical intersection may be found by following the energy gap between the two states to a value close to zero. This method was used to analyze the photochemistry of a large number of molecules [9,37]. An example that lends itself to facile comparison with our method is the photochemistry of butadiene. [38,116]. Figure 46 summarizes their finding in a simple scheme. The conical intersection is presented as a tetraradical, which may stabilize by forming three types of bonds, depending on the electron pairs that form the bonds. This scheme is directly comparable to Figure 36. The upper reaction is an intraanchor reaction, the middle two are identical to those shown in Figure 36 and discussed in Section VI.A1. The reaction shown at the bottom of the scheme (trans–cis isomerization around one of the double bonds) would require, in our method, a different loop than the other three.

Wilsey and Houk [65] used the RBO method to find conical intersections in several olefins, including 1,4-cyclohexadiene (CHDN). This was done by using chemical reactions coordinates to identify the gradient difference vector $\mathbf{x}_1$ and the non-adiabatic vector $\mathbf{x}_2$: When $\mathbf{x}_1$ corresponds to the motion leading to a trimethylene radical and $\mathbf{x}_2$ to the formation of the isomeric [1,3]-shift product, the H/allyl conical intersection is found (Section VI.B). When $\mathbf{x}_1$ corresponds to the motion leading to a C–C bond cleavage to form a diradical and $\mathbf{x}_2$ to the formation of the isomeric [1,3]-shift product, the vinyl–allyl conical intersection is found. This chemically oriented method is similar in spirit to our approach. Their strategy allows the finding of several conical intersections, by defining the gradient difference vector along a desired coordinate. Implicit in their approach is the idea that several conical intersections may be photochemically active for a specific system.

Köppel et al. [111] considered the so-called symmetry-allowed conical intersections. These intersections result from the allowed crossing of potential energy functions of two electronic states, which transform according to different irreducible representations of an appropriate symmetry group along a certain reaction coordinate. A further deformation, which lowers the symmetry, leads to interactions of the two states in first order in the displacement, converting the crossing to a conical intersection. The search for a conical interaction in this case is greatly helped by symmetry considerations, as has been recently demonstrated for malonaldehyde, pyrrol and chlorobenzene [7].

Worth and Cederbaum [100], propose to facilitate the search for finding a conical intersection if the two states have different symmetries: If they cross along a totally symmetric nuclear coordinate, then the crossing point is a conical intersection. Even this simplifying criterion leaves open a large number of possibilities in any real system. Therefore, Worth and Cederbaum base their search on large scale nuclear motions that have been identified experimentally to be important in the evolution of the system after photoexcitation.

A major motivation for the study of conical intersections is their assumed importance in the dynamics of photoexcited molecules. Molecular dynamics methods are often used for this purpose, based on available potential energy surfaces [118–121]. We briefly survey some methods designed to deal with relatively large molecules (>5 atoms). Several authors combine the potential energy surface calculations with dynamic simulations. A relatively straightforward approach is illustrated by the work of Ohmine and co-workers [6,122]. Ab initio calculations of the ground and excited potential surfaces of polyatomic molecules (ethylene and butadiene) were performed. Several specific nuclear motions were chosen to inspect their importance in inducing curve crossing. These included torsion, around C=C and C–C bonds, bending, stretching and hydrogen-atom migration. The ab initio potentials were parametrized into an analytic form in order to solve the dynamic equations of motion. In this way, Ohmine was able to show that hydrogen migration is important in the radiationless decay of ethylene.

Köppel co-workers [121] developed a method that uses accurately calculated potential surfaces for dynamic simulations. The initially calculated adiabatic potential surfaces are diabatized, to avoid singularities that hamper dynamic calculations. Once the diabatic potential surfaces are determined, along with the off-diagonal coupling constants between them,wave-packet dynamics can be performed. The off-diagonal coupling constants are large along the conical intersection seam. The method was demonstrated for a triatomic molecules (O_3, H_2S, NO_2), but is claimed to be suitable for larger systems [121].

Martinez and Ben-Nun [10,123] proposed a combined approach, in which the potential surfaces and the dynamics are treated on equal footing. In their AIMS method, ab initio quantum chemistry and nonadiabatic quantum dynamics are united. The electronic and nuclear Schrödinger equations are solved simultaneously, that is, the electronic structure problem is solved "on the fly" as dictated by the quantum mechanical nuclear dynamics. This method has recently been applied to ethylene. It was found that electronic excitation leads to an excited state that favors a twist of the two methylene groups to a perpendicular geometry, as well as pyramidalization of one of them. The combined effect lowers the energy of the excited state, and leads to several symmetry allowed crossings, the final one being to the ground state.

Our qualitative approach [40,41], which is based on the phase change theorem of Longuet-Higgins, considers spin pairing as the principal factor for locating conical intersections. We consider transitions from the first excited state to the ground-state, and form the loop on the ground-state surface. A given spin-paired system (anchor) may support many nuclear configurations on the ground state surface, but only one of them is usually at an energy minimum. As in all other methods mentioned above, the task is to find the two coordinates defining the loop that surrounds the conical intersection. We use for this purpose the reaction coordinates connecting the chosen anchors: A pair of reaction coordinates is sought, of which one is phase preserving (p) and the other phase inverting (i). There are many such pairs in a polyatomic molecule, which may be sorted out systematically. For any specific product, the reaction coordinate leading from the starting material is a natural choice. The phase change associated with this coordinate is well defined (either phase preserving or inverting). The second coordinate may be chosen from among all other reactions of the reactant, which may be found by considering all possible electron re-pairings. In practice, experiment and chemical intuition are used to facilitate and shorten the search. The three anchors of the loop, which are A—the reactant, B—the desired product, and C—another product, must form a phase inverting loop. This means that either all three reactions ($A \rightarrow B$, $B \rightarrow C$, and $C \rightarrow A$) are phase inverting, or that only one of them is. The loops formed are designated as $i^3$and ip^2, respectively. As shown in Section VI, the method is readily combined with high level quantum calculations for polyatomic systems.

Bornemann and Klessinger [124] used this approach in the analysis of the photoreactions of 2H-azirines. They implemented the method by calculating the ground- and excited-state geometries using the CASSCF method, starting from previous MNDOC–CI results [125]. The coordinates along which the conical intersection was searched for were determined by the phase change rule. The main application of the method described in this paper is expected to be in the analysis of the photochemistry of large systems. The initial location of the conical intersection is not dependent on a numerical algorithm, but on basic principles. The assignment of spin pairing is a relatively straightforward job, and can be done systematically. In principle, the method relies mostly on ground-state species, whose properties are either experimentally accessible, or may be computed. The numerical application, briefly described in Section VI, is still being developed. Methods specializing in the calculations of ground-state properties are notably efficient and accurate. Currently, the properties of both the ground and the excited states need to be computed. In principle, as all conical intersections leading to the ground state are "points" on the ground-state surface, their properties may be calculated by methods specialized for this state.

VIII. IMPACT ON MOLECULAR PHOTOCHEMISTRY AND FUTURE OUTLOOK

Conical intersections are important in molecular photochemistry, according to the current consensus, which is based on the combination of experimental and theoretical data. In this chapter, we tried to show that the location and approximate structure of conical intersections may be deduced by simple considerations of the changes in spin-pairing accompanying a reaction. We have also shown how these ideas may be put to practical computational application.

Chemists have developed several simple rules and methods that have helped to predict the course of photochemical reactions. In this section, we summarize some of these ideas and discuss their relation to the conical intersection model.

In distinction with thermal reactions, photochemical ones involve at least two potential surfaces. Attempts to understand them may be divided into two categories: those treating the ground and excited states separately, and those considering the coupling between the two as an essential ingredient. The first category analyzes photochemical reactions in terms of a kinetic mechanism, and seeks transition states and intermediates, in a manner that is commonly used for thermal reactions [126–128]. The rate-determining step, as a rule, is considered to be on the excited-state surface and the mechanism for the return from the excited to the ground state is not specified—it is assumed that the system will somehow find its way down. This approach, which views the photochemical process as a sequence of elementary reactions, each proceeding on a single reaction coordinate, may be termed the *single coordinate model*. In the second category, the coupling between the excited and the ground state is of central importance. This is *the two coordinate model*, in which concerted motion along *two* coordinates is essential for understanding the photochemical process. One coordinate carries the reactant to the product, the other couples the excited state with the ground state. The situation is identical to the case of the Jahn–Teller effect, where motions along two coordinates are equally important in removing the degeneracy.

A corollary of the single coordinate model is the idea that the photochemical process can be analyzed by a correlation diagram based on a single coordinate. Both MO [19,52] and VB [30,31,39] models have been proposed, and were widely used. The MO-based orbital symmetry conservation rules offered by Woodward an Hoffmann (WH) [52], gained special popularity. These rules, though not based on a solid theoretical basis, were of paramount importance in systematizing organic reactions. They are easy to apply and work very well for thermal reactions [53,54]. For photochemical reactions, the correlation between an occupied MO and an unoccupied one (in the ground state) is considered. In that sense, the WH rules assume that the reaction proceeds on the excited-state surface and results in an electronically excited product. It is usually accepted, that the rules for photochemical reactions and thermal ones are mutually exclusive: reactions that are "forbidden" thermally, are "allowed" photochemically.

The VB model considers the entire system, rather than molecular orbitals; At its heart is the passage of the system from one bonding mode to another [17,18,31,32,45,47] (in our nomenclature, change in spin pairing). This model was considered by some authors to be superior to the MO approach for thermal reactions, as it accounts naturally for the energy barrier found in thermal reactions [44]. As pointed out by Oosterhoff and co-workers [47] for the example of butadiene ring closure, the WH approach considers an antisymmetric excited state, while their model shows that a symmetric excited state is important. They show that for this state, the disrotatory ring-closure mode has an energy minimum along the reaction coordinate on the excited-state potential surface. The system will therefore move along this route, rather than the conrotatory one (preferred on the ground state. In both the WH and the Oosterhoff models, the reaction is taking place on the excited-state surface, and the return to the ground state is assumed to occur somehow (in Oosterhoff's theory, near the minimum). Thus, both models, based on essentially two-state paradigm, do not consider explicitly the equivalent importance of two coordinates.

The concept of biradicals and biradicaloids was often used in attempts to account for the mechanism of photochemical reactions [2,20,129–131]. A biradical (or diradical) may be defined as [132] an even-electron molecule that has one bond less than the number permitted by the standard rules of valence.

Biradical or biradicaloid minima are introduced in simple MO theory in cases where a pair of nearly degenerate approximately nonbonding orbitals is occupied with a total of only two electrons in the ground state [2, p. 187; 30, 133]. The approximate degeneracy of the two MOs can usually occur only at certain nuclear geometries, referred to as biradicaloid geometries. In both VB and MO theories, the spacing between the ground and excited state surfaces of biradicals are small, so that these species are considered as efficient funnels for radiationless transitions [2,133,134]. Salem and co-workers used biradical structures to construct correlation diagrams to analyze photochemical reactions. Their method differs from the WH approach in two respects [30, p. 136]. First, they use resonance structures, rather than the WH orbital configurations; Second, they chose as a symmetry element a reaction plane that does not cut through bonds being made. In distinction with Oosterhoff, Salem explicitly considered the direct passage from an excited reactant to a ground state product. His correlation diagrams are essentially represented by Figure 5—they are plotted against a phase preserving coordinate. As discussed there, it corresponds to a cut across the conical intersection along the phase preserving coordinate. Likewise, the WH and Oosterhoff correlation diagrams are a cut across the conical intersection along a phase-inverting coordinate (see Fig. 6). The conical intersection model uses both for the full account of photochemical reactions.

Biradicals were central building blocks in the construction of MO-based correlation diagrams, in which orbital symmetry was assumed to be preserved (as in the WH treatment) [129,135,136]. For example, the H_4 system was studied by Gerhartz et al. [34] in some detail. The biradical approach leads to a 3×3 CI treatment, which allows the construction of correlation diagrams along certain reaction coordinates. The perfect tetrahedral geometry was recognized as a touching "point" of S_1 and S_0. It was also admitted that neither the 3×3 CI approach, nor the simple VB one are satisfactory: In MO, more than three configurations are required, in VB, ionic terms in addition to covalent ones [129].

The biradical model suggests a connection between the single coordinate model, emphasizing reactions on a single energy surface, and the two-coordinate model, in which the coupling between states is important.

These couplings between the states were introduced as perturbations that mix the ground and excited state (the γ and δ parameters [2,136]). For example, the electronegativity difference between orbitals is parametrized by δ, their energy difference, as a function of the nuclear configuration. In biradicals such as twisted ethylene, even a weak polarizing perturbation δ causes an essentially complete uncoupling of the two zwitterionic configurations (in VB language). This leads to a highly polar character of S_1, near the perpendicular configuration, a phenomenon known as "sudden polarization" [82,137].

By using these parameters, the single coordinate scheme becomes, in practice, a two-coordinate one, the coordinate along which the δ parameter changes serving as the second one. Comparison with the spin-pairing model shows that the biradicaloid model reaches the same conclusions, using essentially perturbation methods (via the parameters γ and δ). For example, in the case of butadiene photocyclization, the γ perturbation leads to cyclobutene, and the δ perturbation, to bicyclobutane. On the other hand, no perturbations are required in the spin-pairing model. It postulates two independent nuclear coordinates from the outset, and finds the asymmetric conical intersection in a conceptually simple way.

Robb and Bernardi and their co-workers [44] used the Evans VB-based model as the starting point for their analysis. Their model is based on the coordinate leading from the reactant to the product (in our notation—the phase-inverting coordinate). However, they recognized the importance of the coupling between the ground and excited state, and introduced specifically a coordinate that leads to the minimization of the energy gap between the two (the gradient difference coordinate [Eq. (25)]. This coordinate was found computationally, but the physical meaning assigned to it was not specified.

The method outlined in this chapter continues the trend proposed by [44], and uses the phase-change concept to provide the physical meaning. It acknowledges the fact that two spin-paired species only are not sufficient to account for a photochemical reaction. A third one is required. The third anchor provides a simple interpretation to the second coordinate, used in the Bernardi–Robb model, as well as in the biradicaloid one. It leads to the same qualitative results as the orbital correlation method and the biradical hypotheses, but it does not introduce ad hoc assumptions. The conical intersection is not a biradical in the simple sense, but the fact that electrons can be re-paired in its neighborhood in several different ways, makes it appear as if the electrons are independent. However, in general these "biradicals" are not intermediates, and cannot be trapped or observed directly by spectroscopic methods. This accounts for the general failure to isolate many proposed "intermediate" biradicals. The model also explains naturally how apparently photochemically "forbidden" reaction products are obtained in practice. The results obtained are based on the phase change rule, which is a direct consequence of Pauli's principle. The method can be used to make simple qualitative applications, enabling the prediction of photochemical products and their stereospecificity. On the other hand, it can be used to guide high-level quantum chemical calculations on fairly complex systems. It can therefore be used by both experimentalists and computational chemists.

APPENDIX A. PHASE INVERTING REACTIONS

I. THE MODEL (see 28)

Consider a molecule consisting of more than three atoms, with an even number of valence electrons, $2n$ ($n \geq 2$). The basic assumption of the model is that the system will tend to form as many valence electron pairs between the atoms as possible. Each pair of electrons that are spin paired, form a bond between the atoms from which they originated. If the pair happens to be situated on a single atom, the pair is counted as a nonbonding pair (as in the lone pair of ammonia).

In the reaction, compound A transforms to B; the total number of electron pairs is preserved, but at least four electrons are assumed to change spin partners.

The task is now to calculate the structure and energy of the system in the transition state between A and B. Its wave function is assumed to be constructed from a linear combination of the two. It is convenient to use VB terminology for this purpose. Let the wave function of A be denoted by a VB function $|A\rangle$ and that of B by $|B\rangle$.

The $|A\rangle$ wave functions is written in the standard fashion:

$$\varphi_A = \sum_p \epsilon_p \mathrm{P}\ 1(1)2(2)\cdots 2n(2n)[\alpha(1)\beta(2) - \beta(1)\alpha(2)][\alpha(3)\beta(4) - \beta(3)\alpha(4)\cdots[\alpha(2n-1)\beta(2n) - \beta(2n-1)\alpha(2n)] \tag{A.1}$$

We use a short hand notation:

$$\phi_A = |A\rangle = (1\bar{2} - \bar{1}2)(3\bar{4} - \bar{3}4)\cdots(2n-1\overline{2n} - \overline{2n-1}2n) \tag{A.2}$$

With ϕ_A containing a normalization factor and all permutations over the atomic orbital wave functions i $(1 = 1, 2, \ldots 2n)$. Likewise, if all electron pairs were exchanged in a cyclic manner, the product wave function, ϕ_B, has the form:

$$|B\rangle = (1\overline{2n} - \bar{1}2n)(2n-1\overline{2n-2} - \overline{2n-1}2n-2)\cdots(3\bar{2} - \bar{3}2) \tag{A.3}$$

If the exchange was done in a different pattern, an corresponding expression would result. If only part of the pairs were re-paired, some of the factors of Eq. (A.3) stay put, and some change. In that case, one can always transform $|B\rangle$ to a form $|B'\rangle$ where all the unchanged factors are grouped together. Each transposition multiplies the determinant by -1. Therefore, if an even number of transpositions are needed, the signs of $|A\rangle$ and $|B'\rangle$ are equal. If an odd number is required, $|A\rangle$ and $|B'\rangle$ have opposite signs. The actual wave function of the system is constructed from the combination of the two VB structures $|A\rangle$ and $|B\rangle$. Two combinations are possible, an in-phase one $|A\rangle + |B\rangle$, and an out-of-phase one $|A\rangle - |B\rangle$. We disregard possible different coefficients, since we are only interested in the sign of the combination. Their energies are given by

$$E^{\pm} = \frac{H_{AA} + H_{BB} \pm 2H_{AB}}{2 \pm 2S_{AB}} \tag{A.4}$$

According to Eq. (A.4), if $H_{AB} < 0$, the ground state will be the in-phase combination, and the out-of-phase one, an excited state. On the other hand, if $H_{AB} > 0$, the ground state will be the out-of-phase combination, while the in-phase one is an excited state. This conclusion is far reaching, since it means that the electronic wave function of the ground state is nonsymmetric in this case, in contrast with common chemical intuition. We show that when an even number of electron pairs is exchanged, this is indeed the case, so that the transition state is the out-of-phase combination.

In order to show this, we have to evaluate the matrix element $H_{AB} = \langle A|H|B\rangle$. We begin by writing out the orbital part of the A wave function explicitly

$$|A\rangle = (1\bar{2} - \bar{1}2)(3\bar{4} - \bar{3}4)\cdots(2n-1\,\overline{2n} - \overline{2n-1}\,2n) = \{1\,\bar{2}\,3\,\bar{4}\cdots 2n-1\overline{2n} - \bar{1}2\,3\,\bar{4}\cdots 2n-1\,\overline{2n} - 1\bar{2}\,\bar{3}\,4\cdots 2n-1\,\overline{2n} + \cdots(-1)^n\bar{1}2\,\bar{3}4\cdots\overline{2n-1}\,2n\} \tag{A.5}$$

The sign of the last term depends on the parity of the system. Note that in the first and last term (in fact, determinants), the spin–orbit functions alternate, while in *all* others there are two pairs of adjacent atoms with the *same* spin functions. We denote the determinants in which the spin functions alternate as the alternant spin functions (ASF), as they turn out to be important reference terms.

Next, assuming that all pairs were exchanged in a cyclic pattern, we transform the VB function $|B\rangle$ [Eq. (A.3)] to a form similar to $|A\rangle$ by making $n-1$ transpositions of the form $(2n,2)$, $(2n-1,3)$, and so on. Each transposition multiplies the function by -1, obtaining

$$\begin{aligned}|B\rangle &= (-1)^{n-1}\{1\,\bar{2}\,3\,\bar{4}\cdots 2n-1\,\overline{2n} - 1\,2\,\bar{3}\,\bar{4}\cdots 2n-1\,\overline{2n} - 1\,\bar{2}\,3\,4\,\bar{5}\,\bar{6}\cdots 2n-1\,\overline{2n} + \cdots + (-1)^n\bar{1}\,2\,3\,\bar{4}\cdots\overline{2n-1}\,2n\} \\ &= (-1)^{n-1}\{1\,\bar{2}\,3\,\bar{4}\cdots 2n-1\,\overline{2n}\} + \cdots + (-1)^{2n-1}\{\bar{1}\,2\,\bar{3}\,4\cdots\overline{2n-1}\,2n\}\end{aligned} \tag{A.6}$$

It is evident that the only determinants that appear in both $|A\rangle$ and $|B\rangle$ are the ASFs.

The cross-term $\langle A|H|B\rangle$ in Eq. (A.4) can now be evaluated. This term may be written as (omitting the normalization constant):

$$H_{AB,CL} = (-1)^{n-1}2\{Q + K_{12} + K_{23} + \cdots K_{ii+1} + \cdots K_{2nl} + \text{higher exchange integrals}\} \tag{A.7}$$

Where the Coulomb integral Q is

$$\begin{aligned} Q &= \langle 1\ \bar{2}\ 3\ \bar{4} \cdots 2n-1\ \overline{2n}|H|1\ \bar{2}\ 3\ \bar{4} \cdots 2n-1\ \overline{2n}\rangle \\ &= \langle \bar{1}\ 2\ \bar{3} \cdots \overline{2n-1}\ 2n|H|\bar{1}\ 2\ \bar{3} \cdots \overline{2n-1}\ 2n\rangle \end{aligned}$$

The exchange integrals K_{ij} contain terms such as $\langle i, i+1|g|i+1, i\rangle + 2S_{i,i+1}$ $\langle i|h|i+1\rangle$ [138]. The second term, representing the attractive interaction between two nuclei and the electronic overlap charge between them, is the dominant one and completely outweighs the first repulsive term. $K_{i,i+1}$ therefore has the same sign as the Coulomb integral Q. (For a similar derivation, see [139].) In Eq. (A.7), $H_{\mathrm{AB,CL}}$ is the cross-term obtained by classic VB theory [140], in which only contributions from electron pairwise transposition permutations were considered. Bonding in these systems is due mainly to the exchange integrals $K_{i,i+1}$ between orbitals in the same cycle [138]. Pauling [140] showed that the most important contributions are due to neighboring orbitals, justifying the neglect of the smaller terms in Eq. (A.7).

Because of the orthogonality of the spin functions, and since we assume no spin–orbit coupling, only the first and last terms in Eqs. (A.5) and (A.6) will contribute to the Coulomb integral in $H_{\mathrm{AB,CL}}$. The Coulomb integrals together with the exchange ones between neighbors, K_{ii+1}, are larger than all other terms and determine which will be the ground state. The sign of their contributions is determined as follows:

When n is odd, the first ASF in both $|\mathrm{A}\rangle$ and $|\mathrm{B}\rangle$ is positive, while the second is negative. The two resulting Coulomb integrals are equal contributing together $2Q$.

When n is even, the two terms in $|\mathrm{A}\rangle$ have equal signs, as in $|\mathrm{B}\rangle$, but the sign in $|\mathrm{B}\rangle$ is opposite to that in $|\mathrm{A}\rangle$. Therefore, the total contribution to the energy is $-2Q$.

Since Q is negative, and $H_{\mathrm{AB,CL}}$ for the ground state must be a negative sign, it follows that the ground state for the odd parity case is the in-phase combination, while for the even parity case, the out-of-phase wave function is the ground state.

This classical VB picture, using only pairwise electron transpositions in the permutations, in which the spins of the two electrons of every bond are paired, is sometimes termed "the perfect pairing approximation" [138]. Figure 7 shows a schematic representation of the different contributions to the in-phase and out-of-phase combinations. Note that the energies of the twin excited states shown in Figure 7 can also be calculated from Eq. (A.7). Their destabilization with respect to the $H_{\mathrm{AA}} + H_{\mathrm{BB}}$ reference due to the cross-term will be larger than the stabilization of the ground state, due to the different contributions of the overlap integral in the denominator of Eq. (A.4).

If only part of the electron pairs were exchanged in the reaction (the usual case), the argument follows similar lines. The important factor is the parity of the exchanges: for an even number of electron pairs exchanged, the out-of-phase combination is always of lower energy than the in-phase one. Therefore, the transition state is phase inverting. For an odd number of electron pairs exchanged, the reaction is phase preserving.

Acknowledgments

We thank Professor S. Shaik, Professor B. Dick, Professor L. S. Cederbaum, and Dr. W. Fuss for many enlightening discussions and suggestions. This research was supported by The Israel Science Foundation founded by The Israel Academy of Sciences and Humanities and partially by The VolkswagenStiftung. The Farkas Center for Light Induced Processes is supported by the Minerva Gesellschaft mbH.

References

1. E. Teller, *J. Phys. Chem.* **41**, 109 (1937).
2. M. Klessinger and J. Michl, *Excited States and Photochemistry of Organic Molecules*, VCH, New York, 1995.
3. W. Fuss, S. Lochbrunner, A. M. Müller, T. Schikarski, W. E. Schmid, and S. A. Trushin *Chem. Phys.* **232**, 161 (1998).
4. E. W.-G. Diau, S. De Feyter, and A. H. Zewail *J. Chem. Phys.* **110**, 9785 (1999).
5. J. M. Metsdagh, J. P. Visicot, M. Elhanine, and B. Soep, *J. Chem. Phys.* **113**, 237 (2000).
6. I. Ohmine, *J. Chem. Phys.* **83**, 2348 (1985).
7. A. L. Sobolewski and W. Domcke, *Chem. Phys.* **259**, 181 (2001).
8. M. Klessinger, *Angew. Chem. Int. Ed. Engl.* **34**, 549 (1995).
9. F. Bernardi, M. A. Robb, and M. Olivucci, *Chem. Soc. Rev.* **25**, 321 (1996).
10. M. Ben-Nun and T. J. Martinez,, *Chem. Phys.* **259**, 237 (2001).
11. D. R. Yarkoni, *Rev. Modern Phys.* **68**, 985 (1996).
12. G. Herzberg and H. C. Longuet-Higgins, *Discuss. Faraday Soc.* **35**, 77 (1963).
13. H. C. Longuet-Higgins, *Proc. Roy. Soc London Ser. A* **344**, 147 (1975).
14. G. N. Lewis, *J. Am. Chem. Soc.* **38**, 762 (1916).
15. L. Pauling, *The Nature of the Chemical Bond*, Cornell University Press, Ithaca, N.Y., 1940.
16. R. P. Feynmann, M. Leighton, and M. Sands, *The Feynmann Lectures in Physics*, Vol. III, Addison-Wesley, Reading, MA, 1965, Chapters 9 and 10.
17. M. G. Evans and E. Warhurst, *Trans. Faraday Soc.* **34**, 614 (1938).
18. M. G. Evans, *Trans. Faraday Soc.* **35**, 824 (1939).
19. M. J. S. Dewar, *The Molecular Orbital Theory of Organic Chemistry*, McGraw-Hill, New York, 1969.
20. H. E. Zimmerman, *J. Am. Chem. Soc.* **88**, 1564,1566 (1966).
21. H. E. Zimmerman, *Acc. Chem. Res.* **4**, 272 (1971).
22. W. A. III, Goddard, *J. Am. Chem. Soc.* **94**, 793 (1972).
23. G. Levin and W. A. III, Goddard, *J. Am. Chem. Soc.* **97**, 1649 (1975).
24. A. F. Voter and W. A. III, Goddard, *Chem. Phys.* **57**, 253 (1981).
25. S. Zilberg and Y. Haas, *J. Photochem. Photobiol.* **144**, 221 (2001).

26. S. S. Shaik, *J. Am. Chem. Soc.* **103**, 3692 (1981). S. Shaik and P. C. Hiberty, *Adv. Quantum. Chem.* **26**, 99 (1995).
27. S. Shaik, S. Zilberg, and Y. Haas, *Acc. Chem. Res.* **29**, 211 (1996).
28. S. Zilberg and Y. Haas, *Int. J. Quant. Chem.* **71**, 133 (1999).
29. E. Hückel, *Z. Phys.* **70**, 204 (1931).
30. L. Salem, *Electrons in Chemical Reactions: First Principles*, John Wiley & Sons, Inc., New York, 1982.
31. W. J. van der Hart, J. J. Mulder, and L. J. Oosterhoff, *J. Am. Chem. Soc.* **94**, 5724 (1972).
32. J. J. Mulder and L. J. Oosterhoff, *Chem Commun.* **305**, 307 (1970).
33. S. S. Xantheas, G. J. Atchity, S. T. Elbert, and K. Ruedenberg, *J. Chem. Phys.* **94**, 8054 (1991).
34. W. Gerhartz, R. D. Poshusta, and J. Michl, *J. Am. Chem. Soc.* **98**, 6427 (1977); **99**, 4263 (1976).
35. A. J. Stone, *Proc. R. Soc Lond. Ser. A* **351**, 141 (1976).
36. S. Glasstone, K. J. Laidler, and H. Eyring, *The Theory of Rate Processes*, McGraw-Hill, New York, 1941.
37. For a review of the method, see M. A. Robb, M. Garavelli, M. Olivucci, and F. Bernardi, in *Reviews in Computational Chemistry*, K. B. Lipkovwitz and D. B. Boyd, eds., Wiley-VCH, New York, 2000, Vol. 15, pp. 87–212.
38. P. Celani, M. Garavelli, S. Ottani, F. Bernardi, M. A. Robb, and M. Olivucci, *J. Am. Chem. Soc.* **117**, 11584 (1995).
39. A. F. Voter and W. A., Goddard, *J. Am. Chem. Soc.* **108**, 2830 (1986).
40. S. Zilberg and Y. Haas, *Eur. J. Chem.* **5**, 1755 (1999).
41. S. Zilberg and Y. Haas, *Chem. Phys.* **259**, 249 (2000).
42. M. J. S. Dewar, *Tetrahedron Suppl.* **8**, Part 1, 75 (1966).
43. M. J. S. Dewar, *Angew. Chem. Int. Ed. Engl.* **10**, 761 (1971).
44. M. A. Robb and F. Bernardi, in J. Bertran and I. G. Csizmadia, eds., *New Theoretical Concepts for Understanding Organic Reactions*, Kluwer, N.Y., 1989, p. 101.
45. E. Heilbronner, *Tetrahedron Lett.* 1923 (1964).
46. D. P. Craig, *J. Chem. Soc.* **997**, (1959).
47. W. T. A. M. van der Lugt, and L. J. Oosterhoff, *J. Am. Chem. Soc.* **91**, 6042 (1969); *Chem. Commun.* 1235 (1968); *Mol. Phys.* **18**, 177 (1970).
48. W. Fuss, Y. Haas, and S. Zilberg, *Chem. Phys.* **259**, 273 (2000).
49. S. Zilberg and Y. Haas, *J. Phys. Chem. A* **103**, 2364 (1999).
50. S. C. Wright, D. L. Cooper, J. Gerratt, and M. Raimondi, *J. Phys. Chem.* **96**, 7943 (1992); P. B. Karadakov, D. L. Cooper, J. Gerratt, and M. Raimondi, *J. Phys. Chem.* **99**, 10186 (1995).
51. S. Zilberg, Y. Haas, D. Danovich, and S. Shaik, *Angew. Chem. Int. Ed. Engl.* **37**, 1394 (1998).
52. R. B. Woodward and R. Hoffmann, *Angew. Chem. Int. Ed. Engl.* **8**, 781 (1969).
53. H. C. Longuet-Higgins and E. W. Abramson, *J. Am. Chem. Soc.* **87**, 2045 (1965).
54. D. M. Silver, *J. Am. Chem. Soc.* **96**, 5959 (1974).
55. H. E. Zimmerman, *Acc. Chem. Res.* **5**, 393 (1972).
56. F. Merlet, S. D. Peyerimhoff, T. J. Bunker, and S. Shih, *J. Am. Chem. Soc.* **96**, 959 (1974); M. Yamaguchi, *J. Mol. Struct. (THEOCHEM)* **365**, 145 (1996).
57. J. J. C. Mulder, *Nouv. J. Chim.* **4**, 283 (1980).
58. G. J. Collin, *Adv. Photochem.* **14**, 135 (1987).

59. R. Hoffmann, R. W. Alder, and C. F. Wolicox, *J. Am. Chem. Soc.* **92**, 4992 (1970); M. S. Gordon and M. W. Schmidt, *J. Am. Chem. Soc.* **115**, 7486 (1993).
60. R. Srinivasan, L. S. White, A. R. Rossi, and G. A. Epling, *J. Am. Chem. Soc.* **103**, 7299 (1981).
61. E. V. Cromwell, D.-J. Liu, M. J. J. Vrakking, A. H, Kung, and Y. T. Lee, *J. Chem. Phys.* **92**, 3230 (1990).
62. E. V. Cromwell, D.-J. Liu, M. J. J. Vrakking, A. H. Kung, and Y. T. Lee, *J. Chem. Phys.* **95**, 297 (1991).
63. A. Kumar, P. K. Chowdhury, K. V. S. R. Rao, and J. P. Mittal, *Chem. Phys. Lett.* **182**, 165 (1991); A. Kumar, P. D. Naik, R. D. Sainy, and J. P. Mittal, *Chem. Phys. Lett.* **309**, 191 (1999).
64. S. De Feyter, E. W.- G. Diau, and A. H. Zewail, *PCC* **2**, 877 (2000).
65. S. Wilsey and K. N. Houk, *J. Am. Chem. Soc.* **122**, 2651 (2000).
66. S. S. Hixson, P. S. Mariano, and H. E. Zimmerman, *Chem. Rev.* **73**, 531 (1973).
67. R. J. Rico, M. Page, and C. Doubleday, Jr., *J. Am. Chem. Soc.* **114**, 1131 (1992).
68. M. J. S. Dewar, *Adv. Chem. Phys.* **8**, 121 (1965).
69. H. A. Jahn and E. Teller, *Proc. R. Soc. London Ser. A* **161**, 220 (1937).
70. I. B. Bersuker, *Chem. Rev.* **101**, 1067 (2001).
71. I. B. Bersuker and V. Z. Pollinger, *Vibronic Interactions in Molecules and Crystals*, Springer, New York, 1996.
72. W. T. Borden and E. R. Davidson, *J. Am. Chem. Soc.* **101**, 3771 (1979).
73. B. E. Applegate, T. A. Miller, and T. A. Barckholz, *J. Chem. Phys.* **114**, 4855 (2001).
74. G. R. Liebling and H. M. McConnel, *J. Chem. Phys.* **42**, 3931 (1965).
75. H. Fischer and J. N. Murrell, *Theor. Chim. Acta (Berl.)* **1**, 463 (1967).
76. B. E. Applegate, A. J. Bezant, and T. A. Miller, *J. Chem. Phys.* **114**, 4869 (2001).
77. E. P. F. Lee and T. G. Wright, *PCCP* **1**, 219 (1999).
78. J. W. Zwanziger and E. R. Grant, *J. Chem. Phys.* **87**, 2954 (1987).
79. B. A. Williams and T. A. Cool, *J. Chem Phys.* **94**, 6358 (1991).
80. L. Serrano-Anders, M. Merchan, I. Nebot-Gil, R. Lindh, and B. O. Roos, *J. Chem. Phys.* **98**, 3151 (1993).
81. We are indebted to Professor T. Martinez for a discussion on this point.
82. V. Bonacic-Koutecky, P. Bruckmann, P. Hiberty, J. Koutecky, C. Leforestier, and L. Salem, *Angew. Chem. Int. Ed. Engl.* **14**, 575 (1975).
83. Y. Haas and U. Samuni, *Prog. React. Kinet.* **23**, 211 (1998).
84. Y. Inoue, *Chem. Rev.* **92**, 741 (1992).
85. M. Squillacote and T. C. Semple, *J. Am. Chem. Soc.* **112**, 5546 (1990).
86. G. Maier, *Angew. Chem. Int. Ed. Engl.* **27**, 309 (1988).
87. We are indebted to Professor Bernhard Dick for drawing our attention to these results, and for an enlightening discussion.
88. Y. Inoui, S. Takamuku, and H. Sakurai, *J. Chem. Soc, Perkin Trans* **2**, 1635 (1977).
89. F. Bernardi, M. Olivucci, M. A. Robb, and G. Tonachini, *J. Am. Chem. Soc.* **114**, 5805 (1992).
90. A. Gilbert and J. Baggott, *Essentials of Molecular Photochemisty*, Blackwell, London, 1991.
91. E. A. Halevy, *Orbital Symmetry and Reaction Mechanisms: The OCAMS View*, Springer, Berlin, 1992.

92. I. J. Palmer, I. N. Ragazos, F. Bernardi, M. Olivucci, M. A. Robb, and G. Tonachini, *J. Am. Chem. Soc.* **115**, 673 (1993); A. L. Sobolewski, C. Woywood, and W. Domcke, *J. Chem. Phys.* **98**, 5627 (1993); J. Dreyer and M. Klessinger, *Eur. J. Chem.* **2**, 335 (1996).
93. M. I. McCarthy, P. Rosmus, H.-J. Werner, P. Botschwina, and V. Vaida, *J. Chem. Phys.* **86**, 6693 (1987).
94. D. R. Eaton, *J. Am. Chem. Soc.* **90**, 4272 (1968); T. H. Whitesides, *J. Am. Chem. Soc.* **91**, 2395 (1969).
95. F. Scandola, O. Traverno, V. Balzani, G. L. Zucchini, and V. Carassity, *Inorg. Chem. Acta.* **1**, 76 (1967).
96. J. K. Burdett, *Inorg. Chem.* **15**, 212 (1976).
97. H. E. Zimmerman, and H. Iwamura, *J. Am. Chem. Soc.* **92**, 2015 (1970).
98. S. Zilberg and Y. Haas, The photochemistry of 1,4-cyclohexadiene in solution and in the gas phase: Conical intersections and the origin of the "helicopter-type" motion of H_2 photogenerated in the isolated molecule," *PCCP* **4**, 34 (2002).
99. M. Merchan, L. Serrano-Andres, L. S. Slater, B. O. Roos, R. McDiarmid, and X. Xing, *J. Phys. Chem. A* **103**, 5468 (1999).
100. G. A. Worth and L. S. Cederbaum, *Chem. Phys. Lett.* **338**, 219 (2001).
101. M. S. Berry, *Proc. R. Soc. London, Ser. A.* **392**, 45 (1984).
102. S. Adhukari and G. D. Billing, *Chem. Phys.* **259**, 149 (2000).
103. A. Kuppermann, in *Dynamics of Molecules and Chemical Reactions*, R. E. Wyatt and J. Z. H. Zhang, eds., Marcel Dekker, New York, 1996.
104. Y. Aharonov and D. Bohm, *Phys. Rev.* **115**, 485 (1959).
105. C. A. Mead, *Rev. Mod. Phys.* **64**, 51 (1992).
106. D. G. Truhlar, *J. Chem. Phys.* **70**, 2282 (1979).
107. G. J. Atchity, K. Ruedenberg, and A. Nanayakkara, *Theor. Chim. Acc.* **96**, 195 (1997).
108. D. R. Yarkoni, *Acc. Chem. Res.* **31**, 511 (1998).
109. D. R. Yarkoni, *Theor. Chem. Acc.* **98**, 197 (1997).
110. C. A. Mead, *J. Chem. Phys.* **78**, 807 (1983).
111. H. Köppel, W. Domcke, and L. S. Cederbaum, *Adv. Chem. Phys.* **57**, 59 (1984).
112. M. Baer, *Chem. Phys.* **259**, 123 (2000).
113. A. Mebel, M. Baer, and S. H. Lin, *J. Chem. Phys.* **112**, 10703 (2000).
114. F. Bernardi, M. Olivucci, and M. A. Robb, *J. Am. Chem. Soc.* **114**, 1606 (1992).
115. M. J. Bearpark, M. A. Robb, and H. B. Schlegel, *Chem. Phys. Lett.* **223**, 269 (1994).
116. M. Olivucci, F. Bernardi, and M. A. Robb, *J. Am. Chem. Soc.* **116**, 2034 (1994).
117. M. Olivucci, I. N. Ragazos, F. Bernardi, and M. A. Robb, *J. Am. Chem. Soc.* **115**, 3710 (1993).
118. W. Domcke and G. Stock, *Adv. Chem. Phys.* **100**, 1 (1997).
119. F. Santoro, C. Petrongolo, G. Granucci, and M. Persico, *Chem. Phys.* **259**, 193 (2000).
120. A. Ferretti, A. Lami, and G. Villani, *Chem. Phys.* **259**, 201 (2000).
121. S. Mahapatra, H. Köppel, L. S. Cederbaum, P. Stampfuss, and W. Wenzel, *Chem. Phys.* **259**, 211 (2000).
122. M. Ito and I. Ohmine, *J. Chem. Phys.* **106**, 3159 (1997).
123. M. Ben-Nun, J. Quenneville, and T. M. Martinez, *J. Phys. Chem. A* **104**, 5161 (2000).
124. C. Bornemann and M. Klessinger, *Chem. Phys.* **259**, 263 (2000).

125. R. Izzo and M. Klessinger, *J. Comput. Chem.* **21**, 52 (2000).
126. E. Havinga and J. Cornelisse, *Pure Appl. Chem.* **47**, 1 (1976).
127. N. J. Turro, *Modern Molecular Photochemistry*, Bejamin/Cummings, Menlo Park, CA, 1978.
128. D. I. Schuster, G. Lem, and N. A. Kaprinidis, *Chem. Rev.* **93**, 3 (1991).
129. J. Michl, *Photochem. Photobiol.* **25**, 141 (1977).
130. L. Salem and C. Rowland, *Angew. Chem. Int. Ed. Engl.* **11**, 92 (1972).
131. L. Salem, *Science*. **191**, 822 (1976).
132. J. A. Berson, *Acc. Chem. Res.* **11**, 466 (1978).
133. J. Michl, *Mol. Photochem.* **4**, 243–257 (1972).
134. J. Michl, *Top. Curr. Chem.* **46**, 1 (1974).
135. J. Michl, in "Photochemical Reactions: Correlation Diagrams and Energy Barriers," G. Klopman, ed., *Chemical Reactivity and Reaction Paths*, John Wiley & Sons, Inc., New York, 1974.
136. J. Michl and V. Bonacic-Koutecky, *Electronic Aspects of Organic Photochemistry*, John Wiley & sons; Inc. New York, 1990.
137. L. Salem, *J. Am. Chem. Soc.* **96**, 3486 (1974).
138. R. McWeeny and B. T. Sutcliffe, *Methods of Molecular Quantum Mechanics*, Academic Press, 1969, Chap. 6.
139. H. Eyring, J. Walter, and G. E. Kimball, *Quantum Chemistry*, John Wiley & Sons, Inc., New York, 1944, Chap. 13.
140. L. Pauling, *J. Chem. Phys. 1*, 280 (1933).

THE CRUDE BORN–OPPENHEIMER ADIABATIC APPROXIMATION OF MOLECULAR POTENTIAL ENERGIES

K. K. LIANG, J. C. JIANG, V. V. KISLOV, A. M. MEBEL, and S. H. LIN

Institute of Atomic and Molecular Sciences,
Academia Sinica, Taipei, Taiwan, ROC

M. HAYASHI

Center for Condensed Matter Sciences,
National Taiwan University, Taipei, Taiwan, ROC

CONTENTS

The Role of Degenerate States in Chemistry: A Special Volume of Advances in Chemical Physics, Volume 124, Edited by Michael Baer and Gert Due Billing. Series Editors I. Prigogine and Stuart A. Rice.
ISBN 0-471-43817-0.

I. INTRODUCTION

The introduction of the conventional Born–Oppenheimer (BO) approximation introduces the concept of electronic potential energy surface (PES), which lays the foundation of the majority of our concepts about molecular systems. However, the crossing of two adiabatic PES is also a consequence of such an adiabatic approximation. There has been much research done in an attempt to remove the singularity brought about by this crossing of multi-dimensional surfaces, namely, the conical intersections. Recently, characterization of conical intersection in molecules and the role played by conical intersection in femtosecond processes have attracted considerable attention [1–4]. The conical intersection is conventionally determined by the use of the adiabatic approximation. There are a number of so-called adiabatic approximations for the time-independent quantum mechanical treatment of molecules [5–13]. The most prominent of them are the BO approximation and the Born–Huang (BH) approximation. This latter name of BH approximation was suggested by Ballhausen and Hansen, but the theory was actually formulated by Born himself. It has also been described as the BO correction, the variational BO approximation and the Born–Handy formula. First, these approximations all start with the separation of the total molecular Hamiltonian into terms of different magnitude. Second, it is very common that, while trying to sort out terms of different magnitude, attempts were given to argue that the crossing terms coupling the momenta of the various atoms in the molecule are negligible after proper transformations. Ballhausen and Hansen made a very instructive comment saying that [8] "The effect of these cross-terms is to correlate the internal motion, so to speak, in such a way that the linear momentum as well as the angular momentum of the entire molecule stay constant." It is worth noting that actually these cross-terms are not only crucial for keeping the momentum constant, but also are important for keeping the energy constant. This point has not been proved rigorously, but it can be understood by noticing that the concept of electronic potential energy is a direct consequence of the adiabatic approximation. The negligence of the nuclear kinetic energy term and the fixation of the nuclear coordinates (and thus the frozening of the dependence of the cross-terms on the nuclear coordinates) are the causes of the dependence of the electronic energy on the nuclear configuration.

The separation of the electronic degrees of freedom from nuclear motions through adiabatic approximation has brought success to the ab initio quantum chemistry computations, but it is also the reason why we are confronted with the very difficult problem of potential energy crossing, in particular, the conical intersections. There may be other approaches, however, in which the energies of the states depend neither functionally nor parametrically on the nuclear configuration, and hence no crossing of energy levels may occur. If an approach like this can be developed, and if it is computationally tractable, then it may be a good method complement to the contemporary quantum chemistry packages for treating the cases in which potential energy surfaces crossing may happen in the traditional approach.

An alternative approximation scheme, also proposed by Born and Oppenheimer [5–7], employed the straightforward perturbation method. To tell the difference between these two different BO approximation, we call the latter the crude BOA (CBOA). A main purpose of this chapter is to study the original BO approximation, which is often referred to as the crude BO approximation and to develop this approximation into a practical method for computing potential energy surfaces of molecules.

In this chapter, we demonstrate the approach of the CBOA, and show that to carry out different orders of perturbation, the ability to calculate the matrix elements of the derivatives of Coulomb interaction with respect to nuclear coordinates is essential. Therefore, we studied the case of the diatomic molecule, and here we demonstrate the basic skill of computing the relevant matrix elements in Gaussian basis sets. The formulas for diatomic molecules, up to the second derivatives of the Coulomb interaction, are shown here to demonstrate that some basic techniques can be developed to carry out the calculation of the matrix elements of even higher derivatives. The formulas obtained may be complicated. First, they are shown to be nonsingular. Second, the Gaussian basis set with angular momentum can be dealt with in similar ways. Third, they are expressed as multiple finite sums of certain simple functions, of order up to the angular momentum of the basis functions, and thus they can be computed efficiently and accurately. We show the application of this approach on the H_2 molecule. The calculated equilibrium position and force constant seem to be reasonable. To obtain more reliable results, we have to employ a larger basis set to higher orders of perturbation to calculate the equilibrium geometry and wave functions.

II. CRUDE BORN–OPPENHEIMER APPROXIMATION

The theory discussed in this section is based on the work of Born and others [5,7]. However, some of the approaches that are not suitable for our need are modified, and proper notations are adopted accordingly.

For a molecular system, we shall separate the total Hamiltonian into three parts:

$$\hat{H} = \hat{T}_e + V(\mathbf{r}, \mathbf{R}) + \hat{T}_N \tag{1}$$

The $\hat{T}$ operators are the usual kinetic energy operators, and the potential energy $V(\mathbf{r}, \mathbf{R})$ includes all of the Coulomb interactions:

$$V(\mathbf{r}, \mathbf{R}) = \frac{e^2}{2}\sum_{i \neq j} \frac{1}{r_{ij}} - e^2 \sum_{A,i} \frac{Z_A}{r_{A,i}} + \frac{e^2}{2}\sum_{A \neq B} \frac{Z_A Z_B}{R_{AB}} \tag{2}$$

Let us consider the simplified Hamiltonian in which the nuclear kinetic energy term is neglected. This also implies that the nuclei are fixed at a certain configuration, and the Hamiltonian describes only the electronic degrees of freedom. This electronic Hamiltonian is

$$\hat{H}_0(\mathbf{r};\mathbf{R}) = \hat{\boldsymbol{T}}_e + V(\mathbf{r};\mathbf{R}) \tag{3}$$

and the complete adiabatic electronic problem to solve is

$$\hat{H}_0|\phi_n\rangle = U_n|\phi_n\rangle \tag{4}$$

Note that the last term in expression (2) of V does not depend on electronic coordinates, and therefore it may be neglected in $\hat{H}_0$. The adiabatic Hamiltonian still depends parametrically on $\mathbf{R}$, and so is the electronic wave funcion $|\phi\rangle$. If we expand the nuclear coordinates or some of the nuclear coordinates with respect to a given configuration, that is, if we define

$$\mathbf{R} = \mathbf{R}_0 + \kappa\boldsymbol{\varrho} \tag{5}$$

where κ is a natural perturbation parameter that will be described later, then we shall expand the Hamiltonian in powers of κ as

$$\hat{H}_0 = \hat{H}_0^{(0)} + \kappa\hat{H}_0^{(1)} + \kappa^2\hat{H}_0^{(1)} + \cdots \tag{6}$$

Here

$$\hat{H}_0^{(0)} = \hat{H}_0(\mathbf{r};\mathbf{R}_0) \tag{7}$$

$$\hat{H}_0^{(1)} = \sum_i \left(\frac{\partial \hat{H}_0}{\partial R_i}\right)_{\mathbf{R}=\mathbf{R}_0} \varrho_i \tag{8}$$

$$\hat{H}_0^{(2)} = \frac{1}{2}\sum_{i,j} \left(\frac{\partial^2 \hat{H}_0}{\partial R_i \partial R_j}\right)_{\mathbf{R}=\mathbf{R}_0} \varrho_i\varrho_j \tag{9}$$

$$\vdots$$

Note that the electronic kinetic energy operator does not depend on the nuclear configuration explicitly. Therefore, we can conclude that

$$\hat{H}_0^{(1)} = \sum_i \left(\frac{\partial V}{\partial R_i}\right)_{\mathbf{R}=\mathbf{R}_0} \varrho_i \tag{10}$$

$$\hat{H}_0^{(2)} = \frac{1}{2}\sum_{i,j} \left(\frac{\partial^2 V}{\partial R_i \partial R_j}\right)_{\mathbf{R}=\mathbf{R}_0} \varrho_i\varrho_j \tag{11}$$

$$\vdots$$

The electronic wave functions and electronic energies are also expanded

$$|\phi\rangle = |\phi^{(0)}\rangle + \kappa\phi^{(1)} + \kappa^2|\phi^{(2)}\rangle + \cdots \tag{12}$$

$$U_n = U_n^{(0)} + \kappa U_n^{(1)} + \kappa^2 U_n^{(2)} + \cdots \tag{13}$$

In the following, it shall always be assumed that the zeroth-order solution is known, that is, we have a complete set of eigenvalues and wave functions, labeled by the electronic quantum number n, which satisfy

$$\hat{H}_0^{(0)}|\phi_n^{(0)}\rangle = U_n^{(0)}|\phi_n^{(0)}\rangle \tag{14}$$

Next, we shall consider how the nuclear kinetic energy is taken into consideration perturbatively. The natural perturbation index κ is chosen to be

$$\kappa = \sqrt[4]{\frac{m}{M_0}} \tag{15}$$

where m is the electron mass and M_0 is some quantity to do with the mass of nuclei. In rectangular coordinates, $\hat{T}_N$ can be written as

$$\begin{aligned}\hat{T}_N &= -\sum_i \frac{\hbar^2}{2m} \times \frac{m}{M_0} \times \frac{M_0}{M_i}\left(\frac{\partial^2}{\partial X_i^2} + \frac{\partial^2}{\partial Y_i^2} + \frac{\partial^2}{\partial Z_i^2}\right) \\ &= -\kappa^4 \sum_i \frac{\hbar^2}{2m} \times \frac{M_0}{M_i}\left(\frac{\partial^2}{\partial X_i^2} + \frac{\partial^2}{\partial Y_i^2} + \frac{\partial^2}{\partial Z_i^2}\right) \\ &\equiv \kappa^4 \hat{H}_1\end{aligned} \tag{16}$$

Since the nuclear coordinates are expanded according to Eq. (5), we can write the derivatives in the kinetic energy expression as

$$\frac{\partial^2}{\partial R_i^2} = \frac{1}{\kappa^2}\frac{\partial^2}{\partial \varrho_i^2} \tag{17}$$

Thus, we have

$$\kappa^4 \hat{H}_1 = \kappa^2 \hat{H}_1^{(2)} \tag{18}$$

where in $\hat{H}_1^{(2)}$, all of the derivatives with respect to nuclear coordinates R are replaced by derivatives with respect to ϱ, while the rest of the expression remains unchanged. In this case, the total Hamiltonian can be expanded into power series of κ as

$$\hat{H} = \hat{H}_0^{(0)} + \kappa \hat{H}_0^{(1)} + \kappa^2\left(\hat{H}_0^{(2)} + \hat{H}_1^{(2)}\right) + \cdots \tag{19}$$

The total Schrödinger equation is

$$\hat{H}|\psi_n\rangle = E_n|\psi_n\rangle \tag{20}$$

where the energy and the wave function are also to be expanded into power series of κ

$$|\psi_n\rangle = |\psi_n^{(0)}\rangle + \kappa|\psi_n^{(1)}\rangle + \kappa^2|\psi_n^{(2)}\rangle + \cdots \tag{21}$$

$$E_n = E_n^{(0)} + \kappa E_n^{(1)} + \kappa^2 E_n^{(2)} + \cdots \tag{22}$$

In the zeroth-order approximation,

$$(\hat{H}_0^{(0)} - U_n^{(0)})|\phi_n^{(0)}\rangle = 0 \tag{23}$$

$$(\hat{H}_0^{(0)} - E_n^{(0)})|\psi_n^{(0)}\rangle = 0 \tag{24}$$

Since $\hat{H}_0^{(0)}$ is an operator on electronic degrees of freedom only, it can be summarized that

$$E_n^{(0)} = U_n^{(0)} \tag{25}$$

$$|\psi_n^{(0)}\rangle = |\chi_n^{(0)}\rangle|\phi_n^{(0)}\rangle \tag{26}$$

Here, $|\chi_n^{(0)}\rangle$ is an arbitrary function of nuclear coordinates. It cannot be determined from Eq. (24) alone, but has to be determined from higher ordered perturbation equations.

In the first-order approximation, we find

$$(\hat{H}_0^{(0)} - U_n^{(0)})|\phi_n^{(1)}\rangle = -(\hat{H}_0^{(1)} - U_n^{(1)})|\phi_n^{(0)}\rangle \tag{27}$$

$$(\hat{H}_0^{(0)} - U_n^{(0)})|\psi_n^{(1)}\rangle = -(\hat{H}_0^{(1)} - E_n^{(1)})|\psi_n^{(0)}\rangle \tag{28}$$

The electronic wave functions can be found to be

$$|\phi_n^{(1)}\rangle = \sum_{m \neq n} \frac{\langle \phi_m^{(0)}|\hat{H}_0^{(1)}|\phi_n^{(0)}\rangle}{U_n^{(0)} - U_m^{(0)}} |\phi_m^{(0)}\rangle \tag{29}$$

The total energy, on the other hand, can be shown to follow:

$$E_n^{(1)} = U_n^{(1)} = 0 \tag{30}$$

This requires that the initially chosen $\mathbf{R}_0$ be the equilibrium configuration of this electronic level. Also, we reach the conclusion that the wave function will be of the form

$$|\psi_n^{(1)}\rangle = |\chi_n^{(0)}\psi_n^{(1)}\rangle + |\chi_n^{(1)}\psi_n^{(0)}\rangle \tag{31}$$

$|\chi_n^{(1)}\rangle$ also has to be determined later.

In the second-order approximation, we have

$$(\hat{H}_0^{(0)} - U_n^{(0)})|\psi_n^{(2)}\rangle = -\hat{H}_0^{(1)}|\psi_n^{(1)}\rangle - (\hat{H}_0^{(2)} + \hat{H}_1^{(2)} - E_n^{(2)})|\psi_n^{(0)}\rangle \tag{32}$$

$$(\hat{H}_0^{(0)} - U_n^{(0)})|\phi_n^{(2)}\rangle = -\hat{H}_0^{(1)}|\phi_n^{(1)}\rangle - (\hat{H}_0^{(2)} - U_n^{(2)})|\phi_n^{(0)}\rangle \tag{33}$$

It can be shown that [7]

$$|\phi_n^{(2)}\rangle = C_{nn}^{(2)}|\phi_n^{(0)}\rangle + \sum_{m\neq n} C_{nm}^{(2)}|\phi_m^{(0)}\rangle \tag{34}$$

where

$$C_{nm}^{(2)} = \frac{1}{U_n^{(0)} - U_m^{(0)}}\left[\sum_{k\neq n}\frac{\langle\phi_m^{(0)}|\hat{H}_0^{(1)}|\phi_k^{(0)}\rangle\langle\phi_k^{(0)}|\hat{H}_0|\phi_n^{(0)}\rangle}{U_n^{(0)} - U_k^{(0)}} + \langle\phi_m^{(0)}|\hat{H}_0^{(2)}|\phi_n^{(0)}\rangle\right] \tag{35}$$

and

$$C_{nn}^{(2)} = \frac{1}{2}\sum_{k\neq n}\left|\frac{\langle\phi_k^{(0)}|\hat{H}_0|\phi_n^{(0)}\rangle}{U_n^{(0)} - U_k^{(0)}}\right|^2 \tag{36}$$

The full wave function and the electronic potential energy are

$$|\psi_n^{(2)}\rangle = |\chi_n^{(0)}\phi_n^{(2)}\rangle + |\chi_n^{(1)}\phi_n^{(1)}\rangle + |\chi_n^{(2)}\phi_n^{(0)}\rangle \tag{37}$$

$$U_n^{(2)} = \langle\phi_n^{(0)}|\hat{H}_0^{(2)}|\phi_n^{(0)}\rangle + \sum_{m\neq n}\frac{|\langle\phi_n^{(0)}|\hat{H}_0^{(1)}|\phi_m^{(0)}\rangle|^2}{U_n^{(0)} - U_m^{(0)}} \tag{38}$$

Furthermore, we obtain the equation of motion of the zeroth-order nuclear wave function:

$$(\hat{H}_1^{(2)} + U_n^{(2)} - E_n^{(2)})|\chi_n^{(0)}\rangle = 0 \tag{39}$$

We can only determine $E_n^{(2)}$ and $|\chi_n^{(0)}\rangle$ up to now. Later, we shall demonstrate that this equation is just the equations of motion of harmonic nuclear vibrations. The set of eigenstates of Eq. (43) can be written as $\{|\chi_{n\mathbf{v}}\rangle\}$, symbolizing that they are the vibrational modes of the nth electronic level, where $\mathbf{v} = (v_1, v_2, \ldots, v_N)$ if $\boldsymbol{\varrho}$ is N dimensional, and v_i is the vibrational quantum number of the ith mode.

In the third-order approximation, the equations are

$$(\hat{H}_0^{(0)} - U_n^{(0)})|\psi_n^{(3)}\rangle = -\hat{H}_0^{(1)}|\psi_n^{(2)}\rangle - (\hat{H}_0^{(2)} + \hat{H}_1^{(2)} - E_n^{(2)})|\psi_n^{(1)}\rangle - (\hat{H}_0^{(3)} - E_n^{(3)})|\psi_n^{(0)}\rangle \tag{40}$$

$$(\hat{H}_0^{(0)} - U_n^{(0)})|\phi_n^{(3)}\rangle = -\hat{H}_0^{(1)}|\phi_n^{(2)}\rangle - (\hat{H}_0^{(2)} - U_n^{(2)})|\phi_n^{(1)}\rangle - (\hat{H}_0^{(3)} - U_n^{(3)})|\phi_n^{(0)}\rangle \tag{41}$$

The electronic wave functions and potential energy can be determined in ways similar to those done in the first and second order. Here we wish to emphasize that, the full wave function in this order is

$$|\psi_n^{(3)}\rangle = |\chi_n^{(0)}\phi_n^{(3)}\rangle + |\chi_n^{(1)}\phi_n^{(2)}\rangle + |\chi_n^{(2)}\phi_n^{(1)}\rangle + |\chi_n^{(3)}\phi_n^{(0)}\rangle + |f_n^{(3)}\rangle \tag{42}$$

where $|f_n^{(3)}\rangle$ satisfies

$$(\hat{H}_0^{(0)} - U_n^{(0)})|f_n^{(3)}\rangle = \frac{\hbar^2}{m}\sum_{\alpha,i}\left(\frac{M_0}{M_\alpha}\right)\left(\frac{\partial}{\partial\varrho_{\alpha,i}}|\chi_n^{(0)}\rangle\right)\left(\frac{\partial}{\partial\varrho_{\alpha,i}}|\phi_n^{(1)}\rangle\right) \tag{43}$$

This means that the electronic and nuclear wave functions cannot be separated anymore, and therefore the adiabatic approximation cannot be applied beyond the second-order perturbation.

In the following, we shall demonstrate techniques for calculating the electronic potential energy terms up to the second order. For simplicity, we shall study the case of H_2 molecule, the simplest multi-electron diatomic molecule.

III. HYDROGEN MOLECULE: HAMILTONIAN

Consider a diatomic molecule as shown in Figure 1. The nuclear kinetic energy is expressed as

$$\hat{T}_N = -\frac{\hbar^2}{2}\left[\frac{1}{M_1}\left(\frac{\partial^2}{\partial X_1^2} + \frac{\partial^2}{\partial Y_1^2} + \frac{\partial^2}{\partial Z_1^2}\right) + \frac{1}{M_2}\left(\frac{\partial^2}{\partial X_2^2} + \frac{\partial^2}{\partial Y_2^2} + \frac{\partial^2}{\partial Z_2^2}\right)\right] \tag{44}$$

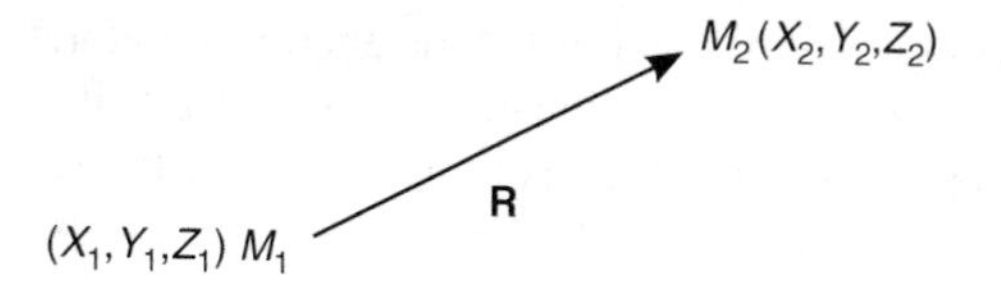

Figure 1. A model two-atom molecule.

Transferring into the center-of-mass coordinates, where

$$\mathbf{R}_0 = (X_0, Y_0, Z_0) = \left(\frac{M_1 X_1 + M_2 X_2}{M_1 + M_2}, \frac{M_1 Y_1 + M_2 Y_2}{M_1 + M_2}, \frac{M_1 Z_1 + M_2 Z_2}{M_1 + M_2} \right) \tag{45}$$

$$\mathbf{R} = (X, Y, Z) = (X_2 - X_1, Y_2 - Y_1, Z_2 - Z_1) \tag{46}$$

where $\mathbf{R}_0$ is the coordinate of the center of mass, one can rewrite the nuclear kinetic energy:

$$\begin{aligned} \hat{T}_N &= \frac{-\hbar^2}{2} \left[\left(\frac{1}{M_1 + M_2} \right) \nabla_0^2 + \frac{M_1 + M_2}{M_1 M_2} \nabla_{\mathbf{R}}^2 \right] \\ &= \frac{-\hbar^2}{2(M_1 + M_2)} \left[\nabla_0^2 + \frac{(M_1 + M_2)^2}{M_1 M_2} \nabla_{\mathbf{R}}^2 \right] \\ &= -\kappa^4 \frac{\hbar^2}{2m} \left\{ \nabla_0^2 + \frac{\mu}{R^2} \frac{\partial}{\partial R} \left(R^2 \frac{\partial}{\partial R} \right) + \frac{\mu}{R^2} \nabla_\Omega^2 \right\} \end{aligned} \tag{47}$$

Here, we have defined

$$\kappa \equiv \sqrt[4]{\frac{m}{M_1 + M_2}} \qquad (\approx 0.1285 \text{ for } H_2) \tag{48}$$

$$\mu \equiv \frac{(M_1 + M_2)^2}{M_1 M_2} \qquad (= 4 \text{ for } H_2) \tag{49}$$

$$\mathbf{R} \equiv (R, \mathbf{\Omega}) \equiv (R, \theta, \phi) \tag{50}$$

$$\nabla_0^2 = \left(\frac{\partial^2}{\partial X_0^2} + \frac{\partial^2}{\partial Y_0^2} + \frac{\partial^2}{\partial Z_0^2} \right) \tag{51}$$

$$\nabla_{\mathbf{\Omega}}^2 \equiv \frac{1}{\sin\theta} \frac{\partial}{\partial\theta} \left(\sin\theta \frac{\partial}{\partial\theta} \right) + \frac{1}{\sin^2\theta} \frac{\partial^2}{\partial\phi^2} \tag{52}$$

Although all of the nuclear coordinates participate in this kinetic energy operator, and in our previous discussions, all of the nuclear coordinates are expanded, with respect to an equivalent position, in power series of the parameter κ, here in the specific case of a diatomic molecule, we found that only the R coordinate seems to have an equilibrium position in the molecular fixed coordinates. This means that actually we only have to, or we can only, expand the R coordinate, but not the other coordinates, in the way that

$$R = R_0 + \kappa\varrho \tag{53}$$

By replacing $\partial/\partial R$ with $\partial/\kappa\partial\varrho$ we have

$$\hat{T}_N = -\kappa^2 \frac{\hbar^2}{2m}\frac{\mu}{\varrho^2}\frac{\partial}{\partial\varrho}\left(\varrho^2\frac{\partial}{\partial\varrho}\right) - \kappa^4\frac{\hbar^2}{2m}\left\{\nabla_0^2 + \frac{\mu}{R^2}\nabla_{\Omega}^2\right\} \tag{54}$$

In other words, for calculating the second-order energy (the vibrational energy), we only have to keep the term to do with the interatomic distance. The other terms, then, will enter the total Schrödinger equation in higher orders.

Now let us use a more specific coordinate system shown in Figure 2 for H_2 molecule. The z axis is taken to be along the internuclei vector $\mathbf{R}$. Adapting this coordinate-system, the consequence is that only the z coordinate of the nucleus are to be expanded in powers of κ. That is, we define

$$Z = R_0 + \kappa\varrho \tag{55}$$

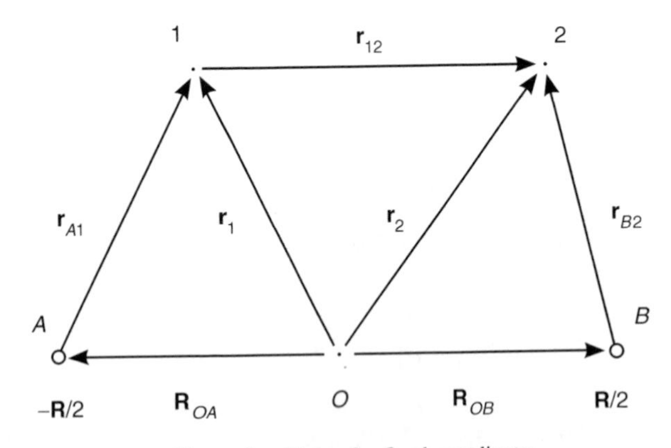

Figure 2. Molecular-fixed coordinates.

Due to our choice of the coordinates, Z is actually identical with R. Therefore, we rewrite the following definitions:

$$\hat{H}_1^{(2)} = -\frac{\hbar^2 \mu}{2m} \frac{1}{\varrho^2} \frac{\partial}{\partial \varrho} \varrho^2 \frac{\partial}{\partial \varrho} \tag{56}$$

$$\hat{H}_0^{(1)} = \left(\frac{\partial V}{\partial R}\right)_{R=R_0} \varrho \tag{57}$$

$$\hat{H}_0^{(2)} = \frac{1}{2}\left(\frac{\partial^2 V}{\partial R^2}\right)_{R=R_0} \varrho^2 \tag{58}$$

$$U_n^{(2)} = \left\{\frac{1}{2}\langle \phi_n^{(0)} | \left(\frac{\partial^2 V}{\partial R^2}\right)_{R=R_0} | \phi_n^{(0)} \rangle + \sum_{m \neq n} \frac{|\langle \phi_n^{(0)} | \left(\frac{\partial V}{\partial R}\right)_{R=R_0} | \phi_m^{(0)} \rangle|^2}{U_n^{(0)} - U_m^{(0)}}\right\} \varrho^2 \tag{59}$$

Thus, the equation we are going to solve for the zeroth-order nuclear motion is

$$\left\{-\frac{\hbar^2 \mu}{2m} \frac{1}{\varrho^2} \frac{\partial}{\partial \varrho} \varrho^2 \frac{\partial}{\partial \varrho} + \left[\frac{1}{2}\langle \phi_n^{(0)} | \left(\frac{\partial^2 V}{\partial Z^2}\right)_0 | \phi_n^{(0)} \rangle + \sum_{m \neq n} \frac{|\langle \phi_n^{(0)} | \left(\frac{\partial V}{\partial Z}\right)_0 | \phi_m^{(0)} \rangle|^2}{U_n^{(0)} - U_m^{(0)}}\right] \varrho^2 \right\} |\chi_n^{(0)}\rangle$$
$$= E_n^{(2)} |\chi_n^{(0)}\rangle \tag{60}$$

Equation (60) is a standard equation of motion of a harmonic oscillator. It can be easily solved as long as the term corresponding to the force constant can be evaluated. To do that with Eq. (66), we need to know how to calculate the matrix elements $\langle \phi_n^{(0)} | (\partial^2 V / \partial Z^2)_0 | \phi_n^{(0)} \rangle$ and $\langle \phi_n^{(0)} | (\partial V / \partial Z)_0 | \phi_m^{(0)} \rangle$ for the given molecular basis set $\{|\phi_n^{(0)}\rangle\}$.

In modern quantum chemistry packages, one can obtain molecular basis set at the optimized geometry, in which the wave functions of the molecular basis are expanded in terms of a set of orthogonal Gaussian basis set. Therefore, we need to derive efficient formulas for calculating the above-mentioned matrix elements, between Gaussian functions of the first and second derivatives of the Coulomb potential terms, especially the second derivative term that is not available in quantum chemistry packages. Section IV is devoted to the evaluation of these matrix elements.

In the work of King, Dupuis, and Rys [15,16], the matrix elements of the Coulomb interaction term in Gaussian basis set were evaluated by solving the differential equations satisfied by these matrix elements. Thus, the Coulomb matrix elements are expressed in the form of the Rys' polynomials. The potential problem of this method is that to obtain the matrix elements of the higher derivatives of Coulomb interactions, we need to solve more complicated differential equations numerically. Great effort has to be taken to ensure that the differential equation solver can solve such differential equations stably, and to make sure that the result is accurate. In this work, we have carried out the integrals explicitly up to the second-order derivatives of the Coulomb interaction. After the lengthy derivations, it should be clear that the results we obtain, in the form of a finite series (with number of terms <12 for the extreme case in which h orbitals are taken into the basis set) of error functions and exponential functions, are simple to calculate, and the numerical accuracy is high, thanks to the existing numerical libraries providing accurate computation of error function.

The starting point of our technique is to make use of the equality

$$\frac{1}{r} = \frac{1}{\sqrt{\pi}} \int_{-\infty}^{\infty} e^{-t^2 r^2} dt \tag{61}$$

By using the geometry defined in Figure 2, the coordinates of the electrons with respect to the nucleus will be written as

$$\frac{1}{r_{Ai}} = (x_i, y_i, z_i + R/2) = \frac{1}{\sqrt{\pi}} \int_{-\infty}^{\infty} e^{-t^2\left[x_i^2+y_i^2+(z_i+R/2)^2\right]} dt \tag{62}$$

$$\frac{1}{r_{Bi}} = (x_i, y_i, z_i - R/2) = \frac{1}{\sqrt{\pi}} \int_{-\infty}^{\infty} e^{-t^2\left[x_i^2+y_i^2+(z_i-R/2)^2\right]} dt \tag{63}$$

Thus, it can be seen that the first-derivatives w. r. t. R are

$$\frac{\partial}{\partial R} \frac{1}{r_{Ai}} = \frac{1}{\sqrt{\pi}} \int_{-\infty}^{\infty} -t^2 z_{Ai} e^{-t^2\left[x_i^2+y_i^2+(z_i+R/2)^2\right]} dt \tag{64}$$

$$\frac{\partial}{\partial R} \frac{1}{r_{Bi}} = \frac{1}{\sqrt{\pi}} \int_{-\infty}^{\infty} t^2 z_{Bi} e^{-t^2\left[x_i^2+y_i^2+(z_i-R/2)^2\right]} dt \tag{65}$$

and the second derivatives are

$$\frac{\partial^2}{\partial R^2} \frac{1}{r_{Ai}} = \frac{1}{\sqrt{\pi}} \int_{-\infty}^{\infty} \left(-\frac{t^2}{2} + t^4 z_{Ai}^2\right) e^{-t^2\left[x_i^2+y_i^2+(z_i+R/2)^2\right]} dt \tag{66}$$

$$\frac{\partial^2}{\partial R^2} \frac{1}{r_{Bi}} = \frac{1}{\sqrt{\pi}} \int_{-\infty}^{\infty} \left(-\frac{t^2}{2} + t^4 z_{Bi}^2\right) e^{-t^2\left[x_i^2+y_i^2+(z_i-R/2)^2\right]} dt \tag{67}$$

The matrix elements of these derivatives are to be evaluated with R equal to its equilibrium value R_0. However, to keep the notation simple, we shall still write R in place of R_0 in later text unless ambiguity may occur.

With the following discussions, it can also be seen that higher order derivatives can be evaluated with similar technique.

IV. MATRIX ELEMENTS OF ANGULAR-MOMENTUM-ADOPTED GAUSSIAN FUNCTIONS

In this work, we shall use the notation

$$|\gamma_\alpha(i)\rangle = \left(\frac{2\alpha}{\pi}\right)^{3/4} e^{-\alpha(\mathbf{r}_i-\mathbf{R}_\alpha)^2} \tag{68}$$

to represent a normalized Gaussian wave function. Here, α is the label to specify the nuclei, which will be either A or B in our cases, and i is the label to specify the electron 1 or 2. In the context in which the label of electron is immaterial, we shall drop the i index.

To incorporate the angular dependence of a basis function into Gaussian orbitals, either spherical harmonics or integer powers of the Cartesian coordinates have to be included. We shall discuss the latter case, in which a primitive basis function takes the form

$$|\eta\rangle = N_{n,l,m,\alpha}\, x^n y^l z^m e^{-\alpha r^2} \tag{69}$$

This type of basis functions is frequently used in popular quantum chemistry packages. We shall discuss the way to evaluate different kinds of matrix elements in this basis set that are often used in quantum chemistry calculation.

A. Normalization Factor

$$\begin{aligned}\langle\eta|\eta\rangle &= N_{n,l,m,\alpha}^2 \int d\tau\, x^{2n} y^{2l} z^{2m} e^{-2\alpha r^2} \\ &= N_{n,l,m,\alpha}^2 \int_{-\infty}^{\infty} dx\, x^{2n} e^{-2\alpha x^2} \int_{-\infty}^{\infty} dy\, y^{2l} e^{-2\alpha y^2} \int_{-\infty}^{\infty} dz\, z^{2m} e^{-2\alpha z^2} \\ &= N_{n,l,m,\alpha}^2 I_x I_y I_z\end{aligned} \tag{70}$$

By using the integral formula in Appendix A, we can obtain the proper expressions for I_x, I_y, and I_z, so that

$$\langle\eta|\eta\rangle = N_{n,l,m,\alpha}^2 \frac{(2n)!(2l)!(2m)!}{n!l!m!(8\alpha)^{(n+l+m)}} \left(\frac{\pi}{2\alpha}\right)^{3/2} \tag{71}$$

Therefore, the normalization factor is

$$N_{n,l,m,\alpha} = \left[\frac{n!l!m!(8\alpha)^{(n+l+m)}}{(2n)!(2l)!(2m)!} \left(\frac{2\alpha}{\pi}\right)^{3/2}\right]^{1/2} \tag{72}$$

B. The Overlap Integrals

Next, we consider the simple overlap integral of two such basis functions with different powers of Cartesian coordinates and different Gaussian width, centered at different points. Let nuclei 1 locate at the origin, and let nuclei 2 locate at $-\mathbf{R}$, then

$$|\eta_1\rangle = N_{n_1,l_1,m_1,\alpha_1} x^{n_1} y^{l_1} z^{m_1} e^{-\alpha_1 (x^2+y^2+z^2)} \tag{73}$$

$$|\eta_2\rangle = N_{n_2,l_2,m_2,\alpha_2} (x+X)^{n_2} (y+Y)^{l_2} (z+Z)^{m_2} e^{-\alpha_2 \left[(x+X)^2+(y+Y)^2+(z+Z)^2\right]} \tag{74}$$

It can be observed that we can separate the overlap integral into the product of three independent spatial integrals

$$\langle \eta_1 | \eta_2 \rangle = N_{n_1,l_1,m_1,\alpha_1} N_{n_2,l_2,m_2,\alpha_2} \times I_x I_y I_z \tag{75}$$

$$I_x = \int_{-\infty}^{\infty} dx\, x^{n_1} (x+X)^{n_2} e^{-\alpha_1 x^2} e^{-\alpha_2 (x+X)^2} \tag{76}$$

$$I_y = \int_{-\infty}^{\infty} dy\, y^{l_1} (y+Y)^{l_2} e^{-\alpha_1 y^2} e^{-\alpha_2 (y+Y)^2} \tag{77}$$

$$I_z = \int_{-\infty}^{\infty} dz\, z^{m_1} (z+Z)^{m_2} e^{-\alpha_1 z^2} e^{-\alpha_2 (z+Z)^2} \tag{78}$$

Obviously, we do not need to calculate all three integrals. We shall calculate I_x and apply the formula to the other integrals.

$$\begin{aligned} I_x &= \int_{-\infty}^{\infty} dx\, x^{n_1} (x+X)^{n_2} e^{-(\alpha_1+\alpha_2)x^2 - 2\alpha_2 X x - \alpha_2 X^2} \\ &= e^{-\alpha_2 X^2} \int_{-\infty}^{\infty} dx\, x^{n_1} (x+X)^{n_2} e^{-(\alpha_1+\alpha_2)[x+\alpha_2 X/(\alpha_1+\alpha_2)]^2 + \alpha_2^2 X^2/(\alpha_1+\alpha_2)} \\ &= e^{-\alpha_1\alpha_2 X^2/(\alpha_1+\alpha_2)} \int_{-\infty}^{\infty} dx\, x^{n_1} (x+X)^{n_2} e^{-(\alpha_1+\alpha_2)x'^2} \end{aligned} \tag{79}$$

where $x' = x + \alpha_2 X/(\alpha_1+\alpha_2)$. Therefore $x = x' - \alpha_2 X/(\alpha_1+\alpha_2)$, and $x + X = x' + \alpha_1 X/(\alpha_1+\alpha_2)$. Thus, by changing the dummy variables, we have

$$\begin{aligned} I_x &= e^{-\alpha_1\alpha_2 X^2/(\alpha_1+\alpha_2)} \int_{-\infty}^{\infty} dx \left(x - \frac{\alpha_2 X}{\alpha_1+\alpha_2}\right)^{n_1} \left(x + \frac{\alpha_1 X}{\alpha_1+\alpha_2}\right)^{n_2} e^{-(\alpha_1+\alpha_2)x^2} \\ &= e^{-\alpha_1\alpha_2 X^2/(\alpha_1+\alpha_2)} \sum_{n=0}^{n_1} \sum_{m=0}^{n_2} \frac{n_1!}{(n_1-n)!n!} \frac{n_2!}{(n_2-m)!m!} \left(\frac{-\alpha_2}{\alpha_1+\alpha_2}\right)^{n_1-n} \left(\frac{\alpha_1}{\alpha_1+\alpha_2}\right)^{n_2-m} \\ &\quad \times X^{n_1+n_2-n-m} \times \int_{-\infty}^{\infty} dx\, x^{n+m} e^{-(\alpha_1+\alpha_2)x^2} \end{aligned} \tag{80}$$

According to the results shown in Appendix A, the integral in Eq. (80) is nonvanishing only when $(n+m)$ is even. Consequently, I_x is a polynomial of order n_1+n_2 in X. If n_1+n_2 is even, I_x is an even polynomial; If n_1+n_2 is odd, I_x is an odd polynomial. Therefore, we can write the result as

$$I_x = \sqrt{\frac{\pi}{\alpha_1+\alpha_2}} e^{-\frac{\alpha_1\alpha_2}{\alpha_1+\alpha_2}X^2} X^{n_1+n_2} \\ \times \sum_{n=0}^{n_1}\sum_{m=0}^{n_2} \frac{n_1!n_2!(n+m)!}{(n_1-n)!n!(n_2-m)!m!\{n+m/2\}!} \frac{1+(-1)^{n+m}}{2} \\ \times [4(\alpha_1+\alpha_2)]^{-(n+m)/2} X^{-n-m} \tag{81}$$

I_y and I_z are obtained in exactly the same way.

C. Interaction Terms with the Nuclei

Next, we shall consider four kinds of integrals. The first is the expectation value of the Coulomb potential by one nucleus for one of the primitive basis function centered at that nucleus. The second is the expectation value of the Coulomb potential by one nucleus for one of the primitive basis function centered at a different point (usually another nucleus). Then, we will consider the matrix element of a Coulomb term between two primitive basis functions at different centers. The third case is when one basis function is centered at the nucleus considered. The fourth case is when both basis functions are not centered at that nucleus. By that we mean, for two Gaussian basis functions defined in Eqs. (73) and (74), we are calculating

$$\langle \eta_1|\frac{1}{r}|\eta_1\rangle = \frac{N_1^2}{\sqrt{\pi}}\int_{-\infty}^{\infty} dt \int d\tau\, x^{2n_1} y^{2l_1} z^{2m_1} e^{-2\alpha_1 r^2} e^{-t^2r^2}$$

$$\langle \eta_2|\frac{1}{r}|\eta_2\rangle = \frac{N_2^2}{\sqrt{\pi}}\int_{-\infty}^{\infty} dt \int d\tau\, (x+X)^{2n_1}(y+Y)^{2l_1}(z+Z)^{2m_1} e^{-2\alpha_2|\mathbf{r}+\mathbf{R}|^2} e^{-t^2r^2}$$

$$\langle \eta_1|\frac{1}{r}|\eta_2\rangle = \frac{N_1N_2}{\sqrt{\pi}}\int_{-\infty}^{\infty} dt \int d\tau\, x^{n_1}(x+X)^{n_2} y^{l_1}(y+Y)^{l_2} z^{m_1}(z+Z)^{m_2} \\ \times e^{-\alpha_1 r^2} e^{-\alpha_2|\mathbf{r}+\mathbf{R}|^2} e^{-t^2r^2}$$

$$\langle \eta_1|\frac{1}{|\mathbf{r}+\mathbf{R}'|}|\eta_2\rangle = \frac{N_1N_2}{\sqrt{\pi}}\int_{-\infty}^{\infty} dt \int d\tau\, x^{n_1}(x+X)^{n_2} y^{l_1}(y+Y)^{l_2} z^{m_1}(z+Z)^{m_2} \\ \times e^{-\alpha_1 r^2} e^{-\alpha_2|\mathbf{r}+\mathbf{R}|^2} e^{-t^2|\mathbf{r}+\mathbf{R}'|^2}$$

N_1 and N_2 are the normalization constants. Let us consider

$$
\begin{aligned}
I_1 &= \int_{\infty}^{\infty} dt \int d\tau\, x^{2n} y^{2l} z^{2m} e^{-2\alpha r^2} e^{-2t^2 r^2} \\
&= \int_{\infty}^{\infty} dt \int_{\infty}^{\infty} dx\, x^{2n} e^{-(2\alpha+t^2)x^2} \int_{\infty}^{\infty} dy\, y^{2l} e^{-(2\alpha+t^2)y^2} \int_{\infty}^{\infty} dz\, z^{2m} e^{-(2\alpha+t^2)z^2} \\
&= \pi^{3/2} \left[\frac{(2n)!}{n!4^n}\right] \left[\frac{(2l)!}{l!4^l}\right] \left[\frac{(2m)!}{m!4^m}\right] \int_{-\infty}^{\infty} dt \left(2\alpha + t^2\right)^{-(n+l+m+3/2)} \\
&= \pi^{3/2} \left[\frac{(2n)!}{n!4^n}\right] \left[\frac{(2l)!}{l!4^l}\right] \left[\frac{(2m)!}{m!4^m}\right] (2\alpha)^{-(n+l+m+3/2)} \int_{-\infty}^{\infty} dt \left(1 + \frac{t^2}{2\alpha}\right)^{-(n+l+m+3/2)} \\
&= \pi^{3/2} \left[\frac{(2n)!}{n!4^n}\right] \left[\frac{(2l)!}{l!4^l}\right] \left[\frac{(2m)!}{m!4^m}\right] (2\alpha)^{-(n+l+m+1)} \int_{-\infty}^{\infty} d\left(\frac{t}{\sqrt{2\alpha}}\right) \left(1 + \frac{t^2}{2\alpha}\right)^{-(n+l+m+3/2)} \\
&= \pi^{3/2} \left[\frac{(2n)!}{n!4^n}\right] \left[\frac{(2l)!}{l!4^l}\right] \left[\frac{(2m)!}{m!4^m}\right] (2\alpha)^{-(n+l+m+1)} \int_{-\infty}^{\infty} dt \left(1 + t^2\right)^{-(n+l+m+3/2)} \qquad (82)
\end{aligned}
$$

By making use of Eq. (A.12), we find

$$
\begin{aligned}
I_1 &= \pi^{3/2} \left[\frac{(2n)!}{n!4^n}\right] \left[\frac{(2l)!}{l!4^l}\right] \left[\frac{(2m)!}{m!4^m}\right] (2\alpha)^{-(n+l+m+1)} \frac{2[(n+l+m)!]^2 4^{n+l+m}}{(2n+2l+2m+1)!} \\
&= \sqrt{2\alpha} \frac{(2n)!(2l)!(2m)!}{n!l!m!(8\alpha)^{n+l+m}} \left(\frac{\pi}{2\alpha}\right)^{3/2} \frac{2[(n+l+m)!]^2 4^{n+l+m}}{(2n+2l+2m+1)!} \\
&= \sqrt{2\alpha} N_{n,l,m,\alpha}^{-2} \frac{2[(n+l+m)!]^2 4^{n+l+m}}{(2n+2l+2m+1)!} \qquad (83)
\end{aligned}
$$

Therefore

$$
\langle \eta_1 | \frac{1}{r} | \eta_1 \rangle = \frac{2[(n_1+l_1+m_1)!]^2 4^{n_1+l_1+m_1}}{(2n_1+2l_1+2m_1+1)!} \sqrt{\frac{2\alpha_1}{\pi}} \qquad (84)
$$

There are two kinds of integrals involving two different centers. The first case is

$$
\begin{aligned}
\langle \eta_2 | \frac{1}{r} | \eta_2 \rangle &= \frac{N_2^2}{\sqrt{\pi}} \int_{-\infty}^{\infty} dt \int d\tau (x+X)^{2n_2} (y+Y)^{2l_2} (z+Z)^{2m_2} e^{-2\alpha_2 |\mathbf{r}+\mathbf{R}|^2} e^{-t^2 r^2} \\
&= \frac{N_2^2}{\sqrt{\pi}} \int_{-\infty}^{\infty} dt \int_{-\infty}^{\infty} dx (x+X)^{2n_2} e^{-2\alpha_2 (x+X)^2} e^{-t^2 x^2} \\
&\quad \times \int_{-\infty}^{\infty} dy (y+Y)^{2l_2} e^{-2\alpha_2 (y+Y)^2} e^{-t^2 y^2} \int_{-\infty}^{\infty} dz (z+Z)^{2m_2} e^{-2\alpha_2 (z+Z)^2} e^{-t^2 z^2}
\end{aligned}
\qquad (85)
$$

Consider the integral

$$
\begin{aligned}
I_2 &= \int_{-\infty}^{\infty} dt \int_{-\infty}^{\infty} dx\, x^{2n} e^{-2\alpha x^2} e^{-t^2\left(x^2-2Xx+X^2\right)} \\
&\quad \times \int_{-\infty}^{\infty} dy\, y^{2l} e^{-2\alpha y^2} e^{-t^2\left(y^2-2Yy+Y^2\right)} \int_{-\infty}^{\infty} dz\, z^{2m} e^{-2\alpha z^2} e^{-t^2\left(z^2-2Zz+Z^2\right)} \\
&= \int_{-\infty}^{\infty} dt\, e^{-t^2R^2} \int_{-\infty}^{\infty} dx\, x^{2n} e^{-\left(2\alpha+t^2\right)x^2+2t^2Xx} \\
&\quad \times \int_{-\infty}^{\infty} dy\, y^{2l} e^{-\left(2\alpha+t^2\right)y^2+2t^2Yy} \int_{-\infty}^{\infty} dz\, z^{2m} e^{-\left(2\alpha+t^2\right)z^2+2t^2Zz} \\
&= \int_{-\infty}^{\infty} dt\, e^{-t^2R^2} \int_{-\infty}^{\infty} dx\, x^{2n} e^{-\left(2\alpha+t^2\right)\left[x-t^2X/(2\alpha+t^2)\right]^2} e^{t^4X^2/(2\alpha+t^2)} \\
&\quad \times \int_{-\infty}^{\infty} dy\, y^{2l} e^{-\left(2\alpha+t^2\right)\left[y-t^2Y/(2\alpha+t^2)\right]^2} e^{t^4Y^2/(2\alpha+t^2)} \\
&\quad \times \int_{-\infty}^{\infty} dz\, z^{2m} e^{-\left(2\alpha+t^2\right)\left[z-t^2Z/(2\alpha+t^2)\right]^2} e^{t^4Z^2/(2\alpha+t^2)} \\
&= \int_{-\infty}^{\infty} dt\, e^{-2\alpha t^2/(2\alpha+t^2)R^2} \int_{-\infty}^{\infty} dx \left(x+\frac{t^2X}{2\alpha+t^2}\right)^{2n} e^{-\left(2\alpha+t^2\right)x^2} \\
&\quad \times \int_{-\infty}^{\infty} dy \left(y+\frac{t^2Y}{2\alpha+t^2}\right)^{2l} e^{-\left(2\alpha+t^2\right)y^2} \int_{-\infty}^{\infty} dz \left(z+\frac{t^2Z}{2\alpha+t^2}\right)^{2m} e^{-\left(2\alpha+t^2\right)z^2} \\
&= \int_{-\infty}^{\infty} dt\, e^{-2\alpha t^2R^2/(2\alpha+t^2)} \left[\sum_{\nu=0}^{2n} \frac{(2n)!}{(2n-\nu)!\nu!} \left(\frac{t^2X}{2\alpha+t^2}\right)^{2n-\nu} \int_{-\infty}^{\infty} x^{\nu} e^{-\left(2\alpha+t^2\right)x^2} dx\right] \\
&\quad \times \left[\sum_{\lambda=0}^{2l} \frac{(2l)!}{(2l-\lambda)!\lambda!} \left(\frac{t^2Y}{2\alpha+t^2}\right)^{2l-\lambda} \int_{-\infty}^{\infty} y^{\lambda} e^{-\left(2\alpha+t^2\right)y^2} dy\right] \\
&\quad \times \left[\sum_{\mu=0}^{2m} \frac{(2m)!}{(2m-\mu)!\mu!} \left(\frac{t^2Z}{2\alpha+t^2}\right)^{2m-\mu} \int_{-\infty}^{\infty} z^{\mu} e^{-\left(2\alpha+t^2\right)z^2} dz\right] \\
&= \int_{-\infty}^{\infty} dt\, e^{-2\alpha t^2R^2/(2\alpha+t^2)} \left[\sum_{\nu=0}^{n} \frac{(2n)!}{(2n-2\nu)!(2\nu)!} \left(\frac{t^2X}{2\alpha+t^2}\right)^{2n-2\nu} \int_{-\infty}^{\infty} x^{2\nu} e^{-\left(2\alpha+t^2\right)x^2} dx\right] \\
&\quad \times \left[\sum_{\lambda=0}^{l} \frac{(2l)!}{(2l-2\lambda)!(2\lambda)!} \left(\frac{t^2Y}{2\alpha+t^2}\right)^{2l-2\lambda} \int_{-\infty}^{\infty} y^{2\lambda} e^{-\left(2\alpha+t^2\right)y^2} dy\right] \\
&\quad \times \left[\sum_{\mu=0}^{m} \frac{(2m)!}{(2m-2\mu)!(2\mu)!} \left(\frac{t^2Z}{2\alpha+t^2}\right)^{2m-2\mu} \int_{-\infty}^{\infty} z^{2\mu} e^{-\left(2\alpha+t^2\right)z^2} dz\right] \\
&= \int_{-\infty}^{\infty} dt\, e^{-2\alpha t^2R^2/(2\alpha+t^2)}
\end{aligned}
$$

$$\times\left[\sum_{\nu=0}^{n}\frac{(2n)!}{(2n-2\nu)!(2\nu)!}\left(\frac{t^2X}{2\alpha+t^2}\right)^{2n-2\nu}\frac{(2\nu)!}{4^\nu\nu!(2\alpha+t^2)^\nu}\sqrt{\frac{\pi}{2\alpha+t^2}}\right]$$
$$\times\left[\sum_{\lambda=0}^{l}\frac{(2l)!}{(2l-2\lambda)!(2\lambda)!}\left(\frac{t^2Y}{2\alpha+t^2}\right)^{2l-2\lambda}\frac{(2\lambda)!}{4^\lambda\lambda!(2\alpha+t^2)^\lambda}\sqrt{\frac{\pi}{2\alpha+t^2}}\right]$$
$$\times\left[\sum_{\mu=0}^{m}\frac{(2m)!}{(2m-2\mu)!(2\mu)!}\left(\frac{t^2Z}{2\alpha+t^2}\right)^{2m-2\mu}\frac{(2\mu)!}{4^\mu\mu!(2\alpha+t^2)^\mu}\sqrt{\frac{\pi}{2\alpha+t^2}}\right]$$
$$=2\sum_{\nu=0}^{n}\sum_{\lambda=0}^{l}\sum_{\mu=0}^{m}\frac{(2n)!(2l)!(2m)!\pi^{3/2}X^{2n-2\nu}Y^{2l-2\lambda}Z^{2m-2\mu}}{(2n-2\nu)!\nu!(2l-2\lambda)!\lambda!(2m-2\mu)!\mu!4^{\nu+\lambda+\mu}}$$
$$\times\int_0^\infty dt\, e^{-2\alpha t^2R^2/(2\alpha+t^2)}\frac{1}{(2\alpha+t^2)^{\nu+\lambda+\mu+3/2}}\left(\frac{t^2}{2\alpha+t^2}\right)^{2(n+l+m-\nu-\lambda-\mu)} \tag{86}$$

Now, let

$$\xi\equiv\frac{t}{\sqrt{2\alpha+t^2}} \tag{87}$$

Since

$$\frac{d\xi}{dt}=\frac{2\alpha}{(2\alpha+t^2)^{3/2}} \tag{88}$$

we find

$$dt=\frac{(2\alpha+t^2)^{3/2}}{2\alpha}d\xi \tag{89}$$

Also, noticing that when $t=0$, $\xi=0$, and $\xi=1$ when $t=\infty$, and

$$\frac{2\alpha}{2\alpha+t^2}=1-\xi^2 \tag{90}$$

we have

$$I_2=\frac{1}{\alpha}\sum_{\nu=0}^{n}\sum_{\lambda=0}^{l}\sum_{\mu=0}^{m}\frac{(2n)!(2l)!(2m)!\pi^{3/2}X^{2n-2\nu}Y^{2l-2\lambda}Z^{2m-2\mu}}{(2n-2\nu)!\nu!(2l-2\lambda)!\lambda!(2m-2\mu)!\mu!4^{\nu+\lambda+\mu}}$$
$$\times\int_0^1 d\xi\, e^{-2\alpha R^2\xi^2}\frac{(1-\xi^2)^{\nu+\lambda+\mu}}{(2\alpha)^{\nu+\lambda+\mu}}\xi^{2[2(n+l+m-\nu-\lambda-\mu)]}$$
$$=\frac{1}{\alpha}\sum_{\nu=0}^{n}\sum_{\lambda=0}^{l}\sum_{\mu=0}^{m}\frac{(2n)!(2l)!(2m)!\pi^{3/2}X^{2n-2\nu}Y^{2l-2\lambda}Z^{2m-2\mu}}{(2n-2\nu)!\nu!(2l-2\lambda)!\lambda!(2m-2\mu)!\mu!(8\alpha)^{\nu+\lambda+\mu}}$$
$$\times\sum_{k=0}^{\nu+\lambda+\mu}\frac{(\nu+\lambda+\mu)!}{(\nu+\lambda+\mu-k)!k!}(-1)^k\int_0^1 d\xi\,\xi^{2[2(n+l+m-\nu-\lambda-\mu)+k]}e^{-2\alpha R^2\xi^2} \tag{91}$$

According to Eq. (A.16),

$$\int_0^1 d\xi\, \xi^{2s} e^{-2\alpha R^2 \xi^2} = \frac{(2s)!}{s!(8\alpha R^2)^s}\left[\frac{1}{2}\sqrt{\frac{\pi}{2\alpha R^2}}\,\mathrm{erf}\left(\sqrt{2\alpha R^2}\right) - e^{-2\alpha R^2}\sum_{j=0}^{s-1}\frac{j!}{(2j+1)!}\left(8\alpha R^2\right)^j\right] \tag{92}$$

where $s = 2(n + l + m - \nu - \lambda - \mu) + k$. Also,

$$\begin{aligned} I_2 &= \frac{1}{\alpha}\sum_{\nu=0}^{n}\sum_{\lambda=0}^{l}\sum_{\mu=0}^{m}\frac{(2n)!(2l)!(2m)!\pi^{3/2}X^{2n-2\nu}Y^{2l-2\lambda}Z^{2m-2\mu}}{(2n-2\nu)!\nu!(2l-2\lambda)!\lambda!(2m-2\mu)!\mu!(8\alpha)^{\nu+\lambda+\mu}} \\ &\quad\times\sum_{k=0}^{\nu+\lambda+\mu}\frac{(\nu+\lambda+\mu)!}{(\nu+\lambda+\mu-k)!k!}(-1)^k\int_0^1 d\xi\,\xi^{2[2(n+l+m-\nu-\lambda-\mu)+k]}e^{-2\alpha R^2\xi^2} \\ &= (2\alpha)^{3/2}\frac{N_{n,l,m,\alpha}^{-2}}{\alpha}\sum_{\nu=0}^{n}\sum_{\lambda=0}^{l}\sum_{\mu=0}^{m}\frac{n!l!m!(8\alpha)^{n+l+m-\nu-\lambda-\mu}X^{2n-2\nu}Y^{2l-2\lambda}Z^{2m-2\mu}}{(2n-2\nu)!\nu!(2l-2\lambda)!\lambda!(2m-2\mu)!\mu!} \\ &\quad\times\sum_{k=0}^{\nu+\lambda+\mu}\frac{(\nu+\lambda+\mu)!}{(\nu+\lambda+\mu-k)!k!}(-1)^k\int_0^1 d\xi\,\xi^{2[2(n+l+m-\nu-\lambda-\mu)+k]}e^{-2\alpha R^2\xi^2} \end{aligned} \tag{93}$$

Therefore

$$\begin{aligned} \langle\eta_2|\frac{1}{r}|\eta_2\rangle &= \sqrt{\frac{8\alpha_2}{\pi}}\sum_{\nu=0}^{n_2}\sum_{\lambda=0}^{l_2}\sum_{\mu=0}^{m_2} \\ &\quad\times\frac{n_2!l_2!m_2!(8\alpha_2)^{n_2+l_2+m_2-\nu-\lambda-\mu}X^{2n_2-2\nu}Y^{2l_2-2\lambda}Z^{2m_2-2\mu}}{(2n_2-2\nu)!\nu!(2l_2-2\lambda)!\lambda!(2m_2-2\mu)!\mu!}\sum_{k=0}^{\nu+\lambda+\mu} \\ &\quad\times\frac{(\nu+\lambda+\mu)!}{(\nu+\lambda+\mu-k)!k!}(-1)^k\int_0^1 d\xi\,\xi^{2[2(n_2+l_2+m_2-\nu-\lambda-\mu)+k]}e^{-2\alpha_2R^2\xi^2} \end{aligned} \tag{94}$$

Next, we consider

$$\begin{aligned} &\langle\eta_1|\frac{1}{r}|\eta_2\rangle \\ &= \frac{N_1N_2}{\sqrt{\pi}}\int_{-\infty}^{\infty}dt\int d\tau\, x^{n_1}(x+X)^{n_2}y^{l_1}(y+Y)^{l_2}z^{m_1}(z+Z)^{m_2}e^{-\alpha_1r^2-\alpha_2|\mathbf{r}+\mathbf{R}|^2-t^2r^2} \end{aligned}$$

$$
\begin{aligned}
&= \frac{N_1 N_2}{\sqrt{\pi}} \int_{-\infty}^{\infty} dt \int_{-\infty}^{\infty} dx\, x^{n_1} (x+X)^{n_2} e^{-\left(\alpha_1+\alpha_2+t^2\right)x^2 - 2\alpha_2 X x - \alpha_2 X^2} \\
&\quad \times \int_{-\infty}^{\infty} dy\, y^{l_1} (y+Y)^{l_2} e^{-\left(\alpha_1+\alpha_2+t^2\right)y^2 - 2\alpha_2 Y y - \alpha_2 Y^2} \\
&\quad \times \int_{-\infty}^{\infty} dz\, z^{m_1} (z+Z)^{m_2} e^{-\left(\alpha_1+\alpha_2+t^2\right)z^2 - 2\alpha_2 Z z - \alpha_2 Z^2} \\
&= \frac{N_1 N_2}{\sqrt{\pi}} \int_{-\infty}^{\infty} dt\, e^{-\alpha_2 R^2} \int_{-\infty}^{\infty} dx\, x^{n_1} (x+X)^{n_2} e^{-\left(\alpha_1+\alpha_2+t^2\right)\left[x+\alpha_2 X/(\alpha_1+\alpha_2+t^2)\right]^2 + \alpha_2^2 X^2/(\alpha_1+\alpha_2+t^2)} \\
&\quad \times \int_{-\infty}^{\infty} dy\, y^{l_1} (y+Y)^{l_2} e^{-\left(\alpha_1+\alpha_2+t^2\right)\left[y+\alpha_2 Y/(\alpha_1+\alpha_2+t^2)\right]^2 + \alpha_2^2 Y^2/(\alpha_1+\alpha_2+t^2)} \\
&\quad \times \int_{-\infty}^{\infty} dz\, z^{m_1} (z+Z)^{m_2} e^{-\left(\alpha_1+\alpha_2+t^2\right)\left[z+\alpha_2 Z/(\alpha_1+\alpha_2+t^2)\right]^2 + \alpha_2^2 Z^2/(\alpha_1+\alpha_2+t^2)} \\
&= \frac{N_1 N_2}{\sqrt{\pi}} \int_{-\infty}^{\infty} dt\, e^{-\alpha_2\left(\alpha_1+t^2\right)R^2/(\alpha_1+\alpha_2+t^2)} \int_{-\infty}^{\infty} dx\, x^{n_1} (x+X)^{n_2} e^{-\left(\alpha_1+\alpha_2+t^2\right)\left[x+\alpha_2 X/(\alpha_1+\alpha_2+t^2)\right]^2} \\
&\quad \times \int_{-\infty}^{\infty} dy\, y^{l_1} (y+Y)^{l_2} e^{-\left(\alpha_1+\alpha_2+t^2\right)\left[y+\alpha_2 Y/(\alpha_1+\alpha_2+t^2)\right]^2} \\
&\quad \times \int_{-\infty}^{\infty} dz\, z^{m_1} (z+Z)^{m_2} e^{-\left(\alpha_1+\alpha_2+t^2\right)\left[z+\alpha_2 Z/(\alpha_1+\alpha_2+t^2)\right]^2} \\
&= \frac{N_1 N_2}{\sqrt{\pi}} \int_{-\infty}^{\infty} dt\, e^{-\left[\alpha_2\left(\alpha_1+t^2\right)/(\alpha_1+\alpha_2+t^2)\right]R^2} \\
&\quad \times \int_{-\infty}^{\infty} dx \left(x - \frac{\alpha_2 X}{\alpha_1+\alpha_2+t^2}\right)^{n_1} \left(x + \frac{(\alpha_1+t^2)X}{\alpha_1+\alpha_2+t^2}\right)^{n_2} e^{-\left(\alpha_1+\alpha_2+t^2\right)x^2} \\
&\quad \times \int_{-\infty}^{\infty} dy \left(y - \frac{\alpha_2 Y}{\alpha_1+\alpha_2+t^2}\right)^{l_1} \left(y + \frac{(\alpha_1+t^2)Y}{\alpha_1+\alpha_2+t^2}\right)^{l_2} e^{-\left(\alpha_1+\alpha_2+t^2\right)y^2} \\
&\quad \times \int_{-\infty}^{\infty} dz \left(z - \frac{\alpha_2 Z}{\alpha_1+\alpha_2+t^2}\right)^{m_1} \left(z + \frac{(\alpha_1+t^2)Z}{\alpha_1+\alpha_2+t^2}\right)^{m_2} e^{-\left(\alpha_1+\alpha_2+t^2\right)z^2}
\end{aligned} \tag{95}
$$

The three integrals of the Cartesian coordinates have the same form. Take the integral with respect to x as, for example,

$$
\begin{aligned}
&\int_{-\infty}^{\infty} dx \left(x - \frac{\alpha_2 X}{\alpha_1+\alpha_2+t^2}\right)^{n_1} \left(x + \frac{(\alpha_1+t^2)X}{\alpha_1+\alpha_2+t^2}\right)^{n_2} e^{-\left(\alpha_1+\alpha_2+t^2\right)x^2} \\
&= \sum_{\nu_1=0}^{n_1} \sum_{\nu_2=0}^{n_2} \frac{n_1! n_2!}{\nu_1!(n_1-\nu_1)!\nu_2!(n_2-\nu_2)!} \left(\frac{-\alpha_2 X}{\alpha_1+\alpha_2+t^2}\right)^{n_1-\nu_1} \left(\frac{(\alpha_1+t^2)X}{\alpha_1+\alpha_2+t^2}\right)^{n_2-\nu_2} \\
&\quad \times \int_{-\infty}^{\infty} dx\, x^{\nu_1+\nu_2} e^{-\left(\alpha_1+\alpha_2+t^2\right)x^2}
\end{aligned} \tag{96}
$$

According to Eqs. (A.2) and (A.6), this integral is not zero only if $\nu_1 + \nu_2$ is an even integer. That is,

$$\int_{-\infty}^{\infty} dx \left(x - \frac{\alpha_2 X}{\alpha_1 + \alpha_2 + t^2} \right)^{n_1} \left(x + \frac{(\alpha_1 + t^2)X}{\alpha_1 + \alpha_2 + t^2} \right)^{n_2} e^{-(\alpha_1+\alpha_2+t^2)x^2}$$
$$= \sum_{\nu_1=0}^{n_1} \sum_{\nu_2=0}^{n_2} \frac{n_1! n_2! X^{n_1+n_2-(\nu_1+\nu_2)}}{\nu_1!(n_1-\nu_1)!\nu_2!(n_2-\nu_2)!} \left(\frac{-\alpha_2}{\alpha_1+\alpha_2+t^2} \right)^{n_1-\nu_1} \left(\frac{\alpha_1+t^2}{\alpha_1+\alpha_2+t^2} \right)^{n_2-\nu_2}$$
$$\times \left(\frac{1+(-1)^{\nu_1+\nu_2}}{2} \right) \frac{(\nu_1+\nu_2)!\sqrt{\pi}}{2^{\nu_1+\nu_2}[(\nu_1+\nu_2)/2]!(\alpha_1+\alpha_2+t^2)^{(\nu_1+\nu_2+1)/2}} \tag{97}$$

Manipulating all three integrals in similar way, we have

$$\langle \eta_1 | \frac{1}{r} | \eta_2 \rangle$$
$$= \pi N_1 N_2 \sum_{\nu_1=0}^{n_1} \sum_{\nu_2=0}^{n_2} \sum_{\lambda_1=0}^{l_1} \sum_{\lambda_2=0}^{l_2} \sum_{\mu_1=0}^{m_1} \sum_{\mu_2=0}^{m_2} \left(\frac{1+(-1)^{\nu_1+\nu_2}}{2} \right)$$
$$\times \left(\frac{1+(-1)^{\lambda_1+\lambda_2}}{2} \right) \left(\frac{1+(-1)^{\mu_1+\mu_2}}{2} \right)$$
$$\times C_{\nu_1}^{n_1} C_{\nu_2}^{n_2} C_{\lambda_1}^{l_1} C_{\lambda_2}^{l_2} C_{\mu_1}^{m_1} C_{\mu_2}^{m_2} \frac{(\nu_1+\nu_2)!(\lambda_1+\lambda_2)!(\mu_1+\mu_2)!(-\alpha_2)^{n_1+l_1+m_1-\nu_1-\lambda_1-\mu_1}}{2^{\nu_1+\nu_2+\lambda_1+\lambda_2+\mu_1+\mu_2} \left(\frac{\nu_1+\nu_2}{2}\right)! \left(\frac{\lambda_1+\lambda_2}{2}\right)! \left(\frac{\mu_1+\mu_2}{2}\right)!}$$
$$\times X^{n_1+n_2-\nu_1-\nu_2} Y^{l_1+l_2-\lambda_1-\lambda_2} Z^{m_1+m_2-\mu_1-\mu_2} \int_{-\infty}^{\infty} dt\, e^{[-\alpha_2(\alpha_1+t^2)/(\alpha_1+\alpha_2+t^2)]R^2}$$
$$\times (\alpha_1+t^2)^{n_2+l_2+m_2-\nu_2-\lambda_2-\mu_2}$$
$$\times (\alpha_1+\alpha_2+t^2)^{-[n_1+n_2+l_1+l_2+m_1+m_2-(\nu_1+\nu_2+\lambda_1+\lambda_2+\mu_1+\mu_2)/2+3/2]} \tag{98}$$

where $C_\nu^n = n!/\{(n-\nu)!\nu!\}$. The remaining integral of t is

$$I = \int_{-\infty}^{\infty} dt \frac{[(\alpha_1+t^2)/(\alpha_1+\alpha_2+t^2)]^{n_2+l_2+m_2-\nu_2-\lambda_2-\mu_2}}{(\alpha_1+\alpha_2+t^2)^{n_1+l_1+m_1-(\nu_1-\nu_2+\lambda_1-\lambda_2+\mu_1-\mu_2)/2+3/2}} e^{[-\alpha_2(\alpha_1+t^2)/(\alpha_1+\alpha_2+t^2)]R^2}$$
$$= 2\int_{0}^{\infty} dt \frac{[(\alpha_1+t^2)/(\alpha_1+\alpha_2+t^2)]^{n_2+l_2+m_2-\nu_2-\lambda_2-\mu_2}}{(\alpha_1+\alpha_2+t^2)^{n_1+l_1+m_1-(\nu_1-\nu_2+\lambda_1-\lambda_2+\mu_1-\mu_2)/2+3/2}} e^{[-\alpha_2(\alpha_1+t^2)/(\alpha_1+\alpha_2+t^2)]R^2} \tag{99}$$

Introducing the new variable ξ

$$\xi = \frac{t}{\sqrt{\alpha_1+\alpha_2+t^2}} \tag{100}$$

By considering the limits of integration, we find that when $t = 0$, $\xi = 0$, and when $t = \infty$, $\xi = 1$. Also,

$$\frac{d\xi}{dt} = \frac{\alpha_1 + \alpha_2}{(\alpha_1 + \alpha_2 + t^2)^{3/2}} \tag{101}$$

$$dt = \frac{(\alpha_1 + \alpha_2 + t^2)^{3/2}}{\alpha_1 + \alpha_2} d\xi \tag{102}$$

$$\frac{1}{\alpha_1 + \alpha_2 + t^2} = \frac{1 - \xi^2}{\alpha_1 + \alpha_2} \tag{103}$$

$$\frac{\alpha_1 + t^2}{\alpha_1 + \alpha_2 + t^2} = \frac{\alpha_1 + \alpha_2 \xi^2}{\alpha_1 + \alpha_2} \tag{104}$$

Therefore

$$I = \frac{2}{\alpha_1 + \alpha_2} \int_0^1 d\xi \, e^{-\alpha_2(\alpha_1+\alpha_2)\xi^2 R^2/(\alpha_1+\alpha_2)} \left(\frac{\alpha_1 + \alpha_2 \xi^2}{\alpha_1 + \alpha_2} \right)^{n_2+l_2+m_2-\nu_2-\lambda_2-\mu_2} \times \left(\frac{1 - \xi^2}{\alpha_1 + \alpha_2} \right)^{n_1+l_1+m_1-(\nu_1-\nu_2+\lambda_1-\lambda_2+\mu_1-\mu_2)/2} \tag{105}$$

Now, let $K_1 = n_1 + l_1 + m_1 - (\nu_1 - \nu_2 + \lambda_1 - \lambda_2 + \mu_1 - \mu_2)/2$, $K_2 = n_2 + l_2 + m_2 - \nu_2 - \lambda_2 - \mu_2$. Notice that K_2 is an integer, and $K_1 + K_2 = n_1 + n_2 + l_1 + l_2 + m_1 + m_2 - (\nu_1 + \nu_2 + \lambda_1 + \lambda_2 + \mu_1 + \mu_2)/2$ is also an integer, therefore K_1 must be an integer. Thus,

$$\begin{aligned} I &= \frac{2e^{-\alpha_1\alpha_2 R^2/(\alpha_1+\alpha_2)}}{(\alpha_1 + \alpha_2)^{K_1+K_2+1}} \int_0^1 d\xi \, e^{-\alpha_2^2 R^2/\alpha_1+\alpha_2\xi^2} (1 - \xi^2)^{K_1} (\alpha_1 + \alpha_2 \xi^2)^{K_2} \\ &= \frac{2e^{-\alpha_1\alpha_2 R^2/(\alpha_1+\alpha_2)}}{(\alpha_1 + \alpha_2)^{K_1+K_2+1}} \sum_{k_1=0}^{K_1} \sum_{k_2=0}^{K_2} C_{k_1}^{K_1} C_{k_2}^{K_2} (-1)^{k_1} \alpha_1^{K_2-k_2} \alpha_2^{k_2} \int_0^1 d\xi \, \xi^{2(k_1+k_2)} e^{-\alpha_2^2 R^2 \xi^2/(\alpha_1+\alpha_2)} \end{aligned} \tag{106}$$

Again, making use of Eq. (A.16), we obtain

$$\begin{aligned} \int_0^1 d\xi \, \xi^{2(k_1+k_2)} e^{-\alpha_2^2 R^2 \xi^2/(\alpha_1+\alpha_2)} &= \frac{(2k_1 + 2k_2)!}{(k_1 + k_2)! \{4\alpha_2^2 R^2/(\alpha_1 + \alpha_2)\}^{k_1+k_2}} \\ &\times \left\{ \frac{1}{2} \frac{\sqrt{(\alpha_1 + \alpha_2)\pi}}{\alpha_2 R} \operatorname{erf}\left(\frac{\alpha_2 R}{\sqrt{\alpha_1 + \alpha_2}} \right) - e^{-\alpha_2^2 R^2/(\alpha_1+\alpha_2)} \right. \\ &\times \left. \sum_{k=0}^{k_1+k_2-1} \frac{k!}{(2k + 1)!} \left(\frac{4\alpha_2^2 R^2}{\alpha_1 + \alpha_2} \right)^k \right\} \end{aligned} \tag{107}$$

By using the notation defined in Eq. (A.13), we find

$$I = \frac{2e^{-\alpha_1\alpha_2 R^2/(\alpha_1+\alpha_2)}}{(\alpha_1+\alpha_2)^{K_1+K_2+1}} \sum_{k_1=0}^{K_1}\sum_{k_2=0}^{K_2} C_{k_1}^{K_1} C_{k_2}^{K_2} (-1)^{k_1} \alpha_1^{K_2-k_2} \alpha_2^{k_2} J_{k_1+k_2}\left(\frac{\alpha_2^2 R^2}{\alpha_1+\alpha_2}\right) \quad (108)$$

Further, defining this integral as

$$K_{K_1,K_2}(R;\alpha_1,\alpha_2) \equiv \int_{-\infty}^{\infty} dt \frac{(\alpha_1+t^2)^{K_2}}{(\alpha_1+\alpha_2+t^2)^{K_1+K_2}} e^{-[\alpha_2(\alpha_1+t^2)/(\alpha_1+\alpha_2+t^2)]R^2} \quad (109)$$

we find

$$\begin{aligned}
&\langle \eta_1 | \frac{1}{r} | \eta_2 \rangle \\
&= \pi N_1 N_2 \sum_{\nu_1=0}^{n_1}\sum_{\nu_2=0}^{n_2}\sum_{\lambda_1=0}^{l_1}\sum_{\lambda_2=0}^{l_2}\sum_{\mu_1=0}^{m_1}\sum_{\mu_2=0}^{m_2} \left(\frac{1+(-1)^{\nu_1+\nu_2}}{2}\right)\left(\frac{1+(-1)^{\lambda_1+\lambda_2}}{2}\right) \\
&\quad \times \left(\frac{1+(-1)^{\mu_1+\mu_2}}{2}\right) C_{\nu_1}^{n_1} C_{\nu_2}^{n_2} C_{\lambda_1}^{l_1} C_{\lambda_2}^{l_2} C_{\mu_1}^{m_1} C_{\mu_2}^{m_2} \\
&\quad \times \frac{(\nu_1+\nu_2)!(\lambda_1+\lambda_2)!(\mu_1+\mu_2)!(-\alpha_2)^{n_1+l_1+m_1-\nu_1-\lambda_1-\mu_1}}{2^{\nu_1+\nu_2+\lambda_1+\lambda_2+\mu_1+\mu_2}\left(\frac{\nu_1+\nu_2}{2}\right)!\left(\frac{\lambda_1+\lambda_2}{2}\right)!\left(\frac{\mu_1+\mu_2}{2}\right)!} \\
&\quad \times X^{n_1+n_2-\nu_1-\nu_2} Y^{l_1+l_2-\lambda_1-\lambda_2} Z^{m_1+m_2-\mu_1-\mu_2} K_{K_1,K_2}(R;\alpha_1,\alpha_2) \qquad (110)
\end{aligned}$$

with K_1 and K_2 defined as above.

In the case of hydrogen molecule, the term $\langle \eta_1 | 1/|\mathbf{r}+\mathbf{R}'|^2 | \eta_2 \rangle$, which involves three centers, does not show up in the calculation. We will not discuss this integral in the present work.

D. Derivatives of the Coulomb Potential

In fact, the Coulomb integrals discussed in Section IV.C are available in contemporary quantum chemistry packages. We do not really need to develop our own method to calculate them. However, it is necessary to master the algebra so that we can calculate the matrix elements of the derivatives of the Coulomb potential. In the following, we shall demonstrate the evaluation of these matrix elements.

Since the derivative is taken with respect to the nuclear coordinate, it is important to choose the convenient coordinates. Earlier, we assigned the origin on one of the nuclei. Now, we will assign the origin on the middle point of the two nuclei. The geometry is shown in Figure 3. Furthermore, the z axis is taken to be along $\mathbf{R}$. That is, the coordinates of the position of the nuclei A is

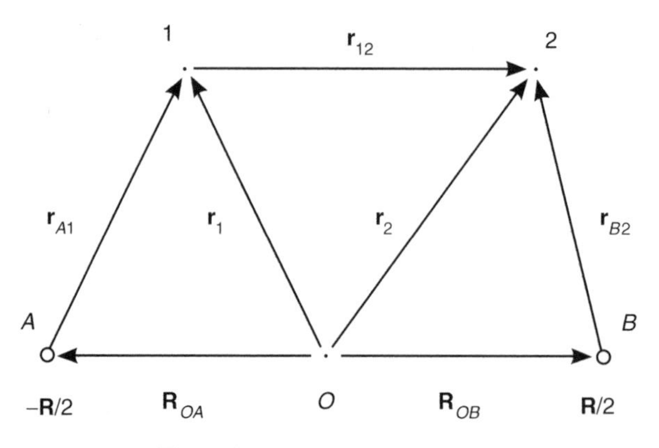

Figure 3. Molecular-fixed coordinates.

$\mathbf{R}_A = (0, 0, -R/2)$, and similarly $\mathbf{R}_B = (0, 0, R/2)$. Later, to calculate the Coulomb interaction terms, we will be dealing with the following terms:

$$\frac{1}{r_{A1}} = \frac{1}{\sqrt{\pi}} \int_{-\infty}^{\infty} dt\, e^{-t^2 r_{A1}^2} = \frac{1}{\sqrt{\pi}} \int_{-\infty}^{\infty} dt\, e^{-t^2 x_1^2 - t^2 y_1^2 - t^2 (z_1 + R/2)^2} \tag{111}$$

$$\frac{1}{r_{B1}} = \frac{1}{\sqrt{\pi}} \int_{-\infty}^{\infty} dt\, e^{-t^2 r_{B1}^2} = \frac{1}{\sqrt{\pi}} \int_{-\infty}^{\infty} dt\, e^{-t^2 x_1^2 - t^2 y_1^2 - t^2 (z_1 - R/2)^2} \tag{112}$$

The quantities $1/r_{A2}$ and $1/r_{B2}$ are defined similarly. In such cases, the first derivatives are

$$\left(\frac{\partial}{\partial R} \frac{1}{r_{Ai}} \right)_0 = \frac{1}{\sqrt{\pi}} \int_{-\infty}^{\infty} dt (-t^2)(z_i + R_0/2) e^{-t^2 r_{Ai}^2} \tag{113}$$

$$\left(\frac{\partial}{\partial R} \frac{1}{r_{Bi}} \right)_0 = \frac{1}{\sqrt{\pi}} \int_{-\infty}^{\infty} dt\, t^2 (z_i - R_0/2) e^{-t^2 r_{Bi}^2} \tag{114}$$

The second derivatives are

$$\left(\frac{\partial^2}{\partial R^2} \frac{1}{r_{Ai}} \right)_0 = \frac{1}{\sqrt{\pi}} \int_{-\infty}^{\infty} dt \left[-\frac{t^2}{2} + t^4 \left(z_i + \frac{R_0}{2} \right)^2 \right] e^{-t^2 r_{Ai}^2} \tag{115}$$

$$\left(\frac{\partial^2}{\partial R^2} \frac{1}{r_{Bi}} \right)_0 = \frac{1}{\sqrt{\pi}} \int_{-\infty}^{\infty} dt \left[-\frac{t^2}{2} + t^4 \left(z_i - \frac{R_0}{2} \right)^2 \right] e^{-t^2 r_{Bi}^2} \tag{116}$$

Also, the notations of the wave functions are to be changed. We shall denote the Gaussian function centered at nucleus A as $|\eta_A\rangle$, and the function centered at nucleus B as $|\eta_B\rangle$.

1. First-Order Derivatives

$$
\begin{aligned}
&\langle \eta_A | \left(\frac{\partial}{\partial R} \frac{1}{r_{A1}} \right)_0 | \eta_A \rangle \\
&= \frac{N_A^2}{\sqrt{\pi}} \int_{-\infty}^{\infty} dt(-t^2) \int d\tau_1 \, x_1^{2n} y_1^{2l} (z_1 + R_0/2)^{2m+1} e^{\left(-2\alpha_1 - t^2\right) r_{A1}^2} \\
&= \frac{N_A^2}{\sqrt{\pi}} \int_{-\infty}^{\infty} dt(-t^2) \int_{-\infty}^{\infty} dx_1 \, x_1^{2n} e^{\left(-2\alpha_1 - t^2\right) x_1^2} \int_{-\infty}^{\infty} dy_1 \, y_1^{2l} e^{\left(-2\alpha_1 - t^2\right) y_1^2} \\
&\times \int_{-\infty}^{\infty} dz_1 \, z_{A1}^{2m+1} e^{\left(-2\alpha_1 - t^2\right) z_{A1}^2}
\end{aligned} \tag{117}
$$

Obviously, this matrix element is zero due to the integral over z. Similarly, we know that

$$
\langle \eta_B | \left(\frac{\partial}{\partial R} \frac{1}{r_{B1}} \right)_0 | \eta_B \rangle = 0 \tag{118}
$$

$$
\langle \eta_A | \left(\frac{\partial}{\partial R} \frac{1}{r_{A2}} \right)_0 | \eta_A \rangle = 0 \tag{119}
$$

$$
\langle \eta_B | \left(\frac{\partial}{\partial R} \frac{1}{r_{B2}} \right)_0 | \eta_B \rangle = 0 \tag{120}
$$

Then, we consider

$$
\langle \eta_A | \left(\frac{\partial}{\partial R} \frac{1}{r_{B1}} \right)_0 | \eta_A \rangle = \frac{N_A^2}{\sqrt{\pi}} \int_{-\infty}^{\infty} dt(-t^2) \int d\tau_1 \, x_1^{2n} y_1^{2l} z_{A1}^{2m} z_{B1} e^{-2\alpha_A r_{A1}^2} e^{-t^2 r_{B1}^2} \tag{121}
$$

By introducing a new variable for the z coordinate $z = z_{A1}$, since $z_{B1} = z - R$, we find that

$$
\begin{aligned}
&\langle \eta_A | \left(\frac{\partial}{\partial R} \frac{1}{r_{B1}} \right)_0 | \eta_A \rangle \\
&= \frac{N_A^2}{\sqrt{\pi}} \int_{-\infty}^{\infty} dt(-t^2) \int_{-\infty}^{\infty} dx_1 \, x_1^{2n} e^{-\left(2\alpha_A + t^2\right) x_1^2} \int_{-\infty}^{\infty} dy_1 \, y_1^{2l} e^{-\left(2\alpha_A + t^2\right) y_1^2} \\
&\times \int_{-\infty}^{\infty} dz \, z^{2m} (z - R) e^{-\left(2\alpha_A + t^2\right) z^2 + 2t^2 R z - t^2 R^2}
\end{aligned} \tag{122}
$$

According to Eqs. (A.2) and (A.6), the integral over x_1 and y_1 can be easily carried out, but the integral over z has to be manipulated.

$$
\begin{aligned}
I_z &= \int_{-\infty}^{\infty} dz\, z^{2m}(z-R)e^{-\left(2\alpha_A+t^2\right)z^2+2t^2Rz-t^2R^2} \\
&= e^{-2\alpha_A t^2R^2/\left(2\alpha_A+t^2\right)} \int_{-\infty}^{\infty} dz\, z^{2m}(z-R)e^{-\left(2\alpha_A+t^2\right)\left[z-t^2R/\left(2\alpha_A+t^2\right)\right]^2} \\
&= e^{-2\alpha_A t^2R^2/\left(2\alpha_A+t^2\right)} \int_{-\infty}^{\infty} dz \left(z+\frac{t^2R}{2\alpha_A+t^2}\right)^{2m}\left(z-\frac{2\alpha_A R}{2\alpha_A+t^2}\right)e^{-\left(2\alpha_A+t^2\right)z^2}
\end{aligned}
\tag{123}
$$

We shall expand the polynomial of z. But recalling that only terms of the even power of z do not vanish, we can write the expansion in the following form:

$$
\begin{aligned}
I_z &= e^{-2\alpha_A t^2R^2/\left(2\alpha_A+t^2\right)}\left\{\frac{-2\alpha_A R}{2\alpha_A+t^2}\sum_{\mu=0}^{m} C_{2\mu}^{2m}\left(\frac{t^2R}{2\alpha_A+t^2}\right)^{2m-2\mu}\int_{-\infty}^{\infty} dz\, z^{2\mu}e^{-\left(2\alpha_A+t^2\right)z^2}\right. \\
&\quad \left. +\sum_{\mu=1}^{m} C_{2\mu-1}^{2m}\left(\frac{t^2R}{2\alpha_A+t^2}\right)^{2m-2\mu+1}\int_{-\infty}^{\infty} dz\, z^{2\mu}e^{-\left(2\alpha_A+t^2\right)z^2}\right\} \\
&= e^{-2\alpha_A t^2R^2/\left(2\alpha_A+t^2\right)}\left\{\frac{-2\alpha_A R}{2\alpha_A+t^2}\left(\frac{t^2R}{2\alpha_A+t^2}\right)^{2m}\sqrt{\frac{\pi}{2\alpha_A+t^2}}\right. \\
&\quad +\sum_{\mu=1}^{m} C_{2\mu}^{2m}\left(\frac{t^2R}{2\alpha_A+t^2}\right)^{2m-2\mu}\left(\frac{2\mu}{2m-2\mu+1}\frac{t^2R}{2\alpha_A+t^2}-\frac{2\alpha_A R}{2\alpha_A+t^2}\right) \\
&\quad \left.\times\int_{-\infty}^{\infty} dz\, z^{2\mu}e^{-\left(2\alpha_A+t^2\right)z^2}\right\}
\end{aligned}
\tag{124}
$$

By inserting the expression for the integral over z, we find

$$
\begin{aligned}
I_z &= \sqrt{\frac{\pi}{2\alpha_A+t^2}}\,e^{-2\alpha_A t^2R^2/\left(2\alpha_A+t^2\right)}\left\{-\frac{2\alpha_R}{2\alpha_A+t^2}\left(\frac{t^2R}{2\alpha_A+t^2}\right)^{2m}\right. \\
&\quad +\sum_{\mu=1}^{m} C_{2\mu}^{2m}\left(\frac{t^2R}{2\alpha_A+t^2}\right)^{2m-2\mu}\left(\frac{2\mu}{2m-2\mu+1}\frac{t^2R}{2\alpha_A+t^2}-\frac{2\alpha_A R}{2\alpha_A+t^2}\right)\frac{(2\mu)!}{2^{2\mu}\mu!} \\
&\quad \left.\times\left(\frac{1}{2\alpha_A+t^2}\right)^{\mu}\right\} \\
&= \sqrt{\frac{\pi}{2\alpha_A+t^2}}\,e^{-2\alpha_A t^2R^2/\left(2\alpha_A+t^2\right)}\left\{\sum_{\mu=0}^{m} C_{2\mu}^{2m}\frac{(2\mu)!}{2^{2\mu}\mu!}\right. \\
&\quad \left.\times\left(\frac{2\mu}{2m-2\mu+1}\frac{t^2R}{2\alpha_A+t^2}-\frac{2\alpha_A R}{2\alpha_A+t^2}\right)\left(\frac{t^2R}{2\alpha_A+t^2}\right)^{2m-2\mu}\left(\frac{1}{2\alpha_A+t^2}\right)^{\mu}\right\}
\end{aligned}
\tag{125}
$$

Thus, we obtain

$$\langle \eta_A | \left(\frac{\partial}{\partial R} \frac{1}{r_{B1}} \right)_0 | \eta_A \rangle$$
$$= \pi N_A^2 \sum_{\mu=0}^{m} C_{2\mu}^{2m} \frac{(2n)!(2l)!(2\mu)!}{4^{n+l+\mu} n! l! \mu!} \int_{-\infty}^{\infty} dt \left(\frac{1}{2\alpha_A + t^2} \right)^{n+l+\mu-1+3/2} \left(\frac{t^2 R}{2\alpha_A + t^2} \right)^{2m-2\mu+1}$$
$$\times \left(\frac{2\alpha_A}{2\alpha_A + t^2} - \frac{2\mu}{2m - 2\mu + 1} \frac{t^2}{2\alpha_A + t^2} \right) e^{-2\alpha_A t^2 R^2/(2\alpha_A + t^2)} \tag{126}$$

By introducing the new variable

$$\xi \equiv \frac{t}{\sqrt{2\alpha_A + t^2}} \tag{127}$$

we have

$$\frac{t^2}{2\alpha_A + t^2} = \xi^2 \tag{128}$$

$$\frac{1}{2\alpha_A + t^2} = \frac{1 - \xi^2}{2\alpha_A} \tag{129}$$

$$dt = \left(2\alpha_A + t^2\right)^{3/2} d\xi \tag{130}$$

Therefore,

$$\langle \eta_A | \left(\frac{\partial}{\partial R} \frac{1}{r_{B1}} \right)_0 | \eta_A \rangle$$
$$= 2\pi N_A^2 \sum_{\mu=0}^{m} C_{2\mu}^{2m} \frac{(2n)!(2l)!(2\mu)!}{4^{n+l+\mu} n! l! \mu! - 1} \left(\frac{1}{2\alpha_A} \right)^{n+l+\mu-1} R^{2m-2\mu+1}$$
$$\times \int_0^1 d\xi \left(1 - \xi^2\right)^{n+l+\mu-1} \left(\xi^2\right)^{2m-2\mu+1} \left(1 - \frac{2m+1}{2m - 2\mu + 1} \xi^2 \right) e^{-2\alpha_A R^2 \xi^2}$$
$$= 2\pi N_A^2 \sum_{\mu=0}^{m} C_{2\mu}^{2m} \frac{(2n)!(2l)!(2\mu)!}{4^{n+l+\mu} n! l! \mu!} \left(\frac{1}{2\alpha_A} \right)^{n+l+2m-\mu+1} R^{2m-2\mu+1}$$
$$\times \left[2\alpha_A \int_0^1 d\xi \left(1 - \xi^2\right)^{n+l+\mu-1} \left(2\alpha_A \xi^2\right)^{2m-2\mu+1} e^{-2\alpha_A R^2 \xi^2} \right.$$
$$\left. - \frac{2m+1}{2m - 2\mu + 1} \int_0^1 d\xi \left(1 - \xi^2\right)^{n+l+\mu-1} \left(2\alpha_A \xi^2\right)^{2m-2\mu+2} e^{-2\alpha_A R^2 \xi^2} \right] \tag{131}$$

According to Eq. (106), let us define the integral

$$
\begin{aligned}
X_{K_1,K_2,\alpha_1,\alpha_2}(R) &\equiv \int_0^1 d\xi \left(1-\xi^2\right)^{K_1}\left(\alpha_1+\alpha_2\xi^2\right)^{K_2} e^{-\alpha_2^2R^2\xi^2/(\alpha_1+\alpha_2)} \\
&= \sum_{k_1=0}^{K_1}\sum_{k_2=0}^{K_2} C_{k_1}^{K_1} C_{k_2}^{K_2}(-1)^{k_1}\alpha_1^{K_2-k_2}\alpha_2^{k_2}\int_0^1 d\xi\, \xi^{2(k_1+k_2)} e^{-\alpha_2^2R^2\xi^2/(\alpha_1+\alpha_2)} \\
&= \sum_{k_1=0}^{K_1}\sum_{k_2=0}^{K_2} C_{k_1}^{K_1} C_{k_2}^{K_2}(-1)^{k_1}\alpha_1^{K_2-k_2}\alpha_2^{k_2} J_{k_1+k_2}\left(\frac{\alpha_2^2}{\alpha_1+\alpha_2}R^2\right) \qquad (132)
\end{aligned}
$$

With Eq. (131), letting $\alpha_1 = 0$, $\alpha_2 = 2\alpha_A$, we find that

$$
\begin{aligned}
&\langle\eta_A|\left(\frac{\partial}{\partial R}\frac{1}{r_{B1}}\right)_0|\eta_A\rangle \\
&= -2\pi N_A^2 \sum_{\mu=0}^{m} C_{2\mu}^{2m}\frac{(2n)!(2l)!(2\mu)!}{4^{n+l+\mu}n!l!\mu!-1}\left(\frac{1}{2\alpha_A}\right)^{n+l+2m-\mu+1} R^{2m-2\mu+1} \\
&\times\left[2\alpha_A X_{n+l+\mu-1,2m-2\mu+1,0,2\alpha_A}(R) - \frac{2m+1}{2m-2\mu+1}X_{n+l+\mu-1,2m-2\mu+2,0,2\alpha_A}(R)\right] \qquad (133)
\end{aligned}
$$

The matrix element between Gaussian functions at different centers is in general of the form

$$
\begin{aligned}
&\langle\eta_A|\left(\frac{\partial}{\partial R}\frac{1}{r_{A1}}\right)_0|\eta_B\rangle \\
&= \frac{N_AN_B}{\sqrt{\pi}}\int_{-\infty}^{\infty} dt(-t^2)\int d\tau\, x_{A1}^{n_A}x_{B1}^{n_B}y_{A1}^{l_A}y_{B1}^{l_B}z_{A1}^{m_A+1}z_{B1}^{m_B}e^{-\alpha_Ar_{A1}^2-t^2r_{A1}^2-\alpha_Br_{B1}^2} \qquad (134)
\end{aligned}
$$

Since we can use the relations $x_{A1} = x_{B1} = x_1 \equiv x$, $y_{A1} = y_{B1} = y_1 \equiv y$, $z_{A1} = z_{B1} + R$, and let $z \equiv z_{B1}$, we have

$$
\begin{aligned}
&\langle\eta_A|\left(\frac{\partial}{\partial R}\frac{1}{r_{A1}}\right)_0|\eta_B\rangle \\
&= \frac{N_AN_B}{\sqrt{\pi}}\int_{-\infty}^{\infty} dt(-t^2)\int_{-\infty}^{\infty} dx\, x^{n_A+n_B}e^{-\left(\alpha_A+\alpha_B+t^2\right)x^2}\int_{-\infty}^{\infty} dy\, y^{l_A+l_B}e^{-\left(\alpha_A+\alpha_B+t^2\right)y^2} \\
&\times\int_{-\infty}^{\infty} dz\,(z+R)^{m_A+1}z^{m_B}e^{-\left(\alpha_A+\alpha_B+t^2\right)z^2-2\left(\alpha_A+t^2\right)Rz-\left(\alpha_A+t^2\right)R^2}
\end{aligned}
$$

$$
\begin{aligned}
&= \frac{N_A N_B}{\sqrt{\pi}} \int_{-\infty}^{\infty} dt(-t^2) e^{-(\alpha_A + t^2)R^2} \int_{-\infty}^{\infty} dx\, x^{n_A+n_B} e^{-(\alpha_A+\alpha_B+t^2)x^2} \\
&\quad \times \int_{-\infty}^{\infty} dy\, y^{l_A+l_B} e^{-(\alpha_A+\alpha_B+t^2)y^2} \\
&\quad \times \int_{-\infty}^{\infty} dz\, (z+R)^{m_A+1} z^{m_B} e^{-(\alpha_A+\alpha_B+t^2)z^2 - 2(\alpha_A+t^2)Rz} \\
&= \frac{N_A N_B}{\sqrt{\pi}} \left(\frac{1+(-1)^{n_A+n_B}}{2}\right)\left(\frac{1+(-1)^{l_A+l_B}}{2}\right) \frac{(n_A+n_B)!(l_A+l_B)!}{2^{n_A+n_B+l_A+l_B}\left(\frac{n_A+n_B}{2}\right)!\left(\frac{l_A+l_B}{2}\right)!} \\
&\quad \times \int_{-\infty}^{\infty} dt(-t^2) e^{-(\alpha_A+t^2)R^2} \left(\frac{\pi}{\alpha_A+\alpha_B+t^2}\right)\left(\frac{1}{\alpha_A+\alpha_B+t^2}\right)^{(n_A+n_B+l_A+l_B)/2} \\
&\quad \times \int_{-\infty}^{\infty} dz (z+R)^{m_A+1} z^{m_B} e^{-(\alpha_A+\alpha_B+t^2)\left[z+(\alpha_A+t^2)/(\alpha_A+\alpha_B+t^2)\right]^2 + (\alpha_A+t^2)^2 R^2/(\alpha_A+\alpha_B+t^2)} \\
&= \frac{N_A N_B}{\sqrt{\pi}} \left(\frac{1+(-1)^{n_A+n_B}}{2}\right)\left(\frac{1+(-1)^{l_A+l_B}}{2}\right) \frac{(n_A+n_B)!(l_A+l_B)!}{2^{n_A+n_B+l_A+l_B}\left(\frac{n_A+n_B}{2}\right)!\left(\frac{l_A+l_B}{2}\right)!} \\
&\quad \times \int_{-\infty}^{\infty} dt(-t^2) e^{-\alpha_B(\alpha_A+t^2)R^2/(\alpha_A+\alpha_B+t^2)} \left(\frac{\pi}{\alpha_A+\alpha_B+t^2}\right)\left(\frac{1}{\alpha_A+\alpha_B+t^2}\right)^{(n_A+n_B+l_A+l_B)/2} \\
&\quad \times \int_{-\infty}^{\infty} dz \left(z + \frac{\alpha_B}{\alpha_A+\alpha_B+t^2} R\right)^{m_A+1} \left(z - \frac{\alpha_A+t^2}{\alpha_A+\alpha_B+t^2} R\right)^{m_B} e^{-(\alpha_A+\alpha_B+t^2)z^2}
\end{aligned} \tag{135}
$$

Note that

$$
\begin{aligned}
&\int_{-\infty}^{\infty} dz \left(z + \frac{\alpha_B}{\alpha_A+\alpha_B+t^2} R\right)^{m_A+1} \left(z - \frac{\alpha_A+t^2}{\alpha_A+\alpha_B+t^2} R\right)^{m_B} e^{-(\alpha_A+\alpha_B+t^2)z^2} \\
&= \sum_{\mu_A=0}^{m_B+1} \sum_{\mu_B=0}^{m_B} C_{\mu_A}^{m_A+1} C_{\mu_B}^{m_B} \left(\frac{\alpha_B R}{\alpha_A+\alpha_B+t^2}\right)^{m_A-\mu_A+1} \left[\frac{-(\alpha_A+t^2)R}{\alpha_A+\alpha_B+t^2}\right]^{m_B-\mu_B} \\
&\quad \times \int_{-\infty}^{\infty} dz\, z^{\mu_A+\mu_B} e^{-(\alpha_A+\alpha_B+t^2)z^2} \\
&= \sum_{\mu_A=0}^{m_A+1} \sum_{\mu_B=0}^{m_B} C_{\mu_A}^{m_A+1} C_{\mu_B}^{m_B} \left(\frac{\alpha_B R}{\alpha_A+\alpha_B+t^2}\right)^{m_A-\mu_A+1} \left[\frac{-(\alpha_A+t^2)R}{\alpha_A+\alpha_B+t^2}\right]^{m_B-\mu_B} \\
&\quad \times \left(\frac{1+(-1)^{\mu_A+\mu_B}}{2}\right) \sqrt{\frac{\pi}{\alpha_A+\alpha_B+t^2}} \frac{(\mu_A+\mu_B)!}{2^{\mu_A+\mu_B}(\mu_A+\mu_B/2)!} \left(\frac{1}{\alpha_A+\alpha_B+t^2}\right)^{(\mu_A+\mu_B)/2}
\end{aligned} \tag{136}
$$

Therefore

$$
\begin{aligned}
&\langle \eta_A | \left(\frac{\partial}{\partial R} \frac{1}{r_{A1}} \right)_0 | \eta_B \rangle \\
&\quad = -2\pi N_A N_B \sum_{\mu_A=0}^{m_A+1} \sum_{\mu_B=0}^{m_B} \left(\frac{1+(-1)^{n_A+n_B}}{2} \right) \left(\frac{1+(-1)^{l_A+l_B}}{2} \right) \left(\frac{1+(-1)^{\mu_A+\mu_B}}{2} \right) \\
&\quad \times C_{\mu_A}^{m_A+1} C_{\mu_B}^{m_B} \frac{(n_A+n_B)!(l_A+l_B)!(\mu_A+\mu_B)!}{2^{n_A+n_B+l_A+l_B+\mu_A+\mu_B} \left(\frac{n_A+n_B}{2}\right)! \left(\frac{l_A+l_B}{2}\right)! \left(\frac{\mu_A+\mu_B}{2}\right)!} R^{m_A+m_B-\mu_A-\mu_B+1} \\
&\quad \times \int_0^\infty \frac{dt}{(\alpha_A+\alpha_B+t^2)^{3/2}} \alpha_B^{m_A-\mu_A+1} \frac{t^2}{\alpha_A+\alpha_B+t^2} \left(\frac{\alpha_B}{\alpha_A+\alpha_B+t^2} - 1 \right)^{m_B-\mu_B} \\
&\quad \times \left(\frac{1}{\alpha_A+\alpha_B+t^2} \right)^{(n_A+n_B+l_A+l_B+2m_A-\mu_A+\mu_B)/2} e^{-\alpha_B\left(\alpha_A+t^2\right)R^2/(\alpha_A+\alpha_B+t^2)} \qquad (137)
\end{aligned}
$$

By introducing the new variable $\xi = t/\sqrt{\alpha_A+\alpha_B+t^2}$ as usual, we have

$$
\begin{aligned}
&\langle \eta_A | \left(\frac{\partial}{\partial R} \frac{1}{r_{A1}} \right)_0 | \eta_B \rangle \\
&= -2\pi N_A N_B \sum_{\mu_A=0}^{m_A+1} \sum_{\mu_B=0}^{m_B} \left(\frac{1+(-1)^{n_A+n_B}}{2} \right) \left(\frac{1+(-1)^{l_A+l_B}}{2} \right) \left(\frac{1+(-1)^{\mu_A+\mu_B}}{2} \right) \\
&\quad \times C_{\mu_A}^{m_A+1} C_{\mu_B}^{m_B} \frac{(n_A+n_B)!(l_A+l_B)!(\mu_A+\mu_B)!}{2^{n_A+n_B+l_A+l_B+\mu_A+\mu_B} \left(\frac{n_A+n_B}{2}\right)! \left(\frac{l_A+l_B}{2}\right)! \left(\frac{\mu_A+\mu_B}{2}\right)!} R^{m_A+m_B-\mu_A-\mu_B+1} \\
&\quad \times \alpha_B^{m_A-\mu_A+1} (-1)^{m_B-\mu_B} \left(\frac{1}{\alpha_A+\alpha_B} \right)^{(n_A+n_B+l_A+l_B+2m_A+2m_B-\mu_A-\mu_B)/2} e^{-\alpha_A\alpha_B R^2/(\alpha_A+\alpha_B)} \\
&\quad \times \int_0^1 d\xi\, \xi^2 \left(1-\xi^2\right)^{(n_A+n_B+l_A+l_B+2m_A-\mu_A+\mu_B)/2} \left(\alpha_A+\alpha_B\xi^2\right)^{m_B-\mu_B} e^{-\alpha_B^2 R^2 \xi^2/(\alpha_A+\alpha_B)}
\end{aligned}
\qquad (138)
$$

The integral over ξ in Eq. (138) is discussed in the Appendix. From Eq. (A.18), we can find the expression for this integral. Inserting it into Eq. (138), we have

$$
\begin{aligned}
&\langle \eta_A | \left(\frac{\partial}{\partial R} \frac{1}{r_{A1}} \right)_0 | \eta_B \rangle \\
&\quad = -2\pi N_A N_B \sum_{\mu_A=0}^{m_A+1} \sum_{\mu_B=0}^{m_B} \left(\frac{1+(-1)^{n_A+n_B}}{2} \right) \left(\frac{1+(-1)^{l_A+l_B}}{2} \right) \left(\frac{1+(-1)^{\mu_A+\mu_B}}{2} \right)
\end{aligned}
$$

$$\times C_{\mu_A}^{m_A+1} C_{\mu_B}^{m_B} \frac{(n_A+n_B)!(l_A+l_B)!(\mu_A+\mu_B)!}{2^{n_A+n_B+l_A+l_B+\mu_A+\mu_B}\left(\frac{n_A+n_B}{2}\right)!\left(\frac{l_A+l_B}{2}\right)!\left(\frac{\mu_A+\mu_B}{2}\right)!} R^{m_A+m_B-\mu_A-\mu_B+1}$$

$$\times \alpha_B^{m_A-\mu_A+1}(-1)^{m_B-\mu_B}\left(\frac{1}{\alpha_A+\alpha_B}\right)^{(n_A+n_B+l_A+l_B+2m_A+2m_B-\mu_A-\mu_B)/2} e^{-\alpha_A\alpha_B R^2/(\alpha_A+\alpha_B)}$$

$$\times I_\xi\left(1, \frac{n_A+n_B+l_A+l_B+2m_A-\mu_A+\mu_B}{2}, m_B-\mu_B, \alpha_A, \alpha_B, R\right) \tag{139}$$

2. *Second-Order Derivatives*

To calculate the matrix elements of second-order derivatives, we have

$$\begin{aligned}&\langle \eta_A | \left(\frac{\partial^2}{\partial R^2}\frac{1}{r_{A1}}\right)_0 | \eta_A \rangle \\ &= \frac{N_A^2}{\sqrt{\pi}} \int_{-\infty}^{\infty} dt \int d\tau \left(t^4 z_{A1}^2 - \frac{t^2}{2}\right) x_{A1}^{2n} y_{A1}^{2l} z_{A1}^{2m} e^{-\left(2\alpha_A+t^2\right) r_{A1}^2}\end{aligned} \tag{140}$$

Neglecting the subscripts of the coordinates for simplicity, one obtains

$$\begin{aligned}&\langle \eta_A | \left(\frac{\partial^2}{\partial R^2}\frac{1}{r_{A1}}\right)_0 | \eta_A \rangle \\
&= \frac{N_A^2}{\sqrt{\pi}}\left[\int_{-\infty}^{\infty} dt\, t^4 \int_{-\infty}^{\infty} dx\, x^{2n} e^{-\left(2\alpha_A+t^2\right)x^2} \int_{-\infty}^{\infty} dy\, y^{2l} e^{-\left(2\alpha_A+t^2\right)y^2} \int_{-\infty}^{\infty} dz\, z^{2m+2} e^{-\left(2\alpha_A+t^2\right)z^2}\right. \\
&\quad \left. - \int_{-\infty}^{\infty} dt \frac{t^2}{2} \int_{-\infty}^{\infty} dx\, x^{2n} e^{-\left(2\alpha_A+t^2\right)x^2} \int_{-\infty}^{\infty} dy\, y^{2l} e^{-\left(2\alpha_A+t^2\right)y^2} \int_{-\infty}^{\infty} dz\, z^{2m} e^{-\left(2\alpha_A+t^2\right)z^2}\right] \\
&= \frac{N_A^2}{\sqrt{\pi}}\left[\frac{(2n)!(2l)!(2m+2)!}{2^{2n+2l+2m+2} n!l!(m+1)!}\int_{-\infty}^{\infty} dt\, t^4 \left(\frac{\pi}{2\alpha_A+t^2}\right)^{3/2}\left(\frac{1}{2\alpha_A+t^2}\right)^{n+l+m+1}\right. \\
&\quad \left. - \frac{(2n)!(2l)!(2m)}{2^{2n+2l+2m} n!l!m!}\int_{-\infty}^{\infty} dt \frac{t^2}{2}\left(\frac{\pi}{2\alpha_A+t^2}\right)^{3/2}\left(\frac{1}{2\alpha_A+t^2}\right)^{n+l+m}\right] \\
&= \pi N_A^2 \frac{(2n)!(2l)!(2m)!}{2^{2n+2l+2m} n!l!m!}\left[(2m+1)\int_0^{\infty} \frac{dt}{(2\alpha_A+t^2)^{3/2}}\, t^4 \left(\frac{1}{2\alpha_A+t^2}\right)^{n+l+m+1}\right. \\
&\quad \left. - \int_0^{\infty} \frac{dt}{(2\alpha_A+t^2)^{3/2}}\, t^2 \left(\frac{1}{2\alpha_A+t^2}\right)^{n+l+m}\right] \\
&= \pi N_A^2 \frac{(2n)!(2l)!(2m)!}{2^{2n+2l+2m} n!l!m!}\left[2m\int_0^{\infty} \frac{dt}{(2\alpha_A+t^2)^{3/2}}\left(\frac{t^2}{2\alpha_A+t^2}\right)^2\left(\frac{1}{2\alpha_A+t^2}\right)^{n+l+m-1}\right. \\
&\quad \left. - \int_0^{\infty} dt \frac{2\alpha_A}{(2\alpha_A+t^2)^{3/2}} \frac{t^2}{2\alpha_A+t^2}\left(\frac{1}{2\alpha_A+t^2}\right)^{n+l+m}\right]\end{aligned} \tag{141}$$

It has to be noted that, when reduced to the pure Gaussian case, that is, when $n+l+m=0$, m must always be 0, and

$$\langle\eta_A|\left(\frac{\partial^2}{\partial R^2}\frac{1}{r_{A1}}\right)_0|\eta_A\rangle = -\pi N_A^2 \int_0^\infty dt \frac{2\alpha_A}{(2\alpha_A+t^2)^{3/2}} \frac{t^2}{2\alpha_A+t^2} \tag{142}$$

By introducing a new variable

$$\xi = \frac{t}{\sqrt{2\alpha_A+t^2}} \tag{143}$$

$$d\xi = \frac{2\alpha_A dt}{(2\alpha_A+t^2)^{3/2}} \tag{144}$$

we find that

$$\begin{aligned}\langle\eta_A|\left(\frac{\partial^2}{\partial R^2}\frac{1}{r_{A1}}\right)_0|\eta_A\rangle &= -\pi N_A^2 \int_0^1 \xi^2\, d\xi \\ &= -\frac{\pi N_A^2}{3}\end{aligned} \tag{145}$$

In the general cases where $n+l+m>0$, it is guaranteed that $n+l+m-1\geq 0$. Therefore, by using the same new variable, we find

$$\begin{aligned}&\langle\eta_A|\left(\frac{\partial^2}{\partial R^2}\frac{1}{r_{A1}}\right)_0|\eta_A\rangle \\ &= \pi N_A^2 \frac{(2n)!(2l)!(2m)!}{(8\alpha_A)^{n+l+m} n!l!m!}\left[2m\int_0^1 d\xi\, \xi^4 (1-\xi^2)^{n+l+m-1} - \int_0^1 d\xi\, \xi^2 (1-\xi^2)^{n+l+m}\right]\end{aligned} \tag{146}$$

where we have made use of the fact that

$$1-\xi^2 = \frac{2\alpha_A}{2\alpha_A+t^2} \tag{147}$$

By noticing that

$$\begin{aligned}\int_0^1 d\xi\, \xi^{2n}(1-\xi^2)^m &= \sum_{\mu=0}^{m} C_\mu^m (-1)^\mu \int_0^1 d\xi\, \xi^{2(n+\mu)} \\ &= \sum_{\mu=0}^{m} C_\mu^m \frac{(-1)^\mu}{2(n+\mu)+1}\end{aligned} \tag{148}$$

we find that

$$\langle \eta_A | \left(\frac{\partial^2}{\partial R^2} \frac{1}{r_{A1}} \right)_0 | \eta_A \rangle$$

$$= \pi N_A^2 \frac{(2n)!(2l)!(2m)!}{2^{2n+2l+2m} n! l! m!} \left[2m \sum_{\mu=0}^{n+l+m-1} C_\mu^{n+l+m-1} \frac{(-1)^\mu}{2(\mu+2)+1} - \sum_{\mu=0}^{n+l+m} C_\mu^{n+l+m} \frac{(-1)^\mu}{2(\mu+1)+1} \right] \tag{149}$$

In the second case,

$$\langle \eta_A | \left(\frac{\partial^2}{\partial R^2} \frac{1}{r_{B1}} \right)_0 | \eta_A \rangle = \frac{N_A^2}{\sqrt{\pi}} \int_{-\infty}^{\infty} dt \int d\tau x_{A1}^{2n} y_{A1}^{2l} z_{A1}^{2m} \left(t^4 z_{B1}^2 - \frac{t^2}{2} \right) e^{-2\alpha_A r_{A1}^2 - t^2 r_{B1}^2} \tag{150}$$

By using the relations: $x_{B1} = x_{A1}$, $y_{B1} = y_{A1}$, and $z_{B1} = z_{A1} - R$, we shall simplify Eq. (150) by letting $x = x_{A1}$, $y = y_{A1}$, and $z = z_{A1}$. Thus

$$\langle \eta_A | \left(\frac{\partial^2}{\partial R^2} \frac{1}{r_{B1}} \right)_0 | \eta_A \rangle$$

$$= \frac{N_A^2}{\sqrt{\pi}} \int_{-\infty}^{\infty} dt \int_{-\infty}^{\infty} dx\, x^{2n} e^{-(2\alpha_A + t^2) x^2} \int_{-\infty}^{\infty} dy\, y^{2l} e^{-(2\alpha_A + t^2) y^2} \times \int_{-\infty}^{\infty} dz\, z^{2m} \left[t^4 (z-R)^2 - \frac{t^2}{2} \right] e^{-(2\alpha_A + t^2) z^2 + 2t^2 R z - t^2 R^2}$$

$$= \frac{N_A^2}{\sqrt{\pi}} \frac{(2n)!(2l)!}{2^{2n+2l} n! l!} \int_{-\infty}^{\infty} dt \left(\frac{\pi}{2\alpha_A + t^2} \right) \left(\frac{1}{2\alpha_A + t^2} \right)^{n+l} e^{-2\alpha_A t^2 R^2 / (2\alpha_A + t^2)} \times \int_{-\infty}^{\infty} dz\, z^{2m} \left[t^4 (z-R)^2 - \frac{t^2}{2} \right] e^{-(2\alpha_A + t^2) [z - t^2 R / (2\alpha_A + t^2)]^2} \tag{151}$$

By making a change of the variable $z' = z - t^2 R / (2\alpha_A + t^2)$ and by ignoring the prime sign in the dummy index results in

$$\langle \eta_A | \left(\frac{\partial^2}{\partial R^2} \frac{1}{r_{B1}} \right)_0 | \eta_A \rangle$$

$$= \frac{N_A^2}{\sqrt{\pi}} \frac{(2n)!(2l)!}{2^{2n+2l} n! l!} \int_{-\infty}^{\infty} dt \left(\frac{\pi}{2\alpha_A + t^2} \right) \left(\frac{1}{2\alpha_A + t^2} \right)^{n+l} e^{-2\alpha_A t^2 R^2 / (2\alpha_A + t^2)} \times \int_{-\infty}^{\infty} dz \left(z + \frac{t^2 R}{2\alpha_A + t^2} \right)^{2m} \left[t^4 \left(z - \frac{2\alpha_A R}{2\alpha_A + t^2} \right)^2 - \frac{t^2}{2} \right] e^{-(2\alpha_A + t^2) z^2} \tag{152}$$

Note that

$$
\begin{aligned}
I_z &\equiv \int_{-\infty}^{\infty} dz \left(z + \frac{t^2 R}{2\alpha_A + t^2} \right)^{2m} \left[t^4 \left(z - \frac{2\alpha_A R}{2\alpha_A + t^2} \right)^2 - \frac{t^2}{2} \right] e^{-\left(2\alpha_A + t^2\right) z^2} \\
&= \int_{-\infty}^{\infty} dz \left(z + \frac{t^2 R}{2\alpha_A + t^2} \right)^{2m} \left[t^4 z^2 - t^4 \frac{4\alpha_A R}{2\alpha_A + t^2} z + t^4 \left(\frac{2\alpha_A R}{2\alpha_A + t^2} \right)^2 - \frac{t^2}{2} \right] e^{-\left(2\alpha_A + t^2\right) z^2} \\
&= t^4 \sum_{\mu=0}^{m} C_{2\mu}^{2m} \left(\frac{t^2 R}{2\alpha_A + t^2} \right)^{2m-2\mu} \int_{-\infty}^{\infty} dz\, z^{2(\mu+1)} e^{-\left(2\alpha_A + t^2\right) z^2} \\
&\quad - t^4 \frac{4\alpha_A R}{2\alpha_A + t^2} \sum_{\mu=0}^{m-1} C_{2\mu+1}^{2m} \left(\frac{t^2 R}{2\alpha_A + t^2} \right)^{2m-2\mu-1} \int_{-\infty}^{\infty} dz\, z^{2(\mu+1)} e^{-\left(2\alpha_A + t^2\right) z^2} \\
&\quad + \left[t^4 \left(\frac{2\alpha_A R}{2\alpha_A + t^2} \right)^2 - \frac{t^2}{2} \right] \sum_{\mu=0}^{m} C_{2\mu}^{2m} \left(\frac{t^2 R}{2\alpha_A + t^2} \right)^{2m-2\mu} \int_{-\infty}^{\infty} dz\, z^{2\mu} e^{-\left(2\alpha_A + t^2\right) z^2} \\
&= \sqrt{\frac{\pi}{2\alpha_A + t^2}} \left\{ t^4 \sum_{\mu=0}^{m} C_{2\mu}^{2m} \left(\frac{t^2 R}{2\alpha_A + t^2} \right)^{2m-2\mu} \frac{(2\mu+2)!}{2^{2\mu+2}(\mu+1)!} \left(\frac{1}{2\alpha_A + t^2} \right)^{\mu+1} \right. \\
&\quad - t^4 \frac{4\alpha_A R}{2\alpha_A + t^2} \sum_{\mu=0}^{m-1} C_{2\mu+1}^{2m} \left(\frac{t^2 R}{2\alpha_A + t^2} \right)^{2m-2\mu-1} \frac{(2\mu+2)!}{2^{2\mu+2}(\mu+1)!} \left(\frac{1}{2\alpha_A + t^2} \right)^{\mu+1} \\
&\quad \left. + \left[t^4 \left(\frac{2\alpha_A R}{2\alpha_A + t^2} \right)^2 - \frac{t^2}{2} \right] \sum_{\mu=0}^{m} C_{2\mu}^{2m} \left(\frac{t^2 R}{2\alpha_A + t^2} \right)^{2m-2\mu} \frac{(2\mu)!}{2^{2\mu}\mu!} \left(\frac{1}{2\alpha_A + t^2} \right)^{\mu} \right\} \\
&= \sqrt{\frac{\pi}{2\alpha_A + t^2}} \left\{ t^2 \sum_{\mu=0}^{m} C_{2\mu}^{2m} \left(\frac{t^2 R}{2\alpha_A + t^2} \right)^{2m-2\mu} \frac{(2\mu)!}{2^{2\mu}\mu!} \left(\frac{1}{2\alpha_A + t^2} \right)^{\mu} \frac{2\mu+1}{2} \left(\frac{t^2}{2\alpha_A + t^2} \right) \right. \\
&\quad - 4\alpha_A \sum_{\mu=0}^{m-1} C_{2\mu}^{2m} \left(\frac{t^2 R}{2\alpha_A + t^2} \right)^{2m-2\mu} \frac{(2\mu)!}{2^{2\mu}\mu!} \left(\frac{1}{2\alpha_A + t^2} \right)^{\mu} (m-\mu) \left(\frac{t^2}{2\alpha_A + t^2} \right) \\
&\quad \left. + \left[t^4 \left(\frac{2\alpha_A R}{2\alpha_A + t^2} \right)^2 - \frac{t^2}{2} \right] \sum_{\mu=0}^{m} C_{2\mu}^{2m} \left(\frac{t^2 R}{2\alpha_A + t^2} \right)^{2m-2\mu} \frac{(2\mu)!}{2^{2\mu}\mu!} \left(\frac{1}{2\alpha_A + t^2} \right)^{\mu} \right\}
\end{aligned}
\tag{153}
$$

In the second summation, we find that, since $m - \mu = 0$ when $\mu = m$, we can safely extend the upper limit to m. Thus,

$$
\begin{aligned}
I_z = \sqrt{\frac{\pi}{2\alpha_A + t^2}} &\left\{ t^2 \sum_{\mu=0}^{m} C_{2\mu}^{2m} \left(\frac{t^2 R}{2\alpha_A + t^2} \right)^{2m-2\mu} \frac{(2\mu)!}{2^{2\mu}\mu!} \left(\frac{1}{2\alpha_A + t^2} \right)^{\mu} \frac{2\mu+1}{2} \left(\frac{t^2}{2\alpha_A + t^2} \right) \right. \\
&\quad - 4\alpha_A \sum_{\mu=0}^{m-1} C_{2\mu}^{2m} \left(\frac{t^2 R}{2\alpha_A + t^2} \right)^{2m-2\mu} \frac{(2\mu)!}{2^{2\mu}\mu!} \left(\frac{1}{2\alpha_A + t^2} \right)^{\mu} (m-\mu) \left(\frac{t^2}{2\alpha_A + t^2} \right)
\end{aligned}
$$

$$+\left[t^4\left(\frac{2\alpha_A R}{2\alpha_A+t^2}\right)^2-\frac{t^2}{2}\right]\sum_{\mu=0}^{m}C_{2\mu}^{2m}\left(\frac{t^2R}{2\alpha_A+t^2}\right)^{2m-2\mu}\frac{(2\mu)!}{2^{2\mu}\mu!}\left(\frac{1}{2\alpha_A+t^2}\right)^{\mu}\Bigg\}$$

$$=\sqrt{\frac{\pi}{2\alpha_A+t^2}}\sum_{\mu=0}^{m}C_{2\mu}^{2m}\left(\frac{t^2R}{2\alpha_A+t^2}\right)^{2m-2\mu}\frac{(2\mu)!}{2^{2\mu}\mu!}\left(\frac{1}{2\alpha_A+t^2}\right)^{\mu}$$

$$\times\left[\frac{2\mu+1}{2}t^2\left(\frac{t^2}{2\alpha_A+t^2}\right)-4\alpha_A(m-\mu)\left(\frac{t^2}{2\alpha_A+t^2}\right)+t^4\left(\frac{2\alpha_A R}{2\alpha_A+t^2}\right)^2-\frac{t^2}{2}\right]$$

$$=\sqrt{\frac{\pi}{2\alpha_A+t^2}}\sum_{\mu=0}^{m}C_{2\mu}^{2m}\left(\frac{t^2R}{2\alpha_A+t^2}\right)^{2m-2\mu}\frac{(2\mu)!}{2^{2\mu}\mu!}\left(\frac{1}{2\alpha_A+t^2}\right)^{\mu}$$

$$\times\left\{4\alpha_A^2R^2\left(\frac{t^2}{2\alpha_A+t^2}\right)^2+\left[\mu t^2-4\alpha_A(m-\mu)-\alpha_A\right]\frac{t^2}{2\alpha_A+t^2}\right\}\tag{154}$$

By inserting Eq. (154) into Eq. (152), we found that

$$\langle\eta_A|\left(\frac{\partial^2}{\partial R^2}\frac{1}{r_{B1}}\right)_0|\eta_A\rangle$$

$$=2\pi N_A^2\frac{(2n)!(2l)!}{2^{2n+2l}n!l!}\sum_{\mu=0}^{m}C_{2\mu}^{2m}\frac{(2\mu)!}{2^{2\mu}\mu!}\int_0^{\infty}\frac{dt}{(2\alpha_A+t^2)^{3/2}}\left(\frac{1}{2\alpha_A+t^2}\right)^{n+l+\mu}\left(\frac{t^2R}{2\alpha_A+t^2}\right)^{2m-2\mu}$$

$$\times\left\{4\alpha_A^2R^2\left(\frac{t^2}{2\alpha_A+t^2}\right)^2+\left[\mu t^2-4\alpha_A(m-\mu)-\alpha_A\right]\frac{t^2}{2\alpha_A+t^2}\right\}e^{-2\alpha_A t^2R^2/(2\alpha_A+t^2)}\tag{155}$$

By introducing the new variable as in Eq. (143), one obtains

$$\langle\eta_A|\left(\frac{\partial^2}{\partial R^2}\frac{1}{r_{B1}}\right)_0|\eta_A\rangle$$

$$=2\pi N_A^2\frac{(2n)!(2l)!}{(8\alpha_A)^{n+l}n!l!}\sum_{\mu=0}^{m}C_{2\mu}^{2m}\frac{(2\mu)!}{(8\alpha_A)^{\mu}\mu!}R^{2m-2\mu}\int_0^1 d\xi\left(1-\xi^2\right)^{n+l+\mu}\left(\xi^2\right)^{2m-2\mu}$$

$$\times\left\{4\alpha_A^2R^2\left(\xi^2\right)^2+\left[2\mu\alpha_A\frac{\xi^2}{1-\xi^2}-4\alpha_A(m-\mu)-\alpha_A\right]\xi^2\right\}e^{-2\alpha_A R^2\xi^2}$$

$$=2\pi N_A^2\frac{(2n)!(2l)!}{(8\alpha_A)^{n+l}n!l!}\sum_{\mu=0}^{m}C_{2\mu}^{2m}\frac{(2\mu)!}{(8\alpha_A)^{\mu}\mu!}R^{2m-2\mu}$$

$$\times\left\{4\alpha_A^2R^2\int_0^1 d\xi\left(1-\xi^2\right)^{n+l+\mu}\left(\xi^2\right)^{2m-2\mu+2}e^{-2\alpha_A R^2\xi^2}\right.$$

$$+2\mu\alpha_A\int_0^1 d\xi\left(1-\xi^2\right)^{n+l+\mu-1}\left(\xi^2\right)^{2m-2\mu+2}e^{-2\alpha_A R^2\xi^2}$$

$$\left.-\alpha_A[4(m-\mu)+1]\int_0^1 d\xi\left(1-\xi^2\right)^{n+l+\mu}\left(\xi^2\right)^{2m-2\mu+1}e^{-2\alpha_A R^2\xi^2}\right\}\tag{156}$$

The second integral in Eq. (155) seemed to be singular when $n + l + \mu = 0$. However, in this case, μ must be zero, and consequently this term will never contribute to the final result for being suppressed by the prefactor. With the definition in Eq. (132), we can write

$$\begin{aligned}
&\langle \eta_A | \left(\frac{\partial^2}{\partial R^2} \frac{1}{r_{B1}} \right)_0 | \eta_A \rangle \\
&= 2\pi N_A^2 \frac{(2n)!(2l)!}{(8\alpha_A)^{n+l} n! l!} \sum_{\mu=0}^{m} C_{2\mu}^{2m} \frac{(2\mu)!}{(8\alpha_A)^{\mu} \mu!} R^{2m-2\mu} \\
&\times \left\{ 4\alpha_A^2 R^2 X_{n+l+\mu, 2m-2\mu+2, 0, 2\alpha_A}(R) / (2\alpha_A)^{2m-2\mu+2} \right. \\
&+ 2\mu\alpha_A X_{n+l+\mu-1, 2m-2\mu+2, 0, 2\alpha_A}(R) / (2\alpha_A)^{2m-2\mu+2} \\
&\left. - \alpha_A [4(m-\mu) + 1] X_{n+l+\mu, 2m-2\mu+1, 0, 2\alpha_A}(R) / (2\alpha_A)^{2m-2\mu+1} \right\} \qquad (157)
\end{aligned}$$

The last kind of second-order derivative considered is of the following form:

$$\begin{aligned}
&\langle \eta_A | \left(\frac{\partial^2}{\partial R^2} \frac{1}{r_{A1}} \right)_0 | \eta_B \rangle \\
&= \frac{N_A N_B}{\sqrt{\pi}} \int_{-\infty}^{\infty} dt \int d\tau \, x_{A1}^{n_A} x_{B1}^{n_B} y_{A1}^{l_A} y_{B1}^{l_B} z_{A1}^{m_A} z_{B1}^{m_B} \left(t^4 z_{A1}^2 - \frac{t^2}{2} \right) e^{-(\alpha_A + t^2) r_{A1}^2} e^{-\alpha_B r_{B1}^2} \qquad (158)
\end{aligned}$$

With the specific geometry, we can let $x_{A1} = x_{B1} = x$, $y_{A1} = y_{B1} = y$, $z_{A1} = z$, and $z_{B1} = z - R$. After changing these variables, we obtain

$$\begin{aligned}
&\langle \eta_A | \left(\frac{\partial^2}{\partial R^2} \frac{1}{r_{A1}} \right)_0 | \eta_B \rangle \\
&= \frac{N_A N_B}{\sqrt{\pi}} \int_{-\infty}^{\infty} dt \int_{-\infty}^{\infty} dx \, x^{n_A+n_B} e^{-(\alpha_A+\alpha_B+t^2)x^2} \int_{-\infty}^{\infty} dy \, y^{l_A+l_B} e^{-(\alpha_A+\alpha_B+t^2)y^2} \\
&\times \int_{-\infty}^{\infty} dz \, z^{m_A} (z-R)^{m_B} \left(t^4 z_{A1}^2 - \frac{t^2}{2} \right) e^{-(\alpha_A+\alpha_B+t^2)z^2 + 2\alpha_B R z - \alpha_B R^2} \\
&= \frac{N_A N_B}{\sqrt{\pi}} \int_{-\infty}^{\infty} dt \left(\frac{1 + (-1)^{n_A+n_B}}{2} \right) \left(\frac{1 + (-1)^{l_A+l_B}}{2} \right) \frac{(n_A+n_B)(l_A+l_B)}{2^{n_A+n_B+l_A+l_B} \left(\frac{n_A+n_B}{2}\right)! \left(\frac{l_A+l_B}{2}\right)!} \\
&\times \left(\frac{\pi}{\alpha_A + \alpha_B + t^2} \right) \left(\frac{1}{\alpha_A + \alpha_B + t^2} \right)^{(n_A+n_B+l_A+l_B)/2} e^{-\alpha_B(\alpha+t^2)/(\alpha_A+\alpha_B+t^2)R^2} \times I_z
\end{aligned}$$

(159)

where

$$I_z = \int_{-\infty}^{\infty} dz \left(z + \frac{\alpha_B}{\alpha_A + \alpha_B + t^2} R \right)^{m_A} \left(z - \frac{\alpha_A + t^2}{\alpha_A + \alpha_B + t^2} R \right)^{m_B}$$

$$\times \left[t^4 \left(z + \frac{\alpha_B}{\alpha_A + \alpha_B + t^2} \right)^2 - \frac{t^2}{2} \right] e^{-(\alpha_A + \alpha_B + t^2) z^2}$$

$$= \sum_{\mu_A=0}^{m_A} \sum_{\mu_B=0}^{m_B} C_{\mu_A}^{m_A} C_{\mu_B}^{m_B} \left(\frac{\alpha_B R}{\alpha_A + \alpha_B + t^2} \right)^{m_A - \mu_A} \left(-\frac{\alpha_A + t^2}{\alpha_A + \alpha_B + t^2} R \right)^{m_B - \mu_B}$$

$$\times \left\{ t^4 \int_{-\infty}^{\infty} dz\, z^{\mu_A + \mu_B + 2} e^{-(\alpha_A + \alpha_B + t^2) z^2} + \frac{2\alpha_B R t^4}{\alpha_A + \alpha_B + t^2} \int_{-\infty}^{\infty} dz\, z^{\mu_A + \mu_B + 1} e^{-(\alpha_A + \alpha_B + t^2) z^2} \right.$$

$$\left. + \left[\left(\frac{t^2 \alpha_B R}{\alpha_A + \alpha_B + t^2} \right)^2 - \frac{t^2}{2} \right] \int_{-\infty}^{\infty} dz\, z^{\mu_A + \mu_B} e^{-(\alpha_A + \alpha_B + t^2) z^2} \right\}$$

$$= \sum_{\mu_A=0}^{m_A} \sum_{\mu_B=0}^{m_B} C_{\mu_A}^{m_A} C_{\mu_B}^{m_B} \left(\frac{\alpha_B R}{\alpha_A + \alpha_B + t^2} \right)^{m_A - \mu_A} \left(-\frac{\alpha_A + t^2}{\alpha_A + \alpha_B + t^2} R \right)^{m_B - \mu_B} \sqrt{\frac{\pi}{\alpha_A + \alpha_B + t^2}}$$

$$\times \left\{ \left(\frac{1 + (-1)^{\mu_A + \mu_B}}{2} \right) t^4 \frac{(\mu_A + \mu_B + 2)!}{2^{\mu_A + \mu_B + 2} \left(\frac{\mu_A + \mu_B + 2}{2} \right)!} \left(\frac{1}{\alpha_A + \alpha_B + t^2} \right)^{(\mu_A + \mu_B + 2)/2} \right.$$

$$+ \left(\frac{1 + (-1)^{\mu_A + \mu_B}}{2} \right) \left[\left(\frac{t^2 \alpha_B R}{\alpha_A + \alpha_B + t^2} \right)^2 - \frac{t^2}{2} \right] \frac{(\mu_A + \mu_B)!}{2^{\mu_A + \mu_B} \left(\frac{\mu_A + \mu_B}{2} \right)!} \left(\frac{1}{\alpha_A + \alpha_B + t^2} \right)^{(\mu_A + \mu_B)/2}$$

$$\left. + \left(\frac{1 - (-1)^{\mu_A + \mu_B}}{2} \right) \frac{2\alpha_B R t^4}{\alpha_A + \alpha_B + t^2} \frac{(\mu_A + \mu_B + 1)!}{2^{\mu_A + \mu_B + 1} \left(\frac{\mu_A + \mu_B + 1}{2} \right)!} \left(\frac{1}{\alpha_A + \alpha_B + t^2} \right)^{(\mu_A + \mu_B + 1)/2} \right\} \tag{160}$$

Note that the first two integrals in Eq. (160) have nonvanishing prefactors when $\mu_A + \mu_B$ is even, while the last integral has nonvanishing prefactor when $\mu_A + \mu_B$ is odd. They do not contribute in the final form at the same time. By inserting I_z into Eq. (159), the familiar forms of integral over t as in the calculation of other matrix elements appear. Again, by changing into the variable $\xi = t/\sqrt{\alpha_A + \alpha_B + t^2}$ and integrating, it can be shown that

$$\langle \eta_A | \left(\frac{\partial^2}{\partial R^2} \frac{1}{r_{A1}} \right)_0 | \eta_B \rangle$$

$$= 2\pi N_A N_B \left(\frac{1 + (-1)^{n_A + n_B}}{2} \right) \left(\frac{1 + (-1)^{l_A + l_B}}{2} \right) \frac{(n_A + n_B)!(l_A + l_B)!}{2^{n_A + n_B + l_A + l_B} \left(\frac{n_A + n_B}{2} \right)! \left(\frac{l_A + l_B}{2} \right)!}$$

$$\times \sum_{\mu_A=0}^{m_A} \sum_{\mu_B=0}^{\mu_A} C_{\mu_A}^{m_A} C_{\mu_A}^{m_A} e^{-\frac{\alpha_A \alpha_B}{\alpha_A + \alpha_B} R^2} \alpha_B^{m_A - \mu_A} (-1)^{m_B - \mu_B} R^{m_A + m_B - \mu_A - \mu_B}$$

$$\times \left(\frac{1}{\alpha_A+\alpha_B}\right)^{(n_A+n_B+l_A+l_B+2m_A+2m_B-\mu_A-\mu_B)/2} \left\{ \left(\frac{1+(-1)^{\mu_A+\mu_B}}{2}\right) \frac{(\mu_A+\mu_B)!}{2^{\mu_A+\mu_B}\left(\frac{\mu_A+\mu_B}{2}\right)!} \right.$$
$$\times \left[\frac{\mu_A+\mu_B+1}{2} \int_0^1 d\xi\, (\xi^2)^2 (1-\xi^2)^{(n_A+n_B+l_A+l_B+2m_A-\mu_A+\mu_B)/2-1} \right.$$
$$\times \left(\alpha_A+\alpha_B\xi^2\right)^{m_B-\mu_B} e^{-\alpha_B^2 R^2 \xi^2/(\alpha_A+\alpha_B)}$$
$$+ \frac{\alpha_B^2 R^2}{\alpha_A+\alpha_B} \int_0^1 d\xi\, (\xi^2)^2 (1-\xi^2)^{(n_A+n_B+l_A+l_B+2m_A-\mu_A+\mu_B)/2}$$
$$\times \left(\alpha_A+\alpha_B\xi^2\right)^{m_B-\mu_B} e^{-\alpha_B^2 R^2 \xi^2/(\alpha_A+\alpha_B)} - \int_0^1 d\xi\, \xi^2 (1-\xi^2)^{(n_A+n_B+l_A+l_B+2m_A-\mu_A+\mu_B)/2-1}$$
$$\left. \times \left(\alpha_A+\alpha_B\xi^2\right)^{m_B-\mu_B} e^{-\alpha_B^2 R^2 \xi^2/(\alpha_A+\alpha_B)} \right] + \left(\frac{1+(-1)^{\mu_A+\mu_B}}{2}\right) \frac{(\mu_A+\mu_B+1)!}{2^{\mu_A+\mu_B+1}\left(\frac{\mu_A+\mu_B+1}{2}\right)!}$$
$$\times \frac{2\alpha_B R}{\sqrt{\alpha_A+\alpha_B}} \int_0^1 d\xi\, (\xi^2)^2 (1-\xi^2)^{(n_A+n_B+l_A+l_B+2m_A-\mu_A+\mu_B-1)/2}$$
$$\left. \times \left(\alpha_A+\alpha_B\xi^2\right)^{m_B-\mu_B} e^{-\alpha_B^2 R^2 \xi^2/(\alpha_A+\alpha_B)} \right\} \tag{161}$$

These integrals over ξ are discussed in the appendix. Inserting the formulas for these integrals obtained from Eq. (A.18) into Eq. (161), we will obtain an expression for computation.

V. HYDROGEN MOLECULE: MINIMUM BASIS SET CALCULATION

To calculate the matrix elements for H_2 in the minimal basis set, we approximate the Slater $1s$ orbital with a Gaussian function. That is, we replace the $1s$ radial wave function

$$\phi_S(\mathbf{r}) = \left(\frac{\zeta^3}{\pi}\right)^{1/2} e^{-\zeta r} \tag{162}$$

with a Gaussian function $|\gamma\rangle$ defined earlier.

It can be shown that

$$S_{AB} = \langle \gamma_A(1)|\gamma_B(1)\rangle = \left(\frac{2\sqrt{\alpha_A\alpha_B}}{\alpha_A+\alpha_B}\right)^{3/2} e^{-\alpha_A\alpha_B R^2/(\alpha_A+\alpha_B)} \tag{163}$$

$$\langle \gamma_A(1)\gamma_A(2)|\frac{1}{r_{12}}|\gamma_A(1)\gamma_A(2)\rangle = \frac{2}{\sqrt{\pi}}\sqrt{\alpha_A} \tag{164}$$

$$\langle \gamma_B(1)\gamma_B(2)|\frac{1}{r_{12}}|\gamma_B(1)\gamma_B(2)\rangle = \frac{2}{\sqrt{\pi}}\sqrt{\alpha_B} \tag{165}$$

$$\langle \gamma_A(1)\gamma_A(2)|\frac{1}{r_{12}}|\gamma_B(1)\gamma_B(2)\rangle = \langle \gamma_B(1)\gamma_B(2)|\frac{1}{r_{12}}|\gamma_A(1)\gamma_A(2)\rangle$$

$$= \frac{2}{\sqrt{\pi}}\sqrt{\frac{\alpha_A+\alpha_B}{2}}\left[\frac{4\alpha_A\alpha_B}{(\alpha_A+\alpha_B)^2}\right]^{3/2} e^{-(\alpha_A^2+\alpha_B^2)R^2/(\alpha_A+\alpha_B)R^2} \tag{166}$$

$$\langle \gamma_A(1)\gamma_B(2)|\frac{1}{r_{12}}|\gamma_A(1)\gamma_B(2)\rangle = \langle \gamma_B(1)\gamma_A(2)|\frac{1}{r_{12}}|\gamma_B(1)\gamma_A(2)\rangle$$

$$= \frac{1}{R}\mathrm{erf}\left(\sqrt{\frac{2\alpha_A\alpha_B}{\alpha_A+\alpha_B}}R\right) \tag{167}$$

$$\langle \gamma_A(1)\gamma_B(2)|\frac{1}{r_{12}}|\gamma_B(1)\gamma_A(2)\rangle = \langle \gamma_B(1)\gamma_A(2)|\frac{1}{r_{12}}|\gamma_A(1)\gamma_B(2)\rangle$$

$$= \frac{2}{\sqrt{\pi}}\sqrt{\frac{\alpha_A+\alpha_B}{2}}\left[\frac{4\alpha_A\alpha_B}{(\alpha_A+\alpha_B)^2}\right]^{3/2} e^{-\alpha_A\alpha_B R^2/(\alpha_A+\alpha_B)} \tag{168}$$

and

$$\langle \gamma_A(1)|\frac{1}{r_{B1}}|\gamma_A(1)\rangle = \frac{1}{R}\mathrm{erf}\left(\sqrt{2\alpha_A}\,R\right) \tag{169}$$

$$\langle \gamma_B(1)|\frac{1}{r_{A1}}|\gamma_B(1)\rangle = \frac{1}{R}\mathrm{erf}\left(\sqrt{2\alpha_B}\,R\right) \tag{170}$$

$$\langle \gamma_A(1)|\frac{1}{r_{A1}}|\gamma_B(1)\rangle = \frac{(4\alpha_A\alpha_B)^{3/4}}{\alpha_B\sqrt{\alpha_A+\alpha_B}} e^{-\alpha_A\alpha_B R^2/(\alpha_A+\alpha_B)} \frac{1}{R}\mathrm{erf}\left(\frac{\alpha_B}{\sqrt{\alpha_A+\alpha_B}}R\right) \tag{171}$$

$$\langle \gamma_A(1)|\frac{1}{r_{B1}}|\gamma_B(1)\rangle = \frac{(4\alpha_A\alpha_B)^{3/4}}{\alpha_A\sqrt{\alpha_A+\alpha_B}} e^{-\alpha_A\alpha_B R^2/(\alpha_A+\alpha_B)} \frac{1}{R}\mathrm{erf}\left(\frac{\alpha_A}{\sqrt{\alpha_A+\alpha_B}}R\right) \tag{172}$$

Now we can calculate the ground-state energy of H_2. Here, we only use one basis function, the $1s$ atomic orbital of hydrogen. By symmetry consideration, we know that the wave function of the H_2 ground state is

$$|\Psi_g\rangle = \|\overset{+}{\sigma}\overset{-}{\sigma}|\rangle \tag{173}$$

where

$$|\sigma\rangle = \frac{1}{\sqrt{2(1+S_{AB})}}\left(|\chi_{1s,A}\rangle + |\chi_{1s,B}\rangle\right) \tag{174}$$

and

$$|\chi_{1s,A}\rangle = \sqrt{\frac{\zeta^3}{\pi}} e^{-\zeta|\mathbf{r}-\mathbf{R}_A|} \tag{175}$$

Here, we shall replace $\chi_{1s,A}$ with a single Gaussian wave function $|\gamma_A(i)\rangle$ as defined earlier. That is, we have used the approximation

$$|\sigma(i)\rangle = \frac{1}{\sqrt{2(1+S_{AB})}}(|\gamma_A(i)\rangle + |\gamma_B(i)\rangle) \tag{176}$$

For H_2, let us write down the zeroth-order electronic Hamiltonian (in atomic unit):

$$\hat{H}_0^{(0)} = \left(-\frac{\nabla_1^2}{2} - \frac{1}{r_{A1}} - \frac{1}{r_{B1}}\right) + \left(-\frac{\nabla_2^2}{2} - \frac{1}{r_{A2}} - \frac{1}{r_{B2}}\right) + \frac{1}{r_{12}} + \frac{1}{R} \tag{177}$$

Let

$$\hat{h}_0 \equiv \frac{1}{R}, \qquad \hat{h}_2 \equiv \frac{1}{r_{12}}$$

$$\hat{h}_1 \equiv \hat{h}_1(1) + \hat{h}_1(2) = \left(-\frac{\nabla_1^2}{2} - \frac{1}{r_{A1}} - \frac{1}{r_{B1}}\right) + \left(-\frac{\nabla_2^2}{2} - \frac{1}{r_{A2}} - \frac{1}{r_{B2}}\right)$$

we have

$$\begin{aligned}
&\langle\Psi_g|\hat{H}_0^{(0)}|\Psi_g\rangle \\
&= \frac{1}{2}\langle\overset{+}{\sigma}(1)\,\bar{\sigma}(2) - \bar{\sigma}(1)\,\overset{+}{\sigma}(2)|\hat{H}_0^{(0)}|\overset{+}{\sigma}(1)\,\bar{\sigma}(2) - \bar{\sigma}(1)\,\overset{+}{\sigma}(2)\rangle \\
&= \frac{1}{2}\Big[\langle\overset{+}{\sigma}(1)\,\bar{\sigma}(2)|\hat{H}_0^{(0)}|\,\overset{+}{\sigma}(1)\,\bar{\sigma}(2)\rangle + \langle\bar{\sigma}(1)\,\overset{+}{\sigma}(2)|\hat{H}_0^{(0)}|\bar{\sigma}(1)\,\overset{+}{\sigma}(2)\rangle \\
&\quad - \langle\overset{+}{\sigma}(1)\,\bar{\sigma}(2)|\hat{H}_0^{(0)}|\bar{\sigma}(1)\,\overset{+}{\sigma}(2)\rangle - \langle\bar{\sigma}(1)\,\overset{+}{\sigma}(2)|\hat{H}_0^{(0)}|\,\overset{+}{\sigma}(1)\,\bar{\sigma}(2)\rangle\Big]
\end{aligned} \tag{178}$$

The last two terms in (178) with negative signs vanish after integrating out the spin part, and, consequently,

$$\begin{aligned}
\langle\Psi_g|\hat{H}_0^{(0)}|\Psi_g\rangle &= \langle\sigma(1)\sigma(2)|\left[\hat{h}_0 + \hat{h}_1(1) + \hat{h}_1(2) + \hat{h}_2\right]|\sigma(1)\sigma(2)\rangle \\
&= \hat{h}_0 + 2\langle\sigma(1)|h1(1)|\sigma(1)\rangle + \langle\sigma(1)\sigma(2)|\hat{h}_2|\sigma(1)\sigma(2)\rangle
\end{aligned} \tag{179}$$

Next, we expand them into the atomic orbitals

$$\begin{aligned}\langle\sigma(1)|\hat{h}_1(1)|\sigma(1)\rangle &= \frac{1}{2(1+S_{AB})}\langle\gamma_A(1)+\gamma_B(1)|\hat{h}_1(1)|\gamma_A(1)+\gamma_B(1)\rangle \\ &= \frac{1}{2(1+S_{AB})}\left[\langle\gamma_A|\hat{h}_1|\gamma_A\rangle+\langle\gamma_B|\hat{h}_1|\gamma_B\rangle+2\langle\gamma_A|\hat{h}_1|\gamma_B\rangle\right]\end{aligned} \tag{180}$$

where

$$\langle\gamma_A|\hat{h}_1|\gamma_A\rangle = E_g(H) - \langle\gamma_A|\frac{1}{r_{B1}}|\gamma_A\rangle \tag{181}$$

$$\langle\gamma_B|\hat{h}_1|\gamma_B\rangle = E_g(H) - \langle\gamma_B|\frac{1}{r_{A1}}|\gamma_B\rangle \tag{182}$$

$$\langle\gamma_A|\hat{h}_1|\gamma_B\rangle = E_g(H)S_{AB} - \langle\gamma_A|\frac{1}{r_{A1}}|\gamma_B\rangle \tag{183}$$

Therefore,

$$\langle\sigma(1)|\hat{h}_1(1)|\sigma(1)\rangle = E_g(H) - \frac{1}{2(1+S_{AB})}\left[\langle\gamma_A|\frac{1}{r_{B1}}|\gamma_A\rangle+\langle\gamma_B|\frac{1}{r_{A1}}|\gamma_B\rangle+2\langle\gamma_A|\frac{1}{r_{A1}}|\gamma_B\rangle\right]$$

The quantity $\langle\sigma(1)\sigma(2)|1/r_{12}|\sigma(1)\sigma(2)\rangle$ can also be expanded but we do not show the final result here. In Figure 4, we show the calculated result of the quantity $\langle\Psi_g|\hat{H}_0^{(0)}|\Psi_g\rangle - 2E_g(H)$. We have simply taken $\zeta = 1$. The corresponding equilibrium position is marked in the figures, which is 0.9112 Å. This value is not realistic, however, the magnitude and the feature of the PES are reasonable. We believe that the more realistic value can be obtained by using larger basis set.

To obtain the force constant for constructing the equation of motion of the nuclear motion in the second–order perturbation, we need to know about the excited states, too. With the minimal basis set, the only excited-state spatial orbital for one electron is

$$|\sigma*\rangle = \frac{1}{\sqrt{2(1-S_{AB})}}\left(|\chi_{1s,A}\rangle - |\chi_{1s,B}\rangle\right) \tag{184}$$

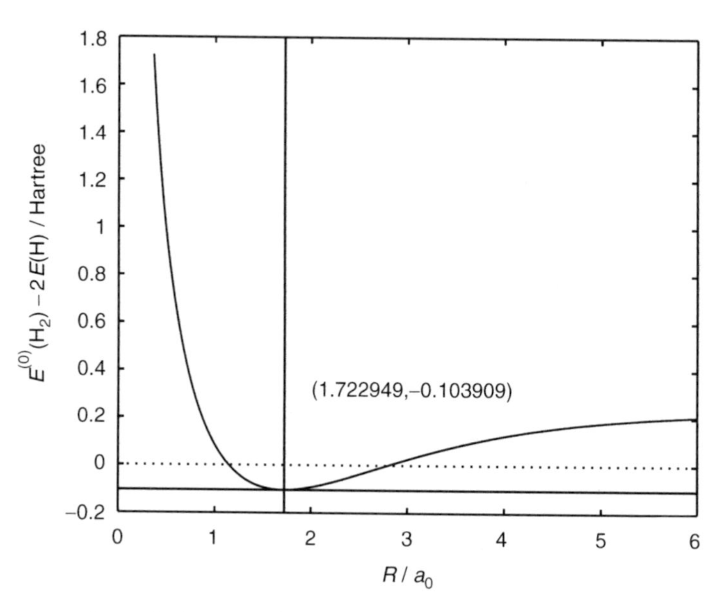

Figure 4. The ground-state potential energy surface with $\zeta = 1$.

With symmetry considerations, we can write down all of the possible spin orbitals:

$$|\Psi_1\rangle = \|\overset{+}{\sigma}\,\overset{+}{\sigma}{}^*|\rangle \tag{185}$$

$$|\Psi_2\rangle = \|\overset{-}{\sigma}\,\overset{-}{\sigma}{}^*|\rangle \tag{186}$$

$$|\Psi_3\rangle = \frac{1}{\sqrt{2}}\left(\|\overset{+}{\sigma}\,\overset{-}{\sigma}{}^*|\rangle + \|\overset{+}{\sigma}{}^*\,\overset{-}{\sigma}|\rangle\right) \tag{187}$$

$$|\Psi_4\rangle = \frac{1}{\sqrt{2}}\left(\|\overset{+}{\sigma}\,\overset{-}{\sigma}{}^*|\rangle - \|\overset{+}{\sigma}{}^*\,\overset{-}{\sigma}|\rangle\right) \tag{188}$$

$$|\Psi_5\rangle = \|\overset{+}{\sigma}{}^*\,\overset{-}{\sigma}{}^*|\rangle \tag{189}$$

Nonvanishing matrix element only exists between $|\Psi_3\rangle$ and the ground state for any one-electron operator $\hat{O}_1$ that does not involve spin. Obviously, $\langle\Psi_1|\hat{O}_1|\Psi_g\rangle$ and $\langle\Psi_2|\hat{O}_1|\Psi_g\rangle$ are zero after integrating over the spin part. It is then clear that all of the triplet states will result in vanishing matrix elements. Therefore $\langle\Psi_4|\hat{O}_1|\Psi_g\rangle$ is also zero. For the doubly excited singlet state, however, we have $\langle\Psi_5|\hat{O}_1|\Psi_g\rangle = 0$ because both spatial orbitals are different in $|\Psi_g\rangle$ and $|\Psi_5\rangle$. The only excited state that we need to calculate, therefore, is the singly excited singlet state $|\Psi_e\rangle = |\Psi_3\rangle$. Although these can be easily demonstrated, we shall neglect the algebra here.

Consequently, the term $U_n^{(2)}$ is reduced to

$$U_g^{(2)} = \langle \Psi_g | \hat{H}_0^{(2)} | \Psi_g \rangle + \frac{|\langle \Psi_e | \hat{H}_0^{(1)} | \Psi_g \rangle|^2}{U_g^{(0)} - U_e^{(0)}} \tag{190}$$

where we have changed the subscripts to g and e for convenience. We have to calculate $U_e^{(0)}$, $\langle \Psi_g | \hat{H}_0^{(2)} | \Psi_g \rangle$, and $\langle \Psi_e | \hat{H}_0^{(1)} | \Psi_g \rangle$ to obtain $U_g^{(2)}$. The zeroth-order excited state energy $U_e^{(0)}$ is

$$\begin{aligned} U_e^{(0)} &= \langle \Psi_e | \hat{H}_0^{(0)} | \Psi_e \rangle \\ &= \frac{1}{2} \langle (|\overset{+}{\sigma} \overset{-}{\sigma}{}^*| + |\overset{+}{\sigma}{}^* \overset{-}{\sigma}|) | \hat{H}_0^{(0)} | (|\overset{+}{\sigma} \overset{-}{\sigma}{}^*| + |\overset{+}{\sigma}{}^* \overset{-}{\sigma}|) \rangle \\ &= \frac{1}{2} (\langle |\overset{+}{\sigma} \overset{-}{\sigma}{}^*| | \hat{H}_0^{(0)} | |\overset{+}{\sigma} \overset{-}{\sigma}{}^*| \rangle + \langle |\overset{+}{\sigma} \overset{-}{\sigma}{}^*| | \hat{H}_0^{(0)} | |\overset{+}{\sigma}{}^* \overset{-}{\sigma}| \rangle \\ &\quad + \langle |\overset{+}{\sigma}{}^* \overset{-}{\sigma}| | \hat{H}_0^{(0)} | |\overset{+}{\sigma} \overset{-}{\sigma}{}^*| \rangle + \langle |\overset{+}{\sigma}{}^* \overset{-}{\sigma}| | \hat{H}_0^{(0)} | |\overset{+}{\sigma}{}^* \overset{-}{\sigma}| \rangle) \end{aligned} \tag{191}$$

Expanding the Slater determinants and integrating out the spin part and collecting terms that are the same under exchange of electron indices, we have

$$\begin{aligned} U_e^{(0)} &= \langle \sigma\sigma^* | \hat{H}_0^{(0)} | \sigma\sigma^* \rangle + \langle \sigma\sigma^* | \hat{H}_0^{(0)} | \sigma^*\sigma \rangle \\ &= \frac{1}{R} + \langle \sigma | \hat{h}_1 | \sigma \rangle + \langle \sigma^* | \hat{h}_1 | \sigma^* \rangle + \langle \sigma\sigma^* | \hat{h}_2 | \sigma\sigma^* \rangle + \langle \sigma\sigma^* | \hat{h}_2 | \sigma^*\sigma \rangle \end{aligned} \tag{192}$$

By expanding the spatial orbitals into atomic orbitals and manipulating them properly, we have

$$\langle \sigma | \hat{h}_1 | \sigma \rangle = E_g(\mathrm{H}) - \frac{\langle \chi_A | \frac{1}{r_{B1}} | \chi_A \rangle + \langle \chi_B | \frac{1}{r_{A1}} | \chi_B \rangle + 2 \langle \chi_A | \frac{1}{r_{A1}} | \chi_B \rangle}{2(1 + S_{AB})} \tag{193}$$

$$\langle \sigma^* | \hat{h}_1 | \sigma^* \rangle = E_g(\mathrm{H}) - \frac{\langle \chi_A | \frac{1}{r_{B1}} | \chi_A \rangle + \langle \chi_B | \frac{1}{r_{A1}} | \chi_B \rangle - 2 \langle \chi_A | \frac{1}{r_{A1}} | \chi_B \rangle}{2(1 - S_{AB})} \tag{194}$$

$$\begin{aligned} &\langle \sigma\sigma^* | \hat{h}_2 | \sigma\sigma^* \rangle + \langle \sigma\sigma^* | \hat{h}_2 | \sigma^*\sigma \rangle \\ &= \frac{1}{2(1 - S_{AB}^2)} \Bigg(\langle \chi_A\chi_A | \frac{1}{r_{12}} | \chi_A\chi_A \rangle - \langle \chi_A\chi_A | \frac{1}{r_{12}} | \chi_A\chi_B \rangle - \langle \chi_A\chi_B | \frac{1}{r_{12}} | \chi_A\chi_A \rangle \\ &\quad + \langle \chi_A\chi_B | \frac{1}{r_{12}} | \chi_A\chi_B \rangle + \langle \chi_B\chi_A | \frac{1}{r_{12}} | \chi_A\chi_A \rangle - \langle \chi_B\chi_A | \frac{1}{r_{12}} | \chi_A\chi_B \rangle \\ &\quad - \langle \chi_B\chi_B | \frac{1}{r_{12}} | \chi_A\chi_A \rangle + \langle \chi_B\chi_B | \frac{1}{r_{12}} | \chi_A\chi_B \rangle \Bigg) \end{aligned} \tag{195}$$

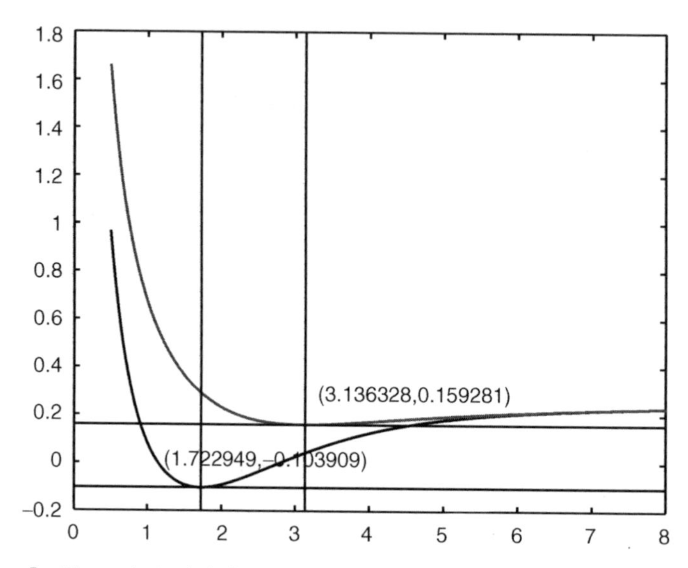

Figure 5. The vertical axis is the potential energy minus $2E_g$ (H) in hartree, and the x axis is R in a_0. Both $U_g^{(0)}$ and $U_e^{(0}$ were plotted for comparison.

Putting them all together, and letting $\zeta = 1$, we can obtain the excited-state potential energy surface (see Fig. 5). We wish to emphasize again that we calculate the energy for a range of the nuclear coordinate because we need to find the minima. In crude BOA, further properties of the molecules, such as vibrational mode frequency (determined by the force constant), anharmonicity, and so on, can be calculated in higher order perturbation, instead of being extracted from the PES curve.

In Figure 5, it can be seen that there is a minimum in the excited state as well. In more realistic calculations, such minimum was not observed. Note that the crude BOA is based on expanding the total wave function in terms of the basis functions obtained at the equilibrium position. In the expression, it seems that we have to find the equilibrium position for each electronic level. This is not practical because if we choose different center of expansion for different electronic levels, we will have to calculate a lot of matrix elements with different centers. Further, this might be impossible because there will be excited states that do not have an equilibrium geometry. Therefore, we chose to expand everything in terms of the basis function at the equilibrium position of the ground electronic state. The zeroth-order electronic energy $U_e^{(0)}$ is also calculated under the equilibrium geometry of the ground state. In other words, the vertical energy difference is used in Eq. (190).

It can be shown that

$$\langle\Psi_e|\left(\frac{\partial V}{\partial R}\right)_0|\Psi_g\rangle = \frac{1}{\sqrt{2(1-S_{AB}^2)}}\left[\langle\chi_A|\left(\frac{\partial}{\partial R}\frac{1}{r_{B1}}\right)_0|\chi_A\rangle - \langle\chi_B|\left(\frac{\partial}{\partial R}\frac{1}{r_{A1}}\right)_0|\chi_B\rangle + \langle\chi_A|\left(\frac{\partial}{\partial R}\frac{1}{r_{A1}}\right)_0|\chi_A\rangle - \langle\chi_B|\left(\frac{\partial}{\partial R}\frac{1}{r_{B1}}\right)_0|\chi_B\rangle\right] \tag{196}$$

but both $\langle\chi_A|\left(\frac{\partial}{\partial R}\frac{1}{r_{A1}}\right)_0|\chi_A\rangle$ and $\langle\chi_B|\left(\frac{\partial}{\partial R}\frac{1}{r_{B1}}\right)_0|\chi_B\rangle$ are zero, and

$$\langle\Psi_g|\left(\frac{\partial^2 V}{\partial R^2}\right)_0|\Psi_g\rangle = \frac{2}{R^3} - \frac{1}{1+S_{AB}}\left[\langle\chi_A|\left(\frac{\partial^2}{\partial R^2}\frac{1}{r_{A1}}\right)_0|\chi_A\rangle + 2\langle\chi_A|\left(\frac{\partial^2}{\partial R^2}\frac{1}{r_{A1}}\right)_0|\chi_B\rangle + \langle\chi_B|\left(\frac{\partial^2}{\partial R^2}\frac{1}{r_{A1}}\right)_0|\chi_B\rangle + \langle\chi_A|\left(\frac{\partial^2}{\partial R^2}\frac{1}{r_{B1}}\right)_0|\chi_A\rangle + 2\langle\chi_A|\left(\frac{\partial^2}{\partial R^2}\frac{1}{r_{B1}}\right)_0|\chi_B\rangle + \langle\chi_B|\left(\frac{\partial^2}{\partial R^2}\frac{1}{r_{B1}}\right)_0|\chi_B\rangle\right] \tag{197}$$

By replacing all of the atomic orbitals with Gaussian functions, we calculate the matrix elements

$$\langle\gamma_A|\left(\frac{\partial}{\partial R}\frac{1}{r_{B1}}\right)_0|\gamma_A\rangle = \frac{1}{R^2}\left[\frac{1}{2}\operatorname{erf}\left(\sqrt{2\alpha_A}\,R\right) - \frac{\sqrt{2\alpha_A}\,R}{\sqrt{\pi}}e^{-2\alpha_A R^2}\right] \tag{198}$$

$$\langle\gamma_B|\left(\frac{\partial}{\partial R}\frac{1}{r_{A1}}\right)_0|\gamma_B\rangle = \frac{1}{R^2}\left[\frac{1}{2}\operatorname{erf}\left(\sqrt{2\alpha_B}\,R\right) - \frac{\sqrt{2\alpha_B}\,R}{\sqrt{\pi}}e^{-2\alpha_B R^2}\right] \tag{199}$$

$$\langle\gamma_A|\left(\frac{\partial^2}{\partial R^2}\frac{1}{r_{A1}}\right)_0|\gamma_A\rangle = -\frac{(2\alpha_A)^{3/2}}{3\sqrt{\pi}} \tag{200}$$

$$\langle\gamma_B|\left(\frac{\partial^2}{\partial R^2}\frac{1}{r_{B1}}\right)_0|\gamma_B\rangle = -\frac{(2\alpha_B)^{3/2}}{3\sqrt{\pi}} \tag{201}$$

$$\langle\gamma_A|\left(\frac{\partial^2}{\partial R^2}\frac{1}{r_{A1}}\right)_0|\gamma_B\rangle = \frac{1}{\sqrt{\pi}}(4\alpha_A\alpha_B)^{3/4}\frac{e^{-\alpha_A\alpha_B R^2/(\alpha_A+\alpha_B)}}{\left(\frac{\alpha_B^2}{\alpha_A+\alpha_B}R^2\right)^{3/2}}\left[\frac{\sqrt{\pi}}{2}\operatorname{erf}\left(\sqrt{\frac{\alpha_B^2}{\alpha_A+\alpha_B}}R\right) - \sqrt{\frac{\alpha_B^2}{\alpha_A+\alpha_B}}Re^{-\frac{\alpha_B^2}{\alpha_A+\alpha_B}R^2} - \left(\frac{\alpha_B^2}{\alpha_A+\alpha_B}R^2\right)^{3/2}e^{-\alpha_B^2R^2/\alpha_A+\alpha_B}\right] \tag{202}$$

$$\langle \gamma_A | \left(\frac{\partial^2}{\partial R^2} \frac{1}{r_{B1}} \right)_0 | \gamma_A \rangle = \frac{1}{\sqrt{\pi} R^3} \left[\frac{\sqrt{\pi}}{2} \operatorname{erf}\left(\sqrt{2\alpha_A} R \right) - \sqrt{2\alpha_A} R e^{-2\alpha_A R^2} - (2\alpha_A R^2)^{3/2} e^{-2\alpha_A R^2} \right] \tag{203}$$

We found that

$$U_g^{(2)} = 0.179874 \tag{204}$$

in the atomic unit. This is the force constant of the oscillator. To demonstrate that this number agrees reasonably with that extracted from the potential energy curve obtained in other quantum chemical calculation, we wish to show that the parabolic curve defined by

$$U = U_g^{(2)} (R - R_0)^2 + U_0 \tag{205}$$

matches the shape of the potential energy curves obtained in other calculations near the bottom of the potential. For this purpose, we chose to compare our result to that from the simple MO calculation done by Slater [14]. This comparison is shown in Figure 6. The equilibrium position is shifted to coincide with that calculated by Slater. Our force constant value appears to be reasonable.

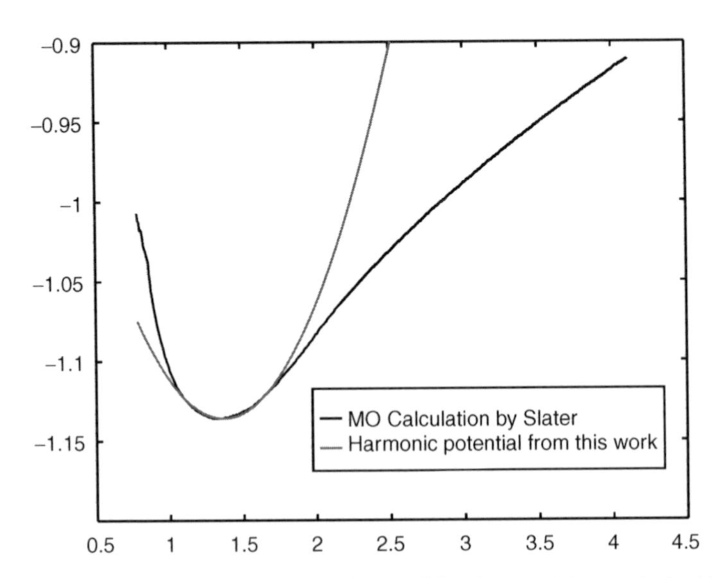

Figure 6. Matching the calculated harmonic potential to the potential curve obtained by Slater with simple MO theory.

VI. CONCLUSIONS

In the crude BO approximation, the problem of PES crossing can be avoided. However, the price is to pay. First, there will be degeneracy instead of crossing that we would encounter; second, all of the molecular properties have to be obtained by carrying out the perturbation calculation, which involves the computation of a huge number of matrix elements in realistic cases; third, since the expansion is around one nuclear configuration, the speed of convergence of the perturbation series might be a problem when the nuclear motion is significant.

Nevertheless, the examination of the applicability of the crude BO approximation can start now because we have worked out basic methods to compute the matrix elements. With the advances in the capacity of computers, the test of these methods can be done in lower and lower cost. In this work, we have obtained the formulas and shown their applications for the simple cases, but workers interested in using these matrix elements in their work would find that it is not difficult to extend our results to higher order derivatives of Coulomb interaction, or the cases of more-than-two-atom molecules.

APPENDIX A: USEFUL INTEGRALS

First, it will be very useful to remember that

$$\begin{aligned}
I_0 &\equiv \int_{-\infty}^{\infty} e^{-ax^2} dx \\
I_0^2 &= \int_{r=0}^{\infty}\int_{\theta=0}^{2\pi} e^{-ar^2} r\,dr\,d\theta = 2\pi \int_0^{\infty} e^{-ar^2} r\,dr = \pi \int_0^{\infty} e^{-ar^2} dr^2 \\
&= -\frac{\pi}{a}\int_1^0 de^{-ar^2} = \frac{\pi}{a} \\
I_0 &= \sqrt{\frac{\pi}{a}}
\end{aligned} \tag{A.1}$$

With a positive integer n, we find

$$I_{2n-1} \equiv \int_{-\infty}^{\infty} dx\, x^{2n-1} e^{-ax^2} = 0 \tag{A.2}$$

$$\begin{aligned}
I_{2n} &\equiv \int_{-\infty}^{\infty} dx\, x^{2n} e^{-ax^2} = 2\int_0^{\infty} dx\, x^{2n} e^{-ax^2} = \int_0^{\infty} x^{2n-1} e^{-ax^2} dx^2 \\
&= -\frac{1}{a}\int_{x=0}^{\infty} x^{2n-1} de^{-ax^2} = -\frac{1}{a}\left[x^{2n-1} e^{-ax^2}\Big|_{x=0}^{\infty} - (2n-1)\int_0^{\infty} dx\, x^{2(n-1)} e^{-ax^2} \right] \\
&= \frac{2n-1}{a}\int_0^{\infty} dx\, x^{2(n-1)} e^{-ax^2} = \frac{2n-1}{2a} I_{2(n-1)}
\end{aligned} \tag{A.3}$$

By doing this iteratively, we can show that

$$I_{2n} = \frac{(2n-1)!!}{2^n a^n} I_0 = \frac{(2n-1)!!}{2^n a^n} \sqrt{\frac{\pi}{a}} \tag{A.4}$$

Note that

$$(2n-1)!! = \frac{(2n)!}{(2n)!!} = \frac{(2n)!}{2^n n!} \tag{A.5}$$

we have

$$I_{2n} = \frac{(2n)!}{2^{2n} n! a^n} \sqrt{\frac{\pi}{a}} \tag{A.6}$$

The following integral is often encountered

$$I_{2\nu+1} \equiv \int_{-\infty}^{\infty} dt \left(1 + t^2\right)^{-(\nu+3/2)} \tag{A.7}$$

Letting $t = \tan\theta$, and noticing that $dt = d\theta/\cos^2\theta$, and $(1+t^2)^{-(\nu+3/2)} = (\cos\theta)^{2\nu+3}$, we obtain

$$\begin{aligned} I_{2\nu+1} &= \int_{\theta=-\pi/2}^{\pi/2} (\cos\theta)^{2\nu+1} d\theta \\ &= \int_{\theta=-\pi/2}^{\pi/2} (\cos\theta)^{2\nu} d\sin\theta \\ &= \sin\theta(\cos\theta)^{2\nu}\Big|_{\theta=-\pi/2}^{\pi/2} + 2\nu \int_{\theta=-\pi/2}^{\pi/2} (\cos\theta)^{2\nu-1} \sin^2\theta \, d\theta \\ &= -2\nu \int_{\theta=-\pi/2}^{\pi/2} (\cos\theta)^{2\nu+1} d\theta + 2\nu \int_{\theta=-\pi/2}^{\pi/2} (\cos\theta)^{2\nu-1} d\theta \qquad \text{(A.8)} \\ &= -2\nu I_{2\nu+1} + 2\nu I_{2\nu-1} \qquad \text{(A.9)} \end{aligned}$$

Therefore,

$$I_{2\nu+1} = \frac{2\nu}{2\nu+1} I_{2\nu-1} \tag{A.10}$$

Consequently,

$$\begin{aligned} I_{2\nu+1} &= \frac{2\nu}{2\nu+1} I_{2\nu-1} \\ I_{2\nu-1} &= \frac{2\nu-2}{2\nu-1} I_{2\nu-3} \\ &\vdots \\ I_3 &= \frac{2}{3} I_1 \end{aligned}$$

and

$$\begin{aligned} I_1 &= \int_{\theta=-\pi/2}^{\pi/2} \cos\theta \, d\theta \\ &= \int_{\theta=-\pi/2}^{\pi/2} d\sin\theta \\ &= 2 \end{aligned} \tag{A.11}$$

Thus

$$\begin{aligned} I_{2\nu+1} &= \frac{(2\nu)!!}{(2\nu+1)!!} I_1 \\ &= 2\frac{(2\nu)!!(2\nu)!!}{(2\nu+1)!} \\ &= 2\frac{4^{\nu}(\nu!)^2}{(2\nu+1)!} \end{aligned} \tag{A.12}$$

Next, we shall consider the integral basically corresponding to the Rys' polynomial problem [15,16]. Letting

$$J_n(a) = \int_0^1 dx\, x^{2n} e^{-ax^2} \tag{A.13}$$

we find that

$$\begin{aligned} J_n &= \frac{1}{2}\int_0^1 x^{2n-1} e^{-ax^2} dx^2 = -\frac{1}{2a}\int_{x=0}^1 x^{2n-1} de^{-ax^2} \\ &= -\frac{1}{2a}\left[x^{2n-1} e^{-ax^2}\Big|_{x=0}^1 - (2n-1)\int_0^1 x^{2(n-1)} e^{-ax^2} dx \right] \\ &= \frac{-e^{-a}}{2a} + \frac{2n-1}{2a}\int_0^1 x^{2(n-1)} e^{-ax^2} dx \\ &= \frac{-e^{-a}}{2a} + \frac{2n-1}{2a} J_{n-1} \end{aligned} \tag{A.14}$$

We can write down the iterative formula

$$J_n = \frac{-e^{-a}}{2a} + \frac{(2n-1)!!}{2a(2n-3)!!} J_{n-1}$$
$$\frac{(2n-1)!!}{2a(2n-3)!!} J_{n-1} = \frac{(2n-1)!!}{2a(2n-3)!!} \frac{-e^{-a}}{2a} + \frac{(2n-1)!!}{(2a)^2(2n-5)!!} J_{n-2}$$
$$\frac{(2n-1)!!}{(2a)^2(2n-5)!!} J_{n-2} = \frac{(2n-1)!!}{(2a)^2(2n-5)!!} \frac{-e^{-a}}{2a} + \frac{(2n-1)!!}{(2a)^3(2n-7)!!} J_{n-3}$$
$$\vdots$$
$$\frac{(2n-1)!!}{(2a)^{n-1}(1)!!} J_1 = \frac{(2n-1)!!}{(2a)^{n-1}(1)!!} \frac{-e^{-a}}{2a} + \frac{(2n-1)!!}{(2a)^n} J_0$$

Summing up all these formula and noticing that

$$J_0 = \int_0^1 e^{-ax^2} dx = \frac{\sqrt{\pi}}{2\sqrt{a}} \operatorname{erf}\left(\sqrt{a}\right) \tag{A.15}$$

we found

$$\begin{aligned} J_n(a) &= \frac{(2n-1)!!}{(2a)^n} \frac{\sqrt{\pi}}{2\sqrt{a}} \operatorname{erf}\left(\sqrt{a}\right) - \frac{(2n-1)!!e^{-a}}{(2a)^n} \left[\frac{1}{1!!} + \frac{2a}{3!!} + \cdots + \frac{(2a)^{n-1}}{(2n-1)!!} \right] \\ &= \frac{(2n-1)!!}{(2a)^n} \frac{\sqrt{\pi}}{2\sqrt{a}} \operatorname{erf}\left(\sqrt{a}\right) - \frac{(2n-1)!!e^{-a}}{(2a)^n} \sum_{k=0}^{n-1} \frac{(2a)^k}{(2k+1)!!} \\ &= \frac{(2n)!}{n!(4a)^n} \frac{\sqrt{\pi}}{2\sqrt{a}} \operatorname{erf}\left(\sqrt{a}\right) - \frac{(2n)!e^{-a}}{n!(4a)^n} \sum_{k=0}^{n-1} \frac{k!(4a)^k}{(2k+1)!} \\ &= \frac{(2n)!}{n!(4a)^n} \left[\frac{1}{2} \sqrt{\frac{\pi}{a}} erf\left(\sqrt{a}\right) - e^{-a} \sum_{k=0}^{n-1} \frac{k!}{(2k+1)!} (4a)^k \right] \end{aligned} \tag{A.16}$$

Finally, we discuss the integral of the form

$$\begin{aligned} & I_\xi(n_1, n_2, n_3, \alpha_A, \alpha_B, R) \\ & \quad \equiv \int_0^1 d\xi\, \xi^{2n_1} \left(1 - \xi^2\right)^{n_2} \left(\alpha_A + \alpha_B \xi^2\right)^{n_3} e^{-\frac{\alpha_B^2 R^2}{\alpha_A + \alpha_B}} \end{aligned} \tag{A.17}$$

which is a more general form of Eq. (131). The modification is simple:

$$
\begin{aligned}
I_{\xi}&(n_1, n_2, n_3, \alpha_A, \alpha_B, R) \\
&= \sum_{\nu_2=0}^{n_2} \sum_{\nu_3=0}^{n_3} C_{\nu_2}^{n_2} C_{\nu_3}^{n_3} (-1)^{\nu_2} \alpha_A^{n3-\nu_3} \alpha_B^{\nu_3} \int_0^1 d\xi \, \xi^{2(n_1+\nu_2+\nu_3)} e^{-\alpha_2^2 R^2 \xi^2/(\alpha_1+\alpha_2)} \\
&= \sum_{\nu_2=0}^{n_2} \sum_{\nu_3=0}^{n_3} C_{\nu_2}^{n_2} C_{\nu_3}^{n_3} (-1)^{\nu_2} \alpha_A^{n_3-\nu_3} \alpha_B^{\nu_3} J_{n_1+\nu_2+\nu_3} \left(\frac{\alpha_B^2}{\alpha_A + \alpha_B} R^2 \right) \qquad \text{(A.18)}
\end{aligned}
$$

Acknowledgments

This work was supported by NSC (Taiwan) and Academia Sinica.

References

1. C. Woywod, W. Domcke, A. L. Sobolewski, and H.-J. Werner, *J. Chem. Phys.* **100**, 1400 (1994).
2. A. M. Mebel, M. Baer, and S. H. Lin, *J. Chem. Phys.* **112**, 10703 (2000).
3. M. Baer, S. H. Lin, A. Alijah, S. Adhikari, and G. D. Billing, *Phys. Rev.* **62A**, 2506 (2000).
4. S. Krempl, M. Winterstetter, H. Plöhn, and W. Domcke, *J. Chem. Phys.* **100**, 926 (1994).
5. M. Born and R. Oppenheimer, *Ann. Phys.* (Leipzig) **84**, 457 (1927).
6. M. Born, *Nachr. Acad. Wiss. Goeff. Math.-Phys.* **2**, 1 (1951).
7. M. Born and K. Huang, '*Dynamical Theory of Crystal Lattices*,' Oxford University Press, 1954, pp. 166–177, 402–407.
8. C. J. Ballhausen and A. E. Hansen, *Ann. Rev. Phys. Chem.* **23**, 15 (1972).
9. N. C. Handy, Y. Yamaguchi, and H. F. Schaeffer, III, *J. Chem. Phys.* **84**, 4481 (1986).
10. A. Martin, J.-M. Richard, and T. T. Wu, *Phys. Rev.* **46A**, 3697 (1992).
11. W. Cenek and W. Kutzelnigg, *Chem. Phys. Lett.* **266**, 383 (1997).
12. S. Golden, *Mol. Phys.* **93**, 421 (1998).
13. T. Azumi and K. Matsuzaki, *Photochem. Photobiol.* **25**, 315 (1977).
14. *Quantum Theory of Matter*, 2nd ed., John C. Slater, McGraw-Hill, New York 1968, p. 420, Fig. 21-2.
15. M. Dupuis, J. Rys, and H. F. King, *J. Chem. Phys.* **65**, 1 (1976).
16. H. F. King and M. Dupuis, *J. Comput. Phys.* **21**, 144 (1976).

CONICAL INTERSECTIONS AND THE SPIN–ORBIT INTERACTION

SPIRIDOULA MATSIKA and DAVID R. YARKONY

Department of Chemistry, Johns Hopkins University, Baltimore, MD

CONTENTS

The Role of Degenerate States in Chemistry: A Special Volume of Advances in Chemical Physics, Volume 124, Edited by Michael Baer and Gert Due Billing. Series Editors I. Prigogine and Stuart A. Rice.
ISBN 0-471-43817-0.

I. INTRODUCTION

Conical intersections are known to play a key role in nonrelativistic, spin-conserving electronically nonadiabatic processes [1]. If we include the spin–orbit interaction we introduce new nonadiabatic pathways and unexpected complications. By coupling states of different spin-multiplicity the spin–orbit interaction gives rise to spin-nonconserving transitions while making conical intersections out of intersections that otherwise would not be. However, the spin–orbit interaction produces a more subtle but no less significant effect when the molecule in question has an odd number of electrons. Let η be the dimension of the branching space, the space in which the conical topography is evinced. More precisely, the branching space is the smallest space in whose orthogonal complement the degeneracy is lifted only at quadratic or higher order in displacements, if it is lifted at all. The orthogonal complement of the branching space is referred to as the seam. The dimension of the seam is $\vartheta = N^{\text{int}} - \eta$, where N^{int} is the number of internal coordinates. Here $\eta = 2$ in the nonrelativistic case and when the spin–orbit interaction is included, provided the molecule has an even number of electrons. However, for a molecule with an odd number of electrons, an odd electron molecule, the dimension of the branching space is 5, in general, or 3 when the system is restricted to C_s symmetry [2].

To study the ramifications of this change, the locus of the seam should be known. However, locating points on this seam of conical intersection is a challenging undertaking. To better understand the difficulty of the task, consider the history of the accidental same-symmetry intersection. For the nonrelativistic Coulomb Hamiltonian, conical intersections can be classified using the point group symmetry of the intersecting states. Intersections are symmetry-required, accidental symmetry-allowed, or accidental same-symmetry, according to whether the electronic states in question carry a multidimensional irreducible representation, distinct one-dimensional irreducible representations, or the same irreducible representation of the spatial point group. While the existence of same-symmetry conical intersections was firmly established over 70 years ago [3], until approximately a decade ago [4,5], virtually all conical intersections based on ab initio wave functions were determined with the help of symmetry. This is a consequence, in part, of the fact that to locate a single point of conical intersection for a same symmetry intersection a two-dimensional branching plane must be searched, whereas for an accidental symmetry-allowed intersection only an one-dimension search is required. Indeed, it was only in the last decade, after the introduction of efficient algorithms [6,7] for locating same-symmetry intersections, that their true significance began to emerge. The situation for odd electron molecules when the spin–orbit interaction is included in the Hamiltonian is similar, but even more extreme, since a five- (or three-) dimensional branching space must be searched.

Contrary to this gloomy assessment, it is rapidly becoming possible to describe non-adiabatic processes driven by conical intersections, for which the spin–orbit interaction cannot be neglected, on the same footing that has been so useful in the nonrelativistic case. An effective algorithm for locating points of conical intersection for odd electron molecules has been developed [8] and an analytic representation of the energies and derivative couplings in the vicinity of these points of conical intersection has been determined [9,10] based on degenerate perturbation theory [11,12]. These advances, in addition to providing conceptual insights, will lead to a more rigorous approach to nonadiabatic dynamics whose computational utility increases with the size of the spin–orbit interaction.

In this chapter, recent advances in the theory of conical intersections for molecules with an odd number of electrons are reviewed. Section II presents the mathematical basis for these developments, which exploits a degenerate perturbation theory previously used to describe conical intersections in nonrelativistic systems [11,12] and Mead's analysis of the noncrossing rule in molecules with an odd number of electrons [2]. Section III presents numerical illustrations of the ideas developed in Section II. Section IV summarizes and discusses directions for future work.

II. THEORY

A. The Electronic Hamiltonian

In this work, relativistic effects are included in the no-pair or large component only approximation [13]. The total electronic Hamiltonian is $H^{\mathrm{e}}(\mathbf{r};\mathbf{R}) = H^0(\mathbf{r};\mathbf{R}) + H^{\mathrm{so}}(\mathbf{r};\mathbf{R})$, where $H^0(\mathbf{r};\mathbf{R})$ is the nonrelativistic Coulomb Hamiltonian and $H^{\mathrm{so}}(\mathbf{r};\mathbf{R})$ is a spin–orbit Hamiltonian. The relativistic (nonrelativistic) eigenstates, $\Psi_i^e(\Psi_{\mathrm{I}}^0)$, are eigenfunctions of $H^e(\mathbf{r};\mathbf{R})(H^0(\mathbf{r};\mathbf{R}))$. Lower (upper) case letters will be used to denote eigenfunctions H^e (H^0). A point of conical intersection of states i, j of $H^e[I$, J of $H^0]$ will be denoted $\mathbf{R}^{x,ij}[\mathbf{R}^{x,IJ}]$.

B. Time-Reversal Symmetry

Table I summarizes the differences in the dimension of the branching space. The origin of these differences is the behavior of the wave functions under

TABLE I
η the Dimension of Branching Space

No. of e^-	H^0	H^e
Even	2	2
Odd	2	5^a

$^a\eta = 3$ when C_s symmetry is present.

time-reversal symmetry [14,15]. The time-reversal operator, T is an antiunitary operator, that is, $\langle\phi|\psi\rangle^* = \langle T\phi|T\psi\rangle$. In addition $T^2 = +1(-1)$ if the number of electrons is even (odd). For a molecule with an odd number of electrons

$$\langle\phi|T\phi\rangle^* = \langle T\phi|T^2\phi\rangle = -\langle T\phi|\phi\rangle = -\langle\phi|T\phi\rangle^* \qquad \text{so that} \qquad \langle\phi|T\phi\rangle = 0 \tag{1a}$$

that is, ϕ and $T\phi$ are orthogonal and degenerate, since T commutes with H^e. This degeneracy owing to time-reversal symmetry is referred to as Kramers' degeneracy [16]. For a molecule with an even number of electrons, ϕ and $T\phi$ are linearly dependent. With the choice $\phi = T\phi$

$$\langle\phi|H^e\psi\rangle^* = \langle T\phi|TH^e\psi\rangle = \langle T\phi|H^eT\psi\rangle = \langle\phi|H^e\psi\rangle \tag{1b}$$

so that $\langle\phi|H^e\psi\rangle$ is real valued. For an odd electron system

$$\langle\phi|H^eT\psi\rangle^* = \langle T\phi|TH^eT\psi\rangle = \langle T\phi|H^eT^2\psi\rangle = -\langle\psi|H^eT\phi\rangle^* \tag{1c}$$

so that, for example, for $\phi = \psi$, $\langle\phi|H^eT\phi\rangle = 0$.

A set of functions will be referred to as time-reversal adapted, provided that for each ϕ in the set $T\phi$ is also in the set.

We are now in a position to explain the results of Table I. As a consequence of the degeneracy of ϕ and $T\phi$, at a conical intersection there are four degenerate functions Ψ^e_i, Ψ^e_j and $T\Psi^e_i \equiv \Psi^e_{Ti}$, $T\Psi^e_j \equiv \Psi^e_{Tj}$. By using Eq. (1c), an otherwise arbitrary Hermitian matrix in this four function time-reversal adapted basis has the form

$$\mathbf{H}^e = \frac{(H^e_{ii} + H^e_{jj})}{2}\mathbf{I} + \begin{pmatrix} -\Delta H^e_{ji} & H^e_{ij} & 0 & H^e_{iTj} \\ H^{e^*}_{ij} & \Delta H^e_{ji} & -H^e_{iTj} & 0 \\ 0 & -H^{e^*}_{iTj} & -\Delta H^e_{ji} & H^{e^*}_{ij} \\ H^{e^*}_{iTj} & 0 & H^e_{ij} & \Delta H^e_{ji} \end{pmatrix} \tag{2a}$$

The eigenvalues of this matrix are [17]

$$\varepsilon_{\pm}(\mathbf{R}) = \frac{H^e_{ii}(\mathbf{R}) + H^e_{jj}(\mathbf{R})}{2} \pm [\Delta H^e_{ji}(\mathbf{R})^2 + |H^e_{ij}(\mathbf{R})|^2 + |H^e_{iTj}(\mathbf{R})|^2]^{1/2} \tag{2b}$$

each of which is twofold degenerate. Since ΔH^e_{ij} is real valued while H^e_{ij} and H^e_{iTj} are complex valued, the five conditions for degeneracy at $\mathbf{R}^{x,ij}$ are

$$\Delta H^e_{ji}(\mathbf{R}^{x,ij}) = 0 \qquad H^e_{ij}(\mathbf{R}^{x,ij}) = 0 \qquad H^e_{iTj}(\mathbf{R}^{x,ij}) = 0 \tag{3}$$

TABLE II
C_s Double Group

	E	σ	R	σ^3
a'	1	1	1	1
a''	1	-1	1	-1
e'	1	i	-1	$-i$
e''	1	$-i$	-1	i

When C_s symmetry is present, Ψ_k^e and $T\Psi_k^e$ $k = i, j$ can be chosen to transform according to the e' and e'' irreducible representations of the C_s double group (see Table II) so that $H_{iTj}^e(\mathbf{R}) = 0$ by symmetry. In this case, there are only three conditions for a degeneracy $\Delta H_{ji}^e(\mathbf{R}) = 0$ and $H_{ij}^e(\mathbf{R}) = 0$, $\mathbf{H}^e$ is block diagonal

$$\mathbf{H}^e = \frac{(H_{ii}^e + H_{jj}^e)}{2}\mathbf{I} + \begin{pmatrix} -\Delta H_{ji}^e & H_{ij}^e & 0 & 0 \\ H_{ij}^{e^*} & \Delta H_{ji}^e & 0 & 0 \\ & & -\Delta H_{ji}^e & H_{ij}^{e^*} \\ & & H_{ij}^e & \Delta H_{ji}^e \end{pmatrix} \tag{2c}$$

and clearly evinces Kramers' degeneracy. Finally, for the even electron case the $T\Psi_i^e$ are linearly dependent and only one of the diagonal blocks survives,

$$\mathbf{H}^e = \frac{(H_{ii}^e + H_{jj}^e)}{2}\mathbf{I} + \begin{pmatrix} -\Delta H_{ji}^e & H_{ij}^e \\ H_{ij}^e & \Delta H_{ji}^e \end{pmatrix} \tag{2d}$$

and H_{ij}^e is real valued, so only two conditions need be satisfied.

This analysis is heuristic in the sense that the Hilbert spaces in question are in general of large, if not infinite, dimension while we have focused on spaces of dimension four or two. A form of degenerate perturbation theory [3] can be used to demonstrate that the preceding analysis is essentially correct and, to provide the means for locating and characterizing conical intersections.

C. Perturbation Theory

Ψ_i^e is expanded in a basis of time-reversal adapted configuration state functions [8] (TRA–CSFs, ψ^e)

$$\Psi_i^e(\mathbf{r};\mathbf{R}) = \sum_{\alpha=1}^{N^{\text{CSF}}} d_\alpha^i(\mathbf{R})\psi_\alpha^e(\mathbf{r};\mathbf{R}) \tag{4a}$$

The $\mathbf{d}^i \equiv \mathbf{d}^{\mathrm{r},i} + i\mathbf{d}^{\mathrm{i},i}$ are the solution of the electronic Schrödinger equation in the TRA–CSF basis

$$[\mathbf{H}^{e,\mathrm{r}}(\mathbf{R}) + i\mathbf{H}^{e,\mathrm{i}}(\mathbf{R}) - E_k^e(\mathbf{R})]\mathbf{d}^k(\mathbf{R}) = \mathbf{0} \tag{4b}$$

where $\mathbf{H}^{\mathrm{e}} = \mathbf{H}^{\mathrm{e,r}} + i\mathbf{H}^{\mathrm{e,i}}$. Near $\mathbf{R}^{x,ij}$ the eigenvalue problem in Eq. (4) can be simplified with the use of a crude adiabatic basis

$$\Psi_k^c(\mathbf{r};\mathbf{R}) = \sum_{\alpha=1}^{N^{\mathrm{CSF}}} d_\alpha^k(\mathbf{R}^{x,ij})\psi_\alpha^e(\mathbf{r};\mathbf{R}) \tag{5}$$

Expanding $\mathbf{H}^{\mathrm{e}}(\mathbf{R})$ to second order gives

$$\mathbf{H}^{\mathrm{e}}(\mathbf{R}) = \mathbf{H}^e(\mathbf{R}^{x,ij}) + \nabla\mathbf{H}^e(\mathbf{R}^{x,ij})\cdot\delta\mathbf{R} + 1/2\delta\mathbf{R}\cdot\nabla\nabla\mathbf{H}^e(\mathbf{R}^{x,ij})\cdot\delta\mathbf{R} \tag{6}$$

where $\delta\mathbf{R} = \mathbf{R} - \mathbf{R}^{x,ij}$. Reexpressing this result in the crude adiabatic basis gives

$$\tilde{\mathbf{H}}^e(\mathbf{R}+\delta\mathbf{R}) \equiv \mathbf{d}^\dagger(\mathbf{R}^{x,ij})\mathbf{H}^{\mathrm{e}}(\mathbf{R})\mathbf{d}(\mathbf{R}^{x,ij}) \tag{7a}$$

$$\approx \mathbf{d}(\mathbf{R}^{x,ij})^\dagger[\mathbf{H}^e(\mathbf{R}^{x,ij}) + \nabla\mathbf{H}^e(\mathbf{R}^{x,ij})\cdot\delta\mathbf{R} + 1/2\delta\mathbf{R}\cdot\nabla\nabla\mathbf{H}^e(\mathbf{R}^{x,ij})\cdot\delta\mathbf{R}]\mathbf{d}(\mathbf{R}^{x,ij}) \tag{7b}$$

$$\equiv \mathbf{E}^e(\mathbf{R}^{x,ij}) + \underline{\tilde{\mathbf{H}}}^{[1]}\cdot\delta\mathbf{R} + 1/2\delta\mathbf{R}\cdot\underline{\underline{\tilde{\mathbf{H}}}}^{[2]}\cdot\delta\mathbf{R} \tag{7c}$$

where $\dagger$ denotes the complex conjugate transpose,

$$E^e(\mathbf{R}^{x,ij})_{kl} = \delta_{kl}E_l^e(\mathbf{R}^{x,ij}) \tag{7d}$$

a single (double) bar under a quantity denotes a vector (matrix) of matrices, so that

$$\underline{\tilde{\mathbf{H}}}^{[1]} = (\tilde{\mathbf{H}}^{(1),1}, \tilde{\mathbf{H}}^{(1),2}, \ldots, \tilde{\mathbf{H}}^{(1),N^{\mathrm{int}}}) \tag{8a}$$

$$\underline{\underline{\tilde{\mathbf{H}}}}^{[2]} = (\tilde{\mathbf{H}}^{(2),11}, \tilde{\mathbf{H}}^{(2),21}, \tilde{\mathbf{H}}^{(2),31}, \ldots, \tilde{\mathbf{H}}^{(2),N^{\mathrm{int}}N^{\mathrm{int}}}) \tag{8b}$$

with

$$\tilde{\mathbf{H}}_{mn}^{(1),\kappa}(\mathbf{R}) = \mathbf{d}^{m^\dagger}(\mathbf{R}^{x,ij})\left[\frac{\partial}{\partial R_\kappa}\mathbf{H}^e(\mathbf{R})\right]\mathbf{d}^n(\mathbf{R}^{x,ij}) \tag{9a}$$

and

$$\tilde{\mathbf{H}}_{mn}^{(2),\kappa\kappa'}(\mathbf{R}) = \mathbf{d}^{m^\dagger}(\mathbf{R}^{x,ij})\left[\frac{\partial}{\partial R_\kappa \partial R_{\kappa'}}\mathbf{H}^e(\mathbf{R})\right]\mathbf{d}^n(\mathbf{R}^{x,ij}) \tag{9b}$$

The Ψ_k^e are expanded in the crude adiabatic basis

$$\Psi_k^e(\mathbf{r};\mathbf{R}) = \sum_{l\in Q} \xi_l^k(\mathbf{R})\Psi_l^c(\mathbf{r};\mathbf{R}) + \sum_{l\in P} \Xi_l^k(\mathbf{R})\Psi_l^c(\mathbf{r};\mathbf{R}) \tag{10}$$

where Q is spanned by the degenerate functions at $\mathbf{R}^{x,ij}$ and P is its orthogonal complement. To describe the vicinity of a conical intersection we require the first-order contributions in $\delta\mathbf{R}$ to Eq. (4b). To accomplish this, we expand $\mathrm{E}_\mathrm{i}^\mathrm{e}(\mathbf{R})$, $\xi(\mathbf{R})$, $\Xi(\mathbf{R})$ in powers of $\delta\mathbf{R}$, giving

$$\xi^k(\mathbf{R}) = \xi^{(0),k}(\mathbf{R}^{x,ij}) + \underline{\xi}^{(1),k}(\mathbf{R}^{x,ij})\cdot\delta\mathbf{R} + 1/2\delta\mathbf{R}^\dagger\cdot\underline{\underline{\xi}}^{(2),k}(\mathbf{R}^{x,ij})\cdot\delta\mathbf{R} \tag{11a}$$

$$\Xi^k(\mathbf{R}) = \underline{\Xi}^{(1),k}(\mathbf{R}^{x,ij})\cdot\delta\mathbf{R} + 1/2\delta\mathbf{R}^\dagger\cdot\underline{\underline{\Xi}}^{(2),k}(\mathbf{R}^{x,ij})\cdot\delta\mathbf{R} \tag{11b}$$

$$E_k^e(\mathbf{R}) = E_k^e(\mathbf{R}^{x,ij}) + E_k^{e,(1)}(\mathbf{R}) + E_k^{e,(2)}(\mathbf{R}) \tag{11c}$$

In Eq. (11b), we observed that since the crude adiabatic basis is used $\Xi^{(0)k} = 0$, for $k\varepsilon Q$. Therefore the degeneracy is lifted at first order in the Q-space only, which is therefore used to identified the branching space. The first-order result is

$$(\tilde{\underline{\mathbf{H}}}^{[1]}\cdot\delta\mathbf{R} - E_i^{e,(1)}(\mathbf{R}))\xi^{(0),i}(\mathbf{R}^{x,ij}) = \mathbf{0} \tag{12}$$

Equation (12) and the qualifying equalities, Eqs. (8) and (9) are the lynchpin for the remainder of this work.

D. Perturbation Theory, Time-Reversal Symmetry, and Conical Intersections

To procede further, it is essential to distinguish between even and odd electron systems. While Eq. (12) is formally independent of the dimension of Q, in the former case there are two independent degenerate functions at $\mathbf{R}^{x,ij}$, Ψ_i^e and Ψ_j^e and $\tilde{\underline{\mathbf{H}}}^{[1]}$ is symmetric; while in the later case there are four degenerate functions Ψ_i^e, Ψ_j^e and $T\Psi_i^e \equiv \Psi_{Ti}^e$, $T\Psi_j^e \equiv \Psi_{\mathrm{Tj}}^e$, and $\tilde{\underline{\mathbf{H}}}^{[1]}$ is Hermitian. Here we restrict our attention to the later case. The analysis for real-valued case (using H^0) can be found in [12].

For odd electron systems in the absence of spatial symmetry $\tilde{\underline{\mathbf{H}}}^{[1]}\cdot\delta\mathbf{R}$ in Eq. (12) becomes

$$\tilde{\underline{\mathbf{H}}}^{[1]}\cdot\delta\mathbf{R} = \delta\mathbf{R}^\dagger\cdot\mathbf{s}^{ij}\mathbf{I} + \delta\mathbf{R}^\dagger\cdot\begin{pmatrix} -\mathbf{g}^{ij} & \mathbf{h}^{ij} & 0 & \mathbf{h}^{iTj} \\ \mathbf{h}^{ij^*} & \mathbf{g}^{ij} & -\mathbf{h}^{iTj} & 0 \\ 0 & -\mathbf{h}^{iTj^*} & -\mathbf{g}^{ij} & \mathbf{h}^{ij^*} \\ \mathbf{h}^{iTj^*} & 0 & \mathbf{h}^{ij} & \mathbf{g}^{ij} \end{pmatrix} \tag{13a}$$

where

$$2\mathbf{g}^{ij} = \mathbf{g}^i - \mathbf{g}^j, \qquad 2\mathbf{s}^{ij} = \mathbf{g}^i + \mathbf{g}^j \tag{13b}$$

$$\mathbf{h}^{ij}(\mathbf{R}) \equiv \mathbf{d}^i(\mathbf{R}^{x,ij})^\dagger \nabla \mathbf{H}^e(\mathbf{R})\mathrm{d}^j(\mathbf{R}^{x,ij}) \equiv \mathbf{h}^{\mathrm{r},ij}(\mathbf{R}) + i\mathbf{h}^{\mathrm{i},ij}(\mathbf{R}) \tag{13c}$$

$$\mathbf{h}^{iTj}(\mathbf{R}) \equiv \mathbf{d}^i(\mathbf{R}^{x,ij})^\dagger \nabla \mathbf{H}^{\mathrm{e}}(\mathbf{R})\mathbf{d}^{\mathrm{T}j}(\mathbf{R}^{x,ij}) \equiv \mathbf{h}^{\mathrm{r},iTj}(\mathbf{R}) + i\mathbf{h}^{\mathrm{i},iTj}(\mathbf{R}) \tag{13d}$$

and

$$\mathbf{g}^i(\mathbf{R}) \equiv \mathbf{d}^i(\mathbf{R}^0)^\dagger \nabla \mathbf{H}^{\mathrm{e}}(\mathbf{R})\mathbf{d}^i(\mathbf{R}^0) \tag{13e}$$

Equation (13) and definitions (8), (9), and (11) enable a description of the energy near, and the singular part of the derivative coupling at, $\mathbf{R}^{x,ij}$.

E. Conical Intersections: Location

At the conical intersection the $\mathbf{d}^k$, $k = i, j, Ti, Tj$ are defined only up to a unitary transformation among themselves. As a result, for a particular point $\mathbf{R}^0$ in the region where Eq. (13) is justified, $\mathbf{g}^{ij}(\mathbf{R}^{x,ij})$, $\mathbf{h}^{ij}(\mathbf{R}^{x,ij})$, and $\mathbf{h}^{iTj}(\mathbf{R}^{x,ij})$ can be chosen such that $\mathbf{R}^0 - \mathbf{R}^{x,ij}$ is parallel to $\mathbf{g}^{ij}(\mathbf{R}^{x,ij})$. In this case, expanding Eq. (3) about $\mathbf{R}^0$, with $\mathbf{R}^0 + \delta\mathbf{R} = \mathbf{R}^{x,ij}$

$$\begin{aligned} -\Delta E_{ij}(\mathbf{R}^0) &= \mathrm{Re}\,\nabla[(\mathbf{d}^i(\mathbf{R}^0) + \mathbf{d}^j(\mathbf{R}^0))^\dagger \mathbf{H}^{\mathrm{e}}(\mathbf{R})((\mathbf{d}^i(\mathbf{R}^0) - \mathbf{d}^j(\mathbf{R}^0))] \cdot \delta\mathbf{R} \\ &\equiv \nabla V_1(\mathbf{R}) \cdot \delta\mathbf{R} \end{aligned} \tag{14a}$$

$$\begin{aligned} 0 = \mathbf{h}^{ij}(\mathbf{R}^0) \cdot \delta\mathbf{R} &= \nabla[(\mathbf{d}^i(\mathbf{R}^0)^\dagger \mathbf{H}^{\mathrm{e}}(\mathbf{R})\mathbf{d}^j(\mathbf{R}^0)] \cdot \delta\mathbf{R} \\ &\equiv \nabla(V_2 + iV_3) \cdot \delta\mathbf{R} \end{aligned} \tag{14b}$$

$$\begin{aligned} 0 = \mathbf{h}^{iTj}(\mathbf{R}^0) \cdot \delta\mathbf{R} &= \nabla[(\mathbf{d}^i(\mathbf{R}^0)^\dagger \mathbf{H}^e(\mathbf{R})\mathbf{d}^{Tj}(\mathbf{R}^0)] \cdot \delta\mathbf{R} \\ &\equiv \nabla(V_4 + iV_5) \cdot \delta\mathbf{R} \end{aligned} \tag{14c}$$

where

$$\Delta E_{ij}(\mathbf{R}^0) = \mathrm{Re}[(\mathbf{d}^i(\mathbf{R}^0) + \mathbf{d}^j(\mathbf{R}^0))^\dagger \mathbf{H}^{\mathrm{e}}(\mathbf{R}^0)(\mathbf{d}^i(\mathbf{R}^0) - \mathbf{d}^j(\mathbf{R}^0))] = V_1(\mathbf{R}^0) \tag{14d}$$

Equations (14a)–(14d) form the basis for our algorithm for locating conical intersections. However, these equations determine only five (or three when C_s symmetry can be imposed) internal nuclear coordinates. Determination of any remaining internal degrees of freedom requires additional constraints. We employ the approach used in our algorithm for determining seams of conical intersection for the nonrelativistic Hamiltonian [7] where geometrical constraints $K^{\mathrm{i}}(\mathbf{R}) = 0$, and/or minimization of the energy of the crossing, provide the additional conditions. Geometrical constraints can also be used to map out the seam of conical intersection in its full dimensionality. This constrained minimization can be accomplished by minimizing the following Lagrangian [7]:

$$L^{ij}(\mathbf{R}, \boldsymbol{\xi}, \boldsymbol{\lambda}) = E_i^e(\mathbf{R}) + \sum_{i=1}^{5} V^i(\mathbf{R})\xi_i + \sum_{i=1}^{N^{\text{con}}} K^i(\mathbf{R})\lambda_i \tag{15a}$$

where $\boldsymbol{\xi}$ and $\boldsymbol{\lambda}$ are Lagrange multipliers. Expanding L^{ij} through second order yields the Newton–Raphson equations [7]:

$$\begin{bmatrix} \mathbf{Q}^{ij}(\mathbf{R}, \boldsymbol{\xi}, \boldsymbol{\lambda}) & \mathbf{v}(\mathbf{R}) & \mathbf{k}(\mathbf{R}) \\ \mathbf{v}(\mathbf{R})^{\dagger} & \mathbf{0} & \mathbf{0} \\ \mathbf{k}(\mathbf{R})^{\dagger} & \mathbf{0} & \mathbf{0} \end{bmatrix} \begin{bmatrix} \delta\mathbf{R} \\ \delta\boldsymbol{\xi} \\ \delta\boldsymbol{\lambda} \end{bmatrix} = - \begin{bmatrix} \mathbf{g}^i(\mathbf{R}) + \mathbf{v}^{\dagger}\boldsymbol{\xi} + \mathbf{k}^{\dagger}\boldsymbol{\lambda} \\ \mathbf{V}(\mathbf{R}) \\ \mathbf{K}(\mathbf{R}) \end{bmatrix} \tag{15b}$$

where $\nabla V^i \equiv \mathbf{v}^i$ and $\nabla K^i \equiv \mathbf{k}^i$ and $\mathbf{Q}^{ij} \equiv \nabla\nabla L^{ij}$. The gradients $\mathbf{v}$ are more costly to evaluate than their nonrelativistic counterparts. For this reason, it is useful to search along the direction corresponding to $\delta\mathbf{R}$ while ΔE_{ij}^e decreases. This simple extension of an idea from conjugate gradient theory can significantly reduce the computational effort needed to solve Eq. (15b). The performance of Eq. (15b) is discussed in Section III.

F. Conical Intersections: Description

In this section, notions used to describe nonrelativistic conical intersections are extended to the present case. For simplicity, unless otherwise specified we consider the $\eta = 3$ case. The analogous treatment for $\eta = 5$ will be reported in [17].

1. Orthogonal Intersection Adapted Coordinates

At a point of conical intersection, $\mathbf{R}^{x,ij}$, the four degenerate wave functions are defined up to a rotation $\mathbf{U}$, consistent with time-reversal symmetry. As a consequence of this arbitrariness the $\mathbf{g}^{ij}(\mathbf{R})$, $\mathbf{h}^{\mathrm{r},ij}(\mathbf{R})$, $\mathbf{h}^{\mathrm{i},ij}(\mathbf{R})$, $\mathbf{h}^{\mathrm{r},iTj}(\mathbf{R})$, $\mathbf{h}^{\mathrm{i},iTj}(\mathbf{R})$ need not be orthogonal. Orthogonality of $\mathbf{g}^{ij}(\mathbf{R})$, $\mathbf{h}^{\mathrm{r},ij}(\mathbf{R})$, $\mathbf{h}^{\mathrm{i},ij}(\mathbf{R})$, $\mathbf{h}^{\mathrm{r},iTj}(\mathbf{R})$, $\mathbf{h}^{\mathrm{i},iTj}(\mathbf{R})$ greatly simplifies the analysis of Eq. (13). In the nonrelativistic, $\eta = 2$, case, the degenerate Ψ_{K}^0, $\mathrm{K} = I, J$ are defined only up to a one parameter rotation. We used this flexibility to require orthogonality of $\mathbf{g}^{IJ}$ and $\mathbf{h}^{IJ}$. Below we demonstrate how this orthogonality requirement can be extended to the $\eta = 3$ case.

For $\eta = 3$ define the rotated states by

$$(\tilde{\mathbf{d}} \quad T\tilde{\mathbf{d}}) = (\mathbf{d} \quad T\mathbf{d})\mathbf{U} \tag{16a}$$

where

$$\mathbf{d}^{\dagger} = (\mathbf{d}^{i^{\dagger}} \quad \mathbf{d}^{j^{\dagger}}) \qquad \mathrm{U} = \begin{pmatrix} \mathbf{u} & \mathbf{0} \\ \mathbf{0} & \mathbf{u}^{*} \end{pmatrix} \tag{16b}$$

and

$$\mathbf{u} = \begin{pmatrix} e^{i(\alpha+\gamma)/2}\cos\beta/2 & -e^{i(-\alpha+\gamma)/2}\sin\beta/2 \\ e^{-i(-\alpha+\gamma)/2}\sin\beta/2 & e^{-i(\alpha+\gamma)/2}\cos\beta/2 \end{pmatrix} \tag{16c}$$

Then, using Eqs. (16), in Eqs. (13b)–(13d) we deduce

$$\tilde{\mathbf{g}}^{ji} = (-\mathbf{g}^{ji}\cos\beta + \mathbf{h}^{\mathrm{r},ji}\sin\beta\cos\gamma + \mathbf{h}^{\mathrm{i},ij}\sin\beta\sin\gamma) \tag{17a}$$

and

$$\tilde{\mathbf{h}}^{ji} = \tilde{\mathbf{h}}^{\mathrm{r},ji} + i\tilde{\mathbf{h}}^{\mathrm{i},ji} \tag{17b}$$

where

$$\begin{aligned} \tilde{\mathbf{h}}^{\mathrm{r},ji} &= \mathbf{g}^{ji}\sin\beta\cos\alpha + \mathbf{h}^{\mathrm{r},ji}(\cos\beta\cos\gamma\cos\alpha - \sin\gamma\sin\alpha) \\ &\quad + \mathbf{h}^{\mathrm{i},ij}(\cos\beta\sin\gamma\cos\alpha + \cos\gamma\sin\alpha) \end{aligned} \tag{17c}$$

$$\begin{aligned} \tilde{\mathbf{h}}^{\mathrm{i},ji} &= \mathbf{g}^{ji}\sin\beta\sin\alpha - \mathbf{h}^{\mathrm{r},ji}(\cos\beta\cos\gamma\sin\alpha - \sin\gamma\cos\alpha) \\ &\quad - \mathbf{h}^{\mathrm{i},ij}(\cos\beta\cos\gamma\sin\alpha - \cos\gamma\cos\alpha) \end{aligned} \tag{17d}$$

Then the three requirements

$$\nu_1 = \tilde{\mathbf{g}}^{ji} \cdot \tilde{\mathbf{h}}^{\mathrm{r},ji} = 0 \tag{18a}$$

$$\nu_2 = \tilde{\mathbf{g}}^{ji} \cdot \tilde{\mathbf{h}}^{\mathrm{i},ji} = 0 \tag{18b}$$

$$\nu_3 = \tilde{\mathbf{h}}^{\mathrm{r},ji} \cdot \tilde{\mathbf{h}}^{\mathrm{i},ji} = 0 \tag{18c}$$

define α, β, γ. For example, Eqs. (17a) and (17c) with $\alpha = \gamma = 0$ give the nonrelativistic limit for Eq. (18a)

$$\nu_1 = 0 = (-\mathbf{g}^{ji}\cos\beta + \mathbf{h}^{\mathrm{r},ji}\sin\beta) \cdot (\mathbf{g}^{ji}\cos\beta + \mathbf{h}^{\mathrm{r},ji}\sin\beta) \tag{19a}$$

The solution to Eq. (19a) is

$$2\mathbf{g}^{ji} \cdot \mathbf{h}^{\mathrm{r},ji}/(\mathbf{h}^{\mathrm{r},ji} \cdot \mathbf{h}^{\mathrm{r},ji} - \mathbf{g}^{ji} \cdot \mathbf{g}^{ji}) = \tan 2\beta \tag{19b}$$

as obtained previously [18]. Further, choosing $\beta = \pi/2$, $\gamma = 0$, $\alpha = 0$ and $\beta = \pi/2$, $\gamma = 0$, $\alpha = \pi/2$ shows that $\mathbf{g}^{ij}, \mathbf{h}^{\mathrm{r},ij}$, and $\mathbf{h}^{\mathrm{i},ij}$ are interchangeable at a conical intersection. The solution to the nonlinear Eqs. (18a)–(18c) can be obtained numerically using the following Newton–Raphson procedure

$$\nu_i(\mathbf{x_n} + \delta\mathbf{x}) = \mathbf{0} = \nu_i(\mathbf{x}_n) + \nabla\nu_i(\mathbf{x}_n)\cdot\delta\mathbf{x} \qquad i = 1\text{–}3 \tag{20a}$$

so that

$$\mathbf{x}_{n+1} = \mathbf{x}_n - \mathbf{F}(\mathbf{x}_n)^{-1}\nu(\mathbf{x}_n) \tag{20b}$$

where $F_{ji} \equiv (\partial/\partial x_j)\nu_i$ is computed by divided difference and $\mathbf{x} = (\alpha, \beta, \gamma)$. The properties of the solutions of Eq. (20) will be discussed in Section III.

The orthogonal $\mathbf{g}^{ij}, \mathbf{h}^{\mathrm{r},ij}$, and $\mathbf{h}^{\mathrm{i},ij}$ define a set of Cartesian axes $\hat{\mathbf{z}} = \mathbf{g}^{ij}/g^{ij}$, $\hat{\mathbf{x}} = \mathbf{h}^{\mathrm{r},ij}/h^{\mathrm{r},ij}$, $\hat{\mathbf{y}} = \mathbf{h}^{\mathrm{i},ij}/h^{\mathrm{i},ij}$, where $g^{ij} = ||\mathbf{g}^{ij}||$, $h^{\mathrm{r},ij} = ||\mathbf{h}^{\mathrm{r},ij}||$, and $h^{\mathrm{i},ij} = ||\mathbf{h}^{\mathrm{i},ij}||$ that span the branching space. The associated coordinates (x, y, z) are referred to as orthogonal intersection adapted coordinates.

2. *A Transformational Invariant*

In the nonrelativistic case, at a given $\mathbf{R}^{x,IJ}$, the quantity $\mathbf{g}^{IJ} \times \mathbf{h}^{IJ}$ was shown to be invariant under the transformation in Eq. (16), for $\alpha = \gamma = 0$. This invariant, whose value depends on $\mathbf{R}^{x,IJ}$, was used to systematically locate confluences, [18–21], intersection points at which two distinct branches of the conical intersection seam intersect. Here, we show that the scalar triple product, $\mathbf{g}^{ij} \times \mathbf{h}^{\mathrm{r},ij}\cdot\mathbf{h}^{\mathrm{i},ij}$ is the invariant for $\eta = 3$. Since the $\mathbf{g}^{ij}$, $\mathbf{h}^{\mathrm{r},ij}$, and $\mathbf{h}^{\mathrm{i},ij}$ cannot be assumed orthogonal the scalar triple product has the following form (suppressing the ij superscripts)

$$\begin{aligned}\mathbf{g}\times\mathbf{h}^{\mathrm{r}}\cdot\mathbf{h}^{\mathrm{i}} &= (g_x\boldsymbol{i} + g_y\boldsymbol{j} + g_z\boldsymbol{k}) \times (h^{\mathrm{r}}_x\boldsymbol{i} + h^{\mathrm{r}}_y\boldsymbol{j} + h^{\mathrm{r}}_z\boldsymbol{k})\cdot(h^{\mathrm{i}}_x\boldsymbol{i} + h^{\mathrm{i}}_y\boldsymbol{j} + h^{\mathrm{i}}_z\boldsymbol{k}) \\ &= (g_x\mathrm{h}^{\mathrm{r}}_y - g_y\mathrm{h}^{\mathrm{r}}_x)h^{\mathrm{i}}_z + (g_yh^{\mathrm{r}}_z - g_zh^{\mathrm{r}}_y)h^{\mathrm{i}}_x + (g_zh^{\mathrm{r}}_x - g_xh^{\mathrm{r}}_z)h^{\mathrm{i}}_y\end{aligned} \tag{21}$$

To demonstrate the invariance insert, Eqs. (17a), (17c), and (17d) into Eq. (21) giving

$$\begin{aligned}\tilde{I} &= \tilde{\mathbf{g}}\times\tilde{\mathbf{h}}^{r}\cdot\tilde{\mathbf{h}}^{i} \\ &= \{(h^{\mathrm{r}}_xh^{\mathrm{i}}_y - h^{\mathrm{r}}_yh^{\mathrm{i}}_x)(\sin\beta\sin\alpha) + (-g_xh^{\mathrm{i}}_y + g_yh^{\mathrm{i}}_x)(\cos\alpha\sin\gamma + \cos\gamma\cos\beta\sin\alpha) \\ &\quad + (g_xh^{\mathrm{r}}_y - g_yh^{\mathrm{r}}_x)(-\cos\alpha\cos\gamma + \cos\beta\sin\alpha\sin\gamma)\}\{-g_z\sin\beta\sin\alpha \\ &\quad - h^{\mathrm{r}}_z(\sin\gamma\cos\alpha + \cos\beta\sin\alpha\cos\gamma) - h^{\mathrm{i}}_z(\cos\beta\sin\alpha\sin\gamma - \cos\gamma\cos\alpha)\} \\ &\quad + \text{cyclic permutations} \\ &= \{(-h^{\mathrm{r}}_xh^{\mathrm{i}}_y + h^{\mathrm{r}}_yh^{\mathrm{i}}_x)g_z + (h^{\mathrm{r}}_yh^{\mathrm{i}}_z - h^{\mathrm{r}}_zh^{\mathrm{i}}_y)g_x + (h^{\mathrm{i}}_xh^{\mathrm{r}}_z - h^{\mathrm{i}}_zh^{\mathrm{r}}_x)g_y\}\{(\sin^2\beta\sin^2\alpha) \\ &\quad + (\cos\alpha\sin\gamma + \cos\gamma\cos\beta\sin\alpha)^2 + (\cos\beta\sin\alpha\sin\gamma - \cos\gamma\cos\alpha)^2\} \\ &= -\mathbf{h}^{\mathrm{i}}\times\mathbf{h}^{\mathrm{r}}\cdot\mathbf{g} = \mathbf{g}\times\mathbf{h}^{\mathrm{r}}\cdot\mathbf{h}^{i} = I\end{aligned} \tag{22}$$

The use of this invariant is discussed in Section III.

3. *Local Topography: Energy*

The topography of a conical intersection affects the propensity for a nonadiabatic transition. Here, we focus on the essential linear terms. Higher order effects are described in [10]. The local topography can be determined from Eq. (13). For $\eta = 3$, Eq. (13) becomes, in orthgonal intersection adapted coordinates

$$\underline{\tilde{\mathbf{H}}}^{[1]} \cdot \delta\mathbf{R} = (\mathbf{s}^{ij} \cdot \delta\mathbf{R})\mathbf{I} - g^{ij} z \boldsymbol{\sigma}_z + h^{\mathrm{r},ij} x \boldsymbol{\sigma}_x - h^{\mathrm{i},ij} y \boldsymbol{\sigma}_y \tag{23}$$

where $\mathbf{I}$ is a 2×2 unit matrix, and the $\boldsymbol{\sigma}$ are the Pauli matrices.

To determine the eigenfunctions and eigenvalues of Eq. (23), it is convenient to introduce spherical polar coordinates, $x = \rho \cos\phi \sin\theta$, $y = \rho \sin\phi \sin\theta$, and $z = \rho\cos\theta$ and make the definitions

$$h^{\mathrm{i},ij} \sin\phi = h(\phi) \sin\zeta(\phi) \qquad h^{\mathrm{r},ij} \cos\phi = h(\phi)\cos\zeta(\phi) \tag{24a}$$

$$h(\phi)^2 = (h^{\mathrm{i},ij} \sin\phi)^2 + (h^{\mathrm{r},ij}\cos\phi)^2 \qquad \tan\zeta(\phi) = \frac{h^{\mathrm{i},ij}}{h^{\mathrm{r},ij}} \tan\phi \tag{24b}$$

$$h(\phi)\sin\theta = q(\theta,\phi)\sin\lambda(\theta) \qquad g^{ij}\cos\theta = q(\theta,\phi)\cos\lambda(\theta) \tag{24c}$$

$$q(\theta,\phi)^2 = (h(\phi)\sin\theta)^2 + (g^{ij}\cos\theta)^2 \qquad \tan\lambda(\theta,\phi) = \frac{h(\phi)}{g^{ij}}\tan\theta \tag{24d}$$

Then Eq. (23) becomes

$$\underline{\tilde{\mathbf{H}}}^{[1]} \cdot \delta\mathbf{R} = \mathbf{I}(\mathbf{s}^{ij} \cdot \delta\mathbf{R}) + \rho q(\theta, \phi)(-\boldsymbol{\sigma}_z \cos\lambda(\theta, \phi) + \sin\lambda(\theta, \phi)\mathbf{M}(e^{i\zeta(\phi)})\boldsymbol{\sigma}_x) \tag{25}$$

where $\mathbf{M}_{ij}(x) = \delta_{ij}(x\delta_{1j} + x^*\delta_{2j})$. This Hamiltonian can be diagonalized by the transformation

$$\mathbf{U}(\theta, \phi) = \begin{pmatrix} \cos\lambda(\theta, \phi)/2 & \sin\lambda(\theta, \phi)/2 \\ -e^{-i\zeta(\phi)}\sin\lambda(\theta, \phi)/2 & e^{-i\zeta(\phi)}\cos\lambda(\theta, \phi)/2 \end{pmatrix} \tag{26a}$$

that is

$$\mathbf{U}^{\dagger}\underline{\tilde{\mathbf{H}}}^{[1]} \cdot \delta\mathbf{R}\mathbf{U} = (\mathbf{s}^{ij} \cdot \delta\mathbf{R})\mathbf{I} - \rho q(\theta, \phi)\boldsymbol{\sigma}_z \tag{27}$$

From the preceding analysis, it is seen that the coordinate space near $\mathbf{R}^{x,ij}$ can be usefully partitioned into the branching space described in terms of intersection adapted coordinates (ρ, θ, ϕ) or $(\mathrm{x, y, z})$ and its orthogonal complement the seam space spanned by a set of mutually orthonormal set $\mathbf{w}^i$, $i = 4 - N^{\mathrm{int}}$. From Eq. (27), spherical radius ρ is the parameter that lifts the degeneracy linearly in the branching space spanned by $\hat{\mathbf{x}}$, $\hat{\mathbf{y}}$, and $\hat{\mathbf{z}}$.

These results can be simplified by introducing scaled orthogonal intersection adapted coordinates $x' = xh^{r,ij}$, $y' = yh^{i,ij}$, and $z' = zg^{ij}$. In these coordinates, $rh(\phi) = r'h'(\phi') = r'$, $\rho q(\theta, \phi) = \rho' q'(\theta', \phi') = \rho'$, $\zeta(\phi) = \zeta'(\phi') = \phi'$, and $\lambda(\theta, \phi) = \lambda'(\theta', \phi') = \theta'$, where $x^2 + y^2 = r^2$. This is the coordinate system to be used to consider Berry's geometric phase theorem [22].

4. *Local Topography: Conical Parameters*

In the nonrelativistic case, the key linear portion of the double cone is characterized by four conical parameters: a strength parameter $d = \sqrt{[(g^{IJ^2}) + (h^{IJ^2})/2]}$, an asymmetry parameter $\Delta = (g^{IJ^2}) - (h^{IJ^2})/(g^{IJ^2}) + (h^{IJ^2})$, and two tilt parameters s_w^{IJ}/d, $w = x, y$. If $\mathbf{s}^{ij} = 0$ the double cone is vertical. The affect of these parameters on nuclear dynamics has been investigated using time dependent wavepackets [23]. Here the situation is similar but more complicated. In this case, the six parameters, s_w^{ij} $w = x, y, z$, g^{ij}, $h^{r,ij}$, and $h^{i,ij}$ can be used to define a strength parameter, $d = \sqrt{[(g^{ij^2}) + (h^{r,ij^2}) + (h^{i,ij^2})]}$, three tilt parameters s_w^{ij}/d $w = x, y, z$ and two asymmetry parameters, $\Delta_2(\phi) = [(h(\phi)^2 - g^{ij^2})/(h(\phi)^2 + g^{ij^2})$ and $\Delta_1 = (h^{r,ij^2}) - (h^{i,ij^2})/(h^{r,ij^2}) + (h^{i,ij^2})]$. In future work, the affect of these and higher order parameters on nuclear dynamics will be considered.

5. *Derivative Couplings*

By using Eq. (10), the derivative couplings

$$\mathbf{f}^{ij}(\mathbf{R}) = \langle \Psi_i^e(\mathbf{r}; \mathbf{R}) | \nabla \Psi_j^e(\mathbf{r}; \mathbf{R}) \rangle_{\mathbf{r}} \tag{28a}$$

are given by

$$\mathbf{f}^{ij}(\mathbf{R}) = \sum_{l \in Q} \xi_l^i(\mathbf{R}) \nabla \xi_l^j(\mathbf{R}) + \sum_{l \in P} \Xi_l^i(\mathbf{R}) \nabla \Xi_l^j(\mathbf{R}) + {}^{\mathrm{CSF}}\mathbf{f}^{ij}(\mathbf{R}) \tag{28b}$$

where the nonsingular term ${}^{\mathrm{CSF}}\mathbf{f}^{ij}(\mathbf{R})$ is largely negligible near a conical intersection [20]. As in the nonrelativistic case, $\mathbf{f}^{ij}$ is singular at a conical intersection. The singularity appears in the lowest order contribution:

$$\mathbf{f}^{kl,(0)} = \sum_{\alpha \in Q} \tilde{\xi}_\alpha^{(0),k}(\theta, \phi, \mathbf{w}) \nabla \tilde{\xi}_\alpha^{(0),l}(\theta, \phi, \mathbf{w}) \qquad k, l \in i, j \tag{28c}$$

To evaluate $\mathbf{f}^{kl,(0)}$ the $\tilde{\xi}^{(0),w}(\rho = 0, \theta, \phi)$, are required. These are given in terms of $\mathbf{U}(\theta, \phi)$ by

$$(\tilde{\xi}^{(0),i}(\rho = 0, \theta, \phi), \tilde{\xi}^{(0),j}(\rho = 0, \theta, \phi)) \equiv (\xi^{(0),i}(\mathbf{R}^{x,ij}), \xi^{(0),j}(\mathbf{R}^{x,ij}))\mathbf{U}(\theta, \phi) \tag{29}$$

Note that since the eigenfunctions are determined only up to an overall phase, the following transformations also diagonalize $\underline{\tilde{\mathbf{H}}}^{[1]} \cdot \delta\mathbf{R}$:

$$\mathbf{V}(\theta, \phi) = \begin{pmatrix} e^{i\zeta(\phi)/2}\cos\lambda(\theta,\phi)/2 & e^{i\zeta(\phi)/2}\sin\lambda(\theta,\phi)/2 \\ -e^{-i\zeta(\phi)/2}\sin\lambda(\theta,\phi)/2 & e^{-i\zeta(\phi)/2}\cos\lambda(\theta,\phi)/2 \end{pmatrix} \tag{26b}$$

$$\mathbf{B}(\theta, \phi) = \begin{pmatrix} \cos\lambda(\theta,\phi)/2 & e^{i\zeta(\phi)}\sin\lambda(\theta,\phi)/2 \\ -e^{-i\zeta(\phi)}\sin\lambda(\theta,\phi)/2 & \cos\lambda(\theta,\phi)/2 \end{pmatrix} \tag{26c}$$

As discussed in detail in [10], equivalent results are not obtained with these three unitary transformations. A principal difference between the **U**, **V**, and **B** results is the phase of the wave function after being transported around a closed loop C, centered on the z axis parallel to but not in the (x, y) plane. The perturbative wave functions obtained from $\mathbf{U}(\theta, \phi)$ or $\mathbf{B}(\theta, \phi)$ are, as seen from Eq. (26a) or (26c), single-valued when transported around C that is $\langle \Psi_i^e(\mathbf{r}; \mathbf{R}_0) | \Psi_i^e(\mathbf{r}; \mathbf{R}_n) \rangle = 1$, where $\mathbf{R}_0 = \mathbf{R}_\mathrm{n}$ denote the beginning and end of this loop. This is a necessary condition for Berry's geometric phase theorem [22] to hold. On the other hand, the perturbative wave functions obtained from $\mathrm{V}(\theta, \phi)$ in Eq. (26b) are not single valued when transported around C.

U, **V**, and **B** also yield different $\mathbf{f}^{kk}$. By using Eqs. (24a)–(24d) and (26a) the derivative couplings are

$$\mathrm{f}^{ij,(0)} = 1/2(\nabla\lambda) + i(\nabla\zeta)(1/2)\sin\lambda \tag{30a}$$

$$\mathbf{f}^{ii,(0)} = -i(\nabla\zeta)\sin^2\lambda/2 \tag{30b}$$

$$\mathbf{f}^{jj,(0)} = -i(\nabla\zeta)\cos^2\lambda/2 \tag{30c}$$

From Eqs. (30a)–(30c), the singularity in $\mathbf{f}^{kl}$, as the conical intersection is approached, is of order $1/\rho$. Only f_ω^{ij}, $\omega = \theta, \phi$ are singular [10]. As in the nonrelativistic case, knowledge of the singular part of the derivative coupling can be used to construct a local diabatic representation that removes the singularity [10].

If $\mathbf{V}(\theta, \phi)$ had been used in lieu of $\mathbf{U}(\theta, \phi)$, $|\mathbf{f}^{ij,(0)}|$ would be unchanged, but $\mathbf{f}^{kk,(0)}$ becomes $\mathbf{f}^{kk,(0),V}$, which is given by

$$\mathbf{f}^{ii,(0),\mathrm{V}} = \mathbf{f}^{ii,(0)} + i\nabla\zeta/2 = i(\cos\lambda)/2\nabla\zeta \tag{30d}$$

$$\mathbf{f}^{jj,(0),\mathrm{V}} = \mathbf{f}^{jj,(0)} + i\nabla\zeta/2 = -i(\cos\lambda)/2\nabla\zeta \tag{30e}$$

Finally, had $\mathbf{B}(\theta, \phi)$ been used, $\mathbf{f}^{ii,(0)}$ would be unchanged, but

$$\mathbf{f}^{jj,(0),\mathrm{B}} = \mathbf{f}^{jj,(0)} + i\nabla\zeta = i(\nabla\zeta)\sin^2\lambda/2 = -\mathbf{f}^{ii,(0)} \tag{30f}$$

III. NUMERICAL RESULTS

In this section, the spin–orbit interaction is treated in the Breit–Pauli [13,24–26] approximation and incorporated into the Hamiltonian using quasidegenerate perturbation theory [27]. This approach, which is described in [8], is commonly used in nuclear dynamics and is adequate for molecules containing only atoms with atomic numbers no larger than that of Kr.

A. $1,2^2A'$ and $1^2A'$ States of $H_2 + OH$

The nonrelativistic $1,2^2A'$ conical intersection seam in the $H_2 + OH$ supermolecule has been well studied [28–30] because of its role in the non-adiabatic quenching reaction

$$H_2 + OH(A^2\Sigma^+) \rightarrow H_2 + OH(X^2\Pi) \qquad \text{or} \qquad H_2O + H(^2S)$$

The $C_{\infty v}$ portion of this seam is a $^2\Sigma^+ - {}^2\Pi$ symmetry-allowed conical intersection. The character of the seam including spin–orbit coupling can be understood by starting with the degenerate $1^2\Pi$ and $1^2\Sigma^+$ states. Turning on the spin–orbit interaction within the $^2\Pi$ manifold splits the $^2\Pi$ state into a (lower energy) $^2\Pi_{3/2}$ state and (higher energy) $^2\Pi_{1/2}$ state (see Fig. 1). Then with the full spin–orbit interaction included the molecule can distort to make either the upper pair, the $2E'$ and $3E'$ states, or the lower pair, the $1E'$ and $2E'$ states, degenerate. From Figure 1, for $C_{\infty v}$ geometries the $2E' - 3E'$ intersection is a "same symmetry", $\Omega = 1/2,\ 1/2$, intersection, while the $1E' - 2E'$ intersection is a different symmetry $\Omega = 3/2,\ 1/2$ intersection. However, both intersections are conical intersections since both $\Omega = 3/2$ and $\Omega = 1/2$ states decompose into one E' and one E'' (Kramers' doublets) when the molecule is distorted to C_s configurations. Here, we consider the more computationally challenging same

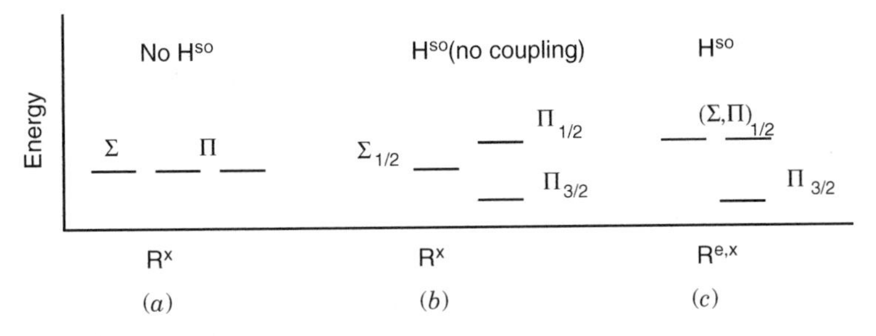

Figure 1. Representation of degenerate states from nonrelativistic components. (*a*) Degenerate zeroth-order states at $\mathbf{R}^x \equiv \mathbf{R}^{x,IJ}$. (*b*) Spin–orbit interaction splits $^2\Pi$ state. (*c*) With full spin–orbit interaction turned on, degeneracy is restored by changing geometry to $\mathbf{R}^{e,x} \equiv \mathbf{R}^{x,ij}$.

symmetry, $2E' - 3E'$, intersection to illustrate the ideas developed in Section 2.F.5. This system provides a stringent test of the numerical procedures since the spin–orbit interaction is relatively modest and the energy splitting changes rapidly in the region of interest.

The nonrelativistic states are described at the first-order configuration interaction level using a six orbital, eight electron, active space with the oxygen $1s$ orbital kept doubly occupied. The molecular orbitals were constructed from a state-averaged multiconfigurational self-consistent field procedure [31] using an extended atomic orbital basis on oxygen and hydrogen. The details of this description can be found in [30].

In the present calculations, the molecule is restricted to C_s symmetry. There are five internal degrees of freedom (the out-of-plane mode is excluded to preserve C_s symmetry). Nuclear configurations will be denoted $\mathbf{R} = (R(H^1\text{—}O), R(O\text{—}H^2), R(H^2\text{—}H^3), \angle H^1OH^2, \angle OH^2H^3)$ corresponding to the arrangement $H^1\text{—}O\text{—}H^2\text{—}H^3$. It will be convenient to refer to the $\mathbf{R}$ by their $R(H^2\text{—}H^3)$ value, writing, $\mathbf{R}(R(H^2\text{—}H^3) = \beta) \equiv \mathbf{R}(\beta) = (R(H^1\text{—}O), R(O\text{—}H^2), R(H^2\text{—}H^3) = \beta, \angle H^1OH^2, \angle OH^2H^3)$. For collinear geometries, the two angles will be suppressed. Equation (14) defines only three internal coordinates. Therefore two additional constraints are needed. These are provided by the value of $R(H^2\text{—}H^3) = \beta$ and/or the energy minimization requirement.

B. Convergence of Eq. (15b)

Figure 2 illustrates the efficacy of Eq. (15), considering convergence, in the absense of geometrical constraints, to a local energy minimum on the relativistic seam of conical intersection. Reported are the relativistic energy separation ΔE^e_{32} and the nonrelativistic energy separation, $\Delta E_{2^2A',1^2A'}$. The search was initiated at the structure indicated on the left-hand side of Figure 2, a point slightly displaced from the nonrelativistic seam. At this point, $\Delta E_{2^2A',1^2A'} \approx 11\,\text{cm}^{-1}$ and $\Delta E^e_{32} \approx 70\,\text{cm}^{-1}$. At the converged structure, achieved after 15 iterations, pictured on the right-hand side, $\Delta E^e_{32} < 0.2\,\text{cm}^{-1}$ while $\Delta E_{2^2A',1^2A'} \approx 70\,\text{cm}^{-1}$. The large changes in ΔE^e_{32} between iterations 8 and 9, and 12 and 13, reflect, in part, the use of the "conjugate gradient" extrapolation noted previously. These results strongly support the utility of the present approach. It is worth noting that once an initial point on a seam is found locating additional points is facilitated by the fact that given an $R^{x,ij}$ corresponding to given K, Eq. (15b) can be used to predict a good starting value for a neighboring $R^{x,ij'}$ corresponding to K′ [32].

C. The Seam: Locus

Further evidence of the efficacy of the algorithm for locating points of conical intersection is provided in Figure 3, which reports additional points on the $2E'$–$3E'$ intersection seam, determined by introducing the geometrical constraint,

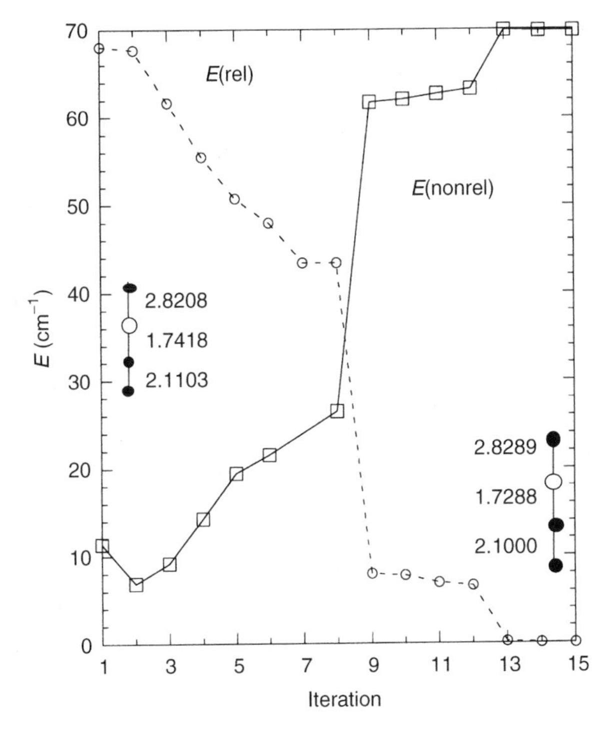

Figure 2. $E(\text{rel}) \equiv \Delta E^e_{32}$ and $E(\text{nonrel}) \equiv \Delta E_{2^2A',1^2A'}$ at each iteration of the solution of Eq. (15b) for $OH + H_2$ using multireference configuration interaction wave functions.

$R(H^2–H^3) = \beta$. Note that $C_{\infty v}$ symmetry was not imposed. The points located on the $2E'$–$3E'$ seam of conical intersection, which are degenerate to $< 1\ \text{cm}^{-1}$, all had $C_{\infty v}$ symmetry. Figure 3 shows, that while the $\eta = 3$ seam is necessarily distinct from the nonrelativistic seam the separation is not large. In future work, it will be interesting to see how this conclusion changes as the magnitude of the spin–orbit interaction increases. Along the nonrelativistic seam the relativistic energy difference, $\Delta E^e_{32}(\mathbf{R}^{x,IJ}(R(H^2–H^3))$, is $\sim 70\ \text{cm}^{-1} > 50\%$ of the $OH(^2\Pi)$ fine structure splitting [33] suggesting that when heavier atoms such a chlorine, where A^{so} is $\sim 780\,\text{cm}^{-1}$ [34], or even bromine or iodine are involved, $\Delta E^e_{ji}(\mathbf{R}^{x,IJ})$, will be much larger, so that nonadiabatic effects may be significantly reduced at the nonrelativistic seam by the inclusion of spin–orbit coupling.

The small $\Delta E^e_{32}(\mathbf{R}^{x,ij}(R(H^2–H^3)) < 1\,\text{cm}^{-1}$ and much larger $\Delta E_{2^2A',1^2A}$ $(\mathbf{R}^{x,ij}(R(H^2–H^3))$, also $\sim 70\,\text{cm}^{-1}$, provide prima facie evidence for a conical intersection of H^e. However, since numerical degeneracies are never exact, an

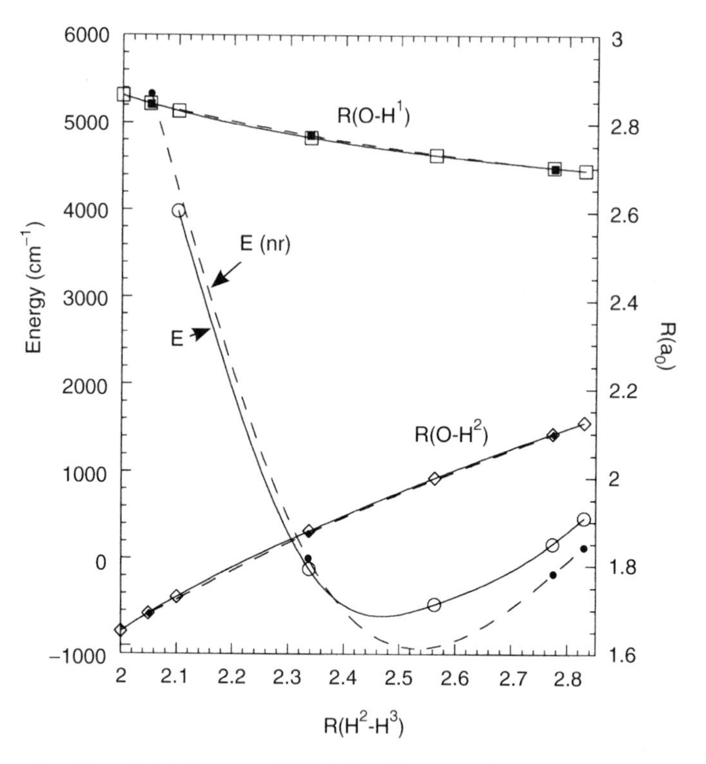

Figure 3. The relativistic seam $\mathbf{R}^{x,ij}(R(\mathrm{H}^2\text{—}\mathrm{H}^3))$: $R(\mathrm{O–H}^1)$, (empty squares), $R(\mathrm{O–H}^2)$ (empty diamonds) $E \equiv E_{2\mathrm{E}'}(R(\mathrm{H}^2\text{—}\mathrm{H}^3))$ (empty circles) on the $2E$–$3E$ seam of conical intersection. Filled markers on $1^2\mathrm{A}' - 2^2\mathrm{A}'$ nonrelativistic seam of conical intersection. The zero of energy is $E(\mathrm{nr}) \equiv E_{1^2\mathrm{A}'}(R^{\mathrm{x,IJ}}(R(\mathrm{H}^2\text{—}\mathrm{H}^3) = 2.336)) = -76.486688$ a.u.

alternative means is required to prove the existence of a conical intersection. The demonstration that $\mathbf{g} \times \mathbf{h}^{\mathrm{r}} \cdot \mathbf{h}^{\mathrm{i}} \neq 0$, that is, that $\mathbf{g}$, $\mathbf{h}^{\mathrm{r}}$, $\mathbf{h}^{\mathrm{i}}$ are linearly independent, serves to confirm the "conical" character near the intersection. It is to the determination of these quantities that we now turn.

D. The Seam: Conical Parameters and the Invariant

The lowest order contributions to the energy are described by the conical parameters g, h^{r}, h^{i}, and s_k, $k = x, y, z$, or by d, $\Delta_i = 1, 2$ and s_k, $k = x, y, z$. Here and below the superscript ij is suppressed when no confusion will result. We also will use the nonrelativistic convention $\mathbf{g}^{\mathrm{ij}}||\hat{\mathbf{x}}$, $h^{\mathrm{r,y}}||\hat{\mathbf{y}}$ and $\mathbf{h}^{\mathrm{i,ij}}||\hat{\mathbf{z}}$, where $||$ is real "is parallel to." These parameters [9] are reported in Figure 4*a* and *b*. Their continuity is attributable to the use of orthogonal intersection adapted coordinates. For comparison, Figure 4*a* and *b* reports the nonrelativistic quantities g^{IJ}, h^{IJ}, and $\mathbf{s}^{IJ}$, respectively. While noting that there is no unique correspondence between the relativistic and nonrelativistic seam points, here the energy optimization in Eq. (15b) allows $R(\text{H}^2\text{—H}^3)$ to define a correspondence. From Figure 4*a* it is seen that for the slowly changing $||\mathbf{g}^{ij}||$, $|(||\mathbf{g}^{IJ}|| - ||\mathbf{g}^{ij}||)|/||\mathbf{g}^{IJ}||$ is small while for the more rapidly varying $\mathbf{h}^{ij}$, $|(||\mathbf{h}^{IJ}|| - ||\mathbf{h}^{r,ij}||)|/||\mathbf{h}^{IJ}||$ is perhaps not unexpectedly, larger but still modest, <0.4.

The invariant $I = \mathbf{g} \times \mathbf{h}^{\mathrm{r}} \cdot \mathbf{h}^{\mathrm{i}}$ is reported in Figure 4*c*, again as a function of $R(\text{H}^2\text{—H}^3)$. Although I decreases as $R(\text{H}^2\text{—H}^3)$ decreases, it is clear that it does not vanish for reasonable values of $R(\text{H}^2\text{—H}^3)$. This confirms the conical character of the intersections reported herein.

The relative magnitudes of $g, h^{\mathrm{r}}, h^{\mathrm{i}}$, and s_i $i = x, y, z$ describe the orientation and shape of the double cone. The relation between these attributes and near

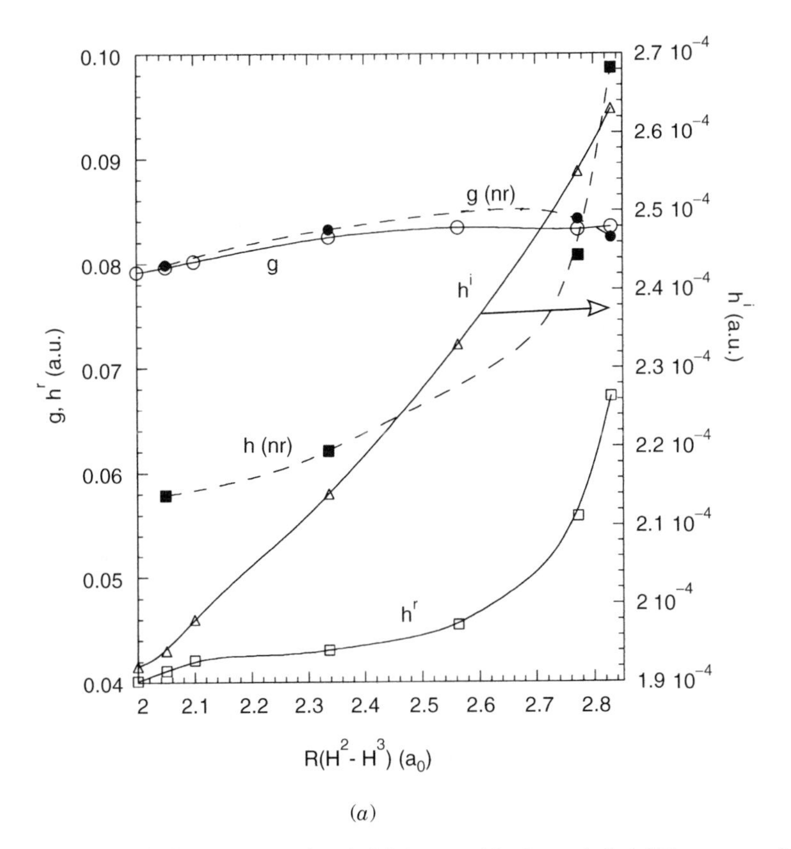

(*a*)

Figure 4. Conical parameters on the relativistic seam. (*a*) g (open circles), h^{r} (open squares), h^{i} (open triangles). The nonrelativistic quantities, g (filled circle) and h(nr) (filled square); (*b*) s_w, w = x (circles), y (squares), z (triangles). Filled (open) markers from nonrelativistic (relativistic) calculations. (*c*) Magnitude of the invariant $I = \mathbf{g} \times \mathbf{h}^{\mathrm{r}} \cdot \mathbf{h}^{\mathrm{i}}$ as a function of $R(\text{H}^2\text{—H}^3)$.

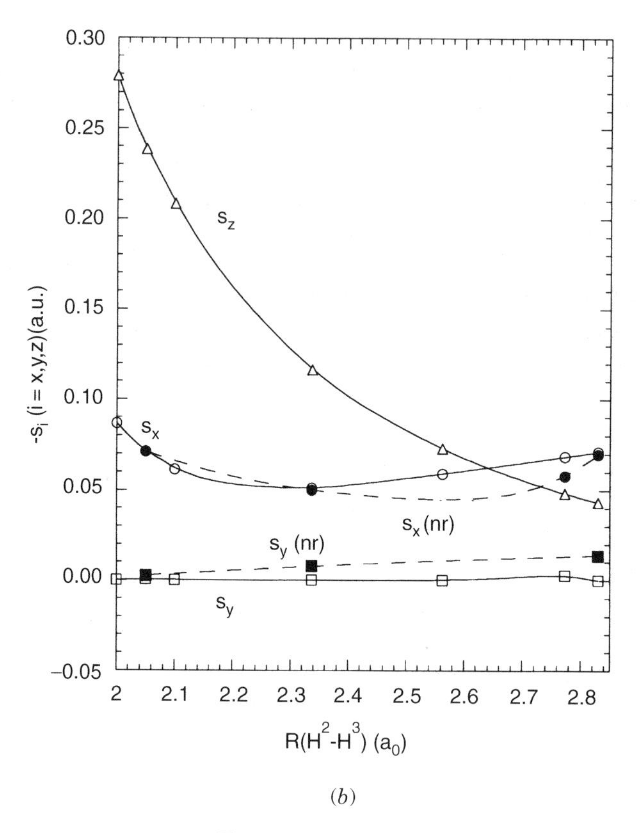

(*b*)

Figure 4 (*Continued*)

conical intersection dynamics has been discussed previously for nonrelativistic conical intersection [23,35]. For the conical intersections encountered here $-s_x < h^{\mathrm{r},ij}(-s_z > h^{\mathrm{i},ij})$ so that the cone is slightly (significantly) tilted in the $x(z)$ direction. Such a topography faciliates the downward (upward) transitions from negative $x(z)$ direction [23]. The linear coupling in the z direction, h^{i}, is, however, quite small so that quadratic terms dominate except very close to the conical intersection. This will limit the role of this coordinate in inducing non-adiabatic transitions. It will be interesting to see how this changes as the size of the spin–orbit interaction increases.

E. Characterizing the Seam: Orthogonal $\mathbf{g}, \mathbf{h}^{\mathrm{r}}$, and $\mathbf{h}^{\mathrm{i}}$

The conical parameters describe the topography of the conical intersection. The directions for $\mathbf{g}$, $\mathbf{h}^{\mathrm{r}}$, and $\mathbf{h}^{\mathrm{i}}$ relate the abstract x, y, z directions to actual molecular

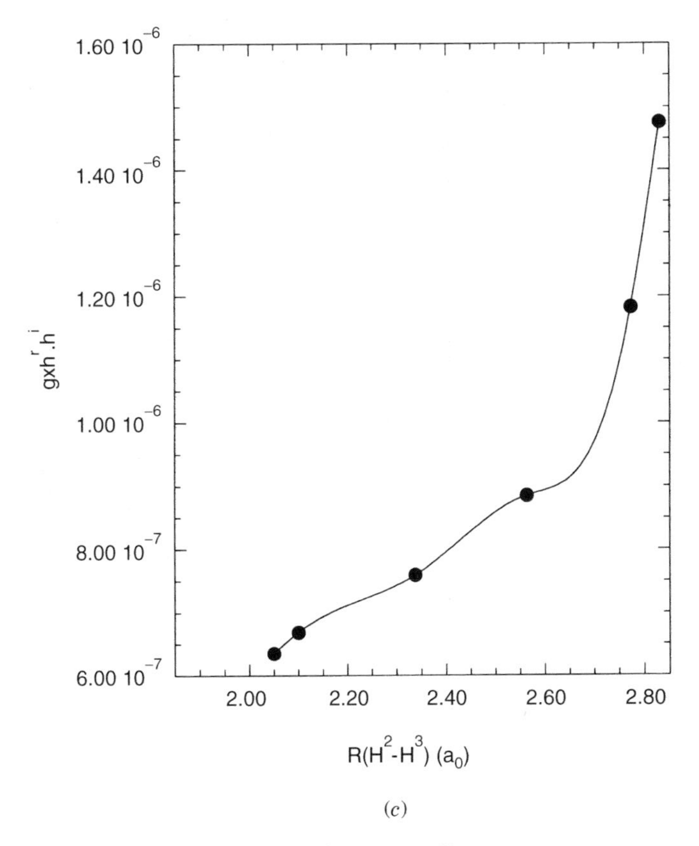

(*c*)

Figure 4 (*Continued*)

displacements. Figure 5 reports $\mathbf{g}$, $\mathbf{h}^{\mathrm{r}}$, and $\mathbf{h}^{\mathrm{i}}$ and their orthogonalized counterparts, $\tilde{\mathbf{g}}$, $\tilde{\mathbf{h}}^{\mathrm{r}}$, and $\tilde{\mathbf{h}}^{\mathrm{i}}$ at $\mathbf{R}^{x,ij}(2.53)$, which is typical. The nascent $\mathbf{g}$, $\mathbf{h}^{\mathrm{r}}$, and $\mathbf{h}^{\mathrm{i}}$ vectors are clearly not symmetry-adapted. The orthogonalization procedure removes this deficiency, with $\mathbf{g}$ and $\mathbf{h}^{\mathrm{i}}$ being symmetry preserving σ displacements and $\mathbf{h}^{\mathrm{r}}$ being a π displacement.

Figure 6 reports $\tilde{\mathbf{g}}^{IJ}$, $\tilde{\mathbf{h}}^{IJ}$ and the seam coordinates $\mathbf{w}^{i}$, $i = 1$–3 for $\mathbf{R}^{x,IJ}(2.336) = (2.774, 1.873, 2.336)$. Comparing Figures 5 and 6 illustrates the general observation that, $\tilde{\mathbf{g}}^{IJ}$ is parallel to $\tilde{\mathbf{g}}^{ij}$, and $\tilde{\mathbf{h}}^{IJ}$ is parallel to $\tilde{\mathbf{h}}^{\mathrm{r},ij}$. These similiarities clearly depend on the use of the orthogonalized vectors. $\tilde{\mathbf{h}}^{\mathrm{i},ij}$ is a linear combination of two nonrelativistic symmetry preserving seam coordinates.

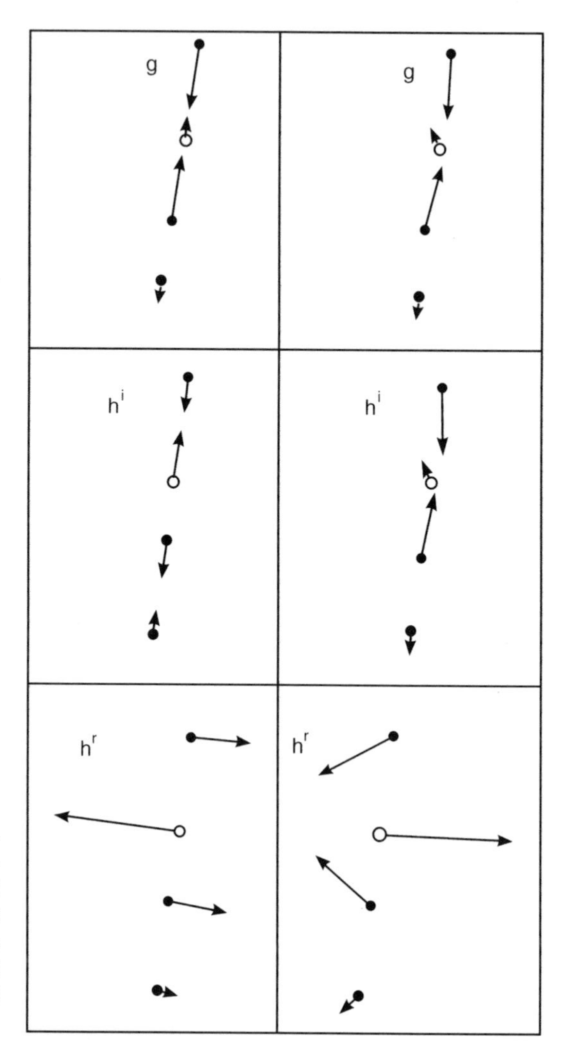

Figure 5. For $\eta = 3$, the vectors $\mathbf{g}$, $\mathbf{h}^r$, and $\mathbf{h}^i$. Nascent (right-hand column) and orthogonalized (left-hand column) results at $\mathbf{R}^{x,ij}(2.53)$. For orthogonal vectors $h^r = 0.0430$, $g = 0.0825$, and $h^i = 0.000233$. Vectors are scaled for visual clarity.

IV. THE FUTURE

In the nonrelativistic case much has been, and continues to be, learned about the outcome of nonadiabatic processes from the locus and topography of seams of conical intersection. It will now be possible to describe nonadiabatic processes driven by conical intersections, for which the spin–orbit interaction cannot be neglected, on the same footing that has been so useful in the nonrelativistic case. This fully adiabatic approach offers both conceptual and potential computational

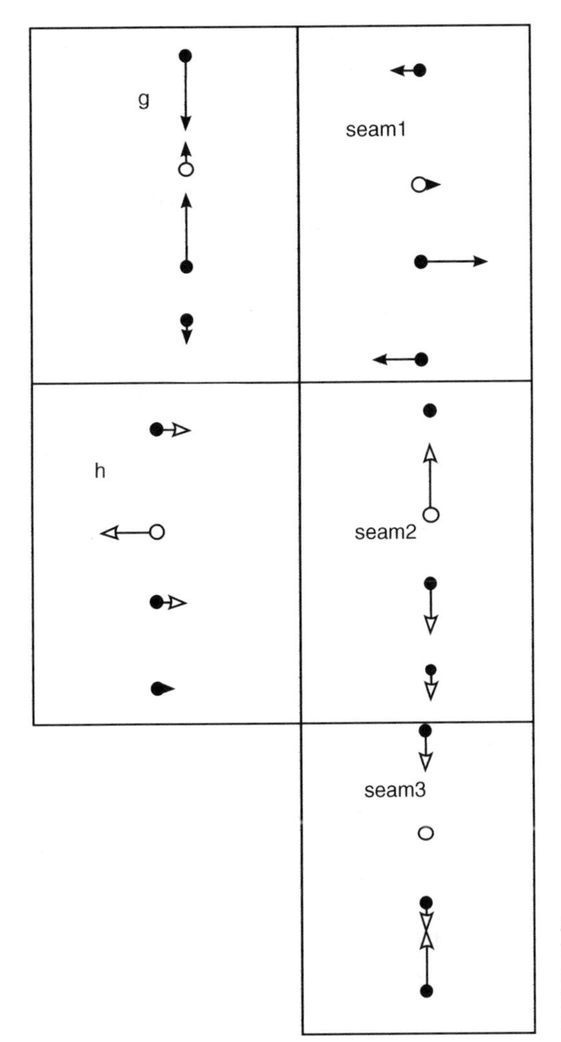

Figure 6. For $\eta = 2$, the orthogonal vectors $\mathbf{g}, \mathbf{h}$ and seam vectors $\mathbf{w}^i$, $i = 1$–3. Orthogonal $\mathbf{g}$ and $\mathbf{h}$ and three seam coordinates at $\mathbf{R}^{x,IJ}(2.33)$. $g = 0.0835$, $h = 0.0514$. Vectors are scaled for visual clarity.

advantages over approaches currently in use [36,37]. It will now be possible to address a key conceptual question: when does the relativistic description with intersections based on H^e, provide a better qualitative picture of a nonadiabatic process than a description with intersections based on H^0. On the computational side, these new capabilities will enable a more rigorous approach whose computational utility increases with the size of the spin–orbit interaction. Diabatic bases deduced from a systematically improvable adiabatic representation will be possible.

At the present time, the solution of the electronic structure problem using full four component wave functions is far from routine [38]. In the future, as progress is made in this area, extension of the present approach to full four component wave functions can be expected.

In summary, the techniques outlined in this work represent the first step on a path that will lead to increased understanding of, and more accurate computational approaches for treating, nonadiabatic processes in which relativistic effects cannot be neglected.

Acknowledgments

This work was supported by the AFOSR, DoE Office of Basic Energy Sciences and the National Science Foundation.

References

1. D. R. Yarkony, *J. Phys. Chem. A* **105**, 6277 (2001).
2. C. A. Mead, *J. Chem. Phys.* **70**, 2276 (1979).
3. J. von Neumann and E. Wigner, *Physik. Z.* **30**, 467 (1929).
4. M. R. Manaa and D. R. Yarkony, *J. Chem. Phys.* **93**, 4473 (1990).
5. S. Xantheas, S. T. Elbert, and K. Ruedenberg, *J. Chem. Phys.* **93**, 7519 (1990).
6. M. J. Bearpark, M. A. Robb, and H. B. Schlegel, *Chem. Phys. Lett.* **223**, 269 (1994).
7. D. R. Yarkony, *Rev. Mod. Phys.* **68**, 985 (1996).
8. S. Matsika and D. R. Yarkony, *J. Chem. Phys.* **115**, 2038 (2001).
9. S. Matsika and D. R. Yarkony, *J. Chem. Phys.* **115**, 5066 (2001).
10. S. Matsika and D. R. Yarkony, *J. Chem. Phys.* **116**, 2825 (2002).
11. C. A. Mead, *J. Chem. Phys.* **78**, 807 (1983).
12. D. R. Yarkony, *J. Phys. Chem. A* **101**, 4263 (1997).
13. B. A. Hess and C. M. Marian, *Relativistic Effects in the Calculation of the Electronic Energy*, in *Computational Molecular Spectroscopy*, P. Jensen and P. Bunker, eds., John Wiley & Sons, Inc., Chichester, UK, 2000, pp. 169–220.
14. L. I. Schiff, *Quantum Mechanics*, McGraw-Hill, New York, 1960.
15. M. Tinkham, *Group Theory and Quantum Mechanics*, McGraw–Hill, New York, 1964.
16. H. Kramers, *Proc. Acad. Sci. Amsterdam* **33**, 959 (1930).
17. S. Matsika and D. R. Yarkony, *J. Phys. Chem A*, accepted for publication (2002).
18. D. R. Yarkony, *J. Chem. Phys.* **112**, 2111 (2000).
19. D. R. Yarkony, *J. Phys. Chem. A* **105**, 2642 (2001).
20. S. Matsika and D. R. Yarkony, *J. Phys. Chem. A.* **106**, 2580 (2002).
21. D. R. Yarkony, *Molec. Phys.* **99**, 1463 (2001).
22. M. V. Berry, *Proc. R. Soc. London Ser. A* **392**, 45 (1984).
23. D. R. Yarkony, *J. Chem. Phys.* **114**, 2601 (2001).
24. H. A. Bethe and E. E. Salpeter, *Quantum Mechanics of One and Two Electron Atoms*, Plenum/Rosetta, New York, 1977.
25. S. R. Langhoff and C. W. Kern, *Molecular Fine Structure*, in *Modern Theoretical Chemistry*, H. F. Schaefer, ed., Plenum, New York, 1977, Vol. 4, p. 381.

26. D. R. Yarkony, *Int. Reviews of Phys. Chem.* **11**, 195 (1992).
27. D. R. Yarkony, *J. Chem. Phys.* **89**, 7324 (1988).
28. M. I. Lester, R. A. Loomis, R. L. Schwartz, and S. P. Walch, *J. Phys. Chem. A* **101**, 9195 (1997).
29. D. R. Yarkony, *J. Chem. Phys.* **111**, 6661 (1999).
30. B. C. Hoffman and D. R. Yarkony, *J. Chem. Phys.* **113**, 10091 (2000).
31. B. H. Lengsfield and D. R. Yarkony, *Nonadiabatic Interactions Between Potential Energy Surfaces: Theory and Applications*, in *State-Selected and State to State Ion-Molecule Reaction Dynamics: Part 2 Theory*, M. Baer and C.-Y. Ng, eds., John Wiley & Sons, Inc., New York, 1992, Vol. 82, pp. 1–71.
32. M. R. Manaa and D. R. Yarkony, *J. Chem. Phys.* **99**, 5251 (1993).
33. D. R. Yarkony, *J. Chem. Phys.* **97**, 1838 (1992).
34. C. E. Moore, *Atomic Energy Levels, Natl. Stand. Ref. Data Ser., Natl. Bur. Stand.*, U.S. GPO, Washington, DC, 1971.
35. G. J. Atchity, S. S. Xantheas, and K. Ruedenberg, *J. Chem. Phys.* **95**, 1862 (1991).
36. A. L. Kaledin, M. C. Heaven, and K. Morokuma, *Chem. Phys. Lett.* **289**, 110 (1998).
37. A. L. Kaledin, M. C. Heaven, and K. Morokuma, *J. Chem. Phys.* **114**, 215 (2001).
38. K. G. Dyall and E. van Lenthe, *J. Chem. Phys.* **111**, 1366 (1999).

RENNER–TELLER EFFECT AND SPIN–ORBIT COUPLING IN TRIATOMIC AND TETRAATOMIC MOLECULES

MILJENKO PERIĆ and SIGRID D. PEYERIMHOFF

Institut für Physikalische und Theoretische Chemie der Universität Bonn, Bonn, Germany

CONTENTS

The Role of Degenerate States in Chemistry: A Special Volume of Advances in Chemical Physics, Volume 124, Edited by Michael Baer and Gert Due Billing. Series Editors I. Prigogine and Stuart A. Rice. ISBN 0-471-43817-0.

I. RENNER OR RENNER–TELLER?

It is not usual to begin a review with the discussion of its title; we find, however, that we have to do that in the present case. In the very representative book *Computational Molecular Spectroscopy* [1], published last year, two papers written by eminent authors and devoted to the same subject, but with different titles, appeared back-to-back; one of them was entitled *The Renner effect* (Jensen and Bunker [2]), the title of the other was *The Renner-Teller effect* (Brown [3]). In his article, Brown explained this controversy in the following way: "The description of the vibrational levels for this pair of nearly degenerate potentials was first investigated in detail by Renner (1934) for a Π electronic state. However, the generalities of the problem had already been discussed by Herzberg and Teller in a preceding paper (1933). It is for this reason that some authors refer to vibronic coupling in a linear molecule as the Renner–Teller effect while the others prefer to call it simply the Renner effect. Both names are used in the literature to describe a single phenomenon. In this chapter, we follow Herzberg (1991) and refer to it as the Renner–Teller or R–T effect."

Formally, there was little reason to give Teller half of the credit for discovering such an important phenomenon. In the very extensive paper [4] referred to by Herzberg [5] as justification for the name "The Renner–Teller effect," there are only a few sentences that discuss in general the consequences of the reduced symmetry in linear molecules upon bending on the reliability of the Born–Oppenheimer approximation [6] and thus on the vibrational structure of molecular spectra. Moreover, the first author of this paper is Herzberg himself. One can, of course, claim that this is the heart of the problem, but we do not believe that such a conclusion required too much intellectual effort of renowned theoreticians/spectroscopists who knew the works by Wigner, Born, and Oppenheimer. On the other hand, seldom has a completely new topic been elaborated on as thoroughly as in Renner's paper [7]. We cannot agree with Jungen and Merer when they state in their excellent review [8] that "Renner's original paper seems not to have received the recognition which we feel it deserves, possibly *because it is in difficult German and in old notation*." We do not find Renner's notation much different from that used today nor his German more difficult than Herzberg's and Teller's; in almost all papers devoted to this subject up to, say, the 1980s the name "Renner effect" had been used. The situation has gradually changed in the last two decades, which should at least partially be ascribed to Herzberg's authority (By the way, Jungen and Merer were among the first to adopt the name "R–T effect.") We join here with those people who accept his recommendation, because he must have known the circumstances better than anybody else.

II. INTRODUCTION

The Renner–Teller effect (in the following text we shall use acronym R–T) arises when the potential energy surface of an electronic state spatially degenerate at the linear molecular geometry (Π, Δ, ...) splits upon bending into two surfaces. From the group-theoretical point of view, this splitting is a consequence of reduction of the axial symmetry $C_{\infty v}$, $D_{\infty h}$ upon bending to C_s, C_v, the latter two point groups possessing only one-dimensional (1D) irreducible representations. At small distortions of linearity the two (nondegenerate) electronic states, corresponding to these potential surfaces, lie close to each other, which manifests itself in a complicated vibrational and rotational structure of the corresponding spectra. The origin of the R–T effect can be interpreted in different terms: [8] (1) as a consequence of the electrostatic interaction between two components of an electronic state with a nonzero angular momentum; (2) as a coupling of two different electronic states through the electronic–rotational Coriolis interaction; (3) from the quantum chemical standpoint, the R–T effect is a consequence of violation of validity of the Born–Oppenheimer approximation.

In his classical paper, Renner explicitly considered only one of several cases that differ from each other from a *quantitative* point of view, namely, that in which the molecular potential energy surfaces for *both* components of a Π state have the minimum at the linear geometry (we shall call this situation "*weak R–T effect*"). The reason for his restriction to Π states was that he realized that the manifestation of the breakdown of the Born–Oppenheimer (BO) approximation seen in the spectral features would be most spectacular just in this case; at the same time this was the only situation for which closed formulas for vibronic (vibrational–electronic) energy levels could be derived by hand (in the framework of the perturbation theory). A generalization of his *theory* to the cases in which one or both of these potential surfaces has/have a minimum at a bent nuclear arrangement and to other degenerate electronic species is (at least from the conceptual point of view) more or less straightforward. It requires, however, other *computational approaches* (numerical, variational) to solve the corresponding equations.

As argued above, it is not Renner who should be blamed for his paper being forgotten for almost 25 years. The reason is that the experimentalists needed this much time to obtain the first spectrum showing the features predicted by him [9,10]. The effect that might have looked exotic in the 1930s has become one of the most usual and best studied spectroscopic problems. A great part of radicals being of crucial interest from an astrophysical/chemical point of view and/or playing as reaction intermediates an important role in various chemical processes have spatially degenerate ground states and thus exhibit the R–T effect. This circumstance determines not only the spectral features of such radicals, but also their behavior in chemical reactions.

A consequence of the continuously growing importance of such systems is the appearance of several excellent reviews devoted to this subject. A comprehensive presentation of the theoretical and experimental situation in this field up to 1974 is given by Duxbury (one of very few "R–T" titles up to that time) [11]. Jungen and Merer's [8] breakthrough paper from 1976 presents various theoretical aspects of the R–T effects in triatomics, particularly the relationship between two apparently opposite, but in fact equivalent, points of views (electrostatic interaction within a degenerate electron state—Coriolis interaction between two separate states). A critical exposure of different models for handling the R–T effect is given by Brown and Jørgensen [12]: It is recommended to everyone who is interested in this subject, particularly to those who really want to understand it completely, although we find this article the most difficult of all mentioned so far. Köppel et al. [13] and Köppel and Domcke [14] present the R–T effect as a particular case of a general multistate and multimode problem. Their models are deliberately so simplified that they enable clear insight into the mechanisms being otherwise obscured. Recently, Brown [3] has given a very clear presentation of the effective Hamiltonian approaches used in experimental investigations of the R–T effect, and a very detailed description of a highly sophisticated theoretical approach has been presented by Jensen et al. [2]. Furthermore, the R–T effect is described in detail in several standard textbooks, for example, by Herzberg [5]. and Bunker and Jensen [15]. Finally, let us mention the reviews by the present authors, which refer to early ab initio studies on triatomic [16,17] and tetraatomic [18] species.

Taking into account that there already exists such a comprehensive literature on the R–T effect, the question naturally arises: Is there a need to write a new review devoted to this subject? Is it, for example, possible to add anything relevant to the review by Brown and Jørgensen [12], to present the matter more clearly than Jungen and Merer [8], or more elegantly than Köppel et al. [13,14]? However, most of the reviews mentioned above are based on the results of their authors and reflect the authors' particular point of view. We feel that a comparison of different approaches is still missing. This is what we try to do in this chapter. Emphasis will be put on results of ab initio methods, but a discussion of approaches based on the use of experimentally derived parameters (potential energy surfaces, equilibrium geometries, etc.) will also be included, because these approaches can be employed equally well (and several of them have been) if the parameters are generated in ab initio calculations.

III. TRIATOMIC MOLECULES

A. Theoretical Treatment

Let us start with a description of the bending vibrations in singlet Σ electronic states of linear triatomic molecules. Already, this simple sentence reflects some of the difficulties with which one is confronted by handling the present subject: The adjective "linear" looks pleonastic in connection with "Σ states," because only linear molecules can have Σ electronic states. On the other hand, we speak about bending of molecules in Σ states. The rigorous meaning of the mentioned sentence is that we consider the bending vibrations of a molecule with linear equilibrium geometry, at which it is in a Σ electronic state; upon bending, the state of the molecule transforms into one of the species classified according to the irreducible representations of the lower symmetry point group corresponding to the bent nuclear arrangement. In this chapter we shall frequently be forced, in order to save time and space, to use similar phrases that may not be strictly correct from a scientific and/or linguistic point of view. Returning now to our problem, we assume the model Hamiltonian in the form

$$H = H_e + T_b + T_r^z \tag{1}$$

where H_e is the electronic Hamiltonian involving the kinetic energy of electrons, their mutual interaction, and the interaction of the electrons with the nuclei. Furthermore, we assume that H_e also includes the nuclear repulsion term. The term T_b is the kinetic energy operator for the bending vibrations of nuclei. It involves the derivatives of the coordinate ρ, defined as a supplement of the bond angle (in radian) and can be written in the form

$$T_b = -\frac{1}{2}\left[T_1(\rho)\frac{\partial^2}{\partial\rho^2} + T_2(\rho)\frac{\partial}{\partial\rho} + T_0(\rho)\right] \tag{2}$$

[Atomic units ($m_e \equiv 1$, $q_e \equiv 1$, $\hbar \equiv 1$) are used throughout this chapter.] The coefficients T_1, T_2, and T_0 are assumed to be in general analytical functions of the bending coordinate ρ. The term T_r^z represent the operator describing the rotation of the molecule around the (principal) axis z corresponding to the smallest moment inertia—this axis coincides at the linear nuclear arrangement with the molecular axis. Now T_r^z can be written in the form

$$T_r^z = A(\rho)R_z^2 \tag{3}$$

where $A = 1/(2I_{zz})$ is the rotational constant and R_z is the z component of the angular momentum of the nuclei. The operator R_z is defined as

$$R_z = -i\frac{\partial}{\partial\phi} \tag{4}$$

where ϕ represents the angle between the instantaneous molecular plane and the space-fixed plane with which it has a common z axis. The model Hamiltonian [Eq. (1)] commutes with R_z and thus the quantum number l corresponding to the latter operator is a good quantum number. It is an integer as any quantum number for projection of a spatial angular momentum. The literature often refers to l as the absolute value of the eigenvalue R_z; we shall consider it here as a signed quantity.

Explicit forms of the coefficients T_i and A depend on the coordinate system employed, the level of approximation applied, and so on. They can be chosen, for example, such that a part of the coupling with other degrees of freedom (typically stretching vibrations) is accounted for. In the space-fixed coordinate system at the infinitesimal bending vibrations, $T_b + T_r^z$ reduces to the kinetic energy operator of a two-dimensional (2D) isotropic harmonic oscillator,

$$\lim_{\rho\to 0} T = T_0 = -\frac{1}{2\mu}\left(\frac{\partial^2}{\partial\rho^2} + \frac{1}{\rho}\frac{\partial}{\partial\rho} + \frac{1}{\rho^2}\frac{\partial^2}{\partial\phi^2}\right) \tag{5}$$

where μ is the corresponding reduced mass. The Hamiltonian [Eq. (1)] does not involve the terms describing the stretching vibrations and the end-over-end rotations. It is supposed that these degrees of freedom can be separated from those considered here.

The wave function corresponding to the Hamiltonian [Eq. (1)] can be assumed in the form

$$\Psi_{l,n} = \psi\frac{1}{\sqrt{2\pi}}e^{il\phi}f^{l,m}(\rho) \tag{6}$$

where Ψ represents the electronic wave function and f is the ρ-dependent part of the nuclear function. Here m is the running index numbering vibrational states corresponding to a particular l state. After integrating over the electronic coordinates and ϕ, the Schrödinger equation corresponding to the Hamiltonian (1) becomes

$$\left[-\frac{1}{2}\left(T_1\frac{\partial^2}{\partial\rho^2} + T_2\frac{\partial}{\partial\rho} - l^2T_3 + T_0\right) + V(\rho)\right]f(\rho) = Ef(\rho) \tag{7}$$

$V(\rho)$ represents the potential energy surface for bending vibrations obtained by solving the electronic Schrödinger equation in the BO approximation; it is two dimensional, but invariant with respect to the coordinate ϕ. In the harmonic approximation, the potential energy function $V(\rho)$ is represented by $1/2\ k\ \rho^2$, where k is the force constant for the bending vibrations, and the kinetic energy operator is then given by Eq. (5) (with $\partial^2/\partial\phi^2$ replaced by $-l^2$). The solutions of the corresponding Schrödinger equation are $E = (\upsilon + 1)\,\omega$, where ω is the harmonic bending frequency, $\omega = \sqrt{(k/\mu)}$ (to simplify the orthography we use for the bending frequency symbol ω instead of more usual ω_2). The corresponding wave functions are

$$R_{\upsilon,l}(\rho) = N_{\upsilon,l}(\sqrt{\lambda}\rho)^{|l|}L_{\upsilon,l}(\sqrt{\lambda}\rho)e^{-1/2\lambda\rho^2} \tag{8}$$

where $L_{\upsilon,l}$ are Laguerre polynomials and

$$\lambda = \sqrt{k\mu} = \mu\omega \tag{9}$$

$R(\rho)$ is thus the harmonic approximation counterpart of the function $f(\rho)$. To simplify the orthography, we use symbol $L_{\upsilon,l}$ instead of the usual $L^{l}_{(\upsilon-l)/2}$. For a given υ, the quantum number l takes values $-\upsilon, -\upsilon + 2, \ldots, 1$ or $0, \ldots, \upsilon - 2, \upsilon$. Since the bending energy (in the harmonic approximation) depends only on the quantum number υ, a vibrational level is thus $\upsilon + 1$ times degenerate.

In the harmonic approximation, the bending vibrational energy scheme for a singlet Σ electronic state corresponds to the upper part of the left-hand side of Figure 1 (given below in Section III.B). Anharmonicity of the bending potential would cause splitting of the levels with the same quantum number υ but different l; while the latter of them remains a good quantum number, because the axial symmetry of the problem is also preserved when the anharmonicity is introduced, the former becomes an approximate good quantum number. In the framework of this model (Σ electronic states, neglected effects of end-over-end rotations, etc.) the bending energy does not depend on the sign of l; this is a reason for the above mentioned widely used convention that assumes l to be a nonnegative integer and for representing the levels corresponding to the same absolute value of l by a single line in Figure 1. Remember, however, that every $|l| \neq 0$ level is twofold degenerate.

Let us see now how the situation is changed if the molecule is a spatially degenerate electronic state. This state has two components that are at linear molecular geometry exactly degenerate with each other. Upon bending, however, the symmetry of the nuclear framework is reduced, consequently having a splitting of the potential surfaces for these component states. At least at small deviation from linearity these two electronic surfaces lie energetically close to each other and the ansatz [Eq. (6)], which assumes the vibronic wave function as a product of a single electronic species with the corresponding vibrational function, ceases to be reliable. Instead, we have to employ the wave function of the form

$$\Psi = \psi_1 f_1(\rho, \phi) + \psi_2 f_2(\rho, \phi) \tag{10}$$

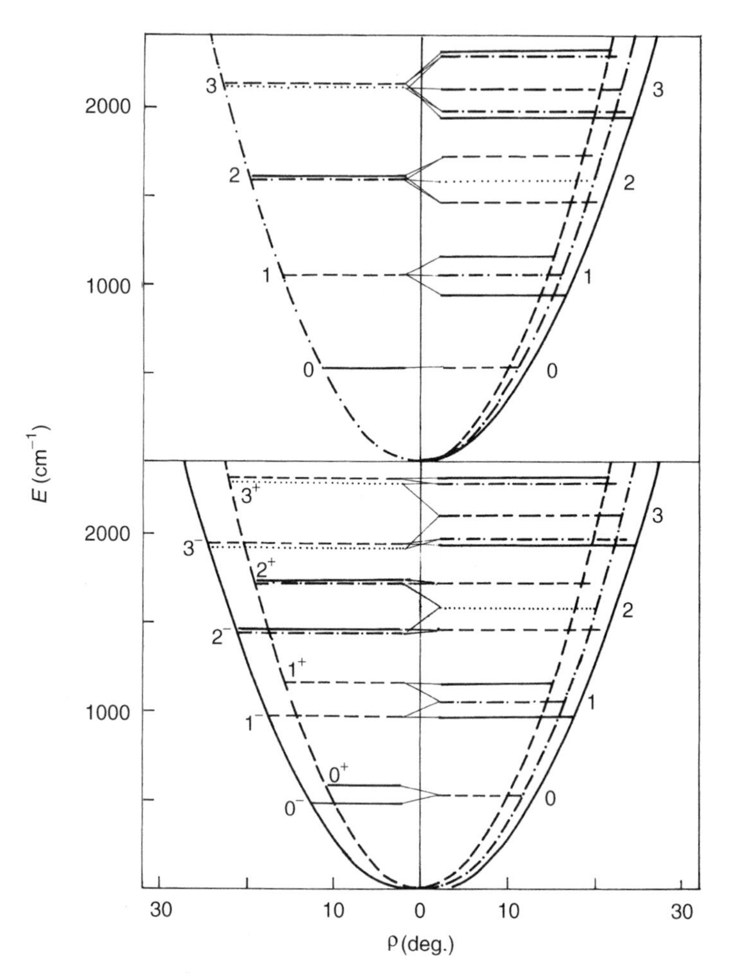

Figure 1. Relationship between bending vibrational structure in Σ electronic states and vibronic strucure in Π states. Left, top: Bending vibrational structure in a Σ electronic state of a triatomic molecule with linear equilibrium geometry. Left, bottom: Bending vibrational structure of two Σ electronic states with the same minimum of bending potential curves. Right: Vibronic structure of a Π electronic state of a molecule with linear equilibrium geometry ($A^3\Pi_u$ state of NCN) [28,29]. Solid curve: Bending potential curve for the lower lying adiabatic electronic state. Dashed curve: Bending potential curve for the upper adiabatic electronic state. Dash–dotted curve: Mean bending potential. Solid horizontal lines: $l = 0$ vibrational levels for Σ electronic states, $K = 0$ vibronic levels for Π electronic state. Dashed lines: $l = 1$ and $K = 1$ levels. Dash–dotted lines: $l = 2$ and $K = 2$ levels. Dotted lines: $l = 3$ and $K = 3$ levels. Vibrational levels are assigned by the quantum numbers υ_2. υ_2^-, and υ_2^+ quantum numbers in the bottom left part of figure denote vibrational levels belonging to the lower and upper Σ electronic state, respectively.

where ψ_1 and ψ_1 are electronic (at the moment not specified more precisely) species and f_1, f_2 are function of the nuclear coordinates.

Now, we have besides the vibrational, the electronic angular momentum; the latter is characterized by the quantum number Λ corresponding to the magnitude of its projection along the molecular axis, L_z. Here we shall consider Λ as a unsigned quantity, that is, for each $\Lambda \neq 0$ state there will be two possible projections of the electronic angular momentum, one corresponding to Λ and the other to $-\Lambda$. The operator L_z can be written in the form

$$L_z = -i\frac{\partial}{\partial\theta} \tag{11}$$

where θ is the coordinate describing collective rotations of the electrons around the z axis. This one-electron approximation was severely criticized by Brown and Jørgensen [12]. Note that, although it will be used in this chapter in several instances in order to achieve a simple presentation of some results, it will never be necessary to apply explicitly expression (11). In handling the R–T effect, one actually needs the (electronic) matrix elements of the operator L_z, which can be derived without assuming that the operator has the form (11). Trying, however, to reconcile the ansatz (11) with justified critics of the authors of [13], we can interpret θ as a coordinate (in general without obvious physical meaning) conjugate to the electronic angular momentum L_z.

The presence of two angular momenta has as a consequence that only their sum, representing the total angular momentum in the case considered, necessary commutes with the Hamiltonian of the system. Thus only the quantum number K, associated with the sum, N_z, of R_z and L_z,

$$N_z = R_z + L_z = -i\left(\frac{\partial}{\partial\phi} + \frac{\partial}{\partial\theta}\right) \tag{12}$$

that is,

$$K = l \pm \Lambda \tag{13}$$

must be a good quantum number, not necessarily l and Λ. It can be easily shown that this is really the case, that is, that l and Λ are generally not even approximately good quantum numbers. The operator for rotation of the nuclei around the z axis can now be written as

$$T_r^z = A(\rho)R_z^2 = A(\rho)(N_z - L_z)^2 \tag{14}$$

Now, we add to (1) the operator describing the spin–orbit (SO) coupling, so that our model Hamiltonian becomes

$$H = H_e + T_b + T_r^z + H_{\text{SO}} \tag{15}$$

We shall consider only the leading part of the spin–orbit operator assumed in the phenomenological form

$$H_{\mathrm{so}} = A_{\mathrm{so}} L_z S_z \tag{16}$$

where A_{so} is the "spin–orbit constant" and S_z the z component of the electronic spin operator. The Hamiltonian [Eq. (15)] commutes with the z component of the spin operator, S_z, and of the total angular momentum, $J_z = N_z + S_z$. Consequently, the corresponding quantum numbers Σ (we choose this symbol in spite of the fact that it is also used to denote $\Lambda = 0$ electronic states and $K = 0$ vibronic species) and $P = K + \Sigma$ are good quantum numbers. In the framework of this model, such also remains the quantum number K. This is not rigorously the case in treatments of the R–T effect at higher levels of sophistication. It can be stated, however, that K is one of the best among "bad" quantum numbers used in molecular spectroscopy. Thus the handling of the R–T effect in the framework of this model can be carried out within a particular K and P (i.e., Σ) subspace separately. In this chapter, we consider K, Σ, and P to be signed; since the vibronic (vibration–electronic) levels with $|K| \neq 0$ (and $|P| \neq 0$) are always doubly degenerate in the framework of this model [one state corresponding to $K = +|K|$ and the other to $K = -|K|$ (and analogously for $P = +|P|$ and $P = -|P|$)], we shall deal, as a rule with nonnegative values for K and P only.

Until now we have implicitly assumed that our problem is formulated in a space-fixed coordinate system. However, electronic wave functions are naturally expressed in the system bound to the molecule; otherwise they generally also depend on the rotational coordinate ϕ. (This is not the case for Σ electronic states, for which the wave functions are invariant with respect to ϕ.) The eigenfunctions of the electronic Hamiltonian, ψ_e^+ and ψ_e^-, computed in the framework of the BO approximation ("adiabatic" electronic wave functions) for two electronic states into which a spatially degenerate state of linear molecule splits upon bending,

$$H_e\psi^+ = V^+\psi^+ \qquad H_e\psi^- = V^-\psi^- \tag{17}$$

are labeled by a *plus* or *minus* according to their behavior with respect to reflection in the molecular plane. The species ψ^+ (of A_1 or B_2 symmetry in the C_{2v} point group, and of A' in C_s) is invariant upon this reflections, and ψ^- (belonging to B_1 or A_2 irreducible representation in the C_{2v} point group, and to A'' in C_s) changes sign. For the sake of convenience, we assume throughout this chapter that the phase factor in ψ^- is i. Both V^+ and V^- are the corresponding adiabatic potentials. Therefore, these wave functions rotate together with the molecular plane. It is thus convenient (although not necessary) to formulate the complete problem in the coordinate system rotating with the molecule in the way that a plane of this frame (we chose it to be xz) coincides with the instantaneous molecular plane at bent nuclear arrangements. We introduce the coordinate transformation

$$\chi = \phi \qquad \alpha = \theta - \phi \tag{18}$$

α is therefore the electronic azimuthal coordinate with respect to the molecular plane. It is easy to prove that the nuclear, electronic, and total (excluding spin) angular momenta defined with respect to the molecule-bound frame are

$$R_z = -i\left(\frac{\partial}{\partial\chi} - \frac{\partial}{\partial\alpha}\right) = N_z - L_z \qquad L_z = -i\frac{\partial}{\partial\alpha} \qquad N_z = -i\frac{\partial}{\partial\chi} \tag{19}$$

Apparently, the most natural choice for the electronic basis functions consist of the adiabatic functions ψ_e^+ and ψ_e^- defined in the molecule-bound frame. By making use of the assumption that K is a good quantum number, we can write the complete vibronic basis in the form

$$\frac{1}{\sqrt{2\pi}} e^{iK\chi}\, \psi_\Sigma^+(\alpha) f^{+,K,n}(\rho) \qquad \frac{1}{\sqrt{2\pi}} e^{iK\chi}\, \psi_\Sigma^-(\alpha) f^{-,K,n}(\rho) \tag{20}$$

The electronic functions are assumed to be eigenfunctions of the spin operator and are thus assigned by the spin quantum number Σ. Their dependence on α, indicated in Eq. (20), should not be taken too seriously; it should just remind us that the functions are defined in the molecule-bound frame, that is, that they do not depend on the nuclear rotational coordinate ϕ. The bending functions f^+ and f^-, corresponding to the electronic species ψ^+ and ψ^-, respectively, are additionally labeled by K and a running index n. The electronic Hamiltonian and the operator T_b are diagonal in the adiabatic electronic basis. The operator R_z couples the functions ψ^+ and ψ^- with each other. Although only the nuclear angular momentum R_z appears explicitly in the model Hamiltonian [Eq. (15)], the electronic angular momentum creeps into its matrix representation as a consequence of the first of relations (19) and the form of the basis functions. The adiabatic electronic functions are, namely, not the eigenfunctions of L_z. Let us denote the electronic matrix elements of the operators L_z and L_z^2 by

$$\begin{aligned} \langle\psi^+(\alpha)|L_z^2|\psi^+(\alpha)\rangle = C^{++} \qquad \langle\psi^-(\alpha)|L_z^2|\psi^-(\alpha)\rangle = C^{--} \\ \langle\psi^+(\alpha)|L_z|\psi^-(\alpha)\rangle = B^{+-} = \langle\psi^-(\alpha)|L_z|\psi^+(\alpha)\rangle \end{aligned} \tag{21}$$

The remaining combinations vanish for symmetry reasons [the operator L_z transforms according to B_1 (A'') irreducible representation]. The nonvanishing of the off-diagonal matrix element B^{+-} is responsible for the coupling of the adiabatic electronic states.

Since the form of the electronic wave functions depends also on the coordinate ρ (in the usual, "parametric" way), the matrix elements (21) are functions of it too. Thus it looks at first sight as if a lot of cumbersome computations of derivatives of the electronic wave functions have to be carried out. In this case, however, nature was merciful: the matrix elements in (21) enter the Hamiltonian matrix weighted with the rotational constant A, which tends to infinity when the molecule reaches linear geometry. This means that only the form of the wave functions, that is, of the matrix elements in (21), in the $\rho \to 0$ limit are really needed. In the above mentioned one-electron approximation

$$\begin{aligned} \lim_{\rho \to 0} \psi^{+}(\alpha) &\simeq \psi_0^{+}(\alpha) = \frac{1}{\sqrt{\pi}} \cos(\Lambda\alpha) \xi_{\Sigma}(\rho_e) \\ \lim_{\rho \to 0} \psi^{-}(\alpha) &\simeq \psi_0^{-}(\alpha) = \frac{i}{\sqrt{\pi}} \sin(\Lambda\alpha) \xi_{\Sigma}(\rho_e) \end{aligned} \tag{22}$$

$\xi_{\Sigma}(\rho_e)$ represents the part of the electronic functions depending on all spatial and spin electronic coordinates except θ. In the lowest order approximation, which is usually sufficiently reliable, it is the same for both electronic species in question. With these electronic functions

$$\begin{aligned} \langle \psi_0^{+}(\alpha) | L_z^2 | \psi_0^{+}(\alpha) \rangle &= \Lambda^2 = \langle \psi_0^{-}(\alpha) | L_z^2 | \psi_0^{-}(\alpha) \rangle \\ \langle \psi_0^{+}(\alpha) | L_z | \psi_0^{-}(\alpha) \rangle &= \Lambda = \langle \psi_0^{-}(\alpha) | L_z | \psi_0^{+}(\alpha) \rangle \end{aligned} \tag{23}$$

Note that the relations (23) are valid also if (22) is questionable. Brown [19] refined the approximation (23) by introducing the "g_K factor," describing the deviation of the mean values for L_z and L_z^2 from integers. Validity of the approximation (23) has been checked by means of explicit ab initio calculations, for example, in [20,21].

After integrating over the electronic coordinates and χ, the model Hamiltonian (15) is represented by the matrix whose elements are

$$\begin{aligned} H^{++} &= V^{+} - \frac{1}{2}\left(T_1 \frac{\partial^2}{\partial \rho^2} + T_2 \frac{\partial}{\partial \rho} + T_0 \right) + A(K^2 + C^{++}) \\ H^{--} &= V^{-} - \frac{1}{2}\left(T_1 \frac{\partial^2}{\partial \rho^2} + T_2 \frac{\partial}{\partial \rho} + T_0 \right) + A(K^2 + C^{--}) \\ H^{+-} &= (-2KA + \Sigma A_{\text{so}}) B^{+-} = H^{-+} \end{aligned} \tag{24}$$

This matrix represents an effective operator that still has to act on the bending functions $f^{+}(\rho)$, $f^{-}(\rho)$. A generalization of (24) to the case when the kinetic energy operator (i.e., the coefficients T_i and A) has a different form in the electronic states ψ^+ and ψ^- can be carried out straightforwardly. In order to have expressions more suitable for discussing the spectroscopic aspects of this problem, we now replace the matrix elements B and C by their asymptotic values (23) and the kinetic energy operator by its small-amplitude counterpart (5) [note, however, that the asymptotic forms of the electronic wave functions, Eq. (25), cannot be employed for calculation of the matrix elements of the electronic operator H_e]:

$$\begin{pmatrix} V^+ - \frac{1}{2\mu}\left(\frac{\partial^2}{\partial\rho^2} + \frac{1}{\rho}\frac{\partial}{\partial\rho} - \frac{K^2+\Lambda^2}{\rho^2}\right) & -\frac{2\Lambda KA}{\rho^2} + \Lambda\Sigma A_{so} \\ -\frac{2\Lambda KA}{\rho^2} + \Lambda\Sigma A_{\rm so} & V^- - \frac{1}{2\mu}\left(\frac{\partial^2}{\partial\rho^2} + \frac{1}{\rho}\frac{\partial}{\partial\rho} - \frac{K^2+\Lambda^2}{\rho^2}\right) \end{pmatrix} \tag{25}$$

The basis consisting of the adiabatic electronic functions (we shall call it "bent basis") has a serious drawback: It leads to appearance of the off-diagonal elements that tend to infinity when the molecule reaches linear geometry (i.e., $\rho \rightarrow 0$). Thus it is convenient to introduce new electronic basis functions by the transformation

$$\psi^{\Lambda}(\theta) = \frac{1}{\sqrt{2}} e^{i\Lambda\phi}(\psi^+ + \psi^-) \qquad \psi^{-\Lambda}(\theta) = \frac{1}{\sqrt{2}} e^{-i\Lambda\phi}(\psi^+ - \psi^-) \tag{26}$$

We write them as $\psi^{\pm\Lambda}(\theta)$ to stress that now we use the space-fixed coordinate frame. We shall call this basis "diabatic," because the functions (26) are not the eigenfunction of the electronic Hamiltonian. The matrix elements of H_e are

$$\begin{aligned} \langle\psi^{\Lambda}(\theta)|H_e|\psi^{\Lambda}(\theta)\rangle &= \frac{V^+ + V^-}{2} = \langle\psi^{-\Lambda}(\theta)|H_e|\psi^{-\Lambda}(\theta)\rangle \\ \langle\psi^{\Lambda}(\theta)|H_e|\psi^{-\Lambda}(\theta)\rangle &= e^{-2i\Lambda\phi}\frac{V^+ - V^-}{2} \\ \langle\psi^{-i\Lambda}(\theta)|H_e|\psi^{\Lambda}(\theta)\rangle &= e^{2i\Lambda\phi}\frac{V^+ - V^-}{2} \end{aligned} \tag{27}$$

We shall call this basis also "linear" because in the one-electron approximation at $\rho \rightarrow 0$ the functions (26) become

$$\begin{aligned} \lim_{\rho\rightarrow 0} \psi^{\Lambda}(\theta) &\simeq \psi_0^{\Lambda}(\theta) = \frac{1}{\sqrt{2\pi}} e^{i\Lambda\theta}\, \xi_{\Sigma}(\rho_e) \\ \lim_{\rho\rightarrow 0} \psi^{-\Lambda}(\theta) &\simeq \psi_0^{-\Lambda}(\theta) = \frac{1}{\sqrt{2\pi}} e^{-i\Lambda\theta}\, \xi_{\Sigma}(\rho_e) \end{aligned} \tag{28}$$

that is, they reduce to the functions describing free rotation of electrons around the molecular axis. These asymptotic forms of the basis functions may be used in computation of the matrix elements of the kinetic energy operator and the spin–orbit part of the model Hamiltonian. The functions (28) are the eigenfunctions of L_z. We can utilize this fact by expanding the rovibrational (rotational–vibrational) part of the complete rovibronic functions in the basis consisting of the eigenfunctions of a 2D harmonic oscillator,

$$\Phi_{\upsilon,l} = \frac{1}{\sqrt{2\pi}} e^{il\phi} R_{\upsilon,l}(\sqrt{\lambda}\rho) \tag{29}$$

where $R_{\upsilon,l}$ are defined by Eq. (8). Applying the operator N_z [Eq. (12)] onto the product of asymptotic electronic basis functions (28) and rovibrational functions (29) one obtains

$$N_z \psi_0^{\Lambda} \Phi_{\upsilon,l} = (l+\Lambda) \psi_0^{\Lambda} \Phi_{\upsilon,l} \qquad N_z \psi_0^{-\Lambda} \Phi_{\upsilon,l} = (l-\Lambda) \psi_0^{-\Lambda} \Phi_{\upsilon,l} \tag{30}$$

Thus, for a particular value of the good quantum number K, the only possible values for l are $K \pm \Lambda$. The matrix representation of the model Hamiltonian in the linear basis, obtained by integrating over the electronic coordinates and ϕ, is thus

$$\begin{aligned} H^{\Lambda,\Lambda} &= \frac{V^+ + V^-}{2} - \frac{1}{2}\left(T_1 \frac{\partial^2}{\partial\rho^2} + T_2 \frac{\partial}{\partial\rho} + T_0\right) \\ &\quad + A\left(K^2 + \frac{C^{++} + C^{--}}{2} - 2KB^{+-}\right) + \Sigma B^{+-} A_{\text{so}} \\ H^{-\Lambda,-\Lambda} &= \frac{V^+ + V^-}{2} - \frac{1}{2}\left(T_1 \frac{\partial^2}{\partial\rho^2} + T_2 \frac{\partial}{\partial\rho} + T_0\right) \\ &\quad + A\left(K^2 + \frac{C^{++} + C^{--}}{2} + 2KB^{+-}\right) - \Sigma B^{+-} A_{\text{so}} \\ H^{\Lambda,-\Lambda} &= \frac{V^+ - V^-}{2} = H^{-\Lambda,\Lambda} \end{aligned} \tag{31}$$

Employing simplifications arising from the use of asymptotic forms of the electronic basis functions and the zeroth-order kinetic energy operator, we obtain

$$\begin{pmatrix} \bar{V} - \frac{1}{2\mu}\left[\frac{\partial^2}{\partial\rho^2} + \frac{1}{\rho}\frac{\partial}{\partial\rho} - \frac{(K-\Lambda)^2}{\rho^2}\right] + \Lambda\Sigma A_{\text{so}} & \dfrac{V^+ - V^-}{2} \\ \dfrac{V^+ - V^-}{2} & \bar{V} - \frac{1}{2\mu}\left[\frac{\partial^2}{\partial\rho^2} + \frac{1}{\rho}\frac{\partial}{\partial\rho} - \frac{(K+\Lambda)^2}{\rho^2}\right] - \Lambda\Sigma A_{\text{so}} \end{pmatrix}$$

$$\bar{V} \equiv \frac{V^+ + V^-}{2} \tag{32}$$

The diagonal elements of the matrix [Eqs. (31) and (32)], actually being an effective operator that acts onto the basis functions $R_{\upsilon,l}$, are diagonal in the quantum number l as well. The factors $\exp(\pm 2i\Lambda\phi)$ [Eqs. (27)] determine the selection rule for the off-diagonal elements of this matrix in the vibrational basis—they couple the basis functions with different l values with one another (i.e., with $l' = l \pm \Lambda$).

The matrices (24) and (31) [or Eqs. (25) and (32)] are equivalent—one can be obtained from another by a unitary transformation. They reflect the two ways of interpreting the R–T effect mentioned in Section II [(2) and (1) respectively].

We employ the general scheme presented above as a starting point in our discussion of various approaches for handling the R–T effect in triatomic molecules. We find it reasonable to classify these approaches into three categories according to the level of sophistication at which various aspects of the problem are handled. We call them (1) minimal models; (2) pragmatic models; (3) benchmark treatments. The criterions for such a classification are given in Table I.

In Table I, 3D stands for three dimensional. The symbol $\rho^{2\Lambda}$ symbol in connection with the bending potentials means that the bending potentials are considered in the lowest order approximation; as already realized by Renner [7], the splitting of the adiabatic potentials has a $\rho^{2\Lambda}$ dependence at small distortions of linearity. With "exact" form of the spin–orbit part of the Hamiltonian we mean the microscopic (i.e., nonphenomenological) many-electron counterpart of, for example, The Breit–Pauli two-electron operator [22] (see also [23]).

Let us stress immediately that "minimal" must not be understood in a pejorative sense: Frequently it is more difficult to develop a simple model

TABLE I

Model	Minimal	Pragmatic	Benchmark
Bent–stretch coupling	Neglected	Indirectly	Full 3D vibrational treatment
b,c Rotations	Separated	Separated	Full vibrational–rotational treatment
Bending potential	$\rho^2\Lambda$	Fully anharmonic	
Kinetic energy	Small amplitude	Large amplitude	Exact
Good quantum numbers	K, P, J	K, P, J	J
$\langle L_z \rangle$, $\langle L_z^2 \rangle$	Integer	Integer	"Exact"
H_{so}	Phenomenologic	Phenomenologic	"Exact"
Other electronic states	Neglected	Neglected	Accounted for
Approach	Perturbative	Numerical/variational	Numerical/variational

incorporating essential features of a phenomenon (as, e.g., those by Köppel et al. [13,14]) than to carry out the most complex computations employing highly sophisticated but straightforward approaches. We refer "benchmark" to those approaches that do not involve any a priori approximation (like, e.g., neglect or indirect handling of the bend–stretch coupling), except of those not playing a significant role in chemical problems (as, e.g., relativistic effects beyond spin–orbit coupling). It should also not be thought that the results of "benchmark" calculations are necessarily more accurate than those achieved in a "pragmatic" handling (as, e.g., in the framework of the approach developed by Jungen and Merer [24–27]). We shall discuss this topic in Sections III.C–III.H. Before doing that, we consider, in Section III.B, the spectroscopic aspects of the R–T effect combined with spin–orbit coupling.

B. Spectroscopic Features

In this section, we briefly discuss spectroscopic consequences of the R–T coupling in triatomic molecules. We shall restrict ourselves to an analysis of the vibronic and spin–orbit structure, determined by the bending vibrational quantum number υ (in the usual spectroscopic notation υ_2) and the vibronic quantum numbers K, P.

1. *Vibronic Coupling in Singlet States of Linear Molecules*

Let us consider a singlet Π electronic species (right-hand side of Fig. 1) and first assume that the magnitude for the splitting of the potential surfaces upon bending is negligible, that is, that the electronic state remains degenerate at small-amplitude bending vibrations (this degeneracy is "accidental," because it does not follow for symmetry reasons), and that the bending potential is harmonic. The bending potential then has the same form as that presented on the left-hand side of Fig. 1 (top), representing a Σ electronic state, but it consists of two potential energy surfaces coinciding with each other. In the Π state, we also have the electronic angular momentum besides the vibrational state. In the (hypothetical) case we consider (no splitting of the potential surfaces), the presence of the additional electronic angular momentum has no effects on the position of vibronic energy levels. That becomes obvious if we look at matrix (32)—its off-diagonal elements vanish in this case. On the other hand, the number of levels is doubled. Furthermore, the presence of two angular momenta has as a consequence that the vibronic levels have to be classified according to the quantum number corresponding to their sum being the only angular momentum that commutes with the Hamiltonian. A simple bookkeeping shows that the lowest lying (nondegenerate) vibrational level of the Σ electronic state, characterized with the quantum numbers $\upsilon = 0$, $l = 0$, correlates with the (doubly degenerate) $\upsilon = 0$, $K = 1$ level of the Π state, that the $\upsilon = 1$, $l = 1$ level of Σ state corresponds, to the manifold of two $K = 0$ and a $K = 2$, $\upsilon = 1$ levels of the Π electronic species and so on.

Now, let us consider a realistic case when the splitting of the potential surfaces is small, but not negligible. These potentials are drawn on the left-hand side of Figure 1, bottom (they correspond quantitatively to the $A^3\Pi_u$ state of NCN [28,29]), and on the right-hand side of both the bottom and the top part of Figure 1. Besides, the mean potential is depicted on the right-hand side, having the same form as the potential on the left-hand side of Figure 1, top, that we considered in the above discussion. All potentials are assumed to be harmonic. On the right-hand side of Figure 1 are also displayed the vibronic energy levels for the $A^3\Pi_u$ state of NCN (corresponding to the value zero of the spin quantum number).

The bottom part of Figure 1 corresponds to the "bent" and the upper part to the "linear" point of view in interpreting the R–T effect. The right-hand side of the upper part shows how the splitting of the potential surfaces affects the positions of vibronic levels; so, for example, the three $\upsilon = 1$ levels now have different energies, and the only degeneracy that remains is that of the two components ($K = 2$ and -2) of the $|K| = 2$ vibronic level. The two $K = 0$ levels differ in their symmetry with respect to reflection in the molecular plane $(+/-)$. The $K = 2$ level has the energy close to that of the unperturbed vibrational level $\upsilon = 1, l = 1$. This is a general characteristic of the lowest lying level ("*unique*" level [8]) for each $K \neq 0$ vibronic series. The explanation is simple: While the other υ, $K \neq 0$ vibronic species result in the first approximation from the interaction between the υ, $l = K - \Lambda$ ($\Lambda = 1$ in the present case) and υ, $l = K + \Lambda$ sublevels, the *unique* levels, for which $K = \upsilon + \Lambda$, correspond exclusively to υ, $l = K - \Lambda$ unperturbed levels, because the υ, $l = K + \Lambda = \upsilon + 2\Lambda$ levels do not exist (remember that $|l| \leq \upsilon$).

The "bent" point of view offers the explanation of one other aspect of the vibronic energy pattern presented. On the left-hand side of the bottom part of Figure 1 are presented the bending levels for two Σ electronic states having the same potential surface as the components of the Π state considered. This situation can be looked upon as a particular case, $\Lambda = 0$, of the matrix representation (25); the coupling between the electronic states vanishes and each of them has its own bending levels with the pattern analogous to that on the left-hand side of Figure 1, top. The difference in the vibronic structure of two Σ and a Π electronic state is caused by the presence of the off-diagonal elements of the matrix (25) in the latter case. However, even in Π electronic states the off-diagonal elements vanish for the particular case $K = 0$ and these vibronic levels belong exclusively to one of the adiabatic electronic states. This is indicated symbolically on the right-hand side of Figure 1 by the corresponding energy level lines matching exactly one of the adiabatic potential curves. The $+/-$ symmetry of a $K = 0$ level is determined by the symmetry of the adiabatic state it belongs to (+ for A_1, B_2, A', − for B_1, A_2, A''). Because for Π electronic states $\Lambda = 1$, these levels coincide exactly with the $l = 1$ (i.e., $K = 1$) levels of one of the Σ electronic species. The existence of vibrational levels not being perturbed by the vibronic coupling is of great importance for spectroscopists, because it enormously facilitates the derivation of the shape of the adiabatic potential surfaces from measured data.

All vibronic levels in spatially degenerate electronic states except $K = 0$ species are more or less shared between both the adiabatic electronic states. The *unique* levels are almost equally shared between two adiabatic electronic states—their energetic position is such as if they belong to the mean adiabatic potential $(V^+ + V^-)/2$. We indicate this on the right-hand side of Figure 1 by the vibronic energy lines matching exactly with the mean potential curve. The other $K \neq 0$ levels belong predominantly to a particular adiabatic electronic state; this is indicated by the lines nearly matching one of the potential curves on the right-hand side of Figure 1.

The situation in singlet Δ electronic states of triatomic molecules with linear equilibrium geometry is presented in Figure 2. This vibronic structure can be interpreted in a completely analogous way as above for Π species. Note that in Δ electronic states there is a single *unique* level for $K = 1$, but for each other $K \neq 0$ series there are *two* levels with a *unique* character.

2. *Combined Vibronic and Spin–Orbit Coupling in Linear Molecules*

Let us see now how the situation is changed in the presence of the spin–orbit coupling. In the central part of Figure 3 are presented the low-lying vibronic levels of the $A^3\Pi_u$ state of NCN, as obtained in recent ab initio calculations [28,29]. On the left and right edge of the figure are displayed the bending vibrational levels corresponding to the case when both the vibronic and spin–orbit coupling are absent.

Going from left to right, we show in the second column the results of calculations in which the spin–orbit constant is set to be $A_{so} = -37$ cm^{-1} (as computed in the mentioned studies) and the vibronic coupling is neglected ($\varepsilon = 0$). The parameter ε is the "Renner parameter," defined for Π electronic states as the ratio of the (quadratic) force constants for the difference and the sum of the adiabatic bending potentials. Inspection of the secular equation (32) shows that in this case there exist three doubly degenerate effective bending potentials, involving the mean electronic energy and the contribution from the spin–orbit part of the Hamiltonian, with the energy spacing equal to $|A_{so}|$: The lowest energy ones correspond to $\Lambda(= 1)$, $\Sigma = 1$ and $-\Lambda$, $\Sigma = -1$; the next two to Λ, $\Sigma = 0$ and $-\Lambda$, $\Sigma = 0$; and the highest energy pair to Λ, $\Sigma = -1$ and $-\Lambda$, $\Sigma = 1$. Each zeroth-order vibrational level with a particular value υ is now split into three levels, each one belonging to one of the effective potentials. These levels are generally degenerate, involving all possible K species with the

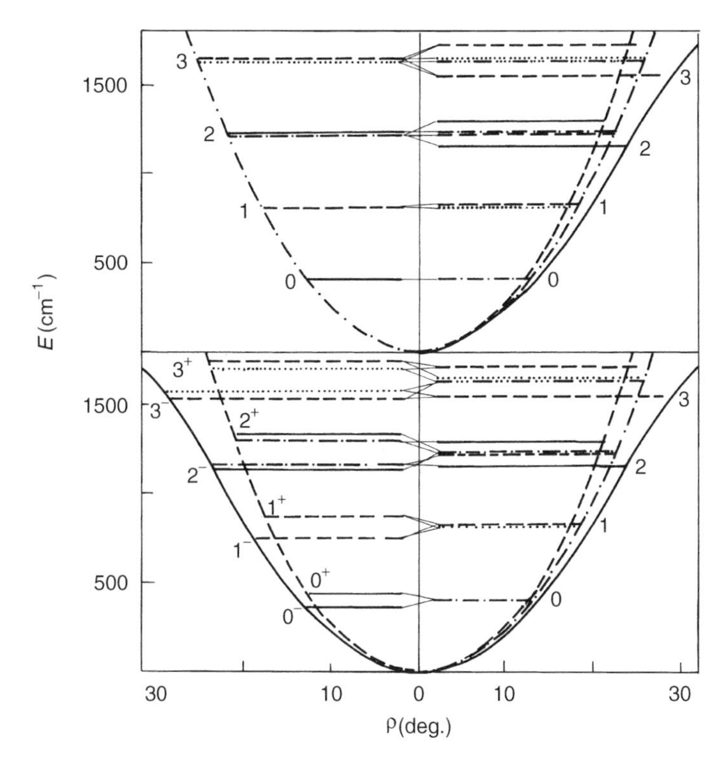

Figure 2. Relationship between bending vibrational structure in Σ electronic states and vibronic structure in Δ states. Left, top: Bending vibrational structure in a Σ electronic state of a triatomic molecule with linear equilibrium geometry. Left, bottom: Bending vibrational structure of two Σ electronic states with the same minimum of bending potential curves. Right: Vibronic structure of a Δ electronic state of a molecule with linear equilibrium geometry. Solid curve: Bending potential curve for the lower lying adiabatic electronic state. Dashed curve: bending potential curve for the upper adiabatic electronic state. Dash–dotted curve: mean bending potential. Bending vibrational structure in a Σ electronic state of a triatomic molecule with linear equilibrium geometry. Solid horizontal lines: $l = 0$ vibrational levels for Σ electronic states, $K = 0$ vibronic levels for Δ electronic state. Dashed horizontal lines: $l = 1$ and $K = 1$ levels. Dash–dotted lines: $l = 2$ and $K = 2$ levels. Dotted lines: $l = 3$ and $K = 3$ levels. Vibrational levels are assigned by the quantum numbers υ_2. The υ_2^- and υ_2^+ quantum numbers at the bottom left part of the figure denote vibrational levels belonging to the lower and upper Σ electronic state, respectively.

combinations of quantum numbers $\pm\Lambda$, Σ associated with the effective potential in question. The exception are the $\upsilon = 0$ levels, being nondegenerate (except for the $\pm|K|$ degeneracy).

Introduction of the vibronic coupling ($\varepsilon \neq 0$) causes removal of the above degeneracy and leads to the general vibronic–spin–orbit pattern presented in the central part of Figure 3. Each vibronic level is characterized by a particular K

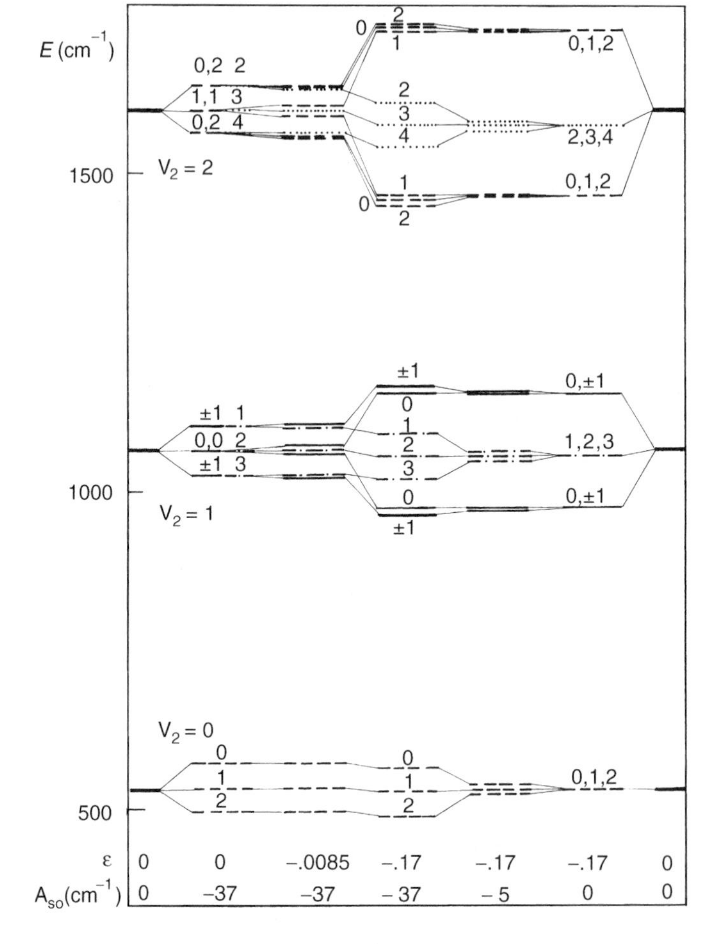

Figure 3. Low-energy vibronic spectrum in a $^3\Pi$ electronic state of a linear triatomic molecule, computed for various values of the Renner parameter ε and spin–orbit constant A_{so} (in cm^{-1}). The spectrum shown in the center of figure ($\varepsilon = -0.17$, $A_{so} = -37\text{cm}^{-1}$) corresponds to the $A^3\Pi_u$ state of NCN [28,29]. The zero on the energy scale represents the minimum of the potential energy surface. Solid lines: $K = 0$ vibronic levels; dashed lines: $K = 1$ levels; dash-dotted lines: $K = 2$ levels; dotted lines: $K = 3$ levels. Spin–vibronic levels are denoted by the value of the corresponding quantum number $P(P = K + \Sigma$; note that Σ is in this case spin quantum number).

and $P = K + \Sigma$ quantum number (i.e., by a particular K, Σ combination). The exception represents the $K = 0$, $P = \pm 1$ levels remaining degenerate with each other. In the case when the vibronic coupling is weak compared to the spin–orbit coupling ($\varepsilon\omega \ll A_{so}$) the coarse structure of the spectrum is determined by the spin–orbit effects. This is illustrated in Figure 3 with the case $\varepsilon = -0.0085$

(corresponding to $\varepsilon\omega = -4.5\ \text{cm}^{-1}$), $A_{\text{so}} = -37\ \text{cm}^{-1}$; note, for example, that the $\upsilon = 1$, $K = 0$ levels are divided into three pairs of close-lying levels, with successive energetic separation between these pairs nearly equal to the value of the spin–orbit constant.

Now, let us consider the rising of the vibronic–spin–orbit structure from the "opposite side," that is, when the vibronic interaction is dominant compared to the spin–orbit coupling (right-hand side part of Figure 3). The structure of the vibronic spectrum for the nonzero value of the Renner parameter (in the concrete case $\varepsilon = -0.17$) and $A_{\text{so}} = 0$ has been disussed above (Fig. 1). The only difference is that each vibronic level is now threefold spin degenerate. When an additional weak spin–orbit interaction is added [Fig. 3, $\varepsilon = -0.17$, $A_{\text{so}} = -5\ \text{cm}^{-1}$ (arbitrary choice)], the spin degeneracy of the vibronic levels is removed, but the energy pattern is quite different from that corresponding to the oposite case of strong spin–orbit and weak vibronic coupling, discussed above. The coarse structure of the spectrum is the same as in the case of no spin–orbit coupling, with the latter interaction causing a relatively small additional splitting of vibronic levels. This splitting is maximally pronounced in *unique* levels, where it is nearly equal to the value of A_{so}, and almost negligible in non*unique* levels. This is a consequence of the composition of the corresponding wave functions. In the first approximation, the magnitude of the spin–orbit splitting is given by $A_{\text{so}}\ \langle L_z \rangle$, where $\langle L_z \rangle$ is the mean value of the electronic angular momentum operator in the vibronic state considered. While the *unique* level belongs almost exclusively to a single *diabatic* electronic state $(+\Lambda)$ and thus $\langle L_z \rangle$ is nearly equal to Λ for it, the other levels belong predominantly to a particular *adiabatic* electronic species, or, in other words they are nearly equally shared between the diabatic states $+\Lambda$ and $-\Lambda$ with the result that $\langle L_z \rangle$ is close to zero in them. Each $K = 0$ level, being threefold spin degenerate at $A_{\text{so}} = 0$, splits at nonvanishing spin–orbit coupling into two levels; the single one corresponds to $\Sigma = 0$ (i.e., $K = P = 0$), and the twofold degenerate level involvs $\Sigma \pm 1$ spin states. The latter vibronic states cannot be exactly classified into $+$ and $-$ species according to their behavior upon reflections in symmetry planes.

When the vibronic and spin–orbit coupling are comparably strong, as in the $A^3\Pi_u$ state of NCN (actually, although $\varepsilon\omega$ and A_{so} are in this case of the same magnitude order, the former quantity is roughly by a factor of 3 largen than the latter), the coarse structure of the part of the spectrum corresponding to a particular quantum number υ is characterized by (1) a relatively large energetic separation of the *unique* level from its non*unique* counterparts; (2) relative proximity of non*unique* levels with different K values; (3) relatively large spin–orbit splitting (roughly equal to the magnitude of A_{so}) of the *unique* level; (4) an efficiently quenched spin–orbit splitting in other levels. Figure 4 presents the ab initio computed dependence of the spin–orbit splitting in $K = 1$ and 2 levels on the

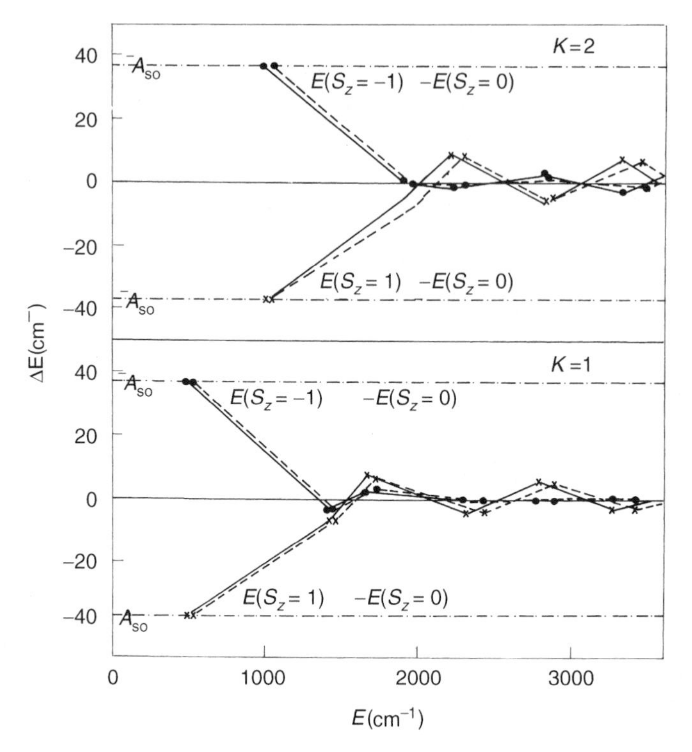

Figure 4. Spin–orbit splitting in $K = 1$ and 2 vibronic levels of the $A^3\Pi_u$ state of NCN. Solid lines connect the results of calculations that employ ab initio computed potential curves [28]. For comparison the results obtained by employing experimentally derived potential curves (dashed lines) [30,31] are also given. Full points represent energy differences between $P = K - 1$ and $P = K$ spin levels, and crosses are differences between $P = K + 1$ and $P = K$ levels.

value of the bending quantum number υ. It shows a "sawtooth" pattern, reflecting the erratic change of the vibronic mean value for the electronic angular momentum from one level to the other, typical for such systems (see also [26]).

The above results of ab initio calculation for the $A^3\Pi_u$ state of NCN (completed by those employing hypothetical values for ε and A_{so}) correspond to the schematic presentation of the effect of combined vibronic and spin–orbit couplings onto the spectral structure of $^3\Pi$ states of linear triatomics, carried out by Hougen [32] and reproduced in Herzberg's book (see Fig. 9 and accompanied discussion in [5]). A more detailed insight can be achieved by inspecting the perturbative formulae given in Appendix A.

The vibronic structure of a $^3\Delta$ electronic state at variable strengths of the vibronic and spin–orbit coupling is presented in Figure 5. The splitting of the

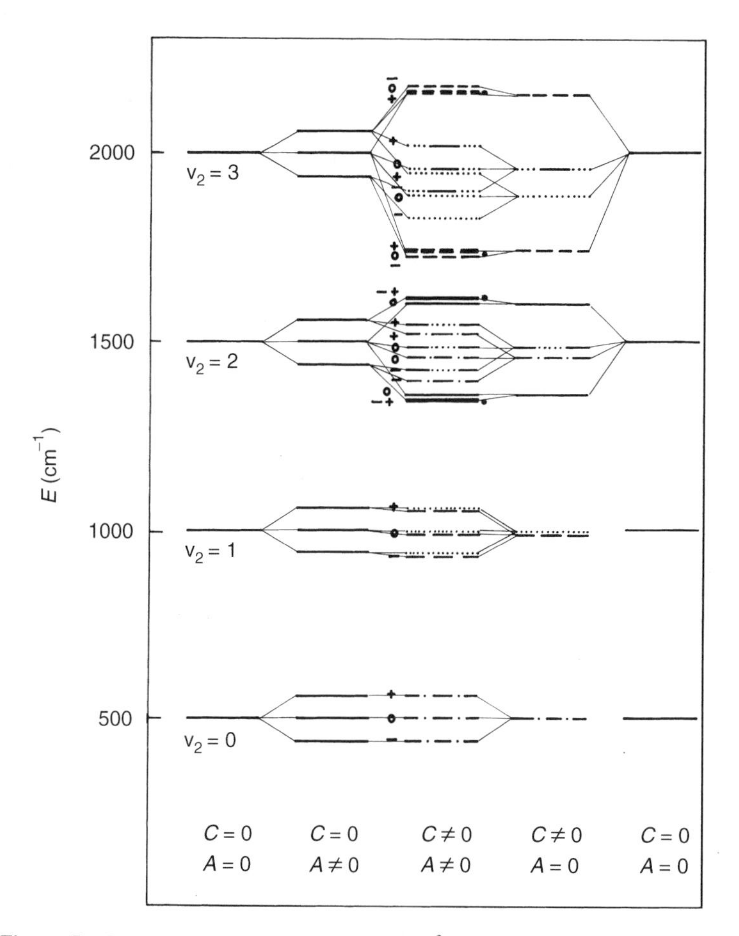

Figure 5. Low-energy vibronic spectrum in a $^3\Delta$ electronic state of a linear triatomic molecule. The parameter c determines the magnitude of splitting of adiabatic bending potential curves, A_{so} is the spin–orbit coupling constant, which is assumed to be positive. The zero on the energy scale represents the minimum of the potential energy surface. ——— : $K = 0$ vibronic levels; - - - - - - : $K = 1$ levels; — · — : $K = 2$ levels; .... :$K = 3$ levels; __ ... __ : $K = 4$ levels; ... __ ... : $K = 5$ levels. Spin-vibronic levels are denoted by a minus ($\Sigma = -1$) , zero ($\Sigma = 0$), or plus ($\Sigma = +1$). Note that Σ is in this case spin quantum number.

adiabatic bending potential curves is assumed in to be in the form $V^+ - V^- = c\rho^4$. Note that in Δ states the maximal splitting of the vibronic levels upon spin–orbit coupling (taking place in *unique* levels) is $2A_{so}$. The vibronic structure of the $^3\Delta$ electronic state is not considered explicitly in Herzberg's book.

3. *Renner–Teller Effect in Nonlinear Molecules*

Now, we discuss briefly the situation when one or both of the adiabatic electronic states has/have nonlinear equilibrium geometry. In Figures 6 and 7 we show two characteristic examples, the $X^2\Pi_u$ state of BH_2 and NH_2, respectively. The BH_2 potential curves are the result of ab initio calculations of the present authors [33,34], and those for NH_2 are taken from [25].

Let us first list the common characteristics of both systems. First at all, it is clear that in these case the "bent point of view," as defined above, is a logical starting point in the discussion, particularly for the NH_2 case. Except in the neighborhood of the joint point of the potential curves at linear geometry, the energy separation of the adiabatic electronic states is so large that the reliability of the BO approximation is in a large area of molecular geometries not threatened. This is reflected also in the structure of the vibronic secular equation (25): For large value of the bending coordinate ρ, the off-diagonal elements

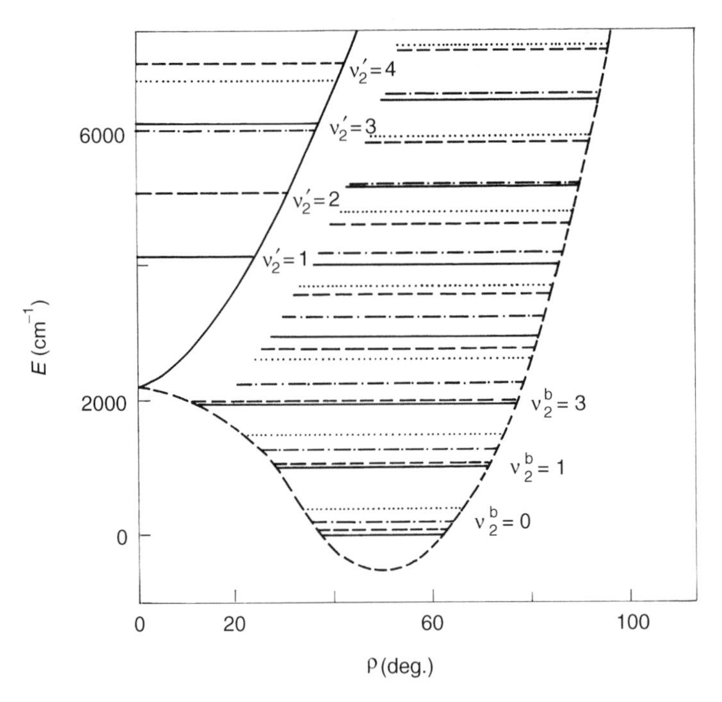

Figure 6. Bending potential curves for the X^2A_1, A^2B_1 electronic system of BH_2 [33,34]. Full hotizontal lines: $K = 0$ vibronic levels; dashed lines: $K = 1$ levels; dash–dotted lines: $K = 2$ levels; dotted lines: $K = 3$ levels. Vibronic levels of the lower electronic state are assigned in "bent" notation, those of the upper state in "linear" notation (see text). Zero on the energy scale corresponds to the energy of the lowest vibronic level.

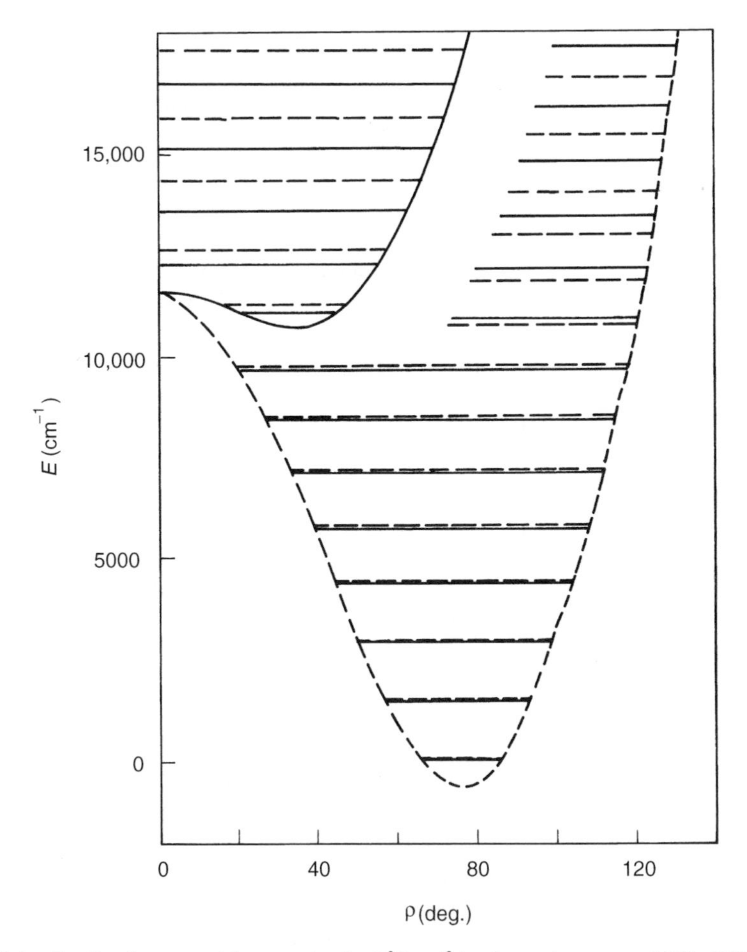

Figire 7. Bending potential curves for the X^2B_1, A^2A_1 electronic system of NH_2 [25]. Full horizontal lines: $K = 0$ vibronic levels; dashed lines: $K = 1$ levels. Zero on the energy scale corresponds to the energy of the lowest vibronic level.

become small compared to their diagonal counterparts. In the energy region below the joint point, the lower electronic state practically does not feel the existence of its higher energy counterpart. While the K vibronic energy pattern in linear molecules (see Section III.1) is analogous to that of a slightly perturbed 2D harmonic oscillator, with the levels of a particular K series associated with *either* even *or* odd vibrational quantum numbers υ (which means that the spacing between two neighboring K levels is 2ω), the K structure of the lower electronic states of BH_2 and NH_2 in the energy region below the barrier to linearity is a typical rotational one: For each vibrational level there are sublevels corresponding to all possible K quantum numbers, with the spacing between successive sublevels being much smaller than that between neighboring vibrational levels, and increasing roughly quadratically with an increase of K. This structure corresponds to the Hamiltonian consisting of the sum of operators for a one-dimensional (1D) harmonic oscillator and a 1D free rotator with the moment of inertia $I = 1/\mu\rho_e^2$, where ρ_e represents the equilibrium value of the bending coordinate ρ for the state in question. The gradual transformation of the Hamiltonian we use is easy to explain: First, if we change the volume element from $\rho d\rho$ (as asummed above) into $d\rho$, the term in the kinetic energy operator involving the first derivative of ρ [for the following qualitative analysis it suffices to consider its simple form given by Eq. (5)] disappears; in the region around the equilibrium geometry of a bent state the molecule vibrates with small amplitudes so that we can replace ρ in the expression $1/\rho^2$ by ρ_e. Thus the complete kinetic energy operator becomes of the form $T = -1/2\mu\, \partial^2/\partial\rho^2 - 1/(\mu\rho_e^2)\; \partial^2/\partial\phi^2$, while the potential, in the quadratic approximation, is $\frac{1}{2}k\,(\rho - \rho_e)^2$. The spin–orbit splitting of the levels far below the barrier to linearity is small and regular (Hund's case b).

With increasing quantum number υ, that is, by approaching the barrier to linearity, the spacing between vibrational levels generally diminishes, reaching minimum in the vicinity of the barrier, as first observed by Dixon [35]. At the same time, the K pattern becomes irregular, reflecting the gradual change of the rotational structure of the bent molecule into the linear molecule vibrational structure. The most spectacular manifestation of this is that at the energy close to that of the joint point of two adiabatic potentials, the $K = 1$ level of the lower electronic state falls below its $K = 0$ counterpart corresponding to the same quantum number υ (see also [8] and [24]).

A consequence of the reordering of vibronic levels described above is the ambiguity concerning the definition of the bending vibrational frequency and the corresponding quantum number υ (one should not wonder about that; such problems are always possible when one uses "bad" quantum numbers; more seriously, the quantum numbers not exactly determining the eigenvalues of the complete Hamiltonian of the system considered). In a bent molecule, the vibrational frequency is (nearly) equal to the separation of two neighboring levels with the same rotational quantum number. In a linear molecule this energy difference is divided by 2. What about the energy region around the barrier? Formally, the problem could be solved by accepting either the former or the latter definition and by employing it consequently. We can label each series of K levels beginning from the lowest one with the running number υ. This is what is called "bent notation" (υ_2^b in Fig. 6). However, if we do that, we have the situation that above the barrier to linearity the energy difference between the vibronic levels with the same quantum number υ (and different K) is comparable to the bending frequency. But the main role of a quantum number associated with the Hamiltonian is just to give at least rough information about the energy of the level labeled by it. That is the reason that besides this "bent notation" the "linear notation" is also employed, the latter one assuming a 2D harmonic oscillator as the zeroth-order problem. In Figure 6, we label in this way (υ_2^l) the levels of the upper electronic state. Note that the problem with two notations, none of them being universally satisfactory, is a consequence of large-amplitude bending vibrations the in course of which a molecule with bent equilibrium geometry passes though the linear nuclear configuration, and thus appears also in electronic states (Σ) not exhibiting the R–T effect. The relationship between the vibrational quantum numbers is simple, $\upsilon_2^{\text{lin}} = 2\,\upsilon_2^b + |l|$ (in order to avoid confusion with the quantum number l we use here superscript "lin" for "linear"), but unfortunately they are many papers in which it has not been explicitly quoted which notation has been meant.

In the upper electronic state, the BH_2 molecule has linear equilibrium geometry. This has as a consequence a peculiar vibronic structure: The lowest lying level belonging to this state is a $K = 0$ species with the energy of 2ω (linear notation). That is so, because it correlates with the lowest lying $l = 1$ ($\upsilon^{\text{lin}} = 1$) vibrational level. This fact has been overlooked in early ab initio studies on BH_2 [36], which has contributed to the misstatement that the theory has failed to reproduce the experimental findings reliably [37] in this apparently simple case (with only seven electrons, BH_2 is the smallest polyatomic molecule whose spectrum has been analyzed in detail thus far). On the other hand, the upper electronic state of NH_2 shows a typical example of quasilinearity, with a single vibrational level lying below the joint point of adiabatic electronic states at linear geometry.

An important difference between the weak and strong R–T effect is that in the latter case there are no levels with clearly pronounced *unique* character. Instead, for each $K \neq 0$ value there are several levels in the energy region around the joint point of the potential surfaces that share among themselves the *unique* character, in the sense, for example, that their spin–orbit splitting, while being much smaller than the value of the spin–orbit coupling constant, is appreciably larger than that for the levels that belong predominantly to a particular adiabatic electronic state. This is reflected in Figure 6 by missing of the $\upsilon_2 = 0$, $K = 1$ vibronic level attributed to the upper electronic state.

As mentioned above, the R–T coupling in the systems with one or both adiabatic electronic states having bent equilibrium geometry is in general pronounced only in the neighborhood of the joint point of the potential surfaces. The exception represents the cases when a vibrational level belonging to the lower potential surface lies accidentally close to a counterpart of the upper electronic state. Note that "pragmatic" approaches that use pure ab initio calculated potential surfaces (as, e.g., those employed by the present authors) generally fail to describe such Fermi-type (local) interactions reliably, because the restricted accuracy of the potential surfaces (errors of typically several hundred reciprocal centimeters) and neglect or approximate handling of some interactions (e.g., bend-stretch) consequently have inherent under- or over-estimating of the energy gap between such levels [16,17].

C. Choice of Hamiltonian

We find it convenient to reverse the historical ordering and to start with (nearly) exact nonrelativistic vibration–rotation Hamiltonians for triatomic molecules. From the point of view of molecular spectroscopy, the optimal Hamiltonian is that which maximally decouples from each other vibrational and rotational motions (as well different vibrational modes from one another). It is obtained by employing a molecule-bound frame that takes over the rotations of the complete molecule as much as possible. Ideally, the only remaining motion observable in this system would be displacements of the nuclei with respect to one another, that is, molecular vibrations. It is well known, however, that such a program can be realized only approximately by introducing the Eckart conditions [38].

$$\sum_{A=1}^{n} m_A(\mathbf{r}_A^0 \times \mathbf{r}_A) = 0 \tag{33}$$

where m_A represents the mass of the nucleus A, $\mathbf{r}_A$ is the instantaneous position, and $\mathbf{r}_A^0$ its equilibrium position in the moving frame. The parameter n is the number of nuclei. The fulfillment of condition (33) ensures decoupling of vibrations from rotations (only) at infinitesimal vibrations. The corresponding quantum mechanical vibration–rotation Hamiltonian in terms of the (vibrational) normal coordinates was derived by Wilson et al. [39,40], and simplified by Watson [41,42]; we shall refer to it as the EWW Hamiltonian. It has (for molecules with nonlinear equilibrium geometry) the form

$$H = \frac{1}{2}\sum_{\alpha,\beta} \mu_{\alpha\beta}(J_\alpha - p_\alpha - L_\alpha - S_\alpha)(J_\alpha - p_\alpha - L_\alpha - S_\alpha) + \frac{1}{2}\sum_{r}^{3n-6} P_r^2 - \frac{1}{8}\sum_{\alpha} \mu_{\alpha\alpha} + V \tag{34}$$

where $\mu_{\alpha\beta}^{-1}$ (α, $\beta = x$, y, or z) represents the element of a matrix that is nearly equal to the instantaneous moment of inertia matrix; J_α, p_α, L_α, S_α are the components of the total, vibrational, electronic, and spin angular momentum operator respectively; P_r is the momentum conjugate with the normal coordinates Q_r; and V is the potential for vibrational motion expressed in terms of Q_r. The natural handling of the corresponding Schrödinger equation is to expand $\mu_{\alpha\beta}$ and V in Taylor series in Q_r about the equilibrium geometry and to apply the perturbation theory.

Use of the Hamiltonian where the vibrational and rotational motions, as well as different vibrational modes, are maximally decoupled from one another should be expected to enable the most consequent realization of the main idea followed in a theoretical, particularly ab initio treatment of the R–T effect, namely, to separate the variables in the Schrödinger equation as much as possible from one another before the actual computations are started. However, while the separation of the nuclear from electronic coordinates in the framework of the BO approximation is practically always carried out (at least as the starting point), the EWW scheme in not the only possible and often not even the most convenient for the treatment of the vibration–rotation problem. First, let us stress that the Eckart's approach assumes the existence of a relatively deep minimum on the potential surface, being well separated from all other (local) minima. Second, the EWW Hamiltonian has the serious drawback that it has different forms for molecules with nonlinear and linear equilibrium geometry. Both of these facts cause difficulties in the description of quasilinear molecules. Furthermore, the nature of interaction between certain vibrational and rotational modes is sometimes such that they have to be treated simultaneously, as, for example, between the bending vibrations and the z-axis rotations in the R–T effect. It should also be mentioned that the form of the normal coordinates is not known in advance and that the transformation of the potential, normally computed as a function of some internal coordinates into the series in normal coordinates is complicated by the fact that the transformation from internal into normal coordinates is nonlinear for noninfinitesimal vibrations [43]. Application of the EWW formalism, being predestinated for a perturbative treatment, loses much of its attractiveness if one intends to solve the complete (or a part) of the vibration–rotation problem by a variational approach. Therefore, it is often advantageous to abandon the Eckart constraints and the use of normal coordinates in order to gain additional flexibility that can be utilized for designing more convenient forms for the particular parts of the Hamiltonian (e.g., to avoid some unpleasant singularities in the vibrational Hamiltonian [44]); furthermore, this enables derivation of explicit analytical expressions for the vibrational–rotational Hamiltonian, as will be shown below. Although some of the disadvantages mentioned above have disappeared in the last few years (so Estes and Secrest [45] derived a unified variant of the Watson's Hamiltonian valid for both nonlinear and linear molecules—the present authors are not able to judge whether it is correct; Wei and Carrington [46] recently presented an exact Eckart-embedded kinetic energy operator in bond coordinates for triatomic molecules; for a review of other various possibilities see, e.g., [47]), for the reasons mentioned above the complete EWW Hamiltonian seems to never have been used in handling the R–T effect.

An alternative form of exact nonrelativistic vibration–rotation Hamiltonian for triatomic molecules (ABC) is that used by Handy, Carter (HC), and co-workers [48–51]. It is given in terms of the geometrically defined vibrational coordinates, we denote by r_1, r_2, and ϑ. The parameter r_1 is the instantaneous value of the distance between the nuclei A and B, r_2 the distance between C and B, and ϑ is the instantaneous ABC bond angle ($\vartheta = \pi - \rho$; this symbol should be not confused with the electronic angular coordinate θ). The molecule-fixed axes are defined such that the molecule lies in the zx plane with the x axis (being parallel to the axis) bisecting the ABC angle and the A nucleus lying in the positive zx quadrant. Thus the z axis coincides with the molecular axis at linear nuclear arrangements. The apparently arbitrary choice of the coordinate axes might look strange; it is not motivated by physical reasons, that is, it does not worry about the strength of the coupling between different motion modes. This becomes understandable in terms of a statement by HC: "... we are not interested in least squares fit procedures for the identification of spectra, and it is possible to label our states from the energies we obtain ..." [50]. The origin of this Hamiltonian is not quite clear: In the first paper in which they used it [48], HC claimed that they had taken it over from Carney et al. [52] (and transformed to the volume element $\sin\vartheta\, dr_1\, dr_2\, d\vartheta$, in our notation) The latter authors also mention Lai and Hagstrom, but the trace is lost somewhere in the library of Indiana University [53]. It is, however, of little importance who the actual author of this Hamiltonian is, because a derivation of such a vibration–rotation Hamiltonian for triatomic molecules employing the chain rule for transformation of the derivatives in Cartesian space-fixed coordinates into curvilinear internal coordinates, using the Podolsky approach [54] (see also [55]) for transforming the classical Hamiltonian in curvilinear coordinate into its quantum-mechanical counterpart, or a combination of both methods (see, e.g., [56]) is not a serious problem (even if it is carried out without the use of computer algebra). The real contribution of HC was to show how to solve the corresponding Schrödinger equation, particularly in the presence of vibronic coupling.

The (kinetic energy part of the) Hamiltonian we are speaking about can be written in the following form:

$$T = T_V + T_{VR} \tag{35}$$

with

$$\begin{aligned} T_V = &-\frac{1}{2\mu_1}\frac{\partial}{\partial r_1^2} - \frac{1}{2\mu_2}\frac{\partial}{\partial r_2^2} - \frac{\cos\vartheta}{m_B}\frac{\partial^2}{\partial r_1 \partial r_2} \\ &-\frac{1}{4}\left(\frac{1}{\mu_1 r_1^2} + \frac{1}{\mu_2 r_2^2} - \frac{2\cos\vartheta}{m_B r_1 r_2}\right)\left(\frac{\partial}{\partial\vartheta^2} + \cot\vartheta\frac{\partial}{\partial\vartheta}\right) \\ &-\frac{1}{4}\left(\frac{\partial}{\partial\vartheta^2} + \cot\vartheta\frac{\partial}{\partial\vartheta}\right)\left(\frac{1}{\mu_1 r_1^2} + \frac{1}{\mu_2 r_2^2} - \frac{2\cos\vartheta}{m_B r_1 r_2}\right) \\ &+\frac{1}{m_B}\left(\frac{1}{r_1}\frac{\partial}{\partial r_2} + \frac{1}{\partial r_2}\frac{\partial}{\partial r_1}\right)\left(\sin\vartheta\frac{\partial}{\partial\vartheta} + \cos\vartheta\right) \end{aligned} \tag{36}$$

and

$$
\begin{aligned}
T_{VR} = {} & \frac{1}{8\cos^2\frac{\vartheta}{2}}\left(\frac{1}{\mu_1 r_1^2}+\frac{1}{\mu_2 r_2^2}+\frac{2}{m_B r_1 r_2}\right)\left(\hat{J}_z+\hat{L}_z+\hat{S}_z\right)^2 \\
& + \frac{1}{8\sin^2\frac{\vartheta}{2}}\left(\frac{1}{\mu_1 r_1^2}+\frac{1}{\mu_2 r_2^2}-\frac{2}{m_B r_1 r_2}\right)\left(\hat{J}_x+\hat{L}_x+\hat{S}_x\right)^2 \\
& + \frac{1}{8}\left(\frac{1}{\mu_1 r_1^2}+\frac{1}{\mu_2 r_2^2}+\frac{2\cos\vartheta}{m_B r_1 r_2}\right)\left(\hat{J}_y+\hat{L}_y+\hat{S}_y\right)^2 \\
& - \frac{1}{4\sin\vartheta}\left(\frac{1}{\mu_1 r_1^2}-\frac{1}{\mu_2 r_2^2}\right)\left[\hat{J}_z+\hat{L}_z+\hat{S}_z, \hat{J}_x+\hat{L}_x+\hat{S}_x\right]_+ \\
& - \frac{i}{2}\left[\left(\frac{1}{\mu_1 r_1^2}-\frac{1}{\mu_2 r_2^2}\right)\left(\frac{1}{2}\cot\vartheta+\frac{\partial}{\partial\vartheta}\right)+\frac{\sin\vartheta}{m_B}\left(\frac{1}{r_2}\frac{\partial}{\partial r_1}-\frac{1}{r_1}\frac{\partial}{\partial r_2}\right)\right] \\
& \times \left(\hat{J}_y+\hat{L}_y+\hat{S}_y\right)
\end{aligned}
\tag{37}
$$

with

$$
\mu_1 = \frac{m_A m_B}{m_A + m_B} \qquad \mu_2 = \frac{m_B m_C}{m_B + m_C} \tag{38}
$$

In Eq. (37) $-\boldsymbol{J}$, $\boldsymbol{L}$, and $\boldsymbol{S}$ are the total, electronic, and spin angular momentum, respectively, all of them defined with respect to the molecule fixed-coordinate axes. In order to avoid problems with the anomalous commutation relations for the components of the total angular momentum, it is following Van Vleck [57] replaced in (37) by its counterpart with a negative sign. (Note that Handy and co-workers prefer to change the sign of $\boldsymbol{J}$ rather than signs of the internal momenta $\boldsymbol{L}$ and $\boldsymbol{S}$, as suggested in the original paper by Van Vleck; for a discussion of this matter see [58] and [59].

The approaches for treating the vibration–rotation problem employing two types of Hamiltonians as described above lead to considerable computational requirements in larger molecules (already for a tetraatomic molecule the potential surface depends on six vibrational coordinates) and/or if some of the vibrational modes are characterized by large amplitudes. For this reason, another strategy is also applied. The vibrational modes are divided into two classes: to the first belong those that are accompanied by relatively small displacements of the nuclei and can easily be handled perturbationally (or even in the harmonic approximation); the second class is build by the vibrational modes for which a large-amplitude handling is necessary, so that they require a more sophisticated treatment. So, for example, in quasilinear molecules (particularly triatomics and tetraatomics, considered in the present study) the bending vibration(s) play(s) a special role. Quite frequently in electronic spectra one observes relatively long progressions in the bending mode, because the equilibrium angles could have very different values in various electronic states, in contrast to the bond lengths, which generally vary less from state to state. This makes a large-amplitude treatment for the bending motion imperative, while the stretching motions can usually be assumed to occur with small amplitudes. Another peculiarity of the bending mode is its relation to the rotations: When a molecule during the bending vibrations approaches linear geometry, a gradual transformation of one of its rotational degrees of freedom into the degenerate bending vibration takes place. Both of these facts play crucial roles in the R–T effect. Such situations have motivated a number of authors to develop various methods in which the emphasis is placed on an accurate treatment of the effective bending problem, while the other nuclear modes are treated more or less conventionally. Among those, the greatest popularity enjoys the approach introduced for triatomic molecules by Hougen, Bunker and Johns (HBJ) [60] and developed further by Bunker and his co-workers, particularly Jensen [61–67].

In the formalism of Bunker et al., the so-called reference configuration plays the role that has the equilibrium configuration in the EWW scheme. It is characterized by fixed bond lengths and the variable valence angle, and is chosen such that the bending motion at each value of the bending coordinate ρ is minimally coupled with the stretching vibrations. This is achieved by introducing in addition to (33) (with the equilibrium configuration replaced by reference configuration) the Sayvetz [68] condition,

$$\sum_{\alpha,i} m_i(\alpha_i - \alpha_i^{\mathrm{ref}}) \frac{\partial \alpha_i}{\partial \rho} = 0 \tag{39}$$

where ref stands for "reference." In analogy with the EWW scheme, the molecular potential and the elements of the μ-matrix are expanded around the reference configuration into Taylor series in the two stretching normal coordinates, with the coefficients depending on ρ,

$$\begin{aligned} \mu_{\alpha\beta} &= \mu_{\alpha\beta}^{\mathrm{ref}} - \sum_{\gamma,\delta,r} \mu_{\alpha\gamma}^{\mathrm{ref}} a_r^{\gamma\delta} \mu_{\delta\beta}^{\mathrm{ref}} Q_r \\ V &= V_0(\rho) + \sum_r V_r Q_r + \frac{1}{2}\sum_r \lambda_r Q_r^2 + \frac{1}{6}\sum_{r,s,t} V_{rst} Q_r Q_s Q_t + \cdots \end{aligned} \tag{40}$$

[The appearance of the (normally small) linear term in V is a consequence of the use of reference, instead of equilibrium configuration]. Because the stretching vibrational displacements are of small amplitude, the series in Eqs. (40) should converge quickly. The zeroth-order Hamiltonian is obtained by neglecting all but the leading terms in these expansions, $\mu_{\alpha\beta}^{\mathrm{ref}}$ and $V_0(\rho) + 1/2\sum \lambda_r Q_r^2$ and has the form

$$H^0 = \frac{1}{2}\sum_{\alpha,\beta} \mu^{\text{ref}}_{\alpha\beta} J_\alpha J_\beta + \frac{1}{2}\mu_{\rho\rho} J^2_{\rho\rho} + V_0(\rho) \tag{41}$$

where $J_\rho = -\mathrm{i}\,\partial/\partial\rho$ and $\mu_{\rho\rho}$ is the inverse of the "reduced mass" for the large amplitude bending vibrations. The Hamiltonian (41) is an essentially 1D operator (in ρ). The coupling between ρ and the other degrees of freedom can be taken into account perturbationally [62,64], or indirectly as in the framework of the semirigid beneder model by Bunker and Landsberg [63], by allowing the bond lengths to vary smoothly with changes in ρ an by correcting correspondingly the potential $V_0(\rho)$.

D. Minimal Models

In his classical paper, Renner [7] first explained the physical background of the vibronic coupling in triatomic molecules. He concluded that the splitting of the bending potential curves at small distortions of linearity has to depend on $\rho^{2\Lambda}$, being thus mostly pronounced in Π electronic state. Renner developed the system of two coupled Schrödinger equations and solved it for Π states in the harmonic approximation by means of the perturbation theory.

For a long time, Renner's paper had been unique. There are two main reasons for this: There have been no reliable experimental results that could have confirmed Renner's predictions and, on the other hand, this work has been carried out so thoroughly that there have been no reasons to try to improve it. The situation changed almost 25 years later, but it changed completely. In 1957–1958 Dressler and Ramsay [9,10] carried out a detailed vibrational and rotational analysis of the absorption spectrum of NH_2 and attributed it to an electronic transition from the ground state in which the molecule is bent, to an exited state in which the molecule "vibrates about a linear configuration." They stated that "The excited state exhibits a previously unobserved and complex pattern of vibronic and rotational energy levels. The vibronic structure of this pattern... may be understood if it is assumed that the combining states are derived from a hypothetical Π state. ... The large splittings observed are due to an interaction between electronic and vibrational motion of the type predicted by Herzberg and Teller (1933) and discussed in detail by Renner" [10]. Thus the experimental data, which Renner vainly had expected to find in the CO_2 spectrum, were eventually available and, on the other hand, the type of the splitting of the potential surfaces was not that which Renner had considered. This situation motivated Pople and Longuet-Higgins [69] to extend Renner's work to the case when in the lower Renner–Teller component state the molecule has a bent equlibrium geometry, and in the upper state it is linear. Before skipping to this work, let us make, however, two small comments concerning the above quotations. Note that Dressler and Ramsay did not explicitly state that the equilibrium geometry of the upper state is linear; they found that the vibrational structure clearly showed that NH_2 behaves as a linear molecule above the (0, 3, 0) level, which was the lowest observed one. They did not exclude the possibility of a small potential maximum at the linear geometry. The second point is that the theoretical work by Pople and Longuet-Higgins paralleled the Dressler–Ramsay's analysis of the spectrum, so that the results of the former work were utilized to interpret the experimental findings.

Pople and Longuet-Higgins (PL–H) developed a model they described by the words: "We shall adopt a simplified model... (which) incorporates the essential features of the situation in, for example, the NH_2 radical; that is to say, the resulting equations of motion are mathematically equivalent to those obtained by Renner from more sophisticated premisses" [69]. They restricted the treatment to three degrees of freedom represented by the coordinates θ, ρ, and ϕ (in our notation). The coordinate θ was defined as the angular distance of the odd electron around the molecular axis, as measured from a fixed plane. The PL–H stated that it is "more properly regarded as the coordinate conjugate to the axial momentum of all the electrons, but the simple interpretation is physically more illuminating." The model Hamiltonian was assumed in the form $H_0 + H'$, where H_0 is the Hamiltonian describing the molecule in the absence of the vibronic coupling and H' the term responsible for the R–T effect. The operator H_0 was assumed as the Hamiltonian for a 2D harmonic oscillator with the eigenfunctions of the form

$$|\upsilon\, l \pm \Lambda\rangle = e^{\pm i\Lambda\theta} e^{il\phi} R_{\upsilon,l}(\rho) \tag{42}$$

corresponding to the asymptotic form of the "linear basis" functions we defined above (with the normalization factor absorbed in $R_{\upsilon,l}$). The coupling term H' was expanded into a series being a symmetric function on the relative angular electronic coordinate $\alpha = \theta - \phi$,

$$H' = V_0(\rho) + V_1(\rho)[e^{i(\theta-\phi)} + e^{-i(\theta-\phi)}] + V_2(\rho)[e^{2i(\theta-\phi)} + e^{-2i(\theta-\phi)}] + \cdots \tag{43}$$

where $V_m(\rho)$ was assumed to be of the order ρ^m at $\rho \to 0$. The matrix representation of the perturbation (43) in the electronic basis $|\pm\Lambda\rangle = 1/\sqrt{(2\pi)}$ exp $(\pm\, i\, \Lambda\, \theta)$, with $\Lambda = 1$ (for Π electronic state) is

$$\begin{pmatrix} V_0 & V_2 e^{-2i\phi} \\ V_2 e^{2i\phi} & V_0 \end{pmatrix} \tag{44}$$

that is, only the constant and quadratic term from the expansion (43) contribute to it. Diagonalization of the matrix (44) leads to the first-order energies

$V^+ = V_0 + V_2$ and $V^- = V_0 - V_2$ (actually, effective operators acting onto functions of ρ and ϕ), corresponding to the zeroth-order vibronic functions of the form $\cos(\theta - \phi)$ and $\sin(\theta - \phi)$, respectively. PL–H computed the vibronic spectrum of NH_2 by carrying out some additional transformations (they found it to be convenient to take the unperturbed situation to be one in which the bending potential coincided with that of the upper electronic state, which was supposed to be linear) and simplifications (the potential curve for the lower adiabatic electronic state was assumed to be of quartic order in ρ, the vibronic wave functions for the upper electronic state were assumed to be represented by sums and differences of pairs of the basis functions with the same quantum number υ and $l = K \pm \Lambda$) to keep the problem tractable by means of simple perturbation theory; they are, however, of no direct concern to us. By deriving the parameters entering their model from the experimental data for unperturbed $K = 0$ vibronic levels, PL–H succeeded in achieving a near coincidence between theoretical and experimental results for all vibronic species. They concluded their paper with the statement that their theory was "limited by a number of severe approximations." Note that Brown and Jørgensen's opinion about this model is not very favorable, but for other reasons [12].

At this place, we make a chronological jump to comment on the paper by Dixon [35], representing a direct continuation of the story about the NH_2 spectrum. This work was motivated by new experimental results (unpublished at that time) by Ramsay et al. for low-lying vibronic levels of the upper electronic state. They indicated, particularly in connection with the predictions based on the Walsh's rules [70], that the upper state could have a slightly nonlinear equilibrium geometry. Thus Dixon allowed both bending curves to have a double minima. These potentials were approximated by a combination of quadratic functions and Gaussian functions. The model was otherwise equivalent to that of PL–H (both of them neglect the coupling of the bending vibrations with the stretching modes and end-over-end rotations, as well as the spin–orbit coupling, and employ the zeroth-order kinetic energy operator). The vibronic problem was solved by diagonalization of a truncated infinite Hamiltonian matrix using a computer. It was found that the molecule is in the upper electronic state quasilinear (equilibrium bond angle of 144°), with only a single $K = 0$ level lying below the barrier to linearity. Dixon's results significantly improved the agreement between theory and experiment.

The first theoretical handling of the weak R–T combined with the spin–orbit coupling was carried out by Pople [71]. It represents a generalization of the perturbative approaches by Renner and PL–H. The basis functions are assumed as products of (42) with the eigenfunctions of the spin operator corresponding to values $\Sigma = \pm 1/2$. The spin–orbit contribution to the model Hamiltonian was taken in the phenomenological form (16). It was assumed that both interactions are small compared to the bending vibrational frequency and that both the vibronic quantum number K and the spin quantum number Σ are conserved. The main conclusions following from this study were: the positions of $K = 0$ levels are shifted with respect to the case when the spin–orbit coupling is neglected and their $+/-$ classification ceases to be precise; the lowest $K \neq 0$ levels are due to the spin–orbit coupling split into pairs of levels, separated from each other by the value of the spin–orbit coupling constant; all other $K \neq 0$ vibronic levels are only slightly split through the spin–orbit interaction. Pople's study concerns doublet electronic states, but it can be modified straightforwardly to other multiplicities. Unfortunately, formulas (3.7) and (3.8) in the original work are erroneous as observed by Hougen, who corrected them in a subsequent study [72] (see also [5]).

The expressions for the rotational energy levels (i.e., also involving the end-over-end rotations, not considered in the previous works) of linear triatomic molecules in doublet and triplet Π electronic states that take into account a spin orbit interaction and a vibronic coupling were derived in two milestone studies by Hougen [72,32]. In them, the "isomorfic Hamiltonian" was introduced, which has later been widely used in treating linear molecules (see, e.g., [55]).

The first handling of the R–T effect in a Δ electronic state of a triatomic was carried out experimentally and theoretically by Merer and Travis [73]. It concerned the $A^2\Delta$ state of CCN. These authors derived the second-order perturbative formulas for the combined effect of the weak vibronic and spin–orbit couplings. The splitting of the bending potential curves due to the R–T interaction was assumed to involve a single term being of fourth order in the bending coordinate; on the other hand, the mean adiabatic potential was assumed to be harmonic. Curiously enough, also in this case the perturbative formulas printed in the original reference were not correct (caused by a trivial error of a factor of 4 concerning the norm of the basis functions used, see [12]).

E. Pragmatic Models

We shall now comment in some detail on the approach developed by Barrow, Dixon, and Duxbury (BDD) [74], historically the first one of those we classify in the category of "pragmatic" ones. BDD extracted from the complete vibration–rotation Hamiltonian the terms describing the bending vibrations and the z-axis rotations. They employed the operator derived by Freed and Lombardi (FL) [75], differing from Eqs. (35)–(37) in the choice of the molecule-bound coordinate system. In the FL's Hamiltonian, the axes of the moving system are attached to the instantaneous principal moments of inertia of the molecule, being the optimal choice for handling the molecular rotations. In the case of symmetric triatomics (ABA) undergoing infinitesimal stretching vibrations, the axes of the HC's molecule-bound frame [48,50] coincide with those preferred by FL (note that the last term in the FL's operator H_3 should be multiplied by 2 to become equal to the correspond term in the HC Hamiltonian).

The first form of the BDD Hamiltonian corresponds to the bending plus z-rotation part of the FL Hamiltionian for symmetric (ABA) molecules, supposed to undergo infinitesimal stretching vibrations. BDD corrected this operator following the idea by FL that the high-frequency stretching vibrations and the low-frequency bending can be handled in a way analogous to the treatment of electronic and vibrational motions in the framework of the usual BO approximation. Thus they carried out integration over the stretching coordinates (assuming parametric dependence of the stretching vibrational frequencies on the bending coordinate ρ) and incorporated the leading part of the stretch–bend coupling, taken in the second-order perturbation theory, into the bending Hamiltonian. This operator has nearly the same form as the zeroth-order bending Hamiltonian by HBJ [60]. In handling the R–T effect, BDD carried out a contact transformation, chosen to diagonalize all of the 2×2 matrix, representing the effective bending operator, except the nuclear kinetic energy operator. This ansatz unifies the "bent" and "linear" ones described above: for small values of ρ the BDD secular problem reduces to (31), at large ρ values to (24.). The bending potential energy curves were assumed to be that of a harmonic oscillator perturbed by a Lorenzian hump, and the system of coupled R–T equations was solved by the Cooley–Numerov numerical integration technique [76]. The spin–orbit part of the Hamiltonian was assumed in the phenomenological form (16). The end-over-end rotations were handled separately. The B rotational constant was computed as an average value of the expression involving the reciprocal value of the instantaneous principal moment of inertia corresponding to the bisector of the valence angle. An accurate calculation of the C rotational constant is somewhat more complex; because of the Coriolis interaction, the factor multiplying the operator J_y^2 in the expression (37) does not reduce to $1/(2I_{yy}^0)$ even if symmetric triatomics undergoing infinitesimal stretching vibrations are considered. This becomes, however, the case when the adiabatic transformation analogous to that described above (for taking into account the stretch–bend interaction) is applied [75]; for a more refined treatment see the original [74].

The BDD approach has been applied in a number of studies that employ the parameters derived from the experimental findings [77–85]. The approach has been extended by Duxbury an co-workers, particularly Alijah; in its present version, involving the new stretch-bender Hamiltonian [84,85], which follows the idea by HBJ [60], it approaches the methods we tentatively call "benchmark."

The approach developed by Jungen and Merer (JM) [24] is of a similar level of sophistication. The main difference is that JM prefer to remove the coupling between the electronic states by a transformation of the Hamiltonian *matrix* (i.e., vibronic energy matrix), rather that of the Hamiltonian itself. They first calculate the large amplitude bending functions for one of the adiabatic potentials, as if it belonged to a Σ electronic state. These functions are used as the basis for matrix representation of the secular problem given by Eq. (31). In the next step, the off-diagonal elements of the Hamiltonian matrix are minimized by a similarity transformation. The resulting matrix is finally diagonalized.

JM employ the semirigid bender Hamiltonain by Bunker and Landsberg [63]. They also neglect the x, y rotations in handling the vibronic problem, that is, they assume K to be a good quantum number. The spin–orbit operator is taken in the phenomenological form involving only z components of the angular momenta. On the other hand, their approach allows considering the dependence of the mean value of L_z on the bending coordinate. JM applied this approach to calculate the vibronic spectra in the X^2B_1, A^2A_1 ($^2\Pi_u$) state of NH_2 and H_2O^+ [25,26] and the $A^1\Pi_u$ state of C_3 [27]. They used the potential energy curves derived by fitting of experimentally determined positions of $K = 0$ levels, not undergoing R–T coupling. The results of these calculations impressively demonstrated that their approach was able to reproduce reliably not only the positions of all $K \neq 0$ vibronic levels measured, but also very fine effects like erratic pattern of the spin–orbit splitting of these levels and the variation of the rotational constants from level to level.

The approach having been employed by the present authors in their ab initio handling of the R–T effect in a series of triatomic molecules (these results are reviewed in [16,17,21,86]) is not very different from the two described above. We have employed the same kind of the kinetic energy operator for large amplitude bending and the same form of the spin–orbit operator as DBB and JM. The vibronic energy levels and wave functions have been computed variationally, by employing as basis functions either the eigenfunctions of a suitably chosen 2D harmonic oscillator [21], or Fourier series in ρ [86]. All matrix elements appearing in the vibronic secular equations are computed by using simple recurrence formulas. Although our program package allows for handling of the R–T effect along both the "linear" and "bent" formalisms, the great majority of the calculations have been carried out in the framework of the first one.

The use of ab initio computed potentials and other relevant quantities has its advantages, as well as its drawbacks. The greatest advantage is that there are no problems like those that are caused by a shortage or insufficient quality of experimental data. Further, some quantities that are difficult to extract from the experimental findings, like variation of the bond lengths or the mean value for L_z upon bending, are easy to compute. The greatest drawback of a pure ab initio handling of the R–T effect is the limited accuracy of quantities entering the model Hamiltonan, particularly of potential energy surfaces. However, the high accuracy achieved in the handling the R–T effect by using the potential curves and structural parameters derived by fitting the experimental data can be deceptive, because it can be based on cancellation of the errors in the potential and kinetic energy part of the model Hamiltonian. An example is the X^2B_1, A^2A_1 ($^2\Pi_u$) system of H_2O^+: The calculations of JM [25,26] excellently reproduced all the experimental findings, although the potentials they employed were based on an incorrect numbering of the vibronic levels observed, as it was shown in later ab initio studies [87,88]. Let us also note that the discrepancy between the ab initio computed bending potential curves and their counterparts derived by fitting the experimentally observed features must not be automatically ascribed to the inaccuracy of the former. The "experimental" potentials correspond to expressions with a certain number of free parameters that are chosen so that the eigenvalues of the 1D model Hamiltonian (into which these parameters enter) match experimentally observed vibronic levels. In this way, they effectively incorporate all kinds of coupling with the other degrees of freedom and the other electronic states. On the other hand, the ab initio potential curves are well-defined 1D sections of the three-dimensional (3D) potential surfaces computed in the framework of the BO approximation. A part of the coupling with the other modes and states can be indirectly incorporated, but this always represents an approximation. Thus both sets of curves do not represent exactly the same quantity; they obtain the same meaning only if the 1D approach is realistic. This matter has been discussed in detail in [17].

The essential equivalence of all three approaches for handling the R–T effect presented in this section have been demonstrated through the computations in which the same input data have been used, that is, in [78] BDD and JM are compared and in [21] and [86] our approach with these two. The result of the latter two studies showed that JM had exaggerated claiming that for a direct diagonalization of the vibronic matrix "it would be necessary to chose an enormous basis in order to avoid truncation errors" [24]. We were able to reproduce their results by diagonalizing matrices of dimensions < 100. With this observation we do not want to question the general utility of the Hamiltonian- or matrix transformations implemented in the approaches by BDD and JM; in the approaches tailored to lean on experimental findings such a subtle handling is of much more relevance than in the ab initio calculations where in some steps the brute force philosophy can be applied without undesirable consequences.

F. Benchmark Handling

Let us first stress that the program of a "benchmark" handling of the R–T effect, as presented in Table I, represents an idealization; in none of the studies that have been published thus far has it been realized in all points.

The most consequent and the most straightforward realization of such a concept has been carried out by Handy, Carter, and Rosmus (HCR) and their co-workers. The final form of the vibration–rotation Hamiltonian and the handling of the corresponding Schrödinger equation in the absence of the vibronic coupling is a result of a short but exciting discussion between HC and Sutcliffe [48,49,89,90]. In variational handling of the R–T effect in singlet electronic states, HC employ the bending–electronic–rotational basis functions of the form [50]

$$\Psi_{\upsilon_2}^{J,K,\Lambda}(\vartheta, \beta, \gamma, r) = P_{n_2}^{|K-\Lambda|}(\cos\vartheta) D_{0K}^{J}(\beta, \gamma)\Phi_e^{\Lambda}(r) \tag{45}$$

(in our notation). Note again that ϑ denotes the bond angle, and not the coordinate conjugate to L_z. The functions $P(\cos\vartheta)$ are associated Legendre polynomials of degree $n_2 = 2\upsilon_2 - |l|$, where $|l| = |K - \Lambda|$. The functions Φ_e are the electronic species of type (26), and $D_{oK}^{J}(\beta, \gamma)$ are the $M = 0$ (end-over-end) rotational wave functions depending on the Euler angles β and γ. The bond stretching expansion functions are expressed in terms of Morse oscillator basis functions. In handling doublet electronic states, this basis is completed by appropriately chosen spin functions [51]. The matrix elements of the Hamiltonian (35)–(37) in this basis are obtained partly analytically and partly by a numerical integration. The only good quantum number assumed in this treatment of the R–T effect is J, corresponding to the total angular momentum of the molecule.

Handy and co-workers are certainly right in claiming that they use "probably the most appropriate general Hamiltonian" [48]. However, in praxis they solve the corresponding Schrödinger equation by making several approximations, some of them being avoided in another treatments: They neglect the geometrical dependence of the mean value of L_z, use the spin–orbit operator in the phenomenological form (16), and assume the spin–orbit coupling constant to be really a constant.

HCR and co-workers carried out a number of studies by employing 3D potential energy surfaces calculated by means of highly sophisticated ab initio approaches [88,91–101]. The results of these computations are in impressive agreement with the corresponding experimental findings. The discrepancies in the order of 100 wavenumbers, as in early ab initio studies [16,17], have been reduced in the HCR studies to only a few wavenumbers. In conclusion of their paper on the X^2B_1, A^2A_1 ($^2\Pi_u$) system of NH_2, Gabriel et al. state: "We believe that the results presented in this paper are near as possible definitive from the theoretician. It has been a major challenge to us for 15 years to be able to compute these properties of NH_2 to such accuracy..." [97].

The excellent agreement of the results of HCR ab initio studies with the corresponding experimental findings clearly shows that the strongest influence on the numerical accuracy of the vibronic levels have effects outside of the R–T effect, that is, primarly the replacement of the effective bending approaches employed in previous works by a full 3D treatment of the vibrational motions (for an analysis of this matter see, e.g., [17]). Let us note, however, that such a high level of accuracy seems never to have been achieved without a slight modification of the ab initio computed potential surfaces (typically, they have been shifted by ~100 wavenumbers). This is at least partially caused by neglect of some fine effects, like for, example, non-adiabatic corrections of the potential surfaces. On this basis, it can be concluded that the HCR' predictions concerning the yet unobserved spectra are somewhat less reliable.

An alternative "benchmark" approach for handling the R–T effect in triatomic molecules has been developed by Jensen and Bunker (JB) and co-workers. It is based on the use of "MORBID" Hamiltonian [66,67,102], a very sophisticated variant of the above described approaches that handle the bending motion in a different way than their stretching counterparts. This method is described in great detail in a recent book [2], so that we restrict ourselves here only to a small comment. It might look anachronistic (this approach postpones that of HCR) to develop a very ambitious approach not employing "probably the most appropriate general Hamiltonian." A justification is given by JB in their book [15]: "However, one disadvantage (...of the approaches like HC's...) is the fact that in practice, many (if not most) interactions between molecular basis states are weak and could be successfully treated by perturbation theory in the form of a contact transformation. In the variational approaches, these weak interactions are treated by direct matrix diagonalization at a high cost of computer time and memory." We cannot judge if this sentence is relevant in the case of triatomics, but it certainly gains weight when the larger molecules are to be handled.

In several papers [102–105] JB presented the results of their calculations on CH_2, CH_2^+, and BH_2. We find their study of the X^2A_1, A^2B_1 state of BH_2 particularly interesting [104], because this system was treated also by Brommer and HCR [94]. The results of JB et al. are of comparable accuracy with those by HCR; like the latter authors, JB were forced to modify their original ab initio potentials slightly to improve the agreement with the available experimental data. An attempt was undertaken to make a direct comparison of both approaches, but it did not lead to a final conclusion, because of difficulties in transforming the HCR potentials into the form they enter within the JB algorithms. Let us note that both works confirmed the conclusion of our old ab initio study [33] that the assignment of the bands observed in the of $A^2B_1 \leftarrow X^2A_1$ absorption spectrum, made by Herzberg and Johns [37], was not correct.

G. Effective Hamiltonians

Another group of approaches for handling the R–T effect are those that employ various forms of "effective Hamiltonians." By applying perturbation theory, it is possible to absorb all relevant interactions into an effective Hamiltonian, which for a particular (e.g., vibronic) molecular level depends on several parameters whose values are determined by fitting available experimental data. These Hamiltonians are widely used to extract from high-resolution [e.g., electron spin resonance (ESR)] spectra precise values for molecular parameters [19,106–111]. Since the main subject of this chapter is ab initio handling of the R–T effect, we refrain from description of the effective Hamiltonian approaches and refer instead to the excellent review published recently by Brown [3].

H. Beyond the Two-State Renner–Teller Effect

In many cases, the two electronic states building an R–T pair are energetically well separated from all other electronic species and the two-state model expressed by the *ansatz* (10) is quite reliable. However, there are also many exceptions from this rule. Even when a spatially degenerate electronic species represents the ground state of the molecule, its interactions with the other species, possibly at nuclear arrangements differing considerably from the equilibrium geometry of the molecule (as found, e.g., in the series of closely related molecules NH_2 [20], PH_2 [112], and SH_2^+ [113]), can take place. When the R–T state is an excited electronic species, the interactions with the other electronic states represents a normal situation. The energetic vicinity of neighboring species has dramatic effects on the vibronic structure within the R–T state. This topic has been investigated by many authors, particulary by Köppel, Domcke, and Cederbaum [13,14,114]. We present here only an example.

The C_2H radical, a species of great astrophysical–chemical interest and an important intermediate in many chemical reactions, has a $^2\Sigma$ ground electronic state and an extremely low-lying $^2\Pi$ excited species. At the equilibrium geometry of the ground state, the energy difference between them is $\sim$4000 cm^{-1}, but it diminishes drastically upon C—C stretching, so that at larger C—C bond lengths the ordering of the states is reversed [115]. A consequence of these facts is a peculiar structure of the vibronic spectrum in both electronic states, which can be understood only in the framework of a coupled three-state electronic problem. This was realized already in early experimental studies carried out by Curl et al. [116]; for an exhaustive literature survey up to 1992 the reader is referred to [117]. The controversies concerning the $X^2\Sigma$, $A^2\Pi$ spectrum of C_2H have motivated a series of ab initio studies on this three-state system, involving computations of its vibronic, spin–orbit and magnetic hyperfine structure [117–124]. The results of these studies contributed to elucidation of a number of measured spectral features and have been used in later experimental works to help in the assignment of the measured data [125–129]. On the other hand, relative simplicity of the theoretical treatment and low computational efforts caused restricted numerical accuracy of the results. Ten years later, Carter et al. [130] published a really impressive ab initio study, which was able to reproduce all available experimental finding quantitatively.

IV. TETRAATOMIC MOLECULES

A. Theoretical Treatment

1. General Remarks

The first evidence of the R–T effect in a tetraatomic molecule was reported by Herzberg in 1963 [131]. Contrary to the situation with triatomics, the first theoretical model (Petelin and Kiselev (PK) [132]) appeared almost 10 years after Herzberg's observation. The theory of the R–T effect in tetraatomic molecules is much more complicated than in triatomics, because of the existence of two bending modes. PK elaborated a perturbative approach for singlet Π electronic states, which described several special coupling cases, the majority of them concerning the situation in which only one bending vibration is excited. Thus most of the perturbative formulas derived were effectively equivalent to their counterparts in the framework of the classical Renner's theory for triatomics. The equations derived by PK have been used by Colin et al. [133] for an analysis of the high-resolution absorption spectrum of acetylene in the 1205–1255-Å region and by several other authors who have studied the structure of spectra of highly exited (Rydberg) states of acetylene [134,135] and the ground state of $C_2H_2^+$ [136–138]. This approach was extended by Tang and Saito [139] and applied to analyze the ground state of HCCS.

The idea of PK was employed by the present authors [140] who developed a variational approach for an ab initio treatment of the Renner–Teller effect in tetraatomic molecules. The approach by PK was extended to handle both Π and Δ electronic states and to treat the bending vibrations beyond the harmonic approximation. The main practical advantage of this method was of course that it enabled us to obtain the term values and the wave functions for all (bending) vibronic levels of interest. It was applied to compute the vibronic structure of two Rydberg-type electronic states of acetylene. Unfortunately, lack of corresponding experimental findings made it impossible to check the reliability of these results. The approach was later extended to take into account the interplay between the vibronic, spin–orbit [141], magnetic hyperfine [142,143] couplings, and the effects of noninfinitesimal bending vibrations [144,145].

For a long time after Herzberg's observation [131], the experimental information on the R–T effect in tetraatomic molecules has been very scarce. The situation has changed, however, in the last decade in which a series of experimental studies on the Rydberg states of acetylene [134,135,146], and of the ground state of the acetylene ion [136–138,147] has been published. This made it possible for us to judge the validity of the ab initio approach proposed [140]. The results of the ab initio computations enabled a very reliable interpretation of all available experimental findings concerning the ground state of $C_2H_2^+$ [141,148]. The method was also applied to predict the vibronic structure of heretofore unobserved spectra of B_2H_2 ($1^1\Delta_g$ electronic state) [149,150] and $B_2H_2^+$ ($X^2\Pi_u$ state) [142,143] and to interpret the spectra of HCCO [145,151] and HCCS [152]. The key points of this approach will be described in Section IV.2. For a more detailed description, the reader is referred to the original references and the reviews [18,153,154].

2. *Hamiltonian*

The complexity of the problem makes it almost imperative to employ every sensible simplification. For this reason, we immediately exclude from consideration the stretching vibrations and the end-over-end rotations. Since we will restrict ourselves to linear and quasilinear tetraatomic molecules (ABCD), this represents a quite acceptable approximation. We are left thus with four nuclear coordinates. We make the following choice: the supplement of the ABC bond angle, ρ_1; the supplement of the BCD bond angle, ρ_2 ($\rho_1 = \rho_2 = 0$ at the linear molecular geometry); the angle between the plane ABC and a space-fixed plane with the common z axis (coinciding with the BC bond length), ϕ_1; the angle between the plane BCD and the same space-fixed plane, ϕ_2. We denote the bond lengths A—B, C—D and B—C by r_1, r_2, and r_3, respectively.

The complete vibration–rotation Hamiltonian for acetylene-like tetraatomic molecules has been derived by Handy et al. by hand [155] and using a computer algebra program [156]. (Note that in both of the mentioned papers there are some minor errors, see also [144,157,158]). Handy uses as bending coordinates ϑ_1 and ϑ_2, connected with those we prefer by the relations $\vartheta_1 = \pi - \rho_1$ and $\vartheta_2 = \pi - \rho_2$, and instead of our ϕ_1 and ϕ_2 the torsional and z-rotational coordinates γ and ϕ, respectively. In [144], from this Hamiltonian was extracted the part involving derivatives with respect to the above mentioned coordinates, and the volume element $\sin\vartheta_1 \sin\vartheta_2 \Pi\, dq_i$ was replaced by $\rho_1\, \rho_2\, \Pi\, dq_i$. This resulted in a two-bending/z-rotation Hamiltonian given by Eqs. (6)–(8) of [144], which allows the treatment of the large-amplitude bending vibrations at constant bond lengths. This kinetic energy operator may look oversimplified. Indeed, some more sophisticated variant can be used, for example, by taking into account the bend–stretch coupling in line with the BDD [74] or HBJ [60,63] approach. Note that there already exist several effective bending operators derived to describe reliably large-amplitude bending in quasilinear tetraatomics [159,160]. We find, however, that there are other type of problems, that are more important for the understanding of spectral features and leave the discussion about optimal kinetic energy operator for a future review. In this chapter, we make a further simpification by considering only the small-amplitude form of the Hamiltonian derived in [144]:

$$\begin{aligned}
T^0 &= T_1^0 + T_2^0 + T_{12}^0 \\
&= -\frac{1}{2}\left(\frac{1}{\mu_1 r_1^2} + \frac{1}{\mu_3 r_3^2} + \frac{2}{m_B r_1 r_3}\right)\left(\frac{\partial^2}{\partial \rho_1^2} + \frac{1}{\rho_1}\frac{\partial}{\partial \rho_1} + \frac{1}{\rho_1^2}\frac{\partial^2}{\partial \phi_1^2}\right) \\
&\quad -\frac{1}{2}\left(\frac{1}{\mu_2 r_2^2} + \frac{1}{\mu_3 r_3^2} + \frac{2}{m_C r_2 r_3}\right)\left(\frac{\partial^2}{\partial \rho_2^2} + \frac{1}{\rho_2}\frac{\partial}{\partial \rho_2} + \frac{1}{\rho_2^2}\frac{\partial^2}{\partial \phi_2^2}\right) \\
&\quad + \left(\frac{1}{\mu_3 r_3^2} + \frac{1}{m_B r_1 r_3} + \frac{1}{m_C r_2 r_3}\right)\left\{\cos(\phi_2 - \phi_1)\left(\frac{\partial^2}{\partial \rho_1 \partial \rho_2} + \frac{1}{\rho_1 \rho_2}\frac{\partial^2}{\partial \phi_1 \partial \phi_2}\right)\right. \\
&\quad \left. + \sin(\phi_2 - \phi_1)\left(\frac{1}{\rho_1}\frac{\partial^2}{\partial \rho_2 \partial \phi_1} - \frac{1}{\rho_2}\frac{\partial^2}{\partial \rho_1 \partial \phi_2}\right)\right\}
\end{aligned} \tag{46}$$

where

$$\mu_1 = \frac{m_A m_B}{m_A + m_B} \qquad \mu_2 = \frac{m_C m_D}{m_C + m_D} \qquad \mu_3 = \frac{m_B m_C}{m_B + m_C} \tag{47}$$

The bond lengths appearing on the right-hand side of Eq. (46) are assumed to take their equilibrium (constant) values.

For symmetric tetraatomic molecules (ABBA), it is convenient to introduce the symmetry coordinates corresponding to the trans and cis bending by the vector relations

$$\boldsymbol{\rho}_T = \frac{\boldsymbol{\rho}_1 - \boldsymbol{\rho}_2}{2} \qquad \boldsymbol{\rho}_C = \frac{\boldsymbol{\rho}_1 + \boldsymbol{\rho}_2}{2} \tag{48}$$

$\boldsymbol{\rho}_1$ and $\boldsymbol{\rho}_2$ represent the displacement vectors of the nuclei A and D (the corresponding polar coordinates are ρ_1, ϕ_1 and ρ_2, ϕ_2, respectively); $\boldsymbol{\rho}_T$ and $\boldsymbol{\rho}_C$ are the displacement vectors and ρ_T, ϕ_T and ρ_C, ϕ_C the corresponding polar coordinates of the terminal nuclei at the (collective) trans-bending and cis-bending vibrations, respectively. As a consequence of the use of these symmetry coordinates the nuclear kinetic energy operator for small-amplitude bending vibrations represents the kinetic energy of two uncoupled 2D harmonic oscillators:

$$\begin{aligned}
T^0 = T_T^0 + T_C^0 &= -\frac{1}{2\mu_T}\left(\frac{\partial^2}{\partial \rho_T^2} + \frac{1}{\rho_T}\frac{\partial}{\partial \rho_T} + \frac{1}{\rho_T^2}\frac{\partial^2}{\partial \phi_T^2}\right) \\
&\quad -\frac{1}{2\mu_C}\left(\frac{\partial^2}{\partial \rho_C^2} + \frac{1}{\rho_C}\frac{\partial}{\partial \rho_C} + \frac{1}{\rho_C^2}\frac{\partial^2}{\partial \phi_C^2}\right)
\end{aligned} \tag{49}$$

where

$$\mu_T = \frac{2mMR^2r^2}{MR^2 + m(R+2r)^2} \qquad \mu_C = \frac{2mMr^2}{M+m} \tag{50}$$

are the reduced masses for the trans and cis bending vibrations respectively. $m \equiv m_A$, $M \equiv m_B$, $r \equiv$ A–B, $R \equiv$ B–B.

3. *Vibronic Problem*

The model Hamiltonian we use is of the form

$$H = H_e + T \tag{51}$$

where H_e represents the electronic operator and T the nuclear kinetic energy operator (46) or (49). The vibronic wave functions describing the situation when two electronic states are coupled with each other are assumed in the same general form as for triatomics,

$$\Psi = \psi_1 f_1 + \psi_2 f_2 \tag{52}$$

where ψ_1 and ψ_2 are electronic species, and f_1, f_2 are functions of nuclear (bending, torsional, and z rotational) coordinates. In [18,153], various possible electronic basis sets for ab initio handling of the R–T effect in tetraatomic molecules were discussed. In the present approach, we employ the basis consisting of the functions

$$\begin{aligned} \psi_1 &= \psi^{\Lambda} = \frac{1}{\sqrt{2}} e^{i\Lambda\tau}(\psi^+ + \psi^-) \\ \psi_2 &= \psi^{-\Lambda} = \frac{1}{\sqrt{2}} e^{-i\Lambda\tau}(\psi^+ - \psi^-) \end{aligned} \tag{53}$$

where ψ^+ and ψ^- represent the solutions of the electronic Schrödinger equation in the framework of the BO approximation (we assume ψ^- to be imaginary):

$$H_e\psi^+ = V^+\psi^+ \qquad H_e\psi^- = V^-\psi^- \tag{54}$$

τ is a "rotational angle," which determines the spatial orientation of the adiabatic electronic functions ψ^+ and ψ^-. In triatomic molecules, this orientation follows directly from symmetry considerations. So, for example, in a Π state one of the electronic wave functions has its maximum in the molecular plane and the other one is perpendicular to it. If a treatment of the R–T effect is carried out employing the space-fixed coordinate system, the angle τ appearing in Eqs. (53)

reduces in triatomics to the angle ϕ between the instantaneous molecular plane and a space-fixed plane with the common z axis. However, at an arbitrary nuclear arrangement, a tetraatomic molecules does not possess any symmetry elements and thus the orientation of its electronic wave functions, that is, the angle τ, cannot be determined on the basis of symmetry considerations. The angle τ in such cases is a function of all nuclear coordinates involved.

The relationship between the angle τ and the nuclear coordinates considered can be derived in the framework of the model analogous to that developed by Pople and Longuet-Higgins [69] for triatomic molecules. Let us consider a tetraatomic molecule in an electronic state that is twofold spatially degenerate at linear nuclear geometry, at which it is described by the Hamiltonian H_0. We represent the molecular Hamiltonian at bent geometries as a sum of H_0 and an additional (electronic) part H', disappearing in the linear limit. We assume the basis electronic wave function of the system in the form given by Eq. (28), we denote now $|\Lambda\rangle$ and $|-\Lambda\rangle$. We suppose that H' consists of two terms, the first, V, having only diagonal elements in this basis (being the same for both basis functions), and the second, W, coupling the basis electroinic states. The matrix representation of H' is thus

$$\begin{pmatrix} \langle\Lambda|V|\Lambda\rangle & \langle\Lambda|W|-\Lambda\rangle \\ \langle-\Lambda|W|\Lambda\rangle & \langle-\Lambda|V|-\Lambda\rangle \end{pmatrix} \equiv \begin{pmatrix} \bar{V} & \bar{W} \\ \bar{W}^* & \bar{V} \end{pmatrix} \tag{55}$$

The first-order energy correction with respect to the unperturbed problem is then

$$V^{\pm} = \bar{V} \pm \sqrt{\bar{W}^*\bar{W}} \tag{56}$$

and the zeroth-order wave functions are

$$\psi^+ = \frac{1}{\sqrt{\pi}} \cos\left[\Lambda(\theta - \tau)\right] \quad \xi_\Sigma(\rho_e) \qquad \psi^- = \frac{i}{\sqrt{\pi}} \sin\left[\Lambda(\theta - \tau)\right] \quad \xi_\Sigma(\rho_e) \tag{57}$$

with

$$\tau = \frac{1}{2\Lambda} \arctan\left\{ i\frac{\bar{W} - \bar{W}^*}{\bar{W} + \bar{W}^*} \right\} \tag{58}$$

Another useful relation is

$$e^{-2i\Lambda\tau} = \frac{\bar{W}}{\sqrt{\bar{W}\bar{W}^*}} \tag{59}$$

The quantities $V^{\pm}$ given by Eq. (56) represent the first-order approximation for the adiabatic bending potentials. If these potentials are known, $\bar{V}$ can be determined as $\bar{V} = (V^+ + V^-)/2$. On the other hand, the potentials $V^\pm$ do not completely determine the quantity $\bar{W}$, but only the product $\bar{W}\bar{W}^*$, $\sqrt{(\bar{W}\bar{W}^*)} = (V^+ - V^-)/2$.

A convenience of electronic basis functions (53) is that they reduce at infinitesimal-amplitude bending to (28) with the same meaning of the angle θ; we may employ these asymptotic forms in the computation of the matrix elements of the kinetic energy operator and in this way avoid the necessity of carrying out calculations of the derivatives of the electronic wave functions with respect to the nuclear coordinates. The electronic part of the Hamiltonian is represented in the basis (53) by

$$\begin{aligned} \langle\psi^\Lambda|H_e|\psi^\Lambda\rangle &= \frac{V^+ + V^-}{2} = \langle\psi^{-\Lambda}|H_e|\psi^{-\Lambda}\rangle \\ \langle\psi^\Lambda|H_e|\psi^{-\Lambda}\rangle &= e^{-2i\Lambda\tau}\frac{V^+ - V^-}{2} \qquad \langle\psi^{-\Lambda}|H_e|\psi^\Lambda\rangle = e^{2i\Lambda\tau}\frac{V^+ - V^-}{2} \end{aligned} \tag{60}$$

The matrix elements (60) represent effective operators that still have to act on the functions of nuclear coordinates. The factors $\exp(\pm 2i\Lambda\tau)$ determine the selection rules for the matrix elements involving the nuclear basis functions.

In a general case, we assume the potential energy part of the Hamiltonian in the form of an expansion involving the terms $\rho_\alpha^m \rho_\beta^n \cos[k(\phi_\beta - \phi_\alpha)]$ (α and β stand for 1 and 2 in the case of ABCD molecules, and for T and C if we consider symmetric ABBA species), subject to certain symmetry constraints (see [144]).

The vibrational part of the molecular wave function may be expanded in the basis consisting of products of the eigenfunctions of two 2D harmonic oscillators with the Hamiltonians $H_\alpha^0 = T_\alpha^0 + 1/2k_\alpha\rho_\alpha^2$ and $H_\beta^0 = T_\beta^0 + 1/2k_\beta\rho_\beta^2$,

$$\Phi_{\upsilon_\alpha,l_\alpha,\upsilon_\beta,l_\beta} = \frac{1}{2\pi} e^{il_\alpha\phi_\alpha} e^{il_\beta\phi_\beta} R_{\upsilon_\alpha,l_\alpha}(\rho_\alpha) R_{\upsilon_\beta,l_\beta}(\rho_\beta) \tag{61}$$

where the functions $R_{\upsilon,l}$ are defined by Eq. (8). It is easy to verify that the model Hamiltonian we use commutes with the projection of the total angular momentum (excluding spin) onto the z axis, N_z,

$$N_z = R_z^\alpha + R_z^\beta + L_z \equiv R_z + L_z \tag{62}$$

where

$$R_z^\alpha = -i\frac{\partial}{\partial\phi_\alpha} \qquad R_z^\beta = -i\frac{\partial}{\partial\phi_\beta} \qquad L_z = -i\frac{\partial}{\partial\theta} \tag{63}$$

It follows that the only possible values for $l_\alpha + l_\beta$ are $K \mp \Lambda$ and the computation of vibronic levels can be carried out for each K block separately. Matrix elements of the electronic operator H_e, diagonal with respect to the electronic basis [first of Eqs. (60)], and the matrix elements of T are diagonal with respect to the quantum number $l = l_\alpha + l_\beta$. The off-diagonal elements of H_e [second and third of Eqs. (60)] connect the basis functions with $l = l_\alpha + l_\beta$ and $l' = l'_\alpha + l'_\beta = l \pm 2\Lambda$.

The spin–orbit coupling in vibronic states is taken into account assuming the spin–orbit part of the Hamiltonian in the phenomenological form is given by Eq. (16). This operator is added to (51) and the total Hamiltonian is diagonalized in the basis of the above basis functions, assumed that the electronic functions are eigenfunctions of the spin operator, with the eigenvalue Σ. The total model Hamiltonian (including spin–orbit operator) commutes with the projection of the total angular momentum on the z axis so that the vibronic secular equation can be solved for each value of the quantum number $P = K + \Sigma$ separately. In the lowest order approximation (A_{so} assumed to be constant) with the electronic basis (53) the spin–orbit contribution to the total model Hamiltonian is diagonal with respect to all quantum numbers (Λ, l_α, υ_α, l_β, and υ_β) labeling the basis functions.

Although the present approach formally does allow for a treatment of large amplitude bending vibrations (when the corresponding kinetic energy operator and the form of the potentials also involving higher order terms in nuclear coordinates is applied) the restrictions of the model imply a more limited viewpoint. The most critical point is the estimate for the angle τ, being related to the low-order approximation of the molecular potentials (this topic is discussed in [145]). Thus we expect that the model in its present form will be reliable at relatively small-amplitude bending vibrations (typically ρ_α, $\rho_\beta \leq 40°$) around linear equilibrium geometry.

In Section IV.A.4, we show what this general model looks like in the case of Π electronic states of symmetric tetraatomic molecules. The situation in Π states of asymmetric tetraatomics is briefly discussed in Section IV.B, where we present the handling of a concrete case, the $X^2\Pi_u$ state of the HCCS radical. For Δ states the reader is referred to original references [18,149,150,153].

4. Π *Electronic States of ABBA Molecules*

In this section, we consider Π electronic state ($\Lambda = 1$) of ABBA type molecules. The additional Hamiltonian H' is of the form

$$H' = V + W = V(\rho_T, \rho_C) + V_T(\rho_T)\lfloor e^{2i(\theta-\phi_T)} + e^{-2i(\theta-\phi_T)}\rfloor + V_C(\rho_C)[e^{2i(\theta-\phi_C)} + e^{-2i(\theta-\phi_C)}] \tag{64}$$

As already noted by PK, [132], the term involving $V_{TC}(\rho_T, \rho_C)\exp[\pm i(2\theta - \phi_T - \phi_C)]$ does not appear for molecules belonging to the $D_{\infty h}$ point group for symmetry reasons.

In the harmonic approximation, V does not involve the cross-term $\sim \rho_T\rho_C$ because ρ_T and ρ_C are the symmetry coordinates. It is thus of the form

$$V = \frac{1}{2}k_T\rho_T^2 + \frac{1}{2}k_C\rho_C^2 \tag{65}$$

In the same approximation,

$$V_T(\rho_T) = \frac{1}{2}\varepsilon_T k_T\rho_T^2 \qquad V_C(\rho_C) = \frac{1}{2}\varepsilon_C k_C\rho_C^2 \tag{66}$$

where ε_T and ε_C are dimensionless parameters analogous to the Renner parameter ε for triatomic molecules. The first-order adiabatic energies [Eq. (56)] are

$$V^{\pm} = V \pm \sqrt{V_T^2 + V_C^2 + 2V_T V_C \cos[2(\phi_T - \phi_C)]} \tag{67}$$

and the angle τ

$$\tau = \frac{1}{2}\arctan\left[\frac{V_T\sin(2\phi_T) + V_C\sin(2\phi_C)}{V_T\cos(2\phi_T) + V_C\cos(2\phi_C)}\right] \tag{68}$$

First, let us note that the adiabatic potentials V^+ and V^- [Eq. (67)], even in the lowest order (harmonic) approximation, depend on the difference of the angles ϕ_T and ϕ_C; this is an essential difference with respect to triatomics where the adiabatic potentials depend only on the radial bending coordinate ρ. The forms of the functions V, V_T, and V_C are determined by the adiabatic potentials via the following relations

$$V = \frac{V^+ + V^-}{2} \qquad \sqrt{V_T^2 + V_C^2 + 2V_T V_C\cos[2(\phi_T - \phi_C)]} = \frac{V^+ - V^-}{2} \tag{69}$$

The second of these relations reduces at $\rho_C = 0$ to

$$V_T = \left(\frac{V^+ - V^-}{2}\right)_{\rho_C = 0} \tag{70}$$

and for $\rho_T = 0$

$$V_C = \left(\frac{V^+ - V^-}{2}\right)_{\rho_T = 0} \tag{71}$$

Thus the function V_T represents one-half of the splitting of the adiabatic potentials computed for pure trans bending, and V_C one-half of the splitting of the cis-bending curves.

By employing the angle τ defined by (68), the perturbative Hamiltonina H' can be formulated in the form completely analogous to the Pople and Longuet-Higgins' *ansatz* [69]:

$$H' = V + V_2 \lfloor e^{2i(\theta-\tau)} + e^{-2i(\theta-\tau)} \rfloor \tag{72}$$

where

$$V_2 = \frac{V^+ - V^-}{2} = \sqrt{V_T^2 + V_C^2 + 2V_T V_C \cos[2(\phi_T - \phi_C)]} \tag{73}$$

Thus the angle τ plays the role analogous to that of the angle ϕ defining the orientation of the instantaneous molecular plane in triatomic molecules. Employing the relations (69) and (59) one obtains

$$e^{\mp 2i\tau} \frac{V^+ - V^-}{2} = V_T e^{\mp 2\phi_T} + V_C e^{\mp 2i\phi_C} \tag{74}$$

Thus in the lowest order approximation the angle τ is eliminated from the off-diagonal matrix elements of H_e [second and third of Eqs. (60)]; it solely determines the selection rules for matrix elements of H_e with respect to nuclear basis functions.

B. An Example: $X^2\Pi$ Electronic State of HCCS

To illustrate the reliability of the above described model for handling the R–T effect and spin–orbit coupling in tetraatomic molecules, we present here the results of calculation of the spectrum for the $X^2\Pi_u$ state of the HCCS radical. In spite of its importance from the astrophysical–chemical point of view and due to its relationship to a number of other radicals of interest involving carbon, hydrogen, sulfur, and oxygen, HCCS has been subject to a relatively small number of experimental and theoretical studies, which all have left a number of questions open. An absorption spectrum detected by Krishnamachari and Venkitachalam [161] in the region 3770 and 4170 Å during the flash photolysis of thiophene, and tentatively assigned to C_4H_3, was latter attributed by Krishnamachari and Ramsay [162] to a $^2\Pi$–$^2\Pi$ electronic transition of HCCS. The complexity of the spectrum, characterized by large spin–orbit splitting in several bands, precluded a definitive determination of the vibrational fundamentals, particularly for the bending modes [163]. The value for the spin–orbit coupling constant of $-185\ \text{cm}^{-1}$ was estimated [164]. An extensive study of the R–T effect in the ground electronic states of HCCS and DCCS was undertaken by Tang and Saito [139]. They derived perturbative formulas for the coupling cases not considered by PK [132] and applied it to interpret the microwave spectra obtained in the frequency range between 160 and 400 GHz. In the simulation of the spectrum they used the Renner parameters ab initio computed by Szalay [165], and concluded that a reliable reproduction of experimental findings was obtained with the value of the spin–orbit coupling constant assumed to be 270 cm^{-1}. This value disagreed with the results of ab initio computations by Goddard ($-360\ \mathrm{cm}^{-1}$) [166] and Szalay and Blaudeau ($-347\ \mathrm{cm}^{-1}$) [167], as well as with the estimate of the previous experimental study by Vrtilek et al. [164].

An extensive ab initio study was thus undertaken with the goal of providing a reliable interpretation of the available experimental findings, and to predict the structure of yet unobserved parts of the HCCS and DCCS long-wavelengths spectra [152]. Potential energy surfaces for the electronic states of the HCCS radical correlating at linear nuclear arrangement with the $X^2\Pi$ state were calculated by means of a large scale multireference configuration-interaction approach (for details see [152]). Particular attention was paid to calculate accurate 3D potential surfaces involving variations of two bending and torsional coordinates that play the central role in vibronic interactions, determining, together with the spin–orbit coupling, the structure of spectra of this radical.

In the lowest order (quadratic) approximation for Π electronic states of asymmetrical (ABCD) tetraatomics, the electronic matrix elements (60) have the forms [18,152,153]:

$$\frac{1}{2}(V^+ + V^-) = \frac{1}{2}k_1\rho_1^2 + \frac{1}{2}k_2\rho_2^2 - K_{12}\rho_1\rho_2\cos(\phi_2 - \phi_1) \tag{75}$$

$$e^{\mp 2i\Lambda\tau}\frac{V^+ - V^-}{2} = \frac{1}{2}\varepsilon_1'\rho_1^2 e^{\mp 2i\Lambda\phi_1} + \frac{1}{2}\varepsilon_2'\rho_2^2 e^{\mp 2i\Lambda\phi_2} + \varepsilon_{12}'\rho_1\rho_2 e^{\mp i\Lambda(\phi_1+\phi_2)} \tag{76}$$

In this approximation, the angle τ does not appear explicitly in the vibronic secular equation. From Eq. (76) it follows that

$$\begin{aligned}\frac{1}{2}(V^+ - V^-)\cos 2\tau &= \frac{1}{2}\varepsilon_1'\rho_1^2\cos 2\phi_1 + \frac{1}{2}\varepsilon_2'\rho_2^2\cos 2\phi_2 + \varepsilon_{12}'\rho_1\rho_2\cos(\phi_1+\phi_2)\\ (V^+ - V^-)^2 &= \varepsilon_1'^2\rho_1^4 + \varepsilon_2'^2\rho_2^4 + 4\varepsilon_{12}'^2\rho_1^2\rho_2^2 + 2\varepsilon_1'\varepsilon_2'\rho_1^2\rho_2^2\cos 2(\phi_2-\phi_1)\\ &\quad + 4(\varepsilon_1'\rho_1^2 + \varepsilon_2'\rho_2^2)\varepsilon_{12}'\rho_1\rho_2\cos(\phi_2-\phi_1)\end{aligned} \tag{77}$$

The expressions (75) and (77) can be used to extract the parameters k_1, k_2, k_{12}, ε_1', ε_2', and ε_{12}' from the mean adiabatic potential and the difference of the adiabatic potentials for two components of the electronic state spatially degenerate at linear molecular geometry.

Thus only *six* parameters are required to determine the potentials entering our model for handling the R–T effect in asymmetric tetraatomic molecules with linear equilibrium geometry in the lowest order approximation, reliable for description of relatively small-amplitude bending (for Π states of ABBA molecules the number of parameters is just four). They can be extracted from few ab initio computed electronic energies for both adiabatic components of the Π electronic states. So, for example, k_1 and ε_1' may be obtained from the values of the electronic energies at the linear geometry and, say, the energies at $\rho_1 = 20^\circ$, $\rho_2 = 0$, because for $\rho_2 = 0$ the expressions (75) and (76)/(77) reduce to $(V^+ + V^-)/\rho_1^2 = k_1$ and $(V^+ - V^-)/\rho_1^2 = \varepsilon_1'$. (Strictly speaking, not even the value at linear geometry, $\rho_1 = \rho_2 = 0$, is necessary). In an analogous way, the values for k_2 and ε_2' can be determined. Having determined the "diagonal" force constant and Renner parameters, one needs only one additional point to estimate the values of k_{12} and ε_{12}'. The simplest way is to use points computed at cis- ($\phi_2 = \phi_1$) or trans-planar ($\phi_2 - \phi_1 = \pi$) geometry: In the former case $(V^+ + V^-)/2 = 1/2k_1\rho_1^2 + 1/2k_2\rho_2^2 - k_{12}\rho_1\rho_2$, in the latter one $(V^+ + V^-)/2 = 1/2k_1\rho_1^2 + 1/2k_2\rho_2^2 + k_{12}\rho_1\rho_2$. In order to get more reliable results, one can, however, fit a somewhat larger sample of ab initio computed energies. Particularly convenient for this purpose are some special cases of formulas (75–77). So, for example, at $\phi_2 = \phi_1$ (cis-planar geometry),

$$V^+ - V^- = \varepsilon_1'\rho_1^2 + \varepsilon_2'\rho_2^2 + 2\varepsilon_{12}'\rho_1\rho_2 \tag{78}$$

which further reduces for $\rho_1 = \rho_2 \equiv \rho$ to $(V^+ - V^-)/\rho^2 = \varepsilon_1' + \varepsilon_2' + 2\varepsilon_{12}'$. For $\phi_2 - \phi_1 = \pi$ (trans-planar geometry),

$$V^+ - V^- = \varepsilon_1'\rho_1^2 + \varepsilon'^2\rho_2^2 - 2\varepsilon_{12}'\rho_1\rho_2 \tag{79}$$

reducing at $\rho_1 = \rho_2 \equiv \rho$ to $(V^+ - V^-)/\rho^2 = \varepsilon_1' + \varepsilon_2' - 2\varepsilon_{12}'$. Summarizing, all the parameters required can be obtained by employing the electronic energies computed at planar geometries.

On the basis of the above analyses, it follows that there is no need to compute multidimensional potential surfaces if one wishes to handle the R–T effect in the framework of the model proposed. In spite of that, such computations were carried out in [152] in order to demonstrate the reliability of the model for handling the R–T effect and to estimate the range in which it can safely be applied in its lowest order (quadratic) approximation. The 3D potential surfaces involving the variation of the bending coordinates ρ_1, ρ_2 and the relative azimuth angle $\gamma = \phi_2 - \phi_1$ were computed for both component of the $X^2\Pi$ state.

Determination of the parameters entering the model Hamiltonian for handling the R–T effect (quadratic force constant for the mean potential and the Renner parameters) was carried out by fitting special forms of the functions [Eqs. (75) and (77)], as described above, and using not more than 10 electronic energies for each of the $X^2\Pi$ component states, computed at cis- and trans-planar geometries. This procedure led to the above mentioned six parameters which, according to the model, suffice to anticipate the form of complete 3D potential surfaces involving variations of the coordinates ρ_1, ρ_2, and γ at relatively small distortions of linearity. The corresponding functions, $1/2(V^+ + V^-)$ and the square root of the second of functions (77), are drawn in Figures 8 and 9. Different symbols in this figures denote the points actually calculated. The agreement between the electronic energies determined by the

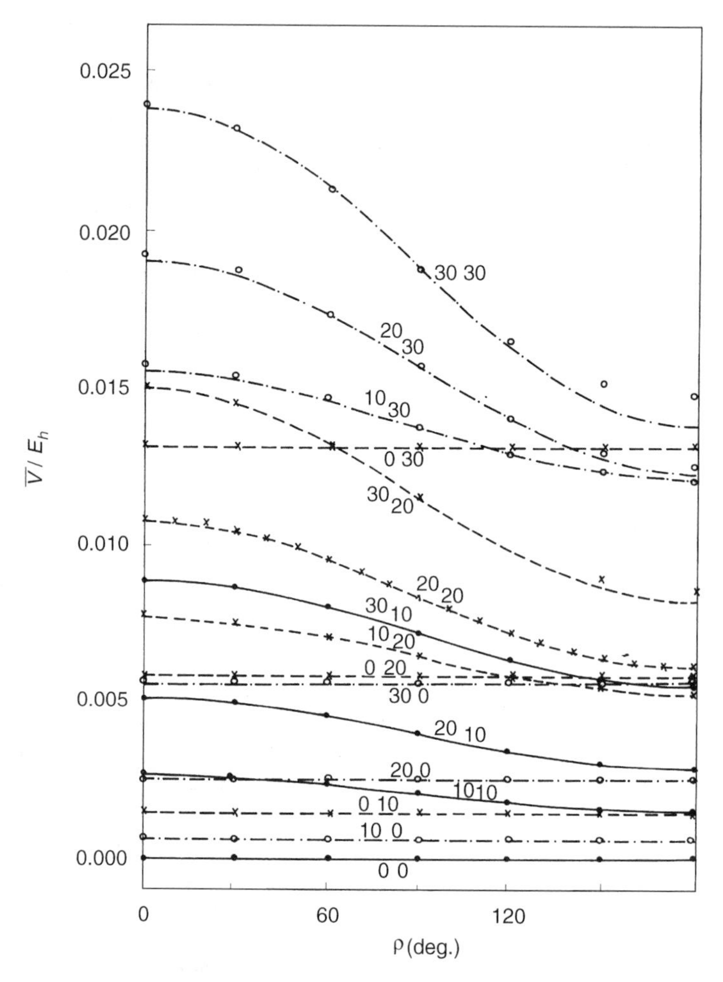

Figure 8. Three-dimensional mean-potential surface for the $X^2\Pi$ state of HCCS, $\bar{V}(\rho_1, \rho_2, \gamma)$, presented in form of its 1D sections. Curves represent the function given by Eq. (75). (with $k_1 = 0.0414$, $k_2 = 0.952$, $k_{12} = 0.0184$) for fixed values of coordinates ρ_1 and ρ_2 (attached at each curve) and variable $\gamma = \phi_2 - \phi_1$. Here $\gamma = 0$ corresponds to cis-planar geometry and $\gamma = \pi$ to trans-planar geometry. Symbols: results of explicit ab initio computations.

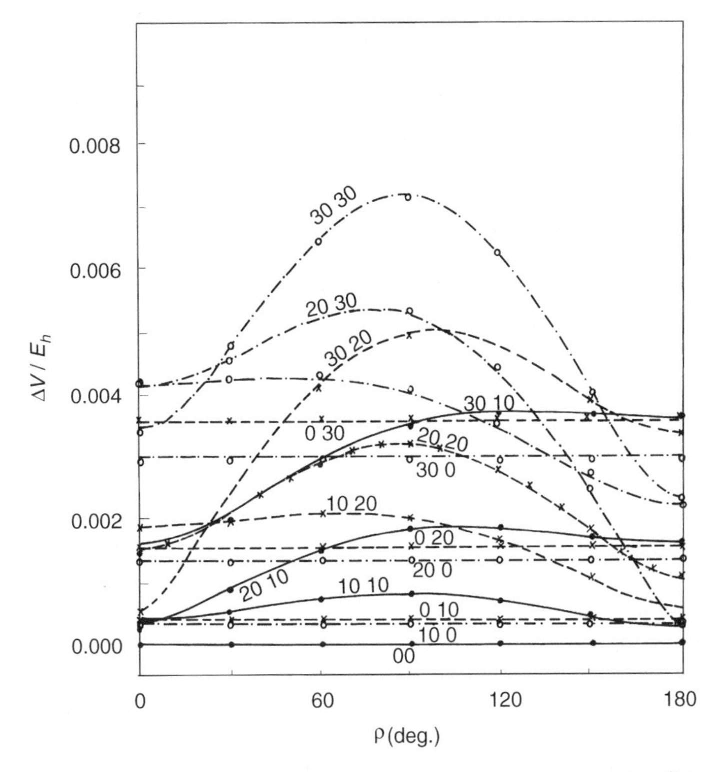

Figure 9. Energy difference (absolute value) between the components of the $X^2\Pi$ electronic state of HCCS as a function of coordinates ρ_1, ρ_2, and γ. Curves represent the square root of the second of functions given by Eq. (77) (with $\varepsilon_1' = -0.011$, $\varepsilon_2' = 0.013$, $\varepsilon_{12}' = 0.005325$) for fixed values of coordinates ρ_1 and ρ_2 (attached at each curve) and variable $\gamma = \phi_2 - \phi_1$. Here $\gamma = 0$ corresponds to cis-planar geometry and $\gamma = \pi$ to trans-planar geometry. Symbols: results of explicit ab initio calculations.

three-parameter formulas (75) and (77) and those actually calculated looks very satisfactory. Some significant discrepancies can be noted only at large deviations from linearity ($\rho_1 = 30°$, $\rho_2 = 30°$), where the harmonic approximation is not expected to be reliable.

For the variational calculations of the vibronic spectrum and the spin–orbit fine structure in the $X^2\Pi$ state of HCCS the basis sets involving the bending functions up to $\upsilon_1 = \upsilon_2 = 11$ with all possible l_1 and l_2 values are used. This leads to the vibronic secular equations with dimensions $\sim$600 for each of the vibronic species considered. The bases of such dimensions ensure full convergence of vibronic energies in the energy range from 0 to 2500 cm^{-1} with respect to the minimum of the potential surface.

Two sets of calculations for the vibronic spectrum were carried out. In the first, the zeroth-order kinetic energy operator given by Eq. (46) is employed. The second set of calculations is performed using the kinetic energy operator for large-amplitude bending with bond lengths kept constant [144]. The results of both sets of calculations do not differ from one another by more than a few wavenumbers (see [152] for a more detailed analysis) and thus we present below only those obtained using the zeroth-order kinetic energy operator.

The results of calculation of the HCCS vibronic spectrum are displayed in Figure 10. Let us skip at this place from the notation used thus far (υ_1, l_1; υ_2, l_2) to the usual spectroscopic one, according to which the indexs 4 and 5 are employed for (predominantly) H–C–C and C–C–S bending modes, respectively. Three sets of vibronic energy levels are given: The first one represent zeroth-order values obtained by neglecting the R–T effect and the spin–orbit coupling—they correspond thus to the mean potential surface for the $X^2\Pi$ electronic state. The second set of vibronic energy values is generated in computations in which the R–T coupling is taken into account, but the spin–orbit coupling is neglected. We present the results for $K = 0, 1, 2,$ and 3 states. The third sample of vibronic energies represents the final results obtained by including both the vibronic and spin–orbit coupling effects. In all cases the composition of the vibronic wave functions in terms of dominating basis functions (denoted by the vibrational quantum numbers υ_4, υ_5) is given. The low-lying vibronic states, computed at different levels of approximation, correlating (approximately) with one another are connected with thin lines.

Since the basis functions employed in variational calculations are *not* expressed in terms of the normal coordinates, the labeling of the vibronic levels by the quantum numbers υ_4 and υ_5 is approximate (so, e.g., the lowest lying $K = 0$ vibronic state, assigned to $\upsilon_4 = 0$, $\upsilon_5 = 1$, contains an appreciable contribution from the basis function with $\upsilon_4 = 1$, $\upsilon_5 = 2$), Harmonic bending frequencies computed in the this study ($\omega_4 = 579$ cm^{-1}, $\omega_5 = 374$ cm^{-1}) are in very good agreement with the corresponding experimental findings (565, 380 cm^{-1}, respectively) [139]. The spin–orbit coupling constant (-261 cm^{-1}) is in excellent agreement with the value extracted by Tang and Saito [139] from the experimental findings.

We do not wish to go into the details of Figure 10. As an illustration of the reliability of the present results we compare, however, in Figure 11, the structure of the measured HCCS spectrum published by Tang and Saito (Fig. 3 of [139]) with the results of the theoretical study. Taking into account the very complex and unusual structure of this kind of spectra, we find the agreement between our ab initio theoretical results and those following from the interpretation of experimental spectra more than satisfactory. While strongly

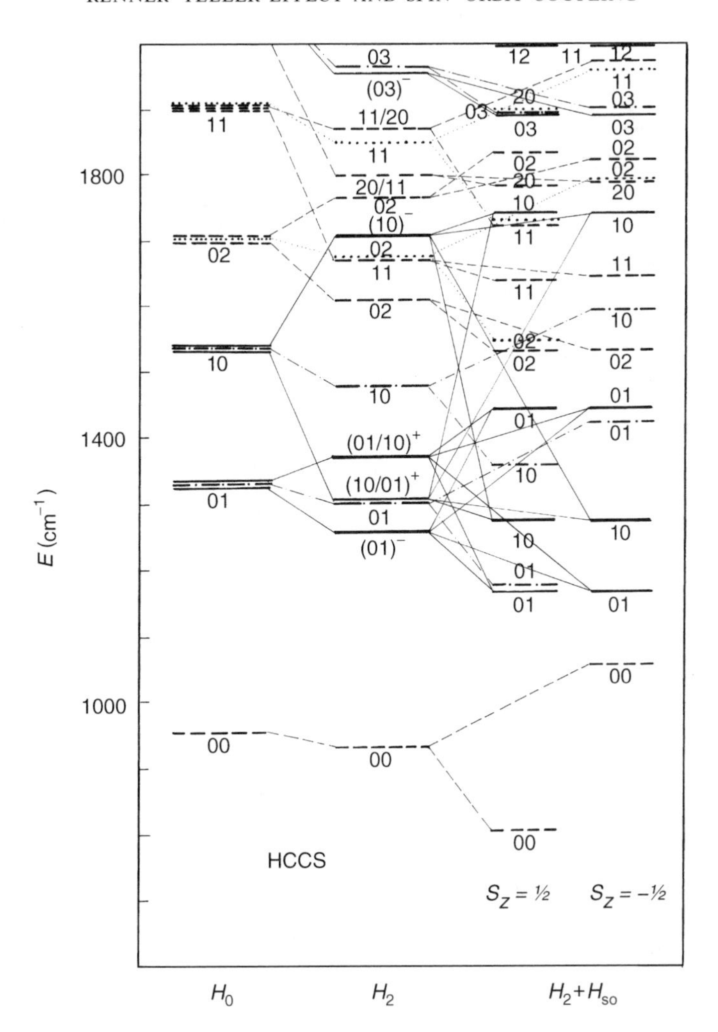

Figure 10. Low-energy vibronic levels in the $X^2\Pi$ state of HCCS computed in various approximations [152]. H_0: zeroth-order approximation (both vibronic and spin–orbit couplings neglected). H_2: vibronic coupling taken into account, spin–orbit interaction neglected. $H_2 + H_{so}$: both vibronic and spin–orbit couplings taken into account. Solid horizontal lines: $K = 0$ vibronic levels; dashed line: $K = 1$; dash–dotted lines: $K = 2$; dotted lines: $K = 3$. Values of the quantum numbers N_4, N_5 of the basis functions dominating the vibronic wave function of the level in question are indicated. Approximate correlation of vibronic states computed in various approximations is indicated by thin lines. In all cases the stretching quantum numbers are assumed to be zero.

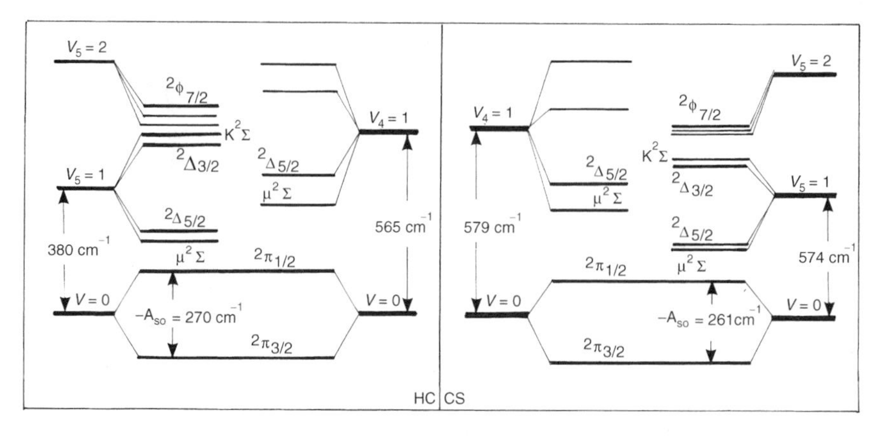

Figure 11. Left: Vibronic structure of the $X^2\Pi$ state of HCCS derived by Tang and Saito from experimental findings (Fig. 3 of [139]). Right: Corresponding results of the ab initio study presented in [152].

supporting the analysis of the experimental findings, carried out by Tang and Saito, the results of the ab initio study offered a number of additional data that could be helpful in future experimental work on this subject.

V. CONCLUSION

In this chapter, we give a review of results of ab initio treatments of the R–T effect and spin–orbit coupling in triatomic and tetraatomic molecules. We have tried to present a comprehensive up to date literature survey and to compare various approaches with one another. Our prime goal was to present the development of ideas that finally resulted in those methods that are nowadays widely used to produce numerical results serving to interpret and to predict features of concrete molecular systems. A consequence of these restrictions is that we only mentioned those model treatments in which the main goal is rather to explain qualitative aspects of the vibronic coupling than to consider particular molecules, and the effective Hamiltonian approaches that do not rely on ab initio computations; we did not even mention several alternative theoretical approaches [168,169] not directly on the main stream we followed in this study. We have tried to document that the handling of the R–T effect by means of modern ab initio techniques not only offers reliable interpretation of experimental findings but also has reached the level of numerical accuracy so that it can compete with high-resolution spectroscopy.

APPENDIX A: PERTURBATIVE HANDLING OF THE RENNER–TELLER EFFECT AND SPIN–ORBIT COUPLING IN Π ELECTRONIC STATES OF TETRAATOMIC MOLECULES

The perturbation theory has for a long time been the leading tool for handling the R–T effect in triatomic molecules (for an overview see, e.g. [8] and [12]). It was also applied in the first theoretical paper on tetraatomic molecules [132] to singlet Π states of both symmetric (ABBA) and asymmetric (ABCD) species. The combined effect of vibronic and spin–orbit coupling in the lowest lying vibronic species was considered by Tang and Saito [139]. Several new perturbative formulas for singlet Π states of symmetric tetraatomic molecules were derived in [153], and a perturbative treatment of $^1\Delta$ electronic states of the same molecular species was presented in [150]. In two recent studies [170,171] the present authors gave the second-order perutbative formulas for the combined effect of vibronic and spin–orbit coupling in Π and Δ electronic states of triatomic and symmetric tetraatomic molecules with linear equilibrium geometry. They were derived employing two schemes for partitioning the model Hamiltonian: In the first, the term descibing the spin–orbit coupling is tretad as perturbation (together with the term responsible for vibronic interaction). In the second approach, reliable for the cases when the spin–orbit coupling constant is comparable in magnitude to the bending frequency, the spin–orbit operator is included into the zeroth-order Hamiltonian. In this chapter, we give only the formula obtained via the first scheme. Moreover, we restrict ourselves to tetraatomic molecules, because the perturbative formulas for triatomics can be looked upon as special cases of their counterparts for tetraatomic species.

The perturbative handling of the R–T effect in molecules with more than three atoms is much more complicated than that for triatomics, because of the presence of more than one bending mode contributing to the vibronic interaction. A consequence is that the perturbative formulas cannot be derived for a general case, because the level of degeneracy of zeroth-order vibronic levels increases rapidly with increasing values for the bending quantum numbers (in the case of triatomics all zeroth-order vibronic levels are either nondegenerate or twofold degenerate).

The present perturbative treatment is carried out in the framework of the minimal model we defined above. All effects that do not crucially influence the vibronic and fine (spin–orbit) structure of spectra are neglected. The kinetic energy operator for infinitesimal vibrations [Eq. (49)] is employed and the bending potential curves are represented by the lowest order (quadratic) polynomial expansions in the bending coordinates. The spin–orbit operator is taken in the phenomenological form [Eq. (16)]. We employ as basis functions the species

$$\Psi_{-\Lambda,\Sigma,K,\upsilon_T,l_T,\upsilon_C,l_C} = \xi_\Sigma(\rho_e)\frac{1}{\sqrt{2\pi}}e^{-i\Lambda\theta}\frac{1}{\sqrt{2\pi}}e^{il_T\phi_T}\frac{1}{\sqrt{2\pi}}e^{il_C\phi_C}R_{\upsilon_T,l_T}(\rho_T)R_{\upsilon_C,l_C}(\rho_C)$$

$$\Psi_{\Lambda,\Sigma,K,\upsilon'_T,l'_T,\upsilon'_C,l'_C} = \xi_\Sigma(\rho_e)\frac{1}{\sqrt{2\pi}}e^{i\Lambda\theta}\frac{1}{\sqrt{2\pi}}e^{il'_T\phi_T}\frac{1}{\sqrt{2\pi}}e^{il'_C\phi_C}R_{\upsilon'_T,l'_T}(\rho_T)R_{\upsilon'_C,l'_C}(\rho_C) \tag{A.1}$$

After integrating over all electronic coordinates except for θ, the electronic operator H_e transforms into the potential for bending vibrations has the form

$$V = \frac{1}{2}k_T\rho_T^2 + \frac{1}{2}k_C\rho_C^2 + \frac{1}{2}\varepsilon_T k_T\rho_T^2[e^{2i(\theta-\phi_T)} + e^{-2i(\theta-\phi_T)}] + \frac{1}{2}\varepsilon_C k_C\rho_C^2[e^{2i(\theta-\phi_C)} + e^{-2i(\theta-\phi_C)}] \tag{A.2}$$

The force constants k_T, k_C and the dimensionless Renner parameters $\varepsilon_T, \varepsilon_C$ are defined by the adiabatic potentials for the components of the Π state at pure trans (V_T^+, V_T^-) and pure cis (V_C^+, V_C^-) bending vibrations,

$$\frac{1}{2}k_T\rho_T^2 = \frac{1}{2}(V_T^+ + V_T^-) \qquad \frac{1}{2}k_C\rho_C^2 = \frac{1}{2}(V_C^+ + V_C^-)$$
$$\frac{1}{2}\varepsilon_T k_T\rho_T^2 = \frac{1}{2}(V_T^+ - V_T^-) \qquad \frac{1}{2}\varepsilon_C k_C\rho_C^2 = \frac{1}{2}(V_C^+ - V_C^-) \tag{A.3}$$

We introduce the dimensionless bending coordinates $q_T \equiv \sqrt{\lambda_T}\rho_T$ and $q_C \equiv \sqrt{\lambda_C}\rho_C$ with $\lambda_T \equiv \sqrt{(k_T\mu_T)} = \mu_T\omega_T$, $\lambda_C \equiv \sqrt{(k_C\mu_C)} = \mu_C\omega_C$, where ω_T and ω_C are the harmonic frequencies for pure trans- and cis-bending vibrations, respectively. After integrating over θ, we obtain the effective Hamiltonian $H = H_0 + H'$, which is employed in the perturbative handling of the R–T effect and the spin–orbit coupling. Its zeroth-order part is of the form

$$H_0 = \left\{-\frac{1}{2}\left(\frac{\partial^2}{\partial q_T^2} + \frac{1}{q_T}\frac{\partial}{\partial q_T} - \frac{l_T^2}{q_T^2}\right) + \frac{1}{2}q_T^2\right\}\omega_T + \left\{-\frac{1}{2}\left(\frac{\partial^2}{\partial q_C^2} + \frac{1}{q_C}\frac{\partial}{\partial q_C} - \frac{l_C^2}{q_C^2}\right) + \frac{1}{2}q_C^2\right\}\omega_C \tag{A.4}$$

The operators $\partial^2/\partial\phi_T^2$ and ∂^2/ϕ_C^2 are replaced in (A.4) by their eigenvalues $-l_T^2$ and $-l_C^2$. The functions $R(\rho_T)$ and $R(\rho_C)$ appearing in (A.1) are the eigenfunctions of the first and the second part of this Hamiltonian, respectively. They are of the form (8). Both l_T and l_C are the quantum numbers for the nuclear angular momentum caused by the 2D trans and cis bending vibrations, respectively. Their algebraic sum with the electronic angular quantum number Λ gives the good (in the framework of the present model) quantum number $K, K = l_T + l_C - \Lambda = l'_T + l'_C + \Lambda$. The sum of K and the spin quantum number Σ results in the quantum number for the projection of the total angular momentum along the molecular axis, $P = K + \Sigma$.

The perturbative part of the effective Hamiltonian is of the form

$$H' = \frac{1}{2}\varepsilon_T\omega_T q_T^2(e^{-2i\phi_T} + e^{2i\phi_T}) + \frac{1}{2}\varepsilon_C\omega_C q_C^2(e^{-2i\phi_C} + e^{2i\phi_C}) \mp \Sigma A_{\mathrm{so}} \quad \text{(A.5)}$$

We introduce for the basis functions the notation

$$|\upsilon_T\ l_T\ \upsilon_C\ l_C\ -\rangle \equiv \frac{1}{\sqrt{2\pi}}e^{il_T\phi_T}\frac{1}{\sqrt{2\pi}}e^{il_C\phi_C}R_{\upsilon_T,l_T}(q_T)R_{\upsilon_C,l_C}(q_C) \quad l_T + l_C = K + \Lambda$$
$$|\upsilon_T\ l_T\ \upsilon_C\ l_C\ +\rangle \equiv \frac{1}{\sqrt{2\pi}}e^{il_T\phi_T}\frac{1}{\sqrt{2\pi}}e^{il_C\phi_C}R_{\upsilon_T,l_T}(q_T)R_{\upsilon_C,l_C}(q_C) \quad l_T + l_C = K - \Lambda$$
$$\text{(A.6)}$$

The zeroth-order Hamiltonian and the spin–orbit part of the perturbation are diagonal with respect to the quantum numbers $K, \Sigma, P, \upsilon_T, l_T, \upsilon_C$ and l_C. The matrix elements $\langle -\ l_C\ \upsilon_C\ l_T\ \upsilon_T|H'|\upsilon'_T\ l'_T\ \upsilon'_C\ l'_C\ +\rangle$ of the remaining part of the perturbative operator can be different from zero only if $l'_T = l_T - 2,\ \upsilon'_T = \upsilon_T,\ \upsilon_T \pm 2,\ l'_C = l_C,\ \upsilon'_C = \upsilon_C$, or $l'_T = l_T,\ \upsilon'_T = \upsilon_T,\ l'_C = l_C - 2,\ \upsilon'_C = \upsilon_C,\ \upsilon_C \pm 2$. These selection rules automatically comprise the *g/u* symmetry restrictions.

As mentioned above, as a consequence of the fact that the degeneracy of the zeroth-order vibronic levels in tetraatomic molecules increases rapidly with increasing value of the bending quantum numbers υ_T and υ_C [it is equal to $2(2\upsilon_T + 1)(2\upsilon_C + 1)$ if spin is neglected] and thus the secular equations to be solved in the framework of the perturbation theory are multidimensional, only several special coupling cases can be handled efficiently. We present here those for which the dimension of the secular equation is effectively not larger than 2; however, some other cases can also be handled, particularly if the spin–orbit coupling is relatively large compared to the separation of the electronic states at relevant bent geometries.

Case Aa $\quad \upsilon_T \neq 0,\ \upsilon_C = 0$

In this case, the situation is essentially equivalent to that for triatomics molecules. (We shall always assume that $\upsilon_T \geq \upsilon_C$; the formulas for the opposite case, $\upsilon_T \leq \upsilon_C$, are obtained from those to be derived by interchanging simply the indexes T and C.) There are two possible cases:

Case Aa1 $\quad K = \upsilon_T + 1 (= K_{\max})$

There is a single nondegenerate zeroth-order vibronic state $|\upsilon_T\, \upsilon_T\, 00+\rangle$. The zeroth-, first-, and second-order formulas read

$$E^{(0)} = (\upsilon_T + 1)\omega_T + \omega_C \tag{A.7}$$

$$E^{(1)} = \Sigma A_{so} \tag{A.8}$$

$$E^{(2)} = \frac{-1}{8} K(K+1)\varepsilon_T^2 \omega_T - \frac{1}{4}\varepsilon_C^2 \omega_C \tag{A.9}$$

Thus the complete expression up to second order is

$$E = (\upsilon_T + 1)\omega_T + \Sigma A_{so} - \frac{1}{8}(\upsilon_T + 1)(\upsilon_T + 2)\varepsilon_T^2 \omega_T + \left(1 - \frac{1}{4}\varepsilon_C^2\right)\omega_C \tag{A.10}$$

For $\omega_T = \omega$, $\varepsilon_T = \varepsilon$, $\upsilon_T = \upsilon$, and $\omega_C = 0$, this formula reduces to its counterparts for *unique* levels of triatomic molecules.

Case Aa2 $\quad K < \upsilon_T + 1$:

The zeroth-order energy level is $|\upsilon_T\, K-1\, 0\, 0\, -\rangle \equiv 14\rangle$ twofold degenerate—the corresponding vibronic states are $|\upsilon_T\ K+1\ 0\ 0\ -\rangle \equiv |1\rangle$ and $|\upsilon_T\ K-1\ 0\ 0\ +\rangle \equiv |2\rangle$. The zeroth-order energy is given by the expression (A.7). The first-order energy correction is

$$E_{1/2}^{(1)} = \mp\frac{1}{2}D \tag{A.11}$$

with

$$D = \sqrt{4\Sigma^2 A_{so}^2 + [(\upsilon_T + 1)^2 - K^2]\varepsilon_T^2\, \omega_T^2} \tag{A.12}$$

The corresponding wave functions are of the form

$$\Psi_1 = c_{11}|1\rangle + c_{12}|2\rangle \qquad \Psi_2 = c_{21}|1\rangle + c_{22}|2\rangle \tag{A.13}$$

where

$$c_{11} = \frac{1}{\sqrt{2}}\sqrt{1 - \frac{B}{D}} \qquad c_{12} = -\frac{1}{\sqrt{2}}\sqrt{\frac{1+B}{D}}$$
$$c_{21} = -c_{12} \qquad c_{22} = c_{11} \tag{A.14}$$

with D given by (A.12) and

$$B = -2\Sigma A_{so} \tag{A.15}$$

The second-order energy correction is

$$E^{(2)}_{1/2} = -\frac{1}{8}(\upsilon_T + 1)\left\{1 \mp \frac{2K\Sigma A_{so}}{D}\right\}\varepsilon_T^2\omega_T - \frac{1}{4}\varepsilon_C^2\omega_C \tag{A.16}$$

Thus the complete energy expression is

$$\begin{aligned} E_{1/2} = {} & (\upsilon_T + 1)(1 - \frac{1}{8}\varepsilon_T^2)\omega_T + (1 - \frac{1}{4}\varepsilon_C^2)\omega_C \\ & \mp \frac{1}{2}\sqrt{4\Sigma^2 A_{so}^2 + [(\upsilon_T + 1)^2 - K^2]\,\varepsilon_T^2\,\omega_T^2} \\ & \pm \frac{K(\upsilon_T + 1)\Sigma A_{so}\varepsilon_T^2\omega_T}{4\sqrt{4\Sigma^2 A_{so}^2 + [(\upsilon_T + 1)^2 - K^2]\,\varepsilon_T^2\,\omega_T^2}} \end{aligned} \tag{A.17}$$

If we put $\omega_T = \omega$, $\varepsilon_T = \varepsilon$, $\upsilon_T = \upsilon$, and $\omega_C = 0$, the energy given by Eq. (A.17) becomes identical to the corresponding expression for triatomic molecules.

Case Ab $\quad K = \upsilon_T + \upsilon_C + 1 = K_{max}$(*unique* levels)

Zeroth-order level, corresponding to the vibronic state $|\upsilon_T\ l_T\ \upsilon_C\ l_C\ +\rangle = |\upsilon_T\ \upsilon_T\ \upsilon_C\ \upsilon_C\ +\rangle$ is nondegenerate. The zeroth, first- and second-order energies are

$$E^{(0)} = (\upsilon_T + 1)\omega_T + (\upsilon_C + 1)\omega_C \tag{A.18}$$

$$E^{(1)} = \Sigma A_{so} \tag{A.19}$$

$$E^{(2)} = -\frac{1}{8}(\upsilon_T + 1)(\upsilon_T + 2)\varepsilon_T^2\omega_T - \frac{1}{8}(\upsilon_C + 1)(\upsilon_C + 2)\varepsilon_C^2\omega_C \tag{A.20}$$

For $\upsilon_C = 0$, the sum of expressions (A.18–A.20) becomes identical to (A.10).

Case Ac $\quad K = \upsilon_T + \upsilon_C - 1 (= K_{max} - 2)\upsilon_T \geq \upsilon_C \neq 0$

Let us use the notation $\upsilon_T \equiv \upsilon$, $\upsilon_C = K - \upsilon + 1$ and

$$\sqrt{\upsilon_T}\varepsilon_T\omega_T \equiv u \qquad \sqrt{\upsilon_C}\varepsilon_C\omega_C \equiv t \tag{A.21}$$

The zeroth-order vibronic level is threefold degenerate, with the wave functions $|\upsilon\ \upsilon\ K-\upsilon+1\ K-\upsilon+|-\rangle \equiv |1\rangle$, $|\upsilon\ \upsilon\ K-\upsilon+1\ K-\upsilon-1\ +\rangle \equiv |2\rangle$ and $|\upsilon\ \upsilon-2\ K-\upsilon+1\ K-\upsilon+1+\rangle \equiv |3\rangle$. The zeroth-order energy is given by Eq. (A.18). The first-order energy corrections are

$$E_2^{(1)} = \Sigma A_{\text{so}} \qquad E_{1/3}^{(1)} = \mp D \tag{A.22}$$

where

$$D = \sqrt{\Sigma^2 A_{\text{so}}^2 + u^2 + t^2} \tag{A.23}$$

The corresponding wave functions are

$$\begin{aligned}
\Psi_2 &= \frac{u}{\sqrt{u^2+t^2}}|2\rangle - \frac{t}{\sqrt{u^2+t^2}}|3\rangle \\
\Psi_{1/3} &= \frac{1}{\sqrt{2}}\sqrt{1 \pm \frac{\Sigma A_{\text{so}}}{D}}|1\rangle \pm \frac{1}{\sqrt{2}}\frac{t}{\sqrt{u^2+t^2}}\sqrt{1 \mp \frac{\Sigma A_{\text{so}}}{D}}|2\rangle \\
&\quad \pm \frac{1}{\sqrt{2}}\frac{u}{\sqrt{u^2+t^2}}\sqrt{1 \mp \frac{\Sigma A_{\text{so}}}{D}}|3\rangle
\end{aligned} \tag{A.24}$$

The second-order energy correction are

$$\begin{aligned}
E_2^{(2)} &= -\frac{1}{8(\upsilon_T \varepsilon_T^2 \omega_T^2 + \upsilon_C \varepsilon_C^2 \omega_C^2)} \\
&\quad \times \{\upsilon_T(\upsilon_T+1)[(\upsilon_T+2)\varepsilon_T^2\omega_T^2 + \upsilon_C\varepsilon_C^2\omega_C^2]\varepsilon_T^2\omega_T \\
&\quad + \upsilon_C(\upsilon_C+1)[\upsilon_T\varepsilon_T^2\omega_T^2 + (\upsilon_C+2)\varepsilon_C^2\omega_C^2]\varepsilon_C^2\omega_C\}
\end{aligned} \tag{A.25}$$

$$\begin{aligned}
E_{1/3}^{(2)} &= -\frac{1}{8}(\upsilon_T+1)\left(1+\frac{t^2}{u^2+t^2}\right)\varepsilon_T^2\omega_T - \frac{1}{8}(\upsilon_C+1)\left(1+\frac{u^2}{u^2+t^2}\right)\varepsilon_C^2\omega_C \\
&\quad \mp \frac{\Sigma A_{\text{so}}}{8D}\left[(\upsilon_T+1)\left(\frac{u^2}{u^2+t^2}-\upsilon_T\right)\varepsilon_T^2\omega_T + (\upsilon_C+1)\left(\frac{t^2}{u^2+t^2}-\upsilon_C\right)\varepsilon_C^2\omega_C\right]
\end{aligned} \tag{A.26}$$

The formulas for $E_{1/3}$ are valid also for the case $\upsilon_C = 0$: They are identical to (A.17) if in the latter, K is replaced by $\upsilon_T - 1$.

The $\Sigma A_{\text{so}} = 0$ limits of the formulas presented in this subsection cover all particular cases (including $\upsilon_T = 1,\ \upsilon_C = 1$, and $\upsilon_T = 2,\ \upsilon_C = 1$) handled in the previous works [18,132,153].

APPENDIX B: PERTURBATIVE HANDLING OF THE RENNER–TELLER EFFECT AND SPIN–ORBIT COUPLING IN Δ ELECTRONIC STATES OF TETRAATOMIC MOLECULES

We restrict ourselves again to symmetric tetraatomic molecules (ABBA) with linear equilibrium geometry. After integrating over electronic spatial and spin coordinates we obtain for Δ electronic states in the lowest order (quartic) approximation the effective model Hamiltonian $H = H_0 + H'$, which zeroth-order part is given by Eq. (A.4) and the perturbative part of it of the form

$$\begin{aligned} H' &= a_T\omega_T q_T^4 + a_C\omega_C q_C^4 + b_0\sqrt{\omega_T\omega_C}q_T^2q_C^2 \\ &+ b_2\sqrt{\omega_T\omega_C}q_T^2q_C^2[e^{2i(\phi_T-\phi_C)} + e^{-2i(\phi_T-\phi_C)}] \\ &+ c_T\omega_T q_T^4[e^{-4i\phi_T} + e^{4i\phi_T}] + c_C\omega_C q_C^4[e^{-4i\phi_C} \\ &+ e^{4i\phi_C}] + c_{TC}\sqrt{\omega_T\omega_C}q_T^2q_C^2[e^{-2i\phi_T}e^{-2i\phi_C} \\ &+ e^{2i\phi_T}e^{2i\phi_C}] \mp 2\Sigma A_{\text{so}} \end{aligned} \tag{B.1}$$

The dimensionless parameters $a_T, \ldots, c_{TC}$ appearing in the last expression are connected with the sums and differences of the adiabatic potentials as shown elsewhere [149,150]. This effective Hamiltonian acts onto the basis functions (A.1) with $\Lambda = 2$.

The zeroth-order Hamiltonian and the spin–orbit part of the perturbation are diagonal with respect to the quantum numbers K, Σ, P, υ_T, l_T, υ_C, and l_C. The terms of H' involving the parameters a_T, a_C, and b_0 are diagonal with respect to both the l_T and l_C quantum numbers, while the b_2 term connects with one another the basis functions with $l'_T = l_T \pm 2$, $l'_C = l_C \mp 2$. The c terms couple with each other the electronic species $-\Lambda$ and Λ. The selection rules for the vibrational quantum numbers are $\upsilon'_{T/C} = \upsilon_{T/C}$, $\upsilon_{T/C} \pm 2$, $\upsilon_{T/C} \pm 4$.

As in the case of Π electronic states of tetraatomic molecules, because of generally high degeneracy of zeroth-order vibronic leves only several particular (but important) coupling cases can be handled efficiently in the framework of the perturbation theory. We consider the following particular cases:

Case Ba $\quad \upsilon_C = 0$

Case Ba1 $\quad K = \upsilon_T + 2$

The zeroth-order vibronic wave function is $|\upsilon_T\ \upsilon_T\ 0\ 0 +\rangle$. The zeroth-order energy is

$$E^{(0)} = (\upsilon_T + 1)\,\omega_T + \omega_C \tag{B.2}$$

The first- and second-order energy corrections are

$$E^{(1)} = (\upsilon_T + 1)(\upsilon_T + 2)a_T\omega_T + 2a_C\omega_C + (\upsilon_T + 1)b_0\sqrt{\omega_T\omega_C} + 2\Sigma A_{\text{so}}$$

$$\begin{aligned} E^{(2)} = & -\frac{1}{2}(\upsilon_T + 1)(\upsilon_T + 2)(4\upsilon_T + 9)a_T^2\omega_T - 9a_C^2\omega_C \\ & -\frac{1}{2}(\upsilon_T + 1)b_0^2\left[(\upsilon_T + 1)\omega_T + \omega_C + \frac{\omega_T\omega_C}{\omega_T + \omega_C}\right] \\ & - b_2^2\left\{4\upsilon_T\omega_T + [(\upsilon_T + 1)(\upsilon_T + 2) + 2]\frac{\omega_T\omega_C}{\omega_T + \omega_C} + \upsilon_T(\upsilon_T - 1)\frac{\omega_T\omega_C}{\omega_C - \omega_T}\right\} \\ & - 2(\upsilon_T + 1)(\upsilon_T + 2)a_T b_0\sqrt{\omega_T\omega_C} - 4(\upsilon_T + 1)a_C b_0\sqrt{\omega_T\omega_C} \\ & -\frac{1}{4}(\upsilon_T + 1)(\upsilon_T + 2)(\upsilon_T + 3)(\upsilon_T + 4)c_T^2\omega_T - 6c_C^2\omega_C \\ & - (\upsilon_T + 1)(\upsilon_T + 2)c_{TC}^2\frac{\omega_T\omega_C}{\omega_T + \omega_C} \end{aligned} \tag{B.3}$$

Case Ba2 $\quad K = \upsilon_T$

The zeroth-order wave function is $|\upsilon_T\ \upsilon_T - 2\ 0\ 0\ +\rangle$

$$E^{(1)} = \upsilon_T(\upsilon_T + 5)a_T\omega_T + 2a_C\omega_C + (\upsilon_T + 1)b_0\sqrt{\omega_T\omega_C} + 2\Sigma A_{\text{so}}$$

$$\begin{aligned} E^{(2)} = & -\frac{1}{2}\upsilon_T(\upsilon_T + 1)(4\upsilon_T + 35)a_T^2\omega_T - 9a_C^2\omega_C \\ & -\frac{1}{2}b_0^2\left[(\upsilon_T + 1)^2\omega_T + (\upsilon_T + 1)\omega_C + \frac{2\upsilon_T\omega_T\omega_C}{\omega_T + \omega_C} + \frac{(\upsilon_T - 1)\omega_T\omega_C}{\omega_C - \omega_T}\right] \\ & - b_2^2\left[(12\upsilon_T - 8)\omega_T + (\upsilon_T^2 + \upsilon_T + 6)\frac{\omega_T\omega_C}{\omega_T + \omega_C}\right. \\ & \left. + (\upsilon_T - 1)(\upsilon_T - 2)\frac{\omega_T\omega_C}{\omega_C - \omega_T}\right] \\ & - 2\upsilon_T(\upsilon_T + 5)a_T b_0\sqrt{\omega_T\omega_C} - 4(\upsilon_T + 1)a_C b_0\sqrt{\omega_T\omega_C} \\ & -\frac{1}{4}\upsilon_T(\upsilon_T + 1)(\upsilon_T + 2)(\upsilon_T + 35)c_T^2\omega_T - 6c_C^2\omega_C \\ & - c_{TC}^2\left[4\upsilon_T\omega_T + \upsilon_T(\upsilon_T + 1)\frac{\omega_T\omega_C}{\omega_C + \omega_T}\right] \end{aligned} \tag{B.4}$$

Case Ba3 $\quad K < \upsilon_T$

The zeroth-order energy level is twofold degenerate. The corresponding vibronic basis functions are $|\upsilon_T\ K{+}2\ 0\ 0\ -\rangle \equiv |1\rangle$ and $|\upsilon_T\ K{-}2\ 0\ 0\ +\rangle \equiv |2\rangle$. The first-order energy correction is

$$E^{(1)}_{1/2} = \frac{1}{2}(3\upsilon_T^2 + 6\upsilon_T - K^2)a_T\omega_T + 2a_C\omega_C + (\upsilon_T + 1)b_0\sqrt{\omega_T\omega_C} \mp \frac{1}{2}D \quad \text{(B.5)}$$

where

$$D = \sqrt{16(Ka_T\omega_T + \Sigma A_{\text{so}})^2 + 9(\upsilon_T^2 - K^2)[(\upsilon_T + 2)^2 - K^2]c_T^2\omega_T^2} \qquad \text{(B.6)}$$

The corresponding vibronic wave functions are of the form (A.13) and (A.14) with D given by (B.6) and

$$B = -4(Ka_T\omega_T + \Sigma A_{\text{so}}) \qquad \text{(B.7)}$$

The second-order energy corrections are of the form

$$\begin{aligned} E_1^{(2)} &= c_{11}^2H_{11} + c_{12}^2H_{22} + 2c_{11}c_{12}H_{12} \\ E_2^{(2)} &= c_{21}^2H_{11} + c_{22}^2H_{22} + 2c_{21}c_{22}H_{12} \end{aligned} \qquad \text{(B.8)}$$

where

$$\begin{aligned} H_{11/22} = &-\frac{1}{4}(\upsilon_T + 1)[17\upsilon_T(\upsilon_T + 2) - 9K(K \pm 4)]a_T^2\omega_T - 9a_C^2\omega_C \\ &-\frac{1}{8}b_0^2\Big\{4(\upsilon_T + 1)^2\omega_T + 4(\upsilon_T + 1)\omega_C \\ &+ [\upsilon_T^2 - (K \pm 2)^2]\frac{\omega_T\omega_C}{\omega_C - \omega_T} + (\upsilon_T \mp K)(\upsilon_T \pm k + 4)\frac{\omega_T\omega_C}{\omega_C + \omega_T}\Big\} \\ &-\frac{1}{2}b_2^2\Big\{4[\upsilon_T(\upsilon_T + 2) - K(K \pm 4) - 4]\omega_T + [\upsilon_T(\upsilon_T - 2) \\ &+ K(K \pm 4) + 4]\frac{\omega_T\omega_C}{\omega_C - \omega_T} + [\upsilon_T(\upsilon_T + 6) + K(K \pm 4) + 12]\frac{\omega_T\omega_C}{\omega_C + \omega_T}\Big\} \\ &- [3\upsilon_T(\upsilon_T + 2) - K(K \pm 4)]a_Tb_0\sqrt{\omega_T\omega_C} - 4(\upsilon_T + 1)a_Cb_0\sqrt{\omega_T\omega_C} \\ &-\frac{1}{8}(\upsilon_T + 1)(2 \mp K)c_T^2\omega_T[17\upsilon_T(\upsilon_T + 2) - 3K(5K \pm 12)] - 6c_C^2\omega_C \\ &-\frac{1}{4}c_{TC}^2\{4(\upsilon_T \mp K)(\upsilon_T \pm K + 2)\omega_T + (\upsilon_T \pm K)(\upsilon_T \pm K + 2)\frac{\omega_T\omega_C}{\omega_C - \omega_T} \\ &+ (\upsilon_T \mp K)(\upsilon_T \mp K + 2)\frac{\omega_T\omega_C}{\omega_C + \omega_T}\Big\} \end{aligned} \qquad \text{(B.9)}$$

and

$$H_{12} = -\sqrt{(\upsilon_T^2 - K^2)[(\upsilon_T + 2)^2 - K^2]}\left\{\frac{17}{2}(\upsilon_T + 1)a_T c_T \omega_T + 3b_0 c_T\sqrt{\omega_T\omega_C} + \frac{1}{2}b_2 c_{TC}\left(4\omega_T + \frac{\omega_T\omega_C}{\omega_C - \omega_T} + \frac{\omega_T\omega_C}{\omega_C + \omega_T}\right)\right\} \quad \text{(B.10)}$$

If we put $\omega_T = \omega,\ \omega_C = 0, a_T = a,\ c_T = c,\ a_C = b_0 = b_2 = c_C = c_{TC} = 0$ all the formulas for case (Ba) reduce to those describing the R–T effect in triatomic molecules.

Case Bb $\quad K = \upsilon_T + \upsilon_C + 2 (= K_{\max})$

The zeroth-order vibronic wave function is $|\upsilon_T\ \upsilon_T\ \upsilon_C\ \upsilon_C\ +\rangle$. The zeroth-order energy is

$$E^{(0)} = (\upsilon_T + 1)\omega_T + (\upsilon_C + 2)\omega_C \quad \text{(B.11)}$$

The first- and second-order energy corrections are

$$E^{(1)} = (\upsilon_T + 1)(\upsilon_T + 2)a_T\omega_T + (\upsilon_C + 1)(\upsilon_C + 2)a_C\omega_C + (\upsilon_T + 1)(\upsilon_C + 1)b_0\sqrt{\omega_T\omega_C} + 2\Sigma A_{so} \quad \text{(B.12)}$$

$$\begin{aligned} E^{(2)} = & -\frac{1}{2}(\upsilon_T + 1)(\upsilon_T + 2)(4\upsilon_T + 9)a_T^2\omega_T - \frac{1}{2}(\upsilon_C + 1)(\upsilon_C + 2)(4\upsilon_C + 9)a_C^2\omega_C \\ & -\frac{1}{2}b_0(\upsilon_T + 1)(\upsilon_C + 1)\left[(\upsilon_T + 1)\omega_T + (\upsilon_C + 1)\omega_C + \frac{\omega_T\omega_C}{\omega_C + \omega_T}\right] \\ & - b_2^2\Big\{2\upsilon_T(\upsilon_C + 1)(\upsilon_C + 2)\omega_T + 2\upsilon_C(\upsilon_T + 1)(\upsilon_T + 2)\omega_C \\ & + (\upsilon_T - \upsilon_C)(2\upsilon_T\upsilon_C + \upsilon_T + \upsilon_C - 1)\frac{\omega_T\omega_C}{\omega_C - \omega_T} \\ & + [(\upsilon_T + 1)(\upsilon_T + 2) + (\upsilon_C + 1)(\upsilon_C + 2)]\frac{\omega_T\omega_C}{\omega_C + \omega_T}\Big\} \\ & - 2(\upsilon_T + 1)(\upsilon_T + 2)(\upsilon_C + 1)a_T b_0\sqrt{\omega_T\omega_C} \\ & - 2(\upsilon_T + 1)(\upsilon_C + 1)(\upsilon_C + 2)a_C b_0\sqrt{\omega_T\omega_C} \\ & -\frac{1}{4}(\upsilon_T + 1)(\upsilon_T + 2)(\upsilon_T + 3)(\upsilon_T + 4)c_T^2\omega_T \\ & -\frac{1}{4}(\upsilon_C + 1)(\upsilon_C + 2)(\upsilon_C + 3)(\upsilon_C + 4)c_C^2\omega_C \\ & -\frac{1}{2}(\upsilon_T + 1)(\upsilon_T + 2)(\upsilon_C + 1)(\upsilon_C + 2)c_{TC}^2\frac{\omega_T\omega_C}{\omega_C + \omega_T} \end{aligned} \quad \text{(B.13)}$$

For $\upsilon_C = 0$, the formulas (B.12) and (B.13) reduce to (B.3).

Case Bc $\quad K = \upsilon_T + \upsilon_C (= K_{\max} - 2)\upsilon_C > 0$

The zeroth-order level is twofold degenerate. The corresponding vibronic basis functions are $|\upsilon_T\ \upsilon_T\ \upsilon_C\ \upsilon_C{-}2\ +\rangle \equiv |1\rangle$ and $|\upsilon_T\ \upsilon_T{-}2\upsilon_C\ \upsilon_C\ +\rangle \equiv |2\rangle$. The zeroth-order energy is (B.11). The first-order energy correction is

$$E^{(1)}_{1/2} = (\upsilon_T^2 + 4\upsilon_T + 1)a_T\omega_T + (\upsilon_C^2 + 4\upsilon_C + 1)a_C\omega_C + (\upsilon_T + 1)(\upsilon_C + 1)b_0\sqrt{\omega_T\omega_C} \mp \frac{1}{2}D + 2\Sigma A_{\text{so}} \tag{B.14}$$

where

$$D = 2\sqrt{[(\upsilon_T - 1)a_T\omega_T - (\upsilon_C - 1)a_C\omega_C]^2 + 16\upsilon_T\upsilon_C b_2^2\omega_T\omega_C} \tag{B.15}$$

The corresponding wave functions have the form (A.13) and (A.14). In the present case, D is given by (B.15) and

$$B = -2(\upsilon_T - 1)a_T\omega_T + 2(\upsilon_C - 1)a_C\omega_C \tag{B.16}$$

The second-order energy corrections have the form (B.8) with

$$\begin{aligned}
H_{11} = &-\frac{1}{2}(\upsilon_T + 1)(\upsilon_T + 2)(4\upsilon_T + 9)a_T^2\omega_T - \frac{1}{2}\upsilon_C(\upsilon_C + 1)(4\upsilon_C + 35)a_C^2\omega_C \\
&- \frac{1}{2}b_0^2(\upsilon_T + 1)\left[(\upsilon_T + 1)(\upsilon_C + 1)\omega_T + (\upsilon_C + 1)^2\omega_C - (\upsilon_C - 1)\frac{\omega_T\omega_C}{\omega_C - \omega_T}\right. \\
&\left. + 2\upsilon_C\frac{\omega_T\omega_C}{\omega_C + \omega_T}\right] - b_2^2\left[2(\upsilon_T^2\upsilon_C - 2\upsilon_T^2 + 7\upsilon_T\upsilon_C - 6\upsilon_T + 6\upsilon_C - 4)\omega_C\right. \\
&+ 2\upsilon_T\upsilon_C(\upsilon_C + 1)\omega_T + (3\upsilon_T^2 + \upsilon_C^2 + 9\upsilon_T + \upsilon_C + 6)\frac{\omega_T\omega_C}{\omega_C + \omega_T} \\
&\left. + (2\upsilon_T^2\upsilon_C - 2\upsilon_T\upsilon_C^2 - \upsilon_T^2 - \upsilon_C^2 + 4\upsilon_T\upsilon_C - 3\upsilon_T + 3\upsilon_C - 2)\frac{\omega_T\omega_C}{\omega_C - \omega_T}\right] \\
&- 2(\upsilon_T + 1)(\upsilon_T + 2)(\upsilon_C + 1)a_T b_0\sqrt{\omega_T\omega_C} - 2\upsilon_C(\upsilon_T + 1)(\upsilon_C + 5)a_C b_0\sqrt{\omega_T\omega_C} \\
&- \frac{1}{4}(\upsilon_T + 1)(\upsilon_T + 2)(\upsilon_T + 3)(\upsilon_T + 4)c_T^2\omega_T - \frac{1}{4}\upsilon_C(\upsilon_C + 1)(\upsilon_C + 2)(\upsilon_C + 35)c_C^2\omega_C \\
&- \frac{1}{2}c_{TC}^2\upsilon_C(\upsilon_T + 1)(\upsilon_T + 2)\left[4\omega_C + (\upsilon_C + 1)\frac{\omega_T\omega_C}{\omega_C + \omega_T}\right]
\end{aligned} \tag{B.17}$$

and

$$H_{12} = -\sqrt{\upsilon_T \upsilon_C}\{12(\upsilon_T + 1)a_T b_2 \sqrt{\omega_T \omega_C} + 12(\upsilon_C + 1)a_C b_2 \sqrt{\omega_T \omega_C}$$
$$+ b_0 b_2 \left[4(\upsilon_T + 1)\omega_T + 4(\upsilon_C + 1)\omega_C + (\upsilon_T + \upsilon_C + 2)\frac{\omega_T \omega_C}{\omega_C + \omega_T}\right.$$
$$\left. + (\upsilon_T - \upsilon_C)\frac{\omega_T \omega_C}{\omega_C - \omega_T}\right] + 4(\upsilon_T + 1)(\upsilon_T + 2)c_T c_{TC}\sqrt{\omega_T \omega_C}$$
$$+ 4(\upsilon_C + 1)(\upsilon_C + 2)c_C c_{TC}\sqrt{\omega_T \omega_C}\} \quad \text{(B.18)}$$

The expression for H_{22} is obtained by interchanging indexes T and C on the right-hand side of Eq. (B.17) ($c_{CT} \equiv c_{TC}$). For $\upsilon_C = 0$, $E_2^{(2)} = H_{22}$ and the second-order energy formula for E_2 reduces to that derived for the case Ba2.

Case Bd $\quad \upsilon_T = 1,\ \upsilon_C = 1,\ K = 0$

In other cases, the zeroth-order vibronic levels are generally more than twofold degenerate and the perturbative handling is much more complicated. An exception is the case $\upsilon_T = 1$, $\upsilon_C = 1$, $K = 0$ with the twofold degenerate zeroth-order level. The basis functions are $|1\ 1\ 1\ 1 -\rangle \equiv |1\rangle$ and $|1\ -1\ 1\ -1 +\rangle \equiv |2\rangle$. The zeroth-order energy is

$$E^{(0)} = 2\omega_T + 2\omega_C \quad \text{(B.19)}$$

The first-order energy correction is

$$E_{1/2}^{(1)} = 6a_T\omega_T + 6a_C\omega_C + 4b_0\sqrt{\omega_T\omega_C} \mp \frac{1}{2}D \quad \text{(B.20)}$$

where

$$D = 4\sqrt{\Sigma^2 A_{so}^2 + 4c_{TC}^2\omega_T\omega_C} \quad \text{(B.21)}$$

The second-order energy correction is

$$E_{1/2}^{(2)} = H_{11} \mp \frac{8c_{TC}\sqrt{\omega_T\omega_C}}{D}H_{12} \quad \text{(B.22)}$$

with

$$
\begin{aligned}
H_{11} = & -39a_T^2\omega_T - 39a_C^2\omega_C - 2b_0^2\left(2\omega_T + 2\omega_C + \frac{\omega_T\omega_C}{\omega_C+\omega_T}\right) \\
& - 12b_2^2\left(\omega_T + \omega_C + \frac{\omega_T\omega_C}{\omega_C+\omega_T}\right) - 24a_Tb_0\sqrt{\omega_T\omega_C} - 24a_Cb_0\sqrt{\omega_T\omega_C} \\
& - 54c_T^2\omega_T - 54c_C^2\omega_C - 2c_{TC}^2\left(2\omega_T + 2\omega_C + \frac{\omega_T\omega_C}{\omega_C+\omega_T}\right)
\end{aligned} \tag{B.23}
$$

$$
\begin{aligned}
H_{12} = & -24(a_Tc_{TC} + a_Cc_{TC} + 2b_2c_T + 2b_2c_C)\sqrt{\omega_T\omega_C} \\
& - 8b_0c_{TC}\omega_T - 8b_0c_{TC}\omega_C - 4b_0c_{TC}\frac{\omega_T\omega_C}{\omega_C+\omega_T}
\end{aligned} \tag{B.24}
$$

Acknowledgments

We wish to thank all our colleagues who collaborated with us in the calculations presented in the calculations presented in this chapter. One of us (M.P.) likes to thank the Deutsche Forschungsgemeinschaft for financial support.

References

1. P. Jensen and P. R. Buenker, eds., *Computational Molecular Spectroscopy*, John Wiley and Sons, Inc., 2000.
2. P. Jensen, G. Osmann, and P. R. Buenker, in *Computational Molecular Spectroscopy*, P. Jensen and R. J. Bunker, eds., John Wiley & Sons, Inc., New York 2000, p. 485.
3. J. M. Brown, in *Computational Molecular Spectroscopy*, P. Jensen and R. J. Bunker, eds., John Wiley & Sons, Inc., New York 2000, p. 517.
4. G. Herzberg and E. Teller, *Z. Phys. Chem.* (B) **21**, 410 (1933).
5. G. Herzberg, *Moleculer Spectra and Molecular Structure III. Electronic Spectra of Polyatomic Molecules*, Van Nostrand, New York, 1967.
6. M. Born and R. Oppenheimer, *Ann. Phys (Leipzig)* **84**, 457 (1927).
7. R. Renner, *Z. Phys.* **92**, 172 (1934).
8. Ch. Jungen and A. J. Merer, in *Molecular Spectroscopy, Modern Research*, K. N. Rao, ed., Vol. 2, Academic Press, New York, 1977, p. 127.
9. K. Dressler and D. A. Ramsay, *J. Chem. Phys.* **27**, 971 (1957).
10. K. Dressler and D. A. Ramsay, *Philos. Trans. R. Soc. Ser. A* **251**, 553 (1958).
11. G. Duxbury, *Molecular Spectroscopy*, Vol. 3, Billing & Sons, Guilford and London, 1975, p. 497.
12. J. M. Brown and F. Jørgensen, in *Advances in Chemical Physics*, I. Prigogine and S. A. Rice, eds., John Wiley & Sons, Inc., New York, 1983, Vol. 52, p. 117.
13. H. Köppel, W. Domcke, and L. S. Cederbaum, *Advances in Chemical Physics*, I. Prigogine and A. C. Rice, eds., John Wiley, New York, 1986, Vol. 67, p. 59.
14. H. Köppell and W. Domcke, *Encyclopedia of Computational Chemistry*, P. V. R. Schleyer et al., eds., Wiley, New York, 1999, p. 3166.

15. P. R. Buenker and P. Jensen, *Molecular Symmetry and Spectroscopy*, NCR Research Press, Ottawa, 1998.
16. M. Perić, S. D. Peyerimhoff, and R. J. Buenker, *Int. Rev. Phys. Chem.* **4**, 85 (1985).
17. M. Perić, B. Engels, and S. D. Peyerimhoff, *Quantum Mechanical Electronic Structure Calculations with Chemical Accuracy*, S. R. Langhoff, ed., Kluwer, Dordrecht, 1995, p. 261.
18. M. Perić, B. Ostojić, and J. Radić-Perić, *Phys. Rep.* **290**, 283 (1997).
19. J. M. Brown, *J. Mol. Spectrosc.* **68**, 412 (1977).
20. R. J. Buenker, M. Perić, S. D. Peyerimhoff, and R. Marian, *Mol. Phys.* **43**, 987 (1981).
21. M. Perić, S. D. Peyerimhoff, and R. J. Buenker, *Mol. Phys.* **49**, 379 (1983).
22. W. Pauli, *Z. Phys.* **43**, 601 (1927); G. Breit, *Phys. Rev.* **34**, 553 (1930).
23. B. A. Hess, C. M. Marian, and S. D. Peyerimhoff, *Advanced Series in Physical Chemistry: Modern Electronic Structure Theory*, C.-Y. Ng and D. R. Yarkony, eds., Singapore: World Scientific, 1995 p. 152; B. A. Hess and C. M. Marian, in *Computational Molecular Spectroscopy*, P. Jensen and P. R. Bunker, eds., John Wiley & Sons, Inc., 2000, p. 169.
24. Ch. Jungen and A. J. Merer, *Mol. Phys.* **40**, 1 (1980).
25. Ch. Jungen, K-E. J. Hallin, and A. J. Merer, *Mol. Phys.* **40**, 25 (1980).
26. Ch. Jungen, K-E. J. Hallin, and A. J. Merer, *Mol. Phys.* **40**, 65 (1980).
27. Ch. Jungen and A. J. Merer, *Mol. Phys.* **40**, 95 (1980).
28. M. Perić, M. Krmar, J. Radić- Perić, and Lj. Stevanović, *J. Mol. Spectrosc.* **208** 271 (2001).
29. M. Perić and M. Krmar, *J. Serb. Chem. Soc.* **66** 613 (2000).
30. S. A. Beaton, Y. Ito, and J. M. Brown, *J. Mol. Spectrosc.* **178**, 99 (1996).
31. S. A. Beaton and J. M. Brown, *J. Mol. Spectrosc.* **183**, 347 (1997).
32. J. T. Hougen, *J. Chem. Phys.* **36**, 1874 (1962).
33. M. Perić, S. D. Peyerimhoff, and R. J. Buenker, *Can. J. Chem.* **59**, 1318 (1981).
34. M. Perić and M. Krmar, *Bull. Soc. Chim. Beograd.* **47**, 43 (1982).
35. R. N. Dixon, *Mol. Phys.* **9**, 357 (1965).
36. C. F. Bender and H. F. Schaefer, III, *J. Mol. Spectrosc.* **37**, 423 (1971); V. Staemmler and M. Jungen, *Chem. Phys. Lett.* **16**, 187 (1972).
37. G. Herzberg and J. W. C. Johns, *Proc. R. Soc. London Ser. A* **298**, 142 (1967).
38. C. Eckart, *Phys. Rev.* **47**, 552 (1935).
39. E. B. Wilson and J. B. Howard, *J. Chem. Phys.* **4**, 269 (1936).
40. E. B. Wilson, E. B. Decius, and P. C. Cross, *Molecular Vibrations*, McGraw-Hill, New York, 1955.
41. J. K. G. Watson, *Mol. Phys.* **15**, 479 (1968).
42. J. K. G. Watson, *Mol. Phys.* **19**, 465 (1970).
43. A. R. Hoy, I. M. Mills, and G. Strey, *Mol. Phys.* **24**, 1265 (1972).
44. J. Tennyson, B. T. Sutcliffe, *J. Mol. Spectrosc.* **101**, 71 (1983).
45. D. Estes and D. Secrest, *Mol. Phys.* **59**, 569 (1986).
46. H. Wei and T. Carrington, Jr., *J. Chem. Phys.* **107**, 2813 (1997); **107**, 9493 (1997).
47. M. Mladenović, *J. Chem. Phys.* **112**, 1070 (2000); **112**, 1082 (2000), **113**, 10524 (2000).
48. S. Carter and N. C. Handy, *Mol. Phys.* **47**, 1445 (1982).
49. S. Carter, N. C. Handy, and B. T. Sutcliffe, *Mol. Phys.* **49**, 745 (1983).
50. S. Carter and N. C. Handy, *Mol. Phys.* **52**, 1367 (1984).

51. S. Carter, N. C. Handy, P. Rosmus, and G. Chambaud, *Mol. Phys.* **71**, 605 (1990).

52. G. D. Carney, L.L Sprandel, and C. W. Kern, *Adv. Chem. Phys.* **37**, 305 (1968).

53. E. K. C. Lai, Master's Thesis, Department of Chemistry, Indiana University, Bloomington, IN, 1975.

54. B. Podolsky, *Phys. Rev.* **32**, 812 (1928).

55. P. R. Bunker and P. Jensen, *Molecular Symmetry and Spectroscopy*, NRC Research Press, Ottawa. 1998.

56. M. Peric, M. Mladenovic, S. D. Peyerimhoff, and R. J. Buenker, *Chem. Phys.* **82**, 317 (1983).

57. J. H. Van Vleck, *Rev. Mod. Phys.* **23**, 213 (1951).

58. J. M. Brown and B. J. Howard, *Mol. Phys.* **31**, 1517 (1976).

59. R. N. Zare, *Angular Momentum*, John Wiley & Sons, Inc., 1988.

60. J. T. Hougen, P. R. Bunker, and J. W. C. Johns, *J. Mol. Spectrosc.* **34**, 136 (1970).

61. P. R. Bunker and J. M. R. Stone, *J. Mol. Spectrosc.* **41**, 310 (1972).

62. A. R. Hoy and P. R. Bunker, *J. Mol. Spectrosc.* **52**, 439 (1974).

63. P. R. Bunker and B. M. Landsberg, *J. Mol. Spectrosc.* **67**, 374 (1977).

64. A. R. Hoy and P. R. Bunker, *J. Mol. Spectrosc.* **74**, 1 (1979).

65. P. Jensen and P. R. Bunker, *J. Mol. Spectrosc.* **118**, 18 (1986).

66. P. Jensen, *J. Mol. Spectrosc.* **128**, 478 (1988).

67. P. Jensen, *J. Chem. Soc. Farady. Trans. 2* **84**, 1315 (1988).

68. A. Sayvetz, *J. Chem. Phys.* **7**, 383 (1939).

69. J. A. Pople and H. C. Longuet-Higgins, *Mol. Phys.* **1**, 372 (1958).

70. A. D. Walsh, *J. Chem. Soc.* 2260, (1953).

71. J. A. Pople, *Mol. Phys.* **3**, 16 (1960).

72. J. T. Hougen, *J. Chem. Phys.* **36**, 519 (1962).

73. A. J. Merer and D. N. Travis, *Can. J. Phys.* **43**, 1795 (1965).

74. R. Barrow, R. N. Dixon, and G. Duxbury, *Mol. Phys.* **27**, 1217 (1974).

75. K. F. Freed and J. R. Lombardi, *J. Chem. Phys.* **45**, 591 (1966).

76. J. W. Cooley, *Math. Comp.* **15**, 363 (1961).

77. I. Dubois, G. Duxbury, and R. N. Dixon, *J. Chem. Soc. Faraday Trans. 2*, **71**, 799 (1975).

78. G. Duxbury and R. N. Dixon, *Mol. Phys.* **43**, 255 (1981).

79. G. Duxbury, *J. Chem. Soc. Faraday Trans.* 2, **78**, 1433 (1982).

80. G. Duxbury and Ch. Jungen, *Mol. Phys.* **63**, 981 (1988).

81. A. Alijah and G. Duxbury, *Mol. Phys.* **70**, 605 (1990).

82. G. Duxbury, A. Alijah, and R. Trieling, *J. Chem. Phys.* **63**, 981 (1993).

83. G. Duxbury, B. McDonald, and A. Alijah, *Mol. Phys.* **89**, 767 (1996).

84. G. Duxbury, B. McDonald, M. Van Gogh, A. Alijah, CH. Jungen, and H. Palivan, *J. Chem. Phys.* **108**, 2336 (1998).

85. G. Duxbury, A. Alijah, B. McDonald, and Ch. Jungen, *J. Chem. Phys.* **108**, 2336 (1998).

86. M. Perić, R. J. Buenker, and S. D. Peyerimhoff, *Mol. Phys.* **59**, 1283 (1986) .

87. W. Reuter, M. Perić, and S. D. Peyerimhoff, *Mol. Phys.* **74**, 569 (1991).

88. M. Brommer, B. Weis, B. Follmeg, P. Rosmus, S. Carter, N. C. Handy, H.-J. Werner, and P. J. Knowles, *J. Chem. Phys.* **98**, 5222 (1993).

89. R. Bartholomae, D. Martin, and B. T. Sutcliffe, *J. Mol. Spectrosc.* **87**, 367 (1981).
90. B. T. Sutcliffe, *Mol. Phys.* **48**, 561 (1983).
91. W. H. Green, Rr., N. C. Handy, P. J. Knowles, and S. Carter, *J. Chem. Phys.* **94**, 118 (1991).
92. M. Brommer, G. Chambaud, E.-A. Reinsch, P. Rosmus, A. Spielfiedel, N. Feautrier, and H.-J. Werner, *J. Chem. Phys.* **94**, 8070 (1991).
93. G. Chambaud and P. Rosmus, *J. Chem. Phys.* **96**, 77 (1992).
94. M. Brommer, P. Rosmus, S. Carter, and N. C. Handy, *Mol. Phys.* **77**, 549 (1992).
95. M. Brommer and P. Rosmus, *J. Chem. Phys.* **98**, 7746 (1993).
96. B. Weis and K. Yamashita. *J. Chem. Phys.* **99**, 9512 (1993).
97. W. Gabriel, G. Chambaud, P. Rosmus, S. Carter, and N. C. Handy, *Mol. Phys.* **81**, 1445 (1994).
98. J. Persson, B. O. Roos, and S. Carter, *Mol. Phys.* **84**, 619 (1995).
99. G. Chambaud, P. Rosmus, M. L. Senent, and P. Palmieri, *Mol. Phys.* **92**, 399 (1997).
100. M. Hochlaf, G. Chambaud, and P. Rosmus, *J. Chem. Phys.* **108**, 4047 (1998).
101. D. Panten, G. Chambaud, P. Rosmus, and P. J. Knowles, *Chem. Phys. Lett.* **311**, 390 (1994).
102. P. Jensen, M. Brumm, W. P. Kraemer, and P. R. Bunker, *J. Mol. Spectrosc.* **171**, 31 (1995).
103. P. Jensen, M. Brumm, W. P. Kraemer, and P. R. Bunker, *J. Mol. Spectrosc.* **172**, 194 (1995).
104. M. Kolbuszewski, P. R. Bunker, W. P. Kraemer, G. Osmann, and P. Jensen, *Mol. Phys.* **88**, 105 (1996).
105. G. Osmann, P. R. Bunker, P. Jensen, and W. P. Kraemer, *Chem. Phys. Lett.* **171**, 31 (1995).
106. A. Carrington, A. R. Fabris, B. J. Howard, and N. J. D. Lucas, *Mol. Phys.* **20**, 961 (1971).
107. P. S. H. Bolman, J. M. Brown, A. Carrinngton, A. Kopp, and D. A. Ramsay, *Proc. R. Soc. London, Ser. A*, **343**, 17 (1975).
108. J. M. Brown, *J. Mol. Spectrosc.* **56**, 159 (1975).
109. J. M. Brown M. Kaise, C. M. L. Kerr, and D. J. Milton, *Mol. Phys.* **36**, 553 (1978).
110. J. M. Brown and F. Jørgensen, *Mol. Phys.* **47**, 1065 (1982).
111. P. Crozet, A. J. Ross, R. Bacis, M. P. Barnes, and J. M. Brown, *J. Mol. Spectrosc.* **172**, 43 (1995).
112. M. Perić, R. J. Buenker, and S. D. Peyerimhoff, *Can. J. Chem.* **57**, 2491 (1979) .
113. P. J. Bruna, G. Hirsch, M. Peri, S. D. Peyerimhoff, and R. J. Buenker, *Mol. Phys.* **40**, 521 (1980).
114. H. Köppel, W. Domcke, and L. S. Cederbaum, *J. Chem. Phys.* **74**, 2945 (1980).
115. S.-K. Shih, S. D. Peyerimhoff, and R. J. Buenker, *J. Mol. Spectrosc.* **64**, 167 (1975); **74**, 124 (1979).
116. P. G. Carrick, A. J. Merer, and R. F. Curl, *J. Chem. Phys.* **78**, 3652 (1983); R. F. Curl, P. G. Carrick, and A. J. Merer, *J. Chem. Phys.* **82**, 3479 (1985).
117. M. Perić, S. D. Peyerimhoff, and R. Buenker, *Z. Phys. D*, **24**, 177 (1992).
118. H. Thümmel, M. Perić, S. D. Peyerimhoff, and R. J. Buenker, *Z. Phys. D*, **13**, 307 (1989).
119. M. Perić, R. Buenker, and S. D. Peyerimhoff, *Mol. Phys.* **71**, 673 (1990).
120. M. Perić, S. D. Peyerimhoff, and R. Buenker, *Mol. Phys.* **71**, 693 (1990).
121. M. Perić, S. D. Peyerimhoff, and R. Buenker, *J. Mol. Spectrosc.* **148**, 180 (1991).
122. M. Perić, W. Reuter, and S. D. Peyerimhoff, *J. Mol. Spectrosc.* **148**, 201 (1991).
123. M. Perić, B. Engels, and S. D. Peyerimhoff, *J. Mol. Spetrosc.* **150**, 56 (1991).
124. M. Perić, B. Engels, and S. D. Peyerimhoff, *J. Mol. Spetrosc.* **150**, 70 (1991).
125. C. Pfelzer, M. Havenith, M. Perić, P. Mürz, and W. Urban, *J. Mol. Spectrosc.* **176**, 28 (1996).

126. C. Schmidt, M. Perić, P. Mürz , M. Wienkoop, M. Havenith, and W. Urban, *J. Mol. Spectrosc.* **190**, 112 (1998).
127. Y.-C. Hsu, P.-R. Wang, M.-C. Yang, D. Papousek, Y.-J. Chen, and W.-Y. Chiang, *Chem. Phys. Lett.* **190**, 507 (1992).
128. Y.-C. Hsu, J. J.-M. Lin, D. Papousek, and Y.-J. Tsai, *J. Chem. Phys.* **98**, 6690 (1993).
129. D. Forney, M. E. Jacox, and W. E. Thompson, *J. Mol. Spectrosc.* **170**, 178 (1995). 32H.
130. S. Carter, N. C. Handy, C. Puzzarini, R. Tarroni, and P. Palmieri, *Mol. Phys.* **98**, 1697 (2000).
131. G. Herzberg, *Discuss. Faraday Soc.* **35**, 7 (1963).
132. A. N. Petelin and A. A. Kiselev, *Int. J. Quantum Chem.* **6**, 701 (1972).
133. R. Colin, M. Herman, and I. Kopp, *Mol. Phys.* **37**, 1397 (1979).
134. M. Takahashi, M. Fujii, and M. Ito, *J. Chem. Phys.* **96**, 6486 (1992).
135. Y. F. Zhu, R. Shedaded, and E. R. Grant, *J. Chem. Phys.* **99**, 5723 (1993).
136. J. E. Reutt, L. S. Wang, J. E. Pollard, D. J. Trevor, Y. T. Lee, and D. A. Shirley, *J. Chem. Phys.* **84**, 3022 (1986).
137. S. T. Pratt, P. M. Dehmer, and J. L. Dehmer, *J. Chem. Phys.* **95**, 6238 (1991).
138. S. T. Pratt, P. M. Dehmer, and J. L. Dehmer, *J. Chem. Phys.* **99**, 6233 (1993).
139. J. Tang and S. Saito, *J. Chem. Phys.* **105**, 8020 (1996).
140. M. Perić, S. D. Peyerimhoff, and R. J. Buenker, *Mol. Phys.* **55**, 1129 (1985).
141. M. Perić, H. Thümmel, C. M. Marian, and S. D. Peyerimhoff, *J. Chem. Phys.* **102**, 7142 (1995).
142. M. Perić, B. Engels, and S. D. Peyerimhoff, *J. Mol. Spectrosc.* **171**, 494 (1995).
143. M. Perić and B. Engels, *J. Mol. Spectrosc.* **174**, 334 (1995).
144. M. Perić, B. Ostojić, B. Schäfer, and B. Engels, *Chem. Phys.* **225**, 63 (1997).
145. B. Schäfer, M. Perić, and B. Engels, *J. Chem. Phys.* **110**, 7802 (1999).
146. M. N. R. Ashfold, B. Tutcher, B. Yang, Z. K. Jin, and L. S. Anderson, *J. Chem. Phys.* **87**, 5105 (1987).
147. M.-F. Jagod, M. Rösslein, C. M. Gabrys, B. D. Rehfuss, F. Scappini, M. W. Crofton, and T. Oka, *J. Chem. Phys.* **97**, 7111 (1992).
148. M. Perić and S. D. Peyerimhoff, *J. Chem. Phys.* **102**, 3685 (1995).
149. M. Perić, C. M. Marian, and B. Engels, *Mol. Phys.* **97**, 731 (1966).
150. M. Perić and B. Ostojić, *Mol. Phys.* **97**, 743 (1966).
151. B. Schäfer-Bung, B. Engels, T. R. Taylor, D. M. Neumark, P. Botschwina, and M. Perić, *J. Chem. Phys.* **115**, 1777 (2001).
152. M. Perić, C. M. Marian, and S. D. Peyerimhoff, *J. Chem. Phys.* **114**, 6086 (1999).
153. M. Perić, B. Ostojić, and B. Engels, *J. Chem. Phys.* **105** , 8569 (1966).
154. M. Perić and S. D. Peyerimhoff, *The Role of Rydberg States in Spectroscopy and Photochemistry*, C. Sandorfy, ed., Kluwer Academic Publishers, Dordrecht/Boston/London, 1999, p. 137.
155. S. Carter and N. C. Handy, *Mol. Phys.* **53**, 1033 (1984).
156. N. C. Handy, *Mol. Phys.* **61**, 207 (1987).
157. M. J. Bramley, W. H. Green, Jr., and N. C. Handy, *Mol. Phys.* **73**, 1183 (1991).
158. S. M. Colwell and N. C. Handy, *Mol. Phys.* **92**, 317 (1997).
159. K. Sarka, *J. Mol. Spectrosc.* **38**, 545 (1971).
160. P. R. Bunker, B. M. Landsberg, and B. P. Winnewisser, *J. Mol. Spectrosc.* **74**, 9 (1979).

161. S. L. N. G. Krishnamachari and T. V. Venkitachalam, *Chem. Phys. Lett.* **55**, 116 (1978).

162. S. L. N. G. Krishnamachari and D. A. Ramsay, *Discuss. Faraday Soc.* **71**, 205 (1981).

163. B. Coquart, *Can. J. Phys.* **63**, 1362 (1985).

164. J. M. Vrtilek, C. A. Gottlieb, E. W. Gottlieb, W. Wang, and P. Thaddeus, *Astrophys. J. Lett.* **398**, L73 (1992).

165. P. G. Szalay, *J. Chem. Phys.* **105**, 2735 (1996).

166. J. D. Goddard, *Chem. Phys. Lett.* **154**, 387 (1989).

167. P. G. Szalay and J.-P. Blaudeau, *J. Chem. Phys.* **106**, 436 (1997).

168. C. Petrongolo, *J. Chem. Phys.* **89**, 1297 (1988).

169. A. Aguilar, M. Gonzales, and L. Poluyanov, *Mol. Phys.* **72**, 193 (1992); **77**, 209 (1992); **81**, 655 (1993).

170. M. Perić and S. D. Peyerimhoff, *J. Mol. Spectrosc.* (2002), in press.

171. M. Perić and S. D. Peyerimhoff, *J. Mol. Spectrosc.* (2002), in press.

PERMUTATIONAL SYMMETRY AND THE ROLE OF NUCLEAR SPIN IN THE VIBRATIONAL SPECTRA OF MOLECULES IN DOUBLY DEGENERATE ELECTRONIC STATES: THE TRIMERS OF ^2S ATOMS

A. J. C. VARANDAS and Z. R. XU

Departamento de Química, Universidade de Coimbra
Coimbra, Portugal

CONTENTS

The Role of Degenerate States in Chemistry: A Special Volume of Advances in Chemical Physics, Volume 124, Edited by Michael Baer and Gert Due Billing. Series Editors I. Prigogine and Stuart A. Rice. ISBN 0-471-43817-0.

I. INTRODUCTION

The full quantum mechanical study of nuclear dynamics in molecules has received considerable attention in recent years. An important example of such developments is the work carried out on the prototypical systems H_3 [1–5] and its isotopic variant HD_2 [5–8], Li_3 [9–12], Na_3 [13,14], and HO_2 [15–18]. In particular, for the alkali metal trimers, the possibility of a conical intersection between the two lowest doublet potential energy surfaces introduces a complication that makes their theoretical study fairly challenging. Thus, alkali metal trimers have recently emerged as ideal systems to study molecular vibronic dynamics, especially the so-called geometric phase (GP) effect [13,19,20] (often referred to as the molecular Aharonov–Bohm effect [19] or Berry's phase effect [21]); for further discussion on this topic see [22–25], and references cited therein. The same features also turn out to be present in the case of HO_2, and their exact treatment assumes even further complexity [18].

For Li_3, Gerber and Schumacher [9] reported the lowest vibrational levels and showed that vibronic coupling is essential to describe the electronic ground state giving rise to the so-called dynamic Jahn–Teller effect. In turn, Mayer and Cederbaum [10] studied the rovibronic coupling in the electronic A system of Li_3. Most recently, Kendrick [14] reported quantum mechanical calculations on the vibrational spectrum of Na_3 using a generalized Born–Oppenheimer treatment. However, a question emerges when we carry out quantum mechanical calculations using a filter diagonalization [26] technique, namely, the efficient minimal residuals (MINRES) filter diagonalization method [11,27] hereafter referred to shortly as MFD. For example, for the vibrational states of the $^1H_3$ electronic ground state, one may compute the full spectrum of the corresponding Hamiltonian, and hence, the problem arises of whether all calculated eigenfunctions are "true" physical molecular vibrational states. We will provide an answer to this question in the following sections of this chapter.

Symmetry considerations have long been known to be of fundamental importance for an understanding of molecular spectra, and generally molecular dynamics [28–30]. Since electrons and nuclei have distinct statistical properties, the total molecular wave function must satisfy appropriate symmetry requirements. Thus, not all calculated states have to be physically acceptable states, and symmetry considerations may allow us to distinguish the "mathematical states" from the "physical states." In this chapter, we discuss the permutational symmetries of the total wave function and its various components for a molecule under the permutation of identical particles. Double group theory will be used as a powerful tool to analyze the molecular states, and an extension of Kramers' theorem [28,30] to its most general form presented. The significant role of nuclear spin will then be emphasized, and some severe consequences will be demonstrated. Thus, the material presented here may be helpful for a detailed understanding of molecular spectra and collisional dynamics.

II. TOTAL MOLECULAR WAVE FUNCTION

The molecular time-independent nonrelativistic Schrödinger equation assumes the form

$$\hat{H}\Omega(\mathbf{R}', \mathbf{i}, \mathbf{r}', \mathbf{s}) = E\Omega(\mathbf{R}', \mathbf{i}, \mathbf{r}', \mathbf{s}) \quad (1)$$

where $\Omega(\mathbf{R}', \mathbf{i}, \mathbf{r}', \mathbf{s})$ is the total molecular wave function, $\hat{H}$ is the total molecular Hamiltonian operator, and E is the total energy; $\mathbf{R}'$ and $\mathbf{r}'$ stand collectively for the nuclear and electronic coordinates in the space-fixed (SF) frame, and $\mathbf{i}$ and $\mathbf{s}$ denote the corresponding nuclear and electronic spin coordinates. For a system consisting of N nuclei and n electrons, there are $3N$ nuclear spacial coordinates and $3n$ electronic ones. In the case of a triatomic molecule, the six nuclear coordinates relative to the center of mass consist of three internal and three external coordinates. The former may be taken as the hyperspherical coordinates [2,31–35] (ρ, θ, ϕ), while the external or orientational coordinates are chosen to be the usual Euler angles (α, β, γ) [36]. As illustrated in Figure 1, these define the orientation of the body-fixed (BF) relative to the SF frames. In the following sections, we will differentiate between these two types of coordinates by expressing $\mathbf{R}' = (\mathbf{R}, \hat{\mathbf{R}})$, where $\mathbf{R} = (\rho, \theta, \phi)$ and $\hat{\mathbf{R}} = (\alpha, \beta, \gamma)$.

In the strictest meaning, the total wave function cannot be separated since there are many kinds of interactions between the nuclear and electronic degrees of freedom (see later). However, for practical purposes, one can separate the total wave function partially or completely, depending on considerations relative to the magnitude of the various interactions. Owing to the uniformity and isotropy of space, the translational and rotational degrees of freedom of an isolated molecule can be described by cyclic coordinates, and can in principle be separated. Note that the separation of the rotational degrees of freedom is not trivial [37].

Consider now the adiabatic approximation [38,39] to the solution of Schrödinger's equation. Such an approximation is based on the fact that the nuclear masses are much larger than the electronic ones and therefore, on average, the nuclei move much more slowly than the electrons. The latter are thus able to follow the nuclear displacements: Their distribution in space is determined by the instantaneous nuclear configuration. To a first approximation, the nuclei may then be regarded as fixed. Accordingly, the total molecular wave function can be divided in two parts: one refers to the electronic wave function $\psi_e(\mathbf{r}, \mathbf{s}; \mathbf{R})$, the other to the nuclear wave function $\chi_{\text{nuc}}(\mathbf{R}', \mathbf{i})$. Regarding the nuclear wave function, it is possible to separate the translational part if the interaction between the translational and the other (rotational and vibrational) nuclear degrees of freedom can be ignored. This case is typical in studies of spectroscopy and collisional dynamics where the measured properties depend on the motions of the interacting species relative to each other but not on the motion of the system as a whole (the space is assumed to be uniform and

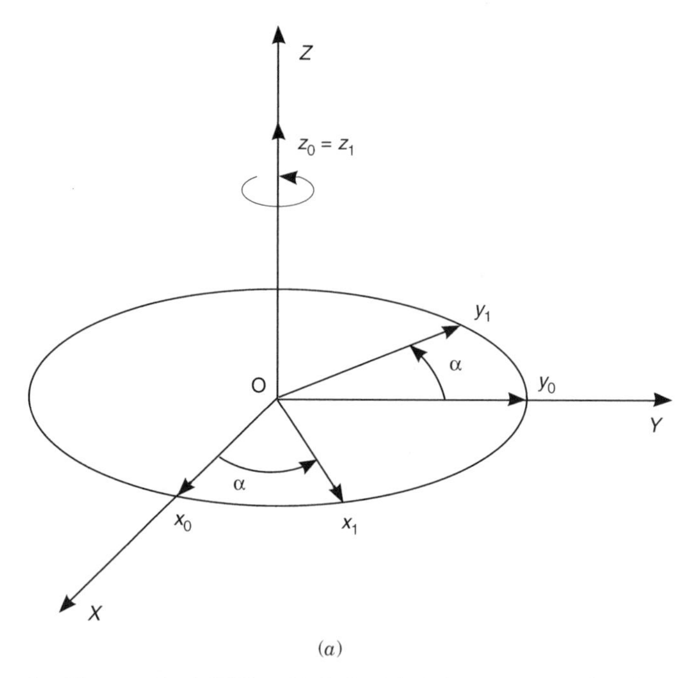

Figure 1. The space-fixed (XYZ) and body-fixed (xyz) frames. Any rotation of the coordinate system (XYZ) to (xyz) may be performed by three successive rotations, denoted by the Euler angles (α, β, γ), about the coordinate axes as follows: (a) rotation about the Z axis through an angle $\alpha(0 \leq \alpha < 2\pi)$, ($b$) rotation about the new y_1 axis through an angle $\beta(0 \leq \beta \leq \pi)$, ($c$) rotation about the new z_2 axis through an angle $\gamma(0 \leq \gamma < 2\pi)$. The relative orientations of the initial and final coordinate axes are shown in panel (d).

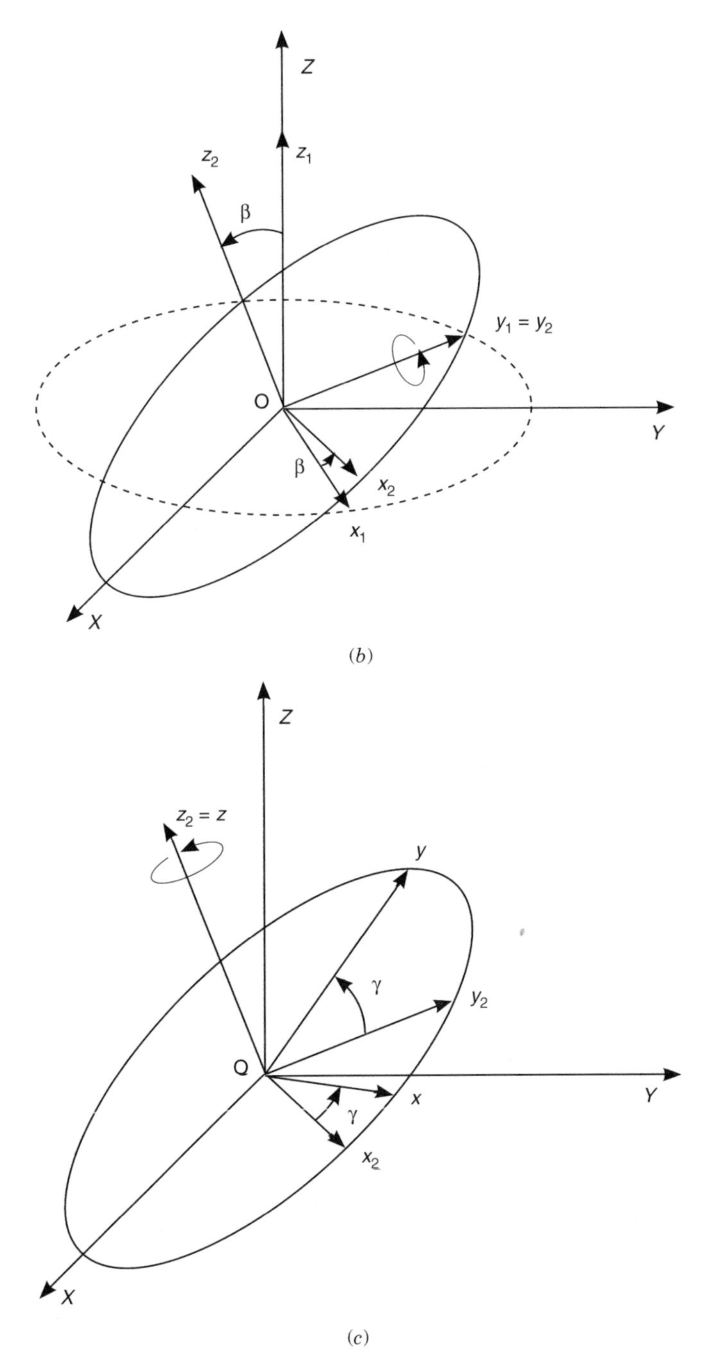

Figure 1 (*Continued*)

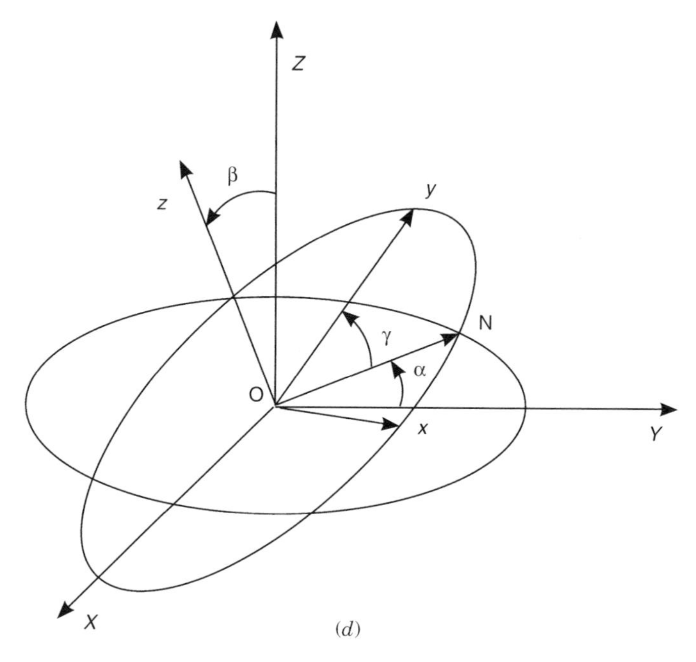

(*d*)

Figure 1 (*Continued*)

isotropic). In this case, after separation of the center of mass motion,[2] the total Hamiltonian operator can be written as

$$\hat{H} = -\frac{\hbar^2}{2\mu}\nabla^2 + \hat{H}_e(\mathbf{r}, \mathbf{s}; \mathbf{R}) \tag{2}$$

where $\hat{H}_e(\mathbf{r}, \mathbf{s}; \mathbf{R})$ is the electronic Hamiltonian that depends parametrically on the $\mathbf{R}$ coordinates. For a triatomic molecule, ∇^2 is the Laplacian with respect to the six nuclear coordinates $\mathbf{R}'$; μ is the three-body reduced mass $\mu = (m_A m_B m_C/M)^{1/2}$; m_A, m_B, and m_C are the masses of nuclei A, B, and C; and $M = m_A + m_B + m_C$. Note that we have separated the nuclear wave functions from the electronic wave functions, and also assumed that the electronic Hamiltonian is written in BF coordinates. Thus, the total molecular wave function can be expanded in the form [39]

$$\Omega(\mathbf{R}', \mathbf{i}, \mathbf{r}, \mathbf{s}) = \sum_n \chi_n(\mathbf{R}', \mathbf{i})\psi_n(\mathbf{r}, \mathbf{s}; \mathbf{R}) \tag{3}$$

[2] The problem of separating the center-of-mass motion in a molecular system is an intricate one that has no implications in the present work; the interested reader is referred to [40] for details.

where $\chi_n(\mathbf{R}', \mathbf{i})$ are the nuclear wave functions, and $\psi_n(\mathbf{r}, \mathbf{s}; \mathbf{R})$ form a complete set of electronic wave functions in BF (the summation should in principle include an integration over the continuum) obtained by solving, for each set of nuclear positions $\mathbf{R}$, the following eigenequation:

$$\hat{H}_e(\mathbf{r}, \mathbf{s}, \mathbf{R})\psi_n(\mathbf{r}; \mathbf{R}) = V_n(\mathbf{R})\psi_n(\mathbf{r}, \mathbf{s}; \mathbf{R}) \tag{4}$$

As usually indicated by the semicolon, both the wave functions and eigenvalues $[V_n(\mathbf{R})]$ depend parametrically on the internal nuclear coordinates.

Substitution of Eq. (3) into the molecular Schrödinger equation leads to a system of coupled equations in a coupled multistate electronic manifold

$$\left\{-\frac{\hbar^2}{2\mu}\left[\nabla^2 + 2\mathbf{F}(\mathbf{R}', \mathbf{i}) \cdot \nabla + \mathbf{G}(\mathbf{R}', \mathbf{i})\right] + \mathbf{V}(\mathbf{R})\right\}\chi(\mathbf{R}', \mathbf{i}) = \mathbf{E}\chi(\mathbf{R}', \mathbf{i}) \tag{5}$$

or, in compact form,

$$\mathbf{H}(\mathbf{R}', \mathbf{i})\chi(\mathbf{R}', \mathbf{i}) = \mathbf{E}\chi(\mathbf{R}', \mathbf{i}) \tag{6}$$

where $\chi(\mathbf{R}', \mathbf{i})$ is a column vector whose components are the nuclear wave functions $\chi_n(\mathbf{R}', \mathbf{i})$ and the matrix elements of $\mathbf{F}(\mathbf{R}', \mathbf{i})$, $\mathbf{G}(\mathbf{R}', \mathbf{i})$, and $\mathbf{V}(\mathbf{R})$ are given by

$$\mathbf{F}_{mn}(\mathbf{R}', \mathbf{i}) = \langle\psi_m(\mathbf{r}, \mathbf{s}; \mathbf{R})|\nabla\psi_n(\mathbf{r}, \mathbf{s}; \mathbf{R})\rangle \tag{7}$$

$$G_{mn}(\mathbf{R}', \mathbf{i}) = \langle\psi_m(\mathbf{r}, \mathbf{s}; \mathbf{R})|\nabla^2\psi_n(\mathbf{r}, \mathbf{s}; \mathbf{R})\rangle \tag{8}$$

$$V_{mn}(\mathbf{R}) = \langle\psi_m(\mathbf{r}, \mathbf{s}; \mathbf{R})|\hat{H}_e(\mathbf{r}, \mathbf{s}, \mathbf{R})|\psi_n(\mathbf{r}, \mathbf{s}; \mathbf{R})\rangle \tag{9}$$

where (and hereafter) the bra–ket notation $\langle|\rangle$ is used to specify integration over the electronic coordinates $\mathbf{r}$ and $\mathbf{s}$ only, and ∇ implies taking the gradient with respect to all the nuclear degrees of freedom. Note that the nonadiabatic coupling terms [of first-order, $\mathbf{F}_{mn}(\mathbf{R}', \mathbf{i})$, and second-order, $G_{mn}(\mathbf{R}', \mathbf{i})$] couple the various electronically adiabatic states, and hence are responsible for electronically nonadiabatic transitions. Note further that in the adiabatic approximation, the matrix formed by the elements $V_{mn} = V_n\delta_{mn}$ is diagonal, whereas the matrix

$$\mathbf{C} = -\frac{\hbar^2}{2\mu}\left[2\mathbf{F}(\mathbf{R}', \mathbf{i}) \cdot \nabla + \mathbf{G}(\mathbf{R}', \mathbf{i})\right] \tag{10}$$

derived from the operator of kinetic energy of the nuclei is nondiagonal.

As is well known, perturbation theory for a single state is different from that for degenerate states. The former leads to the traditional adiabatic approximation whereas, for degenerate or nearly degenerate states, it results into the vibronic Hamiltonian that allows the nonadiabatic mixing of electronic states having the same or close energies. First, let us examine the case of a non-degenerate electronic state $\psi_n(\mathbf{r}, \mathbf{s}; \mathbf{R})$. To first order of degenerate perturbation theory, this is equivalent to considering just the diagonal elements of the Hamiltonian matrix $\mathbf{H}(\mathbf{R}', \mathbf{i})$ or to neglecting all terms in Eq. (3) but the nth one

$$\Omega(\mathbf{R}', \mathbf{i}, \mathbf{r}, \mathbf{s}) = \chi_n(\mathbf{R}', \mathbf{i})\psi_n(\mathbf{r}, \mathbf{s}; \mathbf{R}) \tag{11}$$

Thus, the neglect of the off-diagonal matrix elements allows the change from mixed states of the nuclear subsystem to pure ones. The motion of the nuclei leads only to the deformation of the electronic distribution and not to transitions between different electronic states. In other words, a stationary distribution of electrons is obtained for each instantaneous position of the nuclei, that is, the electrons follow the motion of the nuclei adiabatically. The distribution of the nuclei is described by the wave function $\chi_n(\mathbf{R}', \mathbf{i})$ in the potential $V_{nn} + C_{nn}$, known as the proper adiabatic approximation [41]. The off-diagonal operators C_{mn} in the matrix $\mathbf{C}$, which lead to transitions between the states ψ_n and ψ_m, are called operators of nonadiabaticity and the potential $V_n = V_{nn}(\mathbf{R})$ due to the mean field of all the electrons of the system is called the adiabatic potential.

To obtain the Hamiltonian at zeroth-order of approximation, it is necessary not only to exclude the kinetic energy of the nuclei, but also to assume that the nuclear internal coordinates are frozen at $\mathbf{R} = \mathbf{R}_0$, where $\mathbf{R}_0$ is a certain reference nuclear configuration, for example, the absolute minimum or the conical intersection. Thus, as an initial basis, the states $\psi_n(\mathbf{r}, \mathbf{s}) = \psi_n(\mathbf{r}, \mathbf{s}; \mathbf{R}_0)$ are the eigenfunctions of the Hamiltonian $\hat{H}_e(\mathbf{r}, \mathbf{s}, \mathbf{R}_0)$. Accordingly, instead of Eq. (3), one has

$$\Omega(\mathbf{R}', \mathbf{i}, \mathbf{r}, \mathbf{s}) = \sum_n \chi_n(\mathbf{R}', \mathbf{i})\psi_n(\mathbf{r}, \mathbf{s}) \tag{12}$$

Substitution of Eq. (12) into the Schrödinger equation leads to a system of coupled differential equations similar to Eq. (5), but with the following differences: the potential matrix with elements

$$V_{mn}(\mathbf{R}) = \langle \psi_m(\mathbf{r}) | \hat{H}_e(\mathbf{R}, \mathbf{r}) | \psi_n(\mathbf{r}) \rangle \tag{13}$$

is nondiagonal in the new basis, whereas the matrix of the kinetic energy operator for the nuclei vanishes as the basis functions do not depend on $\mathbf{R}$. For the nondegenerate state case, one has to take into account only the diagonal elements of the Hamiltonian,

$$H_n = -\frac{\hbar^2}{2\mu}\nabla^2 + V_{nn} \tag{14}$$

where V_{nn} plays the role of the potential energy of the nuclei. This is equivalent to looking for solutions of the form in Eq. (11). If a complete basis set is assumed, the eigenvalues of the potential matrix then coincide with the adiabatic potentials V_n from Eq. (4).

This approach is called the Born–Oppenheimer (BO) approximation [38]. It is linked to the proper adiabatic approximation by the unitary transformation of the electronic basis and from this point of view they are equivalent. Of course, such an equivalence is valid only for exact solutions. In fact, the convergence of approximate solutions is different for those two cases. In particular, the BO approximation necessarily involves the expansion of the potential energy in a power series with respect to nuclear displacements from the point $\mathbf{R}_0$, and hence leads to a different convergence when compared with the adiabatic approximation. In some cases though, the weak convergence or even nonvalidity of the BO approximation is caused not by the large contribution of the operator of nonadiabaticity but by the signficant anharmonicity of the potential energy surface, especially in the case of nonrigid molecules where the adiabatic approximation may possibly work. On the other hand, the BO approximation is convenient since it allows the use of symmetry considerations due to the fact that the electronic states $\{\psi_n(\mathbf{r})\}$ form the basis of irreducible representations (IRREPs) of the symmetry group appropriate to the nuclear configuration $\mathbf{R}_0$.

However, for some polyatomic systems, there are electronic states for which the adiabatic and the BO approximations are inapplicable. In such cases, the electronic term energies are very close or coincide at some number (finite or infinite) of points of the configurational space of the nuclei. Such a degeneracy of the terms in the electronic subsystem is usually due to the high symmetry of the nuclear configuration at the point $\mathbf{R}_0$. This is the case, for example, when the potential energy surface shows a conical intersection. As a result of such degeneracies, the BO approximation described in the previous paragraph breaks down. In fact, as first pointed out by Longuet-Higgins and Herzberg [42–44], due to such a conical intersection, a real electronic wave function changes sign when traversing a nuclear path that encircles the degeneracy point. On the other hand, the total electronuclear wave function must be continuous and single-valued, which implies that the nuclear wave function must change sign to compensate that of the electronic wave function. This may be achieved by introducing a geometry-dependent phase factor in the Born–Huang [39] type development as follows:

$$\Omega(\mathbf{R}', \mathbf{i}, \mathbf{r}, \mathbf{s}) = \sum_n \chi_n(\mathbf{R}', \mathbf{i}) e^{iA_n(\mathbf{R})} \psi_n(\mathbf{r}, \mathbf{s}; \mathbf{R}) = \sum_n \chi_n(\mathbf{R}', \mathbf{i}) \tilde{\psi}_n(\mathbf{r}, \mathbf{s}; \mathbf{R}) \quad (15)$$

where $\psi_n(\mathbf{r}, \mathbf{s}; \mathbf{R})$ are the real-valued solutions of Eq. (4), and the $A_n(\mathbf{R})$ are chosen to make $\tilde{\psi}_n(\mathbf{r}, \mathbf{s}; \mathbf{R})$ [and hence $\Omega(\mathbf{R}', \mathbf{i}, \mathbf{r}, \mathbf{s})$] be single valued. Of course,

Eq. (15) may alternatively be written as

$$\Omega(\mathbf{R}', \mathbf{i}, \mathbf{r}, \mathbf{s}) = \sum_n \tilde{\chi}_n(\mathbf{R}', \mathbf{i})\psi_n(\mathbf{r}, \mathbf{s}; \mathbf{R}) \tag{16}$$

where the complex nuclear wave functions $\tilde{\chi}_n(\mathbf{R}', \mathbf{i})$ are now chosen to make $\Omega(\mathbf{R}', \mathbf{i}, \mathbf{r}, \mathbf{s})$ be single valued. Clearly, the $\mathbf{R}$ dependence of $A_n(\mathbf{R})$ must reflect the presence of any conical intersection in accordance with the Berry phase condition, and hence can generally be constructed only after the conical intersections have been located. Although a general approach for determining $A_n(\mathbf{R})$ has been suggested by Kendrick and Mead [45], it remains a nontrivial task. As shown in Appendix A, a simpler approach is possible if one assumes a two-dimensional (2D) Hilbert space model, that is, only two electronic states.

Mead and Truhlar [19,46,47] showed that the ansatz of Eq. (15) leads to the appearance of a vector potential in the nuclear Schrödinger equation. For a X_3-type molecule, the same result can be achieved by multiplying the real nuclear wave function by a complex phase factor such that it changes sign on encircling the conical intersection, and hence makes the resulting complex nuclear wave function single valued [48–50]. Billing and Markovic [51] adopted hyperspherical coordinates within this complex phase factor approach to include such a GP effect in X_3 molecules that have a single D_{3h} conical intersection, since in this coordinate system the GP effect concerns in principle only the ϕ hyperangle. A similar approach has been utilized by the present authors [2] to study the transition state resonances and bound vibrational states of H_3 using a time-dependent wavepacket method. Because all the above methods still use only one electronically adiabatic potential energy surface, they can be said to be based on a generalized BO approximation [52]. Within this spirit, we have also reported [4], following the approach of Baer and Englmann [53] and Xu et al. [5], a generalized BO equation [4] to study the nuclear dynamics in the vicinity of the conical intersection (see Appendix A). Such studies on both the electronically ground doublet state [11] and first-excited doublet state [12] of trimeric hydrogenic systems have shown that the GP effect plays a significant role, and should be taken into account if accurate dynamics results are aimed.

III. GROUP THEORETICAL CONSIDERATIONS

As is well known, when the electronic spin–orbit interaction is small, the total electronic wave function $\psi_e(\mathbf{r}, \mathbf{s}; \mathbf{R})$ can be written[3] as the product of a spatial wave function $\psi'_e(\mathbf{r}; \mathbf{R})$ and a spin function $\psi_{es}(\mathbf{s})$. For this, we can use either

[3] Although for a Slater-type determinant wave function this is true only for two-particle systems, the following discussion is independent of such a restriction.

the SF or the BF coordinate systems. As shown below, it is more appropriate to use BF spin functions, since they will be affected by molecular symmetry operations, and hence must belong to one of the irreducible representations of the symmetry point group of the molecule. For example, for integer spin values, the transformation $\exp(im_S\varphi)$ for rotation by $\varphi = 2\pi$, where $m_S = S, S-1, \ldots, -S$, leads to a retrieval of the spin function to itself [28]. However, when the spin is a half-integer, such a transformation (i.e., a rotation by 2π) will lead to a sign change of the spin function. Indeed, if $\varphi' = \varphi + 2\pi$ and $m_S = l/2$, where l is an integer, one has $\exp(im_S\varphi') = -\exp(im_S\varphi)$. The spin function will then be double valued: A rotation by 2π will not bring the system to its starting point, which can only be achieved through a 4π rotation. A rotation by 2π is therefore a new symmetry element, called R (to denote the corresponding operator we use $\hat{R}$; such a hat notation is also generally used in this work for other operators), with respect to which any spin function may be either symmetric or antisymmetric. As a result there are new symmetry elements $\hat{R}Y_i$, where Y_i stands for any of the original symmetry elements (e.g., C_2, $\sigma, C_3, \ldots$). Such extended point groups are commonly referred to as double groups [28], and we give some examples in Table I; the dashed lines on this table indicate the separation between the traditional point group and its extension. Note that, for twofold axes (C_2) and planes of symmetry (σ), the new elements ($\hat{R}C_2$ and $\hat{R}\sigma$) belong to the same class as the original elements and cause only a doubling of the class. For axes more than twofold and the center of symmetry, they cause a doubling of the number of classes. For example, the class designed $2\hat{C}_3$ of the ordinary point group $\mathbf{D}_{3h}$ has two elements C_3 and C_3^2, while the double group has two extra elements C_3^4 and C_3^5 in the class $2\hat{C}_3^2$, since $\hat{R}C_3 = C_3^4$ and $\hat{R}C_3^2 = C_3^5$. Similarly, one has $\hat{R}S_3 = S_3^5$; for further details, the reader is referred to Herzberg's [28] book.

Note also that on reducing the symmetry of a system, the spin functions for integer spin are resolved by reducing degeneracies [28]. In simple words, this means that by reducing the symmetry, the degenerate spin states in the high-symmetry group split into different states in the lower symmetry group. However, the spin functions for half-integer spin are at most resolved into functions that are still doubly degenerate. Indeed, we may see from Table II that, for integer total electronic spin S or integer total nuclear spin I, on going from $\mathbf{D}_{3h}$ to $\mathbf{C}_{2v}$ the E'' representation transforms to $B_1 + B_2$. Conversely, for the S or I half-integer, the same resolution maintains the E-type degenerate representation. This remaining degeneracy is usually called Kramers' degeneracy to honor the author who first discovered it [28,30]. According to Kramers' theorem, provided that no external magnetic field is present, the degeneracy of a system consisting of an odd number of identical particles with half-integer spin (fermions) is even. This is due to the fact that, as long as no magnetic field is present, there is in all atomic and molecular systems an additional symmetry

TABLE I
Species and Characters of the Extended C_2, C_{2v}, C_{3v}, and D_{3h} Point Groups[a]

$\mathbf{C}_2$	$\hat{I}$	$\hat{C}_2(z)$	$\hat{R}$
A	1	1	1
B	1	−1	1
$E^a_{1/2}$	2	0	−2

$\mathbf{C}_{2v}$	$\hat{I}$	$\hat{C}_2(z)$	$\hat{\sigma}_v(xz)$	$\hat{\sigma}_v(yz)$	$\hat{R}$
A_1	1	1	1	1	1
A_2	1	1	−1	−1	1
B_1	1	−1	1	−1	1
B_2	1	−1	−1	1	1
$E_{1/2}$	2	0	0	0	−2

$\mathbf{C}_{3v}$	$\hat{I}$	$2\hat{C}_3$	$3\hat{\sigma}_v$	$\hat{R}$	$2\hat{C}_3^2$
A_1	1	1	1	1	1
A_2	1	1	−1	1	1
E	2	−1	0	2	−1
$E_{1/2}$	2	1	0	−2	−1
$E_{3/2}$	2	−2	0	−2	2

$\mathbf{D}_{3h}$	$\hat{I}$	$2\hat{S}_3(z)$	$2\hat{C}_3(z)$	$\hat{\sigma}_h$	$3\hat{C}_2$	$3\hat{\sigma}_v$	$\hat{R}$	$2\hat{S}_3^5$	$2\hat{C}_3^2$
A'_1	1	1	1	1	1	1	1	1	1
A'_2	1	1	1	1	−1	−1	1	1	1
A''_1	1	−1	1	−1	1	−1	1	−1	1
A''_2	1	−1	1	−1	−1	1	1	−1	1
E''	2	1	−1	−2	0	0	2	1	−1
E'	2	−1	−1	2	0	0	2	−1	−1
$E_{1/2}$	2	$\sqrt{3}$	1	0	0	0	−2	$-\sqrt{3}$	−1
$E_{3/2}$	2	0	−2	0	0	0	−2	0	2
$E_{5/2}$	2	$-\sqrt{3}$	1	0	0	0	−2	$\sqrt{3}$	−1

[a]As usual, the indices 1/2, 3/2, and 5/2 that appear in the doubly degenerate E representation indicate the values of the projection of the angular momentum vector, $m_J = \pm 1/2, \pm 3/2, \pm 5/2$; see also the text.

element that corresponds to the antilinear time reversal operator $\hat{T}$ (see Appendix B): The evolution of a system (classical or quantum) is invariant when the time t is replaced by $-t$. In fact, an extension of Kramers' theorem will be demonstrated to be valid also for systems with a half-integer total angular momentum quantum number F.

Consider then the total angular momentum $\mathbf{F}$ defined by the vectorial sum of all the angular momenta of the system

$$\mathbf{F} = \mathbf{S} + \mathbf{I} + \mathbf{L} + \mathbf{N} \tag{17}$$

TABLE II
Species of Spin Functions for Some Important Double Groups

S or I	$\mathbf{C}_2$	$\mathbf{C}_{2v}$	$\mathbf{C}_{3v}$	$\mathbf{D}_{3h}$
0	A	A_1	A_1	A_1'
$\frac{1}{2}$	$E_{1/2}$	$E_{1/2}$	$E_{1/2}$	$E_{1/2}$
1	$A + 2B$	$A_2 + B_1 + B_2$	$A_2 + E$	$A_2' + E''$
$\frac{3}{2}$	$2E_{1/2}$	$2E_{1/2}$	$E_{1/2} + E_{3/2}$	$E_{1/2} + E_{3/2}$
2	$3A + 2B$	$2A_1 + A_2 + B_1 + B_2$	$A_1 + 2E$	$A_1' + E' + E''$
$\frac{5}{2}$	$3E_{1/2}$	$3E_{1/2}$	$2E_{1/2} + E_{3/2}$	$E_{1/2} + E_{3/2} + E_{5/2}$

where **S**, **I**, **L**, and **N** are the total electronic spin, nuclear spin, electronic orbital angular momentum, and nuclear orbital angular momentum. For this general case, we can prove that Kramers' theorem still applies since spin and orbital angular momenta must obey the same time-reversal properties (see Appendix C), namely,

$$\hat{T}\hat{S}\hat{T}^{-1} = -\hat{S} \qquad \hat{T}\hat{L}\hat{T}^{-1} = -\hat{L} \tag{18}$$

$$\hat{T}\hat{I}\hat{T}^{-1} = -\hat{I} \qquad \hat{T}\hat{N}\hat{T}^{-1} = -\hat{N} \tag{19}$$

where $\hat{A}$ represents as usual the operator corresponding to the angular momentum (vector) **A**, while the quantum number A defines the eigenvalue $A(A+1)$ of $\hat{A}^2$. Thus,

$$\hat{T}(\hat{S}+\hat{L})\hat{T}^{-1} = -(\hat{S}+\hat{L}) \qquad \hat{T}(\hat{I}+\hat{N})\hat{T}^{-1} = -(\hat{I}+\hat{N}) \tag{20}$$

$$\hat{T}(\hat{S}+\hat{I})\hat{T}^{-1} = -(\hat{S}+\hat{I}) \qquad \hat{T}(\hat{L}+\hat{N})\hat{T}^{-1} = -(\hat{L}+\hat{N}) \tag{21}$$

and hence

$$\hat{T}\hat{F}\hat{T}^{-1} = -\hat{F} \tag{22}$$

Equations (18)–(22) imply that all types of angular momenta have the same time-reversal properties. In fact, it is well known [54,55] that quantities such as energy, coordinates, electric field strength, and so on are invariant under time reversal: The corresponding operators must be time invariant. In turn, the velocity, linear momentum, angular momentum, magnetic field strength, and so on, change sign under time reversal: The corresponding operators must reflect the same property.

Moreover, as also shown in Appendix C for the electronic spin, one has

$$\hat{T}_S^2 = (-\hat{1})^{2S} \tag{23}$$

where hereafter the operator $\hat{T}_A$ stands for a time-reversal operation on the **A** variable. Similarly, for the nuclear spin, one has

$$\hat{T}_I^2 = (-\hat{1})^{2I} \tag{24}$$

Thus,

$$\hat{T}_I^2 \hat{T}_S^2 = (-\hat{1})^{2(I+S)} \tag{25}$$

and finally

$$\hat{T}_F^2 = (-\hat{1})^{2F} \tag{26}$$

since

$$\hat{T}_L^2 = (-\hat{1})^{2L} = \hat{1} \qquad \hat{T}_N^2 = (-\hat{1})^{2N} = \hat{1} \tag{27}$$

Note that $\hat{T}_S$, $\hat{T}_I$, $\hat{T}_L$, and $\hat{T}_N$ operate on the corresponding degrees of freedom, and hence mutually commute. Note especially that L and N always assume integer values.

At this stage, we are ready to prove that Kramers' theorem holds also for the total angular momentum **F**. We will do it by reductio ad absurdum. Then, let $|\Psi_E\rangle$ be the eigenvector of $\hat{H}$ with eigenvalue E,

$$\hat{H}|\Psi_E\rangle = E|\Psi_E\rangle \tag{28}$$

where hereafter we use the Dirac bra–ket notation. Since $\hat{T}$ commutes with $\hat{H}$, one gets

$$\hat{H}\hat{T}|\Psi_E\rangle = E\hat{T}|\Psi_E\rangle \tag{29}$$

Suppose now that the total wave function (state) is nondegenerate, and F is half-integer. From Eq. (29), it then follows

$$\hat{T}|\Psi_E\rangle = c|\Psi_E\rangle \tag{30}$$

where c is a constant, and hence

$$\hat{T}^2|\Psi_E\rangle = \hat{T}c|\Psi_E\rangle = c^*\hat{T}|\Psi_E\rangle = |c|^2|\Psi_E\rangle \tag{31}$$

Note that the third equality in Eq. (31) holds due to the fact that $\hat{T}$ is antiunitary (see Appendix B), and hence can be expressed as

$$\hat{T} = \hat{U}\hat{K} \tag{32}$$

where $\hat{U}$ is an unitary operator and $\hat{K}$ the complex conjugate operator. Clearly, Eq. (31) is in contradiction with the initial hypothesis that $\hat{T}^2 = -\hat{1}$. The eigenstates of a system with half-integer F must therefore be degenerate, as we wished to demonstrate.

We now prove that the degeneracy must be even. For this, we should first demonstrate two Lemmas: (1) $|\Psi_E\rangle$ is orthogonal to $\hat{T}|\Psi_E\rangle$ if $\hat{T}^2 = -\hat{1}$; (2) $\hat{T}|\Psi'_E\rangle$, $|\Psi_E\rangle$ and $\hat{T}|\Psi_E\rangle$ form a set of mutually orthogonal functions, provided that $|\Psi'_E\rangle$ is orthogonal both to $|\Psi_E\rangle$ and $\hat{T}|\Psi_E\rangle$. By first considering Lemma 1, one has

$$\langle\Psi_E|(\hat{T}|\Psi_E\rangle) = -(\langle\Psi_E|\hat{T}^{\dagger 2})(\hat{T}|\Psi_E\rangle) \tag{33}$$

$$= -\left[(\langle\Psi_E|\hat{T}^{\dagger})(\hat{T}^{\dagger}\hat{T}|\Psi_E\rangle)\right]^* \tag{34}$$

$$= -\left[(\langle\Psi_E|\hat{T}^{\dagger})|\Psi_E\rangle\right]^* \tag{35}$$

$$= -\langle\Psi_E|(\hat{T}|\Psi_E\rangle) \tag{36}$$

where the first equality is obtained owing to the fact that $\hat{T}^2 = -\hat{1}$; the second to the fact that $\hat{T}$ is antilinear, and hence obeys the property $\langle\psi|(\hat{T}|\varphi\rangle) = \left[(\langle\psi|\hat{T})|\varphi\rangle\right]^*$ (see Appendix B); the third results from the unitary property of $\hat{T}$ (i.e., $\hat{T}^{\dagger}\hat{T} = \hat{1}$); finally, Eq. (36) follows from applying again the fact that $\hat{T}$ is antilinear, and hence $\langle\varphi|(\hat{T}|\psi\rangle) = \left[(\langle\psi|\hat{T}^{\dagger})|\varphi\rangle\right]^*$. Thus, Eqs. (33)–(36) imply that $\langle\Psi_E|(\hat{T}|\Psi_E\rangle) = 0$, which completes the proof of Lemma 1.

Let us now prove Lemma 2. One has

$$\langle T\Psi'_E|\Psi_E\rangle = (\langle\Psi'_E|\hat{T}^{\dagger})|\Psi_E\rangle \tag{37}$$

$$= \left[\langle\Psi'_E|(\hat{T}^{\dagger}|\Psi_E\rangle)\right]^* \tag{38}$$

$$= \left[\langle\Psi'_E|(\hat{T}^{\dagger 2}\hat{T}|\Psi_E\rangle)\right]^* \tag{39}$$

$$= -\left[\langle\Psi'_E|(\hat{T}|\Psi_E\rangle)\right]^* \tag{40}$$

$$= 0 \tag{41}$$

and

$$\langle T\Psi'_E|(\hat{T}|\Psi_E\rangle) = (\langle\Psi'_E|\hat{T}^{\dagger})(\hat{T}|\Psi_E\rangle) \tag{42}$$

$$= \left[\langle\Psi'_E|(\hat{T}^{\dagger}\hat{T}|\Psi_E\rangle)\right]^* \tag{43}$$

$$= \left[\langle\Psi'_E|\Psi_E\rangle\right]^* \tag{44}$$

$$= 0 \tag{45}$$

which completes the desired proof.

Finally, we can demonstrate that the degeneracy is even when the total angular momentum quantum number F is half-integer, again via a reductio ad absurdum method. Suppose that the degeneracy of the eigenstates is k, then we have k degenerate states $|\Psi_{E,i}\rangle$ $(i = 1, \ldots, k)$, which have in common the same eigenvalue E. One can then form orthogonal pairs of such states such as $|\Psi_E\rangle$ and $\hat{T}|\Psi_E\rangle$. If k is odd, there will be a single state (e.g., $|\phi\rangle$), which has no pair. However, as mentioned above, $\hat{T}|\phi\rangle$ will be orthogonal to all the k states, and $\hat{T}|\phi\rangle$ is nonzero. This implies that the number of total states of the same eigenvalue E is $(k+1)$, which contradicts our initial hypothesis. Thus, we conclude that k must be even, and hence proved the generalized Kramers' theorem for total angular momentum. The implication is that we can use double groups as a powerful means to study the molecular systems including the rotational spectra of molecules. In analyses of the symmetry of the rotational wave function for molecules, the three-dimensional (3D) rotation group $\mathbf{SO}(3)$ will be used.

IV. PERMUTATIONAL SYMMETRY OF TOTAL WAVE FUNCTION

The total Hamiltonian operator $\hat{H}$ must commute with any permutations $\hat{P}_{\mathrm{X}}$ among identical particles (X) due to their indistinguishability. For example, for a system including three types of distinct identical particles (including electrons) like $^7\mathrm{Li}_2\,^6\mathrm{Li}_2$ with a T_d conformation, one must satisfy the following commutative laws:

$$\left[\hat{P}_e,\, \hat{H}\right] = 0 \qquad \left[\hat{P}_{^7\mathrm{Li}},\, \hat{H}\right] = 0 \qquad \left[\hat{P}_{^6\mathrm{Li}},\, \hat{H}\right] = 0 \tag{46}$$

and hence $\hat{P}_{\mathrm{X}}$ (X = e, ^{7}Li, ^{6}Li) are conserved quantities. For a system with N distinct sets of identical particles, there must be N such commutative laws similar to those in Eq. (46), which are relative to the various kinds of permutations; thus, there will be N permutational restrictions on the total wave function $\Omega(\mathbf{R}', \mathbf{i}, \mathbf{r}, \mathbf{s})$. Note from Figures 1$d$ and 2 that under the permutation of two identical nuclei, the hyperspherical coordinates for a three-particle system transform as $(\rho, \theta, \phi, \alpha, \beta, \gamma) \rightarrow (\rho, \theta, -\phi, \alpha, \beta, \gamma + \pi)$.

We will now explain the meaning of the word "identical" used above. Physically, it is meant for particles that possess the same intrinsic attributes, namely, static mass, charge, and spin. If such particles possess the same intrinsic attributes (as many as we know so far), then we refer to them as physically identical. There is also another kind of identity, which is commonly referred to as chemical identity [56]. As discussed in the next paragraph, this is an important concept that must be stressed when discussing the permutational properties of nuclei in molecules.

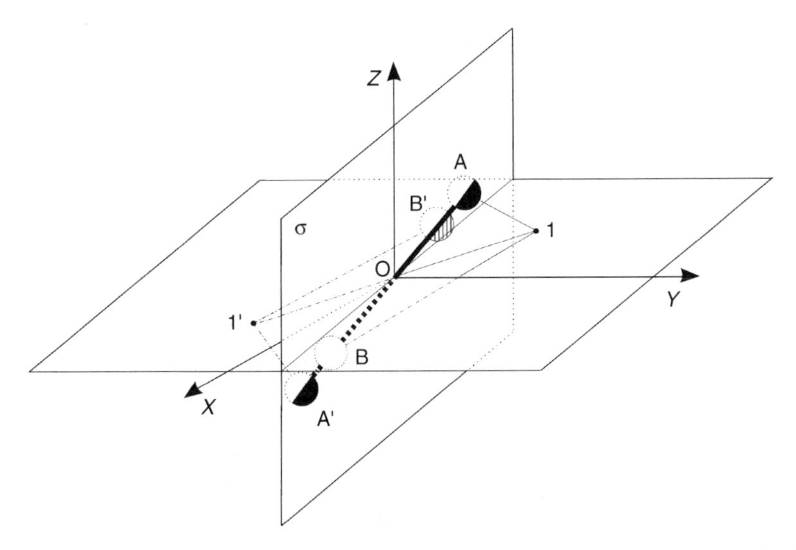

Figure 2. The space-fixed (XYZ) and body-fixed (xyz) frames in a diatomic molecule AB. The nuclei are at A and B, and 1 represents the location of a typical electron. The results of inversions of their SF coordinates are A → A′, B → B′, and 1 → 1′, respectively. After one executes only the reinversion of the electronic SF coordinates, one obtains 1′ → 1. The net effect is then the exchange of the SF nuclear coordinates alone.

Then, let us first examine $^{7}Li_3$ in a D_{3h} structure. In this molecule, the three nuclei not only have the same intrinsic attributes but also have the same molecular environments due to the fact that they are in chemically equivalent positions. Thus, the three nuclei can be exchanged by rotations of the molecule and the permutational symmetry requirement must be satisfied. Now, consider a molecule like methanol (CH_3OH) with four physically identical hydrogens. For any conformation, the methyl hydrogens will be distinguishable from one another through their positions relative to that of the hydroxyl hydrogen. However, the barrier for internal rotation of CH_3 around the CO axis is low, and tunneling from one equivalent conformation to another may occur. Thus, the permutational symmetry requirement must be applied to the methyl hydrogens. A third example is the linear conformation of NNO, with two physically identical nitrogen nuclei. In principle, the permutational symmetry requirement should also be applied to the two ^{14}N nuclei since they are physically identical. However, the two nuclei are placed at different molecular environments, that is, one lies adjacent to the oxygen nucleus while the other does not. Since their permutation involves an extremely high energy process, we may regard such nitrogen nuclei as distinguishable. They can then be said to be chemically nonequivalent [56], and hence not subject to the permutational symmetry requirement in low-energy spectroscopic studies. We should note though that, if their interchange becomes feasible by increasing the energy, the potential energy surface must satisfy the full permutational symmetry requirement dictated by the physical identity alone; this is typically the case in reaction dynamics studies. Another example is the molecule $H^{12}C{\equiv}^{13}CH$ where two hydrogen atoms have nonequivalent chemical environments. Again, although the two hydrogen atoms may be distinguishable from the spectroscopic point of view, the corresponding full potential energy surface must generally be symmetric with respect to their permutation; note that in the BO approximation, the potential energy surface does not depend on the mass of the nuclei, and hence it is the same as for $H^{12}C{\equiv}^{12}CH$. In summary, the permutational symmetry requirement should be applied only to identical particles that are both physically and chemically indistinguishable. In this case, chemical identity implies physically identical particles that have equivalent environments in the molecule and can be brought about by proper rotations of the nuclear framework, or else physically identical particles that may have equivalent chemical environments through some feasible dynamical process. Thus, the concept of chemical identity depends on the energy regime under consideration. Ideally, one should therefore carry out the nuclear dynamics studies using a global potential energy surface [57,58], which has built-in the full permutational symmetry implied by the physical identity of the atoms. Of course, if the equivalent minima are separated from each other by high energy barriers, then it may be an excellent approximation to have just the representation for one of the equivalent minima if one assumes that underbarrier tunneling is negligible. In other words, the concept of chemical identity delimits the nuclei motion to a part of the molecule configuration space. Hereafter, we will refer to identical particles with the above understanding.

Let us examine a special but more practical case where the total molecular Hamiltonian, $\hat{H}$, can be separated to an electronic part, $\hat{H}_e(\mathbf{r}, \mathbf{s}; \mathbf{R}_0)$, as is the case in the usual BO approximation. Consequently, the total molecular wave function $\Omega(\mathbf{R}', \mathbf{i}, \mathbf{r}, \mathbf{s})$ is given by the product of a nuclear wave function $\chi_{\text{nuc}}(\mathbf{R}', \mathbf{i})$ and an electronic wave function $\psi_e(\mathbf{r}, \mathbf{s}; \mathbf{R}_0)$. We may then talk separately about the permutational properties of the subsystem consisting of electrons, and the subsystem(s) formed of identical nuclei. Thus, the following commutative laws $\left[\hat{P}_e, \hat{H}_e\right] = 0$ and $\left[\hat{P}_{\text{X}}, \hat{H}_N\right] = 0$ must be satisfied; $\text{X} = A, B, \ldots$, and all other symbols have their usual meaning.

As pointed out in the previous paragraph, the total wave function of a molecule consists of an electronic and a nuclear parts. The electrons have a different intrinsic nature from nuclei, and hence can be treated separately when one considers the issue of permutational symmetry. First, let us consider the case of electrons. These are fermions with spin $\frac{1}{2}$, and hence the subsystem of electrons obeys the Fermi–Dirac statistics: the total electronic wave function must be antisymmetric under permutation of any two electrons. This requirement implies the Pauli exclusion principle, which states that two electrons cannot occupy the same spin orbital.

Commonly, nuclear dynamics treatment in molecules neglect the interactions between the nuclear spin and the other nuclear and electronic degrees of freedom in the system Hamiltonian. As a result, the eigenenergies become independent of the nuclear spin. In this case, one must impose the requirements that the symmetry properties of the nuclear spin on the total wave function are satisfied, since the nuclei in the molecule have their specific statistical properties. As it is well known, nuclei having zero or integer nuclear spin quantum numbers are bosons and must obey the Bose–Einstein statistics: The nuclear wave function must therefore be symmetric under permutation of any two identical bosonic particles. On the other hand, nuclei having half-integer spin quantum numbers are fermions and must obey the Fermi–Dirac statistics: The nuclear wave function must in this case be antisymmetric with respect to the permutation of any two fermionic nuclei. For example, ^7Li is a fermion with nuclear spin $\frac{3}{2}$, and ^6Li is a boson with nuclear spin 1. Thus, the total wave function of $^7Li_3$ must be antisymmetric under the permutation of the three identical nuclei (note that this involves a three-pair change); see the corresponding $\mathcal{S}_3$ permutational group in Table III. Conversely, the total wave function of $^6Li_3$ must be symmetric under the permutation of the three identical nuclei. In turn, the total wave function of $^7Li_2\,^6Li$ must be antisymmetric under the permutation $\mathcal{S}_2$ of the two identical ^7Li nuclei (see also Table III for the $\mathcal{S}_2$ permutational group). Following the same reasoning, the total wave function of $^6Li_2\,^7Li$ must be symmetric under the permutation of the two identical ^6Li nuclei.

Let us discuss further the permutational symmetry properties of the nuclei subsystem. Since the electronic spatial wave function $\psi_e(\mathbf{r}, \mathbf{s}; \mathbf{R}_0)$ depends parametrically on the nuclear coordinates, and the electronic spacial and spin coordinates are defined in the BF, it follows that one must take into account the effects of the nuclei under the permutations of the identical nuclei. Of course,

TABLE III
Species and Characters of the $\mathcal{S}_2$ and $\mathcal{S}_3$ Permutational Groups

$\mathcal{S}_2$		(1)	(12)
[2]		1	1
$[1^2]$		1	−1
$\mathcal{S}_3$	(1)	2(123)	3(12)
[3]	1	1	1
$[1^3]$	1	1	−1
[21]	2	−1	0

the spin part of the electronic wave function is independent of the permutational properties of the nuclei, which implies that one does not need to take care of the electronic spin wave function when dealing with the permutational properties of the nuclei. However, as it is will be further discussed in Section VIII, the electronic spin **S** will influence the permutational symmetry properties through the total angular momentum **J**. Accordingly, we address in the following sections the consequences of such rules based on the premise that the total wave function $\Omega(\mathbf{R}', \mathbf{i}, \mathbf{r}, \mathbf{s})$ may be written as

$$\Omega(\mathbf{R}', \mathbf{i}, \mathbf{r}, \mathbf{s}) = \psi_e(\mathbf{r}, \mathbf{s}; \mathbf{R}_0)\chi_v(\mathbf{R})\chi_r(\hat{\mathbf{R}})\chi_{ns}(\mathbf{i}) \tag{47}$$

where $\psi_e(\mathbf{r}, \mathbf{s}; \mathbf{R}_0)$, $\chi_v(\mathbf{R})$, $\chi_r(\hat{\mathbf{R}})$, and $\chi_{\text{ns}}(\mathbf{i})$ are the electronic, vibrational, rotational, and nuclear spin functions, respectively.

V. PERMUTATIONAL SYMMETRY OF NUCLEAR SPIN FUNCTION

As discussed before, the nuclear spin functions must belong to one of the irreducible representations of the double group of the molecule. For example, $^6Li_3$ may have nuclear spin quantum numbers $I = 0$, 1, 2, and 3, and hence the permutational symmetries under $\mathcal{S}_3$ will be given according to Table II: A_1' for $I = 0$; $A_2' + E''$ for $I = 1$; $A_1' + E' + E''$ for $I = 2$, and so on. For a molecule with half-integer nuclear spin, all the IRREPs are double-valued due to the Kramers' degeneracy. For example, for $^7Li_3$, the nuclear spin quantum numbers I are half-integer ranging in steps of 1 from $\frac{1}{2}$ up to $\frac{9}{2}$. Thus, from Table II, the permutation symmetries under $\mathcal{S}_3$ will be $E_{1/2}$ for $I = \frac{1}{2}$; $E_{1/2} + E_{3/2}$ for $I = \frac{3}{2}$; $E_{1/2} + E_{3/2} + E_{5/2}$ for $I = \frac{5}{2}$, and so on.

For a nucleus with spin quantum number $I \neq 0$, there are $(2I + 1)$ values of the z component m_I of the spin nuclear angular momentum, with $m_I = -I, -I + 1, \ldots, I - 1, I$. For two such nuclei, the total number of m_I combinations will be $(2I + 1)^2$. Assuming that $\chi_{m_I}(1)$ is the nuclear spin function of nucleus 1 with quantum number m_I, there are $(2I + 1)$ spin functions of the form $\chi_{m_I}(1)\chi_{m_I}(2)$, which are symmetric. Of the remaining $(2I + 1)^2 - (2I + 1)$ ones, one-half can be combined in symmetric states

$$\chi_{ns}^S(1, 2) = \frac{1}{\sqrt{2}}\left[\chi_{m_I}(1)\chi_{m_I'}(2) + \chi_{m_I'}(1)\chi_{m_I}(2)\right] \qquad m_I \neq m_I' \tag{48}$$

and the other one-half in antisymmetric ones

$$\chi_{ns}^A(1, 2) = \frac{1}{\sqrt{2}}\left[\chi_{m_I}(1)\chi_{m_I'}(2) - \chi_{m_I'}(1)\chi_{m_I}(2)\right] \qquad m_I \neq m_I' \tag{49}$$

The total number of symmetric states is then

$$2I + 1 + \frac{1}{2}[(2I + 1)^2 - (2I + 1)] = (2I + 1)(I + 1) \tag{50}$$

and the total number of antisymmetric states is

$$\frac{1}{2}[(2I + 1)^2 - (2I + 1)] = (2I + 1)I \tag{51}$$

In summary, for a homonuclear diatomic molecule there are generally $(2I + 1)(I + 1)$ symmetric and $(2I + 1)I$ antisymmetric nuclear spin functions. For example, from Eqs. (50) and (51), the statistical weights of the symmetric and antisymmetric nuclear spin functions of $^7Li_2$ will be $\frac{5}{8}$ and $\frac{3}{8}$, respectively. This is also true when one considers $^7Li_2\,^6Li$ and $^6Li_2\,^7Li$. For the former, the statistical weights of the symmetric and antisymmetric nuclear spin functions are $\frac{5}{8}$ and $\frac{3}{8}$, respectively; for the latter, they are $\frac{2}{3}$ and $\frac{1}{3}$ in the same order.

For a homonuclear triatomic molecule there are similarly [29] $(2I + 1)(2I + 3)(I + 1)/3$ symmetric, $(2I + 1)(2I - 1)I/3$ antisymmetric, and $(2I + 1)(I + 1)(8I)/3$ degenerate nuclear spin functions. For $^7Li_3$, one therefore has 20 symmetric, 4 antisymmetric, and 40 degenerate nuclear spin functions. The corresponding statistical weights will then be $\frac{5}{16}$, $\frac{1}{16}$, and $\frac{10}{16}$. Following a similar reasoning for $^6Li_3$ one finds 10 symmetric, 1 antisymmetric, and 16 degenerate nuclear spin functions. Thus, the corresponding statistical weights are $\frac{10}{27}$, $\frac{1}{27}$, and $\frac{16}{27}$.

Now, consider a linear polyatomic molecule. If this is of the type $Z \cdots BAAB \cdots Z$ with $D_{\infty h}$ geometry, a $\hat{C}_2$ rotation about an axis perpendicular to the molecular axis at its midpoint will exchange pairs of identical nuclei. If the nuclei $A, B, \ldots, Z$ contain an odd number of fermions, the $\hat{C}_2$ rotation leads to a change of sign of the total wave function; otherwise, it will remain unchanged. By repeated application of the two identical nuclei case, the total number of possible nuclear spin wave functions will be given by $(2I_A + 1)^2 (2I_B + 1)^2 \cdots (2I_Z + 1)^2$, where I_X denotes the nuclear spin of X ($X = A, B, \ldots, Z$). Similarly, the number of possible symmetric and antisymmetric nuclear spin functions can be obtained by an extension of the method used for diatomic molecules. The total number of symmetric states is then given by [56]

$$(2I_A + 1)(2I_B + 1) \cdots (2I_Z + 1)[(2I_A + 1)(2I_B + 1) \cdots (2I_Z + 1) + 1]/2 \tag{52}$$

and the total number of antisymmetric states is [56]

$$(2I_A + 1)(2I_B + 1) \cdots (2I_Z + 1)[(2I_A + 1)(2I_B + 1) \cdots (2I_Z + 1) - 1]/2 \tag{53}$$

which represent the corresponding nuclear statistical weights. For molecules of the form Z$\cdots$BARAB$\cdots$Z with $D_{\infty h}$ symmetry, the corresponding numbers must each be multiplied by [56] $(2I_R + 1)$.

VI. PERMUTATIONAL SYMMETRY OF ELECTRONIC WAVE FUNCTION

In considering the nuclear permutational properties of the total wave function, we must have in mind the corresponding properties of the electronic wave function, since this depends parametrically on the nuclear geometry. The permutational properties of $\psi_e(\mathbf{r}, \mathbf{s}; \mathbf{R}_0)$ under identical-nuclei exchange are determined by those of $\hat{H}_e(\mathbf{r}, \mathbf{s}; \mathbf{R}_0)$, and hence of $V_n(\mathbf{R})$. Since this represents the potential energy of the electrons in the field of the fixed nuclei, it must have the symmetry of the molecule in its nth electronic state. The electronic eigenfunctions for nondegenerate states can therefore only be symmetric or antisymmetric with respect to each symmetry operation that is allowed by the symmetry of the molecule in its equilibrium geometry. For degenerate states, a symmetry operation can only transform an eigenfunction into a linear combination of the degenerate eigenfunctions such that the electron density remains unaltered.

Let us begin by considering a homonuclear diatomic molecule. Clearly, permutation of the nuclei does not affect the internuclear distance, but it does affect the electronic spatial coordinates, since they are defined with respect to the BF axes. To find the effect of interchanging nuclei on the electronic wave function, we invert first the SF coordinates of the nuclei and electrons, and then carry out a second inversion of the SF electronic coordinates alone. The net effect will be the exchange of the SF coordinates of the two nuclei as illustrated in Figure 2. Note that the inversion of the SF does not affect electronic and nuclear spin coordinates. The eigenvalues (± 1) of such an inversion operator indicate the parities of the wave function under consideration. The first inversion (equivalent to a reflection $\hat{\sigma}_v$) leaves the electronic wave functions unchanged (i.e., with even parity) for $\Sigma^+, \Pi^+, \ldots$ electronic states, while it changes their sign (i.e., with odd parity) for $\Sigma^-, \Pi^-, \ldots$ electronic states. This is due to the fact that reflection in the plane containing the nuclei changes (leaves unchanged) the sign of wave function for $-$ $(+)$ states. Because only the Σ^+ and Σ^- have different energies [56], it has often omitted the $\pm$ sign for degenerate states such as Π, Δ ... provided that Λ-type doubling is ignored. The second of the above inversion operations (which inverts back the electronic SF coordinates), inverts the electronic BF coordinates since the nuclei are unaffected by this step, and hence the electronic wave functions can be classified as g or u states according to whether such inversion of the electronic BF coordinates changes or leaves unchanged the sign of the electronic wave function. Thus,

$\Sigma_g^+, \Sigma_u^-, \Pi_g^+, \Pi_u^- \cdots$ electronic states have wave functions that are symmetric under permutation of identical nuclei, whereas $\Sigma_g^-, \Sigma_u^+, \Pi_g^-, \Pi_u^+ \cdots$ electronic states are antisymmetric under such a permutation. The permutational symmetry for linear polyatomics $\mathbf{D}_{\infty h}$ follows similar arguments.

We now consider planar molecules. The electronic wave function is expressed with respect to molecule-fixed axes, which we can take to be the *abc* principal axes of inertia, namely, by taking the coordinates (x, y, z) in Figure 1 coincided with the principal axes (a, b, c). In order to determine the parity of the molecule through inversions in SF, we first rotate all the electrons and nuclei by 180° about the c axis (which is perpendicular to the molecular plane); and then reflect all the electrons in the molecular *ab* plane. The net effect is the inversion of all particles in SF. The first step has no effect on both the electronic and nuclear molecule-fixed coordinates, and has no effect on the electronic wave functions. The second step is a reflection of electronic spatial coordinates in the molecular plane. Note that such a plane is a symmetry plane and the eigenvalues of the corresponding operator $\hat{\sigma}_v$ then determine the parity of the electronic wave function.

In order to determine the permutational symmetry of a nonlinear molecule, one can invoke the permutation group. Here, we give some examples. As is known, the permutation group $\mathcal{S}_3$ is isomorphic to the point group $\mathbf{C}_{3v}$, and $[1^3]$ irreducible representation in the $\mathcal{S}_3$ is antisymmetric with the interchange (12) of any two indentical particles (see Tables I and III). Thus, the A_2 electronic state for a molecule of C_{3v} geometry must be antisymmetric with such interchange of the indentical nuclei. It is obvious that the totally symmetric IRREPs in point groups always correspond to the $[n]$ IRREPs in groups $\mathcal{S}_n$.

Let us focus on the electronic wave function of ground state Li_3, which is known to have B_2 symmetry at its equilibrium geometry in the $\mathbf{C}_{2v}$ point group. In order to determine the permutation symmetry of the B_2 electronic state under the interchange of identical nuclei, we notice a correlation between IRREPs of the group $\mathbf{C}_{2v}$ and those of the group $\mathbf{C}_2$ (see Tables I and III). It is found that B_2 IRREP in the group $\mathbf{C}_{2v}$ is correlated with the B IRREP in the group $\mathbf{C}_2$. In addition, we know that the group $\mathbf{C}_2$ is isomorphic to the permutation group $\mathcal{S}_2$, and the B IRREP in $\mathbf{C}_2$ corresponds to the antisymmetric IRREP $[1^2]$ in $\mathcal{S}_2$. Accordingly, the B_2 electronic state must be antisymmetric with the interchange of the two identical nuclei. In fact, in the B_2 IRREP of $\mathbf{C}_{2v}$, the wave function changes sign under a $\hat{C}_2$ operation. In contrast, the electronic wave function at the lowest point of the conical intersection on the upper sheet of the Li_3 potential energy surface has degenerate character under $\mathcal{S}_3$, since this geometry transforms as E' in the $\mathbf{D}_{3h}$ symmetry point group. Next, consider the isotopomer $^7Li_2\,^6Li$. With the substitution of ^7Li by the isotope ^6Li, the permutational symmetry group of the system has been reduced from $\mathcal{S}_3$ to $\mathcal{S}_2$. Thus, if existing, the spatial degeneracy upon permutation of the nuclei can be

TABLE IV
Resolution of Species of Symmetric Point Groups into Some Point Groups of Lower Symmetry

$\mathbf{K}_h$	$\mathbf{D}_{3h}$	$\mathbf{C}_{3v}$	$\mathbf{C}_{2v}$	$\mathbf{C}_s$
$D_g^0 \equiv S_g$	A_1'	A_1	A_1	A'
$D_u^0 \equiv S_u$	A_1''	A_2	A_2	A''
$D_g^1 \equiv P_g$	$A_2' + E''$	$A_2 + E$	$A_2 + B_1 + B_2$	$A' + 2A''$
$D_u^1 \equiv P_u$	$A_2'' + E'$	$A_1 + E$	$A_1 + B_1 + B_2$	$2A' + A''$
$D_g^2 \equiv D_g$	$A_1' + E' + E''$	$A_1 + 2E$	$2A_1 + A_2 + B_1 + B_2$	$3A' + 2A''$
$D_u^2 \equiv D_u$	$A_1'' + E' + E''$	$A_2 + 2E$	$A_1 + 2A_2 + B_1 + B_2$	$2A' + 3A''$

removed in part or completely, since F is an integer. This is indeed the case for the ground state of $^7Li_2\,^6Li$, with the electronic wave function at the minimum of the lower sheet of the potential energy surface being antisymmetric under $\mathcal{S}_2$. Similarly, the electronic wave function at the lowest point of the conical intersection on the upper sheet of the $^7Li_2\,^6Li$ potential energy surface will be symmetric or antisymmetric[4] under $\mathcal{S}_2$. The spatial degeneracy of the electronic wave function has therefore been removed when resolving $\mathbf{D}_{3h}$ into $\mathbf{C}_{2v}$, since the E' state of the lowest energy structure has been resolved into $A_1 \oplus B_2$ (see Table IV, where the same axis convention as in [28] has been followed: the highest order proper axis always coincides with the z axis.) which correspond to symmetric and antisymmetric wave functions in $\mathcal{S}_2$.

VII. PERMUTATIONAL SYMMETRY OF ROVIBRONIC AND VIBRONIC WAVE FUNCTIONS

Since the total wave function must have the correct symmetry under the permutation of identical nuclei, we can determine the symmetry of the rovibronic wave function from consideration of the corresponding symmetry of the nuclear spin function. We begin by looking at the case of a fermionic system for which the total wave function must be antisymmetric under permutation of any two identical particles. If the nuclear spin function is symmetric then the rovibronic wave function must be antisymmetric; conversely, if the nuclear spin function is antisymmetric, the rovibronic wave function must be symmetric under permutation of any two fermions. Similar considerations apply to bosonic systems: The rovibronic wave function must be symmetric when the nuclear spin function is symmetric, and the rovibronic wave function must be antisymmetric when the nuclear spin function is antisymmetric. This warrants

[4] This and the previous statements can be understood from Tables IX and X, which will be discussed in more detail in subsequent sections.

that the total wave function is totally symmetric under permutation of any two indistinguishable bosons.

As was shown in the preceding discussion (see also Sections VIII and IX), the rovibronic wave functions for a homonuclear diatomic molecule under the permutation of identical nuclei are symmetric for even J rotational quantum numbers in Σ_g^+ and Σ_u^- electronic states; antisymmetric for odd J values in Σ_g^+ and Σ_u^- electronic states; symmetric for odd J values in Σ_g^- and Σ_u^+ electronic states; and antisymmetric for even J values in Σ_g^- and Σ_u^+ electronic states. Note that the vibrational ground state is symmetric under permutation of the two nuclei. The most restrictive result arises therefore when the nuclear spin quantum number of the individual nuclei is 0. In this case, the nuclear spin function is always symmetric with respect to interchange of the identical nuclei, and hence only totally symmetric rovibronic states are allowed since the total wave function must be symmetric for bosonic systems. For example, the ^{12}C nucleus has zero nuclear spin, and hence the rotational levels with odd values of J do not exist for the ground electronic state $(^1\Sigma_g^+)$ of $^{12}C_2$.

Let us now examine the features of the nuclear probability density of a X_3 molecule (X is an ^2S atom) in its electronic ground- and first-excited doublet states. For the lowest vibronic A_1 states, such a nuclear probability density must clearly concentrate[5] at the regions where the potential energy surface itself has A_1 symmetry, which correspond in the case of homonuclear trimeric ^2S systems to the saddle points of the potential energy surface having $^2A_1$ symmetry in $\mathbf{C}_{2v}$. Instead, the nuclear probability density of the lowest A_2 vibronic states will concentrate at regions where the potential energy surface has A_2 symmetry; note that the potential energy surface at the minima has $^2B_2$ symmetry in $\mathbf{C}_{2v}$, and hence A_2 in $\mathcal{S}_3$. Both the A_1 and A_2 probability densities display threefold symmetries on a relaxed triangular plot such as that employed in Figure 3 to represent the Li_3 potential energy surface. Although the nuclear probability density of each component for the E vibronic state has twofold symmetry, their sum must also have threefold symmetry.

VIII. PERMUTATIONAL SYMMETRY OF ROTATIONAL WAVE FUNCTION

The permutational symmetry of the rotational wave function is determined by the rotational angular momentum **J**, which is the resultant of the electronic spin **S**, electronic orbital **L**, and nuclear orbital **N** angular momenta. We will now examine the permutational symmetry of the rotational wave functions. Two important remarks should first be made. The first refers to the $J = 0$ rotational

[5] Of course, for highly excited states, the density is expected to cover wide portions of the molecular configuration space.

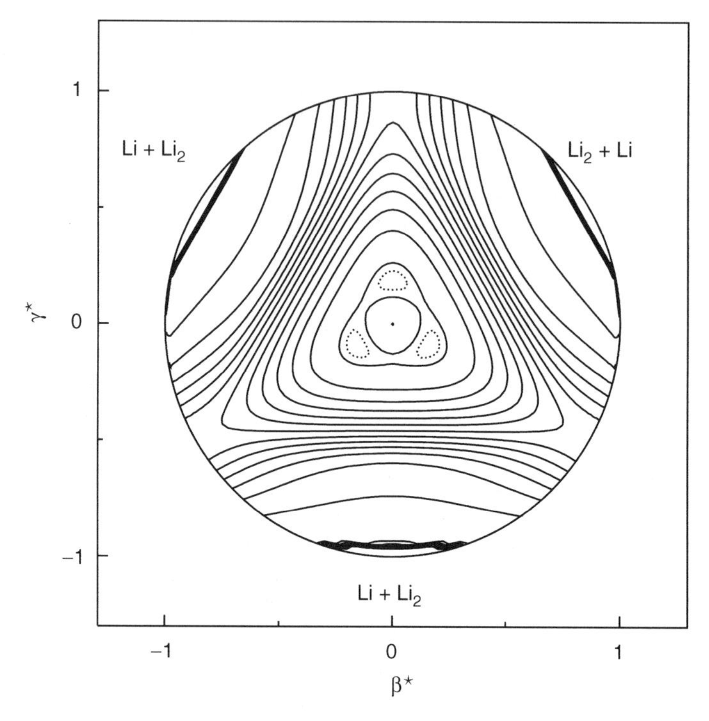

Figure 3. Relaxed triangular plot [68] of the Li_3 ground-state potential energy surface using hyperspherical coordinates. Contours, are given by the expression $E_n(\mathrm{eV}) = -0.56 + 0.045(n-1)$ with $n = 1, 2, \ldots,$ where the dashed line indicates the level $-0.565\,\mathrm{eV}$. The dissociation limit indicated by the dense contouring implies $Li_2(X^1\Sigma_g^+) + Li$.

state, which must be totally symmetric. The second emerges from the extended Kramers' theorem, which imposes half-integer J rotational states to be degenerate. Thus, the lowest rotational state for the electronic ground state of Li_3 corresponds to $J = \frac{1}{2}$, and must be degenerate.

Next, we address some simple cases, begining with homonuclear diatomic molecules in ${}^1\Sigma$ electronic states. The rotational wave functions are in this case the well-known spherical harmonics: for even J values, $\chi_r(\hat{\mathbf{R}})$ is symmetric under permutation of the identical nuclei; for odd J values, $\chi_r(\hat{\mathbf{R}})$ is antisymmetric under the same permutation. A similar statement applies for any $D_{\infty h}$ type molecule.

When the molecule is not in a ${}^1\Sigma$ state there is an interaction between the rotation of the molecule and S and/or L, and the details of coupling the angular momenta are involved. Most nonsinglet molecules with electronic orbital angular momentum $\Lambda = 0$ obey Hund's case (b) coupling. In Case (b), the electronic orbital angular momentum combines with the nuclear orbital angular momentum to give a total angular momentum associated with the quantum number $N = \Lambda, \Lambda + 1, \Lambda + 2, \ldots$. Note that in this case, N is the proper rotational angular momentum. Accordingly, for even N values, $\chi_r(\hat{\mathbf{R}})$ is symmetric under permutation of the identical nuclei; for odd N values, $\chi_r(\hat{\mathbf{R}})$ is antisymmetric under the same permutation. Thus, since the $N = 0, 2, 4, \ldots$ rotational levels for the $^3\Sigma_g^-$ electronic state of $^{16}O_2$ are symmetric with respect to the interchange two ^{16}O nuclei, they cannot exist according to the Pauli principle (see later). For most molecules with electronic orbital angular momentum $\Lambda > 0$ one has Hund's case (a). In Case (a), the axial components of electronic orbital angular momentum combine with the electronic spin angular momentum to give a resultant axial component of total electronic angular momentum associated with the quantum number $\Omega = |\Lambda + \Sigma|$. The resultant angular momentum then combines with the nuclear orbital angular momentum to give a total angular momentum (exclusive of nuclear spin) J, where $J = \Omega, \Omega + 1, \Omega + 2, \ldots$. If J is a half-integer, the rotational levels will be doubly degenerate in the zeroth-order approximation. Yet, the interaction between electronic and rotational motions can lead to the splitting of the degenerate Π electronic level into the nondegenerate Π^+ and Π^- electronic states (so-called Λ-type doubling).

Now, we consider the case of a planar molecule. In general, the rotational wave functions depend on the Eulerian angles (α, β, γ). For planar symmetric tops, the angles α and β give the orientation of the positive direction of the molecular symmetry axis with respect to SF coordinates, while γ is the angle of rotation about the symmetry axis. The angles α and β are unchanged by inversion, while γ is increased by π. The angle γ enters the symmetric top wave function as the factor $\exp(iK\gamma)$, where the quantum number K is the component of the rotational angular momentum J along the molecule-fixed axis. Thus, the inversion multiplies the rotational wave function by $\exp(iK\pi) = (-1)^K$ and its parity will be even for even K and odd for odd K. For planar asymmetric tops, the symmetric top wave functions that occur in the expansion of a given asymmetric top wave function all have either even values of K or odd values of K. A planar symmetric top must be an oblate top. If the asymmetric top level correlates with an oblate top level with K even, then it must be a linear combination of oblate top functions with even K values; similarly for the K-odd correlation. Thus, parities of the asymmetric top wave functions are determined by the values of K. Consider now a homonuclear triatomic molecule, where K is the quantum number for the rotational angular momentum component perpendicular to the plane defined by the three atoms. The rotational contribution to the permutational symmetry [59] is then symmetric if $K = 0$, symmetric or antisymmetric if K is a nonzero integer multiple of three, and degenerate if K is not an integer multiple of three (this includes the half-integer cases referred to above).

We now turn to some well-established cases where severe consequences arise due to the nuclear spin quantum number of the individual nuclei being 0 or $\frac{1}{2}$. Consider the simplest case of spinless nuclei such as $^{16}O_2$. Since the total wave function must be symmetric, it follows that only even rotational states are allowed for the ground vibrational state (this is always symmetric; see Section IX) when the electronic wave function is symmetric; conversely, only odd rotational states are allowed if the electronic wave function is antisymmetric. Since the ground electronic state of $^{16}O_2$ is $^3\Sigma_g^-$, and hence antisymmetric, it then follows that its lowest rotational state must have the rotation quantum number $N = 1$, with one-half of the expected levels being absent in the corresponding Raman spectrum. Similarly, for transitions involving Σ electronic states in homonuclear diatomic molecules with $I = 0$, alternating lines will be missing in the rotational fine structure spectrum. If the nuclei are not identical (e.g., if one is ^{16}O and the other is ^{17}O), the above missing transitions will be restored.

A second well-known example is $^1H_2$. Since the nuclear spin quantum number of ^1H is $\frac{1}{2}$, the total nuclear spin quantum number I can be 0 or 1. When $I = 0$ the nuclear spin function is antisymmetric with respect to the interchange of the two protons. Conversely, the spin functions with $I = 1$ are symmetric under the same operation. Since transitions between the $I = 0$ and $I = 1$ states are forbidden, one may view the hydrogen molecule as consisting of two distinct species: para-H_2 with $I = 0$, and ortho-H_2 with $I = 1$. The electronic ground state of H_2 is a $^1\Sigma_g^+$ state, and hence para-H_2 can only possess rotational states with even or zero J values in order to preserve the antisymmetric nature of the total wave function; on the other hand, ortho-H_2 will have only odd J rotational states. Of course, in statistical equilibrum at room temperature there will be three times as many H_2 molecules in ortho states than in para states. As a result, the alternating lines in the rotational fine structure spectrum show a 3:1 intensity ratio. Such a ratio in the intensities of the rotational fine structure lines for a general homonuclear diatomic with nuclear spin quantum number I is $(I + 1)/I$: for bosons, it represents the relative statistical weight of symmetric to antisymmetric states; for fermions, the relative statistical weight of antisymmetric to symmetric states. Clearly, the nuclear statistical weights of rotational levels will affect the rotational partition function, and hence have implications in various fields such as thermodynamics and reaction kinetics. For linear molecules with identical nuclei that are interchangeable by rotation, the nuclear statistical weights can be calculated by the methods discussed in Section V.

Finally, let us consider molecules with identical nuclei that are subject to $\hat{C}_n$ $(n \geq 2)$ rotations. For C_{2v} molecules in which the $\hat{C}_2$ rotation exchanges two nuclei of half-integer spin, the nuclear statistical weights of the symmetric and antisymmetric rotational levels will be one and three, respectively. For molecules where $\hat{C}_2$ exchanges two spinless nuclei, one-half of the rotational levels (odd or even J values, depending on the vibrational and electronic states) will be missing. For symmetric and spherical tops, there are three or more identical nuclei interchangeable by rotation. For example, for the symmetric top methyl chloride (CH_3Cl), the $K = 0, 3, 6, \ldots$ rotational levels have twice the statistical weight of the $K = 1, 2, 4, \ldots$ rotational levels for the symmetric vibronic states [56]. The rotational levels with different nuclear statistical weights must be summed over separately. However, for sufficiently high temperatures, it may be a good approximation to calculate the rotational partition function by giving to each rotational level an average statistical weight equal to the total spin multiplicity $(2I_{\mathrm{A}} + 1)(2I_{\mathrm{B}} + 1) \ldots (2I_{\mathrm{Z}} + 1)$ divided by a symmetry number that represents the number of different indistinguishable orientations obtained by proper rotations of the nuclear framework. For example, the symmetry numbers of $C_{\infty v}$ and $D_{\infty h}$ molecules are one and two, respectively. For nonlinear molecules, the symmetry number is equal to the order of the rotational subgroup, for example, 12 for both C_6H_6 and CH_4.

IX. PERMUTATIONAL SYMMETRY OF VIBRATIONAL WAVE FUNCTION

We now consider the permutational properties of the vibrational wave function. Similar to the discussion on the permutational symmetry for homonuclear diatomic molecules (and, in general, $D_{\infty h}$ molecules), in order to find the effect of interchanging the nuclei on the vibrational wave function we first invert the SF coordinates of the nuclei and all displacement vectors, and then carry out a back-inversion of the SF displacement vector coordinates alone. The net effect is therefore just the exchange of the SF coordinates of the nuclei. For $D_{\infty h}$ molecules, such an inversion of the nuclear coordinates exchanges all pairs of identical nuclei. Thus, the first inversion leaves the sign of the vibrational wave function unchanged for $\Sigma^+, \Pi^+, \ldots$ vibrational states, while it changes its sign for $\Sigma^-, \Pi^-, \ldots$ vibrational states. The second inversion classifies the vibrational wave functions in g or u according to whether the back-inversion of the displacement vectors leaves the wave function unchanged (the corresponding operator has a $+1$ eigenvalue) or changes its sign (eigenvalue -1). We conclude that the vibrational wave functions are symmetric for $\Sigma_g^+, \Sigma_u^-, \Pi_g^+, \Pi_u^-, \ldots$ vibrational states while being antisymmetric under permutation of identical nuclei for $\Sigma_g^-, \Sigma_u^+, \Pi_g^-, \Pi_u^+, \ldots$ vibrational states. Finally, we note that the ground vibrational states of homonuclear diatomics and, in general, $D_{\infty h}$ molecules are always symmetric under permutation of identical nuclei.

We now consider planar molecules. The electronic wave function is expressed with respect to molecule-fixed axes, which we can take to be the *abc* principal axes of inertia, namely, by taking the coordinates (x, y, z) in Figure 1 coincided with the principal axes (a, b, c). In order to determine the parity of the molecule through inversions in SF, we first rotate all the displacement vectors and and nuclei (in their equilibrum positions) by 180° about the c axis (which is perpendicular to the molecular plane); and then reflect all the displacement vectors in the molecular ab plane. The first step has no effect on both the displacement vectors and the vibrational wave functions. The second step is a reflection of the displacement vectors in the molecular plane. Note that such a plane is a symmetry plane in this case, and the eigenvalues of the corresponding operator then determine the parity of the vibrational wave function. For a molecule with only in-plane vibrational modes, for example, H_2O, the parity of the vibrational wave function is even. For BF_3, there is a out-of-plane mode. The normal coordinate for such an out-of-plane mode is antisymmetric with respect to reflection in the molecular plane. Thus, the parities of the vibrational eigenfunctions of BF_3 are determined by $(-1)^{v_n}$, where v_n is the vibrational quantum number of the out-of-plane mode.

In order to determine the permutational symmetry of a nonlinear molecule, one can invoke the permutation group. The method is similar to that discussed in Section VI. We can see that the A_2 vibrational state for a molecule of C_{3v} geometry must be antisymmetric with an interchange of the two indentical nuclei. Similarly, the totally symmetric IRREPs in the point group always coorespond to $[n]$ IRREPs in groups $\mathcal{S}_n$. Let us consider the vibrational wave function of the electronic ground-state (B_2) Li_3. Noticing that the correlation between IRREPs of the group $\mathbf{C}_{2v}$ and those of the group $\mathbf{C}_2$, and that the group $\mathbf{C}_2$ is isomorphic to the permutation group $\mathcal{S}_2$, one can say the B_2 vibrational state must be antisymmetric with the interchange of the two indentical nuclei. Thus, for $^7Li_2\,^6Li$ and $^6Li_2\,^7Li$, there are symmetric ($A \cong [2]$ IRREP in the $\mathcal{S}_2$ permutation group) and antisymmetric ($B \cong [1^2]$ in $\mathcal{S}_2$) vibrational states that are allowed by symmetry, and hence can be observed spectroscopically.

However, drastic consequences may arise if the nuclear spin is 0 or $\frac{1}{2}$. In these cases, some rovibronic states cannot be observed since they are symmetry forbidden. For example, in the case of $^{12}C^{16}O_2$, the nuclei are spinless and the nuclear spin function is symmetric under permutation of the oxygen nuclei. Since the ground electronic state is Σ_g^+, only even values of J exist for the ground vibrational level $(v_1, v_2^{l_2}, v_3) = (00^00)$, where (v_1, v_2, v_3) are the quantum numbers of symmetric stretching (ν_1), degenerate bending (ν_2), and antisymmetric stretching (ν_3) normal modes, respectively. As usual, l_2 denotes the quantum number for vibrational angular momentum around the internuclear axis of the linear molecule in the degenerate vibrational mode ν_2, which can be shown to assume the values $l_2 = v_2, v_2 - 2, \ldots, 1$ or 0 (see Appendix D). Similarly, the vibrational mode ν_1 is Σ_g^+, and hence the odd J rotational levels of the (10^00) vibrational state are missing, since they are antisymmetric. Following the same reasoning, the even J rotational levels of the (00^01) vibrational state will be missing due to the fact that the vibrational mode ν_3 has Σ_u^+ symmetry. Most severe consequences arise also when the total nuclear spin quantum

TABLE V
The Symmetry Properties of Wave Functions of $^7Li_3$ Electronically Ground State in $\mathcal{S}_3$ Permutation Group

Total	Nuclear Spin	Rovibronic	Vibronic	Electronic[a]	Rotational[b]	Vibrational
A_2	A_1	A_2	A_1	A_2	A_2	A_2
A_2	A_1	A_2	A_2	A_2	A_1	A_1
A_2	A_1	A_2	E	A_2	E	E
A_2	A_2	A_1	A_1	A_2	A_1	A_2
A_2	A_2	A_1	A_2	A_2	A_2	A_1
A_2	A_2	A_1	E	A_2	E	E
A_2	E	E	A_1	A_2	E	A_2
A_2	E	E	A_2	A_2	E	A_1
A_2	E	E	E	A_2	$A_1 \oplus A_2 \oplus E$	E

[a] At minimum of the lower sheet of potential energy surface.
[b] Rotation about the axis perpendicular to the plane of the molecule.

number is $\frac{1}{2}$. For example, for the ground electronic state of $^1H_3$ and $^3H_3$ at $J = 0$, the vibrational states of A_1 symmetry will not be allowed and only the vibrational states of A_2 and E symmetry can be observed.

As discussed above, the permutational symmetry of the total wave function requires the proper combination of its various contributions. These are summarized in Tables V–XII for all isotopomers of Li_3. Note that the conclusions hold provided that the various wave functions have the appropriate symmetries. If, for some reason, one of the components fails to meet such a requirement, then the symmetry of the total wave function will fail too. For example, even if the vibrational wave functions are properly assigned, the total wave

TABLE VI
The Symmetry Properties of Wave Functions of $^7Li_3$ Electronically First-Excited State in $\mathcal{S}_3$ Permutation Group

Total	Nuclear Spin	Rovibronic	Vibronic	Electronic[a]	Rotational[b]	Vibrational
A_2	A_1	A_2	A_1	E	A_2	E
A_2	A_1	A_2	A_2	E	A_1	E
A_2	A_1	A_2	E	E	E	$A_1 \oplus A_2 \oplus E$
A_2	A_2	A_1	A_1	E	A_1	E
A_2	A_2	A_1	A_2	E	A_2	E
A_2	A_2	A_1	E	E	E	$A_1 \oplus A_2 \oplus E$
A_2	E	E	A_1	E	E	E
A_2	E	E	A_2	E	E	E
A_2	E	E	E	E	$A_1 \oplus A_2 \oplus E$	$A_1 \oplus A_2 \oplus E$

[a] At minimum of the conical intersection on the upper sheet of potential energy surface.
[b] Rotation about the axis perpendicular to the plane of the molecule.

TABLE VII
The Symmetry Properties of Wave Functions of $^6Li_3$ Electronically Ground State in $\mathcal{S}_3$ Permutation Group

Total	Nuclear Spin	Rovibronic	Vibronic	Electronic[a]	Rotational[b]	Vibrational
A_1	A_1	A_1	A_1	A_2	A_1	A_2
A_1	A_1	A_1	A_2	A_2	A_2	A_1
A_1	A_1	A_1	E	A_2	E	E
A_1	A_2	A_2	A_1	A_2	A_2	A_2
A_1	A_2	A_2	A_2	A_2	A_1	A_1
A_1	A_2	A_2	E	A_2	E	E
A_1	E	E	A_1	A_2	E	A_2
A_1	E	E	A_2	A_2	E	A_1
A_1	E	E	E	A_2	$A_1 \oplus A_2 \oplus E$	E

[a] At minimum of the lower sheet of potential energy surface.
[b] Rotation about the axis perpendicular to the plane of the molecule.

TABLE VIII
The Symmetry Properties of Wave Functions of $^6Li_3$ Electronically First-Excited State in $\mathcal{S}_3$ Permutation Group

Total	Nuclear Spin	Rovibronic	Vibronic	Electronic[a]	Rotational[b]	Vibrational
A_1	A_1	A_1	A_1	A_2	A_1	A_2
A_1	A_1	A_1	A_2	A_2	A_2	A_1
A_1	A_1	A_1	E	A_2	E	E
A_1	A_2	A_2	A_1	A_2	A_2	A_2
A_1	A_2	A_2	A_2	A_2	A_1	A_1
A_1	A_2	A_2	E	A_2	E	E
A_1	E	E	A_1	A_2	E	A_2
A_1	E	E	A_2	A_2	E	A_1
A_1	E	E	E	A_2	$A_1 \oplus A_2 \oplus E$	E

[a] At minimum of the conical intersection on the upper sheet of potential energy surface.
[b] Rotation about the axis perpendicular to the plane of the molecule.

TABLE IX
The Symmetry Properties of Wave Functions of $^7Li_2\,^6Li$ Electronically Ground State in $\mathcal{S}_2$ Permutation Group

Total	Nuclear Spin	Rovibronic	Vibronic	Electronic[a]	Rotational[b]	Vibrational
B	A	B	A	B	B	B
B	A	B	B	B	A	A
B	B	A	A	B	A	B
B	B	A	B	B	B	A

[a] At minimum of the lower sheet of potential energy surface.
[b] Rotation about the axis through the ^6Li and perpendicular to the $^7Li_2$.

TABLE X
The Symmetry Properties of Wave Functions of $^7Li_2$ ^6Li Electronically First-Excited State in $\mathcal{S}_2$ Permutation Group

Total	Nuclear Spin	Rovibronic	Vibronic	Electronic[a]	Rotational[b]	Vibrational
B	A	B	A	$A \oplus B$	B	$A \oplus B$
B	A	B	B	$A \oplus B$	A	$B \oplus A$
B	B	A	A	$A \oplus B$	A	$A \oplus B$
B	B	A	B	$A \oplus B$	B	$B \oplus A$

[a] At minimum of the conical intersection on the upper sheet of potential energy surface.
[b] Rotation about the axis through the ^6Li and perpendicular to the $^7Li_2$.

TABLE XI
The Symmetry Properties of Wave Functions of $^6Li_2$ ^7Li Electronically Ground State in $\mathcal{S}_2$ Permutation Group

Total	Nuclear Spin	Rovibronic	Vibronic	Electronic[a]	Rotational[b]	Vibrational
A	A	A	A	B	A	B
A	A	A	B	B	B	A
A	B	B	A	B	B	B
A	B	B	B	B	A	A

[a] At minimum of the lower sheet of potential energy surface.
[b] Rotation about the axis through the ^7Li and perpendicular to the $^6Li_2$.

TABLE XII
The Symmetry Properties of Wave Functions of $^6Li_2$ ^7Li Electronically First-Excited State in $\mathcal{S}_2$ Permutation Group

Total	Nuclear Spin	Rovibronic	Vibronic	Electronic[a]	Rotational[b]	Vibrational
A	A	A	A	$A \oplus B$	A	$A \oplus B$
A	A	A	B	$A \oplus B$	B	$B \oplus A$
A	B	B	A	$A \oplus B$	B	$A \oplus B$
A	B	B	B	$A \oplus B$	A	$B \oplus A$

[a] At minimum of the conical intersection on the upper sheet of potential energy surface.
[b] Rotation about the axis through the ^7Li and perpendicular to the $^6Li_2$.

function of systems with conical intersections such as Li_3 may have no physical significance due to failure of the electronic wave function to meet the requirement of single valuedness (i.e., no change of sign when traversing a path that encircles the crossing point). In other words, one needs to include GP effects or treat the dynamics more accurately (e.g., by solving the 2×2 coupled state dynamics problem) in order to warrant the correct symmetry properties of the total wave function. This will be further discussed in Section X.

X. CASE STUDIES: Li_3 AND OTHER ^2S SYSTEMS

A. Potential Energy Surfaces

H_3 (and its isotopomers) and the alkali metal trimers (denoted generally for the homonuclears by X_3, where X is an ^2S atom) are typical Jahn–Teller systems where the two lowest adiabatic potential energy surfaces conically intersect. Since such manifolds of electronic states have recently been discussed [60] in some detail, we review in this section only the diabatic representation of such surfaces and their major topographical details. The relevant 2×2 diabatic potential matrix $\mathbf{W}$ assumes the form

$$\mathbf{W} = \begin{pmatrix} W_{11} & W_{12} \\ W_{21} & W_{22} \end{pmatrix} \tag{54}$$

where $W_{12} = W_{21}$. Specifically, for H_3 and Li_3, two systems that we discuss in detail in this chapter, the matrix elements in Eq. (54) are written as [61]

$$W_{11} = \sum_{i=1}^{3} \mathcal{Q}_i' + X_{\mathrm{EHF}}^{(3)} + \frac{1}{2}(2\mathcal{J}_1' - \mathcal{J}_2' - \mathcal{J}_3') + V_{\mathrm{dc}} \tag{55}$$

$$W_{22} = \sum_{i=1}^{3} \mathcal{Q}_i' + X_{\mathrm{EHF}}^{(3)} - \frac{1}{2}(2\mathcal{J}_1' - \mathcal{J}_2' - \mathcal{J}_3') + V_{\mathrm{dc}} \tag{56}$$

$$W_{12} = W_{21} = \frac{\sqrt{3}}{2}(\mathcal{J}_2' - \mathcal{J}_3') \tag{57}$$

where the $\mathcal{Q}$ and $\mathcal{J}$ values are the well-known Coulomb and exchange integrals that can be obtained semiempirically [62–65] from the lowest singlet and triplet diatomic potential curves, $X_{\mathrm{EHF}}^{(3)}$ is a three-body extended Hartree–Fock type energy, and V_{dc} the total dynamical correlation energy. Note that the prime in the $\mathcal{Q}$ and $\mathcal{J}$ parameters express the fact that such quantities are calculated from the extended Hartree–Fock curves alone [61].

Diagonalization of $\mathbf{W}$ then leads to the two adiabatic surfaces

$$V_{\pm} = \frac{1}{2}[(W_{11} + W_{22}) \pm \sqrt{(W_{11} - W_{22})^2 + 4W_{12}^2}] \tag{58}$$

which may cross when $W_{11} = W_{22}$ and $W_{12} = 0$. For X_3 systems, such a crossing seam is representative of a so-called conical intersection: For a fixed perimeter of the molecule, the crossing seam corresponds to the appex of the double cone defined by the two adiabatic potential energy surfaces V_+ and V_-. Although these become nondegenerate due to the so-called Jahn–Teller effect (see Section X.B), the degeneracy at the locus of conical intersection remains. The location of this crossing seam is defined by the conditions $r_{AB} = r_{BC} = r_{AC}$, where r_{AB}, r_{BC}, and r_{AC} are the interatomic distances. Thus, for homonuclear systems such as Li_3, the conical intersection occurs for D_{3h} symmetries but, for the heteronuclear systems, they may arise at lower symmetries or even do not occur at all [5,66,67]. Clearly, the potential matrix defined in Eq. (54) has the correct asymptotic behavior in the vicinity of the conical intersection (see Appendix E).

In the remainder of this section, we focus on the two lowest doublet states of Li_3. Figures 3 and 4 show relaxed triangular plots [68] of the lower and upper sheets of the Li_3 DMBE III [69,70] potential energy surface using hyperspherical coordinates. Each plot corresponds to a stereographic projection of the

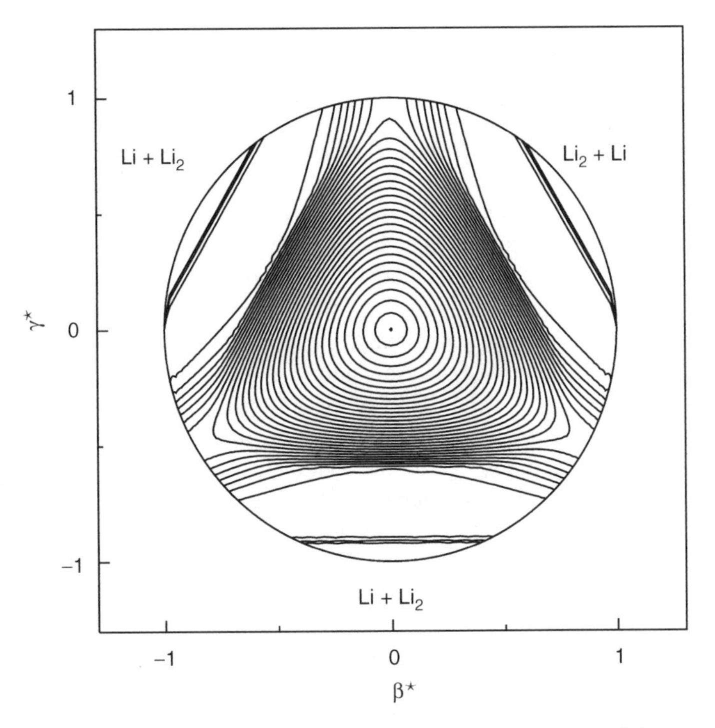

Figure 4. Relaxed triangular plot [68] of the Li_3 first-excited state potential energy surface using hyperspherical coordinates. Contours are given by the expression $E_n(\mathrm{eV}) = -0.56 + 0.045(n-1)$ with $n = 2, 3, \ldots$. The dissociation limit indicated by the dense contouring implies $Li_2(b^3\sum_u^+) + Li$.

surface of an upper one-half of the sphere. The $\beta^{\star}$ coordinate is associated to $\sin(\theta/2)\cos\varphi$, while $\gamma^{\star}$ denotes $\sin(\theta/2)\sin\varphi$. The hyperangle θ runs from zero at the north pole (center of plot) to $\pi/2$ at the equator (the outside circle). The hyperangle φ is measured from the positive $\gamma^{\star}$ axis and grows on going counterclockwise. For the lower sheet, it is noted that the lowest point along the D_{3h} conical intersection seam is located at the origin of the plot and corresponds to an equilateral triangular configuration. As can be seen, the threefold symmetry gives rise to three wells that are equally spaced by 120° intervals around the origin. The minimum energy of the barrier for pseudorotation relative to the bottom of such wells (i.e., the height of the saddle points between the three wells), and the energy of the lowest point along the conical intersection seam are[6] 0.4 meV and 0.0542 eV, respectively. Note that the motion along the hyperradius ρ corresponds to the symmetric stretching mode. Moreover, at the bottom of the well just above the origin, motion along the $\gamma^{\star}$ axis corresponds to the bending mode while motion along the $\beta^{\star}$ axis corresponds to the antisymmetric stretching mode. In addition, motion along the hyperangle φ about the origin corresponds to the pseudorotational motion. Finally, note that the origin in the plot of the upper sheet corresponds to an equilateral triangular geometry.

B. Static Jahn–Teller Effect

Now, we examine the effect of vibronic interactions on the two adiabatic potential energy surfaces of nonlinear molecules that belong to a degenerate electronic state, so-called static Jahn–Teller effect.

For a X_3 molecule in the $\mathbf{D}_{3h}$ symmetry point group, we have a totally symmetric A_1' and a doubly degenerate E' vibrational normal modes in the harmonic-oscillator approximation as illustrated in Figure 5. However, for a real molecule that vibrates anharmonically, we must consider the effects due to anharmonicity. As discussed in Appendix D, we must then use the set of quantum numbers (v_1, v_2, l_2) instead of (v_1, v_{2a}, v_{2b}), and employ the notation $(v_1, v_2^{l_2})$ to label the vibrational levels; l_2 is the vibrational angular momentum quantum number with respect to the symmetry axis. Table XIII gives the assignments of the lowest vibrational levels for Li_3 in the $\mathbf{D}_{3h}$ symmetry point group. From this assignment, one can determine the symmetry of the level. For example, the ground vibrational state $(0, 0^0)$ is A_1' and the level $(1, 1^1)$ is E', since v_1 is totally symmetric and $A_1' \otimes E' = E'$. In turn, for multiply excited degenerate vibrations [28,59], the symmetry A_1' corresponds to $l_2 = 0$, $A_1' \oplus A_2'$ to $l_2 = 3, 6, \ldots$ and E' to $l_2 = 1, 2, 4, 8, \ldots$; note that l_2 is a good quantum number, and hence can be used to specify the symmetry of the vibrational level. For example, the level $(0, 3^1)$ has E' symmetry while $(0, 3^3)$ is $A_1' \oplus A_2'$. Also

[6] Units conversion factors are a.u. of energy $= E_h = 27.211652$ eV $= 4.3598$ aJ $= 2.194746 \times 10^5$ cm^{-1}; a.u. of bond length $= a_0 = 0.529177$ Å $= 0.529177 \times 10^{-10}$ m.

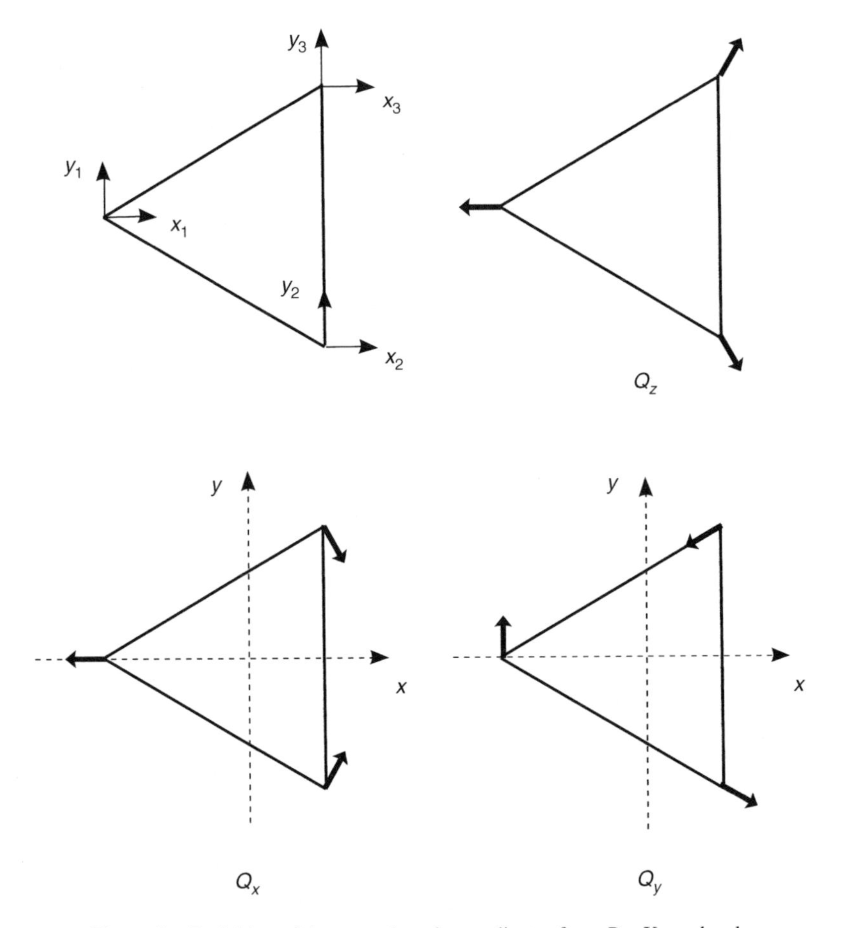

Figure 5. Definition of the normal mode coordinates for a D_{3h} X_3 molecule.

note that anharmonicity splits the degeneracy of the vibrational levels obtained in the harmonic approximation [28].

We now take vibronic interactions into account. In this case, we must determine vibronic states rather than the electronic and vibrational ones. For example, if X_3 in a degenerate E' vibration is singly excited in an E' electronic state, we obtain the vibronic states ${}^{ev}A_1' \oplus {}^{ev}A_2' \oplus {}^{ev}E'$, since ${}^{v}E' \otimes {}^{e}E' = {}^{ev}A_1' \oplus {}^{ev}A_2' \oplus {}^{ev}E'$. If the same vibration is doubly excited (e.g., if $v_2 = 2$, with the symmetric product being $[{}^{v}E' \otimes^{v} E'] = {}^{v}A_1' \oplus {}^{v}E'$: Note that the associated antisymmetric product is ${}^{v}A_2'$), we get the vibronic species $({}^{v}A_1' \oplus {}^{v}E') \otimes {}^{e}E' = {}^{ev}A_1' \oplus {}^{ev}A_2' \oplus 2{}^{ev}E'$. Table XIII shows the symmetries of the lowest 25 vibrational and vibronic states. In turn, the lowest 26 levels calculated for Li_3

TABLE XIII
Vibronic Species of the Vibrational States of Li_3 with Consideration of Geometric Phase Effect[a]

Assignment $(v'_1, v''^{l_2}_2)$ in $\mathbf{D}_{3h}$	Symmetry of Vibrational States	Symmetry of Vibronic States	Assignment[b] (v_1, v_2, v_3) in $\mathbf{C}_{2v}$
$(0, 0^0)$	A'_1	E'	$E'(0,0,0)$
$(0, 1^1)$	E'	$A'_1 \oplus A'_2 \oplus E'$	$A'_2(0,0,0)$, $A'_1(0,0,0)$, $E'(0,0,1)$
$(1, 0^0)$	A'_1	E'	$E'(0,1,0)$
$(1, 1^1)$	E'	$A'_1 \oplus A'_2 \oplus E'$	$A'_2(0,0,1)$, $E'(1,0,0)$, $A'_1(0,0,1)$
$(0, 2^0)$	A'_1	E'	$E'(0,0,2)$
$(0, 2^2)$	E'	$A'_1 \oplus A'_2 \oplus E'$	$A'_2(0,1,0)$, $E'(0,1,1)$, $A'_1(0,1,0)$
$(1, 2^0)$	A'_1	E'	$E'(0,2,0)$
$(1, 2^2)$	E'	$A'_1 \oplus A'_2 \oplus E'$	$A'_2(1,0,0)$, $E'(1,0,1)$, $A'_1(1,0,0)$
$(0, 3^1)$	E'	$A'_1 \oplus A'_2 \oplus E'$	$A'_2(0,0,2)$, $E'(0,0,3)$, $A'_1(0,0,2)$
$(0, 3^3)$	$A'_1 \oplus A'_2$	$2E'$	$E'(1,1,0)$, $E'(0,1,2)$
$(2, 2^0)$	A'_1	E'	$E'(2,0,0)$
$(2, 2^2)$	E'	$A'_1 \oplus A'_2 \oplus E'$	$A'_2(0,1,1)$, $E'(0,2,1)$ $A'_1(0,1,1)$

[a] For simplicity, the left superscripts v and ev are omitted in denoting the vibrational and vibronic states.

[b] In this assignment, we keep the symmetry species of the vibronic state in $\mathbf{D}_{3h}$ but indicate the vibrational quantum numbers for the C_{2v} normal modes. The energy increases from left to right, and up to down.

using the Li_3 double many-body expansion [58,60,71] potential energy surface (DMBE III [69,70]) are shown in Tables XIV and XV.

Now consider the splitting of the potential energy surface for nontotally symmetric (i.e., ${}^vE'$ in $\mathbf{D}_{3h}$) displacements of the nuclei. For such geometries the symmetry is lower, and in general the electronic states become nondegenerate (e.g., ${}^eA_1 \oplus {}^eB_2$ in $\mathbf{C}_{2v}$) instead of being doubly degenerate (${}^eE'$ in $\mathbf{D}_{3h}$). Thus, for displaced positions of the nuclei, we obtain two nondegenerate electronic states of different energy. Jahn and Teller [72] were the first to show that, for a nonlinear molecule, there is always one nontotally symmetric normal mode at least that causes a splitting of the potential energy surface such that the minima do not occur at the most symmetric geometry. They are rather at a certain distance from the most symmetric configuration, with the distance increasing with the magnitude of the vibronic interaction. Consequently, several equivalent minima arise on the potential energy surface for unsymmetric molecular conformations. If the vibronic interaction is strong, a significant amount of vibrational energy may then be required to bring the molecule from one minimum to another, and hence one must regard the molecule as nonsymmetric. Conversely, for weak vibronic interactions, only a small amount of vibrational energy may suffice to make the system flow from one minimum to another. In this case, the molecule may be regarded as symmetric, and the vibronic interaction is treated as a perturbation. Appendix E gives a proof of the Jahn–Teller theorem for a X_3 molecule following Moffitt and Liehr [73].

TABLE XIV
The Losest 26 Energy Levels (in eV) for Ground State Li_3 Without Consideration of the GP Effect

Number	E_n(eV)	Symmetry of Vibration	Assignment (v_1, v_2, v_3) in $\mathbf{C}_{2v}$
1	−0.52816464978	A_1	(0, 0, 0)
2	−0.52086641058	E	(0, 0, 0)
3	−0.50562197957	E	(0, 0, 1)
4	−0.50240784221	A_1	(0, 0, 1)
5	−0.48957138875	A_2	(0, 0, 0)
6	−0.48950339062	E	(0, 0, 2)
7	−0.48710210168	A_1	(0, 1, 0)
8	−0.48054432662	A_1	(1, 0, 0)
9	−0.47611328568	E	(0, 1, 0)
10	−0.47273766106	A_1	(0, 0, 2)
11	−0.47267009384	E	(0, 0, 3)
12	−0.46728845390	E	(1, 0, 0)
13	−0.46008142773	E	(0, 1, 1)
14	−0.45726050697	A_1	(0, 1, 1)
15	−0.45698549489	E	(0, 0, 4)
16	−0.45570955951	A_2	(0, 0, 1)
17	−0.45041884404	A_1	(1, 0, 1)
18	−0.44765547757	E	(1, 0, 1)
19	−0.44399503522	E	(0, 1, 2)
20	−0.44373568206	A_1	(0, 0, 3)
21	−0.44362378636	A_2	(0, 1, 0)
22	−0.43949743434	E	(0, 0, 5)
23	−0.43945450351	A_1	(0, 2, 0)
24	−0.43472274424	A_1	(1, 1, 0)
25	−0.43394121998	E	(1, 0, 2)
26	−0.43100452644	E	(0,2,0)

The treatment of the Jahn–Teller effect for more complicated cases is similar. The general conclusion is that the appearance of a linear term in the off-diagonal matrix elements H_{+-} and H_{-+} leads always to an instability at the most symmetric configuration due to the fact that integrals of the type $\langle \psi_+ | \hat{h}_1^+ | \psi_- \rangle$ do not vanish there when the product $\psi_+^\star \psi_-$ has the same species as a nontotally symmetric vibration (see Appendix E). If Γ is the species of the degenerate electronic wave functions, the species of $\psi_+^\star \psi_-$ will be that of Γ^2, which is the symmetric product of Γ with itself. For example, for a D_{3h} molecule, we have the symmetric product $[{}^eE' \otimes {}^eE'] = {}^eA_1' \oplus {}^eE'$ (the associated antisymmetric product is $\{{}^eE' \otimes {}^eE'\} = {}^eA_2'$), and hence it is the ${}^vE'$ degenerate normal mode that causes the instability since it has the same symmetry as the ${}^eE'$ term which arises from the symmetric product. For a D_{4h} molecule, we have the symmetric product $\left[{}^eE_g \otimes {}^eE_g\right] = {}^eA_{1g} \oplus {}^eB_{1g} \oplus {}^eB_{2g}$

TABLE XV
The Lowest 26 Energy Levels (in eV) for Ground State Li_3 with Consideration of the GP Effect

Number	E_n(eV)	Symmetry of Vibration	Assignment (v_1, v_2, v_3) in $\mathbf{C}_{2v}$
1	−0.52524282512	E	(0, 0, 0)
2	−0.51783314253	A_1	(0, 0, 0)
3	−0.50903972588	A_2	(0, 0, 0)
4	−0.50128197060	E	(0, 0, 1)
5	−0.49188205106	E	(0, 1, 0)
6	−0.48752020977	A_1	(0, 0, 1)
7	−0.48169581419	E	(1, 0, 0)
8	−0.47506737423	E	(0, 0, 2)
9	−0.47429223744	A_2	(0, 0, 1)
10	−0.47257962805	A_1	(0, 1, 0)
11	−0.47109350982	E	(0, 1, 1)
12	−0.46486144408	A_2	(0, 1, 0)
13	−0.45992435398	A_1	(1, 0, 0)
14	−0.45743322309	E	(0, 2, 0)
15	−0.45609295233	A_1	(0, 0, 2)
16	−0.45572118311	E	(1, 0, 1)
17	−0.45380222742	A_2	(1, 0, 0)
18	−0.44725687996	E	(0, 0, 3)
19	−0.44322850657	E	(1, 1, 0)
20	−0.44211312570	A_1	(0, 1, 1)
21	−0.44025040532	E	(0, 1, 2)
22	−0.44004503298	A_2	(0, 0, 2)
23	−0.43625851187	E	(2, 0, 0)
24	−0.43557972347	E	(0, 2, 1)
25	−0.43004385753	A_2	(0, 1, 1)
26	−0.42970907209	A_1	(1, 0, 1)

(note that the associated antisymmetric product is $\{^eE_g \otimes {}^eE_g\} = {}^eA_{2g}$), and hence it is either the $^vB_{1g}$ or $^vB_{2g}$ normal modes that cause the instability, since they have the same nontotally symmetric behavior as the $^eB_{1g}$ and $^eB_{2g}$ terms that arise from the symmetric product.

C. Dynamical Jahn–Teller and Geometric Phase Effects

We begin by discussing the energy levels that arise when a Jahn–Teller instability is present, that is, the dynamic Jahn–Teller effect and the related GP effect. This stems from the observation made by Longuet-Higgins and Herzberg [42–44] that a real-valued electronic wave function changes sign when the nuclear coordinates traverse a closed path encircling a conical intersection. This result has been shown [74] to be valid even for systems that show no symmetry such as LiNaK. In fact, such a geometric phase effect has been rediscovered in a wider context by Berry [21], and hence it is often referred to as the Berry's phase effect (as pointed out in the introduction, a further designation [19] is the molecular Aharonov–Bohm effect since it manifests also in the treatment of a charged particle moving in the presence of a magnetic solenoid).

To be specific, we focus this discussion on studies of the vibrational spectrum of ground state Li_3, which we have carried out using the DMBE III [69,70] Li_3 potential energy surface. All eigenvalues of the system Hamiltonian have been calculated using the MINRES filter diagonalization technique [27]. In turn, the action of the Hamiltonian operator on the nuclear wave function has been evaluated by the spectral transform method in hyperspherical coordinates by using a fast-Fourier transform for ρ and φ and a DVR–FBR transformation for θ^2. In such studies, the GP effect has also been taken into consideration. Thus, six separate sets of calculations have been performed, which include (1) no consideration of GP effect using a basis set of A_1 symmetry; (2) no consideration of GP effect using a basis set of A_2 symmetry; (3) no consideration of GP effect using a basis set of E symmetry; (4) consideration of GP effect using a basis set of A_1 symmetry; (5) consideration of GP effect using a basis set of A_2 symmetry; (6) consideration of GP effect using a basis set of E symmetry. The total number of calculated eigenvalues amounted to 3524 without consideration of GP effect, and 3211 with consideration of GP effect. The full spectra have therefore been calculated, which cover the full range of energies up to the threshold for $Li_2(X^1\Sigma_g^+) + Li$ dissociation. Of the total number of calculated vibrational levels, 953 (920), 750 (817), and 1621 (1474) have been found to belong to A_1, A_2, and E symmetries when GP effects were not (were) taken into consideration. Figures 6–8 show the lowest 40 calculated levels of A_1, A_2, and E symmetries and the corresponding assignments. As one would expect, each vibrational level is associated to three different vibronic levels, for example, the (0,0,0) vibrational level in $\mathbf{C}_{2v}$ is related to the ${}^vA_1'(0,0,0)$, ${}^vA_2'(0,0,0)$, and ${}^vE'(0,0,0)$ levels in $\mathbf{D}_{3h}$. Note that in an obvious correspondence the vibrational states ${}^vA_1'(v_1, v_2, v_3)$, ${}^vA_2'(v_1, v_2, v_3)$, and ${}^vE'(v_1, v_2, v_3)$ are associated to the vibronic states ${}^{ev}A_2'(v_1, v_2, v_3)$, ${}^{ev}A_1'(v_1, v_2, v_3)$, and ${}^{ev}E'(v_1, v_2, v_3)$; see Tables XIII–XV. Note also that the notation (v_1, v_2, v_3) implies that the quantum numbers are associated to the symmetric stretching, bending, and asymmetric streching vibrational modes in $\mathbf{C}_{2v}$. They are of A_1, A_1, and B_2 symmetries, respectively; the correlation between the assignments in $\mathbf{D}_{3h}$ and $\mathbf{C}_{2v}$ is given in Table XIII.

For very small vibronic coupling, the quadratic terms in the power series expansion of the electronic Hamiltonian in normal coordinates (see Appendix E) may be considered to be negligible, and hence the potential energy surface has rotational symmetry but shows no separate minima at the bottom of the moat. In this case, the pair of vibronic levels A_1 and A_2 in $\mathcal{S}_3$ become degenerate by accident, and the D_{3h} quantum numbers (v_1, v_2, l_2) may be used to label the vibronic levels of the X_3 molecule. When the coupling of the

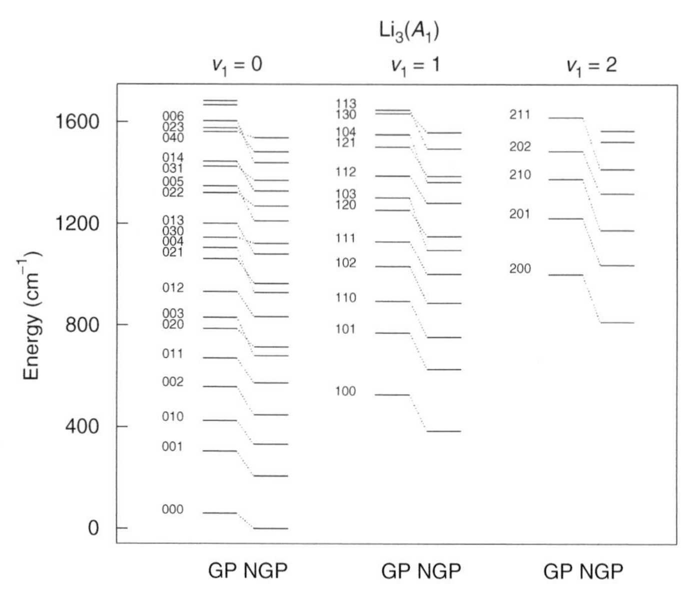

Figure 6. The vibrational levels of the lowest 40 bound states of A_1 symmetry for $^7Li_3$ calculated without consideration and with consideration of GP effect.

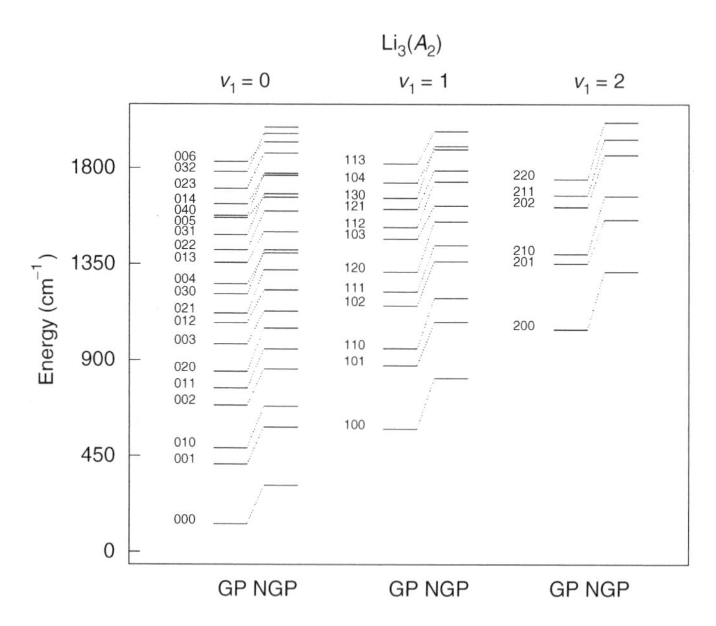

Figure 7. As in Figure 6 for the lowest 40 bound states of A_2 symmetry.

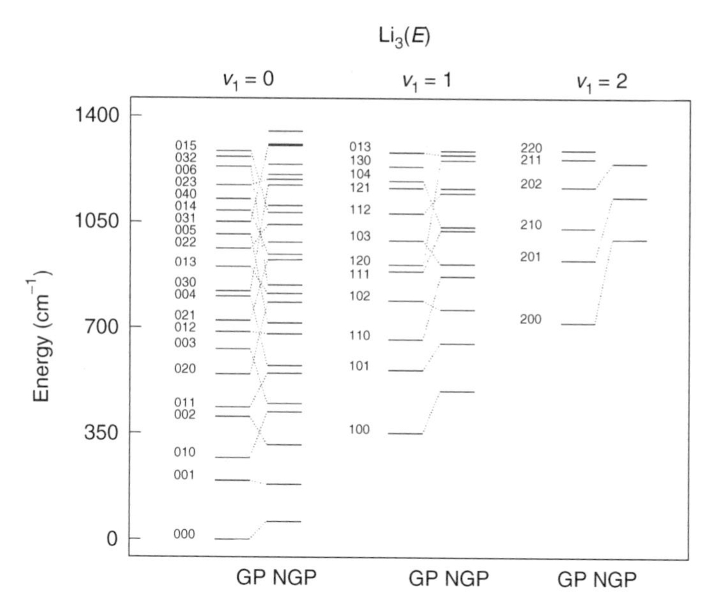

Figure 8. As in Figure 6 for the lowest 40 bound states of E symmetry.

vibrational and electronic motions is strong, (v_1, v_2, l_2) are no longer a good set of quantum numbers. In this case, it is more reasonable to think of the vibronic motion as a result of a vibration of an oscillator with C_{2v} symmetry and a rotation corresponding to the motion on the potential valley around the D_{3h} symmetry axis. Such motion may also be viewed as arising from the circular motion of each Li nucleus around their equilibrium positions in the D_{3h} geometry. As we have noted, there is a vibrational angular momentum on the symmetry axis, and hence the rotation of the molecule as a whole is allowed even when the total angular momentum vanishes; indeed, for $l_2 \neq 0$, the molecule must possess a genuine rotation to compensate the vibrational angular momentum such that $J = 0$.

The spectra of floppy molecules such as Li_3 may therefore have different interpretations. For example, the spectra of specific symmetries have been fitted [11] to within a few percent of error by using the simple vibrational normal mode formula

$$E(v_1, v_2, v_3) = \sum_{i=1}^{3}\left(v_i + \frac{1}{2}\right)\omega_i + \sum_{j \geq i=1}^{3}\left(v_i + \frac{1}{2}\right)\left(v_j + \frac{1}{2}\right)x_{ij} + \sum_{k \geq j \geq i=1}^{3}\left(v_i + \frac{1}{2}\right)\left(v_j + \frac{1}{2}\right)\left(v_k + \frac{1}{2}\right)y_{ijk} \tag{59}$$

where v_i are the vibrational quantum numbers, ω_i are (in energy units) the harmonic frequencies, and x_{ij} and y_{ijk} are higher order anharmonic corrections. Alternatively, Eq. (59) may be interpreted in terms of pseudorotational energies. According to this interpretation, the first and part of the second summations would represent the pseudorotational energy of a $\mathcal{S}_3$ molecule, while the remaining terms would contain the second and higher order coupling pseudorotational terms; the pseudorotational quantum numbers are now defined by $j_i = \left(v_i + \frac{1}{2}\right)$ for $i = 1, 2, 3$. For X_3 molecules, $j_i = \frac{1}{2}, \frac{5}{2}, \frac{7}{2}, \cdots$ will belong to E species, while $j_i = \frac{3}{2}, \frac{9}{2}, \frac{15}{2}, \cdots$ to A_1 or A_2 [28] (see also the discussion later about the quantum number m).

Since there are potential barriers along the pseudorotational path, one must also consider the effects due to tunneling. Each specified level then splits into three owing to such tunneling effects. In fact, now we have distinct zero-point energies for each of the three vibronic states, with their energy differences being determined by the pseudorotational motion that includes the tunneling effects.

The vibronic motion may be described by using the (ρ, θ, φ) coordinates. In particular, the wave functions for the pseudorotational motion along the hyperangle φ that encircles the origin in a X_3 system may assume the form [11]

$$\chi_{pr}(\varphi) = f(\rho, \theta)\exp\left(i\frac{l\varphi}{2}\right)\exp(in\varphi) \qquad n = 0, \pm 1, \pm 2, \ldots \qquad l = 0, 1 \quad (60)$$

where $l = 0(1)$ for the case without (with) consideration of the GP effect. Clearly, they are eigenfunctions of the kinetic energy operator [11] $\hat{K}_\varphi$ with eigenvalues given by

$$m = \left(n + \frac{l}{2}\right)^2 \quad (61)$$

Thus, we can use the approximate quantum number m to label such levels. Moreover, it may be shown [11] that (1) $\sqrt{m}$ is one-half of an integer for the case with consideration of the GP effect, while it is an integer or zero for the case without consideration of the GP effect; (2) the lowest level must have $m = 0$ and be a singlet with A_1 symmetry in $\mathcal{S}_3$ when the GP effect is not taken into consideration, while the first excited level has $m = 1$ and corresponds to a doublet E; conversely, with consideration of the GP effect, the lowest level must have $m = \frac{1}{4}$ and be a doublet with E symmetry in $\mathcal{S}_3$, while the first excited level corresponds to $m = \frac{9}{4}$ and is a singlet A_1. Note that such a reversal in the ordering of the levels was discovered previously by Hancock et al. [59]. Note further that $j_\varphi = \sqrt{m}$ has a meaning similar to the j_i quantum numbers described after Eq. (59). The full set of quantum numbers would then be $(j_\rho, j_\theta, j_\varphi)$, which could be employed for a description of the vibronic motion. The energies of the vibronic levels will then assume the form

$$E(j_\rho, j_\theta, j_\varphi) = \sum_{i=\rho,\theta,\varphi} j_i\omega_i + \sum_{j\geq i} j_i j_j x_{ij} + \sum_{k\geq j\geq i} j_i j_j j_k y_{ijk} \tag{62}$$

where ω_i are the frequencies, $x_{\varphi\varphi}$ is the pseudorotational constant for the φ motion in energy units, and so on for the other x_{ij} and y_{ijk} $(i, j, k \equiv \rho, \theta, \varphi)$ coefficients.

Since Li_3 in its electronic ground doublet state is a very floppy molecule and the vibrational levels are dense, one may expect the vibrational spectrum to be irregular. To understand such behavior, the Li_3 vibrational spectrum has been analyzed statistically with basis on random matrix theory [75]. It has been found [11] that the full spectrum is more regular than each symmetry block per se, which can be understood by recalling that the levels can interact with each other within a symmetry block, while the full spectrum consists of a random superposition of unrelated sequences of energy levels belonging to different symmetries. As discussed in detail elsewhere [11], the spectrum is found to be quasiregular in short range and quasiirregular in long range. It should also be mentioned that the interactions among the levels of the same symmetry result mainly from the so-called Fermi resonances that occur when two or more levels become nearly degenerate and have the same symmetry. Fermi resonances are produced by the anharmonicity of the potential energy surface and make near-degenerate levels (in the harmonic-oscillator approximation) to repel each other. Thus, they originate extra irregularity in the spectrum of a specific symmetry.

One may think that the dynamical Jahn–Teller effect is equivalent to take into consideration the GP effect that arises, for example, due to the conical intersection in X_3-type systems formed from ^2S atoms. We find it pedagogical from our calculations for Li_3 to distinguish three situations. The first corresponds to the calculations in sets (1)–(3) mentioned above, that is, using only one electronic adiabatic BO potential energy surface without consideration of the GP effect. The second refers to the generalized BO treatment in which the GP effect is also considered. Finally, one has the exact (or nearly exact) solution, which is obtained by solving the multistate quantum dynamics (vibronic) problem; see also Section X.D. If one thinks of the first approach as corresponding to include the dynamic Jahn–Teller effect alone, then Table XIV shows that such an effect has a remarkable importance on the vibrational levels as it may be seen by comparing states with equal sets of quantum numbers (v_1, v_2, v_3) in the $\mathbf{C}_{2v}$ point group. In turn, the calculations of sets (4)–(6) would include both the dynamic Jahn–Teller and GP effects. The difference between the above two series of results were then attributable to the GP effect alone; see Figures 6–8. This is found [11] to lead to further shifts of the energy levels, while playing a significant role in the bound vibrational states of Li_3, not only quantitative but also qualitative.

Note that all genuine vibronic degeneracies remain no matter how strong the vibronic interaction is, due to the fact that the permutation–inversion symmetry is unchanged, and hence the potential function retains its original symmetry. For example, the degenerate ${}^{ev}E'$ vibronic levels cannot split even when large amplitude motions are possible. Only the interaction with rotation can produce such a splitting [28].

Next, we discuss the $J = 0$ calculations of bound and pseudobound vibrational states reported elsewhere [12] for Li_3 in its first-excited electronic doublet state. A total of 1944 (1675), 1787 (1732), and 2349 (2387) vibrational states of A_1, A_2, and E symmetries have been computed without (with) consideration of the GP effect up to the $Li_2(b^3\sum_u^+) + Li$ dissociation threshold of -0.0422 eV. Figure 9 shows the energy levels that have been calculated without consideration of the GP effect up to the dissociation threshold of the lower surface, -1.0560 eV, in a total of 41, 16, and 51 levels of A_1, A_2, and E symmetries. Note that they are genuine bound states. On the other hand, the cone states above the dissociation energy of the lower surface are embedded in a continuum, and hence appear as resonances in scattering experiments or long-lived complexes in unimolecular decay experiments. They are therefore pseudobound states or resonance states if the full two-state nonadiabatic problem is considered. The lowest levels of A_1, A_2, and E symmetries lie at -1.4282,

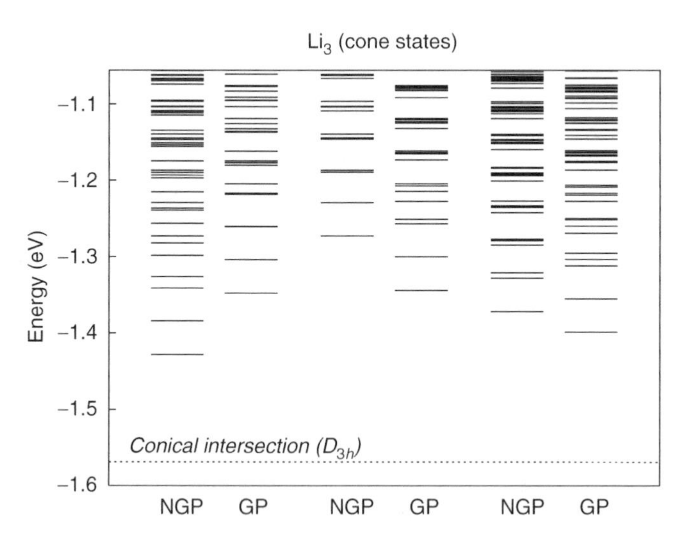

Figure 9. Vibrational levels of the first-excited state Li_3 calculated without consideration (NGP) and with consideration (GP) of geometric phase effect [12].

-1.2729, and $-1.3716\,\mathrm{eV}$, respectively; the minimum energy of the conical intersection seam is $-1.5692\,\mathrm{eV}$. Dynamical Jahn–Teller effects are therefore seen to have a drastic influence on the vibrational levels (resulting in energy shifts of $\sim 0.1\,\mathrm{eV}$). As one would expect, the lowest level in the complete spectrum without consideration of GP effects has A_1 symmetry [11].

The assignment of vibrational quantum numbers to the fundamental modes has been done by inspection of the corresponding nuclear probability density plots. Similarly to the plots obtained for the lower sheet [11], the nuclear probability densities of A_1 and A_2 symmetries show a threefold symmetry, while each component of E symmetry has twofold symmetry about the line $\beta = 0$. The nuclear probability densities of the lowest vibrational states concentrate at narrower regions than the corresponding states for the lower sheet, which reflects the fact that the topographies of the two sheets of the potential energy surface are quite different. Although the lowest 100 levels or so obtained for the lower sheet of the potential energy surface could be fit reasonably well using a C_{2v} normal mode expansion, this is not at all the case for the levels in the upper sheet where only the few lowest levels can be unambigously assigned. Indeed, the higher states are rather anharmonic, and nearly degenerate vibrational states occur frequently. Thus, the normal mode scheme employed for the assignment of the vibrational levels in the lower sheet cannot be used in this case, a difficulty that can be attributed to the conical intersection seam. Note that the five lowest levels of A_1 symmetry have been assigned by ascending order of energy as $(0,0^0)$, $(1,0^0)$, $(2,0^0)$, $(0,1^1)$, and $(3,0^0)$. Note especially that the notation $(v_1, v_2^{l_2})$ is used here (and hereafter) to denote the quantum numbers for the symmetric stretching mode (v_1), the degenerate mode (v_2), and vibrational angular momentum (l_2) [28] referring to D_{3h}. Similarly, the five lowest levels of A_2 symmetry have been assigned by ascending energy to $(0,0^0)$, $(1,0^0)$, $(2,0^0)$, $(0,1^1)$, and $(3,0^0)$. In turn, the four lowest levels of E symmetry are $(0,0^0)$, $(1,0^0)$, $(0,1^1)$, and $(2,0^0)$, again by ascending order in energy. Thus, the fundamental frequencies for the symmetric stretching mode (A_1 symmetry in the D_{3h} point group) are 354.26, 351.01, and $353.34\,\mathrm{cm}^{-1}$ for the A_1, A_2, and E symmetries in the $\mathcal{S}_3$ permutational group, respectively. In turn, the fundamental frequencies for the degenerate mode (E symmetry in the D_{3h} point group) are 824.62, 696.41, and 410.35 cm^{-1} for A_1, A_2, and E symmetries in $\mathcal{S}_3$, respectively.

Similar calculations with consideration of the GP effect have also been reported [12]. A total of 24, 24, and 50 levels of A_1, A_2, and E symmetries have been found below the dissociation threshold of the lower surface, $-1.0560\,\mathrm{eV}$. These are therefore genuine bound states; the cone states lying above such a dissociation threshold are pseudobound states. The lowest levels of A_1, A_2, and E symmetries are found to lie at -1.3475, -1.3438, and $-1.3989\,\mathrm{eV}$, respectively. The notable feature is that the energy levels have been shifted due to the

GP effect, with the shifts being equal to +0.0807, −0.0709, and −0.0273 eV for the A_1, A_2, and E symmetries, respectively. Clearly, such shifts are larger than those obtained in the calculations for the lower adiabatic potential energy surface, namely, +0.0104, −0.0194, and −0.0043 eV (in the above order).

Similar to the case without consideration of the GP effect, the nuclear probability densities of A_1 and A_2 symmetries have threefold symmetry, while each component of E symmetry has twofold symmetry with respect to the line defined by $\beta = 0$. However, the nuclear probability density for the lowest E state has a higher symmetry, being cylindrical with an empty core. This is easyly understand since there is no potential barrier for pseudorotation in the upper sheet. Thus, the nuclear wave function can move freely all the way around the conical intersection. Note that the nuclear probability density vanishes at the conical intersection in the single-surface calculations as first noted by Mead [76] and generally proved by Varandas and Xu [77]. The nuclear probability density of the lowest state of A_1 (A_2) locates at regions where the lower sheet of the potential energy surface has A_2 (A_1) symmetry in $\mathcal{S}_3$. Note also that the A_1 levels are raised up, and the A_2 levels lowered down, while the order of the E levels has been altered by consideration of the GP effect. Such behavior is similar to that encountered for the "trough states" [11].

The five lowest levels of A_1 and A_2 symmetries are found to keep the same order irrespectively of having or not considered the GP effect; by ascending order of energy they are $(0, 0^0)$, $(1, 0^0)$, $(2, 0^0)$, $(0, 1^1)$, and $(3, 0^0)$. However, the GP effect is found to alter the order of the levels $(0, 1^1)$ and $(2, 0^0)$ with E symmetry. The fundamental frequencies for the symmetric stretching mode (A_1 symmetry in the D_{3h} point group) are 352.79, 352.87, and 353.83 cm^{-1} for A_1, A_2, and E symmetries in the $\mathcal{S}_3$ permutational group, respectively. In turn, the fundamental frequencies for the degenerate bending mode (E symmetry in the D_{3h} point group) are 702.44, 748.50, and 769.94 cm^{-1} for A_1, A_2, and E symmetries in $\mathcal{S}_3$, respectively. Thus, no significant changes are observed in the fundamental frequencies of the symmetric stretching modes due to consideration of the GP effect, which implies similar shifts for the two involved eigenenergies. Note that due to the GP effect, the electronic wave function changes sign when traversing a path that encircles the conical intersection. However, the nuclear wave function now has a corresponding change of sign to compensate that in the electronic wave function. The lowest level in the complete spectrum with consideration of the GP effect is then of E symmetry in the $\mathcal{S}_3$ permutation group. As seen, there are considerable differences in the E vibronic wave functions due to inclusion of the GP effect. This can be attributed to having imposed the proper boundary conditions into the wave function and the fact that the E states involve a pseudorotation along the φ coordinate.

We now address the fact that the symmetry of the vibrational modes must be adapted to the nuclear spin multiplicity. Since ^{7}Li is a fermion with nuclear spin

$I = \frac{3}{2}$, the possible nuclear spin multiplicities are 2, 4, 6, 8, and 10. Thus, for the total nuclear wave function to be antisymmetric under permutation of the three identical nuclei, the modes with A_1 and A_2 symmetry in $\mathcal{S}_3$ must be adapted to the nuclear spin multiplicities 2, 4, 6, and 8, and the modes with E symmetry to the nuclear multiplicities 2, 4, 6, 8, and 10. It is well known [29] (see Section V) that there are $(2I+1)(2I+3)(I+1)/3$ symmetric nuclear wave functions, $(2I+1)(2I-1)I/3$ antisymmetric, and $(2I+1)(I+1)(8I)/3$ degenerate. This amounts in the case of $^{7}Li_3$ to 20 symmetric, 4 antisymmetric, and 40 degenerate nuclear spin functions. Thus, the corresponding statistical weights w_{A_1}, w_{A_2}, and w_E will be $\frac{5}{16}$, $\frac{1}{16}$, and $\frac{10}{16}$. Let the nuclear spin weights be $w_{A_1}^{ns}$, $w_{A_2}^{ns}$, and w_E^{ns}, and $w_{A_1}^{v} = w_{A_2}^{v} = w_E^{v} = 1$ the weights associated to the vibrational wave functions. Keeping in mind the direct products $A_1 \otimes E = E$, $A_2 \otimes E = E$, and $E \otimes E = A_1 + A_2 + E$, the statistically averaged frequencies assume the form

$$\omega = (w_{A_1}^{ns} + w_{A_2}^{ns}) w_E^{v} \omega_E + \frac{1}{3} w_E^{ns} (w_{A_1}^{v} \omega_{A_1} + w_{A_2}^{v} \omega_{A_2} + w_E^{v} \omega_E) \tag{63}$$

After simplification, the average fundamental frequencies becomes [12]

$$\omega = \left(1 - \frac{2}{3} w_E^{ns}\right) \omega_E + \frac{w_E^{ns}}{3} (\omega_{A_1} + \omega_{A_2}) \tag{64}$$

where ω stands for ω_1 and ω_2. Equation (64) gives for the symmetric stretching mode 353.04 (353.43) cm^{-1} without (with) consideration of the GP effect. In turn, for the fundamental frequency of the degenerate bending mode, one obtains 556.25 (751.41) cm^{-1} without (with) consideration of the GP effect. Clearly, the fundamental frequency for the symmetric stretching mode has remained almost unaltered due to consideration of the GP effect, while the fundamental frequency for the degenerate bending mode was significantly affected. This is due to the fact that the twofold degenerate bending vibration is a kind of pseudorotation along the φ direction that encircles the conical intersection, and hence is subject to the GP effect. Conversely, consideration of the GP effect is not expected to influence the symmetric stretching vibrational motion along the ρ coordinate much, as indeed verified.

We now compare the results calculated for the fundamental frequency of the symmetric stretching mode with the only available experimental datum [78] of 326 cm^{-1}. The theoretical result is seen to exceed experiment by only ~8.3%. It should be recalled that the $^{7}Li_3$ and $^{6}Li_3$ trimers have for lowest J the values 0 and $\frac{1}{2}$, respectively. Thus, the istopic species $^{6}Li_3$ cannot contribute to the nuclear spin weight in Eq. (64), since the calculations for half-integer J should employ different nuclear spin weights. Note that atomic masses have been used to calculate the vibrational frequencies, which may be justified due to no specification of $^7Li_3$ in [78]. Yet, test calculations have shown that the vibrational levels should not change by >1 or $2\,cm^{-1}$ when $^7Li_3$ is instead considered. In summary, to resolve the remaining discrepancy one requires to refine the potential energy surface and take full consideration of nonadiabatic effects.

As for the trough states, a statistical analysis has been carried out for the calculated cone states [12]. The nearest neighbor spacings are calculated by

$$s_i = \bar{E}_{i+1} - \bar{E}_i \tag{65}$$

where $\bar{E}$ is the unfolded energy. In turn, the mean level spacing is given by

$$\langle s \rangle = \sum_{i=1}^{n} s_i/(n-1) \tag{66}$$

while the second moment assumes the form

$$\langle s^2 \rangle = \sum_{i=1}^{n} s_i^2/(n-1) \tag{67}$$

From these first and second moments, one can calculate the quantity $\langle s^2 \rangle / \langle s \rangle^2$. The results so obtained [12] are given in Table XVI, while Figure 10 shows the corresponding neighbor spacing distributions. It is seen that the level distributions $P(s/\langle s \rangle)$ take values in the interval $0 < P(0) < 1$ for all spectra, which implies in the strict sense of their definitions that they do not obey Wigner, Poisson, or Brody distributions. All spectra seem near Brody type in each block, while the full spectra look like near Poisson type. For each symmetry block, the spectra turn out to be more irregular than the full spectra, irrespectively of taking or not the GP effect into consideration. This is due to the strong interations among states with the same symmetry. However, there is no interation between states with different symmetry, and hence the spectra for each symmetry block are nearly irregular. In turn, the random superposition of unrelated sequences of energy levels leads the full spectra to be more regular.

TABLE XVI
Calculated $\langle s^2 \rangle / \langle s \rangle^2$ for the Vibrational Levels of the First-Excited Electronic Doublet State of Li_3

Method	A_1	A_2	E	Full Spectrum
NGP	1.33	1.34	1.32	1.82
GP	1.32	1.36	1.30	1.75

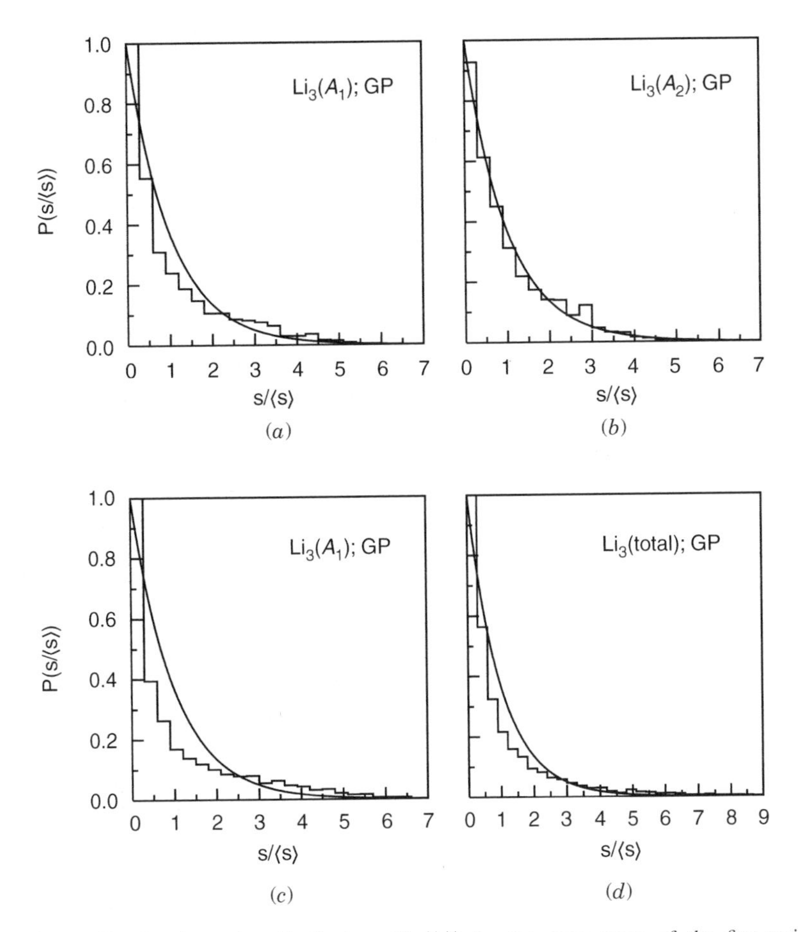

Figure 10. Level spacing distributions $P(s/\langle s\rangle)$ for the cone states of the first-excited electronic doublet state of Li_3 with consideration of GP effects [12]: (*a*) A_1 symmetry; (*b*) A_2 symmetry; (*c*) E symmetry; (*d*) full spectrum. Also shown by the solid lines are the corresponding fits to a Poisson distribution.

Thus, the spectra can be said to be quasiregular in short range and quasiirregular in long range [11].

One should also note that the unfolding procedure makes the average density of levels uniform over the entire energy range. Thus, the difference

$$\Delta_2 = S_2 - \bar{S}_2 \tag{68}$$

where $S_2 = \langle s^2\rangle/\langle s\rangle^2$ is calculated by using the raw spectra, and $\bar{S}_2 = \langle s^2\rangle/\langle s\rangle^2$ is obtained by using the unfolded spectra, represents a measure of the nonuniformity of the spectrum. Such nonuniformity arises from (1) the secular variation of the level spacings that decrease with energy, and (2) the accidental degeneracies and near degeneracies. If such degeneracies can be removed, the uniformity of a spectrum will increase. By this Δ_2 standard, it is found that the spectra (full and those of each symmetry block) are more nonuniform for the cone states than for the trough states. To explain such a finding, one should note the significant differences between the lower and upper adiabatic sheets of the Li_3 potential energy surface, in particular the fact that there is no barrier for pseudorotation on the upper adiabatic sheet. Thus, the pseudorotational motion in the upper sheet is faster than in the lower sheet, which implies a frequency ω_2 for pseudorotation of the degenerate mode considerably larger than that of the breathing mode (ω_1). For example, from our results for the A_1 symmetric vibrational states with consideration of the GP effect, one has $\omega_2 = 702\,\mathrm{cm}^{-1}$, which is nearly twice the value of the frequency $\omega_1 = 353\,\mathrm{cm}^{-1}$ for the symmetric stretching mode. Thus, a lot of levels are expected to gather together leading to frequent accidental degeneracies or near degeneracies. Since in many cases the symmetry requirement for the occurence of Fermi resonances is absent, such degeneracies and near degeneracies cannot be removed. As a result, the first moment $\langle s \rangle$ will be small, and hence S_2 becomes large and leads to an increase in nonuniformity of each symmetry block. Consider, for example, the ideal case where $\omega_2 = 2\omega_1$. Obviously, the states $(2m + n, k^{l_2})$ and $[n, (m + k)^{l'_2}]$ are degenerate in the harmonic oscillator approximation since the energy does not depend on the value of the vibrational angular momentum quantum number. Then, it can be demonstrated by induction that the level $(2m + n, k^{l_2})$ is $[\mathrm{int}(k/2) + \mathrm{int}[(k + m)/2] + 2]$-fold degenerate in the above approximation (this will lead to a near degeneracy in the anharmonic approximation); (n, k) are integers or zero, and m is a nonzero integer. If the above levels have the same symmetry, a Fermi resonance will occur; otherwise, it will be absent. For example, if m is odd there can be no Fermi resonance since the vibrational angular momentum quantum numbers are different. Specifically, for $m = 1$, the $(2 + n, 0^0)$ and $(n, 1^1)$ levels will not suffer a Fermi resonance, with the same being true for the degenerate levels $(2 + n, 1^1)$, $(n, 2^2)$, and $(n, 2^0)$. Instead, for m even, there will be $2[\mathrm{int}(k/2) + 1]$, which will be subject to Fermi resonances. As a specific case, consider $m = 2$: the levels $(4 + n, 0^0)$ and $(n, 2^0)$ will suffer a Fermi resonance, while it will be absent for the $(4 + n, 0^0)$ and $(n, 2^2)$ levels. In the case of m arbitrary but even, the number of levels that cannot be subject to Fermi resonances is given by $[\mathrm{int}[(k + m)/2] - \mathrm{int}(k/2)]$. Clearly, this number increases with increasing m. In summary, at least more than one-half of the degeneracies cannot be removed through Fermi resonances. As a result, the levels will be nonuniformly distributed, leading to an increased nonuniformity of the spectra.

To summarize, the dynamical Jahn–Teller effect is found to be more significant (about one order of magnitute larger) than the GP effect as far as the vibrational states of Li_3 are concerned. Regarding nonadiabatic coupling effects, a comparison of the Li_3 calculations with the only available experimental datum suggests that they play a minor role (at most 0.004 eV, i.e., one order of magnitute smaller than that of the GP effect). Moreover, from the neighbor spacing distributions of the vibrational levels, it has been found that the spectra are quasiregular in short range and quasiirregular in long range, while the Δ_2 standard indicates that the spectra are more nonuniform for the cone states than for the trough ones.

D. Nonadiabatic Coupling Effects

Nonadiabatic coupling between adiabatic (BO) potential energy surfaces leads to a breakdown of the BO approximation. The proper treatment then requires a coupled multistate calculation, which would lead to the exact vibronic levels. For X_3-type systems such as Li_3, we have two adiabatic potential energy surfaces that intersect originating an upper sheet with the shape of an inverted cone, and a lower sheet that looks like a trough. If one carries out a single surface generalized BO calculation (i.e., only with consideration of the GP effect), one obtains either the cone or trough states. For Li_3, the cone states are bound states, although we must distinguish two types of cone states depending on whether they have energies higher than the dissociation energy of the lower surface or not. If they lie above the dissociation threshold, they are imbedded in a continuum and would show up as resonances in reactive scattering or long-lived complexes in studies of unimolecular decomposition. As a result, these states (which are pseudobound states in the single surface calculation) are trully resonance states. If the energies of the cone states are not above the dissociation limit of the lower surface, they are imbedded in the discrete spectrum of the trough states. In this case, they are genuine bound states, and will interact strongly with any other states of the same symmetry. The levels associated to those states, as well as of states that are near the intersection seam, will further shift if nonadiabatic coupling is fully taken into account on the treatment of the problem. A quantitative assessment of such shifts would require the knowledge of the observed frequencies or rigorous nonadiabatic calculations carried out on the same potential energy surface. Unfortunately, such comparisons cannot be done at present.

E. Effects of Electron Spin and Nuclear Spin

For molecules with an even number of electrons, the spin function has only single-valued representations just as the spatial wave function. For these molecules, any degenerate spin–orbit state is unstable in the symmetric conformation since there is always a nontotally symmetric normal coordinate along which the potential energy depends linearly. For example, for an ^3E state of a C_{3v} molecule, the spin function has species $^{es}A_2$ and ^{es}E that upon multiplication by the species ${}^{eo}E$ of the orbital wave function leads to total electronic wave functions of species $({}^{es}A_2 \oplus {}^{es}E) \otimes {}^{eo}E = {}^{e}A_1 \oplus {}^{e}A_2 \oplus 2{}^{e}E$. Thus, the ${}^3E$ state splits into four states of which only two ${}^eE$ states will be unstable. Similarly, for an ${}^3A_1$ state that is orbitally stable, a splitting into ${}^eA_2 \oplus {}^eE$ will occur due to the spin–orbit coupling, of which only the first term is stable (the second, ${}^eE$, is unstable). However, the ${}^eE$ component is only slightly unstable since in an orbitally nondegenerate state the spin–orbit coupling is always very small [28].

The situation is different for an electronic system with an odd number of electrons. By Kramers' theorem, the vibronic coupling cannot remove the degeneracy caused by the half-integer spin S. Moreover, as Jahn [79] has shown, the antisymmetric product of the species of the spin–orbit wave function with itself must have the same species as one of the nontotally symmetric normal vibrations in order to make the Jahn–Teller instability possible. For all axial point groups, the antisymmetric product of any doubly degenerate two-valued representation with itself is totally symmetric; that is, $E_{1/2}, E_{3/2}, \ldots$ states cannot be split by vibronic coupling. Therefore, for all axial point groups when spin–orbit interaction is strong, there is no Jahn–Teller instability. Only a magnetic field such as that connected with a rotation can remove the degeneracy. As follows from the previous discussion, for an ${}^2E$ state, the orbital part of the degeneracy will lead to a Jahn–Teller instability if the spin–orbit coupling is weak. For a strong spin–orbit interaction, ${}^2E$ will split into two states ${}^{eo}E \otimes {}^{es}E_{1/2} = {}^{e}E_{1/2} \oplus {}^{e}E_{3/2}$, with each doublet component remaining doubly degenerate for arbitrary displacements of the nuclei. The general conclusion is that spin–orbit coupling for half-integer spin reduces the instability caused by orbital degeneracy. The above discussion can be applied to the nuclear spin if we consider the nucleus-spin electron–orbit coupling. However, since this coupling is generally smaller than the electron spin–orbit coupling, it may be necessary to take it into account only in special cases, for example, for large values of J.

F. Other Alkali Metal Trimers

In this section, we extend the above discussion to the isotopomers of X_3 systems, where X stands for an alkali metal atom. For the lowest two electronic states, the permutational properties of the electronic wave functions are similar to those of Li_3. Their potential energy surfaces show that the barriers for pseudorotation are very low [80], and we must regard the concerned particles as identical. The ^{23}Na atom has a nuclear spin $\frac{3}{2}$; ^{39}K, ^{40}K, and ^{41}K have nuclear spins $\frac{3}{2}$, 4, and $\frac{3}{2}$; ^{85}Rb and ^{87}Rb have nuclear spins $\frac{5}{2}$ and $\frac{3}{2}$; and ^{133}Cs has a nuclear spin $\frac{7}{2}$. From the above discussion, it then follows that the permutation properties of molecules having individual half-integer spin nuclei will be similar to those involving ^{7}Li; conversely, the permutational properties of molecules consisting of integer spin nuclei will be similar to those containing ^{6}Li.

G. $^1H_3$ and Its Isotopomers

Now, briefly we discuss the molecule $^1H_3$ and its isotopomers, while also highlighting their differences with respect to $^7Li_3$. For the title systems, there have been many investigations, for example, [2,4,5,7,8,59,81], and references cited therein. Similarly to the lithium trimer, H_3 and its isotopomers are important Jahn–Teller systems where the $^vE'$ normal coordinates cause instability and lead to the splitting of the potential energy in the form of a conical intersection between the two involved adiabatic potential energy surfaces ($\tilde{X}^2A'$ and $2^2A'$). Note that intersections of similar kind may occur [81] involving higher states (viz. between the $2^2A'$ and $3^2A'$ electronically adiabatic potential energy surfaces), although this issue of multiple intersections is out of the scope of this work, it will not be discussed here any further. The major difference between the H_3 and Li_3 potential energy surfaces is the fact that the lower sheet of H_3 is of the hat type and can support only resonance states. Moreover, these are mainly located in the collinear saddle-point region that is far from the conical intersection, which may explain why the GP effect calculated on the accurate H_3 DMBE potential energy surface [82] is found [2] to play a minor role on such transition state resonances. Note further that the three equivalent saddle points are located on the outer circle that delimits the potential energy surface when this is viewed as a relaxed triangular plot [68] using hyperspherical coordinates, and are separated from each other by $2\pi/3$. This feature is illustrated in Figure 11, which shows clearly the double-cone shape of the intersection between the $\tilde{X}^2A'$ and $2^2A'$ adiabatic potential energy surfaces near the degeneracy point (this corresponds to the structure with lowest energy along the D_{3h} symmetry line).

As discussed in preceding sections, ^{1}H and ^{3}H have nuclear spin $\frac{1}{2}$, which may have drastic consequences on the vibrational spectra of the corresponding trimeric species. In fact, the nuclear spin functions can only have A_1 (quartet state) and E (doublet) symmetries. Since the total wave function must be antisymmetric, A_1 rovibronic states are therefore not allowed. Thus, for $J = 0$, only resonance states of A_2 and E symmetries exist, with calculated states of A_1 symmetry being purely “mathematical states.” Similarly, only E-symmetric pseudobound states are allowed for $J = 0$. Indeed, even when vibronic coupling is taken into account, only A_2 and E vibronic states have physical significance. Table XVII–XIX summarize the symmetry properties of the wave functions for $^1H_3$ and its isotopomers.

Calculations of bound vibrational levels have been carried out for the first electronically excited state of H_3 with (and without) consideration of the GP effect using the GBO equation; [4,5,53], see Appendix A, Eq. (A.14). The

TABLE XVII
Symmetry Properties of $^1H_3$ and $^3H_3$ Wave Functions in the $\mathcal{S}_3$ Permutation Group

Total	Nuclear Spin[a]	Rovibronic	Rotational[b]	Pseudorotational
A_2	A_1	A_2	A_1	A_2
A_2	A_1	A_2	A_2	A_1
A_2	A_1	A_2	E	E
A_2	E	E	A_1	E
A_2	E	E	A_2	E
A_2	E	E	E	$A_1 \oplus A_2 \oplus E$

[a] Nuclear spin $I = \frac{1}{2}$.
[b] Total angular momentum quantum number $J = 0, 1, 2, \ldots$.

TABLE XVIII
Symmetry Properties of $^2H_3$ Wave Functions in the $\mathcal{S}_3$ Permutation Group

Total	Nuclear Spin[a]	Rovibronic	Rotational[b]	Pseudorotational
A_1	A_1	A_1	E	E
A_1	A_2	A_2	E	E
A_1	E	E	E	$A_1 \oplus A_2 \oplus E$

[a] Nuclear spin $I = 1$.
[b] Total angular momentum quantum number $J = \frac{1}{2}, \frac{3}{2}, \frac{5}{2}, \ldots$.

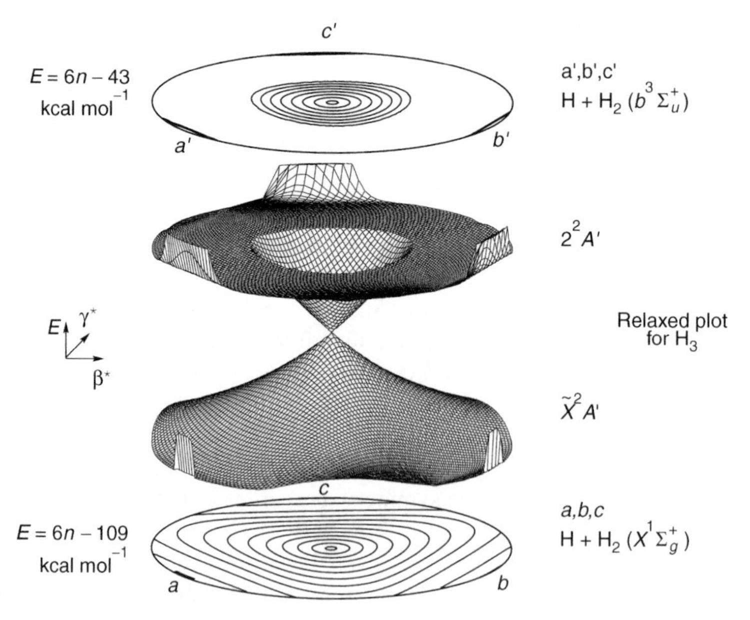

Figure 11. Perspective view [60] of a relaxed triangular plot [68] for the two DMBE adiabatic potential energy surfaces of H_3 using hyperspherical coordinates.

TABLE XIX
Symmetry Properties of $^2H^1H_2$, $^3H^1H_2$, $^2H^3H_2$, and $^1H^3H_2$ Wave Functions in the $\mathcal{S}_3$ Permutation Group

Total	Nuclear Spin[a]	Rovibronic	Rotational[b]	Pseudorotational
B	*A*	*B*	*A*	*B*
B	*A*	*B*	*B*	*A*
B	*B*	*A*	*A*	*A*
B	*B*	*A*	*B*	*B*

[a] Nuclear spins are: $I = \frac{1}{2}$ for ^{1}H and ^{3}H, and $I = 1$ for ^{2}H.
[b] Total angular momentum quantum number $J = \frac{1}{2}, \frac{3}{2}, \frac{5}{2}, \ldots$.

lowest calculated levels are shown graphically in Figure 12. It is seen that the GP effect plays an important role in the spectra of the first excited electronic state of H_3.

Calculations of the GP effect have also been reported for isotopomers of X_3 systems, which we address in the remainder of this section. For such systems, a

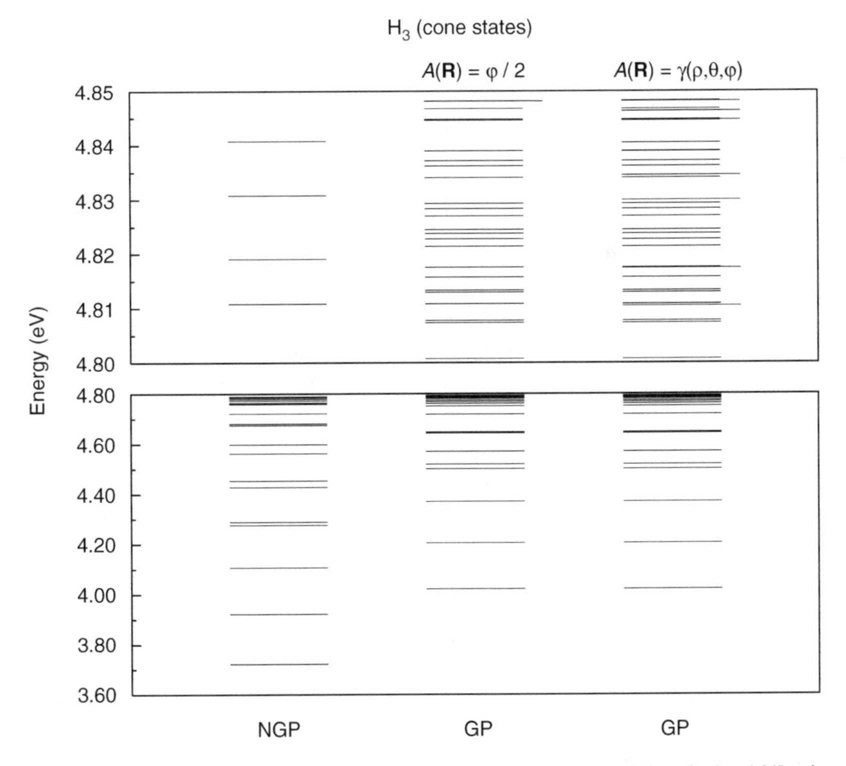

Figure 12. Vibrational levels for the first-excited electronic state of H_3 calculated [4] using: Longuet-Higgins phase $A(\mathbf{R}) = \varphi/2$; Eq. (A.14) with a path-dependent phase $A(\mathbf{R}) = \gamma(\rho, \theta, \varphi)$. The extra levels arising in one calculation but not in the other are indicated by longer line segments.

complication arises due to the mass scaling procedure involved in defining the hyperspherical coordinates. To cope with it, Kuppermann and Wu [83] studied the GP effect in DH_2 using a mass-scaled Jacobi-vectors [84] formula. An alternative approach has been suggested by the present authors, an issue that is discussed next.

The location of the crossing seam (or seam) for an X_3 system is established from the requirement that $r_{AB} = r_{BC} = r_{AC}$, where r_{AB}, r_{BC}, and r_{AC} are the interatomic distances. Since the goal are the the geometric properties produced by this seam, hyperspherical coordinates (ρ, θ, φ) suggest themselves as best suited for this purpose. However, if only two atomic masses are equal, say $m_B = m_C$, the seam is defined [5] by

$$\theta_s = 2 \sin^{-1} \left| \frac{m_B - m_A}{m_B + 2m_A} \right| \tag{69}$$

while φ_s assumes the value π when $m_A > m_B$, and the value zero when $m_A < m_B$. On the other hand, if all three masses are equal, then one has $\theta_s = 0$, and $\varphi_s = 0$ (or π).

The crossing seam in H_3 is therefore characterized by $\theta_s = 0$ [5]. Such a feature warrants that any loop formed by varying φ from 0 to 2π, for fixed values of θ and ρ, will encircle the seam. For HD_2, the equation of the straight line for the seam is defined by ($\theta_s = 0.5048$ rad, $\varphi_s = 0$). Since θ_s is no longer zero, only closed paths (these may be thought of as circular ones, although this does not need to be so) with θ larger than a threshold value (θ_s) will enclose the seam: all others for $\theta < \theta_s$ will not satisfy that requirement, and hence will not show a sign flip on the wave function when this is transported adiabatically along such a closed loop.

A convenient technique to study the sign flip of the wave function is the line-integral approach suggested by Baer [85,86] (an alternative, though more combersome approach, will be to monitor the sign of the wave function along the entire loop [74]). Calculations have been reported [5] using such a line-integral approach for H_3, DH_2, and HD_2 using the 2×2 diabatic DMBE potential energy surface [1]. First, we have shown that the phase obtained by employing the line-integral method is identical (up to a constant) to the mixing angle of the orthogonal transformation that diagonalizes the diabatic potential matrix (see Appendix A). We have also studied this angle numerically along the line formed by fixing the two hyperspherical coordinates ρ and θ and letting φ vary within the interval $[0, 2\pi]$. For H_3, such a line always encircles the seam, and hence the value of the corresponding line integral produces the value π for the geometric (Berry's) phase (of course, deviations may occur if intersections of any of the involved states with higher excited ones are present [81]; this cannot be the case here for H_3 since only two states are considered). In the cases of the two heteronuclear isotopomers, we find similar results but also verify that

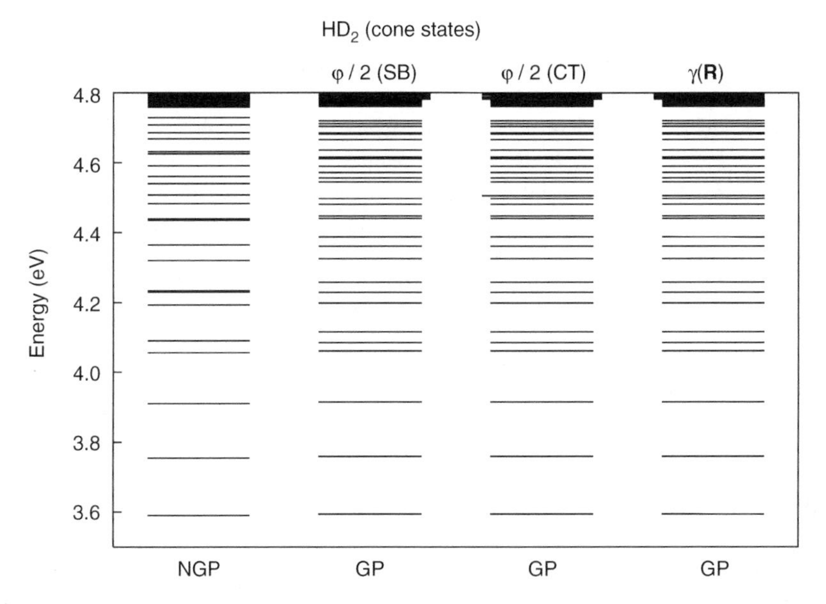

Figure 13. Vibrational levels for the first-excited electronic state of HD_2 calculated [8] using: split basis (SB) technique with $A(\mathbf{R}) = \varphi/2$; coordinate-transformation (CT) treatment with $A(\mathbf{R}) = \varphi/2$; Eq. (A.14) with $A(\mathbf{R}) = \gamma(\rho, \theta, \varphi)$. Shown by the longer line segments are the levels assuming different values in two sets of calculations.

for substantial regions of configuration space where such loops do not encircle the seam, the line integrals yield the value zero for the geometric phase. Such a result may be used to explain why the GP effect has a more remarkable influence on the spectra of H_3 than on its isotopic variants [6], as a set of calculations carried out using the split basis technique [6], the coordinate-transformation approach [8], and the GBO equation (A.14) have consistently demonstrated. Note that the first two approaches employed the traditional Longuet-Higgins phase, $A(\mathbf{R}) = \varphi/2$, whereas Eq. (A.14) uses a path-dependent phase, $A(\mathbf{R}) = \gamma(\rho, \theta, \varphi)$. The lowest levels calculated by such methods are collected in Figure 13, being the reader addressed to the original papers for details concerning the numerical techniques.

XI. CONCLUDING REMARKS

In this chapter, we discussed the permutational symmetry properties of the total molecular wave function and its various components under the exchange of identical particles. We started by noting that most nuclear dynamics treatments carried out so far neglect the interactions between the nuclear spin and the other nuclear and electronic degrees of freedom in the system Hamiltonian. Due to such neglect, one must impose the symmetry properties of the nuclear spin in the total wave function. This requires that the total wave function for identical fermions (bosons) must be antisymmetric (symmetric) under the permutation of the identical particles. Kramers' theorem has been generalized to the case of total angular momentum **F**. It has then been shown that the states of a system with half-integer total angular momentum quantum number must be degenerate. Because molecular double groups are subgroups of the 3D spatial rotation group **SO**(3), the former can be used as a powerful means to study the rotational properties of any molecular system. Based on the permutational symmetry requirements of the total wave function and the extented Kramers' theorem, some severe consequences have then been demonstrated for cases where the nuclear spin quantum number is one-half or zero. The theory has been illustrated by considering in detail the vibrational spectra of the alkali metal trimers where vibronic coupling has been shown to dominate. In this context, we have also reviewed the static and the dynamic Jahn–Teller effects, the GP effect, nonadiabatic coupling, and electron and nuclear spin effects in X_3 (^2S) systems. Although the discussion on $^1H_3$ and its isotopomers has been brief, it has been pointed out that, for $^1H_3$, the A_1 rovibronic states will not be allowed. Thus, for $J = 0$, only resonant states (hat states) of A_2 and E symmetries have physical significance, while only E states are allowed for the pseudobound states (cone states). The implication is that after computing the full spectrum by solving Schrödinger's equation without consideration of nuclear spin effects, one must carefully distinguish the physically allowed solutions from the unphysical (mathematical) ones. For brevity, no attempt has been made to discuss transition selection rules; the interested reader is referred to [28]. Although the material presented in this chapter focused on Li_3 and H_3, the theory that has been reviewed, and in some occasions extended, is general and should be of interest to help understand a wider class of systems.

APPENDIX A: GBO APPROXIMATION AND GEOMETRIC PHASE FOR A MODEL TWO-DIMENSIONAL (2D) HILBERT SPACE

Ignoring all nonadiabatic couplings to higher electronic states, the nuclear motion in a two-state electronic manifold is described explicitly in the adiabatic representation by

$$\left[-\frac{\hbar^2}{2\mu}(\nabla_R^2+\langle\psi_1|\nabla_R^2\psi_1\rangle)+V_1-E\right]\chi_1=\frac{\hbar^2}{2\mu}\left[\langle\psi_1|\nabla_R^2\psi_2\rangle+2\langle\psi_1|\nabla_R\psi_2\rangle\cdot\nabla_R\right]\chi_2 \tag{A.1}$$

$$\left[-\frac{\hbar^2}{2\mu}(\nabla_R^2+\langle\psi_2|\nabla_R^2\psi_2\rangle)+V_2-E\right]\chi_2=\frac{\hbar^2}{2\mu}\left[\langle\psi_2|\nabla_R^2\psi_1\rangle+2\langle\psi_2|\nabla_R\psi_1\rangle\cdot\nabla_R\right]\chi_1 \tag{A.2}$$

where we have supressed obvious coordinate dependences on $\{\psi_i\}$ and $\{\chi_j\}$.

Then, let the two real electronically adiabatic wave functions be written as [4]

$$\psi_1 = \begin{bmatrix} \cos\gamma(\mathbf{R}) \\ \sin\gamma(\mathbf{R}) \end{bmatrix} \qquad \psi_2 = \begin{bmatrix} -\sin\gamma(\mathbf{R}) \\ \cos\gamma(\mathbf{R}) \end{bmatrix} \tag{A.3}$$

where $\gamma(\mathbf{R})$ is the $\mathbf{R}$-dependent mixing angle [5,87–91], which diagonalizes the diabatic potential matrix. By assuming that the electronic basis is orthonormalized, one gets

$$\langle\psi_1|\nabla_R\psi_2\rangle = -\nabla_R\gamma(\mathbf{R}) \tag{A.4}$$

$$\langle\psi_2|\nabla_R\psi_1\rangle = -\langle\psi_1|\nabla_R\psi_2\rangle = \nabla_R\gamma(\mathbf{R}) \tag{A.5}$$

Similarly, one has

$$\langle\psi_I|\nabla_R^2\psi_I\rangle = -\langle\nabla_R\psi_I|\nabla_R\psi_I\rangle = -[\nabla_R\gamma(\mathbf{R})]^2 \tag{A.6}$$

while, for $I \neq J$, one obtains $\langle\nabla_R\psi_I|\nabla_R\psi_J\rangle = 0$. Moreover, one gets

$$\langle\psi_1|\nabla_R^2\psi_2\rangle = \nabla_R\langle\psi_1|\nabla_R\psi_2\rangle - \langle\nabla_R\psi_1|\nabla_R\psi_2\rangle = -\nabla_R^2\gamma(\mathbf{R}) \tag{A.7}$$

and

$$\langle\psi_2|\nabla_R^2\psi_1\rangle = \nabla_R^2\gamma(\mathbf{R}) \tag{A.8}$$

By replacing Eqs. (A.4)–(A.8) into Eqs. (A.1) and (A.2), yields

$$\left\{-\frac{\hbar^2}{2\mu}\left[\nabla_R^2 - (\nabla_R\gamma(\mathbf{R}))^2\right] + V_1 - E\right\}\chi_1$$
$$= -\frac{\hbar^2}{2\mu}\left[\nabla_R^2\gamma(\mathbf{R}) + 2\nabla_R\gamma(\mathbf{R})\cdot\nabla_R\right]\chi_2 \tag{A.9}$$

$$\left\{-\frac{\hbar^2}{2\mu}\left[\nabla_R^2 - (\nabla_R\gamma(\mathbf{R}))^2\right] + V_2 - E\right\}\chi_2$$
$$= \frac{\hbar^2}{2\mu}\left[\nabla_R^2\gamma(\mathbf{R}) + 2\nabla_R\gamma(\mathbf{R})\cdot\nabla_R\right]\chi_1 \tag{A.10}$$

Now, consider a complex nuclear wave function given by a linear combination of the two real nuclear wave functions [42,53],

$$\tilde{\chi} = \frac{1}{\sqrt{2}}(\chi_1 + i\chi_2) \tag{A.11}$$

After some algebraic manipulation, Eqs. (A.1) and (A.2) lead to the following single-surface equations [4]

$$\left\{\frac{\hbar^2}{2\mu}\left[-\nabla_R^2 + (\nabla_R\gamma(\mathbf{R}))^2 + i\nabla_R^2\gamma(\mathbf{R}) + 2i\nabla_R\gamma(\mathbf{R})\cdot\nabla_R\right] + V_1 - E\right\}\tilde{\chi} + \frac{i}{\sqrt{2}}(V_2 - V_1)\chi_2 = 0 \qquad \text{(A.12)}$$

$$\left\{\frac{\hbar^2}{2\mu}\left[-\nabla_R^2 + (\nabla_R\gamma(\mathbf{R}))^2 + i\nabla_R^2\gamma(\mathbf{R}) + 2i\nabla_R\gamma(\mathbf{R})\cdot\nabla_R\right] + V_2 - E\right\}\tilde{\chi} + \frac{i}{\sqrt{2}}(V_1 - V_2)\chi_1 = 0 \qquad \text{(A.13)}$$

By neglecting the second term in Eq. (A.12), one gets

$$\left\{\frac{\hbar^2}{2\mu}\left[-\nabla_R^2 + (\nabla_R\gamma(\mathbf{R}))^2 + i\nabla_R^2\gamma(\mathbf{R}) + 2i\nabla_R\gamma(\mathbf{R})\cdot\nabla_R\right] + V_{1,2} - E\right\}\tilde{\chi} = 0 \qquad \text{(A.14)}$$

which should be valid whenever motion of the nuclei can be confined to the vicinity of the conical intersection, where $V_1 \simeq V_2$. Note that Eq. (A.14) leads to different sets of eigenvalues depending on whether V_1 or V_2 are used, except, of course, when $V_1 = V_2$, which happens only at the conical intersection. Conversely to Baer and Englmann [25,53], our derivation of Eq. (A.14) has not been based on the assumption (so-called BEB approximation in [92] for the lower sheet equation) that χ_2 is small enough everywhere, but on the milder premise that χ_1 and χ_2 are well behaved. Of course, Eq. (A.14) leads to the corresponding BO equations when the derivative coupling elements are constant or zero.

Now, we discuss how the geometric phase is related to the mixing angle in this two-state model. We begin by writing Eq. (A.11) as the gauge transformation

$$\tilde{\chi} = \exp[iA(\mathbf{R})]\chi \qquad \text{(A.15)}$$

where χ is a real wave function, and $A(\mathbf{R})$ is a geometric phase. By analogy, the complex electronic wave function may be written in the form

$$\tilde{\psi} = \exp[iA(\mathbf{R})]\psi \qquad \text{(A.16)}$$

and then

$$\tilde{\psi} = \frac{1}{\sqrt{2}}(\psi_1 + i\psi_2) \qquad \text{(A.17)}$$

where ψ_i is the real-valued electronic wave function corresponding to the wave function χ_i. In the complex plane $\{|\chi_1\rangle, i|\chi_2\rangle\}$, the complex vector state $|\tilde{\chi}\rangle$ will then be characterized by the argument $A(\mathbf{R}) = \arg\tilde{\chi}$ and, similarly, in the complex plane $\{|\psi_1\rangle, i|\psi_2\rangle\}$, $|\tilde{\psi}\rangle$ will satisfy $A(\mathbf{R}) = \arg\tilde{\psi}$.

By using Eq. (A.17), the first derivative coupling for the complex electronic wave function assumes the form [4]

$$\langle\tilde{\psi}|\nabla_R\tilde{\psi}\rangle = i\langle\psi_1|\nabla_R\psi_2\rangle \tag{A.18}$$

where we have used the fact that $\langle\psi_i|\nabla_R\psi_i\rangle = 0$, and $\langle\psi_1|\nabla_R\psi_2\rangle = -\langle\psi_2|\nabla_R\psi_1\rangle$. On the other hand, by noting that ψ is real, one obtains

$$\langle\tilde{\psi}|\nabla_R\tilde{\psi}\rangle = i\nabla_R A(\mathbf{R}) \tag{A.19}$$

A comparison of Eq. (A.18) with Eq. (A.19) then yields

$$\langle\psi_1|\nabla_R\psi_2\rangle = \nabla_R A(\mathbf{R}) \tag{A.20}$$

which shows that, for a set of two electronic states, the derivative coupling is given by the gradient of the geometric phase. A similar result (except for the sign) has been obtained by Baer [25] using a different approach. Note that Eq. (A.20) holds exactly in the vicinity of the crossing seam where the phase $A(\mathbf{R})$ is identical for both sheets. From Eqs. (A.4) and (A.20), one then gets

$$\nabla_R A(\mathbf{R}) = -\nabla_R\gamma(\mathbf{R}) \tag{A.21}$$

which shows that the geometric phase is identical to the mixing angle, except for the sign and a constant term that have no physical implications. Thus, provided that we chose such a constant term to be zero, one has

$$A(\mathbf{R}) = \gamma(\mathbf{R}) \tag{A.22}$$

Note that the mixing angle has the correct sign-change behavior [5]: $\Delta\gamma(\mathbf{R}) = \pi$ for a closed path encircling the crossing seam, and $\Delta\gamma(\mathbf{R}) = 0$ for a closed path that does not encircle it. Thus, the geometric phase defined from Eq. (A.22) displays the correct sign change behavior, while depending on the full set of internal coordinates. Moreover, close to the seam, it shows [5] a behavior similar to the traditional Longuet-Higgins phase of $\varphi/2$, where φ is the pseudorotation angle.

APPENDIX B: ANTILINEAR OPERATORS AND THEIR PROPERTIES

In this appendix, we review some important properties of antilinear operators that are used in the text and Appendix C. Let us then consider an operator $\hat{\mathcal{O}}$ that acts on a state $|\psi\rangle$ to give $\hat{\mathcal{O}}|\psi\rangle$, with the original state being restored after acting twice, that is, $\hat{\mathcal{O}}^2|\psi\rangle = c|\psi\rangle$. Clearly, the space-inversion operator $\hat{I}$ is a well-known example of such an operator. Moreover, as will be shown, the time-reversal $\hat{T}$ and complex conjugate $\hat{K}$ operators provide further examples of such operators.

By the very meaning of a physical state, we must require that

$$\langle \hat{\mathcal{O}}\psi | \hat{\mathcal{O}}\psi \rangle = \langle \psi | \psi \rangle \tag{B.1}$$

This should apply both to linear and antilinear operators, hereafter denoted, respectively, by $\hat{\mathcal{L}}$ and $\hat{\mathcal{A}}$. For a linear operator, the action on a general state $c_1|\psi_1\rangle + c_2|\psi_2\rangle$ is expressed by

$$\hat{\mathcal{L}}(c_1|\psi_1\rangle + c_2|\psi_2\rangle) = c_1\hat{\mathcal{L}}|\psi_1\rangle + c_2\hat{\mathcal{L}}|\psi_2\rangle \tag{B.2}$$

while, for an antilinear operator [93], it assumes the form

$$\hat{\mathcal{A}}(c_1|\psi_1\rangle + c_2|\psi_2\rangle) = c_1^*\hat{\mathcal{A}}|\psi_1\rangle + c_2^*\hat{\mathcal{A}}|\psi_2\rangle \tag{B.3}$$

where c_i $(i = 1, 2)$ are complex numbers. As for linear operators (where we have the definition of Hermitian conjugate of $\hat{\mathcal{L}}|\psi\rangle$ to be $\langle\psi|\hat{\mathcal{L}}^\dagger)$, we define the Hermitian conjugate of $(\hat{\mathcal{A}}|\psi\rangle)$ as $(\langle\psi|\hat{\mathcal{A}}^\dagger)$; recall that the parentheses are necessary. Note that Eq. (B.1) implies that the norm of a vector cannot be altered both for linear and antilinear operators. More generally, a linear operator cannot change the internal product of two vectors, and hence must be unitary. Conversely, antilinear operators change the internal product to its complex conjugate, and hence are called antiunitary [54]. One has

$$\langle \psi | \phi \rangle = \langle \phi | \psi \rangle^* \tag{B.4}$$

which indicates that the unit operator is linear with respect to the ket and antilinear to the bra. In general, a linear operator is linear with respect to the ket and antilinear to the bra. Thus, for an antilinear operator, we have by definition

$$\langle \psi | (\hat{\mathcal{A}} | \phi \rangle) = [(\langle \psi | \hat{\mathcal{A}}) | \phi \rangle]^* \tag{B.5}$$

where the parentheses in $\langle\psi|(\hat{\mathcal{A}}|\phi\rangle)$ indicate the order of the action, with $(\hat{\mathcal{A}}|\cdot)$ being antilinear with respect to both the bra and the ket. In turn, $(\cdot|\hat{\mathcal{A}})$ is linear with respect to both the bra and the ket.

From the definition of Hermitian conjugate and Eq. (B.5), one then gets

$$\langle \psi | (\hat{\mathcal{A}} | \phi \rangle) = \langle \phi | (\hat{\mathcal{A}}^\dagger | \psi \rangle) \tag{B.6}$$

This implies that the Hermitian conjugate of an antilinear operator is also antilinear. It should also be pointed out that the product of two antilinear operators is linear, while the product of a linear and an antilinear operators is antilinear. In general, an antilinear operator may be expressed as a product of a linear and an antilinear operators. From Eq. (B.5), we also have

$$[(\langle\psi|\hat{\mathcal{A}}^{\dagger})(\hat{\mathcal{A}}|\psi\rangle)]^{*} = \langle\psi|(\hat{\mathcal{A}}^{\dagger}\hat{\mathcal{A}}|\psi\rangle) = \langle\psi|\psi\rangle^{*} = \langle\psi|\psi\rangle \tag{B.7}$$

Thus,

$$\hat{\mathcal{A}}^{\dagger}\hat{\mathcal{A}} = \hat{\mathcal{A}}\hat{\mathcal{A}}^{\dagger} = \hat{1} \tag{B.8}$$

a property that is also satisfied by unitary operators.

APPENDIX C: PROOF OF EQS. (18) AND (23)

Let us consider the time evolution of a quantum system, which satisfies the time-dependent Schrödinger equation [55]

$$i\hbar\frac{\partial}{\partial t}|\psi(t)\rangle = \hat{H}|\psi(t)\rangle \tag{C.1}$$

where $\hat{H}$ does not depend explicitly on time. Now, by defining the time-reversal state $|\psi_{\mathrm{rev}}(-t)\rangle$ as

$$|\psi_{\mathrm{rev}}(-t)\rangle = \hat{T}|\psi(t)\rangle \tag{C.2}$$

such a state must satisfy the corresponding time-dependent Schrödinger equation

$$i\hbar\frac{\partial}{\partial(-t)}|\psi_{\mathrm{rev}}(-t)\rangle = \hat{H}|\psi_{\mathrm{rev}}(-t)\rangle \tag{C.3}$$

Now, consider the complex conjugate of Eq. (C.1),

$$i\hbar\frac{\partial}{\partial(-t)}|\psi(t)\rangle^{*} = \hat{H}|\psi(t)\rangle^{*} \tag{C.4}$$

If there is an unitary operator $\hat{U}$ such that

$$\hat{U}\hat{H}\hat{U}^{\dagger} = \hat{H} \tag{C.5}$$

one has, after the action of $\hat{U}$ on Eq. (C.4),

$$i\hbar\frac{\partial}{\partial(-t)}\hat{U}|\psi(t)\rangle^{*} = \hat{U}\hat{H}|\psi(t)\rangle^{*} \tag{C.6}$$

$$= \hat{H}\hat{U}|\psi(t)\rangle^{*} \tag{C.7}$$

By comparing Eq. (C.6) with Eqs. (C.2) and (C.3), the time-reversal operator can be expressed as a product of an unitary and a complex conjugate operators as follows

$$\hat{T} = \hat{U}\hat{K} \tag{C.8}$$

Thus, the time-reversal state can be written as

$$|\psi_{\text{rev}}(-t)\rangle = \hat{T}|\psi(t)\rangle = \hat{U}\hat{K}|\psi(t)\rangle = \hat{U}|\psi(t)\rangle^* \tag{C.9}$$

It is now required for observable quantities that the expectation value of any operator $\hat{\mathcal{O}}$ taken with respect to $|\psi_{\text{rev}}(-t)\rangle$ must be the same as that of the operator $\hat{\mathcal{O}}_{\text{rev}}$ taken relative to $|\psi(t)\rangle$, that is,

$$\langle\psi_{\text{rev}}(-t)|\hat{\mathcal{O}}|\psi_{\text{rev}}(-t)\rangle = \langle\psi(t)|\hat{\mathcal{O}}_{\text{rev}}|\psi(t)\rangle \tag{C.10}$$

Since

$$\langle\psi_{\text{rev}}(-t)|\hat{\mathcal{O}}|\psi_{\text{rev}}(-t)\rangle = \langle\hat{U}\psi(t)|\hat{\mathcal{O}}|\hat{U}\psi(t)\rangle \tag{C.11}$$

$$= \langle\psi(t)|\hat{U}^\dagger\hat{\mathcal{O}}\hat{U}|\psi(t)\rangle \tag{C.12}$$

a comparison with Eq. (C.10), and the fact that any operator associated to an observable quantity is Hermitian, leads to

$$\hat{\mathcal{O}}^T_{\text{rev}} = \hat{U}^\dagger\hat{\mathcal{O}}\hat{U} \tag{C.13}$$

where the superscript T in the time-reversal operator $\hat{\mathcal{O}}^T_{\text{rev}}$ denotes the transpose.

Now, consider the case of spinless particles not subject to external electronic and magnetic fields. We may now choose the unitary operator $\hat{U}$ as the unit operator, that is, $\hat{T} = \hat{K}$. For the coordinate and momentum operators, one then obtains

$$\hat{T}\hat{r}\hat{T}^{-1} = \hat{K}\hat{r}\hat{K}^{-1} = \hat{r} \tag{C.14}$$

$$\hat{T}\hat{p}\hat{T}^{-1} = \hat{K}\hat{p}\hat{K}^{-1} \tag{C.15}$$

$$= \hat{K}(-i\hbar\nabla)\hat{K}^{-1} \tag{C.16}$$

$$= i\hbar\nabla = -\hat{p} \tag{C.17}$$

As a result, the orbital agular momentum operator satisfies the relation

$$\hat{T}\hat{L}\hat{T}^{-1} = \hat{T}(\hat{r}\times\hat{p})\hat{T}^{-1} \tag{C.18}$$

$$= -(\hat{r}\times\hat{p}) = -\hat{L} \tag{C.19}$$

Next, let us address the case of half-spin particles. One has

$$\hat{S} = \hat{S}_x\cdot\mathbf{i} + \hat{S}_y\cdot\mathbf{j} + \hat{S}_z\cdot\mathbf{k} \tag{C.20}$$

where

$$\hat{S}_i = \frac{1}{2}\hbar\sigma_i \tag{C.21}$$

with the matrices σ_i $(i = x, y, z)$ being the Pauli matrices

$$\sigma_x = \begin{pmatrix} 0 & 1 \\ 1 & 0 \end{pmatrix} \qquad \sigma_y = \begin{pmatrix} 0 & -i \\ i & 0 \end{pmatrix} \qquad \sigma_z = \begin{pmatrix} 1 & 0 \\ 0 & -1 \end{pmatrix} \tag{C.22}$$

If we now choose the unitary operator to be real, then it may assume the form

$$\hat{U} = i\sigma_y \tag{C.23}$$

By applying Eq. (C.13) to the spin operators $\hat{S}_i$ and using Eq. (C.22), one then gets after some matrix multiplications

$$\hat{U}^\dagger \hat{S}_x \hat{U} = -\hat{S}_x \qquad \hat{U}^\dagger \hat{S}_y \hat{U} = \hat{S}_y \qquad \hat{U}^\dagger \hat{S}_z \hat{U} = -\hat{S}_z \tag{C.24}$$

We are now ready to prove that

$$\hat{T}\hat{S}\hat{T}^{-1} = -\hat{S} \tag{C.25}$$

since we have

$$\begin{aligned}
\hat{T}\hat{S}_x\hat{T}^{-1} &= i\sigma_y \hat{K}\hat{S}_x\hat{K}\sigma_y^{-1} i^{-1} && \text{(C.26)}\\
&= \sigma_y \hat{K}\hat{S}_x\hat{K}\sigma_y && \text{(C.27)}\\
&= -\sigma_y \hat{K}\hat{S}_x\sigma_y && \text{(C.28)}\\
&= -\frac{1}{2}\hbar\sigma_y\hat{K}\begin{pmatrix} 0 & 1 \\ 1 & 0 \end{pmatrix}\begin{pmatrix} 0 & -i \\ i & 0 \end{pmatrix} && \text{(C.29)}\\
&= -\frac{1}{2}\hbar\sigma_y\hat{K}\begin{pmatrix} i & 0 \\ 0 & -i \end{pmatrix} && \text{(C.30)}\\
&= -\frac{1}{2}\hbar\sigma_y\begin{pmatrix} -i & 0 \\ 0 & i \end{pmatrix} && \text{(C.31)}\\
&= -\frac{1}{2}\hbar\begin{pmatrix} 0 & -i \\ i & 0 \end{pmatrix}\begin{pmatrix} -i & 0 \\ 0 & i \end{pmatrix} && \text{(C.32)}\\
&= -\frac{1}{2}\hbar\begin{pmatrix} 0 & 1 \\ 1 & 0 \end{pmatrix} && \text{(C.33)}\\
&= -\hat{S}_x && \text{(C.34)}
\end{aligned}$$

and, similarly,

$$\hat{T}\hat{S}_y\hat{T}^{-1} = -\hat{S}_y \qquad \hat{T}\hat{S}_z\hat{T}^{-1} = -\hat{S}_z \tag{C.35}$$

Clearly, the above equations and Eq. (C.20) prove Eq. (C.25).

Finally, we demonstrate that

$$\hat{T}^2 = (-\hat{1})^{2S} \tag{C.36}$$

From Eqs. (C.8) and (C.23), we have for $S = \frac{1}{2}$

$$\hat{T}_{1/2}^2 = (i\sigma_y\hat{K})^2 = -\sigma_y^2 = -\hat{1} \tag{C.37}$$

where $\hat{1}$ is the unit operator in a 2×2 vector space. Note that, for spinless particles, we have chosen $\hat{U}$ to be the unit operator in a 1×1 vector space (and hence $\hat{T}_0 = \hat{K}$), which leads to

$$\hat{T}_0^2 = \hat{K}^2 = \hat{1} \tag{C.38}$$

and hence proves Eq. (C.36).

The above discussion is now generalized to arbitrary spin values. First, we note that twice application of the time-reversal operator leads the system back to its original state ψ, that is, $\hat{T}^2\psi = c\psi$. Thus, we have $\hat{T}^2 = c\hat{1}$. Next, consider the following two relations

$$\langle\hat{T}\phi|\hat{T}^2\psi\rangle = (\langle\phi|\hat{T}^\dagger)(\hat{T}^2|\psi\rangle) = \left[\langle\phi|(\hat{T}^\dagger\hat{T}^2|\psi\rangle)\right]^\star = \left[\langle\phi|\hat{T}\psi\rangle\right]^\star = \langle\hat{T}\psi|\phi\rangle \tag{C.39}$$

$$\langle\hat{T}\phi|\hat{T}^2\psi\rangle = c\langle\hat{T}\phi|\psi\rangle \tag{C.40}$$

Thus, we have

$$\langle\hat{T}\psi|\phi\rangle = c\langle\hat{T}\phi|\psi\rangle \tag{C.41}$$

Similarly, we can show that

$$\langle\hat{T}\psi|\hat{T}^2\phi\rangle = \langle\hat{T}\phi|\psi\rangle \tag{C.42}$$

$$\langle\hat{T}\phi|\psi\rangle = c\langle\hat{T}\psi|\phi\rangle \tag{C.43}$$

from Eqs. (C.41) and (C.43) we can obtain

$$\langle\hat{T}\psi|\phi\rangle = c^2\langle\hat{T}\psi|\phi\rangle \tag{C.44}$$

and hence

$$c^2 = 1 \tag{C.45}$$

which proves that $\hat{T}^2 = \pm\hat{1}$. Now, by substituting Eq. (C.8) in $\hat{T}\hat{r}\hat{T}^{-1}$, $\hat{T}\hat{p}\hat{T}^{-1}$, and $\hat{T}\hat{S}\hat{T}^{-1}$, we may show that $\hat{U}$ satisfies equations similar to Eqs. (C.24) and (C.25).

As a first application, consider the case of a single particle with spin quantum number S. The spin functions will then transform according to the IRREPs $D^{(S)}(\alpha)$ of the 3D rotational group $\mathbf{SO}(3)$, where α is the rotational vector, written in the operator form as [36]

$$\hat{D}^{(S)}(\alpha) = \exp\left(-\frac{i}{\hbar}\hat{S}\cdot\alpha\right) \tag{C.46}$$

The spin operator $\hat{S}$ is an irreducible tensor of rank one with the following transformational properties

$$\hat{D}^{(S)}(\alpha)\hat{S}\hat{D}^{(S)}(\alpha)^{-1} = \hat{g}(\alpha)\hat{S} \tag{C.47}$$

where $\hat{g}(\alpha)$ is an operator of $\mathbf{SO}(3)$. Let us then take $\hat{g}(\alpha)$ to be a rotation by π around the y axis. Thus, from Eqs. (C.46) and (C.47), one gets

$$\exp\left(-\frac{i}{\hbar}\hat{S}_y\right)\hat{S}_x\exp\left(\frac{i}{\hbar}\hat{S}_y\right) = -\hat{S}_x \tag{C.48}$$

$$\exp\left(-\frac{i}{\hbar}\hat{S}_y\right)\hat{S}_y\exp\left(\frac{i}{\hbar}\hat{S}_y\right) = \hat{S}_y \tag{C.49}$$

$$\exp\left(-\frac{i}{\hbar}\hat{S}_y\right)\hat{S}_z\exp\left(\frac{i}{\hbar}\hat{S}_y\right) = -\hat{S}_z \tag{C.50}$$

Comparing Eqs. (C.48)–(C.50) with Eq. (C.24), one obtains

$$\hat{U} = \exp\left(-\frac{i}{\hbar}\pi\hat{S}_y\right) \tag{C.51}$$

Since $\hat{S}_y^T = -\hat{S}_y$, we then have

$$\hat{U}^\dagger = \exp\left(-\frac{i}{\hbar}\pi\hat{S}_y^T\right) \tag{C.52}$$

$$= \exp\left(\frac{i}{\hbar}\pi\,\hat{S}_y\right) \tag{C.53}$$

$$= \exp\left(-\frac{i}{\hbar}2\pi\hat{S}_y^T\right)\exp\left(-\frac{i}{\hbar}\pi\hat{S}_y\right) \tag{C.54}$$

$$= \hat{D}^{(S)}(2\pi\mathbf{j})\hat{U} \tag{C.55}$$

$$= (-1)^{2S}\hat{U} \tag{C.56}$$

where $2\pi\mathbf{j}$ indicates a 2π rotation about the y axis. Thus, we have

$$\hat{T} = \exp\left(-\frac{i}{\hbar}\pi\hat{S}_y\right)\hat{K} \tag{C.57}$$

and finally, by comparing with Eq. (C.51), one gets

$$\hat{T}^2 = (-\hat{1})^{2S} \tag{C.58}$$

Finally, for a system of n identical particles, the result is

$$\hat{U} = \exp\left(-\frac{i}{\hbar}\pi\hat{S}_{y,1}\right)\exp\left(-\frac{i}{\hbar}\pi\hat{S}_{y,2}\right)\cdots\exp\left(-\frac{i}{\hbar}\pi\hat{S}_{y,n}\right) \tag{C.59}$$

and hence

$$\hat{T}^2 = (-\hat{1})^{\sum_{i=1}^{n} 2S_i} = (-\hat{1})^{2S} \tag{C.60}$$

APPENDIX D: DEGENERATE AND NEAR-DEGENERATE VIBRATIONAL LEVELS

Here, we discuss the motion of a system of three identical nuclei in the vicinity of the D_{3h} configuration. The conventional coordinates for the in-plane motion are employed, as shown in Figure 5. The normal coordinates (Q_x, Q_y, Q_z), the plane polar coordinates (ρ, φ, z), and the Cartesian displacement coordinates (x_i, y_i, z_i) of the three nuclei $(i = 1, 2, 3)$ are related by [20,94]

$$Q_x = \rho\cos\varphi = \frac{1}{\sqrt{3}}\left\{-x_1 + \left(\frac{1}{2}x_2 + \frac{\sqrt{3}}{2}y_2\right) + \left(\frac{1}{2}x_3 - \frac{\sqrt{3}}{2}y_3\right)\right\} \tag{D.1}$$

$$Q_y = \rho\sin\varphi = \frac{1}{\sqrt{3}}\left\{y_1 + \left(\frac{\sqrt{3}}{2}x_2 - \frac{1}{2}y_2\right) + \left(-\frac{\sqrt{3}}{2}x_3 - \frac{1}{2}y_3\right)\right\} \tag{D.2}$$

$$Q_z = z = \frac{1}{\sqrt{3}}\left\{-x_1 + \left(\frac{1}{2}x_2 - \frac{\sqrt{3}}{2}y_2\right) + \left(\frac{1}{2}x_3 + \frac{\sqrt{3}}{2}y_3\right)\right\} \tag{D.3}$$

where the coordinates (Q_x, Q_y) are the doubly degenerate modes belonging to the E' IRREP in D_{3h}, and Q_z belongs to the A_1' one. Note that Q_x is symmetric with respect to the xz plane, while Q_y is antisymmetric.

The coordinates of interest to us in the following discussion are Q_x and Q_y, which describe the distortion of the molecular triangle from D_{3h} symmetry. In the harmonic-oscillator approximation, the factor in the vibrational wave function due to the two degenerate modes is then (except for a normalization constant and dependence on Q_z) given by

$$\chi(v_1, v_{2a}, v_{2b}) \simeq H_{v_{2a}}(\sqrt{\alpha_2}Q_x)H_{v_{2b}}(\sqrt{\alpha_2}Q_y)\exp[-\alpha_2(Q_x^2+Q_y^2)/2] \quad \text{(D.4)}$$

where $H_{v_{2a}}$ and $H_{v_{2b}}$ are Hermite polynomials of order v_{2a} and v_{2b}, respectively; v_{2a} and v_{2b} are the vibrational quantum numbers, and $\alpha_2 = 2\pi\nu_2/\hbar$, with ν_2 being the frequency of the degenerate mode.

Let us then consider the case where the degenerate mode is doubly excited. In this case, $v_2 = v_{2a} + v_{2b} = 2$ and the corresponding vibrational energy level will be triply degenerate with the associated wave functions being given by

$$\chi_1 = \chi(v_1, 2, 0) \sim 4\alpha_2 Q_x^2 - 2 \quad \text{(D.5)}$$

$$\chi_2 = \chi(v_1, 1, 1) \sim 4\alpha_2 Q_x Q_y \quad \text{(D.6)}$$

$$\chi_3 = \chi(v_1, 0, 2) \sim 4\alpha_2 Q_y^2 - 2 \quad \text{(D.7)}$$

Note that only the polynomial factors have been given, since the exponential parts are identical for all wave functions. Of course, any linear combination of the wave functions in Eqs. (D.5)–(D.7) will still be an eigenfunction of the vibrational Hamiltonian, and hence a possible state. There are three such linearly independent combinations which assume special importance, namely,

$$\chi_1' = \chi_1 - \chi_3 + 2i\chi_2 \sim 4\alpha_2(Q_x^2 - Q_y^2 + 2iQ_xQ_y) \quad \text{(D.8)}$$

$$\chi_2' = \chi_1 + \chi_3 \sim 4\alpha_2(Q_x^2 + Q_y^2) - 4 \quad \text{(D.9)}$$

$$\chi_3' = \chi_1 - \chi_3 - 2i\chi_2 \sim 4\alpha_2(Q_x^2 - Q_y^2 - 2iQ_xQ_y) \quad \text{(D.10)}$$

By using the plane polar coordinates defined in Eq. (D.1), one obtains

$$\chi_1' \simeq 4\alpha_2\rho^2\exp(-\alpha_2\rho^2/2)\exp(2i\varphi) \quad \text{(D.11)}$$

$$\chi_2' \simeq 4(\alpha_2\rho^2 - 1)\exp(-\alpha_2\rho^2/2) \quad \text{(D.12)}$$

$$\chi_3' \simeq 4\alpha_2\rho^2\exp(-\alpha_2\rho^2/2)\exp(-2i\varphi) \quad \text{(D.13)}$$

These new wave functions are eigenfunctions of the z component of the angular momentum $\hat{\pi}_z = -i\hbar\frac{\partial}{\partial\varphi}$ with eigenvalues $m_{v_2} = +2, 0, -2$ in units of $\hbar$. Thus, Eqs. (D.11)–(D.13) represent states in which the vibrational angular momentum of the nuclei about the molecular axis has a definite value. When treating the vibrations as harmonic, there is no reason to prefer them to any other linear combinations that can be obtained from the original basis functions in

Eqs. (D.5)–(D.7). However, when perturbations occur due to anharmonicity, the wave functions in Eqs. (D.11)–(D.13) will provide the correct zeroth-order ones. The quantum numbers v_{2a} and v_{2b} are therefore not physically significant, while v_2 and m_{v_2} or v_2 and $l_2 = |m_{v_2}|$ are. It should also be pointed out that the degeneracy in the vibrational levels will be split due to anharmonicity [28].

Now, consider the general case of a v_2 multiply excited degenerate vibrational level where $v_2 > 2$, which is dealt with by solving the Schrödinger equation for the isotropic $2D$ harmonic oscillator with the Hamiltonian assuming the form [95]

$$\hat{H}_v = \frac{\hbar\sqrt{\lambda}}{2}\left(-\frac{\partial^2}{\partial q_x^2} - \frac{\partial^2}{\partial q_y^2} + q_x^2 + q_y^2\right) \tag{D.14}$$

where we have used the dimensionless normal coordinates $q_i = \sqrt{\alpha_2} Q_i$ $(i = x, y)$, with $\alpha_2 = 2\pi\nu_2/\hbar = \sqrt{\lambda}/\hbar$. The transformation of such a Hamiltonian into polar coordinates leads to

$$\hat{H}_v = -\frac{\hbar\sqrt{\lambda}}{2}\left(\frac{\partial^2}{\partial\rho^2} + \frac{1}{\rho}\frac{\partial}{\partial\rho} + \frac{1}{\rho^2}\frac{\partial^2}{\partial\varphi^2} - \rho^2\right) \tag{D.15}$$

Separation of variables can then be achieved by using

$$\chi_v = R(\rho)\Phi(\varphi) \tag{D.16}$$

where

$$R(\rho) = F(\rho)\exp(-\rho^2/2) \tag{D.17}$$

and

$$F(\rho) = \rho^s \sum_{n=0}^{\infty} a_n \rho^n \tag{D.18}$$

Assuming now that the power series expansion in $F(\rho)$ can be terminated to keep $R(\rho)$ well behaved at large ρ values, it may be shown [95] that

$$\Phi(\varphi) = (2\pi)^{-1/2}\exp(im_{v_2}\varphi) \qquad m_{v_2} = \pm v_2, \pm(v_2 - 2), \ldots, \pm 1 \text{ or } 0 \tag{D.19}$$

$$R(\rho) = N_{v_2 l_2}\rho^{l_2} L_n^{l_2}(\rho^2)\exp(-\rho^2/2) \qquad l_2 = |m_{v_2}| \qquad n = (v_2 + l_2)/2 \tag{D.20}$$

where $L_n^{l_2}(\rho^2)$ are the associated Laguerre polynomials of order n, and the normalization factor assumes the form

$$N_{v_2 l_2} = \sqrt{\frac{2[(v_2 - l_2)/2]!}{\{[(v_2 + l_2)/2]!\}^3}} \tag{D.21}$$

Let us now examine the case of a 3D harmonic oscillator possessing three degenerate normal coordinates (Q_1, Q_2, Q_3), with the degenerate mode being v multiply excited; $v = v_1 + v_2 + v_3$. There are then $(v+1)(v+2)/2$ degenerate vibrational wave functions and energy levels for each value of v, corresponding to the possible different combinations of v_1, v_2, and v_3. It is now convenient to define the polar coordinates (ρ, θ, φ) by the corresponding dimensionless normal coordinates (q_1, q_2, q_3) according to

$$\begin{aligned} q_1 &= \rho \sin\theta \cos\varphi \\ q_2 &= \rho \sin\theta \sin\varphi \\ q_3 &= \rho \cos\theta \end{aligned} \tag{D.22}$$

In such coordinates, the Hamiltonian assumes the form [95]

$$\hat{H}_v = \frac{\hbar\sqrt{\lambda}}{2}\left(-\frac{\partial^2}{\partial q_1^2} - \frac{\partial^2}{\partial q_2^2} - \frac{\partial^2}{\partial q_3^2} + q_1^2 + q_2^2 + q_3^2\right) \tag{D.23}$$

Transformation of the Hamiltonian into polar coordinates then leads to

$$\hat{H}_v = -\frac{\hbar\sqrt{\lambda}}{2}\left\{\frac{1}{\rho^2}\frac{\partial}{\partial\rho}\left(\rho^2\frac{\partial}{\partial\rho}\right) + \frac{1}{\rho^2}\frac{1}{\sin\theta}\frac{\partial}{\partial\theta}\left(\sin\theta\frac{\partial}{\partial\theta}\right) + \frac{1}{\rho^2}\frac{1}{\sin^2\theta}\frac{\partial^2}{\partial\varphi^2} - \rho^2\right\} \tag{D.24}$$

while the vibrational wave equation assumes the form

$$\left\{\frac{1}{\rho^2}\frac{\partial}{\partial\rho}\left(\rho^2\frac{\partial}{\partial\rho}\right) + \frac{1}{\rho^2}\frac{1}{\sin\theta}\frac{\partial}{\partial\theta}\left(\sin\theta\frac{\partial}{\partial\theta}\right) + \frac{1}{\rho^2}\frac{1}{\sin^2\theta}\frac{\partial^2}{\partial\varphi^2} + \left(\frac{2E}{\hbar\sqrt{\lambda}} - \rho^2\right)\right\}\chi_v = 0 \tag{D.25}$$

Separation of variables may now be obtained by using

$$\chi_v = R(\rho)\Theta(\theta)\Phi(\varphi) \tag{D.26}$$

which upon insertion into Eq. (D.25) leads to

$$\left\{\frac{d^2}{d^2\varphi} + m^2\right\}\Phi = 0 \tag{D.27}$$

$$\left\{\frac{1}{\sin\theta}\frac{d}{d\theta}\left(\sin\theta\frac{d}{d\theta}\right) + \left[l(l+1) - \frac{m^2}{\sin^2\theta}\right]\right\}\Theta = 0 \tag{D.28}$$

$$\left\{\frac{1}{\rho^2}\frac{d}{d\rho}\left(\rho^2\frac{d}{d\rho}\right) + \left[\frac{2E}{\hbar\sqrt{\lambda}} - \rho^2 - \frac{l(l+1)}{\rho^2}\right]\right\}R = 0 \tag{D.29}$$

These have as solutions

$$\Phi(\varphi) = (2\pi)^{-1/2}\exp(im\varphi) \tag{D.30}$$

$$\Theta(\theta) = N_{l|m|}P_l^{|m|}(\cos\theta) \tag{D.31}$$

$$R(\rho) = N_{vl}\rho^l L_\tau^{l+1/2}(\rho^2)\exp(-\rho^2/2) \tag{D.32}$$

where

$$l = v, v-2, v-4, \ldots, 1 \text{ or } 0 \tag{D.33}$$

$$m = 0, \pm 1, \pm 2, \ldots, \pm l \tag{D.34}$$

$$\tau = (v + l + 1)/2 \tag{D.35}$$

The functions $P_l^{|m|}$ are associated Legendre polynomials of order $|m|$ and degree l, and $L_\tau^{l+1/2}(\rho^2)$ are associated Laguerre polynomials of degree $(v-1)/2$ in ρ^2. In turn, the normalization factors are found to be

$$N_{l|m|} = \frac{(-1)^l}{2^l l!}\sqrt{\frac{(2l+1)}{2}\frac{(l-|m|)!}{(l+|m|)!}} \tag{D.36}$$

$$N_{vl} = \sqrt{\frac{2[(v-1)/2]!}{\{[(v+l+1)/2]!\}^3}} \tag{D.37}$$

In the configuration space spanned by (q_1, q_2, q_3), we may then define the vibrational angular momentum π through its classical components, that is,

$$\pi_1 = q_2 p_3 - q_3 p_2 \qquad \text{and its (123) cyclic permutations} \tag{D.38}$$

where p_i are the conjugate momenta associated to q_i $(i = 1, 2, 3)$. The operators associated with $\pi^2 = \pi_1^2 + \pi_2^2 + \pi_3^2$ and its projection π_z (denoted M_3 in [95]) along the z axis assume in polar coordinates the form

$$\hat{\pi}^2 = -\hbar^2\left(\frac{\partial^2}{\partial\theta^2} + \frac{\cos\theta}{\sin\theta}\frac{\partial}{\partial\theta} + \frac{1}{\sin^2\theta}\frac{\partial^2}{\partial\varphi^2}\right) \tag{D.39}$$

$$\hat{\pi}_z = -i\hbar\frac{\partial}{\partial\varphi} \tag{D.40}$$

As for the 2D case, it can be shown that χ_v in Eq. (D.26) are eigenfunctions of both $\hat{\pi}^2$ and $\hat{\pi}_z$ defined by

$$\hat{M}^2\chi_v = l(l+1)\hbar^2\chi_v \tag{D.41}$$

$$\hat{M}_z\chi_v = m\hbar\chi_v \tag{D.42}$$

Thus, l and m quantize the vibrational angular momentum and its z component.

So far, we have considered interactions that are degenerate at the harmonic-oscillator level of approximation. For two levels that are nearly degenerate by accident in such an approximation, large perturbations may arise due to anharmonicity that are known as Fermi resonances. It should be noted that Fermi resonances occur only between states of the same symmetry. Thus, they cannot occur between two levels with different values of the vibrational angular momentum quantum number l. As usual, Fermi resonances increase the energy of the upper level while decreasing that of the lower one (in common language, they repel each other). Thus, the spectrum of a specific symmetry tends to be more irregular in the presence of Fermi resonances.

APPENDIX E: ADIABATIC STATES IN THE VICINITY OF A CONICAL INTERSECTION

I. JAHN–TELLER THEOREM

Following Moffitt and Liehr [73], in this appendix we give a proof of the Jahn–Teller theorem for X_3 molecules pertaining to the D_{3h} point group. Let ψ_1 and ψ_2 be the two electronic eigenfunctions that belong to the degenerate electronic states of E' symmetry (denoted ${}^eE'$). The two degenerate normal coordinates are Q_x and Q_y, the former being symmetric and the latter antisymmetric with respect to the xz plane (see Appendix D). Defining complex normal coordinates and electronic eigenfunctions as

$$Q_+ = Q_x + iQ_y = \rho\exp(i\varphi) \tag{E.1}$$

$$Q_- = Q_x - iQ_y = \rho\exp(-i\varphi) \tag{E.2}$$

and

$$\psi_+ = \psi_1 + i\psi_2 \tag{E.3}$$

$$\psi_- = \psi_1 - i\psi_2 \tag{E.4}$$

the electronic energy of the system is in degenerate-state perturbation theory obtained by solving the secular equation

$$\begin{vmatrix} H_{++} - W & H_{+-} \\ H_{-+} & H_{--} - W \end{vmatrix} = 0 \tag{E.5}$$

where the matrix elements are given by

$$H_{++} = \langle \psi_+ | \hat{H}_e | \psi_+ \rangle \qquad H_{--} = \langle \psi_- | \hat{H}_e | \psi_- \rangle \tag{E.6}$$

$$H_{+-} = \langle \psi_+ | \hat{H}_e | \psi_- \rangle \qquad H_{-+} = \langle \psi_- | \hat{H}_e | \psi_+ \rangle \tag{E.7}$$

and the integrations are defined with respect to all the electronic coordinates. Then, by developing $\hat{H}_e$ in a power series expansion of the normal coordinates, one gets

$$\hat{H}_e = \hat{h}_0 + \hat{h}_1^+ Q_- + \hat{h}_1^- Q_+ + \hat{h}_2^+ Q_-^2 + \hat{h}_2^- Q_+^2 + \cdots \tag{E.8}$$

where we have considered only the dependence on the degenerate complex normal coordinates Q_+ and Q_-. Substitution of Eq. (E.8) in Eqs. (E.6) and (E.7) gives

$$\begin{aligned} H_{++} = {} & \langle \psi_+ | \hat{h}_0 | \psi_+ \rangle + \langle \psi_+ | \hat{h}_1^+ | \psi_+ \rangle Q_- + \langle \psi_+ | \hat{h}_1^- | \psi_+ \rangle Q_+ \\ & + \langle \psi_+ | \hat{h}_2^+ | \psi_+ \rangle Q_-^2 + \langle \psi_+ | \hat{h}_2^- | \psi_+ \rangle Q_+^2 + \cdots \end{aligned} \tag{E.9}$$

$$\begin{aligned} H_{+-} = {} & \langle \psi_+ | \hat{h}_0 | \psi_- \rangle + \langle \psi_+ | \hat{h}_1^+ | \psi_- \rangle Q_- + \langle \psi_+ | \hat{h}_1^- | \psi_- \rangle Q_+ \\ & + \langle \psi_+ | \hat{h}_2^+ | \psi_- \rangle Q_-^2 + \langle \psi_+ | \hat{h}_2^- | \psi_- \rangle Q_+^2 + \cdots \end{aligned} \tag{E.10}$$

with similarly expressions for H_{--} and H_{-+}.

For a $\hat{C}_3$ rotation, Q_+, ψ_+, and $\psi_-^\star$ are multiplied by $\omega = \exp(2\pi i/3)$ while Q_-, ψ_-, and $\psi_+^\star$ are multiplied by $\omega^* = \exp(-2\pi i/3)$. Since the Hamiltonian must be totally symmetric, it follows that $\hat{h}_1^+$, $\hat{h}_1^-$, $\hat{h}_2^+$, and $\hat{h}_2^-$ are multiplied by ω, ω^*, ω^2, and ω^{*2}, respectively. The integrals in Eqs. (E.9) and (E.10) will then be different from zero only if the integrands are invariant under all symmetry operations allowed by the symmetry point group, in particular under $\hat{C}_3$. It is readily seen that the linear terms in Q_+ and Q_- vanish in H_{++} and H_{--}. In turn, the first term in H_{+-} and H_{-+} vanishes while one of the linear terms (Q_+ for H_{+-}, and Q_- for H_{-+}) does not vanish. Thus, neglecting quadratic (and higher order) terms, one obtains

$$H_{++} = H_{--} = W_0 \qquad H_{+-} = cQ_+ \qquad H_{-+} = cQ_- \tag{E.11}$$

Substitution of Eq. (E.11) into Eq. (E.5), leads to

$$W_\pm = W_0 \pm c\sqrt{Q_+Q_-} = W_0 \pm c\rho \tag{E.12}$$

Clearly, Eq. (E.12) shows that to a first approximation the electronic energy varies linearly with displacements in ρ, increasing for one component state while decreasing for the other. Thus, the potential minimum cannot be at $\rho = 0$. This is the statement of the Jahn–Teller theorem for a X_3 molecule belonging to the $\mathbf{D}_{3h}$ point group.

II. INVARIANT OPERATORS

We follow Thompson and Mead [13] to discuss the behavior of the electronic Hamiltonian, potential energy, and derivative coupling between adiabatic states in the vicinity of the D_{3h} conical intersection. Let $\hat{\mathsf{A}}$ be an operator that transforms only the nuclear coordinates, and $\hat{A}$ be one that acts on the electronic degrees of freedom alone. Clearly, the electronic Hamiltonian satisfies

$$(\hat{\mathsf{A}}\hat{H}_e)\hat{A}\psi = \hat{A}\hat{H}_e\psi \tag{E.13}$$

since, if ψ is an eigenfunction of $\hat{H}_e$, Eq. (E.13) just expresses the fact that $\hat{A}\psi$ is an eigenfunction of the transformed Hamiltonian with the same eigenvalue (for an arbitrary ψ, it also follows upon its expansion in eigenfunctions of $\hat{H}_e$). Thus,

$$(\hat{\mathsf{A}}\hat{H}_e)\hat{A} = \hat{A}\hat{H}_e \tag{E.14}$$

If Eq. (E.14) is satisfied for all elements of some point group G, $\hat{A}$ will be an invariant operator [13] (the Hermitian conjugate as well as the sum and/or product of two invariant operators are also invariant operators). Such an operator can be expanded in the form

$$\hat{H}_e = \sum_{\Gamma\gamma s} \hat{h}_{\Gamma\gamma s} Q_{\Gamma\gamma s} \tag{E.15}$$

where $Q_{\Gamma\gamma s}$ is a nuclear coordinate transforming as the γ th component of the Γ IRREP of G, the index s refers to different occurrences of the same IRREP, and

$\hat{h}_{\Gamma\gamma s}$ is an electronic operator that is independent of the nuclear coordinates. The requirement of Eq. (E.14) thus becomes

$$\sum_{\Gamma\gamma\gamma' s}\left[\hat{h}_{\Gamma\gamma s}D^{\Gamma}_{\gamma\gamma'}(A)Q_{\Gamma\gamma' s}\right]\hat{A} = \hat{A}\sum_{\Gamma\gamma' s}\hat{h}_{\Gamma\gamma' s}Q_{\Gamma\gamma' s} \tag{E.16}$$

This holds independently of the values of the coordinates, and hence term by term for each $Q_{\Gamma\gamma s}$. One has,

$$\sum_{\gamma}\hat{h}_{\Gamma\gamma s}D^{\Gamma}_{\gamma\gamma'}(A)\hat{A} = \hat{A}\hat{h}_{\Gamma\gamma' s} \tag{E.17}$$

or, equivalently,

$$\hat{A}\hat{h}_{\Gamma\gamma' s}\hat{A}^{-1} = \sum_{\gamma}\hat{h}_{\Gamma\gamma s}D^{\Gamma}_{\gamma\gamma'}(A) \tag{E.18}$$

Thus, the operators $\hat{h}_{\Gamma\gamma' s}$ transform under $\hat{A}\cdots\hat{A}^{-1}$ according to the Γ IRREP of G.

Now, consider the subgroup $\mathbf{C}_{3v}$ of $\mathbf{D}_{3h}$ (since no out-of-plane bending is possible for a triatomic system, and hence the subgroup $\mathbf{C}_{3v}$ may be used for the discussion). Then, Eq. (E.15) contains only four symmetry types of electronic operators: $\hat{h}_{A_1}$, $\hat{h}_{A_2}$, $\hat{h}_x$, and $\hat{h}_y$. The direct product decompositions for $\mathbf{C}_{3v}$ may then be shown (see Table 57 of [28]) to assume the form

$$u_{A_1}v_{A_1} \propto A_1 \qquad u_{A_1}v_{A_2} \propto A_2 \qquad u_{A_1}v_x \propto x \qquad u_{A_1}v_y \propto y \tag{E.19}$$

$$u_{A_2}v_{A_2} \propto A_1 \qquad u_{A_2}v_x \propto y \qquad u_{A_2}v_y \propto -x \tag{E.20}$$

$$u_xv_x + u_yv_y \propto A_1 \qquad u_xv_y - u_yv_x \propto A_2 \tag{E.21}$$

$$-u_xv_x + u_yv_y \propto x \qquad u_xv_y + u_yv_x \propto y \tag{E.22}$$

where the symbol $\propto$ means transforms under $\mathbf{C}_{3v}$ as, and v_x and v_y are arbitrary functions with the transformation properties of the corresponding subscripts (similarly for A_1 and A_2). For example, for $\langle E|\hat{h}_{A_2}|E\rangle$, one has

$$\langle x|\hat{h}_{A_2}|y\rangle = \langle x|u_{A_2}v_y\rangle = -1 \tag{E.23}$$

$$\langle x|\hat{h}_{A_2}|x\rangle = \langle x|u_{A_2}v_x\rangle = 0 \tag{E.24}$$

and so on. Similarly, for $\langle E|\hat{h}_x|E\rangle$, one gets

$$\langle x|\hat{h}_x|y\rangle = \langle x|u_xv_y\rangle = 0 \tag{E.25}$$

$$\langle x|\hat{h}_x|x\rangle = \langle x|u_xv_x\rangle = -1 \tag{E.26}$$

Thus, the nonzero submatrices are

$$\langle A_1|\hat{h}_{A_1}|A_1\rangle = 1 \qquad \langle A_2|\hat{h}_{A_1}|A_2\rangle = 1 \tag{E.27}$$

$$\langle E|\hat{h}_{A_1}|E\rangle = \begin{pmatrix} 1 & 0 \\ 0 & 1 \end{pmatrix} \tag{E.28}$$

$$\langle A_1|\hat{h}_{A_2}|A_2\rangle = 1 \qquad \langle A_2|\hat{h}_{A_2}|A_1\rangle = 1 \tag{E.29}$$

$$\langle E|\hat{h}_{A_2}|E\rangle = \begin{pmatrix} 0 & -1 \\ 1 & 0 \end{pmatrix} \tag{E.30}$$

$$\langle A_1|\hat{h}_x|E\rangle = (1 \quad 0) \qquad \langle A_1|\hat{h}_y|E\rangle = (0 \quad 1) \tag{E.31}$$

$$\langle A_2|\hat{h}_x|E\rangle = (0 \quad 1) \qquad \langle A_2|\hat{h}_y|E\rangle = (-1 \quad 0) \tag{E.32}$$

$$\langle E|\hat{h}_x|A_1\rangle = \begin{pmatrix} 1 \\ 0 \end{pmatrix} \qquad \langle E|\hat{h}_y|A_1\rangle = \begin{pmatrix} 0 \\ 1 \end{pmatrix} \tag{E.33}$$

$$\langle E|\hat{h}_x|A_2\rangle = \begin{pmatrix} 0 \\ 1 \end{pmatrix} \qquad \langle E|\hat{h}_y|A_2\rangle = \begin{pmatrix} -1 \\ 0 \end{pmatrix} \tag{E.34}$$

$$\langle E|\hat{h}_x|E\rangle = \begin{pmatrix} -1 & 0 \\ 0 & 1 \end{pmatrix} \qquad \langle E|\hat{h}_y|E\rangle = \begin{pmatrix} 0 & 1 \\ 1 & 0 \end{pmatrix} \tag{E.35}$$

III. FUNCTIONAL FORM OF THE ENERGY

Since the potential matrix $\mathbf{W}$ is invariant and restricted to E space, it has the form

$$\mathbf{W} = W_{A_1}\begin{pmatrix} 1 & 0 \\ 0 & 1 \end{pmatrix} + W_{A_2}\begin{pmatrix} 0 & -1 \\ 1 & 0 \end{pmatrix} + W_x\begin{pmatrix} -1 & 0 \\ 0 & 1 \end{pmatrix} + W_y\begin{pmatrix} 0 & 1 \\ 1 & 0 \end{pmatrix} \tag{E.36}$$

where W_{A_1}, and so on, are functions of the nuclear coordinates transforming under $\mathbf{C}_{3v}$ as indicated by their subscripts. On the other hand $\mathbf{W}$ must be Hermitian, and in our case can be real, from which it follows that $W_{A_2} = 0$. The energies that reduce to the degenerate pair at the reference configuration are then just the eigenvalues of $\mathbf{W}$:

$$W_{\pm} = W_{A_1} \pm W_R \tag{E.37}$$

where

$$W_R = \sqrt{W_x^2 + W_y^2} \tag{E.38}$$

To all orders, the general form for functions with transformation properties u_{A_1}, and so on, may be shown [13] to be

$$u_{A_1} = f_1\left[z; \rho^2, \rho^3 \cos(3\varphi)\right] \tag{E.39}$$

$$u_{A_2} = \rho^3 \sin(3\varphi) f_2\left[z; \rho^2, \rho^3 (\cos 3\varphi)\right] \tag{E.40}$$

$$u_x = \rho \cos\varphi f_3\left[z; \rho^2, \rho^3 \cos(3\varphi)\right] + \rho^2 \cos(2\varphi) f_4\left[z; \rho^2, \rho^3 \cos(3\varphi)\right] \tag{E.41}$$

$$u_y = \rho \sin\varphi f_3\left[z; \rho^2, \rho^3 \cos(3\varphi)\right] - \rho^2 \sin(2\varphi) f_4\left[z; \rho^2, \rho^3 \cos(3\varphi)\right] \tag{E.42}$$

where f_i $(i = 1, 2, 3, 4)$ are functions formally representable as a double power series in their arguments other than z, with the coefficients being constant or functions of z. From Eq. (E.37)–(E.42), one then obtains

$$W_{A_1} = W_{A_1}\left[z; \rho^2, \rho^3 \cos(3\varphi)\right] \tag{E.43}$$

$$W_R = \rho\sqrt{f^2 + \rho^2 g^2 + 2\rho fg \cos(3\varphi)} \tag{E.44}$$

$$\rightarrow \rho w\left[z; \rho^2, \rho \cos(3\varphi)\right] \qquad (\rho \rightarrow 0) \tag{E.45}$$

where $f = f(z; \rho^2, \rho^3 \cos(3\varphi))$, $g = g(z; \rho^2, \rho^3 \cos(3\varphi))$, and w are formally analytic functions of their arguments. These equations define the correct behavior of the potential energy in the vicinity of the conical intersection, and hence may be valuable in delineating fitting forms, as it was the case for the H_3 DMBE potential energy surface [82].

Acknowledgments

We thank Dr. Alexander Alijah for helpful discussions. This work was supported by Fundação para a Ciência e Tecnologia, Portugal.

References

1. A. J. C. Varandas and H. G. Yu, *Chem. Phys. Lett.* **259**, 336 (1996).
2. A. J. C. Varandas and H. G. Yu, *J. Chem. Soc. Faraday Trans.* **93**, 819 (1997).
3. S. Mahapatra and H. Köppel, *J. Chem. Phys.* **109**, 1721 (1998).
4. A. J. C. Varandas and Z. R. Xu, *J. Chem. Phys.* **112**, 2121 (2000).
5. Z. R. Xu, M. Baer, and A. J. C. Varandas, *J. Chem. Phys.* **112**, 2746 (2000).
6. Z. R. Xu and A. J. C. Varandas, *Int. J. Quantum Chem.* **80**, 454 (2000).
7. Z. R. Xu and A. J. C. Varandas, *Int. J. Quantum Chem.* **83**, 279 (2001).
8. Z. R. Xu and A. J. C. Varandas, J. Phys. Chem. **105**, 2246 (2001).
9. W. H. Gerber and E. Schumacher, *J. Chem. Phys.* **69**, 1692 (1978).
10. M. Mayer and L. S. Cederbaum, *J. Chem. Phys.* **105**, 4938 (1996).
11. A. J. C. Varandas, H. G. Yu, and Z. R. Xu, *Mol. Phys.* **96**, 1193 (1999).
12. A. J. C. Varandas and Z. R. Xu, *Int. J. Quantum Chem.* **75**, 89 (1999).
13. T. C. Thompson and C. A. Mead, *J. Chem. Phys.* **82**, 2408 (1985).

14. B. K. Kendrick, *Phys. Rev. Lett.* **79**, 2431 (1997).
15. D. H. Zhang and J. Z. H. Zhang, *J. Chem. Phys.* **101**, 3671 (1994).
16. V. A. Mandelshtam, T. P. Grozdanov, and H. S. Taylor, *J. Chem. Phys.* **103**, 10074 (1995).
17. A. J. Dobbyn, M. Stumpf, H.-M. Keller, W. L. Hase, and R. Shinke, *J. Chem. Phys.* **103**, 9947 (1995).
18. B. K. Kendrick and R. T Pack, *J. Chem. Phys.* **106**, 3519 (1997).
19. C. A. Mead, *Chem. Phys.* **49**, 23 (1980).
20. T. C. Thompson, D. G. Truhlar, and C. A. Mead, *J. Chem. Phys.* **82**, 2392 (1985).
21. M. V. Berry, *Proc. R. Soc. London, Ser. A* **392**, 45 (1984).
22. C. A. Mead, *Rev. Mod. Phys.* **64**, 51 (1992).
23. D. R. Yarkony, *Rev. Mod. Phys.* **68**, 985 (1996).
24. D. R. Yarkony, *J. Phys. Chem.* **100**, 18612 (1996).
25. M. Baer, *J. Chem. Phys.* **107**, 2694 (1997).
26. D. Neuhauser, *J. Chem. Phys.* **100**, 9272 (1994).
27. H. G. Yu and S. C. Smith, *Ber. Bunsenges. Phys. Chem.* **101**, 400 (1997).
28. G. Herzberg, *Molecular Spectra and Molecular Structure. III. Electronic Spectra and Electronic Structure of Polyatomic Molecules*, van Nostrand, New York, 1966.
29. D. M. Dennison, *Rev. Mod. Phys.* **3**, 280 (1931).
30. E. P. Wigner, *Group Theory and Its Applications to the Quantum Mechanics of Atomic Spectra*, Academic Press, New York, 1959.
31. R. C. Whitten and F. T. Smith, *J. Math. Phys.* **9**, 1103 (1968).
32. B. R. Johnson, *J. Chem. Phys.* **73**, 5051 (1980).
33. B. R. Johnson, *J. Chem. Phys.* **79**, 1906 (1983).
34. B. R. Johnson, *J. Chem. Phys.* **79**, 1916 (1983).
35. J. Zúñiga, A. Bastida, and A. Requena, *J. Chem. Soc. Faraday Trans.* **93**, 1681 (1997).
36. D. A. Varshalovich, A. N. Moskalev, and V. K. Khersonskii, *Quantum theory of angular momentum*, World Scientific, Singapore, 1988.
37. A. Kiselev, *Can. J. Phys.* **56**, 615 (1978).
38. M. Born and J. R. Oppenheimer, *Ann. Phys.* **84**, 457 (1927).
39. M. Born and K. Huang, *Dynamical Theory of Crystal Lattices*, Oxford, London, 1954.
40. E. E. Nikitin and L. Zulicke, *Lecture Notes in Chemistry, Theory of Elementary Processes*, Vol. 8, Springer-Verlag, Heidelberg, 1978.
41. I. B. Bersuker and V. Z. Polinger, *Vibronic Interactions in Molecules and Crystals*, Springer-Verlag, Berlin, 1989.
42. H. C. Longuet-Higgins, *Adv. Spectrosc.* **2**, 429 (1961).
43. G. Herzberg and H. C. Longuet-Higgins, *Faraday Discuss. Chem. Soc.* **35**, 77 (1963).
44. H. C. Longuet-Higgins, *Proc. R. Soc., London Ser. A* **344**, 147 (1975).
45. B. K. Kendrick and C. A. Mead, *J. Chem. Phys.* **102**, 4160 (1995).
46. C. A. Mead and D. G. Truhlar, *J. Chem. Phys.* **70**, 2284 (1979).
47. C. A. Mead, *J. Chem. Phys.* **72**, 3839 (1980).
48. Y.-S. M. Wu, B. Lepetit, and A. Kuppermann, *Chem. Phys. Lett.* **186**, 319 (1991).
49. Y.-S. M. Wu and A. Kuppermann, *Chem. Phys. Lett.* **201**, 178 (1993).
50. A. Kuppermann and Y.-S. M. Wu, *Chem. Phys. Lett.* **213**, 636 (1993).

51. G. D. Billing and N. Marković, *J. Chem. Phys.* **99**, 2674 (1993).
52. B. K. Kendrick and R. T. Pack, *J. Chem. Phys.* **104**, 7475 (1996).
53. M. Baer and R. Englman, *Chem. Phys. Lett.* **265**, 105 (1997).
54. E. Merzbacher, *Quantum Mechanics*, John Wiley & Sons, Inc., New York, 1970.
55. B. G. Levich, V. A. Myanlin, and Y. A. Vdovin, *Theoretical Physics. An Advanced Text*, North-Holland, Amsterdam, 1973, p. 418.
56. I. N. Levine, *Molecular Spectroscopy*, John Wiley & Sons, Inc., New York, 1975.
57. J. N. Murrell, S. Carter, S. C. Farantos, P. Huxley, and A. J. C. Varandas, *Molecular Potential Energy Functions*, John Wiley & Sons, Inc., Chichester, 1984.
58. A. J. C. Varandas, Adv. Chem. Phys. **74**, 255 (1988).
59. G. C. Hancock, C. A. Mead, D. G. Truhlar, and A. J. C. Varandas, *J. Chem. Phys.* **91**, 3492 (1989).
60. A. J. C. Varandas, in *Reaction and Molecular Dynamics*, A. Laganá and A. Riganelli, eds., Springer, Berlin, 2000, Vol. 75 of *Lecture Notes in Chemistry*, p. 33.
61. A. J. C. Varandas, *Int. J. Quantum Chem.* **32**, 563 (1987).
62. F. London, *Z. Electrochem.* **35**, 552 (1929).
63. H. Eyring and M. Polanyi, *Z. Phys. Chem. Abt.* **12**, 279 (1931).
64. S. Sato, *J. Chem. Phys.* **23**, 592 (1955).
65. S. Sato, *J. Chem. Phys.* **23**, 2465 (1955).
66. J. N. Murrell and A. J. C. Varandas, *Mol. Phys.* **57**, 415 (1986).
67. V. M. F. Morais, A. J. C. Varandas, and A. A. C. C. Pais, *Mol. Phys.* **58**, 285 (1986).
68. A. J. C. Varandas, *Chem. Phys. Lett.* **138**, 455 (1987).
69. A. J. C. Varandas and A. A. C. C. Pais, *J. Chem. Soc. Faraday Trans.* **89**, 1511 (1993).
70. A. A. C. C. Pais, R. F. Nalewajski, and A. J. C. Varandas, *J. Chem. Soc. Faraday Trans.* **90**, 1381 (1994).
71. A. J. C. Varandas and A. I. Voronin, *Mol. Phys.* **95**, 497 (1995).
72. H. A. Jahn and E. Teller, *Proc. R. Soc. London, Ser. A* **161**, 220 (1937).
73. W. Moffitt and A. D. Liehr, *Phys. Rev.* **106**, 1195 (1957).
74. A. J. C. Varandas, J. Tennyson, and J. N. Murrell, *Chem. Phys. Lett.* **61**, 431 (1979).
75. M. L. Mehta, *Random Matrices and the Statistical Theory of Energy Levels*, Academic Press, New York, 1967.
76. C. A. Mead, *J. Chem. Phys.* **78**, 807 (1983).
77. A. J. C. Varandas and Z. R. Xu, *Chem. Phys. Lett.* **316**, 248 (2000).
78. J.-P. Wolf, G. Delacrétaz, and L. Wöste, *Phys. Rev. Lett.* **63**, 1946 (1989).
79. H. A. Jahn, *Proc. R. Soc. London Ser. A* **164**, 117 (1938).
80. A. J. C. Varandas and V. M. F. Morais, *Mol. Phys.* **47**, 1241 (1982).
81. R. Abrol, A. Shaw, A. Kuppermann, and D. R. Yarkony, *J. Chem. Phys.* **115**, 4640 (2001).
82. A. J. C. Varandas, F. B. Brown, C. A. Mead, D. G. Truhlar, and N. C. Blais, *J. Chem. Phys.* **86**, 6258 (1987).
83. A. Kuppermann and Y.-S. M. Wu, *Chem. Phys. Lett.* **205**, 577 (1993).
84. A. Kuppermann, *Chem. Phys. Lett.* **32**, 374 (1975).
85. M. Baer, *Chem. Phys. Lett.* **35**, 112 (1975).
86. M. Baer, *Mol. Phys.* **40**, 1011 (1980).
87. A. J. C. Varandas, *J. Chem. Phys.* **107**, 867 (1997).

88. A. J. C. Varandas, A. I. Voronin, and P. J. S. B. Caridade, *J. Chem. Phys.* **108**, 7623 (1998).
89. R. K. Preston and J. C. Tully, *J. Chem. Phys.* **54**, 4297 (1971).
90. D. Grimbert, B. Lassier-Govers, and V. Sidis, *Chem. Phys.* **124**, 187 (1988).
91. F. Gianturco, A. Palma, and F. Schnider, *Chem. Phys.* **137**, 177 (1989).
92. B. K. Kendrick, C. A. Mead, and D. G. Truhlar, *J. Chem. Phys.* **110**, 7594 (1999).
93. E. P. Wigner, *J. Math. Phys.* **1**, 409 (1960).
94. R. N. Porter, R. M. Stevens, and M. Karplus, *J. Chem. Phys.* **49**, 5163 (1968).
95. S. Califano, *Vibrational States*, John Wiley & Sons, Inc., London, 1976.

AUTHOR INDEX

Numbers in parentheses are reference numbers and indicate that the author's work is referred to although his name is not mentioned in the text. Numbers in *italic* show the pages on which the complete references are listed.

SUBJECT INDEX